Essential Reproduction

To the late Professor Sir Bob Edwards (1925–2013)
Nobel Laureate in Physiology or Medicine 2010
who first stimulated and then cultivated my interest
in the science of reproduction

Eighth Edition

Essential Reproduction

Martin H. Johnson
MA, PhD, FRCOG, FRSB, FMedSci, FRS
Emeritus Professor of Reproductive Sciences
Department of Physiology, Development and Neuroscience
University of Cambridge
Fellow of Christ's College, Cambridge
Cambridge, UK
Honorary Academic Fellow of St Paul's College
Sydney, NSW, Australia

WILEY Blackwell

Edition History
Wiley-Blackwell, © Martin Johnson (7e, 2013)

Registered Office(s)
John Wiley & Sons, Inc., 111 River Street, Hoboken, NJ 07030, USA
John Wiley & Sons Ltd, The Atrium, Southern Gate, Chichester, West Sussex, PO19 8SQ, UK

Editorial Office
9600 Garsington Road, Oxford, OX4 2DQ, UK

For details of our global editorial offices, customer services, and more information about Wiley products visit us at www.wiley.com.

Wiley also publishes its books in a variety of electronic formats and by print-on-demand. Some content that appears in standard print versions of this book may not be available in other formats.

Library of Congress Cataloging-in-Publication Data

Names: Johnson, M. H., author.
Title: Essential reproduction / Martin H. Johnson.
Description: Eighth edition. | Hoboken, NJ : Wiley, 2018. | Series: Essentials | Includes bibliographical references and index. |
Identifiers: LCCN 2017051650 (print) | LCCN 2017052067 (ebook) | ISBN 9781119246459 (pdf) | ISBN 9781119246473 (epub) | ISBN 9781119246398 (paperback)
Subjects: | MESH: Reproductive Physiological Phenomena | Mammals–physiology
Classification: LCC QL739.23 (ebook) | LCC QL739.23 (print) | NLM WQ 205 | DDC 573.6/19–dc23
LC record available at https://lccn.loc.gov/2017051650

Cover image: © KTSDESIGN/SCIENCE PHOTO LIBRARY/Gettyimages
Cover design by Wiley

Set in 10/12pt Adobe Garamond Pro by SPi Global, Pondicherry, India
Printed and bound in Singapore by Markono Print Media Pte Ltd

10 9 8 7 6 5 4 3 2 1

Contents

Contributors

Dr Catherine Aiken: clinical vignettes
Professor Peter Braude: Chapters 21 and 22
Professor Dino Giussani: parts of Chapters 16 and 18
Dr Emlyn Parfitt: part of Chapter 23
Dr Andrew Sharkey: parts of Chapters 13 and 16

Preface

There have been many advances in our understanding of the reproductive processes in humans since the seventh edition. Much of this progress has been due to the continuing application to reproductive studies of the expanding range and sensitivity of the techniques of molecular biology, which now allow much more sophisticated descriptions and manipulations of reproductive activities. Advances have also come from the development of live imaging techniques of greater utility. These include imaging of reproductive organs and also of brain function *in situ*, as well as less-invasive long-term imaging of the behaviours of cells and embryos in culture. The advances in medical research on reproduction have been truly spectacular – which, in many parts of this edition, renders potentially dangerous extrapolations from other mammals to humans less necessary.

Major health and social issues continue to place reproduction at the centre of scientific, clinical, political and ethical discourse. The threat posed by the continuing growth in world population to the planet's resources and to our fragile climate continues to present a major challenge. The tragic and unnecessary high maternal mortality rates throughout much of the world are an indictment of our best efforts and intentions to 'do better' – as expressed in the WHO Millennium Aims. On the positive side, progress has been made in managing the effects of infection with human immunodeficiency virus through the cheaper provision of generic drugs as well as in the prevention of infection through the provision of Pre-Exposure Prophylaxis (PrEP) medications. However, it is as yet unclear that transmission rates are coming down, a testament to the importance of understanding sexual behaviours and addressing the impact of gender inequalities. These issues are made more pressing by the increasing resistance of microbes to antibiotics in general and to genitourinary infections in particular, with their implications for individual fecundity. Continuing clinical developments in the field of assisted conception have expanded opportunities for the alleviation or circumvention of sub-fertility, genetic disability and, through stem cells, degenerative disease, but have also ignited new (and old) controversies. Thus, the first birth has been reported of a child born after mitochondrial transfer, the development of a human embryo *in vitro* though to 13 days has been reported, and the first licences have been issued for research on the genetic modification of zygotes. The continued explosion of obesity and the realization that both child and adult health and well-being are affected enduringly by life *in utero* have focused work on pregnancy, the placenta and the neonatal period of care. Finally, we are at last beginning to understand more fully how genetic expression during development interacts with environmental factors to influence complex behavioural phenotypes that include psychiatric disease and antisocial behaviour.

It is clear from all these examples that reproduction reaches into all parts of our lives! Science thus forms just part of this book – albeit at its core. One might have hoped that the advances in scientific knowledge and understanding would helpfully inform prevailing socio-legal discussions, attitudes and values. Sadly, since the last edition, for much of the world, the enlightenment viewpoint based on a cool appraisal of expert evidence has been crushed in a miasma of anti-science rhetoric and antagonism to the expertise underlying it: whether directed against climate change, evolution, embryo research, women's progress, or rights for sexual minorities. It is the duty of all scientists to engage in the public and political debates about these issues, so as to ensure that the evidence is presented accurately and reliably to the public, and, where appropriate, to defend their work against such unenlightened views. This course of action is one that the dedicatee of this book, Professor Sir Bob Edwards, was a pioneer of in promoting public discussion of science – indeed, long before it was fashionable to explain your work to the public, he was criticized roundly by his scientific and medical colleagues for doing just that. Indeed, when belatedly he was awarded the Nobel Prize, he was recognized not just for his unique role in the development of research on human reproduction,but also for the wider role Bob played in reproductive science, medicine and ethics, together with his pioneering role in communicating science to the public. Sadly, since the seventh edition, both Bob and his wife and long-term scientific collaborator, Ruth, have died – a loss for me of my intellectual parents.

Finally, for this edition, the book, having been totally restructured in the previous edition, has been updated in all chapters, whilst hopefully retaining the strong narrative thread to the text. To aid this, more detailed information on deeper, comparative or applied aspects of some topics has been transferred to boxes, tables and figures. In addition, we have added a clinical vignette at the end of each chapter.

The major novel changes to which I draw your attention include the elucidation of the role of Vitamin A in the control of the spermatogenic cycle and wave (Chapter 7), more information about fetal hypoxia and immune reactivity during pregnancy (Chapter 16), and the discovery of more elements in the control of parturition in the primate and human (Chapter 18).

As before, many helpful comments, corrections and letters of advice have been received from readers, students and teachers all over the world. As always, in this eighth edition, now in its 38th year, I have tried to provide for students of reproduction a compact and comprehensive text that carves through the micro-detail of the subject to bring out its theoretical cores, but illustrates it with experiment, information and context.

MHJ, Cambridge 2017

How to use this book

This book represents an integrated approach to the study of reproduction. There can be few subjects that so obviously demand such an approach. During my teaching of reproduction at Cambridge University, the need for a book of this kind was clear to me and my colleagues. I know this volume goes some way towards filling this need because of the many appreciative comments I receive from colleagues at scientific meetings as well as from the Cambridge students.

I have written the book about human reproduction in a comparative context for medical, veterinary and science students at all professional levels. Throughout, I have attempted to draw out the general, fundamental points common to reproductive events in all or most species. However, a great range of variation in the *details* of reproduction is observed amongst different species and, in some respects, very *fundamental* differences are also observed. Where the details differ, I have attempted to indicate this in the numerous tables and figures, rather than clutter the general emphasis and narrative of the text. Where the fundamentals differ, an explicit discussion is given in the text. These fundamental differences should not be ignored. For example, preclinical medical students may consider the control of luteal life in the pig, of parturition in the sheep or of ovarian cyclicity in the rat to be irrelevant to their future interests. However, as a result of extrapolation between species, in the past the human has been treated as a pig, a sheep and a rat (with much discomfort and detriment). If, on finishing this book, the student appreciates the dangers of uncritical extrapolations between species, I will have achieved a major aim.

Science is uncertain and provisional, and this provisionality has been illustrated in several places in this book. I have not tried to give a simple story where a simple story does not exist. Uncertainty can be hard to handle, especially in medicine, but it is a reality that is as important as those informational facts that we think we have certain knowledge of – indeed, knowing the boundary between the certain and the uncertain is perhaps the most important knowledge of all. For you students, this uncertainty also provides future research opportunities!

Confinements of space have unfortunately necessitated the omission of the subject of embryonic development from the text. To give only passing reference to this fascinating subject would be an injustice; to treat it fully would require a text of similar length to the present one. I recommend that the interested student seeks this information elsewhere.

I suggest that you first read through each chapter with only passing reference to tables, boxes and figures. In this way, I hope that you will grasp the essential fundamentals of the subject under discussion. Then re-read the chapter, referring extensively to the tables and figures and their legends, in which much detailed or comparative information is located. Finally, because my approach to reproduction is an integrated one, the book needs to be taken as a whole, as it is more than the sum of its constituent chapters. I hope that you both enjoy and learn from reading it!

Acknowledgements

I owe particular thanks to many people for help at many stages of the preparation of this edition: to present and former students for their interest, stimulation and responsiveness; to my colleagues at Wiley-Blackwell for their help and advice; to the AudioVisual Media team in the Anatomy School at Cambridge for advice and help with photographic illustrations; to the Histology section of the Department of Physiology, Development and Neuroscience, Cambridge, for making available slides for photography; to Professor Adam Balen, Professor Peter Braude, Professor Graham Burton, Professor Dino Giussani, Professor Tomas Hökfelt, Professor Yacoub Khalaf, J. Moeselaar, Dr Tony Plant, Dr J.M. Tanner and Dr Pauline Yahr for allowing me to use their original photographs and data; and to my many colleagues who read and criticized my drafts and encouraged me in the preparation of this edition.

In particular, I am indebted to several friends, colleagues and former students who have collaborated with me in the production of this edition. Thus, my thanks to Dr Catherine Aiken for drafting the clinical vignettes; to Professor Peter Braude for his help with Chapters 21 and 22; to Dr Andrew Sharkey for his help with Chapters 13 and 16; to Dr Emlyn Parfitt for his help with Chapter 23; and to Professor Dino Giussani for his help with Chapters 16 and 18. However, I take full and final responsibility for the contents of the volume.

About the companion website

This book is accompanied by a companion website:

www.essentialreproduction.com

The website includes:

- All figures from the book
- Clinical vignettes and multiple choice questions

Part 1
Introduction

CHAPTER 1

What is reproduction?

Essential Reproduction, Eighth Edition. Martin H. Johnson.

Companion website: www.essentialreproduction.com

The ability to reproduce is a defining feature of all living organisms. Through reproduction, we pass our genes to a new generation. Each new generation in turn reproduces or dies out. The survivors are 'selected', by disease resistance and by successful competition for resources and mates, for their 'fitness' to live and to reproduce. In this way, the gene pool of surviving species is constantly adapting to the prevailing environment to provide the best available 'fit'. Thus, reproduction has been central to our **evolution** as the species *Homo sapiens*.

However, humans transmit more than simply their genes across generations. Humans have evolved high levels of sociability through which **cultures** are formed. Cultural practices are also transmitted across generations, and reproduction itself lies at the very heart of many of our cultural practices and taboos (see Chapters 5, 6, 20 and 23). Human society, by influencing socially and/or medically who survives to reproduce and with whom, is itself now part of the 'selection' process. This pivotal position of reproduction in our culture makes it a sensitive subject for study. Indeed, scientific enquiry into human reproduction was relatively late to the modern research scene and even today can provoke hostility, embarrassment or distress.

In this opening chapter, human reproduction is introduced and contextualized: in relation to other species – **reproductive strategies,** and in relation to time – the **reproductive life cycle**.

Reproductive strategies

Most organisms reproduce **asexually** (or **vegetatively**). For example, many unicellular organisms reproduce themselves **mitotically**, just like the individual cells of our body (Figure 1.1). Mitotic divisions generate two offspring that are genetically identical to each other and to their single parent. Among multicellular organisms, some shed cells or even body parts from which another genetically identical individual can be generated – a process called **regeneration**. Others, including some complex vertebrates such as lizards, reproduce themselves by setting aside a special population of **egg cells** that can differentiate into conceptuses in the absence of a fertilizing spermatozoon. This type of asexual reproduction is called **parthenogenesis**, and generates a completely new organism with the same gene complement as its parent.

Mammals reproduce sexually

Parthenogenesis is simply not an option available to mammals. Thus, although it is possible to activate a mammalian egg (including a human egg) in the complete absence of a spermatozoon,

Figure 1.1 Mitosis and meiosis in human cells. Each human cell contains 23 pairs of homologous chromosomes, making 46 chromosomes in total. Each set of 23 chromosomes is called a **haploid** set. When a cell has two complete sets, it is described as being **diploid**. In this figure, we show at the top a single schematized human cell with just two of the 23 homologous pairs of chromosomes illustrated, each being individually colour-coded. Between divisions, the cell is in **interphase**, during which it grows and duplicates both its **centriole** and the DNA in each of its chromosomes. As a result of the DNA replication, each chromosome consists of two identical **chromatids** joined at the **centromere**. Interphase chromosomes are not readily visible, being long, thin and decondensed (but are shown in this figure in a more condensed form for simplicity of representation).

Lower left panel: In **mitotic prophase**, the two chromatids become distinctly visible under the light microscope as each shortens and thickens by a spiralling contraction; at the end of prophase the **nucleoli and nuclear membrane** break down. In **mitotic metaphase**, microtubules form a **mitotic spindle** between the two centrioles and the chromosomes lie on its **equator**. In **mitotic anaphase**, the centromere of each chromosome splits and the two chromatids in each chromosome migrate to opposite poles of the spindle (**karyokinesis**). During **mitotic telophase** division of the cytoplasm into two daughters (known as **cytokinesis**) along with breakdown of the spindle and the reformation of nuclear membranes and nucleoli occurs, as does the decondensation of chromosomes so that they are no longer visible under the light microscope. Two genetically identical daughter cells now exist where one existed before. Mitosis is a non-sexual or vegetative form of reproduction.

Lower right panel: **Meiosis** involves two sequential divisions. The **first meiotic prophase (prophase 1)** is lengthy and can be divided into several sequential steps: (1) **leptotene** chromosomes are long and thin; (2) during **zygotene**, homologous pairs of chromosomes from each haploid set come to lie side by side along parts of their length; (3) in **pachytene**, chromosomes start to thicken and shorten and become more closely associated in pairs along their entire length at which time **synapsis, crossing over** and **chromatid exchange** take place and nucleoli disappear; (4) in **diplotene** and **diakinesis**, chromosomes shorten further and show evidence of being closely linked to their homologue at the **chiasmata** where crossing over and the reciprocal exchange of DNA sequences have occurred, giving a looped or cross-shaped appearance. In **meiotic metaphase 1**, the nuclear membrane breaks down, and homologous pairs of chromosomes align on the equator of the spindle. In **meiotic anaphase 1**, homologous chromosomes move in opposite directions. In **meiotic telophase 1**, cytokinesis occurs; the nuclear membrane may re-form temporarily, although this does not always happen, yielding two daughter cells each with half the number of chromosomes (only one member of each homologous pair), but each chromosome consisting of two genetically unique chromatids (because of the crossing-over at chiasmata). In the **second meiotic division**, these chromatids then separate much as in mitosis, to yield a total of four haploid offspring from the original cell, each containing only one complete set of chromosomes. Due to chromatid exchange and the random segregation of homologous chromosomes, each haploid cell is genetically unique. At fertilization, two haploid cells will come together to yield a new diploid zygote.

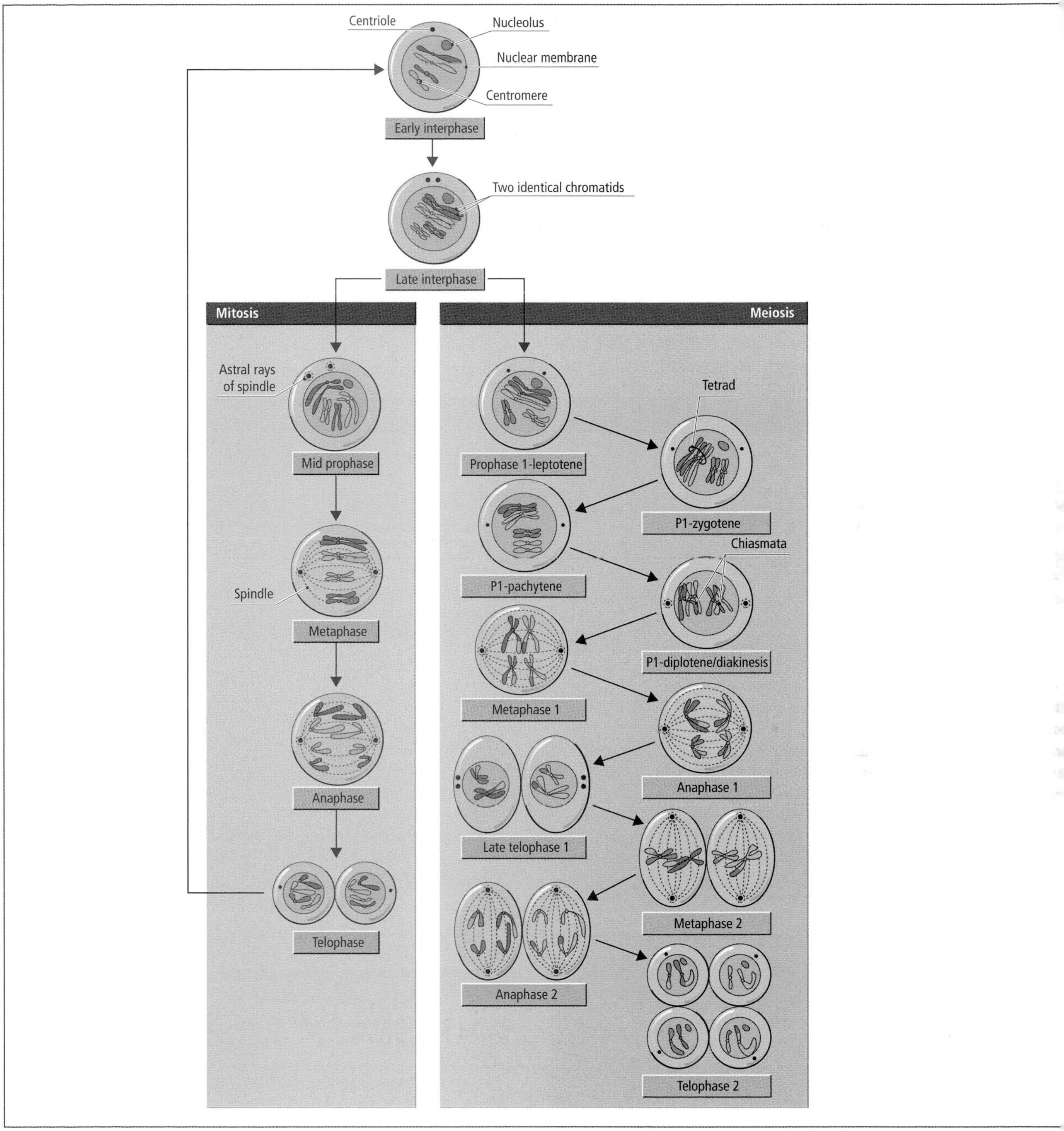

such that it undergoes the early processes of development and may even implant in the uterus, these parthenogenetic conceptuses always fail and die eventually (see page 10 for an explanation as to why this is).

Reproduction in mammals is invariably **sexual**. Sex is defined formally in biology as a process whereby a genetically novel individual is formed as a result of the mixing of genes from two individuals. So, the essential feature of mammalian sexual reproduction is that each new individual receives its chromosomes in two roughly equal portions: half carried in a **male gamete**, the **spermatozoon** (see Chapter 7), and half in a **female gamete**, the **oocyte** (see Chapter 9). These gametes come together at **fertilization** (see Chapter 12) to form the genetically novel **zygote**. In order to reproduce subsequently, the individual formed from that zygote must transmit only half its own chromosomes to the new zygotes of the next generation. In sexually reproducing species, therefore, a special population of **germ cells** is set aside. These cells undergo the division process of **meiosis**, during which the chromosomal content of the germ cells is **reduced by half** and the genetic composition of each chromosome is modified as a result of the **exchange of pieces of homologous chromosomes** (Figure 1.1). The **increased genetic diversity** that is generated within a sexually reproducing population offers a richer and more varied source of material on which natural selection can operate. The population therefore shows greater resilience in the face of environmental challenge. In Chapters 3 and 4, we examine how the two sexes are formed and matured.

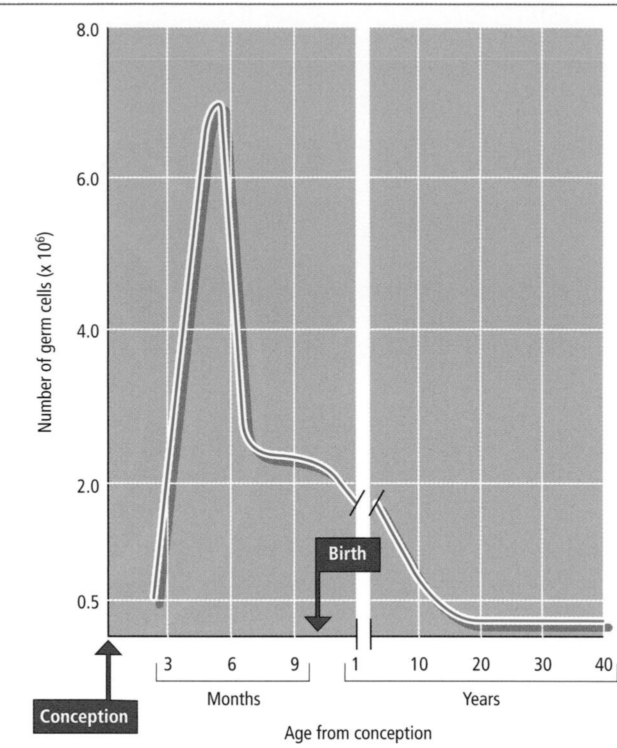

Figure 1.2 Numbers of ovarian germ cells during the life of a human female from conception. Note the steady rise early in fetal life followed by a precipitous decline prior to birth and shortly afterwards. (Source: Drawn from original data from T. Baker.)

Both natural and sexual selection operate in mammals

Asexually reproducing organisms do not need to find a sexual partner. Whether or not they reproduce depends entirely on their survival – **natural selection** operates simply at this level. Sexual reproduction introduces a complication since it involves two individuals. These have to come together and synchronize their egg and sperm production and shedding: spatial and temporal coordination is highly desirable to optimize fertility. The conjunction of two sexes also provides opportunities for mate selection. For successful reproduction, the survival of offspring to sexual maturity is critical and so it is advantageous to share your genes with a mate who has the genes most likely to achieve this success. It is not therefore surprising that mechanisms for recognizing 'fitness' in a sexual partner have evolved, a process called **sexual selection**. However, there is a cost to sexual selection: it involves considerable energy expenditure in locating, attracting and keeping a sexual partner, and also can expose both partners to increased risk of death – from sexual competitors or predators preying on the sexually occupied! So there is an evolutionary trade-off between the pros of sexual selection and the cons of performing it. How sexual selection operates in humans, and the cultural rituals around it, are discussed further in Chapter 6.

Fertilization in mammals is internal

The fertilization of oocytes is achieved in most aquatic and amphibious vertebrates by discharging large numbers of oocytes and spermatozoa into the water – **spawning**. This process of **external fertilization** provides opportunities for the easy predation of eggs and the conceptuses developing from them, and so large numbers are produced to increase the chances of some surviving.

Mammals, in contrast, **fertilize internally** (see Chapter 11). This reproductive strategy **reduces the numbers of eggs shed**, in humans to only one or two at a time, thereby reducing the energy resources invested in egg production. In fact, in an evolutionary hangover, considerable egg wastage still occurs in mammals. Thus, a woman acquires all her eggs when she is herself a fetus, with numbers peaking at around 7 million at 6 months of her fetal life. Thereafter, most eggs die in the ovary during fetal, neonatal and pubertal life (Figure 1.2). Nonetheless, this programmed loss of follicles does conserve energy resources, because it happens before follicle growth has occurred.

Oviparity versus viviparity

Most reptiles and birds, like all mammals, also fertilize internally, but they then lay eggs that contain the developing conceptuses. These eggs, like those shed by externally fertilizing species, must be relatively large, because they have to carry sufficient energy resources to complete the development of young to the point at which they are capable of feeding. In contrast, all mammals are **viviparous**, producing smaller eggs that develop *in vivo* and giving birth to live young, except for the **oviparous** monotremes – the platypus and echidnas. Mammals have evolved not just **fewer eggs**, but also much **smaller eggs** (Figure 1.3).

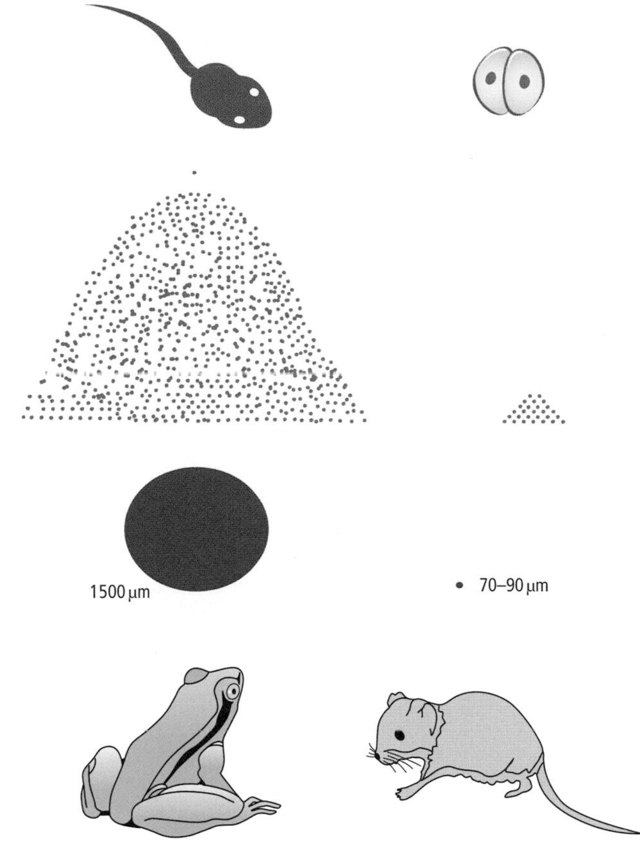

Figure 1.3 Cartoon to show differences between mammalian and frog eggs. From bottom up: frog and mouse; the relative sizes of each egg (diameters); the numbers recoverable from each at a single ovulation; the transition achieved in the first 24 hours after fertilization. Note that human eggs are of very similar size to mouse and elephant eggs, despite giving rise to very different-sized animals; all three are much smaller than frog eggs. This is because frog eggs must carry with them most of what they need to transform into swimming tadpoles that can then feed themselves, something that they do very rapidly! Mammalian eggs in contrast gain their nutrients for growth from the mother: largely through the placenta (see Chapters 14 and 15).

Viviparity also reduces the evolutionary pressure to develop as rapidly as possible, so as to gain the sensory awareness and movement capability helpful for escaping predation. So, in general, mammalian zygotes **develop more slowly** – perhaps most dramatically observed in the mammalian zygote taking 24 hours to divide to just two cells by which time a frog zygote has developed into a swimming tadpole!

A further consequence of viviparous reproduction is that the early growth of the next generation must be nourished within the female genital tract, which is accordingly adapted anatomically and functionally to support this growth. Thus, the **female tract has evolved a dual role** in mammals: it transports spermatozoa to the site of fertilization, and then nourishes the developing conceptus. This dual role imposes complex functional changes on the tract, in which the female genital tract cycles between a preovulatory phase suitable for transmitting sperm to the site of fertilization, and a postovulatory phase in which nurturing of the conceptus predominates, the subject of Chapters 10 and 11. Corresponding changes have evolved in the developing conceptus to optimize its nutrition. These include the development of **specialized membrane systems** and **placentae** for tapping into maternal nutrition in the uterus. These maternal–embryonic interactions are described in Chapters 14 and 15.

Viviparity involves relatively prolonged periods of **gestation or pregnancy**, which make major demands on the **pregnant** female, whose metabolism and physiology are modified to meet the needs of the developing **conceptus, embryo** and **fetus** (see Chapter 15). Pregnancy can often go wrong – a considerable selective cost to the species (see Chapters 16 and 17). Indeed, a major source of evolutionary selective pressure comes at around **parturition** or **birthing** (see Chapter 18).

Parental care

The universal feature unique to all mammals, and through which they are named, is the production of milk from 'mammae' or **nipples** to nurture the neonate (see Chapter 19). **Milk production** is just one aspect of the extended period of **parental care** shown by mammals – a further energy investment in just a few young (see Chapter 20). This parental care takes different forms depending on the relative maturity of the young at birth and the social organization of the species. In mammals that move around in herds, such as sheep and cattle, the young have evolved to be able to walk soon after birth. Where animals are territorial and have dens or nests, the young may be born naked and immature. Parental care is evident in all cases and takes different forms. In higher primates, like ourselves, the young are born very immature and depend for many years on parental and social support, a subject considered in Chapter 20.

Reproduction strategies: summary

Mammals have evolved a **high-investment, low-volume reproductive strategy**. Our reproduction involves sex, the selection of sexual partners, internal fertilization, viviparity and extended parental care. Mammalian eggs and conceptuses are smaller, fewer in number and develop more slowly than those of non-mammals, and have evolved specialized membrane systems for tapping into maternal nutrition. There is heavy parental investment in the relatively few offspring. Whilst these features characterize all eutherian and marsupial mammals, including humans, there is rich variety within the order *Mammalia*. Thus, rodents (such as mice and rats) go for higher volume and faster production of young than do ungulates (such as cattle and sheep). We have already alluded to the wide range of maturity at birth with its consequences for parental care patterns. It is important to keep in mind these variations in the details of reproductive strategy among different mammals, because animal models of reproduction are often used as surrogates for human reproductive enquiries. Even among higher primates there are important reproductive differences. Extrapolation of data and ideas across species can be helpful, but must be undertaken cautiously, and in this book we will be alert throughout to the dangers.

Having identified the reproductive pattern that characterizes mammals, we will now look in a little more detail at time and the reproductive life cycles.

Reproductive life cycles

We have seen that reproduction is central to our lives as mammals. Whether and with whom we mingle our gametes is deeply significant biologically and culturally. At the most reductionist level, each of us can be viewed as a vehicle for our gametes and the chromosomes they transmit to the next generation. It is this distinction between the **germ cells** and the **soma** (or body in which they develop) that is considered here.

The somatic and social life cycles

We are born physically and sexually immature. We then spend the first decade of our life growing and maturing physically and establishing an individual identity. Shortly thereafter, at adolescence, we mature sexually at **puberty** (see Chapter 4). By the early- to mid-teens, we achieve the capacity to produce fertile eggs or sperm (see Chapters 7–12) and, in women, to carry a pregnancy (see Chapters 13–18). This reproductive capacity, or **fecundity**, then characterizes much of our adult life. However, there are **distinct differences between men and women in their life-time fecundity** patterns. Male fecundity, once achieved, persists throughout life, albeit slowly tapering downwards with increasing age. Female fecundity, in contrast, is 'time limited', declining steeply from about 35 years until ending by the **menopause** at around age 50 years (Figure 1.4). This reduction in female fecundity is due to the **loss of quality eggs** (see Chapters 9, 22 and 23), and has been described as a 'public health problem' in Western societies in which many women are delaying having families until well into their 30s. Many do so in the erroneous belief that modern medicine can treat any infertility that emerges should they leave reproduction until later (see Chapter 23). The **social life cycle** has shifted to later in life, while the **somatic life cycle** remains as it always has, the ageing process proceeding apace.

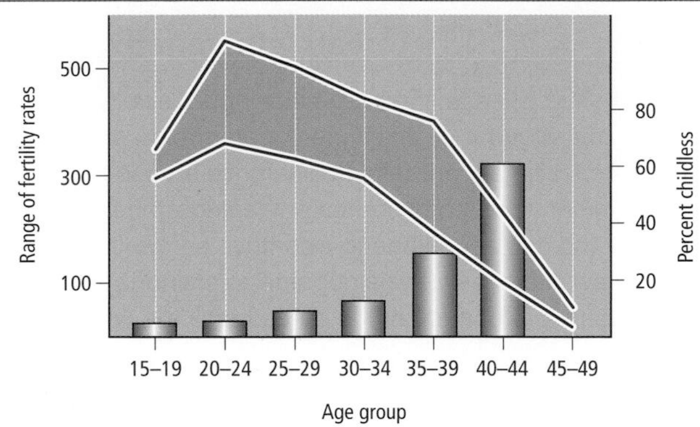

Figure 1.4 Female fecundity is 'time limited'. Rates of fertility (red range) and childlessness (blue bars) by age of woman. The fertility rate data were collected from populations of married women who reported that no efforts were made to limit their fertility. These data approximate to a measure of fecundity by age in humans. Note the steep decline from 35 years. (The range reflects the different circumstances of the populations, drawn from different parts of the world.) The histograms show the proportions of women remaining childless after first marriage at the ages indicated despite continuing attempts to deliver a child. Note again the sharp rise above 35 years, implying a fall in fecundity from this time onwards until the menopause at around age 50 years.

The generative life cycle

The **somatic life cycle** is about the physical transmission of the germ cells across generations, and so we now turn to the germ cells themselves and to the concept of the **generative life cycle** (Figure 1.5). The essential element that sets apart the germ cells from the somatic cells is their potential to give rise to the many tissues of the next generation: their so-called **pluripotency**. Thus, as the **somatic conceptus** develops and grows in size and cell number, so also its complexity increases as muscle, gut, nerve, blood and other types of somatic cell develop. Each cell is specialized for a particular function so that together cells make an effective soma for maximizing the chance of survival and reproductive success. The latter success depends, however, on the subpopulation of persisting **pluripotent cells** within the soma that form the **germ cell lineage**.

What exactly does pluripotent mean? The very early conceptus consists entirely of pluripotent **stem cells** capable of giving rise to a whole organism (Figure 1.5). As somatic cells of different types develop within the embryo, a subpopulation of more pluripotent stem cells can always be detected – diminishing proportionately in number to the developing soma, but traceable throughout until they enter the rudimentary **ovary** or **testis** as the **primordial germ cells** (Figure 1.5; see Chapter 3). These cells will go on to form the oocytes and spermatozoa, one of each of which then combine to provide a new conceptus of pluripotent stem cells, and so the cycle continues through another turn of the generational wheel. In their nature, cycles lack beginnings and ends – unless of course broken by a failure to reproduce.

The concept of this **cycle of pluripotentiality** is central to the generative cycle. Indeed, we can envisage the **somatic cells** – nerve, muscle, skin, gut, etc. – that make up most of our bodies, as being mere vehicles for transmission of our germ cells. So, what is it about this subset of pluripotential cells that makes them so special? After all, every cell in the body – whether a somatic or germ cell – has the same chromosomal and genetic composition (give or take a few atypical cells) and, moreover, all these genes are functionally competent. We know this because you can take a nucleus from an adult somatic cell and inject it into an enucleated oocyte, where it can then direct the formation of a new individual. This process of **somatic cell nuclear transfer (SCNT)** is also known as '**reproductive**

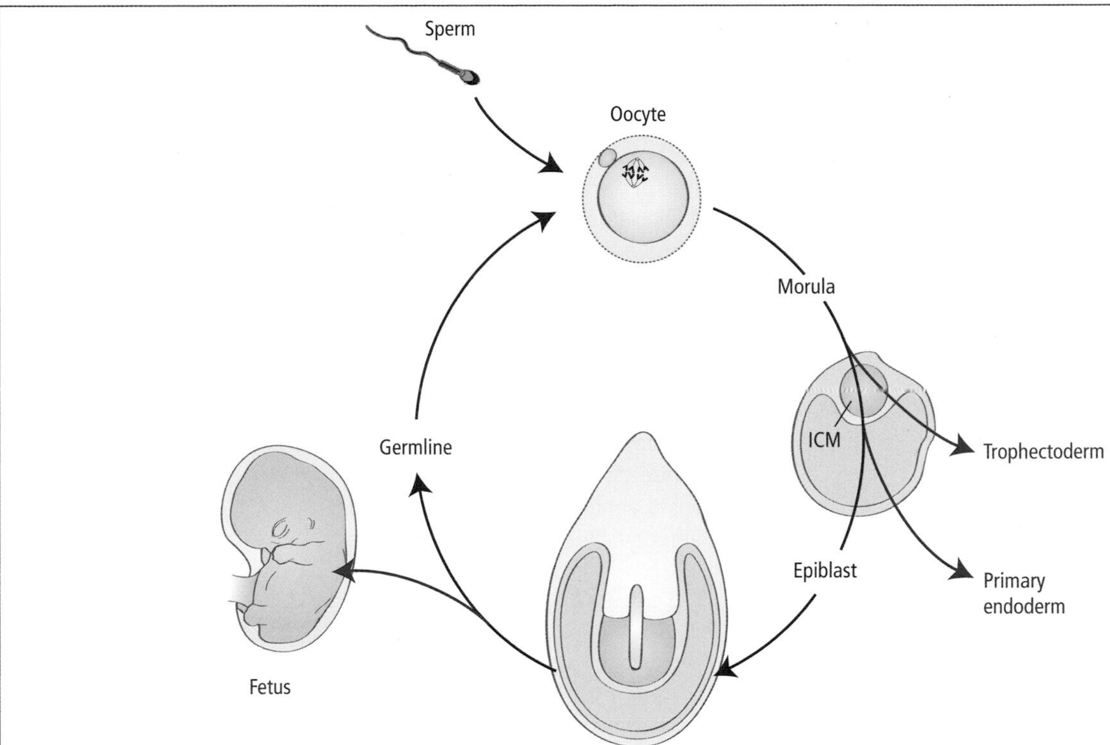

Figure 1.5 The generative life cycle is shown in red and indicates the pluripotent lineage, whereas the somatic life cycle is in blue and ensures the survival and physical transmission of the germ cells across generations. (Source: Adapted from an original idea by Monk (1981) *Differentiation* **19**, 71–76.)

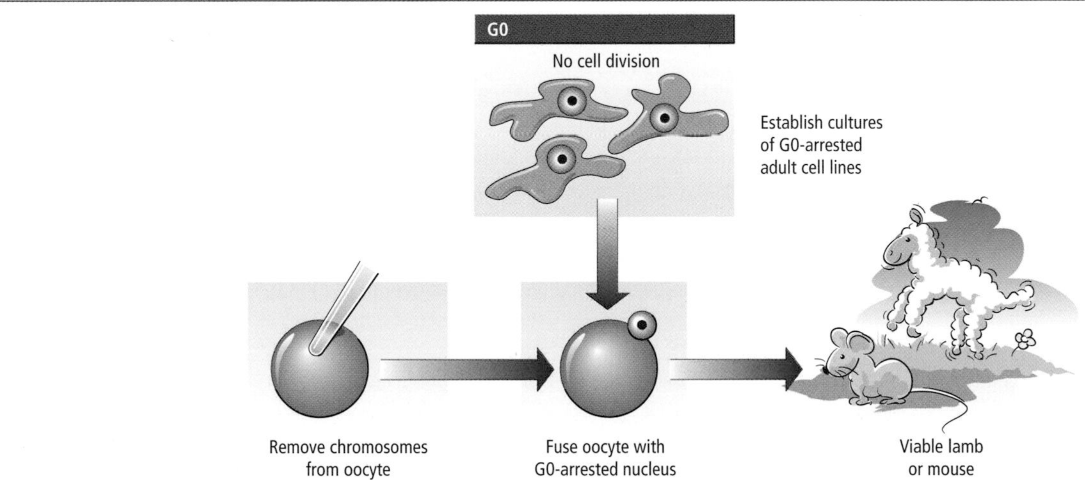

Figure 1.6 Schematic summary of the procedure for somatic cell nuclear transfer (SCNT) in sheep or mice. A differentiated cell is cultured and its division cycle arrested by removal of nutrients (G0 stage). A **karyoplast** (the nucleus with a small amount of cytoplasm and cell membrane surrounding it) is then prepared from the quiescent cell. It is placed next to an unfertilized oocyte from which its own genetic material has been removed by suction. A fusogenic signal is then given. The nucleus and enucleated oocyte fuse and initiate cleavage. The cleaving conceptus is placed into the mouse or ewe uterus and a viable offspring may result. Source: Stewart C. (1997) *Nature*, 385, 769. Adapted with permission from *Nature*.

cloning', and gave rise to the famous sheep, Dolly, who shared all her nuclear genes and chromosomes with the donor (Figure 1.6). Reproductive cloning has now been achieved in many species, including cattle, pigs, goats, cats, dogs, mice and monkeys, so there is every reason to believe it would work in humans too (see Chapters 22 and 23). It essentially defies nature by providing an asexual route to mammalian reproduction. However, it leaves us with a problem. Given that all cells have an identical genetic make-up, how do cells from the pluripotent lineage differ from somatic cells? The answer lies not in genes but in **epigenesis**.

The epigenetic cycle

It is estimated that humans have between 20000 and 25000 genes. Our genes, through their code of DNA triplets, encode proteins and it is proteins that largely make us who we are – **genotype encodes phenotype**. However, only a restricted subset of genes is expressed in any one cell, and that subset is characteristic for each cell type at particular times in its life cycle. Thus, muscle, nerve, skin and gut cells each express different combinations of genes. When these expressing gene sets are studied at the molecular level, they are found, as expected, to be identical in their gene sequences (or genetic codes) to the same genes in different, non-expressing tissues. However, they do differ in three ways:

1. They differ in the pattern of chemical modification to certain cytosine bases in the DNA, lacking methyl groups that are present on non-expressing genes: differences in the **DNA methylation patterns** (Figure 1.7a).
2. They are wrapped up in a distinctive subset of associated proteins called histones that give the chromatin a distinctive looser **euchromatin structure**.

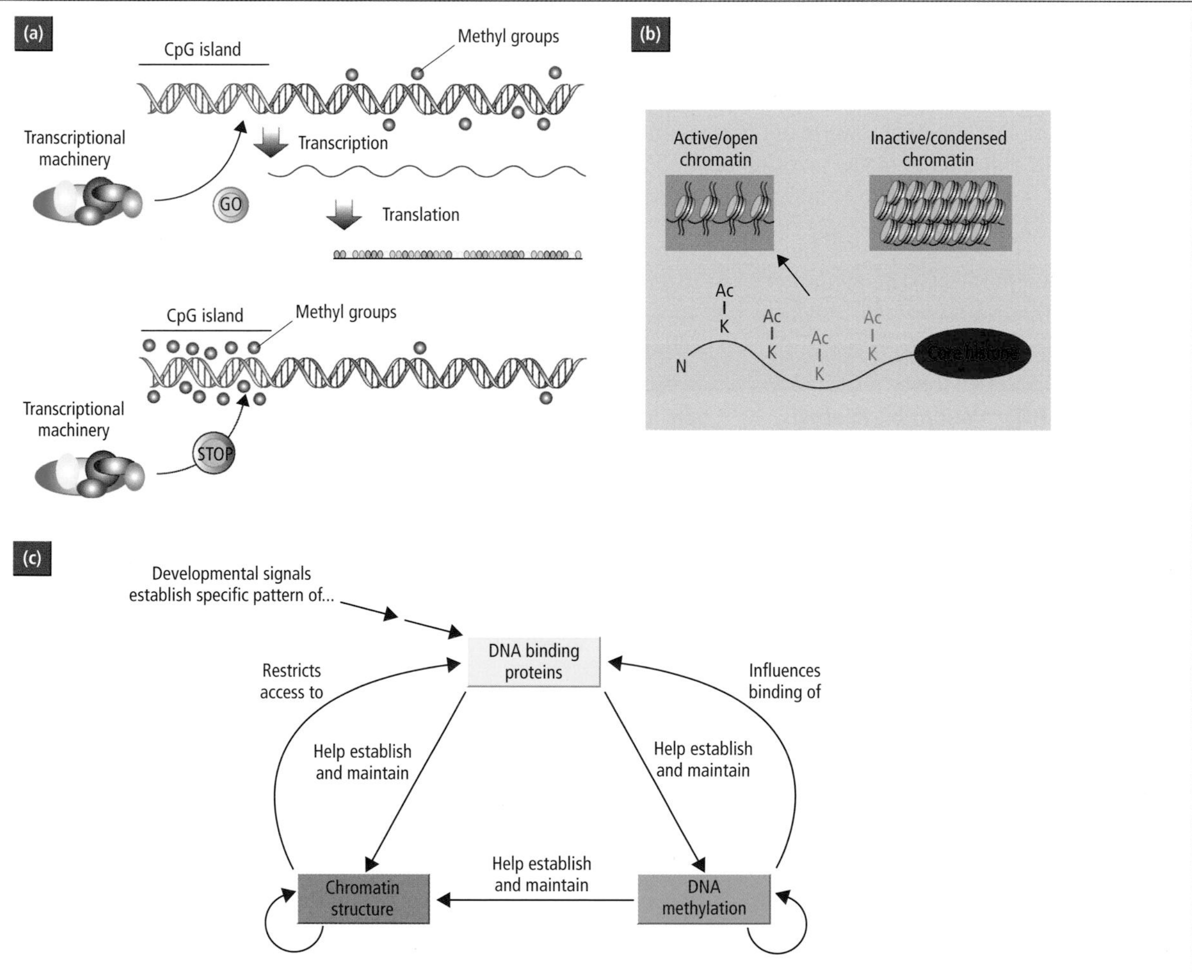

Figure 1.7 Summary of mechanisms available for genetic imprinting. Two sorts of epigenetic modification can leave an imprint. (a) The direct methylation of some, but not all, cytosines (see CpG island in lower part of panel) within the DNA sequence itself. This methylation then blocks access to transcriptional machinery – lower part of (a). Once initiated, this methylation pattern can be copied at each round of DNA replication as long as the maintenance methylase is present, and it is thus heritable through many mitoses. (b) A second sort of epigenetic modification is seen in the histone isotypes present in the chromatin surrounding the promoter region of the genes, as well as in their post-translational modification (e.g. Ac = acetylation). Depending on the chromatin structure, the DNA can be organized in a 'loose' or open state, and so is accessible to the transcriptional machinery and available for expression, or packed tightly and repressed. See also Box 7.3 for discussion of this type of modification during chromatin reorganization during spermatogenesis. (c) Quite complex interactions may occur during development between these two types of epigenetic modification and associated transcriptional proteins. However, these are early days in the science of epigenesis and much remains to be understood (see Chapters 16 and 20 for further discussion of the importance of epigenetic imprinting in health and disease).

3. These histones themselves show characteristic patterns of post-translational modification by acetylation, methylation, etc. (Figure 1.7b).

These three processes are each the result of **epigenetic influences** (from *epi* = outside of the genes), which are so called because they do not affect the genetic code itself (which would be a genetic change), but only the gene organization in ways that affect the **capacity of those genes to be expressed** (Figure 1.7c).

When the cells of the developing early conceptus are examined, the genes in all of its cells are shown to undergo a profound series of changes in their patterns of epigenetic modification. However, those cells that form part of the pluripotent germ cell lineage are characterized by quite distinctive epigenetic patterns from those in non-pluripotent somatic cells. It is clear that this distinctive epigenetic pattern underlies their pluripotency. How this distinctive **cycle of epigenetic patterning** is controlled is the subject of intense study. Understanding 'how' could have profound practical consequences. Thus, these pluripotent cells can now be isolated and persuaded to grow indefinitely *in vitro* as **embryonic stem cells (ESCs)**. ESCs, given their pluripotency, have medical promise as sources of repair for damaged tissues (see Chapters 22 and 23).

Within the developing embryo, these pluripotent stem cells give rise to the germ cells, and as they do so, most of their epigenetic marks are erased – the **epigenetic slate is wiped clean in the germ cells**. Then, during the packaging of the chromosomes in eggs or spermatozoa for transmission to the zygote, new epigenetic marks are placed on the genes. Curiously, the sex of the environment in which the chromosomes are packaged influences whether and how some 100–200 genes are marked. These marks then affect the gene's ability to become transcriptionally active subsequently in the conceptus. This process is called **parental imprinting**, because it leaves a sex-specific imprint on the genes, which is 'remembered' as having been paternally or maternally derived. These **maternal and paternal imprinting processes** mean that, although the oocyte and the spermatozoon each contribute one complete set of chromosomes and genes to the conceptus, each set is not on its own fully competent to direct a complete programme of development. Only when a set of genes from an oocyte is combined with a set of genes from a spermatozoon is a fully functional genetic blueprint achieved. A parthenogenetically activated oocyte lacks access to some crucial genetic information, which, although present in its chromosomes, cannot be accessed because of the maternal imprinting to which it was subjected during oogenesis. In the normal zygote, this information would be provided by genes on the paternally-derived set of chromosomes. **It is parental imprinting that compels us to reproduce sexually** and means that parthenogenesis is not possible in mammals (see also page 6).

Conclusions

This introductory chapter sets the scene and context for all that follows. It does so by setting out some of the key distinguishing features of reproduction in mammals and, in particular, humans. In the rest of this book, we will use human examples wherever possible. However, the advances of molecular genetics have emphasized our close evolutionary relationships to the whole living world and so, despite the clear evidence of the many idiosyncratic features of human reproduction, there are many useful reproductive lessons for us to learn from other species. These are lessons on which we will draw – but cautiously. Having considered the different generative cycles that make up the reproductive life cycles, we now introduce the reproductive body.

Key learning points

- Mammals reproduce sexually through the union of a haploid egg with a haploid spermatozoon.
- Sexual selection operates in mammals and is culturally influenced in humans.
- The mammalian female reproductive tract has a dual role.
- Fertilization is internal and involves spermatozoal transport and fewer eggs being shed.
- Development is viviparous and involves smaller eggs and conceptuses.
- Early development is slower and involves attachment to the mother's uterus that leads ultimately to placental formation.
- Birth marks the beginning of the period of parental care of neonates through milk production: the unique feature of all mammals.
- Overall, mammals have a high-investment, low-volume reproductive strategy.
- The somatic reproductive life cycle involves growth, sexual maturation at puberty, a period of fecundity and then reproductive decline.
- Reproductive decline in women is more marked than in men, occurring earlier, being steeper and resulting in complete loss of fecundity by the menopause.
- The generative life cycle involves the setting aside within the embryo of a population of pluripotent germ cells that develop into the gametes in the ovary and testis.
- The epigenetic life cycle describes the corresponding changes in chromatin and DNA modification that underlie pluripotency.
- Parental imprinting describes the epigenetic marks that identify some genes as being derived from the mother and some from the father, and which also ensure that those genes are not expressed in somatic cells.

Clinical vignette

Why is imprinting important?

A 3-year-old girl who was born by normal delivery at full term was noted to be significantly behind her peers in terms of language development. Her growth was reduced compared to her birth-weight and, at 3 years of age, she was not yet walking confidently. After careful assessment by a paediatrician, global developmental delay was diagnosed. She underwent a series of genetic and functional tests. These revealed a diagnosis of Angelman's syndrome. With specialist therapy, she made some progress in her motor development, but her parents were informed that she would always have severe learning difficulties. Her parents were considering a future pregnancy, but were unclear regarding the risks of a second child being similarly affected.

Angelman's syndrome is a disorder of epigenetic imprinting. The epigenome is a series of modifications to the structure of DNA that determine how it can be transcribed. The gene responsible for Angelman's syndrome is always expressed from the paternal allele, as the maternal allele carries a methyl group at a location that blocks expression. If a mutation is present in the paternal copy of the gene, a normal maternal allele cannot compensate. In this case, the chances of recurrence in a future pregnancy are 50%, because the father is assumed to have one normal copy and one mutated copy of the gene in question. Not only the sequences of genes, but also their expression patterns are vital in determining disease risk.

FURTHER READING

General reading

Bewley S, Davies M, Braude P (2005) Which career first? *British Medical Journal* **331**, 588–589.

Clarke AE (1998) *Disciplining Reproduction: Modernity, American Life Sciences, and 'the Problems of Sex'*. University of California Press, Berkeley, CA.

Darwin C (1871) *The Descent of Man, and Selection in Relation to Sex*. John Murray, London.

Hrady SB (2009) *Mothers and Others: The Evolutionary Origins of Mutual Understanding*. Harvard University Press, Cambridge, MA.

Potts M, Short RV (1999) *Ever since Adam and Eve: The Evolution of Human Sexuality*. Cambridge University Press, Cambridge.

Senner CE (2011) The role of DNA methylation in mammalian development. *Reproductive BioMedicine Online* **22**, 529–535.

Stewart C. (1997) Nuclear transplantation. An udder way of making lambs. *Nature*, **385**, 769.

More advanced reading

Monk M, Salpekar A (2001) Expression of imprinted genes in human preimplantation development. *Molecular Cellular Endocrinology* **183**, (Suppl. 1), S35–40.

Winston NJ, Pickering SJ, Johnson MH, Braude PR (1991) Parthenogenetic activation and development of fresh and aged human oocytes. *Fertility and Sterility* **56**, 904–912.

CHAPTER 2

The infrastructure of reproduction

Essential Reproduction, Eighth Edition. Martin H. Johnson.

Companion website: www.essentialreproduction.com

In this second introductory chapter, we move from a consideration of reproductive strategies and life cycles to consider the reproductive infrastructure. This infrastructure can be broadly divided into the **reproductive hardware** and **reproductive messengers**.

The reproductive hardware

At the core of sexual reproduction lie the creation and fusion of the two types of gamete: the female **oocyte** and the male **spermatozoon**. Their production occurs in distinctive female and male **gonads**: the **ovary** and **testis**. Thus, the human male (Figure 2.1) develops obvious **external genitalia** and a system of internal ducts and glands that conveys the spermatozoa in **seminal fluid** from the testis to the **penis** and thence reproductively to the **vagina** (see Chapters 8 and 11). The female (Figure 2.1) develops less prominent external genitalia (see Chapters 10 and 11), but has an internal system of ducts that accommodate the erect penis and its ejaculated spermatozoa and transport some of the latter through the **cervix** and **uterus** to the **oviduct** (or **fallopian tube**). The oviduct is the site of fertilization, but the fertilized oocyte must then pass back to the uterus to **implant** and **gestate** until delivery through the cervix and vagina (see Chapters 12–18). After birth, the woman's breasts have developed to provide milk for the neonate, whereas the man's breasts remain vestigial (see Chapter 19). Thus, the basic anatomical differences between men and women are rooted in their different reproductive roles.

In addition to the production of different gametes, each gonad also has a distinctive pattern of hormone production, notably of the **sex steroid hormones**. The sex steroids play key roles in both the development (see Chapter 3) and the functioning of the two sexes (Figure 2.1). Thus, they are critically involved in the acquisition of sexual maturity at **puberty** (see Chapter 4). They orchestrate the cyclic changes in fecundity in females (see Chapter 10), and the seasonal changes in fecundity in many male and female mammals (see Chapters 8 and 10). They play key roles in pregnancy and parturition, and in preparing for maternal lactation and parental care (see Chapters 13–19). They can also influence the behaviour of males and females: in many species to ensure that mating will only occur between different sexes at times of maximum fecundity (see Chapters 5 and 6).

However, the sex steroids do not act alone but in conjunction with a range of other hormones as well as with the nervous system. Key hormones amongst these are the **gonadotrophins and prolactin**, large proteinaceous hormones produced by the **pituitary gland**, and some smaller hormones produced by the **hypothalamus – oxytocin, gonadotrophin releasing hormone (GnRH)**, and **prolactin inhibitory factor (PIF)**. Before these various hormones and their actions are introduced, we first explore some key anatomical features of those parts of the nervous system of reproductive interest.

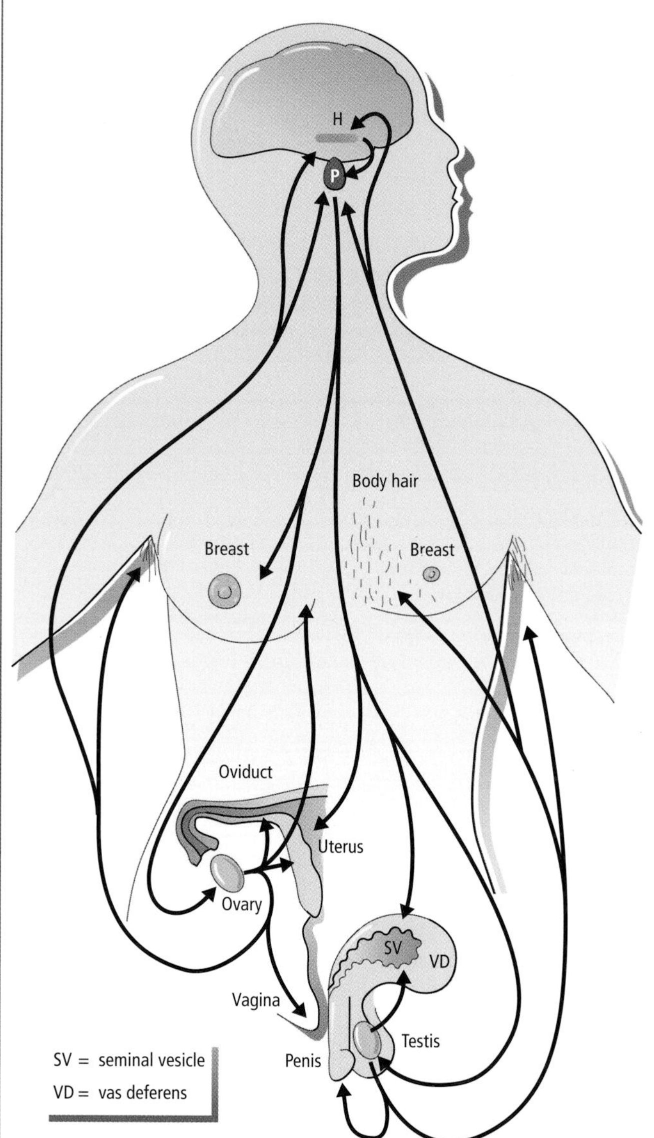

Figure 2.1 A schematic view of the main anatomical structures associated with the reproductive system. The hypothalamic region (H) sits at the base of the brain and is connected to the pituitary gland (P). The pituitary gland communicates hormonally with a range of reproductive organs and tissues including the breasts (see Chapter 19), the uterus and cervix (see Chapters 9, 10 and 18), and the gonads: testis and ovary (see Chapters 4, 7 and 8). The gonads themselves in turn communicate hormonally with the pituitary and the brain (see Chapters 8 and 10 for details). The gonads also influence hormonally multiple sites in the internal and external genitalia (see Chapters 3, 4, 8, 10, 18 and 19).

The brain, hypothalamus and pituitary

In higher primates the central nervous system (CNS), and particularly the **neocortex**, play an important role in reproduction. It does so largely via a much older part of the brain called the **hypothalamus**, which mediates a range of hormonal and environmental influences on reproduction. The hypothalamus in turn acts mostly via the **pituitary gland**.

The pituitary

The pituitary gland, or **hypophysis**, lies at the base of the brain to which it is attached by the **pituitary stalk**, or **infundibulum** (Figure 2.2). It has two main lobes in humans (Figure 2.3). A posterior lobe derived embryologically from a brain outgrowth (the **neurohypophysis**), and a larger anterior lobe derived embryologically from the dorsal pharynx (the **adenohypophysis**). The infundibulum connects the pituitary anatomically and functionally to the overlying hypothalamus in the region of the **median eminence** (Figure 2.3). The adenohypophysis contains several types of hormone-secreting cells, the two of most reproductive importance being the **gonadotrophs** (site of **gonadotrophin** production) and **lactotrophs** (site of **prolactin** production), whilst the neurohypophysis, being nervous in origin, secretes small neuropeptide hormones such as **oxytocin**.

The hypothalamus

The hypothalamus is a relatively small but complex region at the base of the brain, lying between the midbrain and the forebrain (Figure 2.3). Its boundaries are ill-defined, conventionally being described as: (1) superiorly the **hypothalamic sulcus** separating it from the **thalamus**; (2) anteriorly the **lamina terminalis**; and (3) posteriorly a vertical plane immediately behind the **mammillary bodies** (Figure 2.3). It is split symmetrically into left and right halves in the midline by the **third ventricle**, containing cerebrospinal fluid, such that the ventricle's floor and walls are formed by the hypothalamus (Figure 2.4). The hypothalamus has many different functions, only some of which are reproductive. Each function is associated with various **hypothalamic areas** or **nuclei** (Figures 2.3 and 2.4), those particularly concerned with reproductive functions being the **supraoptic, paraventricular, arcuate**, **ventromedial** and **suprachiasmatic nuclei**, and also two less easily defined areas, the **medial anterior hypothalamic** and **medial preoptic areas**.

The hypothalamic connections with the pituitary

The hypothalamic nuclei and areas have **either direct neural or indirect vascular connections with the pituitary gland**. Large neurons (the so-called **magnocellular neurosecretory system**) are located in the supraoptic and paraventricular nuclei (Figures 2.3:green fibres, and Figure 2.5a), and are the site of synthesis of the small peptide hormone, **oxytocin** (plus a second neuropeptide, **vasopressin**). Each hormone is synthesized in a distinctive subset of neurons, and passes along axons projecting from their cell bodies directly to the posterior lobe of the pituitary via the **hypothalamo–hypophyseal tract**, to be stored in the posterior lobe (see Figure 2.12c), before release into the bloodstream.

The anterior pituitary, in contrast, receives its hypothalamic input indirectly by a vascular route (Figure 2.4). A variety of small neuropeptide hormones, including **GnRH**, is synthesized in the **parvocellular** (i.e. small-celled) **neurosecretory system** of the hypothalamus. In primates, parvocellular neurons are clustered mainly in the arcuate nuclei (Figures 2.3: blue fibres, and 2.5b), whilst others are more diffusely located in the medial preoptic and anterior hypothalamic areas (Figures 2.4 and 2.5b).

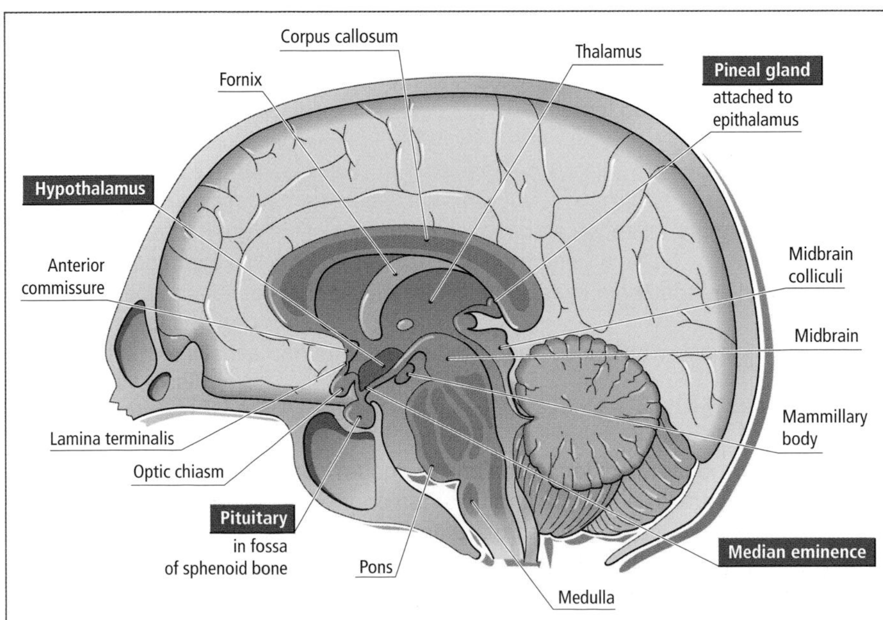

Figure 2.2 A sagittal section through the human brain with the pituitary and pineal glands attached. Note the comparatively small size of the hypothalamus and its rather compressed dimensions ventrally. The pineal is attached by its stalk to the epithalamus (habenula region) and lies above the midbrain colliculi. The third ventricle is a midline slit-like structure, which has been opened up by the midline cut exposing the medial surface of the brain. The thalamus (above) and the hypothalamus (below) form one wall (the right in this view) of the third ventricle, which are better viewed in Figure 2.4.

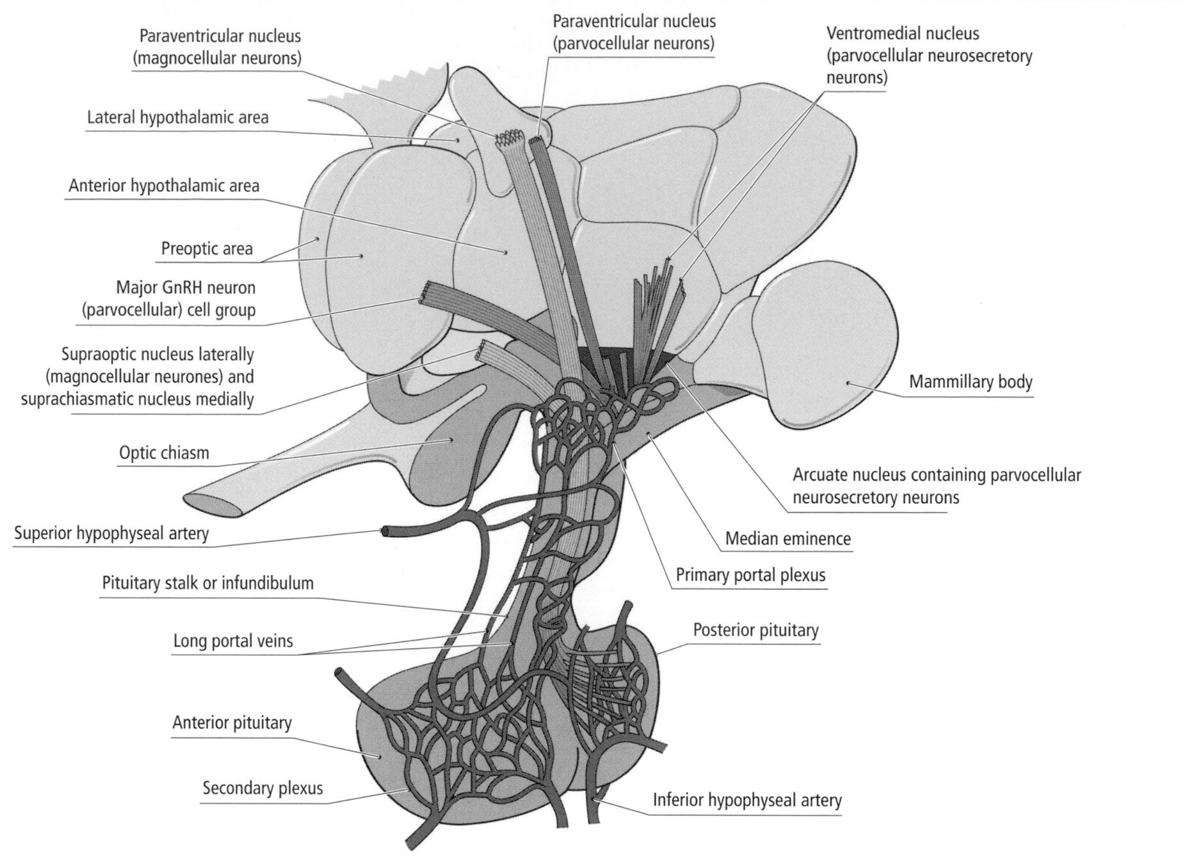

Figure 2.3 A highly schematic and enlarged view of the human hypothalamus and pituitary. Note the portal capillary system (red) derived from the superior hypophyseal artery and running from the median eminence/arcuate nucleus region of the hypothalamus above to the anterior lobe of the pituitary below. The anatomically and functionally well-defined paraventricular, ventromedial and arcuate nuclei contain the cell bodies of parvocellular neurons whose axons (blue) terminate in close association with the portal capillaries. Parvocellular neurons also arise in the less well-defined anterior hypothalamic/preoptic areas continuum and send axons to the portal plexus. Magnocellular neurons (green) are also located in the paraventricular nuclei, with a second group in the supraoptic nuclei, and send axons along the infundibulum to the posterior pituitary. (Source: Heimer L. (1983) *The Human Brain and Spinal Cord*. Reproduced with permission from Springer-Verlag, Berlin.)

Axons of these neurons project to the median eminence (Figure 2.3:blue fibres), where they terminate in the pericapillary space of the **primary portal plexus**. These vessels pass from the hypothalamus to the anterior pituitary, carrying the released neuropeptides in the **portal blood** where they act on the gonadotrophs.

The lactotrophs are also regulated via the portal plexus but quite differently from the gonadotrophs. Thus, unlike other pituitary hormones, prolactin is secreted spontaneously in large amounts when the vascular links between the pituitary and hypothalamus are **disconnected**. This observation means that regulation of secretion is mainly by **inhibition**, and has led to the search for the hypothalamic **prolactin inhibitory factor** (**PIF**). PIF is the catecholamine **dopamine**, and is found in neurons of the arcuate nucleus, the axons of which project to the portal capillaries in the medial and lateral palisade zones of the external layer of the median eminence (Figures 2.3: blue fibres, and 2.6; Box 2.3). It is secreted into the portal blood from the terminals of this **tuberoinfundibular dopamine (TIDA) system** and carried to the lactotrophs, where it depresses prolactin secretion.

Having described the main anatomical and physiological features concerned with reproduction, we can now examine the nature of the key hormones in more detail.

Reproductive messengers

In this volume, you will be introduced to many reproductive messengers, but here we focus on a few key hormones. We start with the sex steroids, introduced above (see page 14), which are the central players in reproduction. Here, a more detailed account of this family is given.

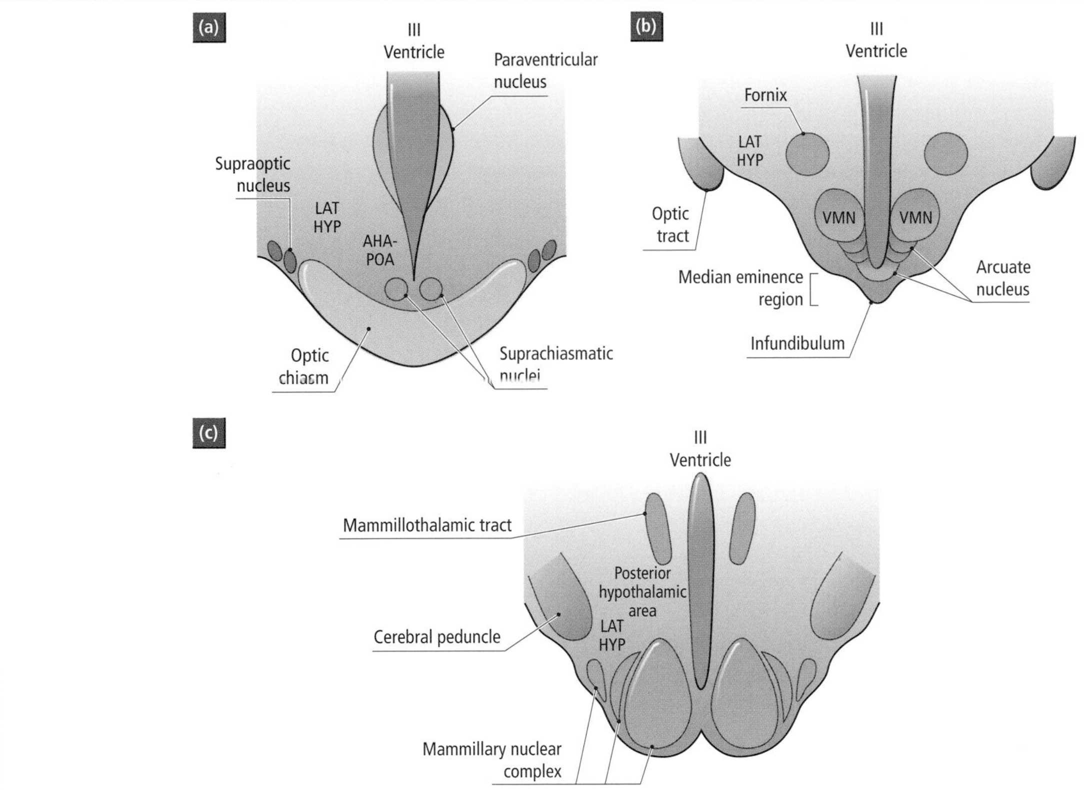

Figure 2.4 Three coronal sections at different anterior–posterior levels of the human hypothalamus (consult Figure 2.3 to construct the planes of sections). (a) Through the optic chiasm, note the third ventricle in the midline flanked by the anterior paraventricular (magnocellular) nuclei, which, together with the laterally placed supraoptic nuclei, synthesize oxytocin and vasopressin. The latter are then transported along the hypothalamo–hypophyseal tract (axons of neurons with cell bodies in these nuclei) to the posterior pituitary. The region of the suprachiasmatic nuclei and the anterior hypothalamic–preoptic area (AHA–POA) are also shown. (b) Through the infundibulum and median eminence, and showing the relationship between the parvocellular arcuate and ventromedial nuclei (VMN). The capillary loops of the portal plexus are found in this region. (c) Through the level of the mammillary bodies and showing the mammillary nuclear complex. The area labelled LAT HYP in all three sections is the lateral hypothalamus, and is composed of many nerve fibres ascending (largely aminergic) from the brainstem and descending from the rostral limbic and olfactory areas. This pathway represents a major input/output system for the more medially placed hypothalamic nuclei.

The sex steroids

These steroids make up a large group of molecules all derived from a common sterol precursor: **cholesterol** (Figure 2.7). There are four main families of steroid: the **progestagens**, **androgens**, **oestrogens** (American spelling estrogens) and **corticosteroids**, of which only the first three are identified as **sex steroids**. In **steroidogenic** tissues, most steroid is synthesized from acetate with cholesterol as an intermediate product (Figure 2.7). The **steroid biosynthetic pathway** is illustrated in Figure 2.8, which also provides a visual framework of the molecular relationships of the different steroid family members. The conversion of cholesterol to **pregnenolone** is the first and rate-limiting step in steroid synthesis, and so an important point of regulation. Although there are three distinctive classes of sex steroid, they are related structurally to each other. Indeed, they can be seen as different generations of a biosynthetic family, the progestagens being 'grandparental' and the androgens being 'parental' to the oestrogens (Figure 2.8). Interconversion from one class of steroid to another is undertaken by a series of enzymes arranged together as a **biosynthetic unit**, taking in substrate and passing the molecule along a production line with little 'leakage' of any intermediates. For example, the enzymes 17α-hydroxylase, 17,20-desmolase, 17-ketosteroid reductase and 3β-hydroxy steroid dehydrogenase would form an enzyme package for the synthesis of testosterone from pregnenolone. This close relationship between the different classes of steroids means that an enzymatic defect at one point in the synthetic pathway may have far-reaching effects, as will be seen in Chapters 3 and 4.

Within each class of sex steroid, there are several natural members. The criteria for membership of each class are

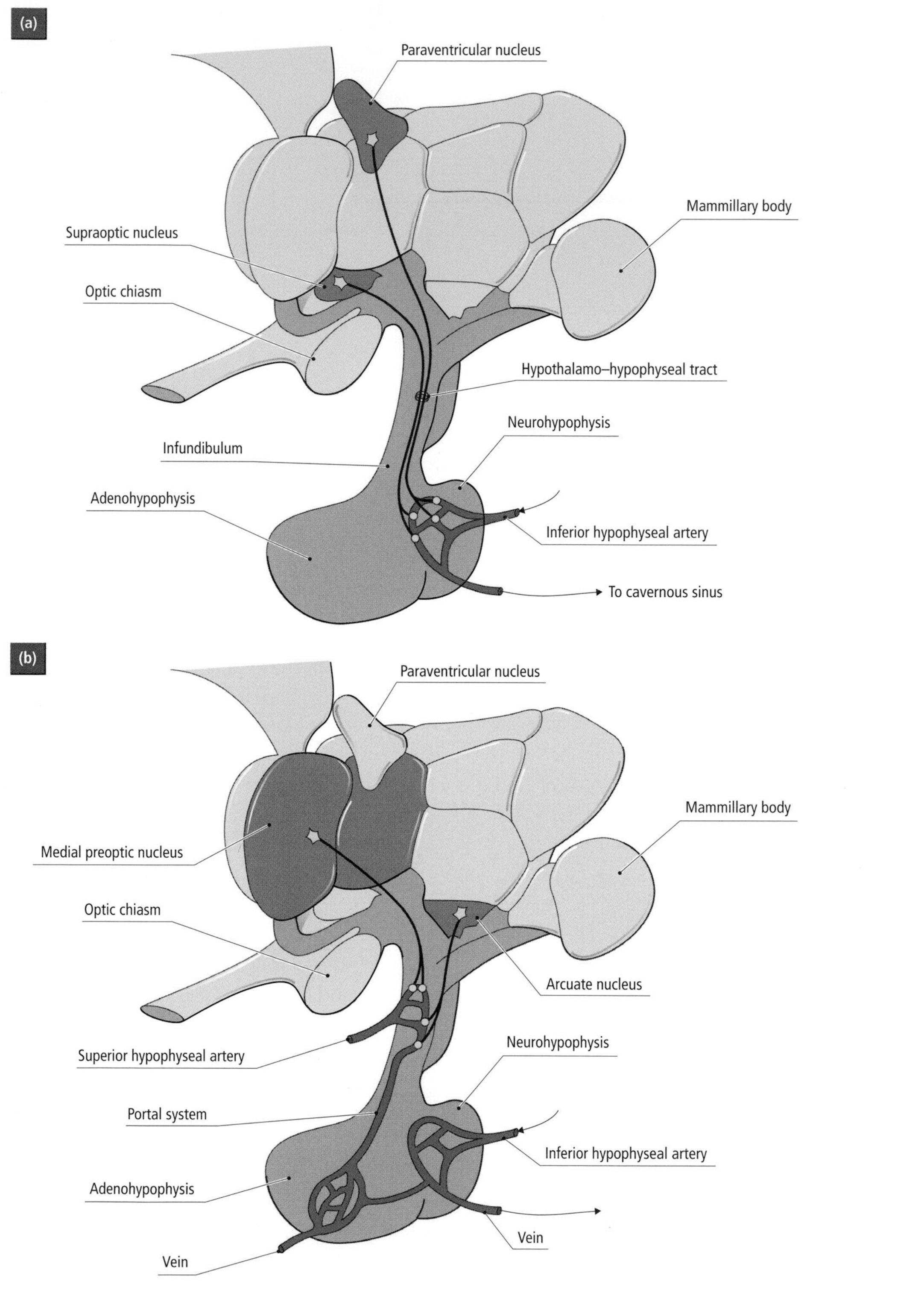

Figure 2.5 (a) Schematic representation of the magnocellular neurosecretory system. Neurons in the supraoptic and paraventricular nuclei send their axons via the hypothalamo–hypophyseal tract and infundibulum to the neurohypophysis, where terminals lie in association with capillary walls, the site of neurosecretion. (b) Schematic representation of the parvocellular GnRH neurosecretory system. Neurons in the medial preoptic and anterior hypothalamic areas, and in the arcuate nucleus send axons down to the portal vessels in the external layer (palisade zone) of the median eminence, where neurosecretion occurs.

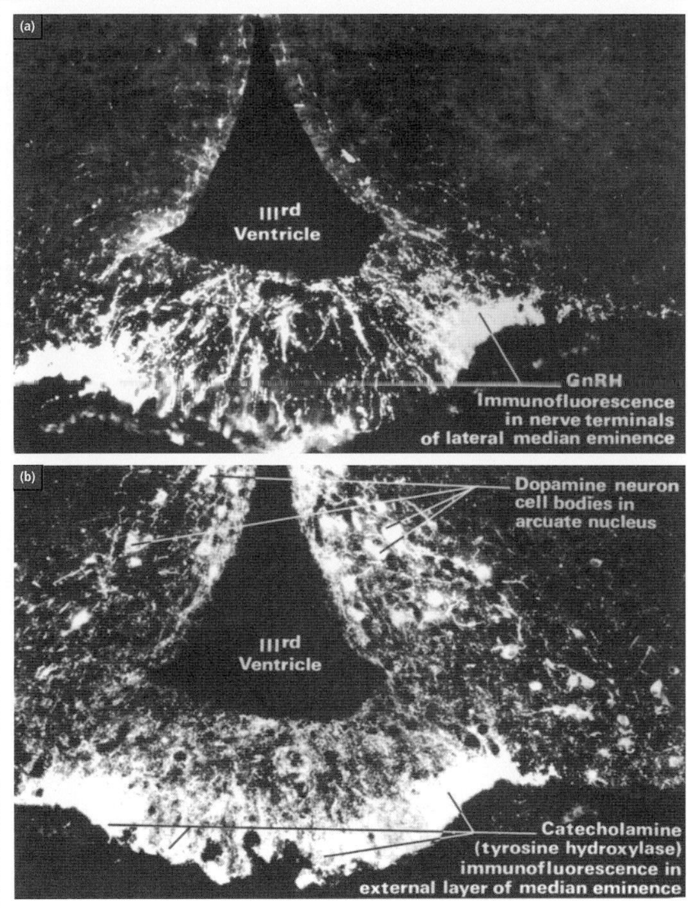

Figure 2.6 (a) GnRH detected by immunofluorescence in the median eminence (middle region). Note the very dense network of GnRH neuron terminals in the lateral region (lateral palisade zone), but no fluorescent cell bodies are visible. (b) Tyrosine hydroxylase immunofluorescence in an adjacent section through the median eminence. The dense terminal fluorescence in the external layer largely represents dopamine (although noradrenaline-containing terminals will also be labelled by this procedure). Large dopamine-containing cell bodies are clearly visible in the arcuate nucleus. (Source: Courtesy of Professor Tomas Hökfelt, Karolinska Institutet, Stockholm, Sweden.)

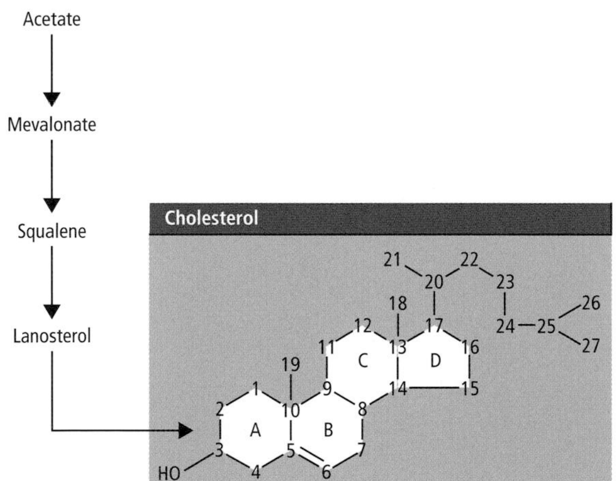

Figure 2.7 Basic structure of the cholesterol molecule. Each of the 27 carbon atoms is assigned a number and each ring a letter. The individual carbon atoms are simply carrying hydrogen atoms unless otherwise indicated (e.g. C-1 and C-19 are $-CH_2-$ and $-CH_3$ residues, whereas C-3 is –CHOH–). Cholesterol is converted to pregnenolone by cleavage of the terminal six carbons, leaving a steroid nucleus of 21 carbons. The conversion occurs on the inner mitochondrial membrane and requires NADPH, oxygen and cytochrome P-450. Pregnenolone is then converted to the sex steroids in the adjacent smooth endoplasmic reticulum.

similarity of chemical structure that then reflects their common functional properties. Tables 2.1–2.3 summarize the principal natural progestagens, androgens and oestrogens, together with some of the functional characteristics of each class. It is often said that progestagens are associated with the preparations for pregnancy and its maintenance, androgens with the development and maintenance of male characteristics and fecundity, and oestrogens with the development and maintenance of female characteristics and fecundity. Although broadly correct, this statement is a simplification, and a number of exceptions will be encountered, e.g. androgens stimulate secondary sex patterns and behaviour in females (see Chapters 5 and 6) and oestrogens stimulate epididymal function in males (see Chapter 11).

Sex steroid receptors

Whether or not a particular steroid affects a tissue, and how it affects it, depend mainly on whether that tissue expresses a steroid **receptor**. The interaction between the two, like a key in a lock, then leads to secondary changes in the receptor (or in a molecule adjacent to it) that alert the cell to the arrival of the hormone. Steroids, being lipid soluble, can pass freely *into* a target cell nucleus to combine with an **intranucleoplasmic receptor**, thereby activating it (Figure 2.9). Activation involves

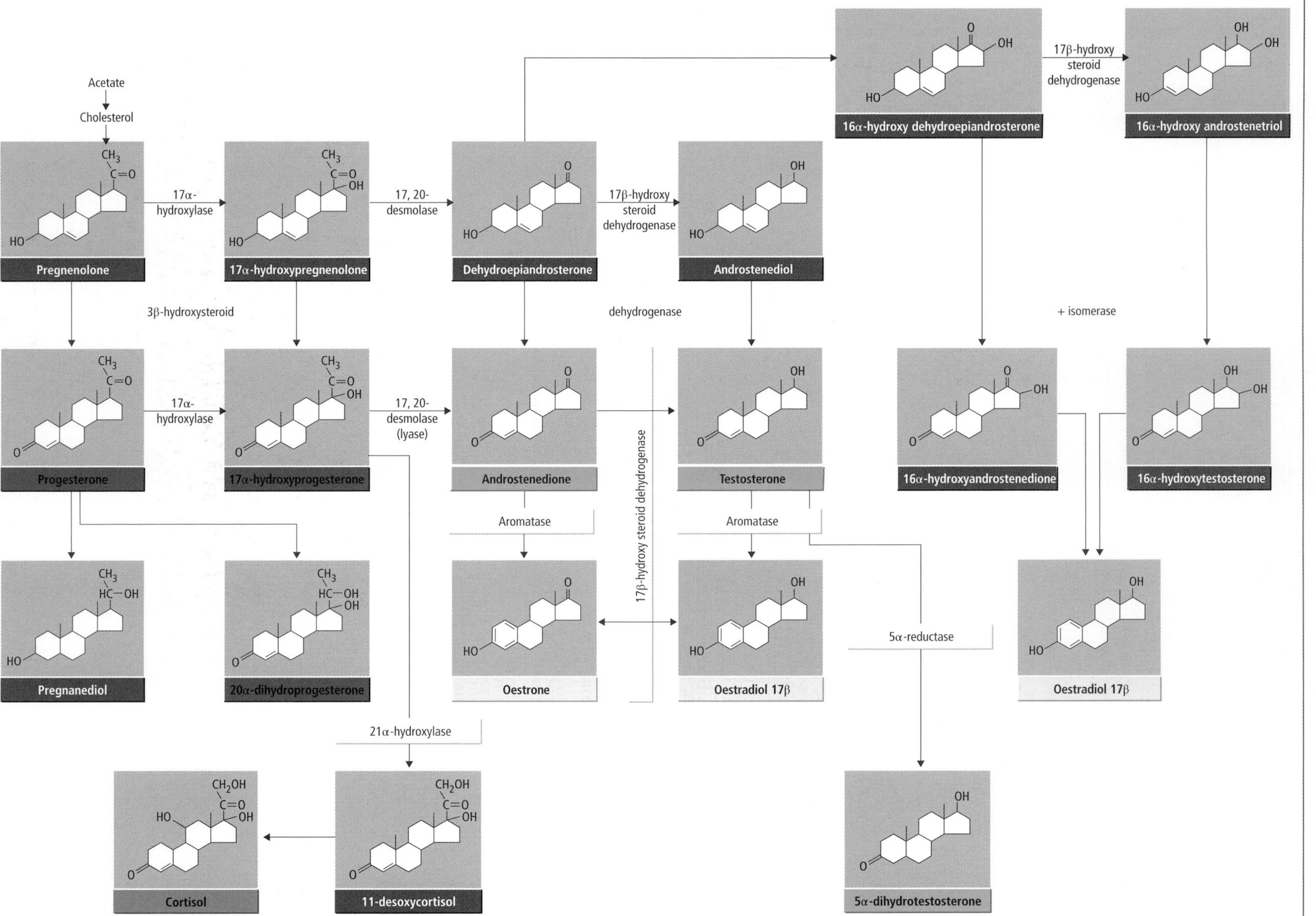

Figure 2.8 Pathway of interconversion of steroids. Some of the principal enzymes involved are indicated, although the scheme is very simplified; many of them have multiple isoforms. Note that extensive interconversions are possible; however, in any given tissue, absence of certain enzymes will mean that only parts of the matrix will be completed. Clearly, the enzymes present will therefore determine which steroids can and cannot be made from which available substrates. The members of the four major classes of sex steroids are shown with distinct colour bars: progestagens, red; androgens, green; oestrogens, yellow; and corticosteroids, blue. Structurally, natural progestagens are characterized by 21 carbons (C_{21} steroids), a double bond between C4 and C5, a β-acetyl at C17 and a β-methyl at C13. Natural androgens are characterized by 19 carbons (C19 steroids); strong androgenic activity is associated with a β-hydroxylated C17 and a ketone structure (–C=O) at C3. Natural oestrogens have 18 carbons (C18 steroids), an aromatized A ring hydroxylated at C3, and a β-hydroxyl group at C17. Note that in steroid terminology the suffixes denote the following: -ol = hydroxyl group; -diol = two hydroxyl groups; -one = ketone group; -dione = two ketone groups. An unsaturated –C–C-link is indicated by -ene; two such links are indicated by -diene. Note also that aromatase denotes 19-hydroxysteroid dehydrogenase + C-10,19 lyase.

Table 2.1 Principal properties of natural progestagens and their receptors*

Progestagens	Relative potency (%)	Key properties	Receptors
Progesterone (P4) 17α-Hydroxyprogesterone (17α-OHP) 20α-Hydroxyprogesterone (20α-dihydroprogesterone or 20α-OHP)	100 40–70 5	1. Prepare uterus to receive conceptus 2. Maintain uterus during pregnancy 3. Stimulate growth of mammary glands but suppress secretion of milk 4. Mild effect on sodium loss via distal convoluted tubule of kidney 5. General mild catabolic effect 6. Regulate secretion of gonadotrophins	Two isoforms PR-A and -B, sharing 780 identical amino acids, but with PR-B having an additional 164 amino acids. Each receptor, on binding P4, transcriptionally activates different genes by binding to sequences in their promoters. The traditional PR is also implicated in the rapid non-transcriptional pathway, via EGF receptor phosphorylation.

* In Tables 2.1–2.3 the relative potencies are only approximate since they vary with species and with the assay used. This variation is due partly to differences in the relative affinity of receptors in different tissues, partly to differences in local enzymic conversions of steroids within tissues and partly to differences in systemic metabolism: see text for discussion of these factors.
Common abbreviations or alternative names encountered in the literature are also recorded in Tables 2.1–2.3.

Table 2.2 Principal properties of natural androgens*

Androgens	Relative potency (%)	Key properties	Receptors
5α-dihydrotestosterone (DHT) Testosterone (T) Androstenedione (A4) Dehydroepiandrosterone (DHEA)	100 50 8 4	1. Induce and maintain differentiation of male somatic tissues 2. Induce secondary sex characters of males (deep voice, body hair, penile growth) and body hair of females 3. Induce and maintain some secondary sex characters of males (accessory sex organs) 4. Support spermatogenesis 5. Influence sexual and aggressive behaviour in males and females 6. Promote protein anabolism, somatic growth and ossification 7. Regulate secretion of gonadotrophins (testosterone) 8. Anticorticosteroid effects (DHEA)	A single androgen receptor isoform, AR, of 918 amino acids, encoded on the X chromosome. However, the gene is highly polymorphic and shows variation in the number of CAG codon repeats in exon 1, the number of which is inversely correlated with both levels of transcriptional activity and receptor expression. Thus, isoforms with shorter repeats are more sensitive to ambient androgen levels. The traditional AR is also thought to act as a surface receptor for a rapid non-classical response to androgens, where it acts by phosphorylating the EGF receptor. However, a G-protein coupled receptor [GPRC6A] has also been implicated in the rapid response to androgens.

* As for Table 2.1.

Table 2.3 Principal properties of natural oestrogens*

Oestrogens	Relative potency (%)	Key properties	Receptors
Oestradiol 17β (estradiol or E_2) Oestriol (estriol or E_3) Oestrone (estrone or E_1)	100 10 1	1. Stimulate secondary sex characters of female. Prepare uterus for spermatozoal transport 2. Increase vascular permeability and tissue oedema 3. Stimulate growth and activity of mammary gland and endometrium 4. Prepare endometrium for progestagen action 5. Mildly anabolic; stimulate calcification 6. Active during pregnancy 7. Regulate secretion of gonadotrophins 8. Associated with sexual behaviour in some species	Two receptors exist, ERα (c.595 amino acids) and ERβ, which has several isoforms arising from splice variants giving a size range of c.480–530 amino acids. Both receptors (but especially ERα) have been implicated in the non-classical pathway, but alternative putative receptors, namely Gq-mER and GPR30/GPER have been described.

* As for Table 2.1.

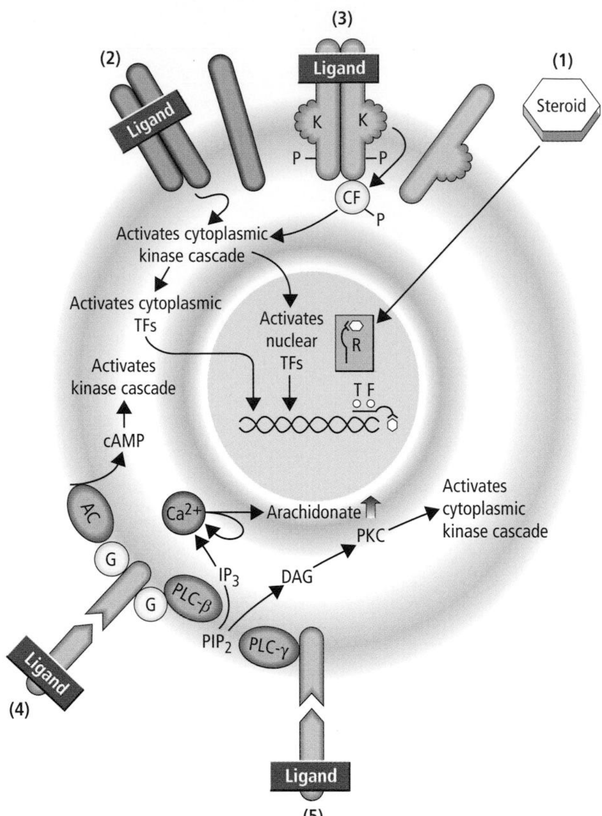

Figure 2.9 Receptor activation and second messengers. Ligands can influence cell function, and ultimately nuclear gene expression, in a variety of ways. Here we summarize in a very simplified form a very complex and rapidly evolving field of study. In doing so we refer to some ligands that will appear later in the text. (1) Steroid hormones, being lipid soluble, pass freely into cell nuclei where they bind receptors (R), displacing associated stabilizing proteins, such as HSP90, leading to receptor activation by phosphorylation; the activated complex can then bind to steroid-specific response elements (SREs) in the DNA and to transcription factors (TF) to activate steroid-specific genes. (2) Some ligands (prolactin, growth hormone, placental lactogen, LIF, leptin, TNFα) cross-link two receptor chains (as hetero- or homo-dimers). The cross-linked complex then activates the Jak-Stat cytosolic kinase cascade, resulting in either the phosphorylation of TFs in the cytoplasm and their translocation to the nucleus, or the translocation of the kinases themselves to the nucleus where they phosphorylate and activate TFs. Hormonal ligands that act in this way can cross-link overlapping spectra of receptor monomers to induce overlapping spectra of downstream cascades, so accounting for some of the redundancy seen amongst peptide ligands. (3) Other ligands, including growth factors of the EGF (EGF and HB-EGF), insulin (insulin, IGF 1 and 2, relaxin) and TGFβ 9 (Amh, activin, inhibin and BMPs) families bind to, and multimerize, receptors that **are themselves kinases** (K). Multimerization leads to kinase activation and both autophosphorylation of the receptor (–P) and phosphorylation of cytoplasmic factors (CF–P), and thereby to a cascade of kinase activity culminating in TF activation. Again, there is some overlap of downstream kinases and their targets both within this class of ligand–receptor interaction and between this class and class 2 above. The EGF and insulin family receptors are tyrosine kinases and are dimerized on ligand binding. Their activation then triggers numerous downstream second messenger pathways including phospholipase Cγ, ras/MAP kinases, and multiple STAT isoforms. The TGFβ family members form a complex with two type II receptors and then recruit two type I receptors, all four receptors having serine/threonine kinase activity. The activated kinases phosphorylate the receptors themselves and the intracellular signalling Smad proteins. The activated Smads translocate to the nucleus to bind with co-factors to response elements in cytokine target gene promoters. (4) Some ligands (LH, FSH, CG, GnRH, oxytocin, arginine vasopressin) bind to receptors which then associate with G proteins (G). G proteins can then act in at least two ways: (i) to stimulate phospholipase Cβ (PLC-β) to hydrolyze phosphatidylinositol phosphate (PIP_2) to 1,4,5-triphosphate (IP_3) and diacylglycerol (DAG), which release Ca^{2+} and activate protein kinase C (PKC) respectively; or (ii) to modulate activity of adenyl cyclase (AC) and so the output of cAMP. (5) Some ligands (EGF) may activate phospholipase Cγ (PLC-γ) directly, without G-protein mediation.

phosphorylation of the receptor and a conformational change. This steroid–receptor complex, but not the steroid or receptor alone, can then bind to specific DNA sequences in the chromatin: the so-called acceptor sites or **steroid response elements (SREs)** specific for each steroid (e.g. ARE, PRE and ERE for androgens, progestagens and oestrogens, respectively). Binding of adjacent transcription factors also occurs, leading to a rapid rise in the activity of RNA polymerase II, production of mRNA species specific for the steroid and, in the continuing presence of the complex, a more general stimulation of nucleolar and transfer RNA synthesis. Steroid action through this so-called '**classical' pathway** has recently been supplemented by evidence of a **non-classical pathway**, in which the steroid action is transduced by a receptor present in the plasma membrane. In general, the intranuclear receptors seem to be involved in mediating the surface actions of steroid signalling, although other distinct receptors may also be involved. The use of the same receptors fits with the observation that the rapid signalling pathway tends to function via a similar intracellular signalling system to the slower classical pathway and so provides an early response either priming or restraining the classical pathway.

Since steroid–receptor interaction depends on a good stereochemical fit between the two molecules, it is not surprising that the general three-dimensional molecular structure of steroid families correlates with their general biological activities. In general, the better the fit, the stronger the effect: hence the explanation for there being '**strong' oestrogens** such as **oestradiol 17β**, and '**weak' oestrogens** such as **oestriol** and **oestrone** (Table 2.3). However, some steroids may bind to a given receptor, but be unable to activate it. This can happen if one part of the steroid is involved in receptor binding, but a second structurally and spatially distinct domain is required for receptor activation. If a molecule is generated that retains only the first domain, it can occupy the receptor without activating it. In so doing, it may prevent other steroids from binding and thus act as an **antagonist**. In this way, high levels of progesterone will compete with dihydrotestosterone to bind the androgen receptor, and so function as an **antiandrogen**.

Pharmacologists have taken advantage of both the blocking and mimicking effects that can operate through receptors to develop drugs for therapeutic use, such as **controlling fertility** or **hormone-dependent tumour growth**. Some that are in clinical or experimental use are summarized in Box 2.1. In addition, some synthetic chemicals (food additives, plastics, pesticides) released into the food chain can bind to hormone receptors and thereby block or activate them inappropriately: so-called **endocrine disruptors**. Most attention has focused on developmental reproductive abnormalities in male fetuses, such as **cryptorchidism** (failure of testis descent) and **hypospadia** (urethral opening not at penis tip), but acute exposure of adult men is also claimed to affect fertility. For example, a number of **pesticides** with steroidal properties have been associated with **lowered sperm counts**, increases in **testicular cancer** and an increased risk of **spontaneous abortion** in partners.

As the tissue specificity of steroid action depends on both the hormone and its receptor being present, hormonal activity can be regulated *either* by controlling the availability of the hormone *or* by controlling the expression of the receptor. The latter control is extremely important in reproduction and can be achieved either by varying the *amount* of receptor available, or the *structure* of the receptor as a result of different types of post-translational modification that affect its stability and how it interacts with the hormone and/or the intracellular second messenger signalling system. Thus, the maturation of the ovarian follicle (see Chapter 9), the changing uterine function during the menstrual cycle (see Chapter 10) and the changes in the female tract at birth (see Chapter 18) are all critically dependent on the acquisition, loss or modification of particular receptors on relevant tissues at appropriate times. So, it is important to note that the **measurement of variation in hormone levels alone will not provide an adequate basis for understanding reproductive function**. The receptor profile must also be known.

Protein and peptide hormones

Although central players in the choreography of reproduction, the sex steroids do not, of course, act alone. They act alongside several other types of hormone, some of the most important of which will be briefly introduced here, the remainder being encountered later in the book.

Several protein hormones are intimately involved in steroid function, of which the most important are the three **gonadotrophic glycoproteins**: **follicle-stimulating hormone (FSH)**, **luteinizing hormone (LH)** and **chorionic gonadotrophin (CG)** (Box 2.2; Table 2.4). FSH and LH are produced in the pituitary gland, while CG is produced in the placenta. LH and CG bind to the same receptors, but FSH has its own receptor. The sex steroids are both regulated by and regulate LH and FSH through feedback pathways that will be described in Chapters 7 and 9. In addition, there is a family of **somatomammotrophic polypeptides**, which exert pervasive effects on tissue growth and function, including effects on the mammary gland. The three main members all evolved from a single ancestoral gene through duplication and modification, and so are structurally and functionally related: **prolactin** (**PRL**; Box 2.3), **placental lactogen** (**PL**; also called **placental somatomammotrophin**) and **growth hormone** (**GH**; also called **somatotrophin**). Each consists of a single polypeptide chain. PRL and PL are particularly concerned with lactation, GH plays a role in puberty, and a placental variant called GH-V is active in pregnancy (Table 2.5).

Box 2.1 Synthetic agonists/antagonists in clinical or experimental use

Historically, the discovery of hormones has tended to take a path that follows roughly this course: first an 'activity' is detected in a tissue, which is then extracted and purified and its structure determined; then, if it is structurally simple enough, it is synthesized. Nowadays even more complex, large proteins can be synthesized after the identification and cloning of the gene encoding it. However, for glycoproteins, such as gonadotrophins, the carbohydrate side chains can prove problematic. Once structural analysis and synthesis are possible, then the synthesis of analogues can be attempted and their properties explored to see whether they function as agonists or antagonists. It was in this way that the oral contraceptive pill was synthesized and tested in the 1950s. Below are listed some of the main antagonists and agonists to key hormones.

Progestagens

1. Derivatives of 19-nortestosterone (testosterone lacking the C19-methyl group attached to C10 and with an ethynyl group –C–CH at C17) that are active as progestagens: norethisterone (also called norethindrone), norethisterone acetate or enantate, norethynodrel/ethynodiol diacetate: converted to norethisterone before becoming active as progestagens, norgestrel (also called levonorgestrel)
2. Derivatives of 17α-hydroxyprogesterone by esterification of the 17-hydroxyl group that have progestagenic activity: medroxyprogesterone acetate, chlormadinone acetate, magestrol acetate, provera
3. So-called third-generation progestins, which have little androgen activity and can be used at much lower doses in contraceptives: desogestrel, gestodene, norgestimate, levonorgestrel, nestorone and etono-gestrel.
4. Antiprogestins: RU486 (mifepristone), onapristone (ZK 98299)

Oestrogens

1. Derivatives of oestradiol 17β having oestrogenic activity: ethinyloestradiol with an ethinyl group at C17α, and mestranol with an ethinyl group at C17α and an –OCH3 group at C3
2. Initially stimulate, then antioestrogenic: clomiphene citrate (clomid)
3. Agonist: diethylstilboestrol
4. Antioestrogens: nafoxidine, tamoxifen, 4-hydroxytamoxifen, MER 25 (nafoxidine binds to the receptor but the complex fails to bind to chromatin; in contrast, the tamoxifen–receptor complex binds to chromatin but is ineffective in stimulating the acceptor site)
5. Aromatase inhibitors: letrozole, anastrozole, exemestane

Androgens

1. 17β-ester derivatives (fat-soluble/injectable agonists): stanozolol, methenolone, boldenone, trenbolone, dromostanolone, various testosterone and nandrolone esters
2. 17α-alky derivatives (oral agonists): oxymetholone, oxandrolone, fluoxymesterone, ethylestrenol, danazol, methyltestosterone, methandrostenolone, anti-androgens; cyproterone and cyproterone acetate

Peptides

GnRH analogues (buserelin, nafarelin, histrelin, goserelin, lupon) used to suppress gonadotrophin output during premature puberty or infertility treatment (see Chapters 4, 10 and 22) or, in conjunction with selective sex steroid replacement, in the control of a range of steroid-dependent pathologies.

LH/FSH/hCG

Genes for α and β chains have been cloned and co-expressed in cell lines to produce bioactive recombinant hormones. Fusing the unique terminal 29 amino acids from the hCG β-chain to extend the β-chain of FSH (to form **corifollitropin alpha**) dramatically increases its stability in blood and thus its bioactivity, and so reduces the number of individual injections required with conventional ovulatory treatments. Removal of N-linked carbohydrate chains of either LH or FSH can lead to receptor occupancy without activation and thereby their function as antagonists (see Box 2.2). In addition, a low molecular weight, orally active agonist of the LH receptor, called **Org 43553** has recently been synthesized and is active in humans and rats where it has a shorter half-life than LH, which may prove useful in reducing the incidence of Ovarian Hyper-Stimulation Syndrome (OHSS).

Box 2.2 Biochemistry of gonadotrophins

Each of the gonadotrophins is a heterodimeric globular protein consisting of two glycosylated polypeptides (α- and β-chains) linked non-covalently. The same α-chain is used in all three hormones, but the β-chain is unique to each, conferring most of the specific functional properties of each hormone. Each polypeptide has complex *O*- or *N*-linked carbohydrate side chains that are unique for each hormone (see Table 2.4). The removal or modification of the carbohydrate side chains does not appear to reduce appreciably binding to target receptors and indeed may even increase it. However, carbohydrate changes can influence hormone biosynthetic and secretion rates (*N*-linked), stability and half-life in the circulation (especially the *O*-linked on tail of HCG), as well as the ability of the hormone to activate receptors after binding (*N*-linked). This latter property means that hormones lacking some carbohydrate chains can occupy receptors non-functionally and thereby act as **antagonists**. This property may be useful therapeutically and may also be important physiologically, because analysis of plasma and urinary gonadotrophins reveals a wide range of molecular sizes for each hormone that reflects modifications to the glycosylation pattern. Moreover, the balance of modifications varies under different conditions. Understanding the full functional importance of the variation and how it is controlled naturally is a current challenge for clinical research.

Recently, the isotypic variants of amino acid sequences of the α and β chains have also been shown to be functionally important. Thus, naturally occurring genetic variants in the population at large have been shown to bind and activate receptors more or less efficiently.

Table 2.4 Properties of human gonadotrophins

	Luteinizing hormone (LH) (also called interstitial cell-stimulating hormone [ICSH]) Molecular mass c. 28 kDa	**Follicle-stimulating hormone (FSH) Molecular mass c. 28 kDa**	**Chorionic gonadotrophin (hCG) Molecular mass c. 37 kDa**
Secreted from	Anterior pituitary gonadotrophs	Anterior pituitary gonadotrophs	Placental trophoblast
Main actions	Leydig cells (see Chapter 8) Thecal cells: antral follicles (see Chapter 9) Granulosa cells: preovulatory follicles (see Chapter 9) Luteal cells: corpus luteum (see Chapter 9)	Sertoli cells (see Chapters 7 and 8) Granulosa cells: follicles (see Chapter 9)	Luteal cells (see Chapters 13 and 14)
Composition	α-chain* (116 amino acids and 2 *N*-linked¹ carbohydrate chains) β-chain† (121 amino acids and one *N*-linked carbohydrate chain)	α-chain* as for LH β-chain† of 111 amino acids and 2 *N*-linked¹ carbohydrate chains	α-chain* as for LH β-chain† of 145 amino acids (a 29 carboxy addition to LHβ) with 2 *N*-linked¹ + *O*-linked¹ carbohydrate chains
Receptor	85–92-kDa glycoprotein; G-protein coupled; adenyl cyclase linked	75-kDa (675 aa) receptor that dimerizes on FSH binding; G-protein coupled; adenyl cyclase linked	As for LH

* The common or backbone subunit chain – encoded in a single gene on chromosome 6q12–q21.
† The subunit chain conferring specificity. FSHβ single gene on chromosome 11p13, having two splice variants (4262 bp). LH/hCGβ gene cluster on chromosome 19q13.32 (LH – 1111 bp; CG – 1467 bp). Four CG genes are expressed, three with low activity. There is 80% homology between the common parts of CG and LH β-chains, but only 36% homology between LH and FSH β-chains.
¹ *O*-linked carbohydrate chains are attached to a serine or threonine residue via *N*-acetylgalactosamine; *N*-linked chains to asparagines via *N*-acetylglucosamine.

Box 2.3 Prolactin and PIF

Prolactin is essential for lactation (see Chapter 19), but otherwise has widespread facilitator effects on the actions of other hormones, such as LH. This it seems to do by modulating the number of ovarian and Leydig cell receptors for LH. Similarly, prolactin increases the uptake of androgen and increases 5α-reductase activity in the prostate, acting synergistically with testosterone, which maintains the prolactin receptors. Prolactin also potentiates the effects of testosterone on the seminal vesicles. It also supports numerous non-reproductive functions, such as regulation of kidney and adrenocorticotrophic activity (these tissues having higher prolactin-binding activity even than mammary tissues). For these reasons, prolactin is often described as a **co-hormone**.

PIF is a catecholamine called **dopamine** (Figure 2.10a). It is synthesized from tyrosine in the arcuate neurons of the hypothalamus and is also the precursor of noradrenaline (norepinephrine), which in turn is the precursor of adrenaline (epinephrine). The lactrotrophs express the D_2-like subtype of dopamine receptors. Binding of dopamine to D_2-like receptors activates coupled G-proteins, resulting in several functional consequences. Acute prolactin release is inhibited, transcription of its RNA is reduced, and lactotroph mitosis is suppressed. The overall outcome is reduced prolactin release. Agonists of dopamine or of D_2 dopamine receptors (**bromocriptine**) suppress prolactin secretion, while dopamine D_2 receptor antagonists (e.g. **haloperidol**, **metoclopramide** and **domperidone**) increase prolactin secretion by direct actions on the lactotrophs.

What regulates the activity of the PIF-secreting TIDA neurons? It appears that the answer to this question is prolactin itself. Increases in circulating prolactin levels result in an increase in dopamine turnover within the arcuate neurons. This increased turnover is related to an increase in tyrosine hydroxylase activity, the rate-limiting enzyme in the intra-neuronal synthesis of dopamine.

Table 2.5 Properties of the human somatomammotrophic polypeptides+

	Prolactin (PRL)* Molecular mass 23 kDa (isoforms 16, 25, 50–60 and 100)	**Placental lactogen** (PL; also called chorionic somatomammotrophin [CSA], L, A and B for three variant forms) Molecular mass 22 kDa**	**Growth hormone*** (GH-N; also called GH1) GH-V (also called GH2 and chorionic somatomammotrophin) a second GH placental gene variant Molecular mass 20 and 22 kDa by alternative splicing**
Secreted by	Anterior pituitary lactotrophs and (in humans) placental decidua and many other tissues	Cytotrophoblast to week 6, then syncytiotrophoblast and invasive mononuclear trophoblast. (In farm animals, binucleate cells)	GH-N: pituitary somatotrophs and various peripheral tissues GH-V: syncytial trophoblast
Main actions	Leydig cells; seminal vesicle and prostate; ovarian follicles; corpus luteum; mammary glands; amnion	Maternal intermediary metabolism; mammary gland; fetal growth	Prenatal and postnatal growth; general follicle and testicular support; puberty; breast development. Often acts via IGF-1
Composition	Polypeptide chain of 199 amino acids (multiple isoforms)*	Non-glycosylated polypeptide chain of 191 amino acids (25% homology to PRL, 85% to GH)**	191 amino acids and acylated, deamidated, and, for GH-V only, glycosylated variants***
Receptor	Binds PRL receptor: long and short forms which are modified to at least seven isoforms	Binds PRL-R (same affinity) and GH-R (affinity 1/2000 < GH)	Binds GH-R and PRL-R (GH-V binds PRL-R > GH-N and GH-R < GH-N)

+ These three are members of a large family that in humans are evolved from a primary GH gene, and include prolactin-like proteins (PLPs), PRL-related proteins (PRPs), proliferins and proliferin-related proteins, many of which are made in the placenta.
* A single 10-kb gene on chromosome 6 with five exons and an extra upstream exon 1a only expressed in decidual tissue.
** Three genes on chromosome 17, in a cluster with the GH genes, encoding CS A, B and L.
*** Two genes on chromosome 17 encoding GH-N and GH-V (differs from 22 kDa variant of GH-N by only 8 amino acids).

(a)

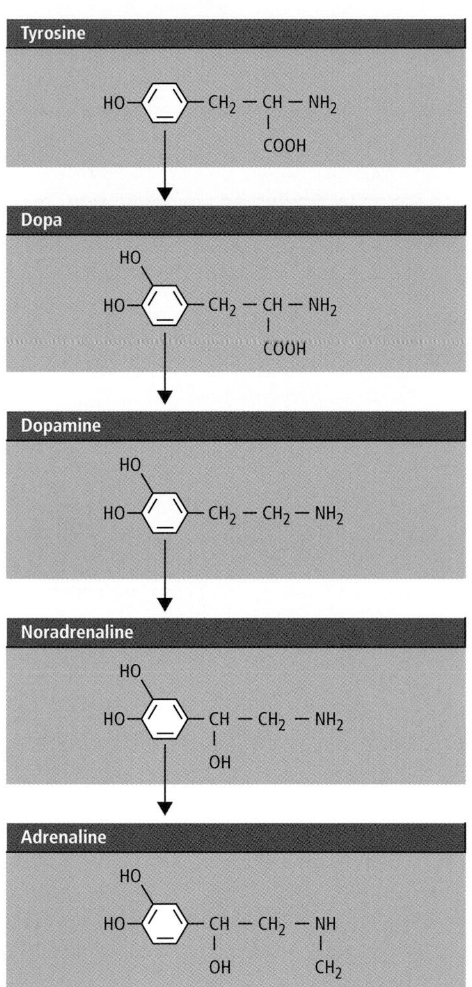

(b)

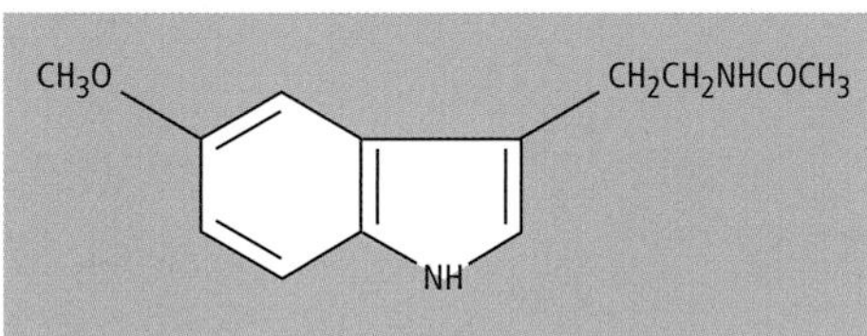

Figure 2.10 (a) Biosynthetic pathway (from top downwards) and molecular structures of the catecholamines: dopamine, noradrenaline and adrenaline. **Dopamine** is released from terminals in the hypothalamus and reaches the anterior pituitary where it is an important modulator of prolactin secretion (see Box 2.3). (b) Structure of the indolamine = melatonin. Melatonin is synthesized from serotonin (5-hydroxytryptamine) in the **pineal gland** and is under environmental control. It is released into the bloodstream and exerts important actions within the hypothalamus in order to regulate reproductive activity in seasonally breeding mammals (see Box 4.3).

Two small **neuropeptides** play important roles in reproduction. Each of them is cleaved from a larger polypeptide precursor just before being secreted. **GnRH** is one member of a large family of peptides produced in the hypothalamus (Figure 2.11). It is involved in the release of LH and FSH from the anterior pituitary. **Oxytocin** (Figure 2.12a, b) is also mainly produced in the hypothalamus, but is released from the posterior pituitary (Figure 2.12c) to act on the uterus (see Chapters 13 and 18) and the mammary gland (see Chapter 19). It is also produced and acts at several other sites.

Proteins and peptides cannot enter cells like lipids, and so use receptors located on the surface of the cell. Binding activates different sorts of transmembrane events (summarized in Figure 2.9). In each of these cases, the **receptor acts as a transducer**, passing on the information on the arrival of the hormonal ligand outside the cell to effect metabolic responses inside the cell. The intracellular molecular systems mediating these responses are called **second messenger systems**. Figure 2.9 does not do justice to the complex interactions of different second messenger systems or to the variety of systems that each hormone–receptor complex activates. Unravelling this complexity is important for an understanding of just how the cells of the body integrate the large variety of extracellular messages that they receive.

General features of reproductive hormones

Before leaving the subject of hormones and their mode of action, there are three general features that it is useful to be aware of when reading this book.

Hormones act over a range of distances

Hormones can act as messengers over variable distances within the body, and a variety of hormonal secretion and transport patterns has evolved to achieve this. For example, most steroids act after secretion into the bloodstream, in which they are distributed rapidly throughout the tissues of the body: the process of **endocrine secretion**, meaning literally 'secreted inwards'. However, not all steroid actions are achieved in this way. Thus, androgens and oestrogens within the adult testis can pass into the fluids within its duct system and be carried downstream to affect locally the epididymis and vas deferens (see Chapter 11): a process of **exocrine secretion**, meaning 'secreted outwards'. Hormones can also act very close to their site of secretion on adjacent cells and tissues. For example, androgens influence the differentiation of the embryo's genital duct system by local diffusion from the fetal testis (see Chapter 3): a process of **paracrine secretion**, meaning acting 'close by'. Indeed, hormones secreted by a cell may act back on the same cell to influence its own function: a process of auto-stimulation described as **autocrine activity**. Finally, some hormones, which in certain circumstances can be released from a cell to act on other

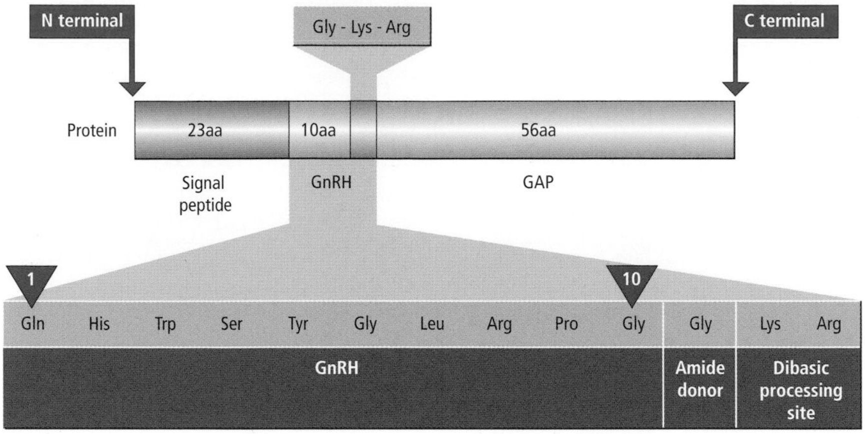

N terminal
Gly - Lys - Arg
C terminal
Protein 23aa 10aa 56aa
Signal peptide
GnRH
GAP
1
10
Gln His Trp Ser Tyr Gly Leu Arg Pro Gly Gly Lys Arg
GnRH
Amide donor
Dibasic processing site

Figure 2.11 GnRH is a decapeptide derived by cleavage from a larger precursor called prepro-GnRH. The upper blue bar shows the structure of the human cDNA for prepro-GnRH (molecular weight 10 000), which comprises the decapeptide GnRH preceded by a signal sequence of 23 amino acids and followed by a Gly–Lys–Arg sequence necessary for enzymatic processing and C-terminal amidation of GnRH. The C-terminal region of the precursor is occupied by a further 56 amino acids that constitute the so-called GnRH-associated peptide or *GAP*, the function of which, if any, is unknown. The amino acid sequence of the GnRH is shown in the lower part of the figure. Recently, a second GnRH gene has been identified (*GnRH-II*) differing at three amino acids from *GnRH-I* (pGlu–His–Trp–Ser–***His***–Gly–***Trp–Tyr***–Pro–Gly). There is a suggestion that it may preferentially release FSH, but the evidence is unclear. (Source: Nikolics K. *et al.* (1983) *Nature* **316**, 511–517. Reproduced with permission of *Nature*.)

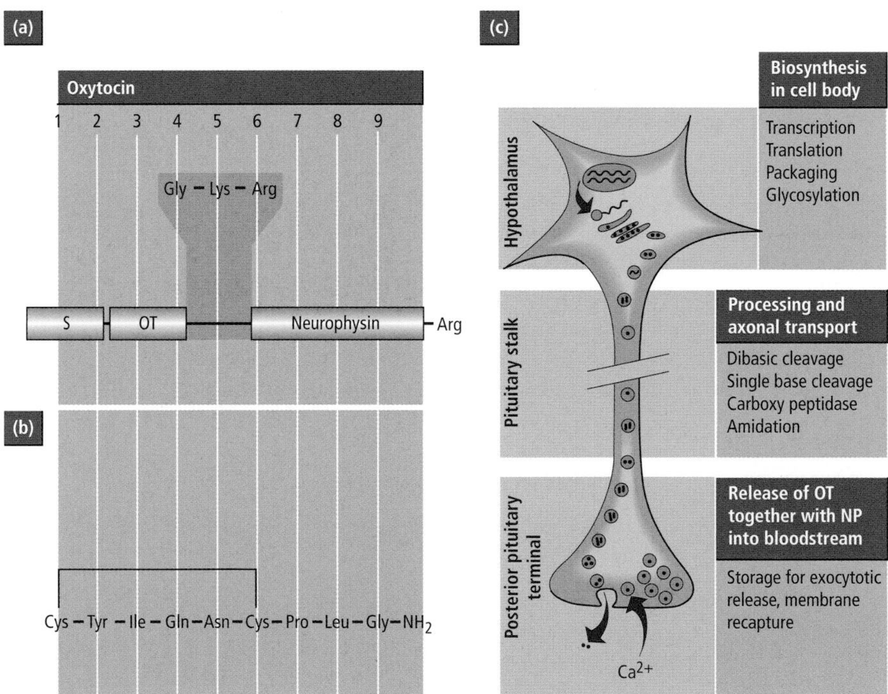

Figure 2.12 Oxytocin, like GnRH, is also derived from a larger precursor (a) consisting of (central yellow bar): a leader sequence (S); the oxytocin nonapeptide sequence (OT); a Gly–Lys–Arg linker sequence, serving the same function as described for GnRH in the legend to Figure 2.11, and a neurophysin sequence. Neurophysin is an intracellular protein able to bind oxytocin, and in this way the hormone is packaged for transport along the axon from the hypothalamus to the terminals of the neurons in the posterior pituitary. Neurophysin is released, along with oxytocin, into the bloodstream, but it has no clear function in the body. (b) The nine amino acids comprising oxytocin contain a hexapeptide 'ring' and a tripeptide 'tail'. (c) A schematized structure of a hypothalamic neurosecretory neuron indicating the cellular location of the various synthetic and processing stages that result in the release of oxytocin (OT) and neurophysin (NP) from neurohypophyseal terminals. (Source: Robinson I.C.A.F. (1986) *Neuroendocrinology* (eds S.L. Lightman and B.J. Everitt), Blackwell, Oxford, pp. 154–176.)

cells at a distance, can sometimes be anchored to the surface of the producing cell, in which case they can act on adjacent cells only: a process of **juxtacrine activity**. These examples indicate that the same hormone can use a variety of routes 'to carry its message'. Whether it acts locally or pervasively will depend not only on the nature and direction of its secretion, but also on the **blood supply** of its secreting tissue and the **solubility** of the hormone itself in the fluids bathing this tissue.

Factors affecting hormone levels in the blood

For endocrine hormones, the blood level is influenced by three factors: **its production rate, its clearance rate** and **the presence of binding proteins**. Many hormones detectable in the blood are produced from a single major source, and so hormone levels will usually reflect the activity of that source. However, some hormones may come from multiple sources, either naturally (e.g. during pregnancy when fetal, placental and/or maternal sources of many hormones are present; see Chapter 14) or pathologically (e.g. tumours commonly secrete chorionic gonadotrophin and placental lactogen). Additionally, other hormones may be produced by a mixture of direct secretion and **interconversion from other hormonal substrates**. For example, 10–30% of oestrone comes from ovaries directly, the remainder being derived by metabolic conversion of ovarian oestradiol 17β and of adrenal androstenedione by peripheral tissues, notably the liver. This distinction between primary secretion and metabolic conversion can be important in interpreting changes in the mean blood levels of a steroid, particularly in pathological conditions.

The rate at which a hormone is removed from the blood is reflected in its **half-life**. An awareness of hormonal half-lives is important in interpreting changes in hormone levels (Table 2.6). Thus, short half-lives mean that hormonal levels are more responsive to secretory changes. Changes in half-life may also inform about the state of the hormone itself (e.g. whether it has been modified to increase its rate of clearance) or about the functional competence of the organs effecting clearance. Removal of hormones from the blood is affected only marginally by utilization in receptor complexes, and mainly by metabolic conversion (e.g. of steroids in the liver) to a range of biologically inactive or less active derivatives.

Binding proteins are also important for interpretation of blood or tissue levels of a hormone. For example, the steroid hormones and their precursors are relatively insoluble in water. In order for them to reach their target cells, they must either **bind to carrier proteins** or **be chemically modified** to increase their solubility in plasma. Thus, most of the cholesterol and sex steroids in the blood are bound with carrier protein molecules in equilibrium, with much lower levels of free steroid in aqueous solution, rather in the way that haemoglobin carries reservoirs of bound oxygen (Table 2.7). Steroids are also rendered more soluble in plasma by conjugation to give **steroid glucosiduronates and sulphates**. In such a state, they show reduced biological activity and must be deconjugated again in target tissues in order to become fully active. In contrast, protein hormones are more freely soluble in the aqueous fluids of the body, but do not readily enter cells, tending to act at their surfaces. Nonetheless, some protein

Table 2.6 Half-lives of some hormones in the blood

Hormone	Half-life
Steroids	2–3 min
Prostaglandins	3–10 min
Inhibin	3–6 min
Gonadotrophins:	
LH	24 h
FSH	30–45 h
CG	48–96 h*
Prolactin and placental lactogen	10–20 min and 24 h (biphasic)

* Stabilized by the *O*-linked carbohydrate chains on tail of β-chain.

Table 2.7 Steroid-binding proteins in human plasma

Binding protein	Percentage of non-conjugated steroids bound*			
	Progestagens	Androgens	Oestrogens†	Cortisol
Albumin	48	32	63	20
Cortisol-binding globulin§	50	1	–	70
Sex steroid-binding globulin	–	66	36	–
Free steroid	2	1	1	10

* Steroids conjugated as sulphates or glucosiduronates bind weakly to albumin only.
† Oestrone and oestriol bind mainly to albumin.
§ Also called transcortin.
Note: Albumin is a low-affinity/high-capacity binding protein whilst the globulins are high-affinity/low-capacity binding proteins. There are sex differences, females in general binding proportionately more androgens/oestrogens to sex steroid-binding globulin than to albumin.

hormones do complex with binding proteins in the blood and/or tissues, and this binding can affect their availability for binding to receptors (Box 2.4).

Factors affecting hormone levels at their site of action

The levels of a hormone in the fluids bathing potential target cells depend on the balance between its arrival and its removal. These in turn depend on the local blood circulation, the nature and proximity of its secreting cells (local and paracrine or distant and endocrine), its circulating blood levels and its stability. A high local blood flow will tend to dissipate autocrine and paracrine secretions but facilitate the local distribution of endocrine secretions. Tissue levels of a hormone are also affected by its local metabolism. Some tissues can take a circulating hormone, transform it enzymatically and then utilize it locally; two important examples of this are given in Box 2.5.

Box 2.4 Protein hormones and binding proteins

The small **growth factors or cytokines insulin and insulin-like growth factor 1 (IGF-1)**, both of which will be encountered later in the book, bind to a number of different **insulin growth factor-binding proteins (IGFBPs)**, and the levels of these proteins modulate both the plasma/tissue concentration and the activity of the cytokines, complicating the interpretation of their actions. Similarly, **follistatin** (a single-chain glycopolypeptide with size variants of 32–39 kDa produced in the ovary) binds to another cytokine that will be encountered later, called **activin**, which modulates its capacity to influence follicular development in the ovary (see Chapter 9). Thus, the activity of a hormone can be regulated by varying the level of its binding protein.

The presence of variable levels of binding proteins can complicate the interpretation of hormone measurements. Thus, when the levels of the binding proteins themselves vary, the bound hormone levels will vary accordingly. For example, cases of **androgenization in women** can be associated with normal androgen levels but reduced levels of sex steroid-binding globulin. In these cases, the proportion of androgen readily available to receptors has risen and androgenization occurs. Conversely, during pregnancy, the levels of steroid-binding proteins rise six-fold (see Chapter 14 for details), leading to a corresponding increase in total, measurable steroid levels, although not of free steroid. It is difficult to determine accurately the ratio of free to bound steroid in blood samples. However, samples of body secretions, such as saliva, contain levels of steroids that reflect primarily the levels of unbound blood steroid available for equilibration. Analysis of saliva for steroid levels is also useful because frequent sampling on an outpatient basis is possible.

Box 2.5 Two examples of local tissue transformation of hormones

5α-reductase

This enzyme is present in many of the androgen target tissues, e.g. the male accessory sex glands and the skin and tissues of the external genitalia. It converts testosterone, the main circulating androgen, to 5α-dihydrotestosterone (Figure 2.8), which has a much higher affinity for the androgen receptor in these target cells, and thus is a much 'stronger' androgen than testosterone. In the prenatal and immature male, in whom both the testicular and blood levels of testosterone are relatively low, testosterone is secreted and converted to 5α-dihydrotestosterone in the tissues possessing 5α-reductase. The low levels of this highly active androgen then exert local effects on the external genitalia, increasing phallic and scrotal size and rendering them clearly distinguishable from the clitoris and labia in females. The importance of this 5α-reductase activity is revealed in people who are genetically deficient in it. Affected male infants have poorly developed male external genitalia, and at birth may be classed as females or intersexes. At puberty, the external genitalia are suddenly exposed to much higher levels of circulating testosterone, and respond with a sudden growth to normal size. This so-called **'penis-at-twelve' syndrome** illustrates vividly the role of local 5α-reductase activity in mediating many of the actions of testosterone (see also Chapter 5).

Androgen aromatization to oestrogen

A second example of local steroid interconversion appears to be even more dramatic. The hypothalamus of male and female rodents is able to take circulating testosterone and convert it to oestradiol 17β by local aromatizing activity (Figure 2.8). The high local levels of oestradiol 17β then interact with a local oestrogen receptor to stimulate activity. Thus, paradoxically, the actions of testosterone to masculinize the neonatal rat brain and to maintain masculine sexual behaviour in adulthood (see Chapters 4 and 5) are accomplished only after aromatization to oestradiol. This phenomenon emphasizes that **the terms 'male' and 'female' hormones should be used with care**.

Conclusions

Although the hormones involved in reproduction are diverse, an understanding of the activities of each of them is helped if two fundamentally important general points are grasped. First, hormonal activity in the body (tissue, fluids or blood) may reflect changing primary secretion, secondary interconversions, metabolic clearance or the levels of binding proteins. Second, hormone activity may also be regulated at the level of the target tissues by altering the level or activity of endogenous receptors, by changing patterns of paracrine stimulation between adjacent tissues or by local hormonal interconversions. These important conclusions will be revisited in the following chapters as we explore reproductive function in more detail.

Key learning points

- The reproductive body centres on anatomical and functional differences between testes and ovaries and their associated reproductive tracts.
- The gonads interact with the pituitary and the brain, especially the hypothalamus.
- The hypothalamus and pituitary are connected by both nervous and vascular routes, involving release of oxytocin, GnRH and PIF.
- The pituitary connects with the gonads and reproductive tissues primarily through the hormones gonadotrophins, prolactin and oxytocin.
- The gonadal sex steroids are the central messengers of reproduction.
- There are three main classes of sex steroid: progestagens, androgens and oestrogens.
- Hormones act as reproductive messengers throughout the body via endocrine, exocrine, paracrine, autocrine and juxtacrine activities.
- Some tissues may appear to be regulated by a hormone, when in fact an adjacent tissue may be the hormone target, which then locally stimulates the adjacent tissue by paracrine activity.
- The blood level of a hormone is influenced by three factors: its production rate, its clearance rate and the presence of binding proteins.
- Hormones may be bound to carrier molecules that stabilize them, neutralize them or increase their total amount in the blood.
- Variation in the levels of carrier molecules can lead to variations in the blood levels of hormones measured.
- Hormones are metabolized as they pass through the body; this may make them more or less active or soluble, or may clear them completely from the blood, giving them a short half-life.
- The delivery of an endocrine hormone to a tissue depends on the blood levels and the blood flow.
- Different kinds of hormones differ in both their chemical structures and their biological properties.
- These two distinguishing properties are interrelated because hormones work through receptors.
- The chemical structure of the hormone determines both how well it fits as a receptor ligand and the effectiveness with which it activates the receptor, although different parts of the hormone molecule may be responsible for each of these properties.
- Receptors are activated when the hormone binds and their activation initiates a downstream cascade of intracellular responses: a second messenger system.
- Most receptors are on the cell surface, but some, those for steroids, for example, are mainly intracellular.
- Some receptors will bind more than one hormone.
- The functional effectiveness with which a hormone binds and activates a receptor determines whether it is a strong- or weak-acting hormone.
- Local target tissues can metabolize hormones before they interact with receptors to increase or decrease their potency locally.
- The type and level of receptors present in a cell or tissue determine the sensitivity of that tissue to different hormones.
- Tissues can regulate their receptors and so control their responsiveness to hormones.
- Where more than one hormone can bind and activate a receptor, they are called agonists.
- Some molecules bind but do not activate receptors: these function as antihormones or antagonists.
- Naturally occurring and manufactured analogues of and antagonists to hormones exist, and may exert toxic or therapeutic effects via their interactions with receptors.

Clinical vignette

The importance of the pituitary gland

A 33-year-old woman was admitted at full term in labour with her first child. Immediately after delivering a healthy baby boy, she had a major haemorrhage, losing over 2 litres of blood due to failure of uterine contraction. After delivery, she was unable to breast-feed due to poor milk supply. She made a slow recovery from her birth experience, which was attributed to excessive blood loss. However, 6 months following delivery she was still experiencing severe fatigue, sweats, weight gain, and had not yet resumed menstruation. Referral to an endocrinology specialist was made, who ran a number of tests including assays of thyroid stimulating hormone (TSH), luteinizing hormone (LH), follicle stimulating hormone (FSH), and adrenocorticotrophic hormone (ACTH). All of the gonadotrophic hormones were found to be at very low levels, and a diagnosis of Sheehan's syndrome was made. After commencing hormonal replacement therapy, her well-being was much improved. She conceived her second child through IVF treatment and had a subsequent uneventful delivery.

The pituitary gland is highly engorged during pregnancy and thus vulnerable to ischaemia from reduced blood pressure, most usually secondary to haemorrhage. The symptoms are wide-ranging and reflect the key role of the pituitary in regulating endocrine function. Lack of TSH causes hypothyroidism, with associated weight gain, lethargy and fatigue. Lack of ACTH causes secondary adrenal insufficiency, which is associated with fatigue and an impaired response to physiological stressors. Lack of gonadotrophin (LH and FSH) production leads to amenorrhoea and infertility. Careful hormone replacement can replicate normal endocrine function and prevent symptoms.

FURTHER READING

General reading

Binart N, Bachelot A, Bouilly J (2010) Impact of prolactin receptor isoforms on reproduction. *Trends in Endocrinology and Metabolism* **21**, 362–368.

De Gendt K, Verhoeven G (2012) Tissue- and cell-specific functions of the androgen receptor revealed through conditional knockout models in mice. *Molecular and Cellular Endocrinology* **352**, 13–25.

Hammes SR, Levin ER (2011) Recent advances in extranuclear steroid receptor actions. *Endocrinology* **152**, 4489–4495.

Hsueh AJ, Bouchard P, Ben-Shlomo I (2005) Hormonology: a genomic perspective on hormonal research. *Journal of Endocrinology* **187**, 333–338.

Micevych PE, Wong AM, Mittelman-Smith AM (2015) Estradiol membrane-initiated signaling and female reproduction. *Comprehensive Physiology* **5**, 1211–1222.

Nilsson S, Mäkelä S, Treuter E, *et al.* (2001) Mechanisms of estrogen action. *Physiological Reviews* **81**, 1535–1565.

Patil M (2014) Gonadotrophins: the future. *Journal of Human Reproductive Sciences* **7**, 236–248.

Pike AC (2006) Lessons learnt from structural studies of the oestrogen receptor. *Best Practice Research in Clinical Endocrinology & Metabolism* **20**, 1–14.

Sharpe RM (2001) Hormones and testis development and the possible adverse effects of environmental chemicals. *Toxicology Letters* **120**, 221–232.

Singh AB, Harris RC (2005) Autocrine, paracrine and juxtacrine signaling by EGFR ligands. *Cellular Signalling* **17**, 1183–1193.

More advanced reading

Cassidy A, Albertazzi P, Lise Nielsen I, *et al.* (2006) Critical review of health effects of soyabean phyto-oestrogens in post-menopausal women. *Proceedings of the Nutrition Society* **65**, 76–92.

Davey RA, Grossmann M (2016) Androgen receptor structure, function and biology: from bench to bedside. *The Clinical Biochemist Reviews* **37**, 3–15.

Fares F (2006) The role of O-linked and N-linked oligosaccharides on the structure-function of glycoprotein hormones: development of agonists and antagonists. *Biochimica Biophysica Acta* **1760**, 560–567.

Forsyth IA, Wallis M (2002) Growth hormone and prolactin – molecular and functional evolution. *Journal of Mammary Gland Biology and Neoplasia* **7**, 291–312.

Fraser LR, Beyret E, Milligan SR, Adeoya-Osiguwa SA (2006) Effects of estrogenic xenobiotics on human and mouse spermatozoa. *Human Reproduction* **21**, 1184–1193.

Freeman ME, Kanyicska B, Lerant A, Nagy G (2000) Prolactin: structure, function and regulation of secretion. *Physiological Reviews* **80**, 1523–1631.

Gromoll J, Simoni M (2005) Genetic complexity of FSH receptor function in relation to ovarian responsiveness. *Trends in Endocrinology and Medicine* **16**, 368–373.

Hull KL, Harvey S (2001) Growth hormone: roles in female reproduction. *Journal of Endocrinology* **168**, 1–23.

Imamov O, Shim GJ, Warner M, Gustafsson JA (2005) Estrogen receptor beta in health and disease. *Biology of Reproduction* **73**, 866–871.

Kahn SM, Hryb DJ, Nakhla AM *et al.*(2002) Sex hormone-binding globulin is synthesized in target cells. *Journal of Endocrinology* **175**, 113–120.

Linzer DIH, Fisher SJ (1999) The placenta and the prolactin family of hormones: regulation of the physiology of pregnancy. *Molecular Endocrinology* **13**, 837–840.

Mangelsdorf DJ, Thummel C, Beato M, et al. (1995) The nuclear receptor superfamily: the second decade. *Cell* **83**, 835–839.

Nagirnaja L, Rulla K, Uuskulaa L, Hallasta P, Grigorovaa M, Laana M (2010) Genomics and genetics of gonadotropin beta-subunit genes: unique *FSHB* and duplicated *LHB/CGB* loci. *Molecular and Cellular Endocrinology* **329**, 4–16.

Peluso JJ (2006) Multiplicity of progesterone's actions and receptors in the mammalian ovary. *Biology of Reproduction* **75**, 2–8.

Penning TM (2003) Hydroxysteroid dehydrogenases and pre-receptor regulation of steroid hormone action. *Human Reproduction Update* **9**, 193–205.

Petrelli G, Mantovani A (2002) Environmental risk factors and male fertility and reproduction. *Contraception* **65**, 297–300.

Reed MJ (2004) Role of enzymes and tissue-specific actions of steroids. *Maturitas* **48** (Suppl. 1), S18–S23.

Rispoli LA, Nett TM (2005) Pituitary gonadotropin-releasing hormone (GnRH) receptor: structure, distribution and regulation of expression. *Animal Reproduction Science* **88**, 57–74.

Themmen APN (2005) An update of the pathophysiology of human gonadotrophin subunit and receptor gene mutations and polymorphisms. *Reproduction* **130**, 263–274.

Toran-Allerand CA (2005) Estrogen and the brain: beyond ERα, ERβ, and 17β-estradiol. *Annals of the New York Academy of Sciences* **1052**, 136–144.

Tung L, Abdel-Hafiz H, Shen T, *et al.* (2006) Progesterone receptors (PR)-B and PR-A regulate transcription by different mechanisms: AF-3 exerts regulatory control over coactivator binding to PR-B. *Molecular Endocrinology* **20**, 2657–2670.

Wehling M (1997) Specific nongenomic actions of steroid hormones. *Annual Reviews in Physiology* **59**, 365–393.

Yong EL, Loy CJ, Sim KS (2003) Androgen receptor gene and male infertility. *Human Reproduction Update* **9**, 1–7.

Part 2
Making women
and men

CHAPTER 3
Sex

Essential Reproduction, Eighth Edition. Martin H. Johnson.

Companion website: www.essentialreproduction.com

We saw in Chapter 1 that in mammals sexual reproduction is obligatory and that for reproductive purposes there are two sexes: female and male. This chapter explores how the two sexes develop. The first point to make is that the genesis of two sexes depends on genetic differences.

The genetic determinant of sex is located on the Y chromosome

Examination of human chromosomes reveals a consistent difference between the sexes in **karyotype** (or pattern of chromosomal morphologies). Thus, the human has 46 chromosomes, **22 pairs of autosomes** and **one pair of sex chromosomes** (Figure 3.1). Human females, and indeed all female mammals, are known as the **homogametic sex** because the two sex chromosomes are **both X chromosomes** and all the post-meiotic oocytes are similar to one another in that they each carry one X chromosome. Conversely, the male is termed the **heterogametic sex**, as his pair of sex chromosomes consists of **one X and one Y chromosome**, so producing two distinct populations of spermatozoa, one bearing an X and the other a Y chromosome (Figure 3.1). Examination of a range of human patients with chromosomal abnormalities has shown that if a Y chromosome is present, then the individual develops testes. If the Y chromosome is absent, ovaries develop. The number of X chromosomes or autosomes present does not affect the primary determination of gonadal sex (Table 3.1). Similar studies on a whole range of other mammals show that Y-chromosome activity determines gonadal sex. Thus, the first step towards sexual dimorphism

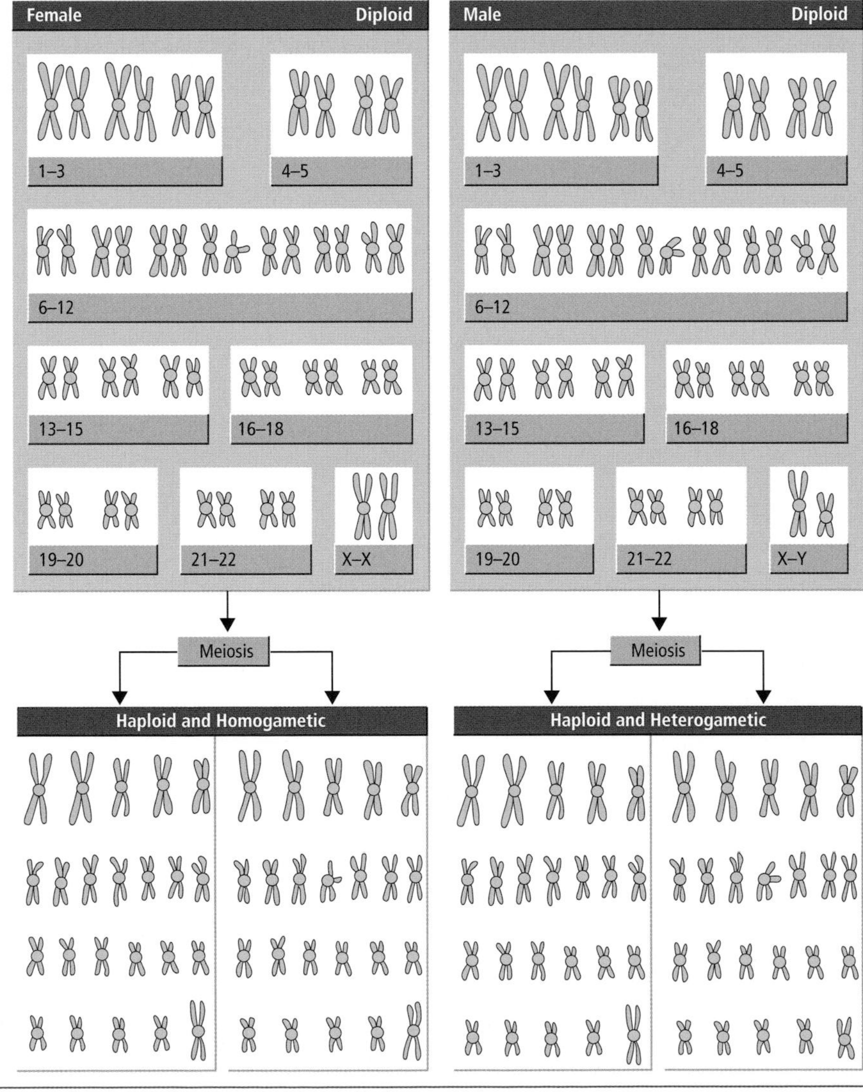

Figure 3.1 Karyotypes of two mitotic human cells: one male and one female. Each cell was placed in colchicine, a drug that arrested them in mitotic metaphase when the chromosomes were condensed and clearly visible (see Figure 1.1). The chromosomes were stained and then classified according to the so-called **'Denver' system**. The 44 autosomes (22 pairs of homologues) are grossly similar in size in each sex, but the pair of sex chromosomes are distinguishable by size, being XX (both large) in the female and XY (one large, one small) in the male. After meiotic division, all four female cells (only two shown) contain one X chromosome : the homogametic sex. In contrast, two of the male cells each contain an X chromosome and two contain a Y chromosome: the heterogametic sex.

Table 3.1 Effect of human chromosome constitution on the development of the gonad

Autosomes	Sex chromosomes	Gonad	Syndrome
44	XO	Ovary	Turner
44	XX	Ovary	Female
44	XXX	Ovary	Super female
44	XY	Testis	Male
44	XXY	Testis	Klinefelter
44	XYY	Testis	Super male
66	XXX	Ovary	Triploid*
66	XXY	Testis	Triploid*
44	XX^{sxr}	Testis	Sex reversed**

* Non-viable.
** An X^{sxr} chromosome carries a small piece of Y chromosome translocated onto the X chromosome: see text.

in mammals is the issuing of an instruction by the Y chromosome saying: '**make a testis**'.

The Y chromosome itself is small. Moreover, most of its DNA is **heterochromatic** (that is, very condensed and incapable of synthesizing RNA). Therefore, the many structural genes required to make an organ as complex as the testis cannot be located on the Y chromosome alone. Indeed, these genes are known to lie on other autosomal chromosomes, and some even lie on the X chromosome. What the Y chromosome contains is a 'switching' or controller gene, which somehow regulates the expression of all these other structural genes by determining whether and when they should become activated. The identity and location on the Y chromosome of this 'make a testis' gene were discovered initially by the study of some rare and atypical individuals.

Clinicians identified a few men with an XX sex chromosomal constitution and women with an XY chromosomal constitution – a situation called **sex reversal**. At first sight, these sex-reversed people appear to contradict all that has been said above (see Table 3.1). However, careful examination of the DNA sequences on the short arm of the Y chromosome of many XY females has revealed either that short pieces of DNA are missing (**chromosomal deletions**) or that there are **mutations** of one or more nucleic acid bases. By comparing the DNA sequences in a large number of such patients, it is possible to find one region of the Y chromosome common to all of them that is affected by deletion or mutation. This region is a likely locus for a testis-determining gene. Supportive evidence comes from many of the XX males, who are found to have **translocations** of small pieces of the Y chromosome to one of their autosomes or X chromosomes. Again, the critical piece of Y chromosome that must be translocated to yield an XX male seems to come from the same region as is damaged in the XY females. This region contains a gene called ***SRY*** (in humans), which stands for **sex-determining region of the Y gene** (see Box 3.1 for gene/protein notations). The gene is located close to the end of the short arm of the human Y chromosome. Genes that encode a conserved sequence of about 88 amino acids (**the Sry box**) have been found in all other mammals examined (except egg-laying monotremes) and are also associated with the development of a testis. In the mouse the gene is called ***Sry*** and also lies on the short arm of the Y chromosome, but nearer to the centromere.

The identification of the mouse homologue was important, because it enabled a critical experimental test of the function of this region of the Y chromosome to be performed. Thus, a region of DNA containing only the *Sry* gene was excised from the Y chromosome and injected into the nuclei of one-cell XX mouse embryos. The excised material can integrate into the chromosomal material of the XX recipient mouse, which now has an extra piece of DNA. If this piece of DNA is functional in issuing the instruction 'make a testis', the XX mouse should develop a testis. This is what happened, strongly supporting the idea that the region containing the controller gene had been identified. This gene encodes an *Sry* protein that binds DNA and localizes to the nucleus. These features might be expected in a controller gene that influences other downstream genes. But how does the *Sry* gene act to cause a testis to be generated?

Box 3.1 Gene/protein notations

Throughout, the book follows the international conventions on gene nomenclature (http:// www.informatics.jax.org/mgihome/nomen/gene.shtml), as exemplified for Sry:

Human genes/mRNAs: *SRY*; human protein: SRY.
Mouse/non-human genes/mRNAs: *Sry*; mouse/non-human protein: Sry.

Where a generic statement is made about mammals in general, the mouse notation is used.

When, where and how does *Sry* act?

The embryonic gonads develop from bipotential precursors that are indistinguishable in males and females. This precursor is derived from common **somatic mesenchymal tissue** precursors called the **genital ridge primordia**. These primordia develop at about 3.5–4.5 weeks in human embryos, on either side of the central dorsal aorta, on the posterior wall of the lower thoracic and upper lumbar region (Figure 3.2b,c). These two mesenchymal knots form the basic matrices of the two gonads. Three waves of ingressing cells then expand this matrix and do so in sex-specific ways to give the final forms of the ovary and testis.

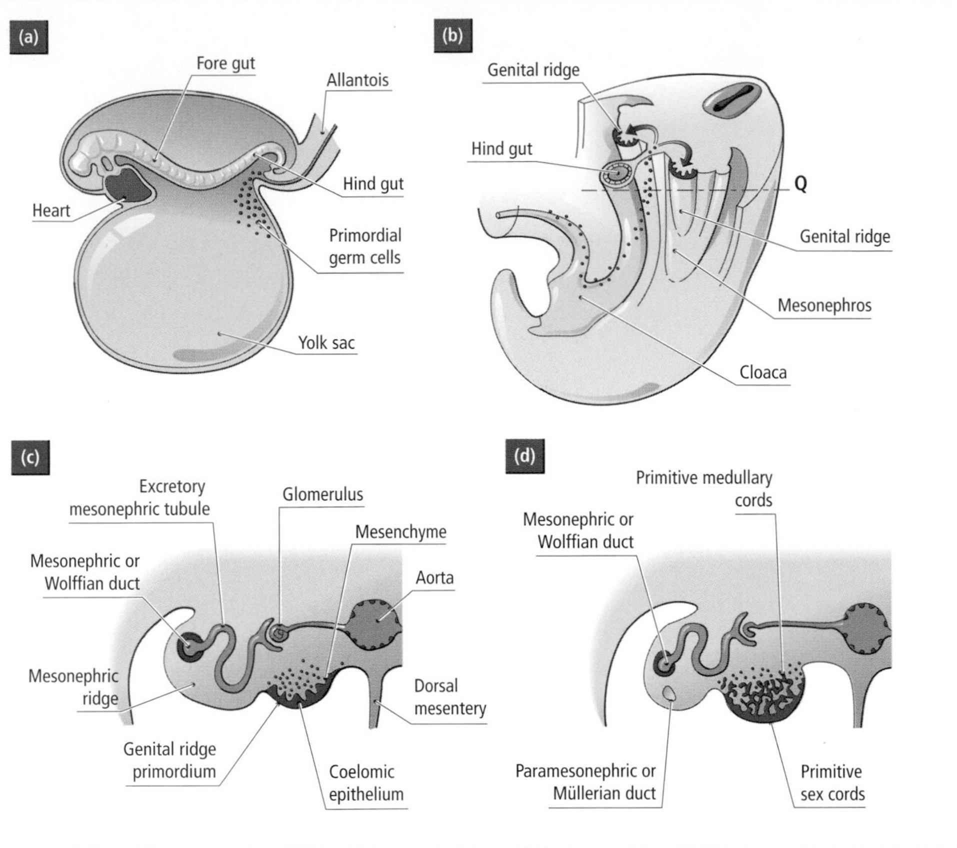

Figure 3.2 A 3-week human embryo showing: (a) the origin of the primordial germ cells and (b) the route of their migration. Section Q is the plane of transverse section through the lumbar region shown at 4 weeks in (c). In (d) the same plane of section is shown at 5 weeks of development: the 'indifferent gonad' stage. (Source: Langman J. (1969) *Medical Embryology*. Williams & Wilkins, Baltimore. Reproduced with permission from LWW.)

Primordial germ cells

One wave of migration consists of the gamete precursors called the **primordial germ cells** (**PGCs**; see page 8). These cells are identifiable first in the human embryo at about 3 weeks as a cluster of about 100 cells in the epithelium of the yolk sac near the base of the developing allantois (Figure 3.2a; Box 3.2). By the 13–20-somite stage, the PGC population, expanded by mitosis, can be observed migrating to the connective tissue of the hind gut and from there into the gut mesentery (Figure 3.2b). From about the 25-somite stage onwards, 30 days or so after fertilization, most PGCs have passed into the region of the developing kidneys, and thence into the adjacent genital ridge primordia. This migration of PGCs is completed by 6 weeks and occurs primarily by amoeboid movement. The genital ridges may produce a chemotactic substance to attract the PGCs, as PGCs co-cultured in a dish with a genital ridge move towards it. Moreover, gonad primordial tissue grafted into abnormal sites within the embryo attracts germ cells to colonize it. However, ***Sry* is *not* expressed in PGCs**.

Primitive sex cords

At about the same time as the PGCs are entering the genital ridges, a second group of cells also migrates in. These cells are derived from the columnar **coelomic** (or **germinal**) **epithelium**

Box 3.2 The origin of primordial germ cells (PGCs)

As we saw in Chapter 1 (pages 8–9) PGCs form part of this cycle and are derived from a subpopulation of pluripotent cells set aside between 2 and 3 weeks of gestation in an embryonic tissue called the **epiblast** (see Figure 13.2). Recent studies in the mouse have located their origin to a particular region of the epiblast adjacent to developing early placental tissue, which is found to transmit paracrine signals to the epiblast that cause this subset of cells to become PGCs. The identity of the inductive signal is established as the cytokine, **bone morphogenetic protein, Bmp4**. The local epiblastic response to Bmp4 is to activate transcriptional regulators and acquire germ cell features. Genetic expression markers of the onset of germ cell competence within the epiblast include an interferon-inducible transmembrane protein (**fragilis**), which then induces the expression of ***Stella***, a gene expressed only in germ cells, but never in somatic cells, and is thus a marker of continuing pluripotency in the mouse.

that overlies the genital ridge mesenchyme. They migrate in as columns called the **primitive sex cords** (Figure 3.2d). In contrast to the PGCs, the further development of these cells *does* depend on whether or not the *Sry* gene is expressed in them. In the developing human male from about 41 days, ***SRY* expression is restricted to the nuclei of cells of the sex cords** peaking at about 44 days. These cells proliferate vigorously and penetrate deep into the medullary mesenchyme, surrounding most of the PGCs to form **testis cords** by 52 days (Figure 3.3a). The same sequence of events (but with a different absolute time scale) occurs in mice. These sex-cord cells will eventually become **Sertoli cells**, the main supporting cell for spermatogenesis. Because *Sry* expression is limited to the precursor Sertoli cells (**pre-Sertolic cells**), it has been suggested that the *Sry* gene actually issues the instruction: '**make a Sertoli cell**'. Now enclosed within the cords, these PGCs are known as **prospermatogonia** and will later give rise to spermatozoa. After Sry performs this vital role in mice, its expression ceases, but not in humans or goats, although its continued lower level expression there may no longer be essential.

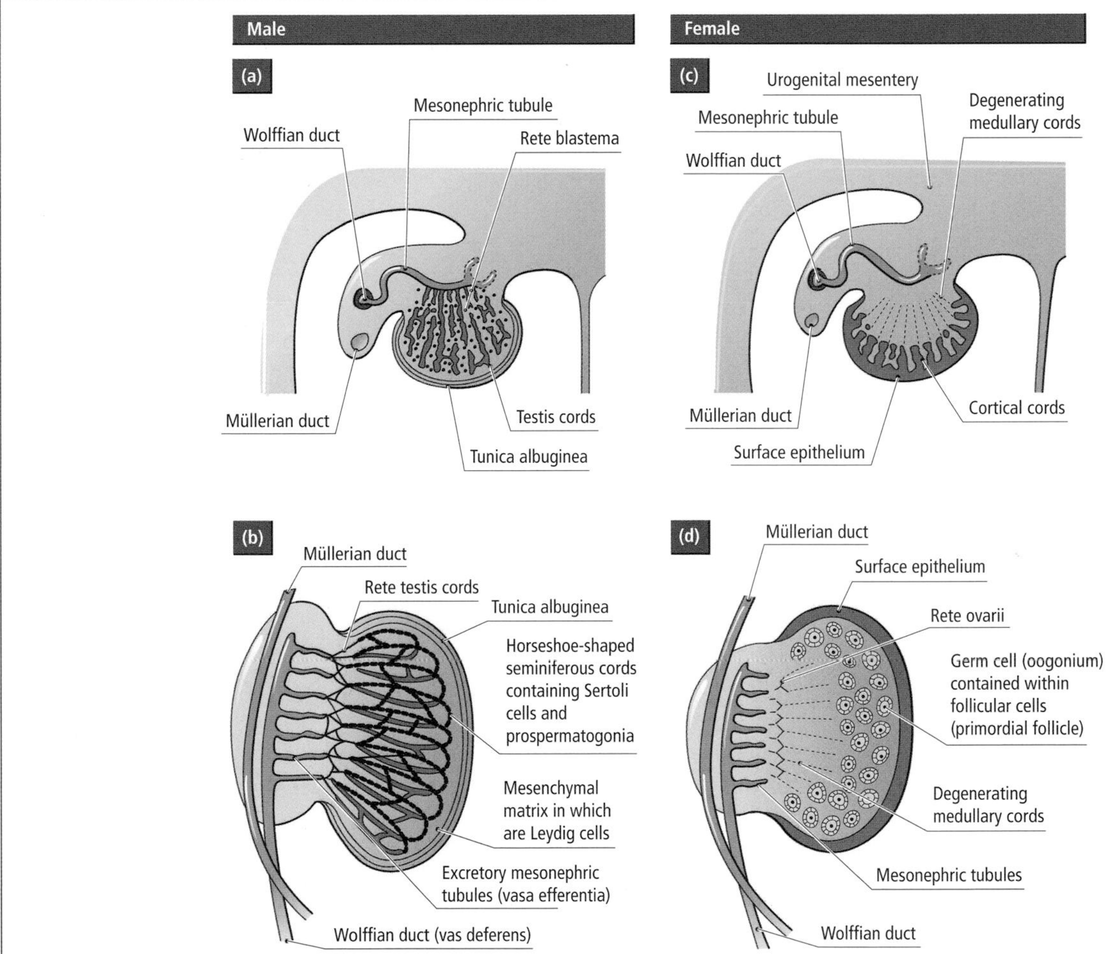

Figure 3.3 Testicular development during (a) week 8 and (b) weeks 16–20 of human embryo–fetal life. (a) The primitive sex cords proliferate into the medulla, establish contact with the mesonephric medullary cords of the rete testis blastema and become separated from the coelomic epithelium by the **tunica albuginea** (fibrous connective tissue), which eventually forms the **testicular capsule**. (b) Note the horseshoe shape of the seminiferous cords and their continuity with the rete testis cords. The vasa efferentia, derived from the mesonephric tubules, connect the seminiferous cords with the Wolffian duct.

Comparable diagrams of ovarian development around (c) week 7 and (d) weeks 20–24 of development. (c) The primitive sex cords are less well organized and more cortical, while medullary mesonephric cords are absent or degenerate. The cortical coelomic epithelial cells condense around the arriving primordial germ cells to yield primordial follicles, as shown in (d). In the absence of medullary cords and a persistent **rete ovarii**, no communication is established with the mesonephric tubules. Hence, in the adult, oocytes are shed from the surface of the ovary and are not transported by tubules to the oviduct (compare with the male, see Chapters 9 and 11). (Source: Langman J. (1969) *Medical Embryology*. Williams & Wilkins, Baltimore. Reproduced with permission from LWW.)

It is important to note that for effective functionality, *Sry* expression must not be delayed. If it is not expressed adequately over a critical 6-hour window of opportunity in the mouse (but presumptively longer in the human), then despite any later rise, either an ovary or a mixed gonad (often called an ovo-testis) develops. The chance of inducing the pre-Sertoli cell has been lost. It seems probable that the timing of *Sry* expression depends in part upon the expression of ***Steroidogenic Factor 1***, or ***Sf1***, which is expressed in both supporting and steroidogenic cell precursors before, during and after *Sry* expression, and which also acts as a cofactor with *Sry*.

Females lack any *Sry* expression, and their sex cords are ill-defined and do not penetrate deeply into the ridge. Instead, the cells condense cortically as small clusters around the PGCs, now called **oogonia**. This clustering initiates formation of the **primordial ovarian follicles** (Figure 3.3c,d). In these follicles the condensing cord cells will give rise to the **granulosa cells** of the primordial follicle, while the oogonia will give rise to **oocytes** (see Chapter 9).

Mesonephric cells

The third wave of migratory cells comes from the **mesonephric primordia**, which lie just lateral to the genital ridges (Figure 3.2c), and, like the sex cords, they show major sex differences. In both sexes, mesonephric cells contribute to the vasculature tissue of the gonad, but do so in a way that is highly sexually dimorphic. Thus, if the inward migration of testicular mesonephric vascular precursor cells is impaired, then the complex intertubular microvascular organization does not develop, and indeed cord formation itself is impaired, suggesting that the vascular elements play a role in this reorganization. It was also thought that mesonephric cells were the source of **Leydig cells**, the eventual main source of androgens, but recent evidence suggests a dual origin for these precursors, some coming from the coelomic epithelium too. Some of the Leydig cell precursors also condense on the developing testis cords and stimulate formation of a basement membrane on which they then sit as non-steroidogenic **myoid cells**, thereby contributing to the formation of the **seminiferous cords**, the forerunners of the adult **seminiferous tubules** (Figure 3.3b). The mesonephric tissue also forms the **rete blastema** or **rete testis cords**, later becoming the **rete testis**, which forms part of the male sperm-exporting duct system (Figure 3.3a,b).

Although contributing to the sexual dimorphism of the gonad, **the mesonephric cells do *not* express *Sry*.** Rather, their inward migration and differentiation are attributable to the action of the pre-Sertoli cells (see next section). In the female, the mesonephric vascular and Leydig cell precursors in males are paralleled by equivalent cells that will eventually form, respectively, blood vessels and condensations around the developing follicles called **thecal cells**.

With these three waves of inward migration completed, the basic patterns of testis and ovary are established. Moreover, the site and time of *Sry* expression are established beyond doubt as being required during a narrow developmental window of opportunity in the Sertoli precursor cells. How does *Sry* achieve this?

Molecular mechanisms underlying *Sry* action

Sry is a DNA-binding protein that acts as a transcription factor (TF); indeed, the Sry domain responsible is highly conserved across different mammals and is mutated in human XY females. Sry also carries two sites that specify nuclear localization, and mutation of either of these sites also renders it non-functional. Thus, Sry must first enter the nucleus and then bind DNA to be active. Where does it bind and with what consequences? It seems that the critically important site of Sry binding is upstream of a *Sry*-related autosomal gene called ***Sox9***, which is also a TF, and which starts to rise at about 41 days in human male when SRY rises, but peaks at 44 days and then plateaus for 18 weeks. As for the production of Sry, Sf1 acts as a cofactor with Sry in inducing the expression from the ***Sox9*** gene. *Sox9* is one member of a large family of genes that all evolved from Sry originally. *Sox9* activation is then self-sustaining since Sox9 protein induces **Sox8**, which then acts with Sox9 to induce **Sox10**, which in turn acts back together with Sf1, to maintain production of both Sox8 and 9, so that their levels remain high even if Sry levels fall: a **positive feedback** action. The importance of SOX9 in humans is seen with mutations that compromise its functionality and result in ovo-testes or even XY ovaries: **campomelic dysplasia syndrome** (so called because of associated skeletal abnormalities). Moreover, genetic duplication of *SOX9*, which can constitutively elevate levels of its protein, results in an XX man with testes. The same phenotypes can be obtained experimentally in mice. These observations demonstrate that, even in the absence of Sry, Sox9 does occupy a key role in the molecular hierarchy of sex determination.

Sox9, possibly with Sox8, then acts to stimulate the expression of downstream genes. So, **Sry determines executive policy, then Sox9 puts the policy into practice**. Now the question becomes, how does the Sox protein complex achieve this managerial function? It does so through a series of genes lower in the hierarchy, of which three are particularly important.

- Sox9 stimulates the production of **prostaglandin D_2 (PGD_2)** by the pre-Sertoli cell. PGD_2 acts locally as a paracrine hormone (see page 27) to drive Sox9 importation into the nuclei of adjacent pre-Sertoli cells. It thereby provides further positive feedback within the developing testis to reinforce all precursor cells along the same path of Sertoli cell differentiation, so **reducing the chance of an ovo-testis forming**.
- Sox9 stimulates (probably indirectly) the production of **fibroblast growth factor 9 (Fgf9)**, which acts as a **chemotactic factor for the third wave of mesonephric cells**. Fgf9 also acts **to promote Sox9 production**: another positive feedback loop. In mice lacking *Fgf9* genes (or its main receptor Fgfr2), the mesonephric migration fails, myoid cells do not form, the testis cords regress and reinforcement

of Sox9 production does not occur. An XY ovary or ovo-testis results.

- Sox9 stimulates the production of **anti-Müllerian hormone** (Amh; or Müllerian inhibitory factor, Mih), the significance of which will become clear below (see page 45).

In addition to these three genes, Sox9, together with Sf1, and possibly Sox8 and Sox10, activates directly or indirectly some 1000 genes that become active in the construction of the testis. These same transcription factors are also responsible for inhibiting the actions of genes involved making an ovary, to which we now turn our attention.

The ovary

The ovary may appear from the foregoing account to take an essentially passive developmental route. However, such a conclusion is simplistic. Thus, in the absence of *Sry* activity, a distinctive female-specific programme of gene expression is set in train and leads to the differentiation of granulosa and theca cells. Key genes include ***Wnt4***, ***Rspo1*** and ***Foxl2***, and their expression in the granulosa/Sertoli cell precursors is suppressed in the presence of Sry and Sox9, both acting with SF1. Wnt4 and Rspo1 work by respectively inducing and stabilizing **β-catenin, which functions as an important anti-testis factor**; indeed, if *β-catenin* is expressed constitutively, then an XX testis forms. Deletions or mutations of *Rspo1* and/or *Wnt4* lead to the development of XX partially or completely masculinized testes, whilst over-expression of *Rspo1* and/or *Wnt4* in XY embryos leads to XY females. Foxl2 seems to take over the ovary-driving function from β-catenin post-developmentally, as if the *Foxl2* gene is inactivated in the granulosa cells of adult female mice, the ovaries can transdifferentiate to testes! Even more strikingly, if the gene encoding a transcription factor called **Dmrtl** is knocked out in the adult XY male, the Sertoli cells start to transdifferentiate into granulosa cells, Foxl2 levels rise and oestrogen synthesis is detected! *Dmrt1* is a gene involved in male sex determination in fish. It is also expressed in both the germ and Sertoli cells of the male fetal mouse testis. However, *Dmrt1*-null mutant mice are born as males with testes, which only later show evidence of transdifferentiation. Moreover, they do so in association with an uncharacteristic decline in Sox9 levels. It seems that, whilst expression of *Dmrt1* is not essential for testis development, it may be essential for its maintenance through a positive effect on Sox9 expression, thereby maintaining the suppression of Foxl2 expression. Thus, the data from *Dmrt1* studies reinforce a molecular model in which the **ratio of Sry/Sox9 to Foxl2/Rspo1/Wnt4/*β*-catenin** is suggested as being important not only for gonad sex determination, but throughout life.

Disorders of gonad development

So far we have seen that **genetic maleness leads to gonadal maleness**, the *Sry* activity on the Y chromosome converting an indifferent gonad into a testis via the actions of Sox9 suppressing *Foxl2*, *Rspo1* and *Wnt4* genes. Given the various positive feedback loops operating in the male and described in the previous sections, this primary step in sexual differentiation is remarkably efficient, and only rarely are individuals found to have both testicular and ovarian tissue. Such individuals are called **primary** (or **true**) **hermaphrodites** and arise in some cases because of the presence of a mixture of XY and XX (or XO) cells, but in others as the outcome of the various rare genetic conditions described above. Similarly rare is **sex reversal**, in which an XX testis or an XY ovary is present.

Sexual differentiation within the gonads

With the establishment of the fetal gonad, the main role of the *Sry* gene in sexual determination is completed. From this point onwards, the gonads themselves assume the pivotal role in directing sexual differentiation both pre- and post-natally.

PGCs and sexual differentiation

Above (see page 39) we saw that the initial decision as to whether to make an ovary or a testis depends on expression of *Sry* in developing pre-Sertoli cells, not in PGCs. However, the impact of this decision on PGCs is profound and immediate, and with very different consequences in each of the developing gonads.

- In the fetal testes, as the PGCs become enclosed in developing sex cords, they cease mitotic proliferation and enter a prolonged period of **mitotic arrest** until puberty (Figure 3.4a), when, as **spermatogonia**, the mitotic cycle is reactivated. In contrast, in the fetal ovary, the PGCs behave very differently. First, all the oogonial germ cells cease dividing mitotically and **enter into their first meiotic division**, thereby becoming **primary oocytes**. This early entry into meiosis depends on the activity of **retinoic acid**, which is synthesized by mesonephric cells in both gonads, but in testes the turnover of retinoic acid is much higher due to the catabolic action of the enzyme CYP26B1 expressed in both Sertoli and Leydig cells. In addition, Fgf9 acts on the male germ cells to prevent the entering meiosis. There is a major consequence for women of this early termination of mitosis: by the time of birth **a woman has all the oocytes within her ovaries that she will ever have** (Figure 1.2). If these oocytes are lost, e.g. by exposure to X-irradiation, they cannot be replaced from stem cells and the woman will be infertile. This situation is distinctly different from that in the male in which the mitotic proliferation of spermatogonial stem cells continues throughout adult reproductive life.
- Having entered meiosis so prematurely, the primary oocytes become enclosed within primordial follicles as a result of the condensation of surrounding granulosa cells derived from invading sex cords (Figure 3.3c). The formation of

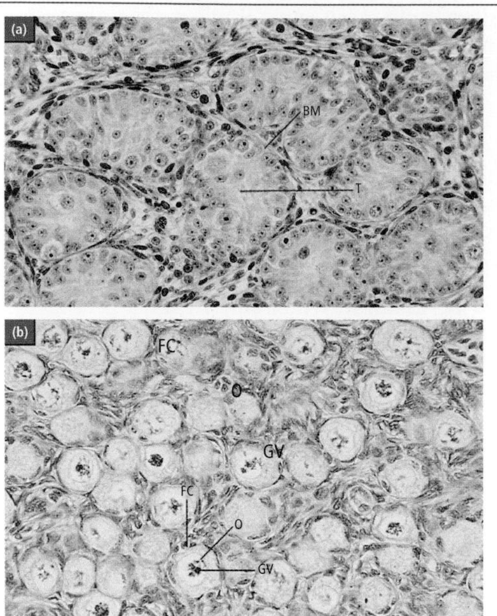

Figure 3.4 Sections through immature (a) testis and (b) ovary, both at the same magnification. Note in (a) that each tubule is surrounded by a basement membrane (BM), and within the tubule there is no lumen (T) and a relatively homogeneous-looking set of cells comprising a very few spermatogonial stem cells and mostly Sertoli cells, as seen here. Note in (b) that each oocyte (O) is relatively large and contains within it a distinctive nucleus (the germinal vesicle, GV). The ooctye is surrounded by a thin layer of follicle granulosa cells (FC) to form the primordial follicle.

the primordial follicles precipitates a second major change: the oocytes abruptly **arrest their progress through first meiotic prophase at diplotene**, their chromosomes still enclosed within a nuclear membrane called the **germinal vesicle** (Figure 3.4b; see also Figure 1.1 for details of meiosis). The oocyte halted at this point in meiosis is said to be at the **dictyate stage** (also called **dictyotene**). The primordial follicle may stay in this arrested meiotic state for up to 50 years in women, with the oocyte metabolically ticking over and waiting for a signal to resume development. The reason for storing oocytes in this extraordinary protracted meiotic prophase is unknown. Although a few follicles may resume development sporadically and incompletely during fetal and neonatal life, regular recruitment of primordial follicles into a pool of growing follicles occurs first at puberty.

- The third remarkable feature occurring in the ovary is the **death of most of the meiotically arrested oocytes** at or around the time of birth, depending on the species (Figure 1.2). The cause of death is unknown and the reasons for it are unclear. The consequence of it is that the stock of female germ cells available for use in adult life is reduced even further.

PGCs and fertility

Once PGCs are clearly differentiating along pathways towards oocytes or spermatozoa, their own genetic composition does assume importance. Thus, women suffering from **Turner syndrome** (see Table 3.1), who have a normal autosomal complement but only one X chromosome, develop an ovary (no *SRY*). Subsequently, however, full oocyte growth requires the activity of both X chromosomes, and the activity of only one X leads to oocyte death. Secondary loss of the follicle cells follows, leading to **ovarian dysgenesis** (abnormal development), and a highly regressed or **streak ovary**.

Conversely, men with **Klinefelter syndrome** (see Table 3.1) have a normal autosomal complement of chromosomes, but possess three sex chromosomes: two X and one Y. Testes form due to expression of *SRY*. However, most of the germ cells die later in life when they enter meiosis. Their death is the result of the activity of two X chromosomes rather than one.

These syndromes provide us with two important pieces of clinical evidence. First, initiation of gonad formation can occur when sex cord cells have only one Y regardless of the number of Xs (testis) or one X (ovary) chromosome. Second, completion of normal gonad development requires that the germ cells have two X chromosomes in an ovary but do not have more than one X chromosome in a testis.

Sexually dimorphic somatic differentiation depends on the endocrine activity of the fetal testis

Endocrine activity in the ovaries is not essential for somatic sexual differentiation during fetal life. In contrast, the fetal testes actively secrete two essential hormones. The **Leydig cells secrete androgens** (maximally between weeks 8 and 24 of pregnancy), and the **Sertoli cells secrete anti-Müllerian hormone** (**Amh**; see Box 3.3) under the influence of Sox9. These two hormones are the main messengers of male sexual differentiation issued by the testis. In their absence, female sexual differentiation occurs. Thus, sexual differentiation must again be actively diverted along the male line, whereas differentiation along the female line again seems to reflect an inherent trend.

The internal genitalia

Unlike the gonad, which develops from a single indifferent but bipotential primordium, **two distinct unipotential primordia** give rise to the male and female **internal genitalia** (see Figures 3.3 and 3.5). These are both located in the mesonephros adjacent to the developing gonad, and are called the **Wolffian** or **mesonephric** (male) and **Müllerian** or **paramesonephric** (female) ducts. For sexual differentiation to proceed, one primordial set must be selected and the other must regress.

Box 3.3 Anti-Müllerian hormone (Amh)

Amh is also called Mis/h/f for Müllerian inhibitory substance/hormone/factor. It is made by fetal, neonatal and prepubertal Sertoli cells, after which its synthesis is down regulated. It is also made by granulosa cells postnatally until the menopause, but *not* in the female fetus. It is a dimeric glycoprotein hormone and a member of the transforming growth factor-β gene family with a molecular mass of 140 kDa. Its gene is located on chromosome 19p13.3. It is activated by a protease that cleaves it into 110- and 25-kDa proteins, of which the smaller is sufficient for activity.

AMH acts through binding to two receptors, both serine/threonine kinases: a specific anti-Müllerian hormone receptor, type II AMHR2 (573 amino acids; molecular mass 82 kDa), the gene for which is located on chromosome 12q13. In addition, like other members of the TGFβ family, AMH also binds to a non-specific type I receptor that is shared with the bone morphogenetic protein family. On binding, a tetramer forms (2 x AMHR2 and 2 x AMHR1), and allows the R2 kinase to phosphorylate serine and threonine residues located near the transmembrane region of RI. This activates the RI kinase to phosphorylate cytoplasmic second messengers called **SMADs**, which accumulate in the nucleus to control gene expression (type 3 mechanism in Figure 2.9). Müllerian duct regression is mediated by apoptosis of epithelial cells and their epithelio-mesenchymal transformation.

Persistent Müllerian duct syndrome (PMDS) is rare, and characterized by the presence of Müllerian derivatives in XY men who are otherwise masculinized normally. Either bilateral cryptorchidism with a pelvic uterus or the descent of one testis dragging the oviduct is observed. Descended testes are normal, but have abnormalities of the excurrent ducts, especially the epididymis and vas deferens. PMDS is caused roughly equally by recessive autosomal mutations of the *Amh* and *Amhr-II* genes. Conversely, **agenesis of the Müllerian duct** is seen in Mayer-Rokitansky-Küster-Hauser (MRKH) syndrome, the cause of which is unknown but is probably due to a mix of genetic and environmental factors.

Further reading

https://ghr.nlm.nih.gov/condition/mayer-rokitansky-kuster-hauser-syndrome#genes

di Clemente N, Belville C (2006) Anti-Müllerian hormone receptor defect. *Best Practice Research in Clinical Endocrinology and Metabolism* **20**, 599–610.

Donahoe PK, Clarke T, Teixeira J, Maheswaran S, MacLaughlin DT (2003) Enhanced purification and production of Müllerian inhibiting substance for therapeutic applications. *Molecular and Cellular Endocrinology* **211**, 37–41.

Visser JA (2003) AMH signaling: from receptor to target gene. *Molecular and Cellular Endocrinology* **211**, 65–73.

In the female, the Wolffian ducts regress spontaneously and the Müllerian ducts persist and develop to give rise to the **oviducts, uterus and cervix and upper vagina** (Figure 3.5). If a female fetus is **castrated** (gonads removed), internal genitalia develop in a typical female pattern. This experiment demonstrates that ovarian activity is not required for development of the female tract.

In the male, the two testicular hormones prevent this spontaneous development of female genitalia. Thus, androgens secreted by the testis actively maintain the Wolffian ducts, which develop into the **epididymis, vas deferens and seminal vesicles**. If androgen secretion by the testes should fail, or be blocked experimentally, then the Wolffian duct system regresses and these organs fail to develop. Conversely, exposure of female fetuses to androgens causes the development of male internal genitalia. Testicular androgens have no influence on the Müllerian duct system, however, and its regression in males is under the control of the second testicular hormone, Amh. Thus, *in vitro* incubation of the primitive internal genitalia of female embryos with Amh provokes abnormal regression of the Müllerian ducts. Null mutations of *Amh* or its receptor, in the human or mouse, result in persistent Müllerian ducts.

The external genitalia

The primordia of the external genitalia, unlike those of the internal genitalia, are bipotential (Figure 3.6). In the female, the **urethral folds and genital swellings** remain separate, thus forming the **labia minora and majora**, while the **genital tubercle** forms the **clitoris** (Figure 3.6). If the ovary is removed, these changes still occur, indicating their independence of ovarian activity. In contrast, androgens secreted from the testes in the male cause the **urethral folds to fuse**, thereby enclosing the **urethral tube** and contributing, together with cells from the genital swelling, to the **shaft of the penis**. Androgens also cause the genital swellings to fuse in the midline, so forming the **scrotum**, and the genital tubercle to expand, so forming the **glans penis** (Figure 3.6). Exposure of female fetuses to androgens will 'masculinize' their external genitalia, while castration or suppression of endogenous androgens in the male results in 'feminized' external genitalia.

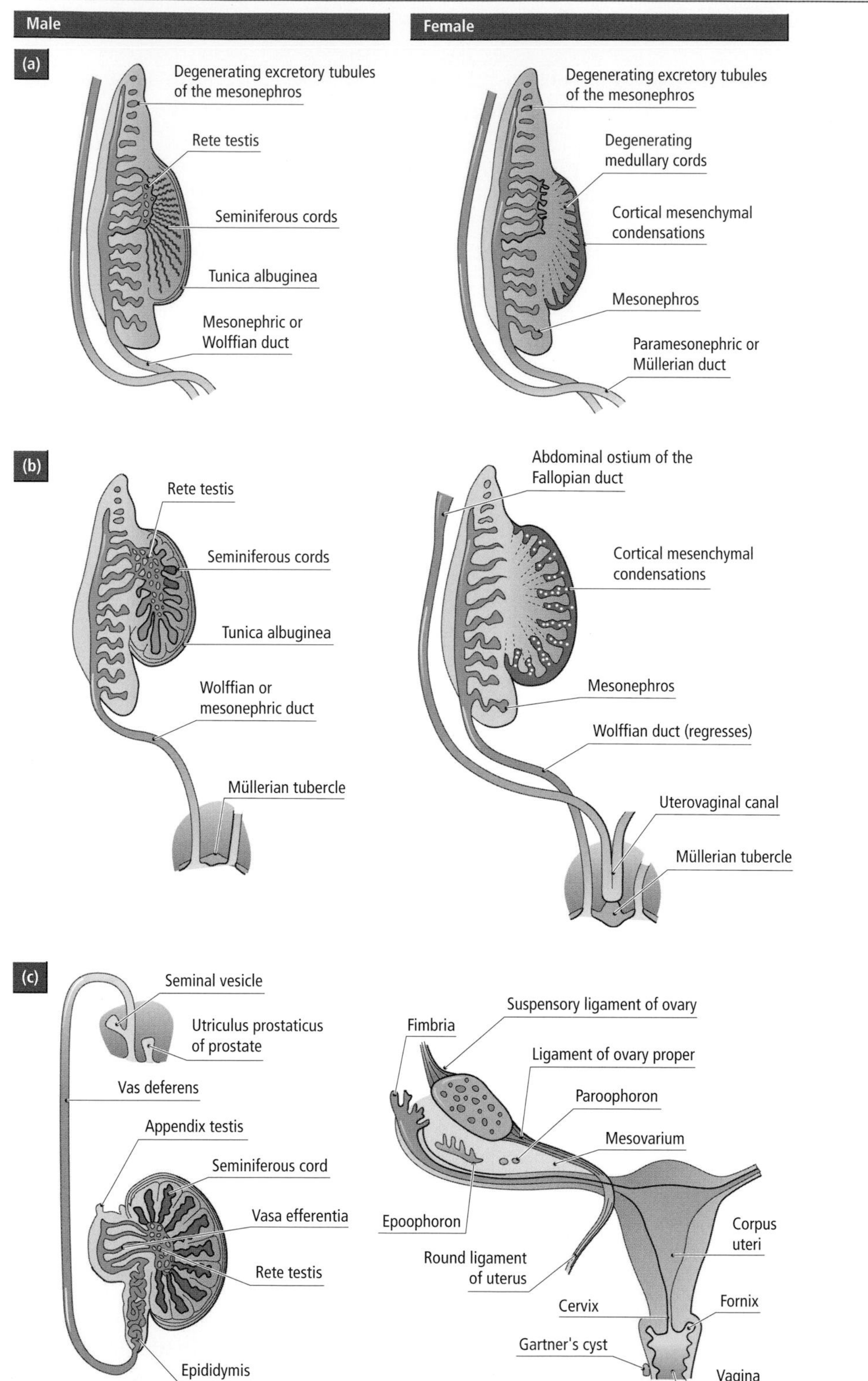

Figure 3.5 Differentiation of the internal genitalia in the human male (left) and female (right) at: (a) week 6; (b) the fourth month; and (c) the time of descent of the testis and ovary. Note the paramesonephric Müllerian and mesonephric Wolffian ducts are present in both sexes early on, the former eventually regressing in the male and persisting in the female, and vice versa. The **appendix testis** and **utriculus prostaticus** in the male, and **epoophoron, paroophoron** and **Gartner's cyst** in the female are thought to be remnants of the degenerated Müllerian and Wolffian ducts, respectively. (Source: Langman J. (1969) *Medical Embryology*. Williams & Wilkins, Baltimore. Reproduced with permission from LWW.)

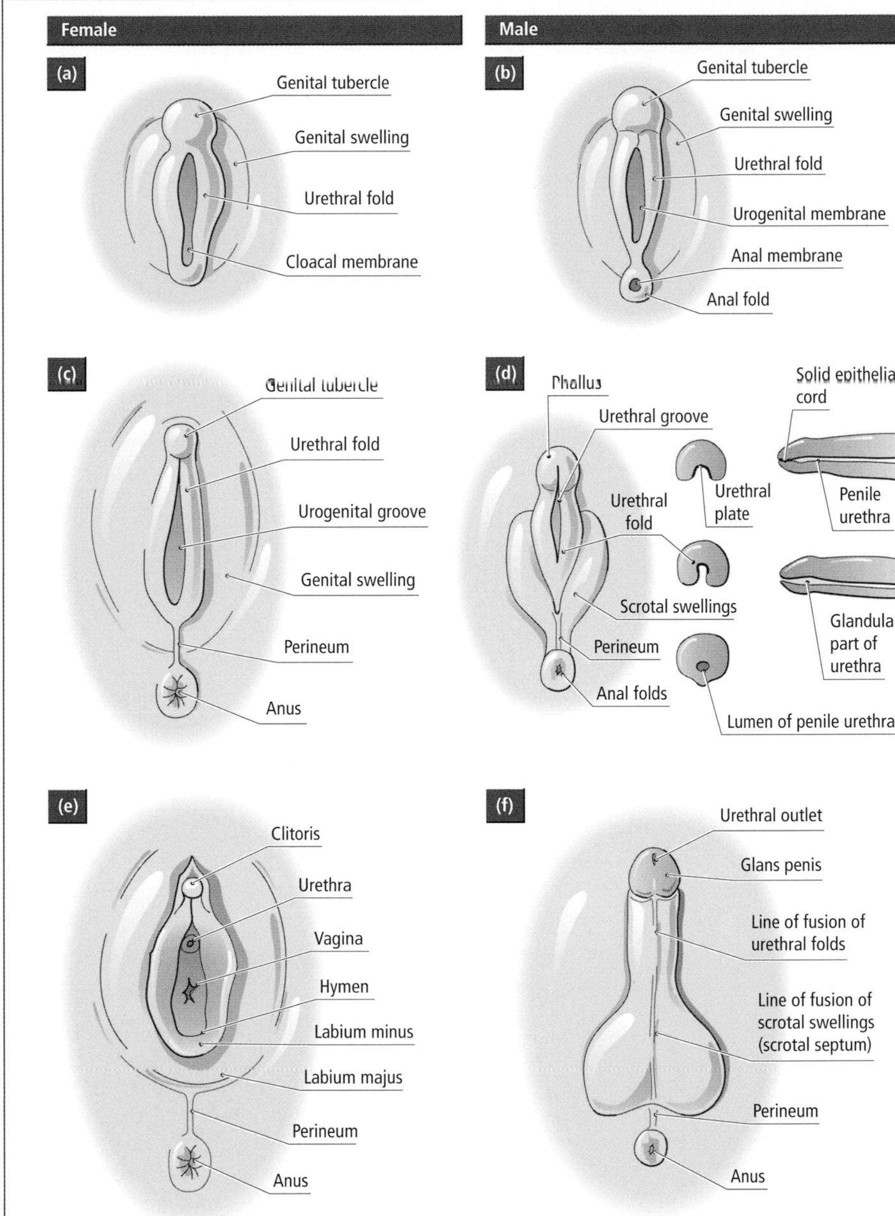

Figure 3.6 Differentiation of the external genitalia in the human female (left) and male (right) from common primordia shown at: (a) 4 weeks; and (b) 6 weeks. (c) In the female, the **labia minora** form from the urethral folds and the genital tubercle elongates to form the **clitoris**. (d) Subsequent changes by the fifth month are more pronounced in the male, with enlargement of the genital tubercle to form the **glans penis** and fusion of the urethral folds to enclose the urethral tube and form the **shaft of the penis** (the genital swellings probably also contribute cells to the shaft). (e) The definitive external genitalia of the female at birth. (f) The definitive external genitalia of the male at birth. (Source: Langman J. (1969) *Medical Embryology*. Williams & Wilkins, Baltimore. Reproduced with permission from LWW.)

Secondary hermaphrodites have genitalia of the sex not expected from their gonads

Failures of endocrine communication between the gonads and the internal and external genital primordia can lead to a dissociation of gonadal and genital sex. Such individuals are called **secondary** (or **pseudo) hermaphrodites**. For example, in the genetic syndrome of **androgen insensitivity syndrome** (**AIS**; also called **testicular feminization** or **Tfm**) the genotype is XY (male), and testes develop normally and secrete androgens and AMH. However, the fetal genitalia are wholly or partially insensitive to the action of androgens because they **lack a fully effective androgen receptor** (see page 19). This deficiency results in complete or partial regression of the androgen-dependent Wolffian ducts and in the development of female external genitalia. Meanwhile, the AMH secreted from the testes exerts its action fully on the Müllerian ducts, which regress. Thus, this genetically male individual, bearing testes and secreting androgens, nonetheless appears fully or partially female with labia, a clitoris and a vagina, but lacks other components of the internal genitalia (Figure 3.7a).

A naturally occurring counterpart to testicular feminization is the genetically based **adrenogenital syndrome** (**AGS**; also called **congenital adrenal hyperplasia or CAH**) in female fetuses, in which the XX female develops ovaries as usual. However, as a result of genetic defects in the corticosteroid synthesizing enzymes, the fetal adrenal glands become

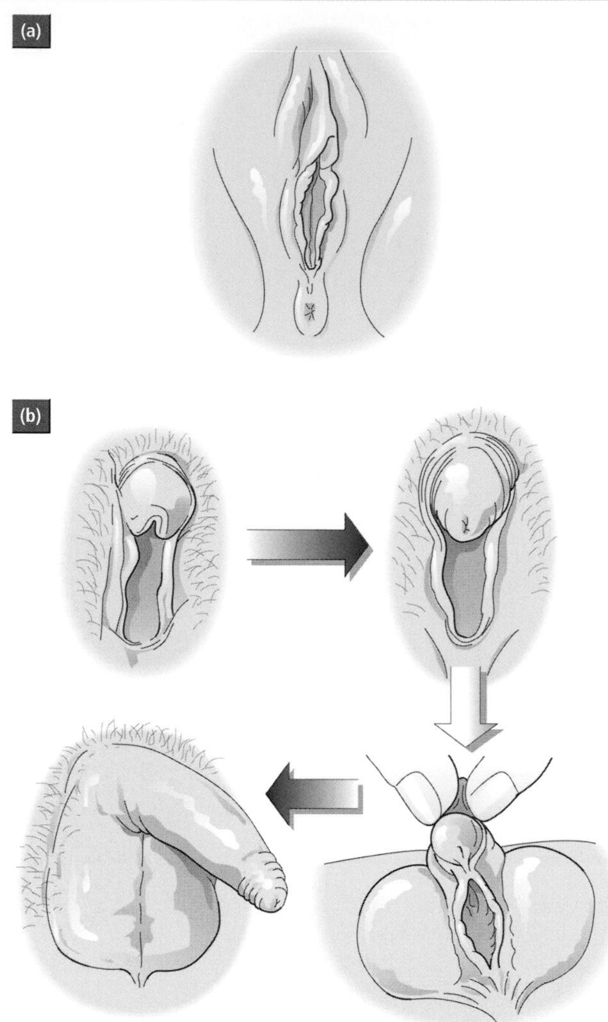

Figure 3.7 (a) External genitalia of an XY adult with complete androgen insensitivity syndrome (testicular feminization). Affected individuals are phenotypic females with normal external genitalia, breasts and a female gender identity. On examination, these women are found to have an XY chromosome constitution, abdominal testes and blood levels of androgens in the male range. Examination of the usual androgen target tissues of these women reveals an absence or deficiency of androgen receptors. (b) The external genitalia from XX girls with adrenogenital syndrome show varying degrees of masculinization, from an enlarged clitoris to development of a small penis and (empty) scrotum. Ovaries are present internally. The adrenal cortex has inappropriately secreted androgens at the expense of glucocorticoids during fetal life and directed development of the genitalia along the male line. For example, not uncommon genetic deficiencies in the fetal adrenal gland are reduced activity of 21α-hydroxylase (which converts 17α-hydroxyprogesterone to 11-desoxycortisol; Figure 2.8) or of 11β-hydroxylase (which converts 11-desoxycortisol to cortisol). In either case, there is a resulting deficiency of corticosteroids, which leads to further compensatory stimulation of the corticosteroid biosynthetic path (congenital adrenal hyperplasia), which in turn leads to the accumulation of high levels of 17α-hydroxyprogesterone. This steroid is then converted by 17,20-desmolase to androgen, which masculinizes female fetuses. Clearly, the more severe cases could lead to sex assignment as a boy or to indecision. Conversely, genetic deficiency of 17,20-desmolase itself results in depressed androgen output, and the failure of male fetuses to masculinize (see Chapter 5, Gender stereotypes and gender identities). (Source for Figure 3.7b: Overzier C. (1963) *Inter-sexuality*. Academic Press, London. Reproduced with permission from Elsevier.)

hyperactive in an attempt to overcome the lack of corticosteroids, and secrete large quantities of precursor steroids, some with strong androgenic activity (see page 16 and Figure 2.8). These androgens stimulate development of the Wolffian ducts, and also cause the external genitalia to develop along the male pattern. The Müllerian system remains, as no AMH has been secreted. Thus, the individual appears partially or even wholly masculinized with a penis and scrotum, but is genetically and gonadally female and possesses the internal genitalia of both sexes (Figure 3.7b).

Individuals with **persistent Müllerian duct syndrome** present as genetic males in whom either AMH production or

Box 3.4 Distribution of genes in four adult mouse tissues that show more than a two-fold difference in expression between males and females

Microarray technology can be used to assess numbers of expressed transcripts in adult tissues taken from males and females. These can be compared for significant differences in expression by sex of origin of tissue, otherwise matched by age, strain of mouse, etc. Significant differences in expression can be selected as >1 fold or higher. Shown as an example are data from Yang *et al.* (2006) in which a stringent level of >2 fold has been selected.

Tissue	Active genes identified	Number expressed dimorphically (%)	Number female biased	Number male biased
Liver	12845	142 (1.1)	61	81
Adipose	16664	215 (0.8)	120	95
Muscle	7367	45 (0.6)	13	32
Brain	4508	6 (0.1)	1	5

Further reading

Burgoyne PS, Arnold AP (2016) A primer on the use of mouse models for identifying direct chromosomal effects that cause sex differences in non-gonadal tissues. *Biology of Sex Differences* **7**, 68.

Gabory A, Attig L, Junien C (2009) Sexual dimorphism in environmental epigenetic programming. *Molecular and Cellular Endocrinology* **304**, 8–18.

Ngun TC, Ghahramani N, Sánchez FJ, Bocklandt S, Vilain E (2010) The genetics of sex differences in brain and behavior. *Frontiers in Neuroendocrinology* **32**, 227–246.

Yang X, Schadt EE, Wang S *et al.* (2006) Tissue-specific expression and regulation of sexually dimorphic genes in mice. *Genome Research* **16**, 995–1004.

responsiveness to it is inadequate. They therefore have testicular androgens that stimulate external genitalia and Wolffian ducts, but retain Müllerian duct structures. These men are thus genetically and gonadally male but possess the internal genitalia of both sexes.

Apart from the problems of immediate clinical management raised by diagnosis of these syndromes, abnormalities of development of the external genitalia can have important long-term consequences. The single most important event in the identification of the sex of the newborn human is examination of the external genitalia. These may be unambiguously male or female, regardless of whether the genetic and gonadal constitutions correspond. They may also be ambiguous as a result of only partial masculinization during fetal life. Sex assignment at birth is one important step that contributes to the development of an individual's **gender identity**, so uncertainty or error at this early stage can have major consequences for an individual's self-perception later in life as a man or a woman. This issue is discussed in Chapter 5.

The non-reproductive somatic tissues are also sexed

The reproductive somatic tissues of the body have been the primary concern of this chapter. However, it is clear that by adulthood most, if not all, somatic tissues of the body are 'sexed'. This is shown most dramatically in microarray studies that reveal many transcript profile differences between male and females (Box 3.4). The mechanisms underlying these many somatic tissue sex differences and the qualitative details of how they impact on the reproductive process will be covered in later chapters. For the moment, it is sufficient to note how pervasive the sexing of tissues is.

Conclusions

In this chapter we have seen that sexual differentiation occurs through a process of steady divergence, which begins with the expression (or not) of a genetic message, which, through a series of positive reinforcements, establishes the structure and nature of the fetal gonad as male, and then extends from the testis via its hormonal secretions to many tissues of the body. Thus, sex may be defined at several levels and by several parameters. Concordance at all levels may be incomplete, and the medical, social and legal consequences of this 'blurring' of a clear, discrete sexual boundary may pose social and/or clinical problems. Some of these will be considered in Chapters 5 and 6. First, however, having defined by a broad set of criteria how the two sexes are established, in the next chapter, we examine how sexual maturity is achieved.

Key learning points

- The developing gonad arises from a unipotential genital primordium of mesenchymal tissue and three invading cell populations: the primordial germ cells, germinal epithelial cells and mesonephric cells.
- Males and females are distinguished simply by the presence or absence of a Y chromosome.
- A single gene called *Sry* on the Y chromosome acts by issuing the instruction 'make a testis'.
- In the absence of *Sry*, an ovary develops.
- The primordial germ cells do not express *Sry* in either sex.
- In the testis, Sertoli cell precursors derived from the male germinal epithelial cells are the sole site of *Sry* expression: so 'make a testis' may be expressed as 'make a Sertoli cell'.
- Sry acts via a second transcription factor called Sox9: expressed only in pre-Sertoli cells.
- Sox 9 then induces Sox8 and Sox10 to form a Sox complex which with Sf1 sustains Sox9 production.
- The granulosa cells of the follicle derive from the female germinal epithelial cells.
- Ovarian expression of *Wnt4*, *Rspo1* (acting via β-catenin) and *Foxl2* in granulosa cells acts to maintain ovarian development but is suppressed by Sox9 in testes.
- The ratio of Sry/Sox9 to *Foxl2/Rspo1/Wnt4/b*-catenin is important for gonad sex determination, and for stability of gonad function throughout life.
- In males, Sox9 activates expression of prostaglandin D2, fibroblastic growth factor 9 (Fgf9) and Müllerian inhibitory hormone (Amh), each of these then acts to sustain Sox9 production.
- Fgf9 also induces mesonephric cells to migrate into the male genital ridge and form myoid cells that are essential for stabilizing testis tubule development, as well as suppressing meiosis in male germ cells.
- The vascular cells of the gonads are derived from ingressing mesonephric cells, as are some of the Leydig cells in the male and thecal cells in the female.
- The embryonic testis makes two main hormones: androgens in Leydig cells and Amh in Sertoli cells.
- The androgens stimulate the Wolffian ducts to make the epididymis, vas deferens and prostate.
- Amh causes the Müllerian ducts to regress.
- In the absence of Amh, the Müllerian ducts become the oviducts, uterus, cervix and upper vagina.
- Common bipotential precursors of the external genitalia are stimulated to become the scrotum and penis by the testicular androgens, which become the labia and clitoris in the absence of androgens.
- The ovary is not required for the development of the prepubertal female, but the testis is required for the development of the prepubertal male.
- A sex reversed individual has an XX testis or an XY ovary.
- Sex reversed individuals are infertile because fertility in females require two X chromosomes, whereas two X chromosomes (whether or not a Y chromosome is present) prevent full maturation of spermatozoa.
- A primary hermaphrodite has both ovarian and testicular tissue.
- A secondary hermaphrodite has internal and/or external genitalia at variance with the sex of their gonads.
- In the fetal/neonatal ovary, the germ cells enter meiosis under the influence of retinoic acid, but then arrest in prophase of first meiosis at the germinal vesicle stage.
- Most female germ cells die around the time of birth.
- Male germ cells also arrest, but in mitosis, having been protected from the meiosis-inducing effect of retinoic acid by its rapid degradation.

Clinical vignette

Congenital adrenal hyperplasia

A newborn baby boy was noted by his mother to be vomiting frequently, sleepy, and disinterested in breast-feeding within a few days of birth. After 10 days with no improvement, she sought advice from her general practitioner (GP), who felt that the baby was drowsy and appeared dehydrated. The GP organized an urgent review by the paediatric team at the local hospital. By the time the baby was reviewed in hospital, he was severely dehydrated, with cool peripheries and a dangerously low blood pressure. Emergency resuscitation with intravenous saline, dextrose, and ionotropes was initiated. The baby's condition stabilized over the next few hours. At the time of initial collapse, his serum sodium level was found to be 110 mmol/L (normal range 133–146 mmol/L) and his serum potassium level was 8.7 mmol/L (normal range 3.7–5.2 mmol/L). On hearing these results, the senior paediatrician immediately undertook a careful physical inspection of the baby. She noted that the baby had a small penis with a urethra emerging from the underside (hypospadias) and that there were no testes in the scrotal sac. The diagnosis of salt-wasting congenital adrenal hyperplasia was suspected, and later confirmed by measurement of 17-hydroxyprogesterone levels. The baby's karyotype was later confirmed as XX.

The adrenal cortex synthesizes glucocorticoids, mineralocorticoids and sex steroids from common inactive precursor forms. A series of specific enzymes are required to catalyze side-chain modifications to convert the precursors into the active forms. If one enzyme is missing, then a particular pathway is blocked, and a corresponding amount of precursor will be shifted down the available pathways. The most common missing enzyme is 21-hydroxylase (21-OH). This enzyme catalyzes the conversion of hydroxyprogesterone or progesterone to 11-deoxycorticosterone, a vital step in the synthesis of both mineralocorticoids and glucocorticoids. Consequently, in the severe form of the enzyme deficiency, the baby will be deficient in both aldosterone (an important mineralocorticoid) and cortisol (an important glucocorticoid). A correspondingly higher amount of sex steroid will be produced. In the absence of aldosterone, this baby has lost high amounts of sodium in the urine, leading to life-threatening dehydration and metabolic acidosis. Virilization of the external genitalia of the female fetus has occurred as a result of high testosterone levels during differentiation of the genital tubercle. This condition requires life-long treatment with replacement of mineralocorticoid and glucocorticoid hormones.

FURTHER READING

General reading

Bowles J, Koopman P (2010) Sex determination in mammalian germ cells: extrinsic versus intrinsic factors. *Reproduction* **139**, 943–958.

Brennan J, Capel B (2004) One tissue, two fates: molecular genetic events that underlie testis versus ovary development. *Nature Reviews in Genetics* **5**, 509–521.

Carrillo AA, Berkovitz GD (2004) Genetic mechanisms that regulate testis determination. *Reviews in Endocrine and Metabolic Disorders* **5**, 77–82.

Crow JF (1994) Advantages of sexual reproduction. *Developmental Genetics* **15**, 205–213.

Dinapoli L, Capel B (2008) SRY and the standoff in sex determination. *Molecular Endocrinolology* **22**, 1–9.

Edson MA, Nagaraja AK, Matzuk MM (2009) The mammalian ovary from genesis to revelation. *Endocrine Reviews* **30**, 624–712.

Kashimada K, Koopman P (2010) Sry: the master switch in mammalian sex determination. *Development* **137**, 3921–3930.

Koopman P (2009) Sex determination: the power of DMRT1. *Trends in Genetics* **25**, 479–481.

Nef S, Vassalli J-D (2009) Complementary pathways in mammalian female sex determination. *Journal of Biology* **8**, 74.

Oktem O, Bulent U (2010) Understanding follicle growth in vivo. *Human Reproduction* **25**, 2944–2954.

Ostrer H (2001a) Identifying genes for male sex determination in humans. *Journal of Experimental Zoology* **290**, 567–573.

Ostrer H (2001b) Sex determination: lessons from families and embryos. *Clinical Genetics* **59**, 207–215.

Piprek RP (2009) Genetic mechanisms underlying male sex determination in mammals. *Journal of Applied Genetics* **50**, 347–360.

Sekido R, Lovell-Badge R (2009) Sex determination and SRY: down to a wink and a nudge? *Trends in Genetics* **25**, 19–29.

Sekido R, Lovell-Badge R (2013) Genetic control of testis development. *Sex Development* **7**, 21–32.

She, Z-Y, Yang W-X (2017) *Sry* and *Sox* genes: how they participate in mammalian sex determination and gonadal development? *Seminars in Cell & Developmental Biology* **63**, 13–22.

Veitia RA (2010) FOXL2 *versus* SOX9: a lifelong 'battle of the sexes'. *BioEssays* **32**, 375–380.

More advanced reading

Adams IR, McLaren A (2002) Sexually dimorphic development of mouse primordial germ cells: switching from oogenesis to spermatogenesis. *Development* **129**, 1155–1164.

Albrecht KH, Eicher EM (2001) Evidence that *Sry* is expressed in pre-Sertoli cells, and Sertoli cells and granulosa cells have a common precursor. *Developmental Biology* **240**, 92–107.

Bashamboo A, Ledig S, Wieacker P *et al.* (2010) New technologies for the identification of novel genetic markers of disorders of sex development (DSD). *Sexual Development* **4**, 213–224.

Bouma GJ, Hudson QJ, Washburn LL, Eicher EM (2009) New candidate genes identified for controlling mouse gonadal sex determination and the early stages of granulosa and Sertoli cell differentiation. *Biology of Reproduction* **82**, 380–389.

Chaboissier M-C, Kobayashi A, Vidal VI *et al.* (2004) Functional analysis of *Sox8* and *Sox9* during sex determination in the mouse. *Development* **131**, 1891–1901.

Colvin JS, Green RP, Schmahl J, Capel B, Ornitz DM (2001) Male-to-female sex reversal in mice lacking fibroblast growth factor 9. *Cell* **104**, 875–889.

Garcia-Ortiz JE, Pelosi E, Omari S *et al.* (2009) Foxl2 functions in sex determination and histogenesis throughout mouse ovary development. *BMC Developmental Biology* **9**, 36.

Hanley NA, Hagan DM, Clement-Jones M *et al.* (2000) *SRY, SOX9,* and *DAX1* expression patterns during human sex determination and gonadal development. *Mechanisms in Development* **91**, 403–407.

Jeanes A, Wilhelm D, Wilson MJ *et al.* (2005) Evaluation of candidate markers for the peritubular myoid cell lineage in the developing mouse testis. *Reproduction* **130**, 509–516.

Koopman P, Gubbay J, Vivian N *et al.* (1991) Male development of chromosomally female mice transgenic for *Sry*. *Nature* **351**, 117–121.

Koopman P, Münsterberg A, Capel B *et al.* (1990) Expression of a candidate sex-determining gene during mouse testis differentiation. *Nature* **348**, 450–452.

Matson CK, Murphy MW, Sarver AL *et al.* (2011) DMRT1 prevents female reprogramming in the postnatal testis. *Nature* **476**, 101–104.

Ohe K, Lalli E, Sassone-Corsi P (2002) A direct role of SRY and SOX proteins in pre-mRNA splicing. *Proceedings of the National Academy of Sciences of the United States of America* **99**, 1146–1151.

Raymond CS, Parker ED, Kettlewell JR *et al.* (1999) A region of human chromosome 9p required for testis development contains two genes related to known sexual regulators. *Human Molecular Genetics* **8**, 989–996.

Sekido R, Lovell-Badge R (2008) Sex determination involves synergistic action of SRY and SF1 on a specific Sox9 enhancer. *Nature* **453**, 930–934.

Sinclair AH, Berta P, Palmer MS *et al.* (1990) A gene from the human sex-determining region encodes a protein with homology to a conserved DNA-binding motif. *Nature* **346**, 240–244.

Spiller CM, Wilhelm D, Koopman P (2010) Retinoblastoma 1 Protein modulates XY germ cell entry into G1/G0 arrest during fetal development in mice. *Biology of Reproduction* **82**, 433–443.

Uhlenhaut N, Jakob S, Anlag K *et al.* (2009) Somatic sex reprogramming of adult ovaries to testes by FOXL2 ablation. *Cell* **139**, 1130–1142.

Vainio S, Heikkilä M, Kispert A, *et al.* (1999) Female development in mammals is regulated by *Wnt-4* signalling. *Nature* **397**, 405–409.

Vidal VP, Chaboissier MC, de Rooij DG, Schedl A (2001) *Sox9* induces testis development in XX transgenic mice. *Nature Genetics* **28**, 216–217.

Wilson JD, George FW, Griffin JE (1981) The hormonal control of sexual development. *Science* **211**, 1278–1284.

Yang X, Schadt EE, Wang S *et al.* (2006) Tissue-specific expression and regulation of sexually dimorphic genes in mice. *Genome Research* **16**, 995–1004.

CHAPTER 4

Sexual maturation

Essential Reproduction, Eighth Edition. Martin H. Johnson.

Companion website: www.essentialreproduction.com

In Chapter 3, we described how the gonads and genitalia develop during embryonic and early fetal life. We saw that the female path of development is taken unless there is intervention via first genetic (*Sry/Sox9*) and then endocrine (androgens/Amh) activities, when a male develops. Now the transition from the sexually immature late fetus to a sexually mature adult is considered. In higher primates, and humans in particular, this period of maturation is remarkably extended, involving prolonged parental care through **infancy** (from birth to weaning), **childhood** (a period of dependence on adults for survival) and the **juvenile** stage (when survival without adults is possible), to the **adolescent** stage. Adolescence marks the transition from sexual immaturity to full adulthood and is very extended in humans, usually defined biologically as beginning at around 12–14 years and ending at around 24–25 when adult cognitive development is completed. This extended period of parental care, and its impact on the well-being of the offspring, are considered in Chapter 20. In this chapter, we are concerned more with the physical changes that convert a reproductively immature fetus to a young adult with the **potential to reproduce sexually**: the so-called state of **fecundity**. The central pivot about which the discussion is structured is **puberty**.

Prepubertal development

Prepubertal development is characterized by slow but steady physical growth, **growth of the brain that plateaus at 6–8 years** after which brain maturation dominates (see page 61), and some small but significant changes to the reproductive organs and tissues. Further sexual divergence of male and female physical phenotypes occurs only at a very slow pace and both internal and external genitalia remain immature, growing slowly in line with general body growth. Aside from transient activity in the neonatal hypothalamic/pituitary/gonadal axis around the time of birth, the prepubertal period of slow growth depends on low and variable levels of gonadal hormones in the male (but not in the female). Despite this relative quiescence, some important reproductive changes do occur over this period.

The testes migrate to a scrotal position

The gonads develop in the upper lumbar region of the embryo, yet by adulthood in most mammalian species, including humans, the testes have by birth descended through the abdominal cavity, and over the pelvic brim through an inguinal canal to arrive in the scrotum (Figure 4.1). Evidence of this extraordinary migration is found in the nerve and blood supplies to the testis, which retain their lumbar origins and pass on an extended course through the abdomen and pelvis to reach their target organ. The transabdominal descent of the testis towards the inguinal canal involves two ligamentous structures: at the superior pole of the testis is the **suspensory ligament**, while inferiorly is the **gubernaculum**, which attaches the testis to the posterior abdominal wall (Figure 4.1). As the fetal male body grows, the suspensory ligament elongates but the gubernaculum does not, and thus in the male the relative position of the testis becomes increasingly caudal or pelvic. Two hormones, both secreted by the developing Leydig cells, are responsible for these male-specific effects. Androgens act on the suspensory ligament, allowing its elongation, while **insulin-like 3** (**Insl3**; Box 4.1) acts on the gubernaculum to mature and stabilize it.

Testicular migration in humans may be arrested developmentally at some point on the migratory route, resulting in one or both testes being non-scrotal, a condition known as **cryptorchidism** (hidden gonad). The consequences of cryptorchidism in men demonstrate that a scrotal position is essential for normal testicular function. Thus, although adult endocrine activity is not affected in any major way, spermatogenesis is arrested, testicular metabolism is abnormal and the risk of testicular tumours increases. These effects can be simulated by prolonged warming of the scrotal testis experimentally or by wearing thick, tight underwear. The normal scrotal testis functions best at temperatures 4–7 °C lower than the abdominal 'core' temperature. Cooling of the testis is improved by **copious sweat glands in the scrotal skin** and by the blood circulatory arrangements in the scrotum. The testicular blood supply (via the **internal spermatic arteries**) is coiled (or even forms a rete in marsupials) and passes through the **spermatic cord** in close association with the draining venous **pampiniform plexus**, which carries peripherally cooled venous blood (Figure 4.2). Therefore, a heat exchange is possible, cooling the arterial and warming the venous blood (see also Box 4.1).

Testicular growth and activity are important for male development

By weeks 16–20 of human fetal life, the testis consists of an outer fibrous capsule, the **tunica albuginea**, enclosing vascularized stromal tissue, which contains condensed Leydig cells and solid seminiferous cords comprised of a basement membrane, Sertoli cells and prospermatogonial germ cells. The seminiferous cords connect to the developing excurrent duct system, comprising cords of the **rete testis**, the **vasa efferentia** and thereby to the **epididymis**.

The Leydig cells in the fetal human testis actively secrete testosterone from at least weeks 8–10 of pregnancy onwards and show a prenatal peak (blood levels of 2 ng/ml around weeks 13–15). Thereafter, blood levels decline and plateau by 5–6 months prenatally at 0.8 ng/ml. This transient **prenatal peak** is a feature of many mammals, although in some (rat and sheep), the peak may approach or span the period of parturition, and only begin its decline postnatally. The males of some primate species, including humans, also show a

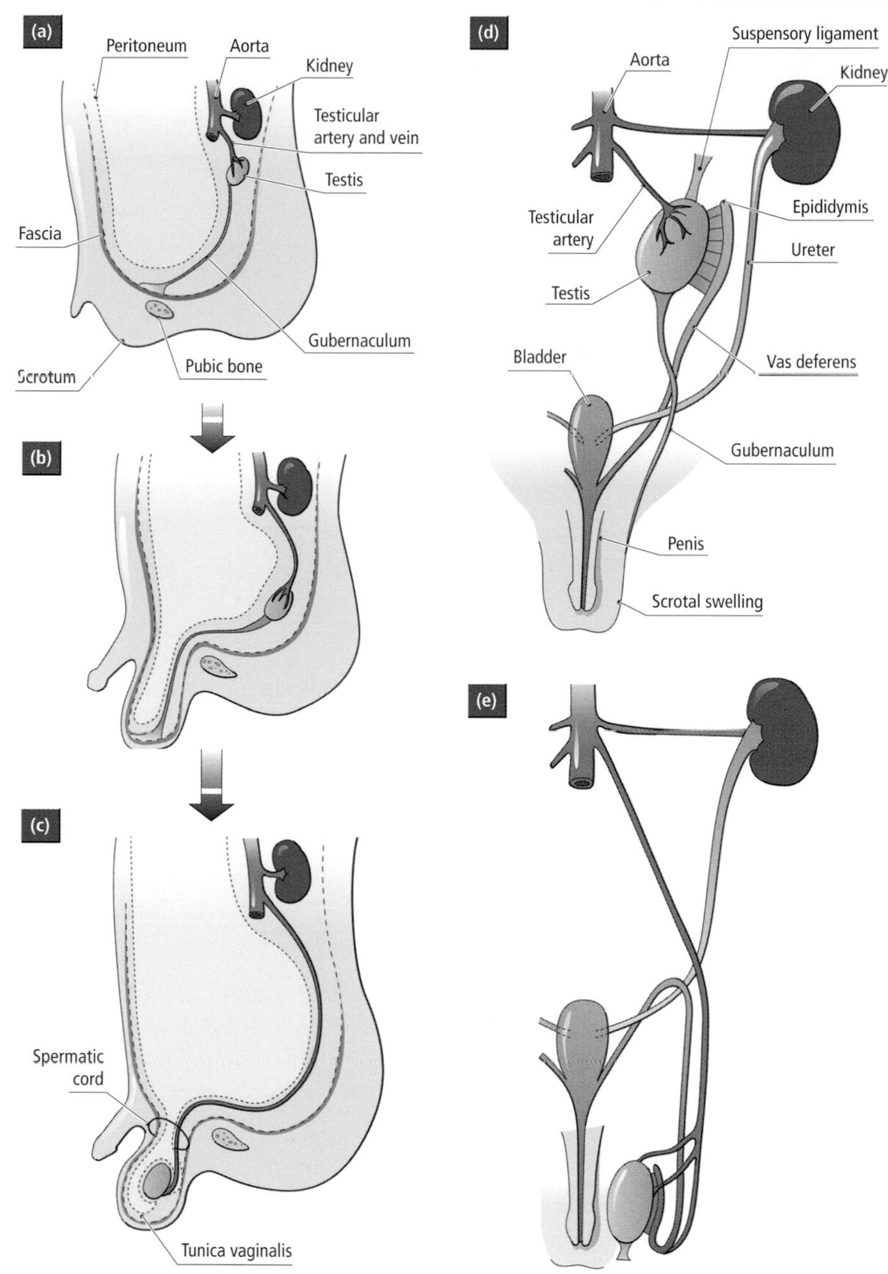

Figure 4.1 (a–c) Parasagittal sections and (d, e) front view of the testis migration through a developing male abdomen. The initial retroperitoneal, abdominal position of the testis shifts pelvically between 10 and 15 weeks (a to b), extending the blood supply (and Wolffian duct derivatives, not shown) as the gubernaculum shortens and the suspensory ligament, shown in (d) connecting the testis to the posterior abdominal wall, lengthens and then regresses (e). A musculofascial layer (b) evaginates into the scrotal swelling accompanied by peritoneal membrane, forming the **processus vaginalis**. Between weeks 25 and 28 of pregnancy in the human (c), the testis migrates over the pubic bone behind the processus vaginalis (which then wraps around it forming a double-layered sac), reaching the scrotum by weeks 35–40. The fascia and peritoneum become closely apposed above the testis, obliterating the peritoneal cavity and leaving only a **tunica vaginalis** (c) around the testis. The fascial layers, the obliterated stem of the processus vaginalis, the vas deferens and testicular vessels and nerves form the **spermatic cord** (c). (d, e) show the extended course ultimately taken by the testicular vessels and vas deferens.

Box 4.1 Evolutionary evidence of testis migration

Comparative biological study of the male testis provides evidence of 'evolutionary cryptorchidism'. Thus:

- In elephants, hyraxes and the monotremes (platypus and echidna), the testes normally do not descend at all from the lumbar site
- In armadillos, whales and dolphins, the testes migrate only part of the route to the rear of the lower abdomen
- In hedgehogs, moles and some seals, they lodge in the inguinal canal
- In most rodents and wild ungulates, they retain mobility in the adult, migrating in and out of the scrotum, to and from inguinal or abdominal retreats.

The human scrotal testis clearly requires a lower ambient temperature for normal function, but this requirement may be a secondary consequence of its scrotal position rather than the original evolutionary drive for its migration. Thus, those species in which testes remain in the abdomen survive, flourish and reproduce despite the high testicular temperature. It remains unclear as to why testes evolved a scrotal location in so many species, given their greater physical vulnerability!

In mammals, such as humans, who have exclusively scrotal testes, a null mutation of the *Insl3* gene results in no Insl3 expression and failure of gubernacular maturation and testicular descent. It would be interesting to look at *Insl3* gene expression in the above non-scrotal species!

The *INSLA3* gene on chromosome 19 encodes a 131-amino acid pro-protein, molecular mass 14 kDa, which is assembled into a heterodimer of A (26 amino acids) and B (37 amino acids) chains. It is a member of the insulin super-family (comprising insulin, insulin-like growth factors [IGFs] I and II, and relaxin, which it most closely resembles, and acts via the relaxin type 2 receptor [RXFP2; see Figure 2.9]).

Recent evidence from the rat has suggested that androgens act not only on the gubernaculum but also on the sexually dimorphic genito-femoral nerve, stimulating it to produce the transmitter calcitonin gene-related peptide or Cgrp, and that this factor is critical in acting on the gubernaculum to ensure full migration to the scrotum. Thus, transection of the nerve prevents migration, and exposure of the gubernaculum *in vitro* to Cgrp leads to increased cell proliferation at the gubernacular tip, but only after exposure of the gubernaculum to androgens. Cgrp also stimulates the cremester muscle as well as the gubernaculm itself to rhythmically contract. Evidence for a similar role for CGRP in humans is at present circumstantial.

postnatal peak in plasma testosterone, underpinned by rises in gonadotrophins, with concentrations reaching 2–3 ng/ml by 3 months postpartum (Figure 4.3), but declining to around 0.5 ng/ml by 3–4 months. A second very modest infantile rise in androgens can occur at 1 year and extend to puberty, when a prepubertal peak in androgen output occurs, reaching levels of about 9 ng/ml. This capacity to secrete testosterone is, as seen in Chapter 3, essential for the establishment of the male phenotype. It is also important for the continuing development of the male phenotype and, in many if not all species, can also influence the development of masculine behaviour patterns (see Chapters 5 and 6). The Sertoli cells continue to produce anti-Müllerian hormone (AMH) throughout fetal life up until puberty, when levels drop sharply (see Box 3.3).

Throughout fetal and early postnatal life, testis size increases slowly but steadily. The prospermatogonial germ cells undergo only limited mitotic proliferation and contribute little to this growth. At puberty there is a sudden increase in testicular size to which all parts of the testis contribute: the solid seminiferous cords canalize to give rise to tubules; the intratubular Sertoli cells increase in number, size and activity; the germ cells resume mitotic activity and begin the process of forming spermatozoa; and endocrine secretion by the intertubular Leydig cells increases sharply. These changes herald the onset of sexual maturity and the development of male fecundity. The details of how the mature testis functions are described in Chapters 7 and 8.

Ovarian growth and activity are not essential for female development

The ovary, unlike the testis, retains its position within the abdominal cavity, attached to the posterior abdominal wall by the **ovarian mesentery or mesovarium**. In humans, it shifts slightly to a pelvic location. The ovary, like the testis, grows slowly but steadily in size during early life. As in the testis, little of this growth is due to the germ cells themselves. However, quite unlike the situation in the testis, the ovarian germ cells are undergoing the major changes described in Chapter 3. Over this prenatal period the output of steroids by the ovary is minimal and, indeed, removal of the ovary at any stage up to puberty does not affect prepubertal development. In some species, including humans, there may be a transitory stimulation of ovarian endocrine activity spanning the period of birth, but this does not appear to be important for female development. As for the male, it is at puberty that gonadal activity resumes to herald the onset of fecundity.

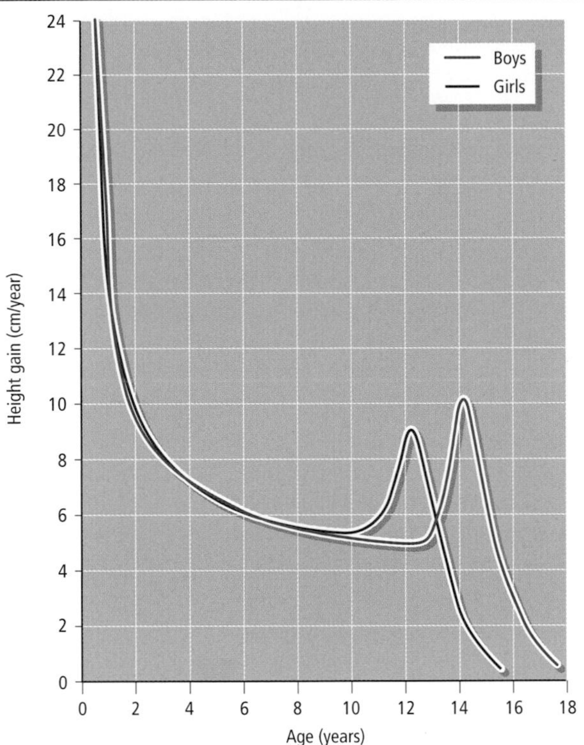

Figure 4.4 Growth velocity curves for boys and girls. Note the later time of 'take-off' in boys, which generally ensures a greater height at the start of the adolescent growth spurt. Also note that average peak height velocity is 9 cm/year for girls and 10 cm/year for boys. (Source: Tanner J.M. (1962) *Growth at Adolescence*. Blackwell Scientific Publications. Reproduced with permission of John Wiley & Sons.)

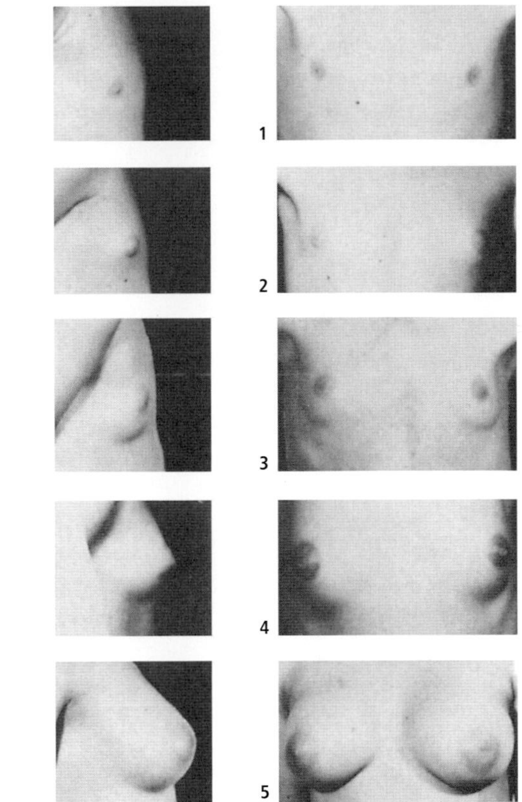

Figure 4.5 Stages of female breast development (regulated by oestrogen): 1, preadolescent stage during which the papilla (nipple) alone is elevated; 2, breast bud stage in which the papilla and breast are elevated as a small mound and the areolar area increases; 3, continued enlargement of the breast and areola, but without separation in their contours; pigmentation increases; 4, further breast enlargement but with the papilla and areola projecting above the breast contour; 5, mature stage in which the areola has become recessed, and forms a smooth contour with the rest of the breast – only the papilla is elevated. Classification of this stage is independent of breast size, which is determined principally by genetic and nutritional factors. (Source: van Wieringen J.C., Wafelbakker F., Verbrugge H.P., de Haas J.H. (1971) Growth Diagrams 1965 Netherlands: Second National Survey on 0–24-year-olds. Reproduced with permission from LWW.)

Reproductive changes

In addition to growth and changes in body composition, development of the **secondary sexual characteristics** occurs at puberty, e.g. **breasts** (Figure 4.5), **genitalia** and **pubic hair** (Figures 4.6 and 4.7), and **beard growth** and **voice change**. In women, ovarian oestrogens regulate growth of the breast and female genitalia, but **androgens secreted by the ovary and the adrenal gland** control the growth of female pubic and axillary hair. Testicular androgens not only control development of the genitalia and body hair in boys, but also, by **enlarging the larynx and laryngeal muscles**, lead to **deepening of the voice**.

These various characteristics develop at different chronological ages in different individuals (Figure 4.8). However, the **sequence** in which the changes occur is quite characteristic for each sex (Figure 4.9). This is important for the clinician, who can make comparisons between individuals, populations and cultures, or detect abnormalities using **staging criteria** (summarized in Figures 4.5–4.7). For example, a boy with advanced penile and pubic hair growth but small testes must have a non-gonadal source of excess androgen, such as congenital adrenal hyperplasia or an adrenal tumour.

Behavioural aspects

Puberty is the time of major physical and reproductive changes. Simply experiencing these can be deeply confusing. However, these changes are compounded by the continuing development of the brain towards a full adult functional capacity involving both emotional and cognitive functions. This process is accompanied by the formation and assertion of an independent adult identity, encompassing fundamental shifts in social relationships with parents, peers and romantic partners, a key component of which is learning to manage one's emotional reactions. All these changes and demands can

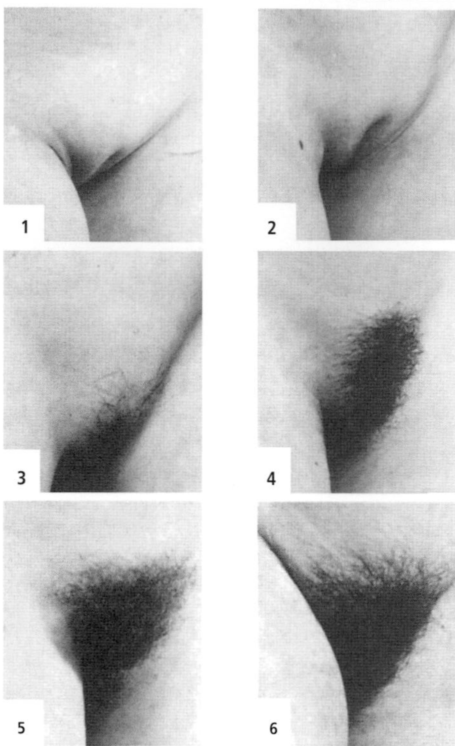

Figure 4.6 Stages of pubic hair development in girls (regulated by adrenal androgen): 1, no pubic hair is visible; 2, sparse growth of long, downy hair which is only slightly curled and situated primarily along the labia; 3, appearance of coarser, curlier and often darker hair; 4, hair spreads to cover labia; 5, hair spreads more over the junction of the pubes and is now adult in type but not quantity; no spread to the medial surface of the thighs; 6, adult stage in which the classical 'inverse triangle' of pubic hair distribution is seen, with additional spread to the medial surface of the thighs. Pubic hair maturation is accompanied by apocrine odour development, and skin oiliness and acne. (Source: van Wieringen J.C., Wafelbakker F., Verbrugge H.P., de Haas J.H. (1971) Growth Diagrams 1965 Netherlands: Second National Survey on 0–24-year-olds. Reproduced with permission from LWW.)

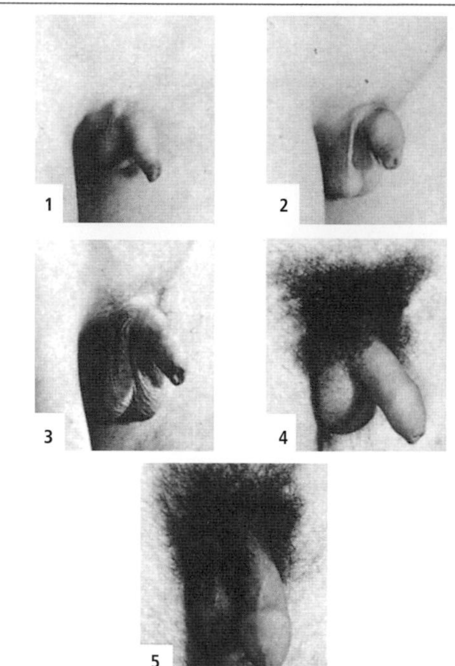

Figure 4.7 Stages of external genitalia development in boys (regulated by androgens): 1, preadolescent stage during which penis, testes and scrotum are of similar size and proportion as in early childhood; 2, scrotum and testes have enlarged; texture of the scrotal skin has also changed and become slightly reddened; pubic hair is at the base of the penis and downy; 3, testes and scrotum have grown further but now the penis has increased in size: first in length and then in breadth; facial hair appears for the first time on the upper lip and cheeks, and pubic hair is longer and more extensively distributed; 4, further enlargement of the testes, scrotum (which has darkened in colour) and penis; the glans penis has now begun to develop; 5, adult stage; facial hair has now extended to lower lip and chin. Hair on the chest, back, abdomen and more of the face is genetically variable and starts appearing 3 or so years after stage 5 is achieved. (Source: van Wieringen J.C., Wafelbakker F., Verbrugge H.P., de Haas J.H. (1971) Growth Diagrams 1965 Netherlands: Second National Survey on 0–24-year-olds. Reproduced with permission from LWW.)

Figure 4.8 Schematic of five boys all aged 14 years illustrating marked individual differences in physical maturation at the same chronological age. (Source: Tanner J.M. (1962) *Growth at Adolescence*. Blackwell Scientific Publications. Reproduced with permission of John Wiley & Sons.)

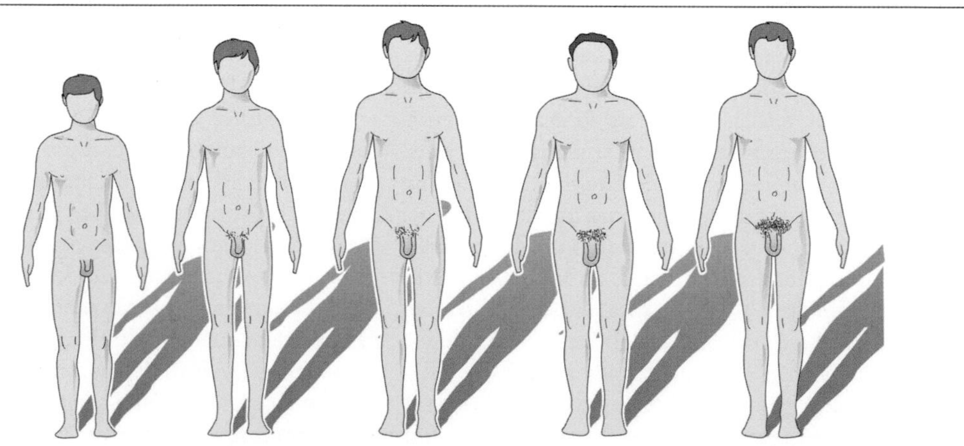

(a)

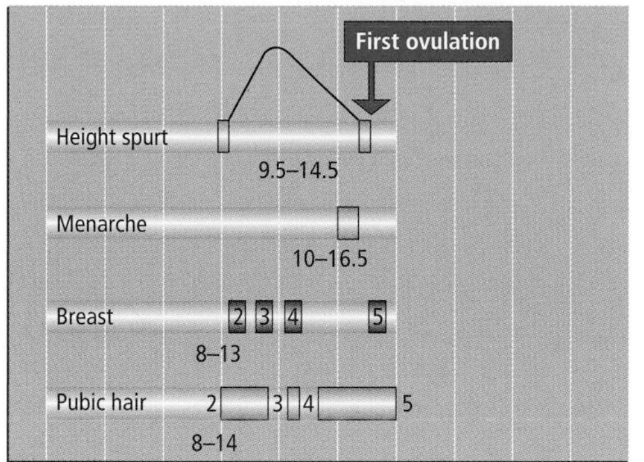

(b)

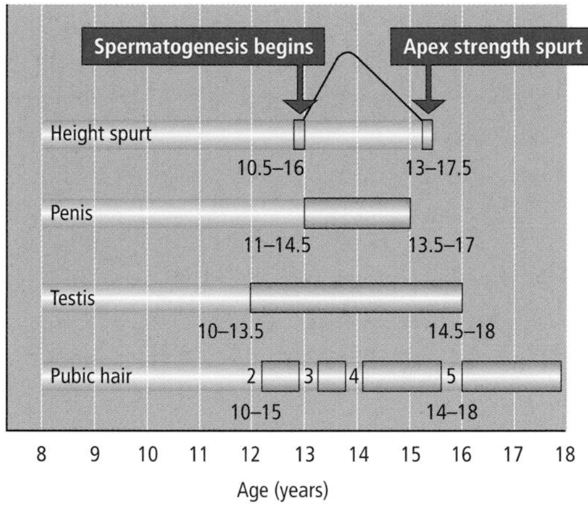

Figure 4.9 Summary of the sequence of events during puberty in (a) girls and (b) boys. The figures below each symbol represent the range of ages within which each event may begin and end. The figures within each symbol refer to the stages illustrated in Figures 4.5–4.7. (Source: Tanner J.M. (1962) *Growth at Adolescence*. Blackwell Scientific Publications. Reproduced with permission of John Wiley & Sons.)

render adolescence **a time of behavioural turmoil and hyperemotionality**. For example, adolescents can engage in more **reward-seeking** and **risky behaviours** often in **conflict with authorities** such as their parents, and in a greater **susceptibility to peer pressure**. Some mental health problems are increased, especially during early adolescence, such as **depression, conflict, mood swings, self-harming and suicidal intent**. The underlying causes of these behavioural manifestations are several, and include awareness of changing social expectations, physicality and sexuality, and the first strong experiences of romantic love, jealousy and rejection externally, whilst internally they are experiencing fluctuating hormone levels, which exert direct and lasting effects on brain functioning, and the asynchronous maturation of those brain regions concerned with emotional and cognitive functions, the latter not achieving full adult status until the mid-20s. Sex differences in brain development are also observed (discussed on pages 79–81).

Whilst most adolescents experience these behavioural changes, the extent to which they become and remain problematic varies with culture and individual circumstance. Thus, adolescence is generally less difficult for individuals who have been in **securely attached relationships since birth** or where there is **little perceived conflict between the 'emergent self' and the surrounding cultural values**. For example, gay adolescents in sexually intolerant cultures experience mental and associated physical health problems disproportionately, especially if they fear they will not be supported by their family (see also page 90).

A distinctive pattern of hormonal changes underlies puberty

Surprisingly, the earliest detectable endocrine change associated with puberty is a progressive increase in the plasma concentration of adrenal androgens, particularly dehydroepiandrosterone (DHEA) and its sulphated variant DHEAS. This rise is called **adrenarche**, occurs only in the great apes and humans and spans years 8–15 of age. Its significance for puberty is unclear. It does not seem to be essential for puberty to occur, but might be a marker for the termination of the period of rapid brain growth often seen as underlying evolutionarily the delayed puberty in great apes.

The **key effectors of pubertal change are the gonadal steroids**. It is now clear that the output of **pituitary gonadotrophins** (see Table 2.4) drives these changes in steroid activity. During the childhood and juvenile stages, gonadotrophin output and blood levels remain very low in both sexes (Figure 4.10a). From puberty onwards, pulses of both follicle stimulating hormone (FSH) and luteinizing hormone (LH) become evident, and mean levels rise gradually to reach adult levels (Figure 4.11b). Initially most gonadotrophin secretion occurs at night during sleep (Figure 4.10b). In late puberty, daytime LH pulses also increase (Figure 4.10c), but are less than those still occurring at night, until finally the adult pattern of higher basal levels is achieved with no pulsing variation through the 24-hour cycle.

In boys, testosterone plasma levels follow the gonadotrophins. Thus, early in puberty, testosterone levels in boys rise at night when LH secretion becomes elevated (Figure 4.10b) followed later by increases during the day, the greatest changes appearing during pubertal stage 2, when testosterone concentrations may change from 0.2 to 2.4 ng/ml. There are smaller increases in plasma testosterone concentrations in girls between pubertal stages 1 and 4.

In girls, oestradiol levels rise consistently from low levels through the stages of puberty to reach the concentrations seen in mature women (see Figure 4.11a). In boys, plasma

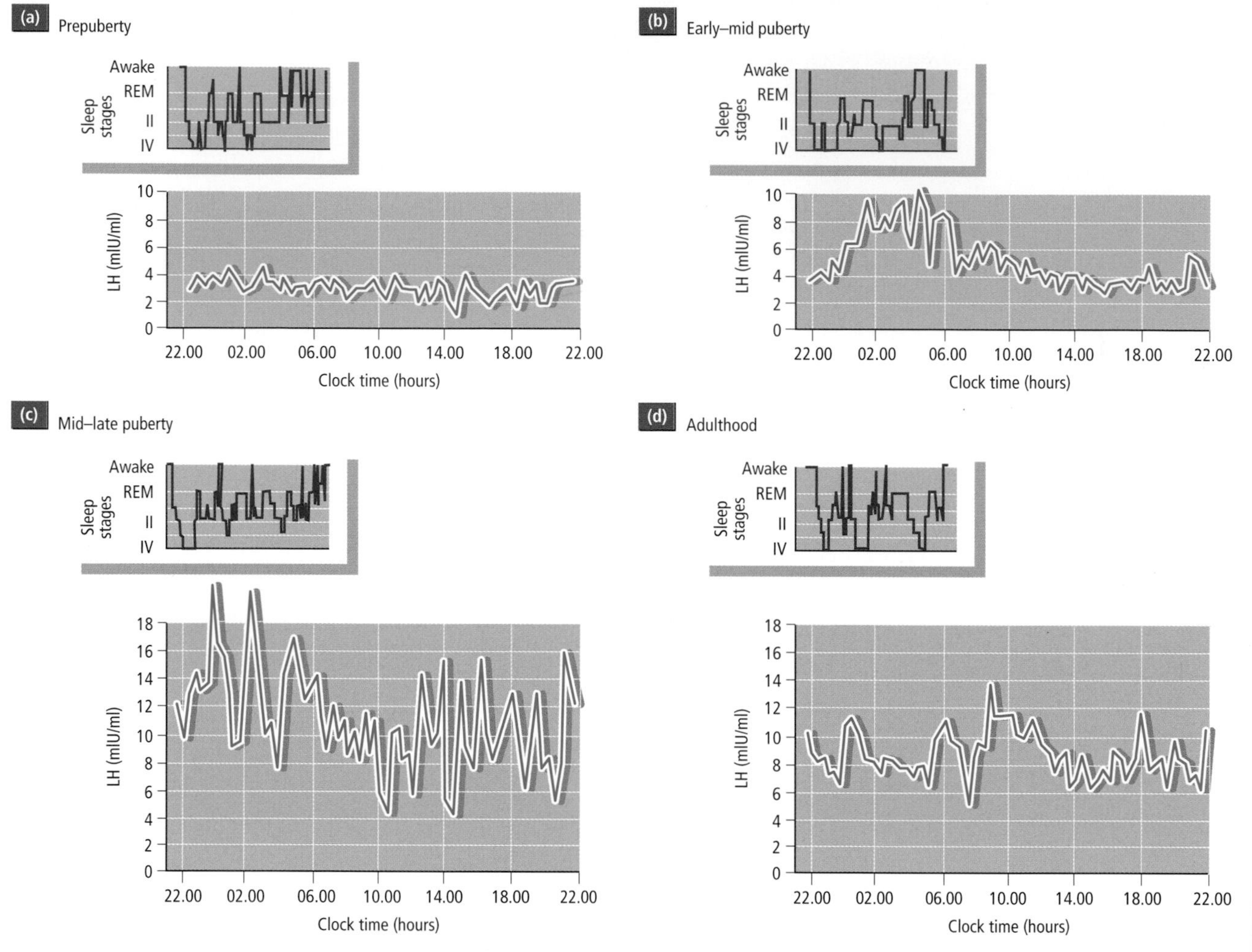

Figure 4.10 Plasma LH concentrations over a 24-hour period in: (a) a prepubertal girl (9 years); (b) an early pubertal boy (15 years); (c) a late pubertal boy (16 years); and (d) a young adult male. The sleep pattern for each nocturnal sleep period is depicted in the top left-hand corner of each graph (REM, rapid eye movement or 'paradoxical' sleep). Note the marked daily rhythm in (b), with sleep-augmented LH secretion, and the overall higher LH concentrations in (d) compared with (a), but no clear daily rhythm in either. Note that prior to the juvenile quiescent phase in higher primates, low frequency pulsing of GnRH and gonadotrophins is observed, but is switched off at the end of the infant phase, and, interestingly, is switched off for longer and more completely in boys than in girls. (Source: Grumbach M. (1978) *Reproductive Endocrinology* (eds S.S.C. Yen & R. Jaffe). WB Saunders & Co, Philadelphia. Reproduced with permission from Elsevier.)

concentrations of oestrone are higher than oestradiol, but both are considerably lower than in girls at comparable stages of puberty. In males, about half the oestradiol is derived from extraglandular aromatization of testosterone, and a quarter, or less, from testicular secretion.

In summary, the peripubertal period is associated with the activation of the gonads (and adrenals), which results in elevated steroid secretion. These events are, in turn, dependent on increased trophic stimulation by FSH and LH. In the next section, the control of these trophic stimuli, the neuropharmacological regions of the central nervous system (CNS) involved, and the nature and timing of the trigger that induces them are examined.

The central nervous system plays a key role in the onset of puberty

The distinctive peripubertal pattern of gonadotrophin output suggests a **central maturational role for the CNS** and, in particular, the hypothalamus at puberty. Is there evidence to support this view?

Hypothalamic GnRH secretion drives pubertal development

In Chapter 2 (see page 28), we introduced the small neuropeptide hormone: GnRH (gonadotrophin releasing hormone; see Figure 2.11). GnRH is released from the hypothalamus,

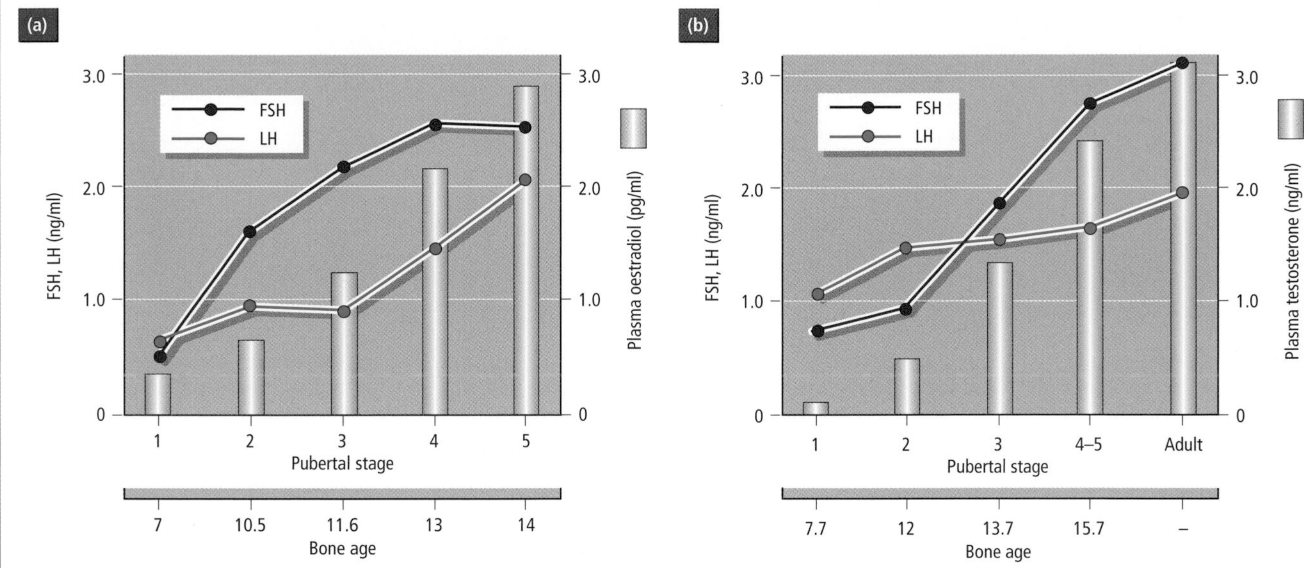

Figure 4.11 Plasma concentrations of gonadotrophins and steroid hormones during various pubertal stages in (a) girls and (b) boys. Bone age is assessed by examining radiographs of the hand, knee and elbow, for comparison with standards of maturation in a normal population. It is an index of physical maturation, and better correlated with the development of secondary sexual characteristics than chronological age. (Source: Weitzmann E.D. (1975) *Recent Progress in Hormone Research*. Academic Press, New York. Reproduced with permission from Elsevier.)

passes via the portal system to the pituitary and there acts on GnRH receptors to release the gonadotrophins that in turn act on the gonads. Histochemical techniques have localized GnRH production to two major subsets of hypothalamic neurons: a diffuse group of neurons within a continuum centred on the **medial preoptic** and adjacent **anterior hypothalamic areas**, and a smaller cluster in the **arcuate nucleus** (see Figure 2.5b). Nerve terminals from these areas containing GnRH are associated with the portal capillaries of the lateral palisade zone of the median eminence (see Figure 2.5b), which is the primary site of GnRH neurosecretion into the portal vessels. Any pathological, experimental or genetic abnormality in GnRH synthesis, storage, release or action results in partial or complete failure of puberty and of gonadal function.

In Figure 4.10c,d, **LH output is seen to be pulsatile**. Concurrent sampling of portal blood to measure GnRH secretion and of peripheral blood to measure gonadotrophins has shown that both pulses occur approximately once an hour (**circhorial**). Thus, each peripheral LH peak reflects an underlying pulsed release of GnRH. This pulsatility is crucial for the effective functioning of GnRH. Thus, where failure of GnRH production or release has occurred, the consequences can be reversed by use of an intravenous infusion pump programmed to deliver exogenous GnRH – but only when delivered in pulses at a frequency approximating that seen naturally. In contrast, continuously infused GnRH does *not* restore function. This critical requirement for pulsatile GnRH secretion arises from the effects that GnRH has on its receptors on the gonadotrophs. Thus, a first GnRH pulse triggers an initial release of stored LH/FSH within minutes and lasts for 30–60 minutes. At the same time, more secretory granules containing LH/FSH move into a zone beneath the plasmalemma. As a consequence of this **self-priming** granule mobilization, a second exposure to GnRH results in a much larger 'primed' release of LH. Exposure to pulsatile GnRH also stimulates gonadotrophin biosynthesis. However, after binding, some GnRH–receptor complexes are internalized to lysosomal structures where degradation of the peptide can occur. So, continuous exposure of gonadotrophs to GnRH results in maintained occupancy of the receptors and is followed eventually by their wholesale internalization and by receptor down-regulation. GnRH can then no longer stimulate pituitary LH and FSH production or secretion.

That an increased pulsatile GnRH secretion provides the mechanism underlying puberty comes from experiments in juvenile female rhesus monkeys, which were provided with an external pump delivering pulses of GnRH intravenously at hourly intervals. As can be seen in Figure 4.12, these pulses lead to rising LH, oestradiol and progesterone outputs that can support premature menstrual cycles. Similar experiments on male juveniles give a comparable male maturational outcome. Clinical studies confirm these experimental results. Thus, precocious puberty may occur in children as young as 2 years of age, often as a result of a CNS tumour activating GnRH pulses prematurely. Conversely, puberty can be delayed or reversed by administering

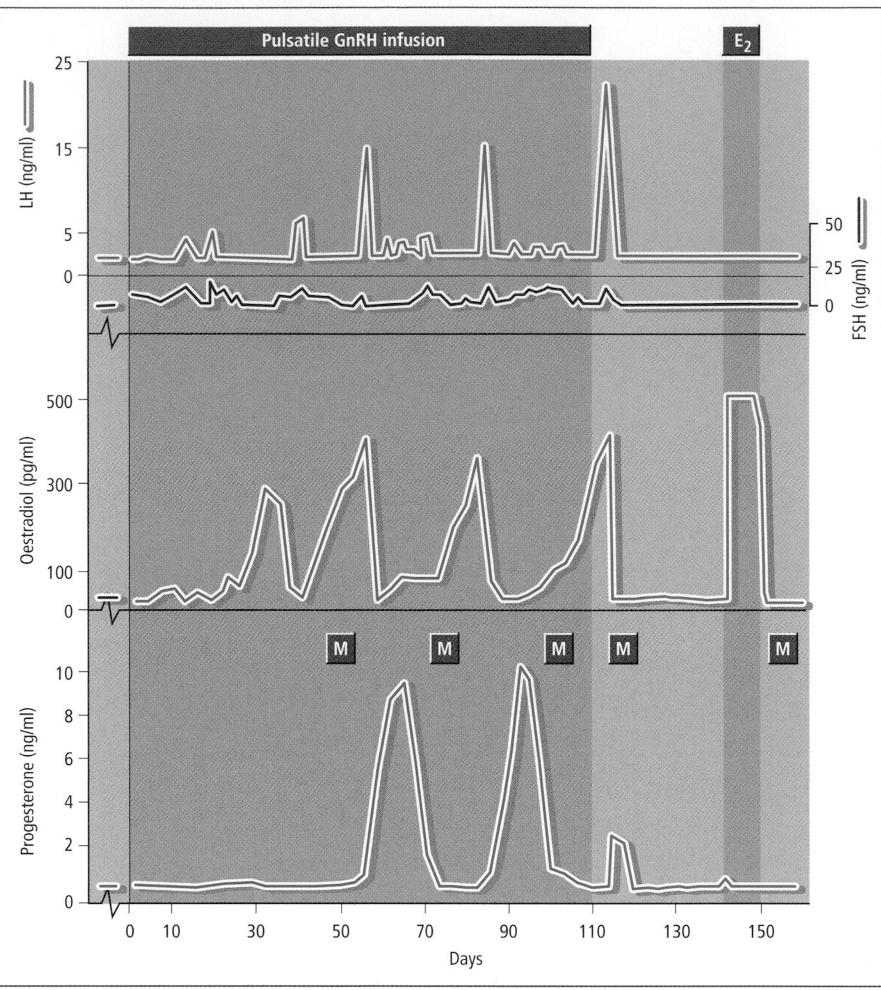

Figure 4.12 Induction of premature menstrual cycles in an immature female rhesus monkey by the infusion of GnRH (1 mg/min for 6 minutes once every hour) shown by the horizontal bar (day 0–110); levels of LH, SH, oestradiol and progesterone were undetectable in blood samples prior to GnRH infusions, but rose to produce a full menstrual cycle each with a period of menstrual flow (M). Cessation of GnRH infusions was followed by prompt re-entry into a non-cyclic, prepubertal state.

GnRH inhibitors or stimulated if overdue by injecting GnRH analogues (see Box 2.1).

These data all suggest that the **most important event in the initiation of puberty is activation of the hypothalamic mechanism, which delivers GnRH pulses to the anterior pituitary**. Once this has occurred, the pituitary and ovary are able to respond instantly. Figure 4.12 also shows rather dramatically that switching off the GnRH pump is followed by re-entry into the immature, prepubertal state. This reversibility indicates that simply exposing the hypothalamus to adult levels of circulating steroids does not contribute to the attainment of a 'mature' pattern of functioning. These experiments provide a convincing demonstration that puberty arises solely as a consequence of a maturational event within the CNS, translated to the pituitary–gonadal system as a stream of GnRH pulses. Thus, the genesis of puberty must therefore be sought within the CNS. Indeed, juvenile prepubertal GnRH neurons are full of RH, suggesting that they are ready to secrete but simply lack a signal. So how is a GnRH secretion restrained or damped during the juvenile years? And what triggers its reactivation at the onset of adolescence?

What neural mechanisms regulate GnRH secretion at puberty initiation?

The neural sites mediating puberty initiation must involve the hypothalamus, given that the GnRH neurons provide the final common effector pathway. Although clinical cases of advanced or delayed puberty can be associated with an underlying neuropathology in the hypothalamus, genetic studies have proved to be more informative. Thus, cases of delayed or absent puberty have been associated with inactivating genetic lesions in a gene encoding the 54-amino acid neuropeptide transmitter, **kisspeptin 1** (***Kiss1***), or in the gene for its receptor ***Kiss1R*** (also known as ***Gpr54***). Additionally, mutations of either the *KISS1* or *KISS1R* genes that render each constitutionally active have been found in children with precocious puberty. In primates (including humans), kisspeptin-expressing neurons are present in the arcuate nuclei with neurons that project to the hypothalamic median eminence, where their terminals associate intimately with GnRH-containing neurons that themselves express the KISS1r receptor. KISS1 is released at this site in pulses that match the pulsed release of GnRH. *Kiss1* gene expression rises, along with the number of Kiss1-positive neurons, as puberty is initiated in female monkeys, and the administration of exogenous kisspeptin to juvenile monkeys activates GnRH output and thereby gonadotrophin secretion, and advances puberty. Kiss1R antagonists block Kiss1's effect on LH and FSH output in female rats. Thus, **kisspeptin-Kissr seems to be an essential part of the pathway controlling GnRH secretion**. However, it is likely that kisspeptin 1 is just one of several neurotransmitters that form a network of balancing regulatory factors controlling the onset of active GnRH production at puberty and its suppression prior to puberty (Box 4.2; Figure 4.13).

How are the neural mechanisms activated?

We now consider what is known of how puberty timing might be triggered via the stimulation of kisspeptin, GnRH and then gonadotrophin pulses to activate the gonads. The factors responsible for triggering and timing the onset of puberty have proved elusive. Is there a clock built into the brain, or is the brain monitoring responsively some parameter of ageing or growth? Figure 4.14 indicates a **secular trend towards an earlier menarche in girls and puberty in boys in Western Europe and the USA from the mid-19th century to the mid-20th century** – although this trend may have plateaued or even reversed over the past 50 years. What factors have changed that might have contributed to the earlier attainment of sexual maturity, and do they give any insight into the mechanisms controlling the initiation of puberty?

Clearly, there may be more than one answer to these questions. Health care and personal health have improved during this time, along with living conditions and socio-economic standards. Experimental and clinical studies have implicated two factors that might explain this trend to earlier puberty, and in the mechanisms underlying pubertal

Box 4.2 Neuropharmacological systems and the regulation of GnRH output

A network of inhibitory and facilitatory neurons is implicated in the initiation of pubertal output of GnRH (as well as in its regulation in the mature adult; see Box 10.3).

Negative effects on GnRH output are exerted by two neurotransmitter systems: **γ-aminobutyric acid (GABA)** present in the preoptic and mediobasal hypothalamic areas (yellow in Figure 4.13) and the **opioid β-endorphin** in a subset of arcuate nucleus neurons that richly innervate the medial preoptic area GnRH-containing neurons (green in Figure 4.13). Whilst the use in juveniles of opioid antagonists such as naloxone has not, under a range of conditions, led to increased GnRH pulsing, reduced GABA activity has been clearly implicated in pubertal activation in females. Thus, microdialysis experiments in the rhesus hypothalamus showed that GABA release became depressed as Kiss1 activity and GnRH output rose. In addition, premature juvenile discharge of GnRH can be achieved either by down-regulating GABA synthesis or by blocking its interaction with its receptor using bicuculline. In humans, treatment of childhood premature gonadarche with a GABA agonist decreased gonadotrophin output and regressed the pubertal changes. Agonist use has also been reported to delay puberty. Thus, release from GABAergic inhibition is implicated in puberty initiation.

Conversely, three main neurotransmitter systems exert positive effects on GnRH output: hypothalamic **glutamate** release in the vicinity of the GnRH cell bodies (but not terminals) is elevated at puberty initiation. GnRH neurons express glutamate receptors and respond electrically to stimulation by glutamate analogues such as NMDA. Moreover, the sustained pulsatile administration of NMDA prepubertally results in increased GnRH pulsing and precocious puberty, and glutamate receptor antagonists interrupt the GnRH pulsing. In addition, **noradrenergic** neurons in the brainstem medulla oblongata (black in Figure 4.13) richly innervate the hypothalamus and synapse directly on both preoptic and median eminence GnRH+ve cell bodies, which express both α and β adrenergic receptors. In adult monkeys, α-adrenergic stimulation facilitates GnRH pulsatility permissively, and after ablation of noradrenergic tone GnRH pulses decline. However, it is less clear that this system is involved in pubertal initiation.

More recently, patients with pubertal failure associated with severe congenital gonadotrophin deficiency have been identified and found to be homozygous for loss-of-function mutations in the gene ***TAC3*** (encoding a **neuropeptide Neurokinin B**) or that for its receptor ***TACR3*** (encoding ***NK3R***). Neurokinin B is co-expressed in hypothalamic arcuate nucleus neurons that also express kisspeptin 1, thereby implicating it in the positive regulation of GnRH output. Interestingly, the Kiss1 neurons also contain a third peptide **dynorphin** (which has led to these neurons being called **KNDy cells**). Dynorphin has opioid activity, and acts to oppose the release of GnRH. Significantly, receptors for both dynorphin and neurokinin B are also found on these KNDy cells, suggesting that an autofeed back loop may be operating. Thus, the KNDy cells may act as pulse generators for the pulsatile release of kisspeptin to the GnRH neurons.

Further reading

Topaloglu AK, Reimann F, Guclu M *et al*. (2009) TAC3 and TACR3 mutations in familial hypogonadotropic hypogonadism reveal a key role for Neurokinin B in the central control of reproduction. *Nature Genetics* **41**, 354–358.

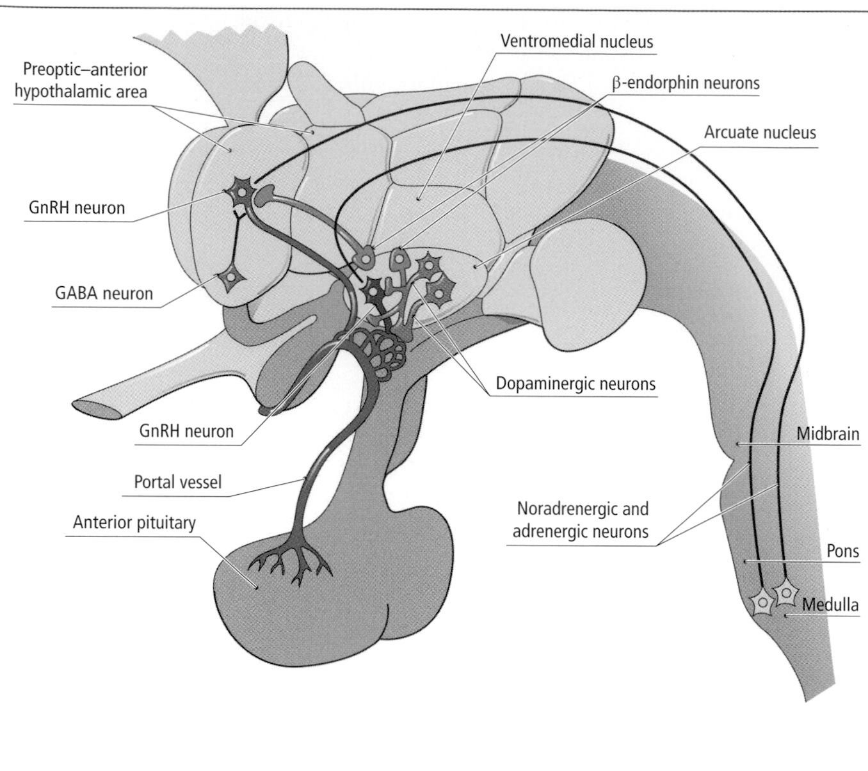

Figure 4.13 Schematic diagram to show some of the postulated neurochemical interactions that may control GnRH secretion. In primates, GnRH neurons lie in the medial preoptic area (blue) and arcuate nucleus (purple) and project to the portal vessels in the median eminence, especially to the lateral palisade zone (see Figures 2.4–2.6). γ-Aminobutyric acid (GABA) neurons (yellow) in the preoptic and mediobasal hypothalamic areas and opioid β-endorphin neurons (green) richly innervate the medial preoptic area GnRH-containing neurons to influence output of GnRH. Dopamine neurons (mauve) in the arcuate nucleus modulate prolactin release (and may affect GnRH output, but this is controversial). Noradrenergic and adrenergic neurons (black) in the medulla oblongata project to the medial anterior hypothalamus and preoptic area, and may also participate in the regulation of GnRH secretion (see Box 4.2). (Source: Moore R.Y. (1978) *Reproductive Endocrinology* (eds S.S.C. Yen & R. Jaffe). WB Saunders & Co, Philadelphia. Reproduced with permission from Elsevier.)

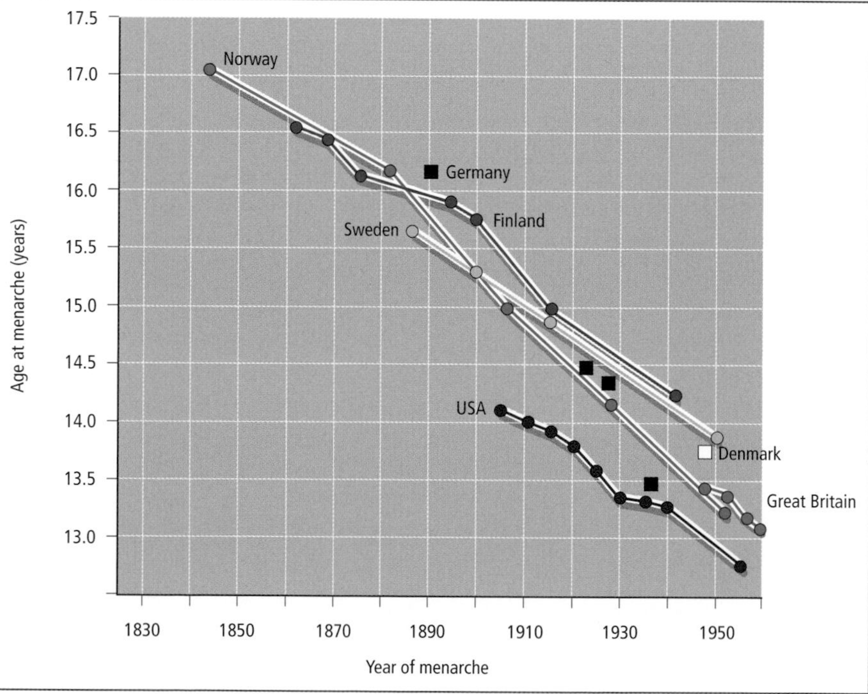

Figure 4.14 Secular trend towards an earlier age at menarche in girls from Western Europe and the USA.

onset. These are photoperiod and nutrition. In many animals **the light:dark cycle (day:night ratio) is known to influence fertility** profoundly (Box 4.3). Western society has used electricity to artificially extend the length of the day, such that daily dark periods may be consistently as short as 7–8 hours all the year round, as if we are in constant 'long days' or a summer photoperiod. Although such changes in day length might advance puberty, there is little direct evidence to support the idea. Nutrition, in contrast, does indeed seem to be important.

Box 4.3 Evidence that light–dark cycles influence reproduction in animals

There are two levels at which control over the timing of reproduction has evolved.

Circadian rhythms control reproductive function in some species

The temporal control of reproductive activity in females is complicated because the production of a viable oocyte is itself a cyclical event that must be matched to other cyclical events occurring within the life of an animal. In a nocturnally active rodent, for example, potential encounters with mates will be restricted to the hours of darkness. This selection pressure has led to the development of a reproductive strategy that ensures that both maximum female fecundity (ovulation) and receptivity to males occur at night. The likelihood of conception is therefore maximized. Reversal of the light–dark cycle causes a 12-hour shift in the timing of ovulation, demonstrating the essential role of information about light in setting the neural timer. The timing mechanism underlying this response to a change in the light–dark cycle lies in the hypothalamus in the **suprachiasmatic nuclei** (see Figure 2.4a). Neurons in this nucleus show an intrinsic activity rhythm of roughly 24 hours, or **circadian periodicity** (circa = approximately, diem = day). This oscillator continues to run even in continuous darkness, and is thus an example of a **self-sustaining biological clock** (or oscillator). However, it can be **entrained to the ambient light–dark cycle**, through photic stimuli which reach the suprachiasmatic nuclei via a direct retinal input, the **retinohypothalamic tract**. Lesions of the suprachiasmatic nuclei disrupt many circadian functions, including ovulation. It is important to realize that this circadian system controls a wide range of other behavioural and endocrine rhythms, both reproductive and non-reproductive, which are held in a very strict, temporal relationship to each other. Female primates have more flexible patterns of reproductive activity than rodents, and both ovulation and mating may occur at any time of day.

Circannual rhythms control reproductive function in seasonal breeders

In seasonal environments, where adverse climate and the availability of food are major determinants of offspring survival, and therefore of the reproductive success of the parents, it is adaptive to ensure that young are born in the spring or early summer. This tight control over birth season, apparent in many domestic and wild species, is achieved by a precise restriction on the **month(s) of fecundity** and hence the timing of conception. In species with short gestation times, such as hamsters and birds, winter is a time of infertility with gonadal development suspended until spring. In species with longer gestation times, such as sheep and deer, the anticipation of spring must begin much earlier and seasonal changes in autumn act as a stimulus to reproductive function. This leads to the dramatic spectacle of the **rut** when animals that have been reproductively quiescent for the entire year suddenly become sexually active. Males may develop pronounced secondary sexual features, such as antlers, become fertile, aggressive and territorial, and spend their whole time engaged in an intense competition for access to females. Females come into heat and actively show interest in males, accepting their attempts to copulate.

In some species, such as deer and ground squirrels, there is good evidence that seasonal cycles are under the control of an endogenous **circannual oscillator**, a biological clock with a period of approximately 1 year. In other species, there is no endogenous rhythmicity and the seasonal rhythms observed in the field are triggered by cyclical stimuli within the environment. Of these, **photoperiod** is by far the most important. This is exemplified in the laboratory, where artificial manipulation of day lengths can be used to drive all of the components of the annual reproductive cycle. For example, exposure of Syrian hamsters to less than 12.5 hours of light per day (pseudowinter) leads to gonadal atrophy and the loss of sexual behaviour. In contrast, these short photoperiods stimulate gonadal activity in species, such as sheep, that normally mate in the autumn. All of these effects are mediated by changes in the frequency of the GnRH pulse generator in the hypothalamus, which then determines the level of secretion of gonadotrophins and steroids.

Photic circannual information obviously has access to the GnRH neurons, but does it use the same pathways that are involved in the circadian control of reproduction? Certainly, the suprachiasmatic nuclei have an important role to play because lesions of these structures completely block photoperiodic sensitivity. However, the pathways involved are not exclusively intrahypothalamic. It is now well recognized that the **pineal gland**, which sits over the dorsal midbrain, attached to the **epithalamus** in the posterior–dorsal third ventricle (see Figure 2.2), is the mediator of photoperiodic time measurement. Removal of the gland or interruption of its sympathetic innervation leaves animals insensitive to changing day length. The primary pineal hormone, **melatonin** (the indolamine: N-acetyl-5-methoxytryptamine; see Figure 2.10b), is synthesized and released into the bloodstream only in the hours of darkness, exhibiting a true circadian rhythmicity driven by the suprachiasmatic nuclei. At night, the circadian signal increases sympathetic activation of the gland, resulting in a dramatic rise in the activity of the enzyme, **N-acetyl transferase**, the rate-limiting step in melatonin biosynthesis.

The crucially important feature of the circadian melatonin signal is that it provides a precise representation of the length of the night, so that as days shorten and nights lengthen in autumn, the duration of the nocturnal melatonin peak is increased. Conversely, after the winter solstice, the photoperiod increases and the duration of the melatonin signal falls. The changing shape of the rhythm of circulating melatonin is detected within the hypothalamus and somehow leads to alterations in GnRH secretion. The melatonin signal is such a powerful regulator of neuroendocrine state that in pinealectomized animals the entire reproductive axis can be turned on or off by repeated nightly administration of programmed infusions of melatonin, which mimic the pattern of its secretion typical of either long or short photoperiods. The same melatonin signal leads to opposite neuroendocrine responses in spring-breeding species, since a progressive decrease in the duration of the night-time melatonin signal results in reproductive activation as days lengthen in the spring. Unsurprisingly, in humans, who reproduce throughout the year, melatonin does not seem to exert effects on reproduction.

Further reading

Srinivasan V, Spence WD, Pandi-Perumal SR, Zakharia R, Bhatnagar KP, Brzezinski A (2009) Melatonin and human reproduction: shedding light on the darkness hormone. *Gynecological Endocrinology* **25**, 779–785.

Body weight appears to be a critical determinant of pubertal activation

In Figure 4.15, it can be seen that although age at menarche changed considerably between the mid-19th and mid-20th century, the **body weight at menarche remained surprisingly constant** at about **47 kg for females**. A comparable observation of a critical weight for **boys of 55 kg** underlying sexual maturation has also been made. Similar constancy is seen in the weight at onset of the adolescent growth spurt. These associative data have suggested a causal relationship between a critical weight and activation of the hypothalamic–pituitary– gonadal axis, and thus the timing of the growth spurt. According to this view, **body weight, or something associated with it, may trigger and therefore time the onset of puberty**. The earlier occurrence of puberty today may therefore be explained by improvements in nutrition, health care and social living conditions. Indeed, examination of cultures, such as the nomadic Lapps, that until recently have not experienced such major improvements in living standards and nutrition, show that between 1870 and 1930 there was little or no trend towards an earlier menarche.

Direct evidence in support of this view is, at first sight, abundant. Moderately obese girls experience an earlier menarche than lean girls. Malnutrition is associated with delayed menarche. Primary amenorrhoea is extremely common in ballet dancers at professional schools who are in the very low range of weight for height and relative fatness for their age. Adolescent girls with the complex syndrome of anorexia nervosa, who have a very low food intake (particularly carbohydrate) and body weight, show primary amenorrhoea and/or delayed puberty. Moreover, the amenorrhoea is associated with a body weight below 47 kg, while in some anorexic girls who begin to feed, the recurrence of menstruation is associated with attainment of a 47-kg body weight.

This impressive array of supportive data also seems like common sense. Thus, the attainment of a body size sufficient to cope with the demands placed on it by adult reproductive activities such as pregnancy (some 50 000 kcal) would be a logical signal for the onset of puberty. However, whether it is simply a total body weight threshold that is monitored is controversial. Thus, the time of onset of puberty might be related causally to other parameters, such as lean body weight, absolute or proportionate body fat, or total body water. Indeed, it is quite **difficult to predict age of menarche from knowledge of an individual's body weight and weight gain**. Second, many anorexic girls who re-attain their critical body weight do not begin having menstrual cycles, although this may simply reflect the fact that anorexia nervosa is a far more complicated syndrome than just a disorder of food intake and consequential weight loss. Third, whilst **obesity in girls seems to advance puberty, in boys it has less clear effects**. Finally, menarche is a rather late event in puberty, and may therefore be much removed from the critical factors that determine the onset of those endocrine changes described above, even if menarche itself is often related to body weight. So, if body weight *per se* is not obviously predictive of puberty onset, what might be and how? Recent studies in molecular endocrinology have explored the triggering signals that might reflect some aspect of body weight.

Hormonal involvement in the activation of puberty?

Of several hormones investigated for a possible role in monitoring some aspect of body growth or metabolism and thereby triggering puberty initiation, a polypeptide, **leptin**, has been implicated (Box 4.4). Leptin is produced by white adipocytes, so its output provides a measure of fat mass. It acts on the hypothalamus to reduce appetite and increase thermogenesis. Leptin is the product of the ***obese* gene**, mutations in which lead to increased food intake and lowered body temperature, in turn leading to obesity. Homozygous null mutant mice are also infertile, a feature reversed by treatment with leptin. So might leptin be involved in the activation of the

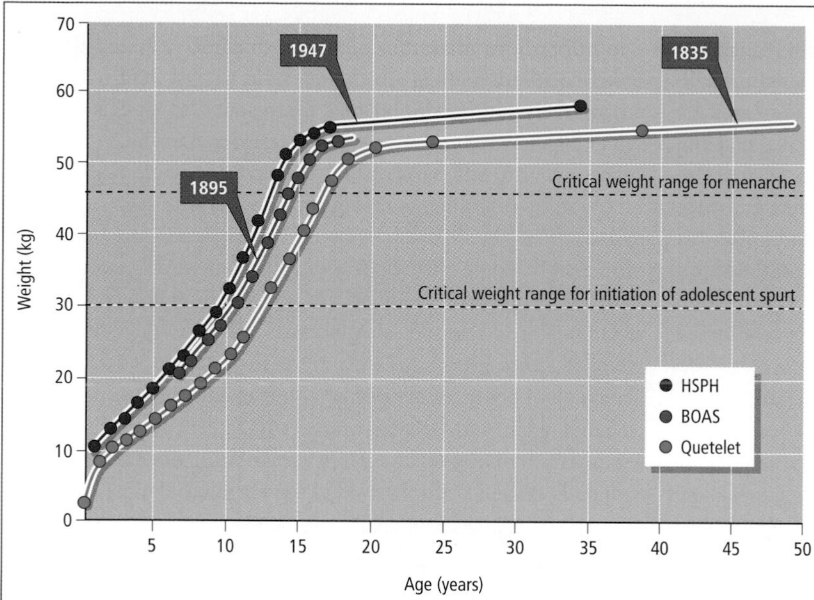

Figure 4.15 Age plotted against body weight in three populations of girls in 1835, 1895 and 1947. Note the constant weights at initiation of the growth spurt (30 kg) and at menarche (47 kg). (Populations: Belgian girls in 1835 from data of Quetelet, 1869; American girls in 1895 adapted from data of BOAS; and in 1947 adapted from data of Reed and Stuart, 1959.)

Box 4.4 Leptin

Leptin (from *leptos* or thin) is a 16-kDa single polypeptide chain protein of 167 amino acids encoded by the *OBESE* (or *OB*) gene on chromosome 7. It is secreted mainly by white adipocytes, its output providing a measure of fat mass. It acts by binding to its receptor called Lep-R (or Ob-R), which has multiple splice forms, of which the long one is the most important (see Figure 2.9, type 3).

Leptin is a member of a large family of peptides called the **fibroblast growth factors** (or **FGFs**). We have already encountered FGF9 in the context of testis differentiation (see page 40), and will encounter other members later, including leukaemia inhibitory factor (LIF; see page 219). All are monomeric single-chain peptides stimulating cells through a type 3 receptor interaction (Figure 2.9).

hypothalamic–pituitary–gonadal axis at puberty? Leptin does indeed rise during puberty (but in girls only) and patients with mutations in the leptin gene or that for its receptor fail to enter puberty. Moreover, this failure can be overcome by treatment with leptin. Infusion of leptin into the median eminence region of the hypothalamus also leads to GnRH release, but not through an action on GnRH neurons, which can secrete GnRH when lacking receptors for leptin. However, leptin is unable to initiate puberty precociously. Thus, it seems unlikely to be a trigger for GnRH secretion, but perhaps does provide some form of permissive or background presence that is essential for a triggering signal to operate – perhaps via kisspeptin 1, the expression of which is down-regulated under conditions of negative energy balance or starvation in pubertal rats, both conditions associated with delayed puberty, the latter being partially prevented by the supply of a Kiss1 agonist, despite the persistence of low body weight. There is some evidence to support the idea that leptin may work via an effect on Kiss1 neurons. Thus, the leptin receptor has been found on a subset of arcuate nucleus cells that express Kiss1, and leptin-deficient mice show depressed levels of *Kiss1* mRNA and of Kiss1 positive neurons. However, these studies were not done on pubertal mice, and the jury is still out on where and how leptin might work.

The thyroid hormones, **thyroxine (T4) and triiodothyronine (T3)**, are also heavily involved in body weight control and thermogenesis, and recently have also been implicated in puberty initiation in primates. Thus, the burst in pulsatile GnRH release at the termination of the juvenile phase of development has been shown in ablate-and-replace studies to depend on a permissive action of thyroid hormones. It is unclear, however, whether this action is mediated indirectly by an effect on somatic development or directly on the hypothalamus (see Box 4.2). It is also unclear whether and how this presumed action might reflect body mass.

Conclusions

Puberty is a crucial reproductive transition. In humans and higher primates there is a particularly prolonged period between birth and puberty initiation, which may mean that the features by which the timing in higher primates is controlled are unique. It is clear that activation of GnRH pulsing is involved and that the CNS via kisspeptin 1 provides the route by which activation is mediated. There is clear evidence linking some aspect of body weight increase and growth to the triggering of puberty, but as yet which aspect and the nature of the link between these two processes have not been identified.

Key learning points

- Male and female fetuses do show limited sexual divergence after they have formed.
- In male fetuses, the migration of the testes to a cooler scrotal position is important for later fertility and health.
- Migration is regulated by androgens and Insl3.
- Puberty is a state of transition (adolescence) between the juvenile and adult states.
- In primates, it is preceded by infancy (from birth to weaning), childhood (a period of dependence on adults for survival), and the juvenile stage (when survival without adults is possible).
- In higher primates, the onset of puberty is considerably delayed postnatally compared with other mammals.
- Puberty involves gonadarche in all mammals and adrenarche in great apes and humans.
- Physical changes include sexually dimorphic growth differences and differences in lean muscle and fat mass.
- Secondary sexual characteristics develop at puberty under the influence of the rising output of gonadal steroids.
- Adrenal androgens stimulate the appearance of pubic and axillary hair in both males and females.
- Characteristic increases in the secretion of GnRH, gonadotrophins and gonadal steroids occur during pubertal activation.
- Activation of pulsatile hypothalamic GnRH secretion is the key and primary event underlying gonadal activation.
- The activation of GnRH pulses is stimulated by Kisspeptin1.
- One of the first endocrine changes to occur is an increase in gonadotrophin secretion at night.
- Lesions in the CNS can be associated with advanced or delayed puberty, but are difficult to interpret mechanistically.
- There is a secular trend towards an earlier age of onset of puberty.
- This secular trend is primarily linked to attainment of a critical body weight at progressively earlier ages during the last century.
- Attainment of a critical body weight, or some related feature such as a critical body fat content, seems to be a key trigger to the onset of puberty.
- Endocrine markers of body development such as leptin are implicated in the onset of puberty, but are unlikely to be causally involved as its trigger.

Clinical vignette

Young athletes at puberty

A 15-year-old girl had trained in gymnastics to a high standard since early childhood. She trained for at least 2 hours six-days-a-week in the gymnasium with her coach and often undertook additional sessions at the weekends. She was been successful in her sport at regional level and under consideration for the national team. Her mother had concerns that her daughter had not menstruated at the age of 15, and took her to see a paediatric endocrinologist (a specialist in hormonal disorders in childhood) for assessment. The endocrinologist undertook a careful history and physical examination. He established that the girl had never menstruated, and that she had not grown significantly over the past year. She was 152 cm tall and weighed 40 kg. Her breasts were Tanner stage 2, and she had scanty axillary and pubic hair growth. She had female external genitalia of normal appearance. Blood tests revealed that her gonadotrophin levels were in the normal childhood range. A pelvic ultrasound scan revealed a small anteverted uterus and normal ovaries. She and her mother were reassured that puberty was underway, and that her development did not fall outside of the normal range. She subsequently had her first menstruation just prior to her 16th birthday.

It is not uncommon for girls who train as athletes during their peri-pubertal years to experience later puberty, particularly if they train for activities where low body mass is desirable. This is most often seen in ballet dancers, gymnasts, and endurance athletes. The precise reason for puberty being later in elite young athletes is not completely clear.

The intense physical training required in such sports is often assumed to be the cause of later puberty in these groups. However, it is important to note that young people are selected into these groups based on their physiques, and that the association between training and later puberty might be a result of their baseline physiology rather than an effect of training per se. Pubertal timing may also be influenced by other factors that commonly surround young athletes, for example, psychological stress, low calorie intake, and familial later puberty. It is notable that the normal range of puberty for female gymnasts overlaps with the upper range of puberty in the general population, and display similar pubertal timing to short, late-maturing individuals.

FURTHER READING

General reading

Burt Solorzano CM, McCartney CR (2010) Obesity and the pubertal transition in girls and boys. *Reproduction* **140**, 399–410.

Campbell B (2011) Adrenarche in comparative perspective. *American Journal of Human Biology* **23**, 44–52.

Conway GS, Jacobs HS (1997) Leptin: a hormone of reproduction. *Human Reproduction* **12**, 633–635.

Cortes ME, Carrera B, Rioseco H *et al.* (2015) The role of kisspeptin in the onset of puberty and in the ovulatory mechanism: a mini-review. *Journal of Pediatric and Adolescent Gynecology* **28**, 286–291.

Dhillo WS, Murphy KG, Bloom SR (2007) The neuroendocrine physiology of kisspeptin in the human. *Reviews in Endocrinology and Metabolic Disorders* **8**, 41–46.

Ebling FJP (2005) The neuroendocrine timing of puberty. *Reproduction* **129**, 675–683

Guyer AE, Silk JS, Nelson EE (2016) The neurobiology of the emotional adolescent: from the inside out. *Neuroscience and Behavioral Reviews* **70**, 74–85.

Herting MM, Sowell ER (2017) Puberty and structural brain development in humans. *Frontiers in Neuroendocrinology.* **44**, 122–137.

Hutson JM, Southwell BR, Li R, Lie G, Ismail K, Harisis G, Chen N (2013) The regulation of testicular descent and the effects of cryptorchidism. *Endocrine Reviews* **34**, 725–752.

Karapanou O, Papadimitriou A (2010) Determinants of menarche. *Reproductive Biology and Endocrinology* **8**, 115.

Millar RP, Roseweir AK, Tello JA *et al.* (2010) Kisspeptin antagonists: unraveling the role of kisspeptin in reproductive physiology. *Brain Research* **1364**, 81–89.

Oakley AE, Clifton DK, Steiner RA (2009) Kisspeptin signaling in the brain. *Endocrine Reviews* **30**, 713–743.

Plant TM (2016) Neuroendocrine control of the onset of puberty. *Frontiers in Neuroendocrinology* **38**, 73–88.

Sanchez-Garrido MA, Tena-Sempere M (2013) Metabolic control of puberty: roles of leptin and kisspeptins. *Hormones and Behavior* **64**, 187–194.

Smith JT, Smith JT, Clifton DK, Steiner RA (2006) Regulation of the neuroendocrine reproductive axis by kisspeptin-GPR54 signaling. *Reproduction* **131**, 623–630.

Walvoord EC (2010) The timing of puberty: is it changing? Does it matter? *Journal of Adolescent Health* **47**, 433–439.

More advanced reading

Anand-Ivell R, Heng K, Hafen B *et al.* (2009) Dynamics of INSL3 peptide expression in the rodent testis. *Biology of Reproduction* **81**, 480–487.

Carel J-C, Lahlou N, Roger M, Chaussain JL (2004) Precocious puberty and statural growth. *Human Reproduction Update* **10**, 135–147.

Clément K, Vaisse C, Lahlou N *et al.* (1998) A mutation in the human leptin receptor gene causes obesity and pituitary dysfunction. *Nature* **392**, 398–401.

De Bellis MD, Keshavan MS, Beers SR *et al.* (2001) Sex differences in brain maturation during childhood and adolescence. *Cerebral Cortex* **11**, 552–557.

Devesa J, Almenglo C, Deveso P (2016) Multiple effects of growth hormone in the body: is it really the hormone for growth? *Clinical Medicine Insights: Endocrinology and Diabetes* **16**, 47–71.

Dobbing J, Sands J (1973) Quantitative growth and development of human brain. *Archives of Disease in Childhood* **48**, 757–767.

Fergani C, Navarro VM (2017) Expanding the role of tachykinins in the neuroendocrine control of reproduction. *Reproduction* **153**, R1–R14.

Frish RE (1972) Weight at menarche: similarity for well-nourished and under-nourished girls at differing ages and evidence for historical constancy. *Pediatrics* **50**, 445–450.

Frisch RE, Revelle R, Cook S (1973) Components of weight at menarche and the initiation of the adolescent growth spurt in girls: estimated total water, lean body weight and fat. *Human Biology* **45**, 469–483.

Hull KL, Harvey S (2001) Growth hormone: roles in female reproduction. *Journal of Endocrinology* **168**, 1–23.

Lehman MN, Merkley CM, Coolen LM, Goodman RL (2010) Anatomy of the kisspeptin neural network in mammals. *Brain Research* **1364**, 81–89.

Mann DR, Plant TM (2010) The role and potential sites of action of thyroid hormone in timing the onset of puberty in male primates. *Brain Research* **1364**, 175–185.

Nef S, Prada LF (1999) Cryptorchidism in mice mutant for *Insl3*. *Nature Genetics* **22**, 295–299.

Reed RB, Stuart HC (1959) Patterns of growth in height and weight from birth to eighteen years of age. *Pediatrics* **24**, 904–921.

Silveira LG, Tusset C, Latronico AC (2010) Impact of mutations in kisspeptin and neurokinin B signalling pathways on human reproduction. *Brain Research* **1364**, 72–80.

Tanner JM (1978) *Foetus into Man; Physical Growth from Conception to Maturity*. Open Books, Wells.

Tanner JM (1986) *Growth at Adolescence*. Blackwell Scientific Publications, Oxford.

Topaloglu AK, Reimann F, Guclu M *et al.* (2009) TAC3 and TACR3 mutations in familial hypogonadotropic hypogonadism reveal a key role for Neurokinin B in the central control of reproduction. *Nature Genetics* **41**, 354–358.

Zimmermann S, Steding G, Emmen JM *et al.* (1999) Targeted disruption of the *Insl3* gene causes bilateral cryptorchidism. *Molecular Endocrinology* **13**, 681–691.

CHAPTER 5

Gender

Essential Reproduction, Eighth Edition. Martin H. Johnson.

Companion website: www.essentialreproduction.com

In Chapters 2 and 3, sex was defined in biological terms as the creation of a genetically unique individual as a result of the equal contribution of chromosomes from two parents: hence, two types of gamete (oocytes and spermatozoa) are produced from two types of gonad (ovary and testis) in two types of individual (female and male). The features by which the two sexes were described and differentiated therefore included their chromosomes, genes, gonads, gametes, hormones and anatomical structures (upper part of Table 5.1). The developmental relationships between these features were explored, and the sources of some of the rare 'errors' of sexual differentiation described. What then is gender and how does it relate to sex? This chapter examines these questions. At the outset, it must be emphasized that there exists considerable variation in the ways that the term gender is used.

Gender is a system of classification based on sex

There is an assumption, broadly universal across cultures and history, that the identification of one of the features defining an individual as male or female could reasonably be expected to predict that all the other features would also be concordantly male or female. Thus, the presence or absence of a penis at birth is taken generally as diagnostic of a male or female, respectively. Of course, discordances can and do exist, and we now understand more of the nature and origin of many of them. Estimates of the incidence of ambiguous external genitalia are understandably problematic, but figures of 0.1–0.2% of babies with major ambiguity and 1–2% with less severe ambiguity have been suggested (Box 5.1). Although small in percentage terms, this amounts to a large number of individuals. The conventional approach to genital ambiguity in modern Euro-American cultures has been to intervene as early in childhood as possible to remove or reduce ambiguity and to assign a clear anatomical and thereby social sex to the baby. An 'intersex' state was not considered acceptable.

However, other cultures have taken a different approach, and accepted intersex individuals, often according them a special social status as a distinctive '**third sex**', e.g. the hijra in India or the berdache among some North American indigenous peoples. A move towards a more flexible approach to the clinical management of genital ambiguity has recently occurred in Euro-American societies, in part through pressure from people who were assigned a 'sex' medically, and in their view inappropriately, as babies (see later).

Box 5.1 How frequently is concordance for chromosomal, gonadal and genital sex absent?

Cause	Estimated frequency/1000 live births
Non-XX females or non-XY males	1.93
Complete or partial androgen insensitivity	0.08
Congenital adrenal hyperplasia	15.08
True hermaphrodites	0.01
Vaginal agenesis	0.17

Source: Data adapted from Blackless *et al.* (2000).

In the UK, the birth certificate has to record the baby's sex as male or female; intersex is not a legal option (Births and Deaths Registration Act of 1953). When confronted with sexual ambiguity, in general one of two courses of action is practised today. The first involves a conservative surgical and endocrinological intervention until the child grows and expresses a gender identity as masculine, feminine or intermediate. Alternatively, a more active intervention is to suppress puberty endocrinologically in order to give more time for the child to make an informed decision about their future before major body changes make surgical intervention more difficult and potentially less successful. The pros and cons of these two treatment options are the subject of hot debate.

Further reading

Blackless M, Charuvastra A, Derryck A, Fausto-Sterling A, Lauzanne K, Lee E (2000) How sexually dimorphic are we? Review and synthesis. *American Journal of Human Biology* **12**, 151–166.

Cohen-Kettanis PT, Delemarre-van de Waal HA, Gooren LJ (2008) The treatment of adolescent transsexuals: changing insights. *Journal of Sex Medicine* **5**, 1892–1897.

Green R (2008) A tale of two conferences. *The Gender Trust News* **73**, 10–11.

The bipolar biological classification of individuals as either male or female is paralleled by a bipolar allocation of many other traits, some of which are summarized in the lower part of Table 5.1 as **gender attributes**. Unlike the features characterizing sex, these attributes are based more on attitudes, expectations, behaviour and roles; some of them may appear contentious or less absolute; many are complex; and many vary in detail or substance with different cultures or, within a culture, over historical time. In Table 5.1, these attributes have been grouped under the broad heading of gender because they are associated with sex, but not obviously, invariably or simply so. Moreover, there is not always any immediately obvious essential or causal relationship between those features listed under sex and those attributes listed under gender. It is for this reason that gender is defined here as a **system of classification based on sex**. In order to distinguish sex from gender, this book uses the **terms male and female to describe sexual features** and the words **masculine and feminine to describe gender attributes**. The nature of the relationship between sex features and gender attributes forms the substance of this chapter. First, some of the gender differences summarized in Table 5.1 will be examined in more detail. Then, the basis of gender differences in behaviour is explored since, collectively, these will influence social interactions and thereby the socially-based gender attributes.

Gender stereotypes and gender identities

These two concepts should be grasped for an understanding of gender.

A gender stereotype is the set of beliefs about what it means to be a man or a woman in a particular society

The gender attributes listed in Table 5.1 constitute the elements of gender stereotypes. Gender stereotypes provide a description that is broadly recognizable as defining what it means to be masculine or feminine in a society. The precise attributes appropriate to each gender will vary from one society to another or in the same society over time. However, social, historical and anthropological studies reveal a remarkable consistency in the extent to which each of those attributes listed recurs with greater or lesser emphasis in the gender stereotypes of a range of different societies. For example, the exclusion of women from public life or from particular social or work roles is more evident in strict Islamic societies or traditional Judaeo-Christian societies than in modern secular societies. However, in the latter societies, such gender stereotyping still persists in that certain roles remain associated strongly with men (e.g. priests) or women (e.g. housewives) even if many of these associations are much weaker than they once were. The behaviour expected of men and women also

Table 5.1 Sex and gender: commonly used distinguishing descriptors

Sexual features	**Male**	**Female**
Chromosome	Y present	Y absent
Gene	*SRY* active in Sertoli cell	*SRY* inactive
Gonad	Testis	Ovary
Gamete	Spermatozoon	Oocyte
Hormone	Androgens, MIH	Low androgens, MIH
External phenotype	Penis, scrotum	Clitoris, labia
Internal phenotype	Vas deferens, prostate, etc.	Oviduct, uterus, vagina
Gendered attributes	**Masculine**	**Feminine**
Inter-gender interactions	Encouraged	Discouraged
Intra-gender interactions	Competitive	Cooperative
Social role	Public, extrovert, outspoken in the workplace, powerful, independent, forceful	Private, domestic, quiet, powerless, care provider
Reproductive role	Disposable and transitory	Essential and enduring
Sexual role	Active, insertive, dominant	Passive, receptive, submissive
Work role	Rule setting and enforcing, leadership, military, artistic, ritualistic and priesthood	Domestic, childrearing, agricultural, creative
Appearance	Characteristic for gender	Characteristic for gender
Temperament and emotion	Competitive, combative aggressive, ambitious, not showing vulnerability	Cooperative, consensual, expressive, empathic
Intellect and skills	Mathematical, spatial, systematizing	Linguistic, people-centred

differs. Rowdy, aggressive behaviour from men is resignedly expected and often excused ('boys will be boys'), whereas the same behaviour from women may be considered 'unlady-like'. The wearing of earrings by men or of trousers by women was until recently in British society very gender astereotypic: there were social rules about what constituted appropriately gendered body decoration and clothing, many of which still linger in today's attitudes and values, albeit much attenuated.

Although it may appear difficult in a society in flux to define the current gender stereotypes in terms acceptable to all, nonetheless there tends to be a normative social view about those elements constituting masculine and feminine behaviour. The cohesiveness of that view can be particularly strong for the members of each generation: a person's peers. In framing a gender stereotype, no claim is being made that this stereotype is true for all or indeed for any female or male. It is rather **a shared cultural belief about what men and women are like**. This social consensus about what it means to be a man or a woman is important for an individual's perceptions of themself and of those around them. It provides a yardstick against which to measure their own masculinity or femininity and that of those whom they meet.

This measuring process is important because those who appear to stray too far from the stereotype are generally regarded negatively, or as curiosities or as rebellious. In societies in which gender is a strong social element, it is **less acceptable for men to appear feminine than for women to appear masculine**, although there are boundaries in both directions. This asymmetry may result from the fact that men tend to be more powerful than women in such societies, and so their attributes are more valued socially. So in societies in which gender stereotypes are being eroded, there tends to be more acceptance of the perceived masculinization of women's stereotypes and more resistance to the feminization of men's stereotypes. However, as economies shift increasingly towards a service function, in which traditionally feminine attributes are more valued, the employment opportunities for traditionally masculine men are reduced and these men become marginalized as their masculine attributes are less valued. A key message from this brief discussion is the strong cultural contingency of gender attributes.

Gender stereotyping provides a short-hand for classifying people by sex

We are presented with a bewildering array of social information. Part of the process of our development as children is to learn how to interpret the world around us. Sex differences are an important part of that world. By learning a gender stereotype, or indeed any other stereotype (ethnicity, race, class, age, employment), we are provided with a social shorthand that enables some rapid preliminary assessments to be made of each individual encountered. **Recognizing someone as male or female allows us to associate the various attributes of gender stereotypes** and thereby conditions our immediate behaviour patterns in ways that are socially appropriate for our and their gender. Of course, this process will tend to reinforce the gender stereotype of the society. It does not, however, preclude later reactions to the individual as an individual. If you doubt the importance of social sketching of this sort, consider your reaction on being introduced to someone whose sex and gender are not immediately obvious. How comfortable are you, and how does it affect your behaviour? Or consider how you react when, in a different culture, you find that the accepted gender stereotypes conflict with those of your own culture: e.g. men cuddling in public or women being excluded from public life? **Humans are social beings and the rules by which societies function are therefore important.**

Gender identity describes the personal concept of 'me as a man or a woman'

We have a social view that there are two genders defined broadly by the gender stereotypes of our society. Each of us is part of that society. It therefore follows that each of us has a view of ourselves as being masculine or feminine and of conforming to a greater or lesser degree to the stereotype. The extent to which each individual feels confident of his or her position within this bipolar gender spectrum is a measure of the strength and security of their gender identity. **Most individuals have gender identities that are fully congruent with their sex**. Thus, most women and men who are physically female and male, respectively, have **strong gender identities**. Some individuals may feel less certain about their gender identities, although they nonetheless identify congruently with their physical sex: they may be said to have **weak gender identities**. A few individuals may feel that their gender identities are totally at variance with their otherwise congruent genetic, gonadal, hormonal and genital sex. Such people are described as being **transsexual** or **transgendered**. Transgendering may occur in either direction: the **male-to-female transgendered** consider themselves to be females with a female gender identity but with otherwise male bodies, whereas the **female-to-male transgendered** feel themselves to be men in an otherwise woman's body. Traditionally, more male-to-female transgendered individuals have been identified than female-to-male, although differential reporting may account for much of this difference. The transgendered may adopt the gender roles of the physically different sex, and some may undergo surgical and hormonal treatments so as to bring their bodies and their bodily functions (their sex) as closely congruent to their gender identity as is possible (**females becoming trans men** and **males becoming trans women**) (Box 5.2). Trans men and women provide us with perhaps the strongest justification for making the distinction between sex and gender.

Gender differences may not be as great as they first appear to be

Intuitively, when looking at the gender attributes in Table 5.1, it is possible simultaneously to recognize the gender stereotypes as familiar while rejecting them as an oversimplification. For example, whereas men in general might not readily express

Box 5.2 The law and trans men and women

The European Union (EU) introduced legislation in Article 19 of the 1997 Treaty of Amsterdam empowering it to 'take appropriate action to combat discrimination based on sex, racial or ethnic origin, religion or belief, disability, age or sexual orientation'. In consequence, the countries of the EU are all required to pass legislation putting this Article into effect in ways that permit recognition of trans men and women in their 'new' identities. Where this has not been done, the European Court has broadly upheld the rights of trans people to be treated fairly. The precise form of the legislation introduced in different countries has varied.

In the UK, the relevant laws are:

- The Equality Act 2010, which makes it unlawful to discriminate on the grounds of sex, race, age, disability, sexual orientation, gender reassignment, marriage, civil partnership or religion or belief.
- The Gender Recognition Act 2004, which provides for a 'gender recognition certificate', which then allows the trans person to apply for a new birth certificate, to marry, etc. In order to qualify for a certificate, 'expert evidence' must be produced to establish the person's gender identity, effectively a gatekeeper role for clinicians. Then, only after 2 years living in one's 'acquired gender' (the terminology the Act uses) can the person get a certificate.

Further reading

Reed R (2005) Transsexuals and European human rights law. *Journal of Homosexuality* **48**, 49–90.

vulnerable emotions through crying and admissions of helplessness, many individual men do express such emotions and show such behaviour. Individual women can be just as competitive and aggressive as men, although overall these attributes are associated much less with women than with men. Many studies have attempted to make objective and quantitative measurements of gender differences, through the use of behavioural and cognitive function tests and the use of questionnaires to address attitudes. For most attributes, the degrees of **variation within populations** of men and of women are so great that **the overlap between men and women is too large** to produce significant differences between the sexes (Box 5.3). Some of these differences increase or decrease significantly with age or with cultural or educational changes. Moreover, rarely if ever do any differences observed have predictive validity: it is not possible from the measurement of a gender attribute in an individual to predict reliably whether that individual is a man or a woman.

There is thus a paradox. Society has a clear concept of what it means to be masculine and feminine. Moreover, most individuals profess a very clear concept of themselves as masculine or feminine and an understanding of what that means for their place in society. Yet both objectively and subjectively it is not possible to sustain a strongly bipolar description of a gendered society. Men and women overlap greatly in the attitudes that they express, in their patterns of behaviour, in their skills and, increasingly, in the roles they adopt. There is more a continuum of attributes than a bipolar segregation. Some societies reflect this reality and are relatively non-gendered, but most societies have a bipolar gendered organization despite the lack of evidence for its inevitability. Why? Presumably such social organization is seen to have advantages, e.g. for the production and raising of children, controlling patterns of inheritance, the division of labour, or the ability to resist external threats. In order to take this discussion further, we will turn to a consideration of how a gendered society might arise.

The brain and behavioural dimorphism

It will be clear from the foregoing discussion that gender is a concept applicable to humans. Does this therefore mean that studies on the origin of sex differences in the behaviour of animals are of no use to us in trying to understand gender differences in humans? We examine this question first for non-primates, and then for non-human primates, before finally considering whether and how this evidence applies to humans.

In animals, hormones condition sex differences in behaviour and brain structure

There is overwhelming evidence that the exposure of animals to **sex hormones during a critical period of early life** is associated with the **sexually dimorphic behaviour** displayed later in adulthood. Thus, the distinctive urination patterns shown by the dog (cocked leg) compared with that of the bitch (squatting) is due to exposure to androgens. The critical exposure period may be in late fetal life (e.g. guinea-pig, sheep) or neonatally (e.g. rat, mouse, hamster). The most intensively studied behaviour patterns are those associated with play patterns and with copulation. For example, during sexual interaction with females, adult male rats show **courtship behaviour** (e.g. pursuing females and anogenitally investigating them), **mounting**, **intromission** and **ejaculation** (Figure 5.1). Conversely, adult females display **soliciting** and **receptive postures**, such as **lordosis** (Figure 5.1). These behaviours are typical for each sex, even if not exclusive to each – males occasionally solicit and even accept mounts from other males, while females in heat will often mount one another.

Treatment of female rats with testosterone during the first 5 days of life increases their display of masculine sexual behaviour in adulthood and reduces their display of feminine patterns. Castration of male rats to remove the influence of androgens during this same critical period has the reverse effects. Thus, '**masculinization**' in the rat (and other non-primates) is accompanied by '**defeminization**'. Androgens influence the development of these behavioural differences by

Box 5.3 Summary of findings from a meta-analysis of studies on sex differences in humans (Shibley Hyde, 2005)

In a range of published studies, 124 traits were analyzed to see whether there were significant differences between populations of men and women.

- For 78% of these traits, there was effectively no difference.
- For 15% of them, there was a moderate population difference. Traits observed more frequently in the male population included spatial perception, mental rotation, physical and verbal aggression, assertiveness, body esteem, sprinting, activity level, self-efficacy of computer use. Traits more frequently observed in women included spelling and language skills, and smiling when aware of being observed.
- For only 6% of traits were large differences observed. Observed more frequently for men were mechanical reasoning, masturbation and permissive attitudes to casual sex, and for women agreeableness.
- Only 2% of traits showed very large sex differences, throw velocity/distance and grip strength being significantly more frequent in males.

Conclusions and qualifications

It is important to note that for all traits there was overlap. Some are clearly related to the anabolic actions of androgens on muscles, and others may be culturally conditioned – expectations from gender stereotypes perhaps influencing attribute acquisition. Where statistically significant sex differences are found, it is important to note that scores for some traits may vary with factors such as age and experience, mood, motivation, practice and ambient hormone levels, and that these may differ for the two sexes.

Overall, what impresses is how similar the two sexes are. Humans do not seem very dimorphic behaviourally!

Gender differences nonetheless are often highlighted

There are two types of reaction to the evidence that men and women show big overlaps in attributes. One reaction is to focus on those statistically significant average differences that are observed and to seek to understand their origins – the 'women are from Venus, men are from Mars' approach. This reaction may also be used to justify the perceived different needs/treatments of men/boys and women/girls in education, health, employment, etc.

The alternative reaction, while accepting that some average differences do exist between the sexes, is to focus on people first and foremost, given the overlap between the sexes and the complex biological and social origins of sex differences. This approach accepts the notion of a less gendered society than hitherto in which people of either sex are freer to flourish without constraint of stereotype. A debate about biological sex differences among scientists illustrates these distinctive reactions (see Further reading).

These two differing approaches take us into political and social theory, and we simply alert readers to read and interpret the evidence base as objectively as possible despite the strong academic and social reactions that discussion of sex differences can evoke.

Further reading

Baron-Cohen S (2003) *The Essential Difference: Male and Female Brains and the Truth about Autism*. Basic Books, New York (takes quite a strong position about innate biological differences between male and female brains, and relates the analysis to the higher incidence of autism among males; acknowledges considerable overlap and sex-atypical patterns).

Barres BA (2006) Does gender matter? *Nature* **442**, 133–136 (questions biological origins of sex differences in scientific success).

Lawrence PA (2006) Men, women and ghosts in science. *PLoS Biol.* **4**, e19, 13–15 (takes the approach that biological sex differences are inevitably a part of our make-up as humans and cannot be ignored).

Shibley Hyde J (2005) The gender similarities hypothesis. *American Psychologist* **60**, 581–592.

effects on the **structure of the developing brain**, generating several neuroanatomical sex differences, especially in the medial amygdala and hypothalamus, in particular the sexually-dimorphic nucleus of the preoptic area (SDN-POA) and the bed nucleus of the stria terminalis (BNST; Figure 5.2). In animals, then, there is clear evidence that hormones lead to neuroanatomical and behavioural changes, in much the same way that they also lead to the development of sexually dimorphic genitalia. Thus, sex differences in brain structure and in behaviour seem to fit into the same conceptual framework of

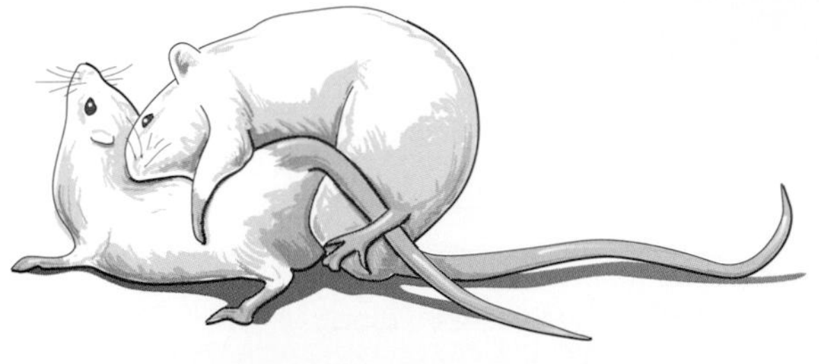

Figure 5.1 Sex-dependent behaviour patterns in male and female rats. Note the immobile lordosis posture shown by the receptive female, which enables the male to mount and achieve intromissions, which will result in ejaculation. Receptivity and lordosis are shown predominantly by females; mounting, intromission and ejaculation patterns of behaviour are shown predominantly by males.

sex, as was described for the gonads, hormones and genitalia in Chapter 3. **There is then no need for a 'gender category' in these species.**

In non-human primates, sex differences in behaviour are influenced by hormonal exposure early in life

To what extent do androgens exert the same effects on the development of sexually dimorphic behaviour in non-human primates? Results from experiments on rhesus monkeys suggest similarities. Young females, exposed to high levels of androgens during fetal life, display levels of sexually dimorphic behaviour in their patterns of childhood play that are intermediate between normal males and females (Figure 5.3). Moreover, although

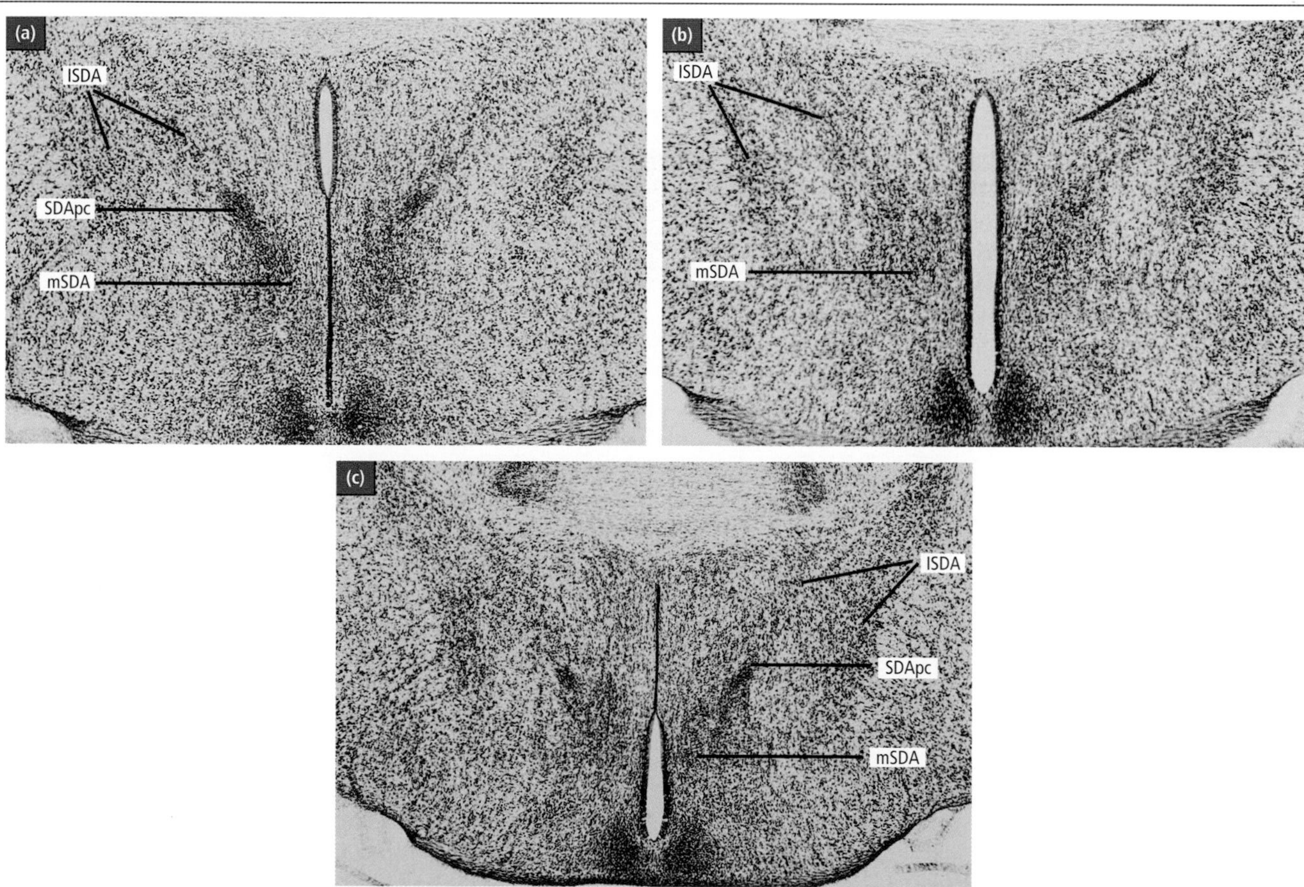

Figure 5.2 Photomicrographs of coronal sections through the preoptic area of three 21-day-old gerbils (*Meriones unguiculatus*). Section (a) is taken from a male, (b) from a female and (c) from a female treated neonatally with androgens (testosterone propionate: 50 mg on the day of birth and 50 mg the next day). The sexually dimorphic area in the gerbil (SDA) can be divided into several regions: medial (mSDA); lateral (lSDA); and pars compacta (SDApc). The SDA differs between males and females in a number of aspects: prominence (not necessarily size); acetylcholinesterase histochemistry; steroid binding; and various other neurochemical characteristics, but most obviously in the presence or absence of the SDApc. Thus, the SDApc is virtually never found in females (compare a with b). Note that in females treated neonatally with testosterone, there is a clear SDApc (compare c with b). These pictures provide clear evidence of the impact of hormones during a critical period of early life on the differentiation of this part of the brain. The medial preoptic area in general is closely involved with the regulation of sexual behaviour (see Chapter 6), and some progress has been made in relating specific aspects of sexual behaviour to subdivisions of the SDA. It is also important to note that such sex differences in the structure of the preoptic area are found in many species, from rats to humans, but the precise details of the dimorphism vary considerably. (Source: Courtesy of Dr Pauline Yahr, University of California, Irvine, USA.)

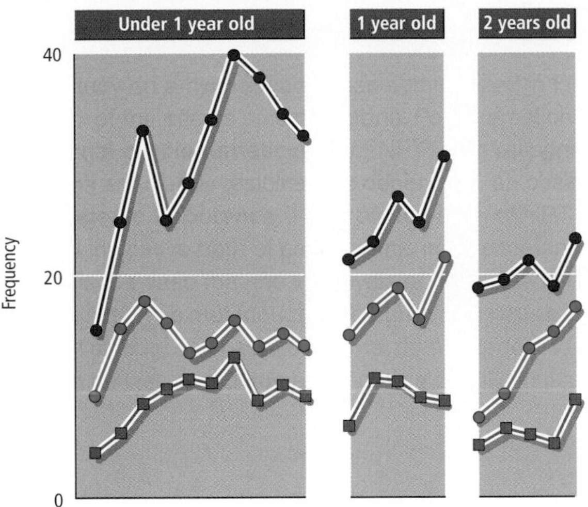

Figure 5.3 Frequency of 'rough-and-tumble play' during the first, second and third years of life of a male rhesus monkey (red circles), female rhesus monkey (blue squares) and female rhesus monkey treated prenatally with androgens (green circles). Note that males display this behaviour at a higher frequency than females and that androgenized females are intermediate.

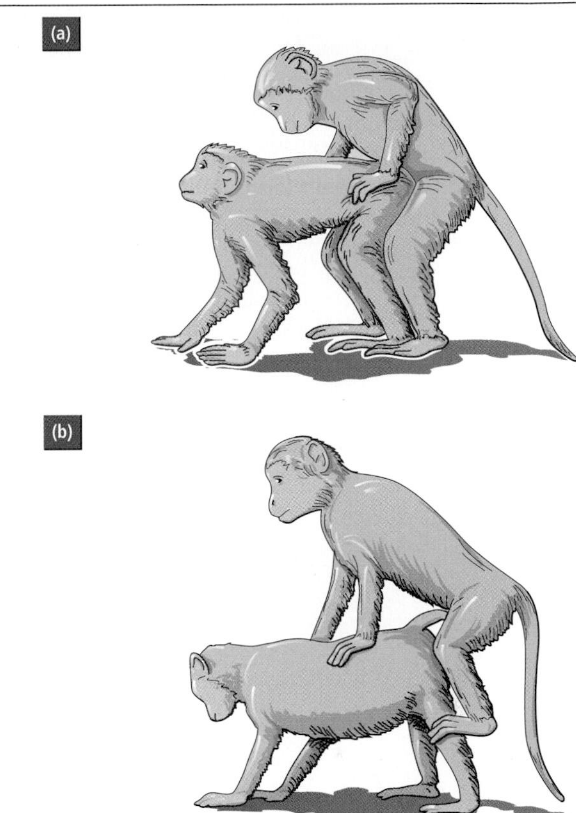

Figure 5.4 Sexually dimorphic patterns of mounting behaviour in young rhesus monkeys. (a) Early in life, both males and females show immature mounts by standing on the cage floor. (b) During development, males show progressively more mature mounts in which they clasp the female's calves so that she supports his weight entirely. Androgenized females display more of the latter type of mature mounts than do untreated females.

both male and female infant monkeys will mount other infants (Figure 5.4a), only males progressively display mounts of a mature pattern (Figure 5.4b). Androgenized females, however, do develop this mature mounting pattern, and as adults they attempt to mount other females at a higher frequency than do non-androgenized females. Thus, neonatal androgenization produces persistent 'masculinization' of behaviour. However, the androgenized female monkeys as adults show normal menstrual cycles and can become pregnant. They must therefore display patterns of adult feminine sexual behaviour at least adequate for them to interact successfully with males, suggesting that they are not totally or permanently 'defeminized'.

These results suggest a **less complete or persistent effect of androgens** on the development of sexually dimorphic behaviour in primates than in non-primates. Why might this be? One explanation is technical, and results from the different timing of the critical androgen-sensitive effect on brain structure. In rats, this occurs neonatally, after genital phenotype is established, so making it easily accessible to selective manipulation. If a critical 'behavioural' period also exists in primates, it occurs during fetal life, and may be prolonged postnatally. Attempts to androgenize primate fetuses *in utero* often lead to abortion if doses of administered androgens are too high. The genitalia also tend to be masculinized, which might affect the subsequent social interactions and learning of the infant. Thus, in primates, **a specific selective effect of androgen on the brain is more difficult to achieve**. However, it is also possible that in non-human primates, the prescriptive hormonal determination of sexually dimorphic behaviour seen in non-primates does not occur. **Androgens may predispose to masculine patterns of behaviour**, but other factors may also influence the degree to which they are expressed.

What about sex differences in brain structure in non-human primates? As for non-primates, a few such differences exist, including some in the hypothalamic region particularly concerned with reproductive and sexual behaviours and in the amygdala concerned with play patterns. These neuroanatomical differences also seem to result from endocrine exposure in fetal life.

In humans, both sex and gender differences in brain structure and function have been described

Not surprisingly, the difficulty in studying non-human primates is exacerbated further when humans are considered. Neuroanatomical studies are understandably limited and often conflicting, relying as they have on postmortem samples with their attendant problems of selection and distortion. Nonetheless, a few consistent sex differences in the structural organization of the brain have been reported, e.g. in the size of a small region of the anterior hypothalamus called the **third interstitial nucleus** (**INAH3**; Figure 5.5). However, the significance of these sex differences for gender identity and attributes is less clear. Claims have been made that the size and organization of **INAH3**, the closely related **INAH1** and the **central bed nucleus of the stria terminalis** (**cBST**) are associated specifically with gender identity as opposed to sex. Thus, all three nuclei are reported as being smaller and with fewer neurons in women than in men, and also smaller in trans women (male-to-female transgendered; Figure 5.6). However, the number of individual brains studied is small, as are the measured gender differences, and there is overlap between genders such that nuclear size is not predictive for gender. It is also not clear when these size differences first appear, what causes them (hormones or usage?) or what they might mean functionally. Until we know more about the time at which brain differences emerge and are able to study more brains from a larger range of individuals with gender or endocrine anomalies, it will be difficult to draw firm conclusions.

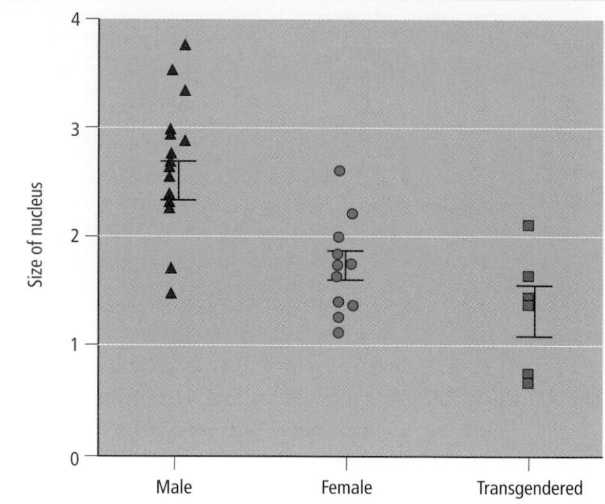

Figure 5.6 The scatter of sizes of the central bed nucleus of the stria terminalis (cBST) in human adult males, females and trans males (male-to-female transgendered). Error bars ± SEM. Note the sex difference and that the trans male nuclear volume is closer to the female pattern. (Source: Data from Zhou JN, Hofman MA, Gooren LJ, Swaab DF (1995) *Nature* **378**, 68–70.)

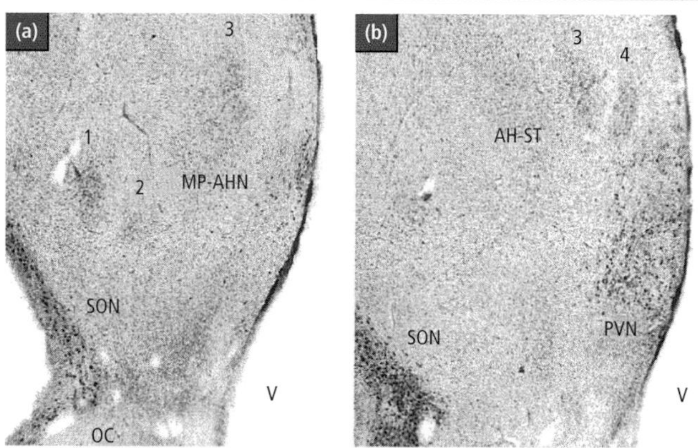

(c)

INAH3 parameter measured	Male brain (n)	Female brain (n)
Volume (mm³)	0.121 (31)	0.073* (34)
Volume/brain weight (mm³/g)	0.087 (31)	0.061* (34)
Neurons/mm³	15730 (21)	16021 (21)
No. neurons	1794 (21)	1123* (21)

Figure 5.5 (a, b) Two coronal sections through the hypothalamus from a presumed heterosexual male to show the four INAH nuclei. In (a) INAH1 and 2 are seen, with INAH3 emerging within the area of the medial preoptic/anterior hypothalamic nucleus (MP-AHN). INAH3 and 4 are prominent in the posterior section (b), close to a cell group comprising the anterior hypothalamic nucleus (AH) and the bed nucleus of the stria terminalis (ST) (PVN, paraventricular nucleus; oc, optic chiasm; SON, supraoptic nucleus; v, third ventricle). (c) A summary of data comparing INAH3 in men and women; * significant differences. The study also compared these data with those for the brains of 14 homosexual men, with values of 0.096 mm³, 0.069 mm³/g, 18792/mm³ and 1831 respectively – none differing significantly from heterosexual men. (Source: Data from Byne *et al*. (2001) *Hormones and Behavior* **40**, 86–92.)

Many of these difficulties inherent in postmortem studies can be overcome by use of **neuroimaging technologies**. These are particularly informative when used in conjunction with developmental longitudinal studies on children. Thus, imaging studies have revealed that as children progress from childhood through puberty, average sex differences in both the sizes of a number of brain structures and the timings of growth patterns are evident (Figure 5.7). Two points about these sorts of data need to be made. First, they are mean data and the considerable variation within each sex means that any single data point is not necessarily predictive of sex. Second, the earlier development of the female brain (some 2–4 years earlier compared with boys) resembles the earlier body growth pattern and puberty onset described in Chapter 4 (see page 58).

Neuroimaging has also contributed substantially to the study of male/female functional differences in the living brain. A proliferation of such studies on adults has shown that there are population sex differences in brain activities in a variety of brain regions, each associated with the performance of particular activities or tasks, including those related to emotion, memory, learning, olfaction, vision, hearing, navigation, processing of faces, and pain perception. Functional evidence has also been reported for sex differences in lateralization, women

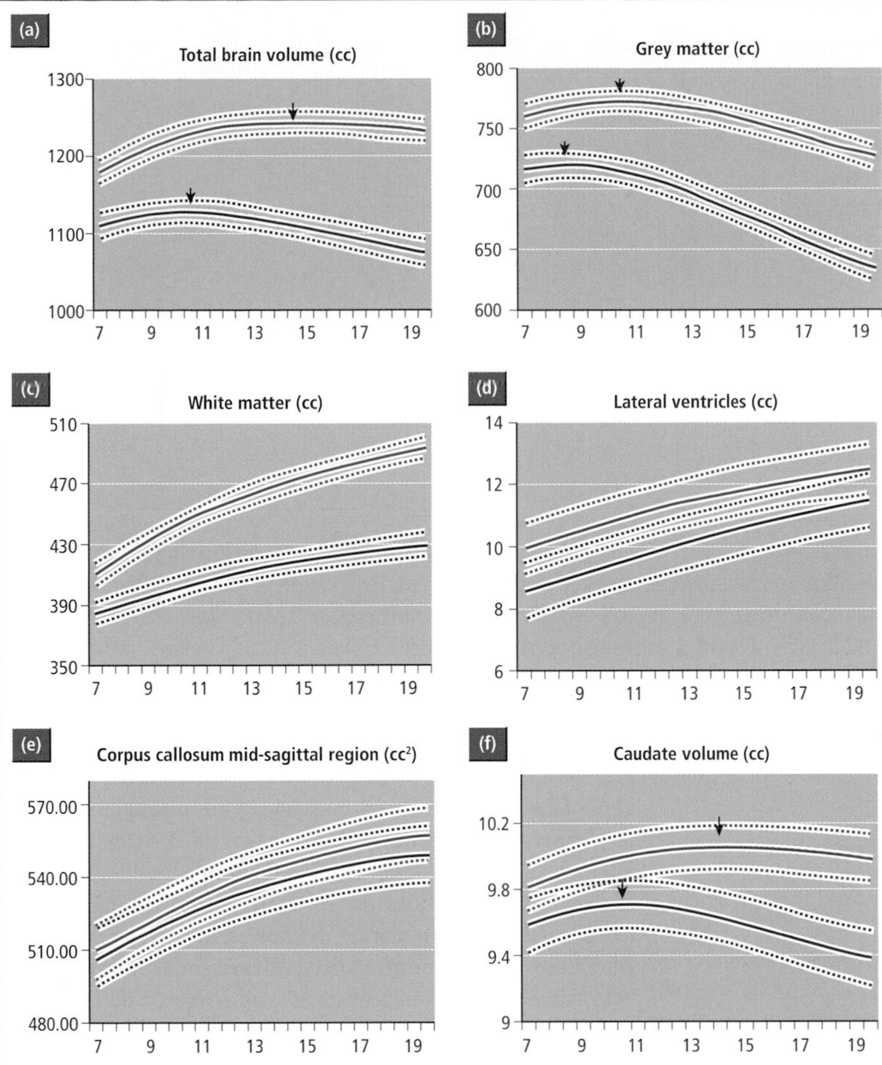

Figure 5.7 Mean values for morphometric volumes of different brain features in developing males (blue) and females (red) between the ages 6 and 20 years (upper and lower 95% confidence intervals shown). Curves a, b, c and f differ significantly in height and shape; curve d differs in height only. (Source: Adapted from Lenroot RK *et al.* (2007) *Neuroimage* **36**, 1065–1073.)

showing more left-lateralized language and emotion processing, whereas men tend to show right-lateralized visuospatial activity. There is also clear evidence for sexual dimorphism in the amygdala, a region of the brain involved in emotional processing, and in the hippocampus, a region involved in memory. Thus, the brain sex differences observed have been much more extensive and marked than might have been anticipated from the behavioural studies reported above. **The brain seems to be more 'sexed' than does behaviour**! Do these findings mean that sex differences in brain and behaviour should be regarded simply as further examples of the sexual dimorphism described for other somatic features in Chapter 3, there being no such thing as gender? Again, studies on trans people provide us with some clues.

Thus, a few studies have applied functional neuroimaging to trans people in an attempt to discriminate sex from gender. In general, for those few areas of brain activity where 'gendered' differences have thus far been reported, the male-to-female trans individuals examined show intermediate activities between those of men and women, e.g. in hypothalamic functions and laterality effects. At best, these observations on trans people can be said to be sex atypical, but the underlying basis for this atypicality is unclear. Overall, more longitudinal functional studies are needed for secure identification of gender-based brain organizational differences.

Given these many sex differences in brain activity and structure reported, as well as the apparently 'gendered' subset of these sex differences that appear to be atypical for sex in trans individuals, can anything be said about causation? There are three sorts of explanation offered for the sex and gender differences in human brain structure and function: (1) they are programmed into development genetically – usually assumed to be mediated mainly via the differential action of androgens in male fetuses; (2) they are socially generated postnatally by the way babies are treated, or (3) they arise from an interaction between both the foregoing two influences. In the next section, the possible genetic and endocrine basis for sex and gender differences is considered.

Might the underlying basis for sex and gender differences in humans be hormonal?

When gene expression patterns are analysed developmentally in animals, many sex differences are observed before the expression of Sry and androgens. These are presumed to be due (and in some cases shown to be) directly or indirectly to differences in the expression of genes on the sex chromosomes and/or of parentally imprinted genes. These expression profiles suggest a role for non-Sry/androgen mediated mechanisms in expressing somatic sex. However, most sex differences appear after Sry/androgen activation.

So what about the relationship between hormones and gendered behaviour in humans? This question has been studied in both adults and children using various of the gendered attributes summarized in Table 5.1. It is important to reemphasize that in humans the two sexes differ quantitatively in gender attributes, with much overlap. The influence of prenatal hormones on subsequent behaviour has been investigated in **genetic females with adrenogenital syndrome** (AGS; see page 43): nature's counterpart to experimental animals treated exogenously with androgens during the critical period of neural differentiation. However, it is important to note that we are not dealing with 'pure' androgen effects in these girls/women, as under- or (therapeutic) over-exposure to corticosteroids is known to affect brain structure and behaviour directly. Studies of girls with mild AGS have revealed increased levels of energy expenditure and athletic interests more characteristic of boys, and a decreased incidence of 'rehearsals' of maternal behaviour and doll-play activities, together with diminished interest in dresses, jewellery and hairstyles. This spectrum of behaviour, termed **tomboyism**, is accepted in Western culture, and provides few if any problems for children so affected. Tomboyism might be thought to be a consequence of the effects of androgens on the fetal brain, rather like the changes in rough-and-tumble play in infant monkeys exposed prenatally to androgens (Figure 5.3). However, as a group, AGS girls had similar feminine gender identities to a group of non-AGS tomboys, and only slightly weaker gender identities than control girls. Further study of AGS females as adults has revealed evidence of enduring behavioural consequences. Thus, they show a continuing level of higher interest in more masculine pursuits, a slightly higher incidence of dissatisfaction with their female gender identity and, in some studies, a substantial increase in lesbianism compared with controls. A subset of AGS women do identify as trans men, but not necessarily the most androgenized. Thus, overall the studies on AGS girls/women do provide support for either androgens or masculinized genitalia contributing to defeminization in humans. However, it is important to note that the AGS girls and women studied will, by definition, span the low to moderate part of the androgenization scale, where genital virilization is incomplete.

How to summarize? It seems clear that there are sex differences and, albeit less securely, even gender difference in brain structure/activity in humans. It is unclear when they arise and what causes them. The functional imaging studies are providing evidence linking sex differences in the brain to many of the gendered differences in behaviour described in Table 5.1. Evidence from animals suggests that hormones can influence brain organization and thereby behaviour, but even in animals, especially in monkeys, there is no rigid and absolute causal relationship between the two. In humans, where there is even greater flexibility and overlap of sex-related behaviour patterns and of gendered attributes, a role for hormones prenatally or neonatally is plausible, but more research is needed to find out the full extent and nature of any influence.

Gender development may form part of social learning in humans

In the discussion earlier, we referred again to the significance of the assignment of sex to a baby as a boy or a girl depending on the presence or absence of a penis. It was pointed out that this assignment might then affect both how the individual saw him- or herself and how the parents and peers viewed and treated the developing child. We now explore this area further in our analysis of gender development.

Babies are treated differently depending on their sex

When individual adults are handed the same baby, having been told variously that it is a girl or a boy, their play and communication with it differ according to their perception of its sex. This sort of study shows that babies of different sexes are likely to be treated differently simply because they are of different sexes. A second example makes an additional point. When adults were shown the same video sequence of a child playing, and some were told that it is a boy and others a girl, their interpretations of its behaviours depended on the sex that they believed it to be. For example, when the child was startled and believed to be a boy, it was perceived more often as being angry, whereas the same startled behaviour, when believed to be that of a girl, was perceived as fearful distress. This sort of study tells us that **expectations about how a male or female baby should behave** can lead adults to interpret the same behaviour very differently. It raises the possibility that some developing behaviours may be **reinforced in gender-stereotypic ways as a result of the expectations of others**. Thus, girls are expected to be softer and more vulnerable, and are played with more gently. They are also expected to be more vocal and socially interactive, and parents spend more time in these sorts of behaviours with girls. Boys in contrast are encouraged to do things, are less directly communicated with and are disciplined or roughly handled more often. In general, boys tend to be much more prone to gender stereotyping in this regard than girls.

These sorts of observations emphasize how important and subtle gender stereotypes are and how they are applied to children from the moment of birth. Indeed, parents seem quite anxious to encourage differences between boys and girls by the types of toys they offer them, the clothes they provide for them, and the activities they encourage and discourage. Rewards and approval are offered when children conform to parental gender stereotypes. These parental gendering activities are particularly marked for the first 2–3 years of a child's life. It is precisely over this period that a child develops its own sense of gender identity. By 2 years children label themselves consistently as male or female, and soon thereafter reliably associate certain sorts of behaviour and activities with males and females. They appear to have both a gender identity and a gender stereotype. By 5 years of age they also seem to realize that gender is fixed and cannot be changed across time or situation: they have a sense of **gender constancy**. Indeed, if 3–6-year-old children are shown a video of another child, their descriptions of its behaviour are very different if they are told that it is a boy than if told it is a girl. They actually seem to be even more rigidly gender stereotyping than their parents when performing the same task! Since children spend a lot of time with one another, they are likely to reinforce gender stereotypes in each other: **peer pressure** in action.

Children are, of course, cognitive beings. They do not simply absorb subconsciously impressions of the world around them, although that does occur. They see and hear what goes on around them in the household, in the media, at school. They see men and women and what they do and do not do. Models of male and female behaviour are provided all around them. So, there may also be a copying element in the development and elaboration of their growing gender identity and the ways in which they express it. However, copying a model implies identification with that model in the first place and so it is likely that **copying is a secondary process** that may relate more to **the expression of a gender identity** than its initial establishment.

Thus, a lot of evidence supports the view that gender stereotypes are applied to babies and children very early in life, and that children also use them and apply them to their world from an early age. The child's environment is thus immersed in gender stereotyping. Does this mean that the way in which babies are treated and gender stereotyped causes their own gender to develop? It is entirely plausible to suggest that at least some gendered patterns of behaviour that develop in boys and girls may be induced differentially by the way in which they are treated by others and as a result of the expectations of others. In effect, the gender stereotype of a society may be 'taught' to its children by the way they are treated. If this were so, it might be suggested that ambiguity on the part of parents about the sex of their child could affect the development of gender identity. Cases of transgendering might be associated with a sexually ambiguous childhood: e.g. parents treating their son more like a girl, clothing him in dresses and not reinforcing 'boyish' activities in play and sport. The evidence for this suggestion is far from clear. Just because a suggestion is plausible, it does not mean that it is true. What is the evidence?

Gendered behaviour by babies may affect the way that they are treated

Above (see page 82), the evidence that exposure to androgens during fetal or neonatal life might influence gendered behaviour was reviewed. The evidence was consistent with there being a possible influence on childhood play patterns. There are indeed a few claims of intrinsic behavioural differences between newborn male and female babies, although these are not yet strong enough to convince. However, it is at least plausible to suggest that just as babies may respond differently to adults who show gender-specific behaviour towards them, so sex differences in baby behaviour might induce different responses in adults. Clearly, the experiments described above, in which adults were (correctly or incorrectly) 'told' the sex of a child and responded in ways typical for the believed gender, cannot be explained in this way. However, in these experiments when a male baby was handled by adults, half of whom thought it was male and the other half of whom thought it was female, and they were then asked about their experiences, there were differences. Thus, where reality and belief were congruent, the adults had felt more 'comfortable' than when there was conflict. This may mean that they were picking up on inconsistencies in the baby's behaviour that conflicted with expectations. What this may be telling us is that adults are sensitive to the baby's behaviour as being boy-like or girl-like. It does not, of course, tell us whether these sensed differences in behaviour were due to hormonal influences on the baby or to previous social learning by it. Here is the core of our dilemma. From the moment of birth, boys and girls are likely to be treated differently, so how can we separate cleanly the effects of hormones from those of learning?

One way to achieve this might be to look at babies who were born boys but 'became' girls postnatally (or rarely vice versa). A single highly influential case was provided in the 1960s by monozygotic male twins: the so-called **Money twins**, named after the clinician who described the case. One twin (John) was genitally damaged at circumcision. This boy was reassigned as a girl (Joan), given genital plastic surgery, provided with hormone therapy and brought up as a girl. Joan was described as having a female gender identity and was taken by John Money as decisive evidence that sex of rearing 'trumped' genetic, endocrine and gonadal sex in the establishment of gender identity. However, it was later found that in adulthood, Joan rejected her female identity, reverted to John, married and had an adopted child. Amazingly, this one unfortunate case study determined paediatric policy on genital ambiguity until the late 1990s (Box 5.4).

Recently, more systematic prospective studies on the development of gender identity in a range of patients with different combinations of genetic, gonadal, genital and rearing sex have begun to appear. These are based on thorough descriptions of each sex variable in relation to the measured outcome of gender identity. At best, conclusions are highly provisional, but suggest that many XY individuals exposed to prenatal androgens but reared as females are likely to declare male

Box 5.4 The Money twins and their impact on paediatric practice

John Money was an eminent sexologist and the John/Joan twin case was highly influential for paediatric policy in cases of sexual ambiguity. The general principle that developed from it was that clinicians alone should decide on a course of sex assignment and then reinforce that decision in all that is said to parents and the child concerned. Thus, early surgical and endocrine interventions were undertaken, and neither parents nor developing child were told about the ambiguity of sex ('just treating a developmental incompleteness'). This well-meaning clinical paternalism was intended to foster clear parental commitment to a clinically agreed sex of rearing, so as to facilitate the child's entrance into a highly gendered society. However, when the outcome of the John/Joan case became known, with Joan rejecting her feminine identity, this policy was undermined and was abandoned by the American Academy of Pediatrics in the late 1990s. Both in the USA and elsewhere, attempts are now underway to develop and review more sensitive policies for the management of sexual ambiguity. These take account of recent outcome studies, such as Reiner's, changing social attitudes, as well as the parents' concerns. These policies encourage minimal early interventions, as far as possible restricted to those necessary for medical reasons and even suppression of puberty (see Box 5.1).

Further reading

Chau P-L, Herring J (2004) Men, women, people: the definition of sex. In: *Sexuality Repositioned: Diversity and the Law* (ed. B Brooks-Gordon *et al.*), pp. 187–214. Hart Publishing, Oxford.

Money J, Ehrhardt A (1973) *Man & Woman/Boy & Girl: The Differentiation and Dimorphism of Gender Identity from Conception to Maturity*. Johns Hopkins University Press, Baltimore.

Reiner WG (2005) Gender identity and sex-of-rearing in children with disorders of sexual differentiation. *Journal of Pediatric Endocrinology & Metabolism* **18**, 549–553.

sexual identities later in life. XX individuals exposed to high prenatal androgens and reared as males are more likely to develop male identities. Such evidence is important if assignment of sex at birth or soon thereafter is to be attempted. Of course, the alternative, as described earlier, is to accept the intersex state as a valid interim and/or long-term option – a situation that would require both social and legal sanction for it to be acceptable to many parents.

These studies indicate that the establishment of gender identity is clearly complex and roles for both the gender of rearing and fetal androgens are likely, perhaps interacting in ways we do not yet understand. In this context, the male transgendered are of particular interest. We need to understand whether their exposure to androgens was atypical and whether they were reared unambiguously as boys according to their genital sex. They develop a feminine gender identity, but why? Understanding how the transgendered develop may throw interesting and important light on the relative roles of hormones and environment in gender development.

Summary

Four potential elements that might contribute towards the establishment of gender have been considered: sex chromosome constitution, hormones, social learning and brain structure. The brain is central since the expression of attitudes and behaviour, which form the basis of social interactions, is the result of neural processes. Both the organization and function of the brain can be influenced by genes, hormones and learning. Hormones affect brain structure and behaviour in non-primates and modulate behaviour sex-dependently in primates, although their impact appears to be less prescriptive. The patterns of usage of neuronal circuits that come from interactions with the environment, including social learning, can affect brain organization and function, so that learning and rehearsal are associated with changes to the 'hard wiring' of the brain. Some of these environmental influences have been related to changes in the epigenetic structure of the chromatin around steroid receptors, providing a possible molecular basis for the developmental plasticity observed. When we consider the development of gender, a clear separation of endocrine and social factors has not been achieved. It might be suggested that such a rigid separation is also impossible, since each may interact with and reinforce the other. Small sex differences in the behaviour of newborn babies may be induced by androgens. These subtle differences may be detected by parents and peers, who then have clear expectations and beliefs derived from gender stereotypes that condition their behaviour towards the baby's actions and anatomy. These interactions tend to amplify small differences into larger ones. Soon the baby/child engages in the process actively. The process is a dynamic one, susceptible and responsive to cultural difference and change, based on the undoubted cognitive flexibility of humans, and well suited to the development of a social mammal.

Gender and reproduction

Gender has been defined and discussed as a system of classifying individuals based on their sex. Sex in mammals, as we saw in Chapter 1 (see pages 4–6), is fundamentally about reproduction and genetics. The process of reproduction involves the bringing together of a male and a female so that their haploid gametes can unite at fertilization. In mammals, one of each type of gamete is obligatory for the successful production of a new individual. Even a brief look at the gender attributes listed in Table 5.1 reveals that many can be related plausibly to the different reproductive roles of males and females. The generally nurturing, emotional, consensual, creative and private attributes of females and the more aggressive, competitive, powerful attributes of males seem well suited to the explicitly reproductive gender roles. Thus, males are essentially disposable. Their only necessary role in reproduction can be briefly discharged, whereas females have an extended essential role. Because of this, females are a precious resource, which, in times of danger, must be given protection if the social group is to survive. A single male could, in principle, provide all the sperm needed for many females. Moreover, all those unnecessary males will be a drain on resources if food is limited: males are costly biologically. They are therefore disposable in war or in risky competition with one another. It is thus tempting to explain gender differences in human societies entirely in terms of their value to the reproductive process. It is also tempting to conclude that the broad similarities between the reproductive roles of animals and humans must mean that gender differences in humans, like sex differences in animals, are the product of an evolutionary process which is, at its heart, genetically programmed and so ultimately genetically determined. These temptations should be resisted.

Undoubtedly the genetic inheritance of humankind exerts powerful effects on us and on our behaviour. However, what distinguishes humans from most other animals is the powerful additional legacy left to us by our culture. Humans are distinguished by our capacity to use information around us, to learn as we grow, to conceptualize and to establish and transmit cultures, including complex language, in ways not open to most animals. This mental flexibility may operate within limits imposed by our genetic inheritance, but it also operates on opportunities presented by that same inheritance. Even a superficial view of the widely different cultural roles that men and women have in different societies, how they are treated, valued and behave, shows the power of cultural inheritance. This is not surprising, given what we saw about how children learn about gender stereotypes from the society around them.

Conclusions

So, in conclusion, reproduction and sex are tightly, inevitably and invariably linked through our biological and genetic inheritance, whereas reproduction and gender are linked more loosely and elastically through our cultural inheritance. This point is made more clearly when we examine the varied functions associated with courtship and coition in humans. That is the subject of Chapter 6.

Key learning points

- Gender is a system of classification based on sex.
- A gender stereotype is a set of social beliefs about what it means to be a man or a woman. It may include appearance, behaviour, role (social, sexual and employment) and emotional and attitudinal attributes. It provides a shorthand for classifying people socially by sex.
- A gender identity is an inner state of awareness of one's own identity as a man or a woman in society. It is usually congruent with one's sex.
- Trans people have a gender identity that is not congruent with their sex.
- Although gender is a bipolar system based on two distinct sexes, when gender differences are measured in populations of men and women, they are not found to be bipolar. Indeed, most attributes show large overlap between genders and few are reliably predictive of gender.
- In animals, hormones condition sex differences in both brain structure and behaviour.
- In humans, there is evidence of differences in brain structure and function between both the sexes and the genders, but these differences are not large and their cause is unknown.
- The hormonal environment of higher primate and human fetuses and neonates has some effects on behaviour in infancy, some of which may persist into adult life. It is not known how these endocrine effects are exerted. It is not clear whether they are exerted directly on the brain, indirectly via effects on, for example, genital anatomy, or by a combination of both routes. The effects are not absolute but quantitative.
- In humans, the newborn baby is treated differently from the moment of its birth according to its perceived gender.
- In humans, the behaviour of a baby is interpreted to mean different things depending on its perceived gender.
- Babies of different sexes seem to show different behaviours quite early in neonatal life, but it is not clear whether the origin of these differences is endocrine, genetic, socially learnt or a mixture of all three.
- Human infants establish a gender identity by 3 years of age and start to develop a gender stereotype shortly thereafter. They develop a sense of gender constancy by 5 years of age. They then develop the expression of their gender identity by copying the sex-stereotyped behaviour that they observe around them.
- Gender stereotypes in most societies include reproductive roles and share attributes relevant to these reproductive roles. However, gender stereotypes are not limited to reproduction and reflect the fact that in humans sexual activity is not exclusively or even primarily a reproductive activity. There appears to be a large element of social learning in the construction of gender stereotypes and identities, and this forms part of the cultural inheritance that is transmitted transgenerationally.

Clinical vignette

Transgender conversion

AB, a 42-year-old XY individual with male primary and secondary sexual characteristics, presented to their general practitioner (GP) requesting help with gender reassignment. AB had been living and self-defining their gender as male, but explained that after several years of adopting a female-gender persona at weekends and away from work they wished to make a permanent transition to live as a female. AB explained that since an early age they had felt 'trapped in the wrong body' and as a child had attracted comments on their tastes in clothing and play from adults and peers. As AB grew up, this aspect of their life came to feel confusing and difficult. AB made conscious efforts during their early adulthood to adopt a gender-role that AB perceived as masculine – qualifying as a mechanic and playing sports to a high level. AB was sexually attracted to both men and women. Despite some early sexual experimentation with partners of both sexes, AB had married a female partner in their early twenties, and had two children. AB described that, as they had become older, (s)he felt less able to suppress feelings of being female and that the pressures of living as 'the wrong gender' had become less tolerable. AB had become depressed and their marriage broke up. Since moving out of the family home AB had begun to experiment with a female persona, including dressing in feminine clothes, using make-up, and adopting a feminine name. Having felt a sense of release and freedom in a female gender role, AB decided to seek help to make a permanent transition. AB had delayed doing so for nearly two years, because of worries about the impacts that the transition might have on relationships with family, colleagues, and particularly the couple's children. The GP was able to offer referral to a specialist gender reassignment service, information about support services for trans-gendered individuals, and general health advice to help cope with the difficulties that AB perceived might arise during the transition.

Gender dysphoria disorder describes the feeling that an individual's phenotypic sex does not 'match' with their gender. The term gender refers to the individual's self-representation as male, female, or any variant of these. The 'dysphoria' experienced by those who are gender-variant often arises more from society's reinforcement of gender-normative traits and stigmatization than from gender variances themselves. It is thus important for doctors and other health-care providers to realize that all individuals have a 'gender-narrative' that runs throughout their lives, from the development of awareness of self in early childhood onwards. For the majority of individuals, this is a straightforward and easily recognized story, but many others have complicated gender-narratives. These may not fit easily into reductionist concepts of binary 'male' and 'female' genders, and may be fluid throughout life. An understanding among doctors of the nuanced concepts of gender variance, and the therapeutic options that can help to resolve gender dysphoria is key to ensuring that individuals can access help.

FURTHER READING

General reading

Auger AP, Auger CJ (2011) Epigenetic turn ons and turn offs: chromatin reorganization and brain differentiation. *Endocrinology* **152**, 349–353.

Cahill L (2006) Why sex matters for neuroscience. *Nature Reviews in Neuroscience* 7, 477–484.

de Vries GJ, Södersten P (2009) Sex differences in the brain: the relation between structure and function. *Hormones and Behavior* **55**, 589–596.

Diamond M (2009) Clinical implications of the organizational and activational effects of hormones. *Hormones and Behavior* **55**, 621–632 (a history of ideas on gender).

Giedd JN, Rapaport JL (2010) Structural MRI of pediatric brain development: what have we learned and where are we going? *Neuron* **67**, 728–734.

Gooren L (2006) The biology of human psychosocial differentiation. *Hormones and Behaviour* **50**, 589–601.

Hines M (2010) Sex-related variation in human behavior and the brain. *Trends in Cognitive Sciences* **14**, 448–456.

McCarthy MM (2008) Estradiol and the developing brain. *Physiological Reviews* **88**, 91–134.

Thornton J, Zehr JL, Loose MD (2009) Effects of prenatal androgens on rhesus monkeys: a model system to explore the organizational hypothesis in primates. *Hormones and Behaviour* **55**, 633–645.

Zhou JN, Hofman MA, Gooren LJ, Swaab DF (1995) A sex difference in the human brain and its relation to transexuality. *Nature* **378**, 6.8–70

More advanced reading (see also Boxes 5.1–5.4)

Berenbaum SA, Bailey JM (2003) Effects on gender identity of prenatal androgens and genital appearance: evidence from girls with Congenital Adrenal Hyperplasia. *Journal of Clinical Endocrinology & Metabolism* **88**, 1102–1106.

Berglund H, Lindstrom P, Dhejne-Helmy C, Savic I (2008) Male-to-female transsexuals show sex-atypical hypothalamus activation when smelling odorous steroids. *Cerebral Cortex* **18**, 1900–1908.

Byne W, Tobet S, Mattiace LA *et al.* (2001) The interstitial nuclei of the human anterior hypothalamus: an investigation of variation with sex, sexual orientation, and HIV status. *Hormones and Behavior* **40**, 86–92.

Chau P-L, Herring J (2004) Men, women, people: the definition of sex. In: *Sexuality Repositioned: Diversity and the Law* (ed. B Brooks-Gordon *et al.*), pp. 187–214. Hart Publishing, Oxford.

Chura LR, Lombardo MV, Ashwin E *et al.* (2010) Organizational effects of fetal testosterone on human corpus callosum size and asymmetry. *Psychoneuroendocrinology* **35**, 122–132.

Deaux K (1985) Sex and gender. *Annual Review of Psychology* **36**, 49–81.

Frisen L, Nordenström A, Falhammar H *et al.* (2009) Gender role behavior, sexuality, and psychosocial adaptation in women with congenital adrenal hyperplasia due to CYP21A2 deficiency. *Journal of Clinical Endocrinology & Metabolism* **94**, 3432–3439.

Gabory A, Attig L, Junien C (2009) Sexual dimorphism in environmental epigenetic programming. *Molecular and Cellular Endocrinology* **304**, 8–18.

Garcia-Falgueras A, Ligtenberg L, Kruijver FP, Swaab DF (2011) Galanin neurons in the intermediate nucleus (InM) of the human hypothalamus in relation to sex, age and gender identity. *Journal of Comparative Neurology* **519**, 3061–3084.

Garcia-Falgueras A, Swaab DF (2008) A sex difference in the hypothalamic uncinate nucleus: relationship to gender identity. *Brain* **131**, 3132–3146.

Gizewski ER, Krause E, Schlamann M *et al.* (2009) Specific cerebral activation due to visual erotic stimuli in male-to-female transsexuals compared with male and female controls: an fMRI study. *Journal of Sex Medicine* **6**, 440–448.

Good CD, Lawrence K, Thomas NS *et al.* (2003) Dosage-sensitive X-linked locus influences the development of amygdala and orbitofrontal cortex, and fear recognition in humans. *Brain* **126**, 2431–2446 (brain sex and chromosomal constitution).

Greenberg JA (2002) Deconstructing binary race and sex categories: a comparison of the multiracial and transgendered experience. *San Diego Law Review* **39**, 917–942.

Hines M, Brook C, Conway GS (2004) Androgen and psychosexual development: core gender identity, sexual orientation and recalled childhood gender role behavior in women and men with congenital adrenal hyperplasia (CAH). *Journal of Sex Research* **41**, 75–81.

Kruijver FPM, Zhou JN, Pool CW, Hofman MA, Gooren LJ, Swaab DF (2000) Male-to-female transsexuals have female neuron numbers in a limbic nucleus. *Journal of Clinical Endocrinology & Metabolism* **85**, 2034–2041.

Lenroot RK, Gogtay N, Greenstein DK, *et al.* (2007). Sexual dimorphism of brain developmental trajectories during childhood and adolescence. *Neuroimage* **36**, 1065–1073.

Luders E, Sánchez FJ, Gaser C *et al.* (2009) Regional gray matter variation in male-to-female transsexualism. *Neuroimage* **46**, 904–907.

McDowell L (2004) Sexuality, desire and embodied performances in the workplace. In: *Sexuality Repositioned: Diversity and the Law* (ed. B Brooks-Gordon *et al.*), pp. 85–108. Hart Publishing, Oxford.

Pasterski VL, Geffner ME, Brain C *et al.* (2005) Prenatal hormones and postnatal socialization by parents as determinants of male-typical toy play in girls with congenital adrenal hyperplasia. *Child Development* **76**, 264–278.

Paus T, Nawaz-Khan I, Leonard G *et al.* (2010) Sexual dimorphism in the adolescent brain: role of testosterone and androgen receptor in global and local volumes of grey and white matter. *Hormones and Behavior* **57**, 63–75.

Savic I, Berglund H, Gulyas B, Roland P (2001) Smelling of odorous sex hormone-like compounds causes sex-differentiated hypothalamic activations in humans. *Neuron* **31**, 661–668.

Schulz KM, Molenda-Figueira HA, Sisk CL (2009) Back to the future: the organizational-activational hypothesis adapted to puberty and adolescence. *Hormones and Behaviour* **55**, 597–604.

van Wingen G, Mattern C, Verkes RJ *et al.* (2010) Testosterone reduces amygdala-orbitofrontal cortex coupling. *Psychoneuroendocrinology* **35**, 105–113.

CHAPTER 6

Sexual selection

Essential Reproduction, Eighth Edition. Martin H. Johnson.

Companion website: www.essentialreproduction.com

Reproduction and sex are tightly linked through our biological and genetic inheritance. The sexual and reproductive partnerships established by our ancestors provided the basis of that inheritance. Sexual selection thus has enduring consequences. However, this does not mean that the selection of partners in contemporary human societies is conditioned simply by biological evolutionary considerations. Gender intervenes, and reproduction is linked to gender through our cultural inheritance. This latter point is made most clearly when we examine the varied functions associated with reproductive pairings in human societies. Production of offspring is obviously one such function. However, erotic pleasure quite distinct and separable from reproduction is another: unlike most animals, women and men can and do mate regardless of their fecundity. Courtship and coition can also serve a wider emotional purpose, involving feelings such as self-worth, security, dependence and power. Courtship and coition also have social and economic functions: when formalized in kinships they establish and perpetuate patterns of inheritance and power in a society. We should also remember that coition has a consequence not always welcomed by humans, but essential for some microorganisms: of transmitting them and the diseases they may cause through a society. These varied and wide-ranging functions and consequences of courtship and coition mean that society tries to control the processes with customs and laws (see Chapter 23). These customs and laws then contribute to the cultural inheritance that we learn as part of the gender stereotype of our society.

The relationship between sex, gender and reproduction raised here will be revisited in many chapters later in this book. Now, however, we will complete our consideration of how men and women are made, by looking at human sexuality and how sexual partners are selected.

Sexuality

Courtship and coition can involve intense and pleasurable sexual fantasies and feelings. This state of sexual excitement in humans is described as the **erotic**. In this book, we reserve the use of the term **sexuality** for this erotic experience and its expression in human lives. This definition is not uniformly agreed and would be considered by some to be controversial and too narrow. Sometimes, sexuality is used to describe all that it means to be a man or a woman, a sort of all-pervasive state that is difficult to distinguish clearly from gender itself. We find this definition too diffuse to be useful. Of course, the erotic and its associations can be very pervasive and, as we will see, not limited simply to the events surrounding courtship and coition. However, at heart, our sexuality is about **inner erotic excitement and fantasy** and its outward **expression in sexual erotic behaviour**. A **sexual individual** is one who is erotically functional: mentally and/or behaviourally. An **asexual individual** lacks erotic experiences and fantasy.

As will be seen, the physiology of erotic arousal itself seems to be similar for men and women, and descriptions of what it is like to be in an erotically aroused state are not gender specific (see Chapter 11 for further discussion). What then distinguishes erotic experiences in different individuals and genders?

Sexuality can be categorized

A commonly used system for classifying sexuality utilizes the object of sexual arousal as its starting point. Examples of such a classification are shown in Table 6.1. Four things are striking about the contents of Table 6.1.

First, there is a wide range of erotically arousing stimuli. It is important to note that they are not necessarily mutually exclusive. For example, a person may be aroused by both men and women (**bisexual**), or by the opposite sex *and* by objects or cross-dressing, or be sadomasochistic with the same and/or opposite sex partner(s). The sexuality of humans is complex. Moreover, the stimuli of erotic arousal may change for an individual with age, experience or social expectation. So this labelling system is imperfectly rigid, and the use of labelling as shorthand can be misleading.

Second, the stimuli are made up of a mixture of objects, people and activities, and in some cases are described in terms of how they are used erotically but in others are not. When we deal with the sexual, there are two levels of description. There is the inner world of conscious arousal, imagination and fantasy: this is usually given the name **sexual identity**, akin to the conceptual inner state of gender identity described in Chapter 5. It is an acknowledgement by a person of their own state of being as a sexual individual. Their own state may or may not fit with the categories used in Table 6.1, although most people will tend to use the labels that society provides for them. Thus, someone might describe themselves as 'asexual', 'bisexual', 'homosexual' or 'heterosexual'; that would be the verbal expression of their sexual identity. This inner world may or may not be expressed through behaviour and sexual attitudes. Society usually has a clear expectation of how people with different sexualities will behave and what their attributes will be, a **sexual stereotype**, and this can often be absorbed as a part of the sexual identity of people.

Third, many of the stimuli clearly have nothing to do with procreative sex, since procreation is impossible or unlikely in the context of arousal by them. This emphasizes the clear **separation of reproductive and erotic activities** that can be observed in humans compared with other mammals in which

Table 6.1 A classification system for sexualities

Object causing arousal	Classification of person aroused	Comments
Nothing	Asexual	Not generally admitted
Person of opposite sex Person of same sex	Heterosexual Homosexual	Social norm in most cultures Acceptable in some forms in many societies; illegal or disapproved of in others
Immature person Inanimate object	Paedophiliac Paraphiliac	Generally unacceptable Acceptable if not causing harm to others or distress to paraphiliac him- or herself
Excrement Wearing clothes of other gender	Coprophiliac Fetishistic transvestite	Generally disapproved of Often confused with transgendered but is not a gender issue; may be accepted or ridiculed
Watching others naked and/or engaged in sex Self-displaying when naked or engaged in sex Receiving or inflicting pain during sex	Voyeur Exhibitionist Sadomasochist	Broadly disapproved of unless 'formalized' or paid for Broadly disapproved of unless 'formalized' or paid for Held to be illegal in Europe

reproductive fecundity and sexual arousal are mostly very closely co-regulated. In this regard, humans resemble their closest evolutionary relatives among the higher primates and especially the much-studied chimpanzees and bonobos. Thus, these species show quite clearly that sexual interactions, both within and between sexes and across age groups, can have a social role in addition to a sexual role. Genital showing and looking, touching and rubbing, erection and mounting are commonly observed between individuals of the same sex, and appear to be seen as pleasurable and reassuring. Such same-sex interactions are often called **socio-sexual** to distinguish them from the **eroto-sexual** interactions between males and females, although it is unclear how real this distinction is. The main point to understand is that sexual stimulation can form part of the social glue for a social species such as ours.

Fourth, the social acceptability of different sexual stimuli varies with the stimuli, the type of person involved and the society in which they are experienced. Heterosexuality, although a social norm in most societies, is circumscribed heavily by restrictions on its expression in many, e.g. within marriage, caste, race or ethnic group or by the relative age differentials of the partners. Homosexual acts between men have been viewed variously as essential, desirable, acceptable, immoral, sinful, illegal and pathological in different cultures and at different times, whereas those between women have been ignored, ridiculed, politicized, accepted, encouraged and celebrated. Similarly, paedophilia has been, and still is, variously defined according to a wide range in the age of sexual consent in different societies, both historical and contemporary. The **social regulation of sexual expression** is usually strict, whether by legal or social sanction. In many cases, it is so strict that individuals will hide or deny any sexual feelings that do not conform to approved sexual stereotypes, or may only express those feelings covertly. This strong social and self-censorship makes research in the area of sexuality very difficult. People may lie, distort or remember selectively in retrospective studies using questionnaires or interviews. Even in prospective studies, the behaviour and attitudes observed and recorded may reflect a strong impact of social expectations, as we will see in the next two sections, which consider how we acquire sexual identities.

The developmental origins of sexuality are complex

Given the wide range of erotic stimuli, it would seem very unlikely that there is a direct genetic basis for our sexualities. It is difficult to see why evolution should have selected genes for fetishistic transvestism! It would be more reasonable to expect that evolution might have selected genes that encouraged sexual arousal in general and by the opposite sex in particular, since that would presumably promote the most effective transmission of those same genes to future generations. So, might there be something qualitatively different about the basis of sexual arousal by people as stimuli as opposed to by objects or situations? There is little clear evidence on this point. Thus, just as there appears to be considerable emancipation of our gender from our genes, such that social learning plays a substantial role, so the same may have happened with our sexuality. There may be evolutionary advantages to flexible and adaptive social and sexual structures that came with this emancipation. Thus, whether or not there are genetic or anatomical correlates and even causes of human sexualities, there seems likely to be a strong element of social learning too. As with gender identity, it is difficult to disentangle the threads.

Twin and familial studies have suggested that there may be a genetic element in the establishment of our sexuality: a finding much trumpeted in the popular press. However, the results are far from decisive. The studies have focused almost exclusively on the question of how male homosexuality is determined. Monozygotic twins are reported to show a higher concordance of homosexuality than same-sex dizygotic twins. However, this finding does not demonstrate a 'gene for sexuality'. Indeed, although some familial studies have suggested that some homosexual men are more likely to carry a particular set of genetic markers on the X chromosome, none of these results has been confirmed in other studies. Caution is required in interpreting these sorts of genetic study of complex behavioural traits. Thus, genetically similar individuals are likely to share common experiences because they have similar characteristics. This might predispose them to responses more likely to lead to development of a particular sexuality. For example, imagine that our sexuality is learnt in early childhood. The way in which it is learnt may depend on the maturation of the nervous system as well as the social surroundings. Imagine a gene or genes that advanced slightly the maturation of one part of the nervous system over another part. That might change the learning pattern and so influence the probability of a particular sexuality developing. This hypothetical scenario is presented to illustrate that, although **there must be a genetic influence on sexuality, this does not mean that there is a genetic cause**. The origins of sexuality are likely to be more complex and multifactorial.

A second line of evidence comes from studies on the structure and activity of the brain. In Chapter 5 (see pages 78–81), examples were given of suggested sexual and gender dimorphisms in brain structure and functional organization in humans, amongst which was INAH3 (see Figure 5.5). Similarly, there are reports that this same nucleus, which differs in men and women, is intermediate in self-declaring homosexual men (see caption to Figure 5.5). However, as for the studies on the brains of trans people, the number of men in the studies is small and the overlap in the size values between homosexual men and non-homosexual men is too great to be predictive. Functional imaging studies have reported sex differences in the anterior hypothalamic (preoptic and ventromedial nuclei) responses to androstadienone, a volatile testosterone derivative present in sweat and presumptively detected by olfaction. Thus, women showed higher activation rates than men, as did homosexual men.

Since in animals some hypothalamic nuclei differ in size as a result of perinatal androgen exposure, is there any evidence linking exposure to androgens to sexual attraction towards women? We have the same problems here that we encountered when considering androgens and gender development. Although women who have adrenogenital syndrome do show a higher incidence of attraction towards other women, most such women have a heterosexual attraction to men. Numerous other women with no evidence of fetal androgen exposure are attracted to women. Conversely, adult homosexual men show no evidence of reduced androgens compared with heterosexual men, and no good evidence for lower fetal androgen levels either (Box 6.1). So, **fetal androgen levels may influence sexual orientation, but are unlikely to be exclusive determinants of it**.

So where does social learning fit? There is some evidence from work on paraphilias and fetishisms relating sexual arousal experiences in early childhood to the stimuli likely to be arousing in the adult: **associative learning**. Children from an early age do show evidence of arousal, such as phallic erection, and seem to derive pleasure from phallic stimulation. It is possible that the coincidence of arousal with an emotionally charged event or object in childhood might lead to the association of eroticism with that event or object in later life. However, the evidence on this point is far from clear, and we simply do not know how we become eroticized to particular stimuli. The question of social learning in the development of sexuality is considered further in the next section.

Social learning, gender and sexuality

Many highly gendered societies, in which heterosexuality is the social norm and homosexuality is disapproved of, place a strong emphasis on the link between gender and heterosexuality. Thus, an integral part of being feminine is to be attracted to men and of being masculine is to be attracted to women. A heterosexual identity thus becomes subsumed into a gender identity such that the two are conflated conceptually. This conflation is evident in the sexual stereotypes of traditional Judaeo-Christian and Islamic societies. Thus, **masculine men** are seen as **sexually dominant, active, insertive and initiating** whereas **feminine women** are **sexually passive, receptive and submissive**. Deviations from these stereotypes are stigmatized; witness the stereotypes of the sexually passive, effeminate and 'unmanned' gay man and the sexually aggressive, masculine and defeminized lesbian. However, in practice, heterosexual individuals show a much wider range of astereotypical sexual behaviours and, as we have already seen, heterosexuality need not be 'pure', but can coexist within an individual with wider sexual interests, such as sadomasochism, paraphilias and bisexuality. Moreover, although some gay men and lesbians may have insecure gender identities as men and women, and may indeed conform to effeminate and butch stereotypes, respectively; many others do not. There are many gay men who are both homosexual and masculine, and lesbians who are both homosexual and feminine. Insecurity of gender identity for homosexuals is, of course, a likely outcome in a society in which sexual and gender stereotypes are conflated and variation from accepted gender and sexual stereotypes is stigmatized. In other societies in which this conflation of gender and sexuality does not occur, there appears little problem in masculine men and feminine women expressing homosexual emotions and behaviour. In this regard, the transgendered are again instructive. Both male-to-female and female-to-male transgendered individuals may find men, women or both sexually arousing. Thus, a transgendered individual of the male sex with

Box 6.1 Fetal androgens and homosexuality in women and men

The problem in demonstrating directly a relationship between fetal androgen exposure and homosexuality in men is that there is no clear idea of the nature of the expected relationship. Is it the level or the timing of androgen exposure, or some other feature such as responsiveness to androgen that is important? So, aside from use of animal models to frame questions – with all the attendant concerns that this approach raises – surrogate and stable measures of androgen exposure in fetuses have been sought.

First, a brief comment on animal models for homosexuality. Contrary to popular opinion, homosexual behaviour is observed naturally and regularly in animals. For example, some 8% of rams exhibit a sexual preference for other males. Moreover, one study has shown that fetal androgen exposure contributes to their homosexual preference and this behaviour is correlated with a smaller size of the sexually dimorphic nucleus in the preoptic area in 'homosexual' male goats.

What about surrogate measures in humans? Two in particular have been advanced: the relative lengths of the second digit finger to the fourth digit finger (the so-called 2D:4D ratio) – claimed in some but not all as smaller in heterosexual men and in congenital adrenal hyperplasia (CAH) woman than in women, although not differing appreciably or consistently in homosexual and heterosexual men, and the auditory evoked potentials (AEPs) of the auditory system – claimed as differing in form and strength between heterosexual women and men plus masculinized women, whereas again homosexual men did not differ appreciably from heterosexual men. Both the AEPs and 2D:4D ratios differ in newborn male and female infants and then persist throughout life. In the absence of prospective studies, measurements from gay women and men are only performed in adulthood, but, because it is considered unlikely that either measure could result from differential usage postnatally, each measure is assumed to result from a fetal activity. Thus, it is claimed that these developmental traits that differ between the sexes are also associated with sexual preference. There are questions about this claim. First, how clear are the data and are they predictive? Second, are they really immune from later variation? Third, do they really result from androgen exposure and if so how?

First, there are significant associations between some features of each trait with adult homosexuality, but they are of variable direction and are not predictive. Thus, only four of 12 features of AEPs differ between homosexual and heterosexual men. Moreover, of the four, two were less like those in women and thus described as 'hyper-masculinized'; the other two being more like those in women and thus 'hypo-masculinized'. For the 2D:4D ratio, homosexual men are *not* in fact more like women and thus cannot be described as 'feminized'; indeed, even the sex differences, although significant at a population level, are not predictive of sex. Second, there is evidence that AEPs can be influenced in adults; e.g. women who are postovulatory or taking oral contraceptives appear to be reversibly masculinized. Third, direct evidence that fetal androgen exposure causes the differences is lacking. For example, studies on animals suggest that oestrogen receptor isotypes may have more influence on the digit ratio than do androgens, which many think are minor influences on the 2D:4D ratio. Thus, these studies suggest that if there is an influence of fetal androgen exposure on sexual orientation in humans, other factors are likely to be of more significance.

Further reading

Balthazart J (2011) Minireview: hormones and human sexual orientation. *Endocrinology* **152**, 2937–2947.

Breedlove SM (2010) Organizational hypothesis: instances of the fingerpost. *Endocrinology* **151**, 4116–4122.

Forstmeier W, Mueller JC, Kempenaers B (2010) A polymorphism in the oestrogen receptor gene explains covariance between digit ratio and mating behaviour. *Royal Society Proceedings: Biological Science* **277**, 3353–3361.

McFadden D (2011) Sexual orientation and the auditory system. *Frontiers in Neuroendocrinology* **32**, 201–213.

Roselli CE, Reddy RC, Kaufman KR (2011) The development of male-oriented behavior in rams. *Frontiers in Neuroendocrinology* **32**, 164–169.

a feminine gender identity may find men sexually attractive, in which case he is homosexual before gender reassignment and she is heterosexual afterwards! The example of trans people emphasizes the importance of **uncoupling sexuality from gender** conceptually, even if in traditional Judaeo-Christian and Islamic cultures they have been conflated socially.

The conflation of sexuality and gender further complicates study of the possible social learning of sexuality. Several retrospective studies suggest that gays and lesbians recall having more ambiguous gender experiences in childhood than do self-defining heterosexual men and women. For example, gay men recalled playing with girls and girls' toys and games, and lesbians recalled being tomboyish. However, retrospective studies suffer from the dangers of selective recall and denial, which, as we saw earlier, is a dangerous possibility in an area as sensitive as this. Prospective studies of children referred to clinicians precisely because they were displaying gender-atypical play patterns have shown that as adults these individuals manifest a higher incidence of homosexuality than control children. It is difficult to know how to interpret these findings, especially as they are based on children so seriously different (or perceived to be so) as to be referred to a gender clinic. Precisely because society, parents and/or the children themselves associate gender and sexual stereotypes, they are more likely to develop in tandem: there is after all the basis of a socially learned element to each. The results do not mean that they must develop in tandem or that having one identity causes a person to have the other (in either direction).

Summary

We do not understand how people acquire a sexual identity. Indeed, our understanding of the complex nature of sexual identity is rudimentary. Our systems for classifying sexuality are at best approximate and still based largely on a historical view of all sexual deviation from a narrowly defined heterosexuality as being pathological or socially undesirable. This is not a helpful starting point for looking at the natural expression of sexuality. There undoubtedly are influences of genes, hormones, brain structure and social learning on how our sexualities develop, but there is little evidence that any one of these actually causes each of us to have a particular sexual identity. The fact that different societies construct different systems of sexual stereotypes and that these become absorbed (internalized) through social learning into each individual's sexual identity implies that social learning must play a large part in the construction of sexuality, perhaps building on or interacting with the various influences of genes and hormones to affect brain function and structure.

Next we consider what we know of the biology underlying sexual interest and arousal as distinct from sexual orientation.

Eliciting male sexual interest and arousal

Stimuli capable of eliciting sexual behaviour surround most of us most of the time, but sexual interaction occurs only sporadically. What determines when these sexual stimuli induce sexual activity? The answers to this question, particularly in primates and humans, are again complex, but if we examine non-primate species we see rather clearly that hormones are very important. A castrated male or female rat, cat or dog does not display sexual activity when placed with a member of the opposite sex. Treatment with the appropriate hormones results in the activation of sexual behaviour: sexual stimuli become effective again in inducing sexual responses. In these species, therefore, hormones increase the probability that sexual stimuli will elicit sexual activity.

In primates, including humans, this strong influence of sex hormones has been modified considerably, such that social, experiential and volitional factors become increasingly important, reflecting the increasing size and complexity of the brain, particularly the neocortex. So, the ways in which hormones affect sexual behaviour in animals and humans are described next, starting with males.

Sexual behaviour in male animals

The clearest example of testosterone's role in male sexual behaviour is seen in seasonally breeding species, such as the red deer. For a large part of the year, the testes are quiescent, awakening as the time for 'rutting' approaches: the output of testosterone and spermatozoa rises, and male sexual behaviour is observed until testis regression occurs. This natural situation can be replicated experimentally in adult rats by castration and its reversal by testosterone treatment (Figure 6.1). However, although plasma testosterone is virtually undetectable within hours of castration, the decline in sexual behaviour takes several weeks, and conversely, after treatment of the castrate with testosterone, sexual behaviour is restored only slowly. Moreover, the longer the interval between castration and the initiation of testosterone replacement, the longer it takes to restore sexual behaviour (Figure 6.2). Additionally, not all elements of masculine sexual behaviour are affected equally by removal of testosterone. Thus, mounting persists for much longer than do penile intromission and ejaculation. Where is testosterone acting and how does it achieve these effects? The brain is one important target, affecting the male's **motivational state directly**. But so too are the genitalia, with testosterone affecting their sensitivity and so **indirectly influencing sexual behaviour**.

These conclusions have been reached as a result of studies in the rat, in which testosterone exerts its behavioural effects

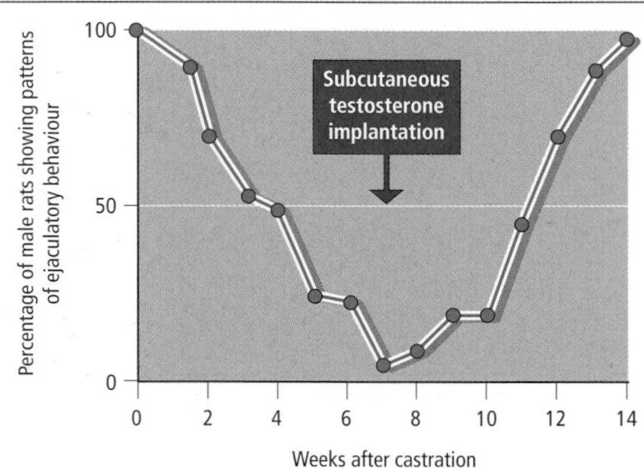

Figure 6.1 Effects of castration and testosterone replacement on patterns of ejaculatory behaviour in the male rat. Note that appreciable levels of the behaviour persist for some weeks after castration and that a comparable time is required for restoration of the behaviour following the subcutaneous implantation of testosterone. (Source: Bermant G. & Davidson J.M. (1974) *Biological Bases of Sexual Behaviour*. Harper & Row, New York. Reproduced with permission from LWW.)

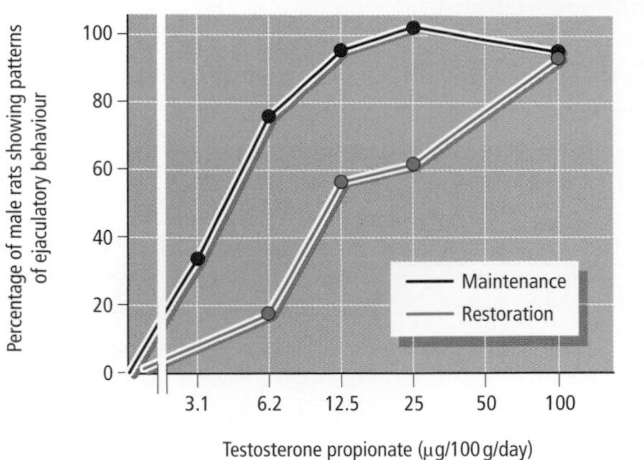

Figure 6.2 Differential testosterone requirements for 'maintaining' ejaculatory behaviour in newly castrated rats and 'restoring' the same behaviour in long-term castrates. Note that more testosterone propionate was required in the latter group; treatment, as in Figure 6.1, was given for several weeks to effect this change. (Source: Bermant G., Davidson J.M. (1974) *Biological Bases of Sexual Behaviour*. Harper & Row, New York. Reproduced with permission from LWW.)

on the brain only after it has been aromatized to oestradiol (see Box 2.5). Thus, after castration, small amounts of oestrogen administered to male rats reverse the motivational deficits, such as reduced mounting behaviour, but regression of seminal vesicles, prostate and cornified spines on the penis persist, as do the attendant deficits in related sexual responses, such as erection and penile intromission. Conversely, when dihydrotestosterone, which cannot be aromatized to oestrogen in the brain, is given to castrated male rats, it has potent stimulatory effects on sex accessory glands, the penis and erectile sexual responses, but little or no effect on mounting behaviour. Only by combining extremely small amounts of oestradiol with dihydrotestosterone can the full restorative effect of testosterone be achieved.

Unfortunately, this divided action of testosterone does not seem to apply in primates, preventing the same dissection of behaviour. However, it is clear that castration of male monkeys is usually followed by a slow reduction in sexual activity after weeks, months or years and rarely to complete absence, but with considerable inter-individual variability. For example, a male monkey with a history of separate sexual interactions with each of two females may, after castration, continue copulating more or less unchanged with one female, but lose his sexual interest completely in the other. Clearly, while testosterone has an important influence on sexual activity in monkeys, its presence is not obligatory and other factors can maintain sexual interest, arousal and behaviour. Moreover, testosterone treatment effectively restores sexual interest to pre-castration levels, but the level of androgen required is much less than pre-treatment levels, suggesting that normally testosterone levels are well above threshold for behavioural effects.

Sexual behaviour in men

A clear influence of testosterone on human sexual behaviour is seen during the transition through puberty. Prepubertal boys do experience sexual arousal and erections, but at lower levels than do pubescing and post-pubescent adolescents as androgens rise (see Chapter 4). Likewise, hypogonadal or agonadal men, in whom low levels of sexual arousal and activity can be treated by testosterone therapy, revert to reduced arousal and activity on androgen withdrawal, but not to cessation (Box 6.2). These situations can be contrasted with the chronic testosterone treatment of intact men, which generally does not result in an increase in sexual activity. Conversely, antiandrogens, such as cyproterone, used clinically to treat intact men with antisocial patterns of sexual behaviour, such as exhibitionism or paedophilia, only reduce, and rarely eliminate, sexual arousal. So, it seems that androgen treatment is only beneficial at stimulating already existing sexual interest and activity where low testosterone levels exist in men (or boys). There is a critical range of low plasma testosterone concentrations over which a clear positive sexual response to chronic treatment can be expected. Once the adult range of testosterone levels is reached, that dose relationship is lost. Moreover, sexual activity can persist in the absence of androgen, which acts to increase its frequency and intensity.

However, even in men with testosterone levels in the adult range, acute elevations in testosterone are associated with acute behavioural responses. Thus, an encounter with a potential sexual partner leads to an acute androgen rise associated with an increased attention to the partner, an uncoupling of the (rational) frontal cortex from the (emotional) amygdala, thereby reducing emotional control and increasing the chance of risky behaviour. However, whether androgens are responsible for or only correlated with this change in behaviour is not clear.

Finally, considerable variation amongst men is observed in their reactions to androgen levels – as was also described for non-human primates. Some of this variation may be related to the many genetic polymorphisms observed in the androgen receptor that affect its sensitivity to ambient androgens (see Table 2.2).

Female sexual behaviour

In mammals other than higher primates, females are only sexually receptive around the time of ovulation when they are said to be '**on heat**' or '**in oestrus**'. Ovariectomy is followed by a prompt and usually complete abolition of both female **receptivity** (they will no longer accept mounts by the male) and **proceptivity** (they no longer 'solicit' male attention). Restoration of these elements of feminine sexual behaviour is

Box 6.2 Androgens and impotence

In the study on a hypogonadal man summarized in Figure 6.3, the frequency of his sexual thoughts was measured as an index of the effects of testosterone. Penile erections, whether spontaneously during rapid eye movement (REM) sleep (**nocturnal penile tumescence** or NPT) or in response to sexual stimuli, have also been measured. Only nocturnal erections are clearly testosterone dependent, an observation that has led to the **differential diagnosis of organic and psychogenic impotence**. Thus, preservation of NPT in men who lose erectile ability during sexual interactions is strongly indicative both of psychogenic impotence and of the likely ineffectiveness of testosterone therapy. Moreover, while it is the case that the low levels of plasma testosterone seen in hypogonadal men are highly correlated with lower frequencies of sexual thoughts and erections induced by sexual fantasy (internal stimuli), erections in response to external stimuli, such as erotic films, do not appear to be so testosterone dependent (Figure 6.4). However, penile rigidity tends to be less and detumescence is more likely as soon as the external stimulus is removed.

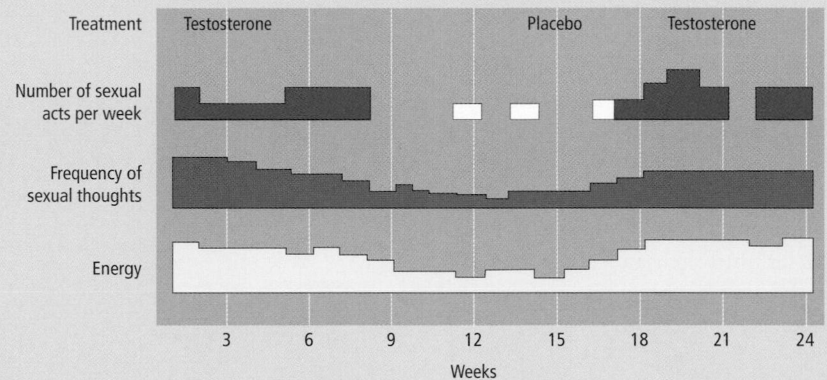

Figure 6.3 The effects of testosterone replacement in a hypogonadal man aged 40 years and castrated 1 year earlier for testicular neoplasia. (Red bars = sexual activity to orgasmic ejaculation; white bars = incomplete sexual activity). Note that about 3 weeks after stopping testosterone treatment, sexual activity, sexual thoughts and energy all declined, but never ceased entirely. There was no response to placebo, but a response within 1 or 2 weeks of restarting testosterone treatment. (Source: Bancroft J. (1983) *Human Sexuality and its Problems*. Churchill Livingstone, Edinburgh. Reproduced with permission from Elsevier.)

As with studies on animals, the nature of the stimulus and the response must be considered carefully when interpreting these data. An interesting issue in studies of sexual behaviour in men is whether erections occur as a result of sexual arousal, or whether they also contribute to its development. Is penile erection a response or a stimulus? Erections may be such an important indicator to men of sexual excitement that a decrease in erectile ability may contribute to a much higher threshold for sexual arousal and hence decreased sexual activity. A clear separation of genital and motivational responsiveness to androgens in hypogonadal men has simply not been achieved, but actions at both sites, as in rats, are probable.

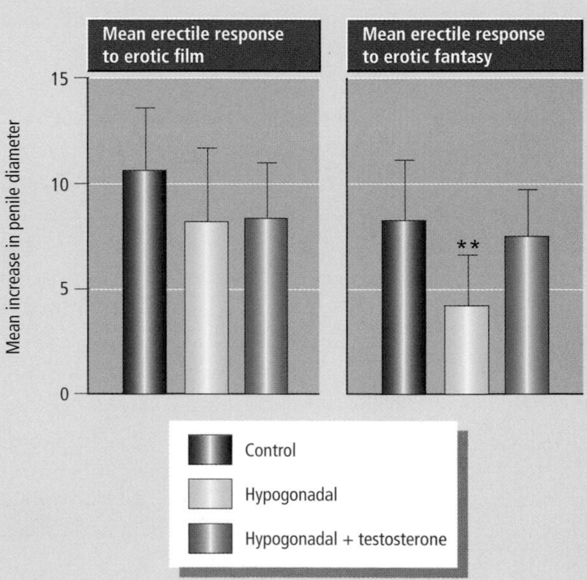

Figure 6.4 Erectile response (measured as increase in penile diameter) to an erotic film and fantasy in hypogonadal men with and without testosterone replacement. The hypogonadal men did not differ from controls in their response to the film, but their response to fantasy was significantly lower than that for controls when they were androgen deficient, but improved with testosterone replacement. The latency of their erectile response was significantly reduced after hormone replacement. $^{**}P < 0.01$. (Source: Bancroft J. (1989) *Human Sexuality and its Problems*, 2nd edn. Churchill Livingstone, Edinburgh. Reproduced with permission from Elsevier.)

achieved equally rapidly by treatment with oestradiol, or in some species, with oestradiol followed by progesterone. So, sexual behaviour in female non-primates seems to be largely dependent on their hormonal state. But again the situation for female primates is more complex.

Oestrogens and monkey sexual behaviour

If sexual interactions between a male and female monkey are observed, they usually follow a cyclic pattern (Figure 6.5a). Moreover, ovariectomy is followed by a reduction in sexual interactions, restored by treating females with oestradiol. However, when the sexual responses of males and females are observed, it is found that it is mainly the male's behaviour that declines markedly after ovariectomy and increases after restorative oestradiol treatment. **Oestradiol changes the sexual 'attractiveness' of females to males, and does so through an effect on the vagina**, where it affects the vaginal odour (Figure 6.5b, D). Thus, male rhesus monkeys rendered reversibly anosmic fail to discriminate between oestrogen-treated and untreated ovariectomized females. They copulate with the latter until their anosmia is reversed, at which point they usually cease promptly until oestrogen treatment of the female (systemic or intravaginal) is resumed.

Analysis of vaginal secretions has revealed the presence of a mixture of simple **aliphatic acids** (acetic, propionic, isobutyric, butyric and isovaleric), the concentrations of which vary during the menstrual cycle. Moreover, they disappear after ovariectomy, but are restored after oestrogen treatment. Vaginal lavages taken from oestrogen-treated female monkeys and placed on the perineum of untreated females stimulate the sexual interest of males, as indeed do vaginal secretions from women and other species of monkey. Not surprisingly, therefore, the same aliphatic acids have been found in human vaginal secretions and they also vary in concentration during the menstrual cycle. Proof of the involvement of the aliphatic acids comes from the observation that a synthetic mixture in the correct proportions placed intravaginally in female monkeys can activate the same sexual interest of males (Figure 6.6). The effectiveness of this mixture can be enhanced by the addition of phenolic compounds (phenylpropanoic and parahydroxy-phenylpropanoic acids), which are also present in vaginal secretions, but are ineffective when applied alone. Thus, a female monkey's vagina produces oestradiol-dependent odour cues. The acids are not a glandular product, but result from microbial action on vaginal secretions, and can be blocked by use of penicillin. Progesterone can decrease the sexual attractiveness of female monkeys and does so by reducing the sex-attractant properties of vaginal secretions. This observation explains the decrease in sexual interaction during the second half of the menstrual cycle. Clearly **vaginal secretions can both 'turn on' and 'turn off' a male's sexual interest in a female**.

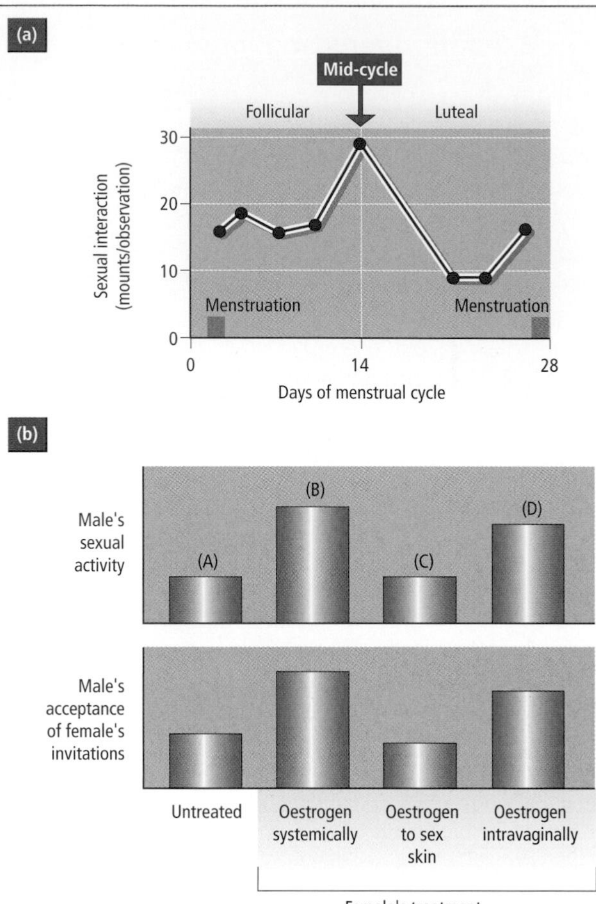

Figure 6.5 (a) Sexual interactions in rhesus monkeys, measured as mounts by the male, between male: female pairs during the menstrual cycle. Note the high levels of interaction, which peak at mid-cycle and fall thereafter. The premenstrual rise is characteristic. This pattern of interaction seems to be more due to fluctuating interest in the female by the male than vice versa, as is revealed more clearly in (b), which summarizes the results of experiments demonstrating that: (A) in ovariectomized females, male sexual activity is low, as is male acceptance of sexual invitations by females; (B) treating females with oestradiol, by subcutaneous injections, increases both these parameters of the males' behaviour; (C) the effect is lost if the oestradiol is smeared, as a cream, on the female's perineum; (D) if it is placed in the vagina, an effect identical to that seen in (B) occurs. These data indicate that oestradiol increases sexual attractiveness by an action on the vagina.

Finally, it must be emphasized that these effects of odours on the sexual activity of males are not all or none. Some males are oblivious to changes in a female's odour; others seem to have their sexual activity completely regulated by them.

Androgens and female monkey sexual behaviour

Oestrogen-treated, ovariectomized female monkeys are both attractive to males and willing to solicit and accept their

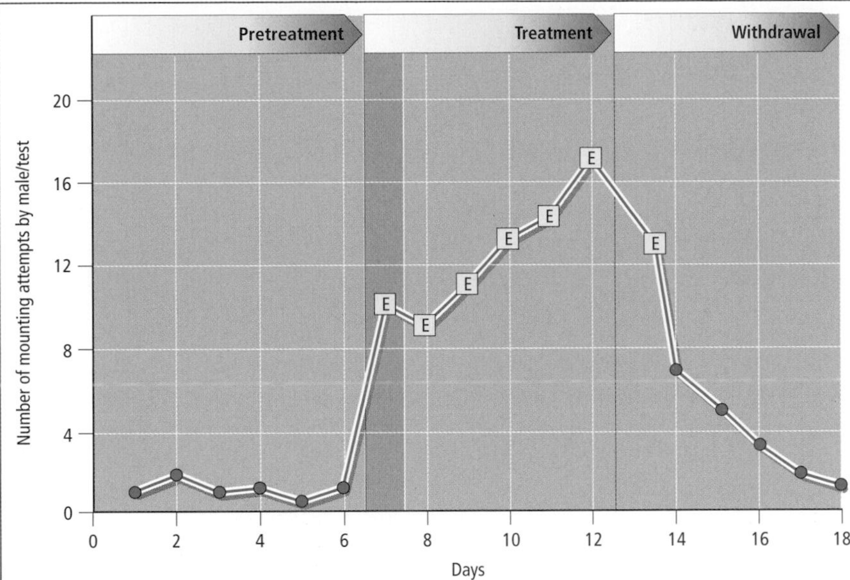

Figure 6.6 Summary of effects on three male rhesus monkeys of a synthetic mixture of aliphatic acids applied to the perineal sexual skin of an ovariectomized female. Mounting behaviour is markedly stimulated during treatment, and ejaculations (E) occur consistently. Withdrawal of the mixture is followed by a reversal of these changes in the male sexual behaviour.

attempts to copulate. However, while oestrogens act to attract males, it is **androgens** that **affect female sexual responsiveness to males**. Thus, if all circulating androgens, including those from the adrenal cortex, are removed or suppressed, a marked reduction in proceptive and receptive behaviours occurs (Figure 6.7), even in the continuing presence of oestradiol. These changes in the sexual behaviour of females are reversed by treatment with testosterone. Although the decreases in sexual behaviour in female monkeys deprived of androgens are considerable, some mounting attempts by males will be accepted, just as androgen-deprived males may continue to display sexual interest in females. It is again the incidence of such acceptances and interest that is reduced. Thus, **the strict dependence of sexual behaviour on steroid hormones seen in female non-primates has no obvious parallel in primates**.

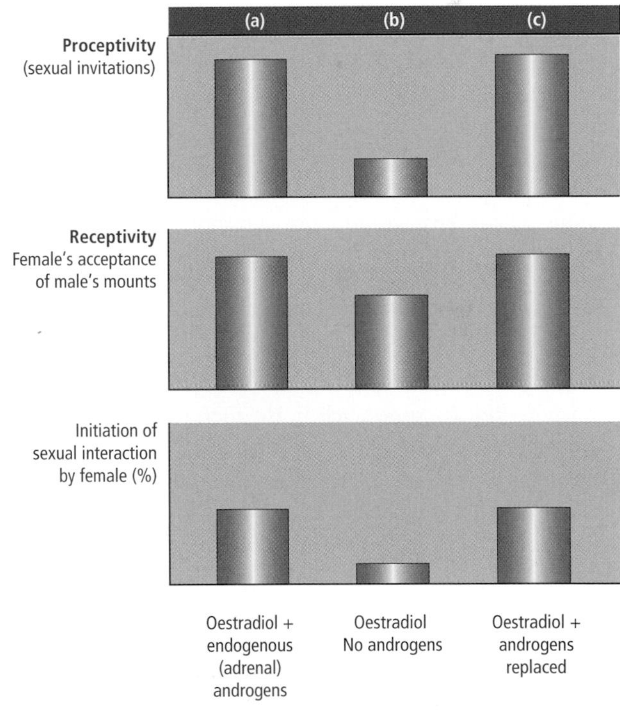

Figure 6.7 A block diagram summarizing the results of experiments on adrenal androgens and sexual behaviours in ovariectomized female rhesus monkeys receiving oestradiol benzoate throughout the experiment to ensure their attractiveness to the males. (a) Controls. (b) Removal of endogenously secreted androgens from the adrenal by suppression with dexamethasone or adrenalectomy with glucocorticoid replacement, caused large decreases in proceptive behaviour and initiation of sexual interaction by the female, and smaller decreases in receptive behaviour. (c) These decreases were reversed by treatment with testosterone propionate or androstenedione.

Steroid hormones in women affect human sexual behaviour

Postmenopausal women, like those who lose their ovaries, do not usually suffer a complete loss of libido. However, oestrogen replacement therapy may be helpful to ensure vaginal lubrication and so prevent a secondary decline in sexual activity due to pain at intercourse. As with men, direct effects of sex steroids on the genitalia in maintaining levels of sexual activity must be taken into account before making assumptions about changes to motivational states.

Several reports have suggested that, as for non-human primates, coital activity varies during the menstrual cycle. For example, a prospective diary study on 68 women aged 25–35 years, all of whom were non-fecund because of intrauterine contraceptive devices (IUCDs) or ligated oviducts, indicated a significant increase in coital activity as the mid-cycle ovulation

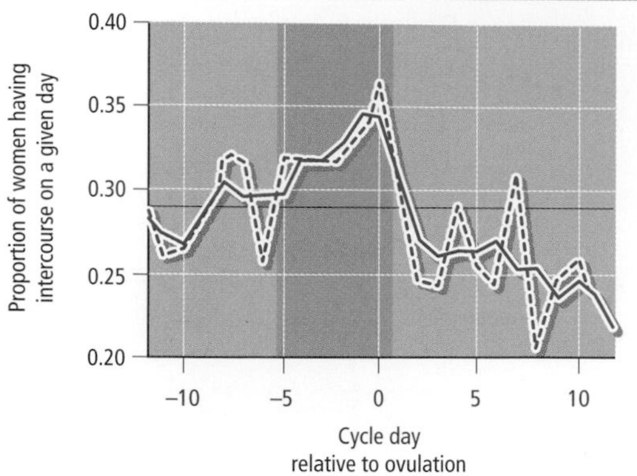

Figure 6.8 Plot of intercourse frequency averaged for 68 women (aged 24–35 years and having ligated tubes or IUCDs) relative to day of menstrual cycle (Day 0 = ovulation; 171 menstrual cycles analyzed). Solid line shows 3-day moving average and dashed line shows mean values for each day. Horizontal line shows mean overall frequency. (Source: Wilcox AJ, Baird DD, Dunson DB *et al.* (2004) *Human Reproduction* **19**, 1539–1543, by permission of Oxford University Press.)

approached (Figure 6.8). Why? Diary records indicate both increased male partner initiation of sexual interaction and increased female libido just before ovulation. This situation resembles that described above in non-human primates. Are similar mechanisms at work?

The male interest could reflect a response to female behaviour or some physical indicator such as pheromones. There is some evidence that pheromones in women attract sexual interest by men, but no direct evidence that vaginal odours are involved. Thus, if men are given women's T-shirts to sniff, then those taken from periovulatory women are associated with higher testosterone levels in the men than are those from perimenstrual women. However, the effects are small, and do not identify which odours, if any, might be involved.

Might the increased libido of women in the peri-ovulatory period reflect the transient rise in androgens observed then (see Figure 10.7 for details)? If women are given acute doses of androgen, then they, like men, show evidence of a characteristic sexual interest and uncoupling of the neocortex and amygdala. Anecdotal evidence from clinical studies on postmenopausal women suggests that androgens may affect sexual activity in some women, but inter-individual variability is again high – possibly in part arising from the androgen receptor polymorphisms described in Table 2.2. Anti-androgen treatment for female acne and hirsutism was followed by reduced sexual satisfaction in most women. Controlled studies of the effects of androgens and oestrogens on sexual behaviour in bilaterally ovariectomized women, in whom androgens fall substantially and immediately, has also provided evidence implicating androgens in sexual interest, fantasy and arousal (Figure 6.9), but not in sexual activity or orgasm. However, other studies have reached different conclusions, although all report a positive impact of androgens on some

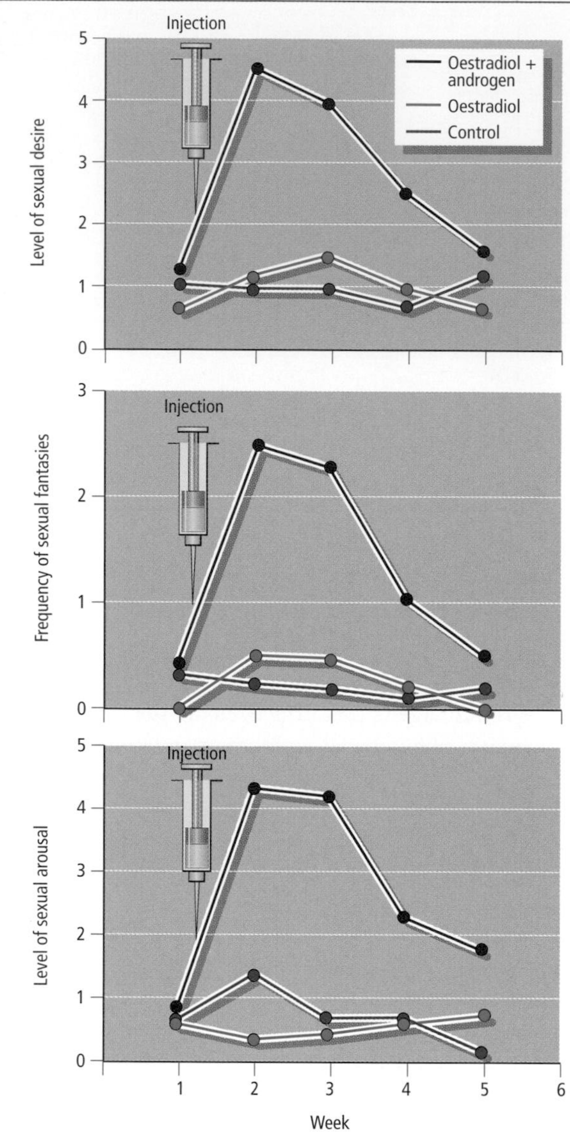

Figure 6.9 The effects on women who had undergone ovariectomy of treatment with oestradiol, oestradiol plus androgen or no hormone. Levels of sexual desire, the frequency of sexual thoughts and level of sexual arousal are all significantly greater in the combined treatment group, indicating the impact of androgenic steroids on sexuality in women.

aspects of sexual function. These data suggest that androgens may play some not-fully-understood role in sexual functioning in women, but still emphasize how other variables are influential.

Summary

This account of steroid hormones and sexual behaviour reveals four important points. First, it is clear that gonadal hormones are not essential for sexual interactions in humans (and other higher primates), and they exert a less dramatic effect than in other mammals. Second, human sexual behaviour is influenced by developmental, social, environmental and emotional factors that may interact with hormones. Third, where hormones do seem to exert an influence, it is important to define the site and mechanism of that influence: genital, behavioural or a mix of both? Finally, there is evidence that ovarian (or adrenal) and testicular hormones may influence sexual behaviour in primates by an action affecting motivation. Now we explore briefly the possible sites of hormone action in the brain where the expression of sexual behaviour might be influenced.

The brain, hormones and sexual behaviour

Testosterone and the male brain

Again, it is important to distinguish motivational and performance aspects of sexual behaviour. Thus, in many non-primate and some primate species, the ability of males to copulate is severely impaired by lesions in the medial preoptic area and adjacent anterior hypothalamus (see Figure 2.3). Thus, male rats with preoptic area lesions show high levels of interest in oestrous females and make repeated, if ill-directed and non-intromitting, attempts to mount. These hypothalamic areas are rich in androgen receptors and the local implantation of testosterone fully restores sexual behaviour in castrated male rats (Figure 6.10). These results argue that this area is of importance in the control of the performance aspect of sexual behaviour, in addition to any peripheral effects of androgens. In contrast, castrated male rats show no interest in females in heat, suggesting a non-hypothalamic site for androgen action underlying the motivational behaviour that precedes copulation. Male rhesus monkeys with lesions in the preoptic area

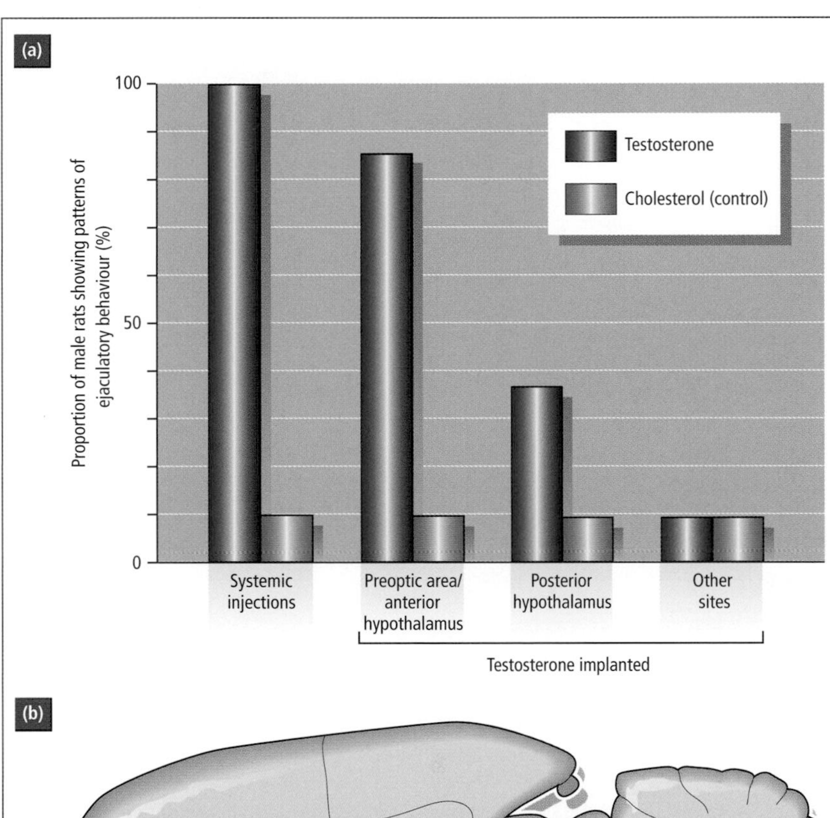

Figure 6.10 (a) The effects of testosterone implanted in the CNS on the sexual behaviour of castrated male rats. Testosterone placed in the preoptic–anterior hypothalamic continuum induced high levels of ejaculatory patterns, comparable to those observed following systemic treatment with the hormone. Testosterone placed in the posterior hypothalamus or elsewhere in the brain had no significant effect. Important controls were provided by the fact that cholesterol had no effect on sexual activity in any site, so arguing for some degree of hormonal specificity. More recently it has been shown that, at least in rats, oestradiol has similar effects. (b) Sites ♂ in the hypothalamus at which testosterone-induced increases in sexual behaviour in male rats. AHA, anterior hypothalamic area; ARC, arcuate nucleus; DMN, dorsomedial nucleus; OT and OCH, optic tract and chiasm; POA, preoptic area; PVN, paraventricular nucleus; SCH, suprachiasmatic nucleus; VMN, ventromedial nucleus.

show a different behavioural dissociation. Thus, they make little attempt to copulate with females, but masturbate to ejaculation at other times. Clearly, the capacity for sexual arousal and the performance of copulatory reflexes are separable neurally in primates as well.

Recent clinical neuroimaging studies on sexually aroused heterosexual men and women have demonstrated the activation of complex cortical and subcortical neural circuits. However, in men, the activation of the amygdala and the hypothalamus was more marked than in women.

Sites of hormone action in the female brain

The androgens that underlie proceptivity in female monkeys seem to act within the anterior hypothalamus. Thus, local implants of testosterone in an area extending from the ventromedial nucleus to the preoptic area reverse the decrease in sexual activity that follows androgen deprivation in female monkeys (Figure 6.11).

Summary

A great deal is known about the strict, controlling role of hormones in the sexual behaviour of non-primate species and about the hypothalamic sites where their actions are exerted. But even in these species, many aspects of the neural regulation of sexual behaviour are incompletely understood. In non-human primates, although the effects of hormones and their sites of action in males and females have been described to some extent, it is clear that gonadal hormones are not the only determinants of sexual activity. Thus, social interactions may profoundly influence the activity of the hypothalamic– pituitary–gonadal axis and hence fertility. The ways in which they can do so are considered next.

Social factors influence sexual behaviour in higher primates

It will be clear from the foregoing discussion that steroids can influence but do not determine sexual behaviour in higher primates. Similarly, it will hopefully be clear that human sexuality is complex and the origins of its complexity are poorly understood. Studies of primates living in social groups can reveal how the social context in which individuals interact can affect their sexual behaviour and also implicate interactions with hormones. Thus, if plasma testosterone levels are measured in male monkeys in the absence of females, all males, whether single or together as a group, have very similar testosterone levels. However, when oestrogen-treated females are introduced to an all-male group, one male becomes dominant, displays sexual activity with the females and is aggressive to subordinate males. Moreover, his plasma testosterone levels rise significantly. None of these changes is seen in the subordinate male(s) (Figure 6.12). These changes in plasma testosterone are due to behavioural interactions, because if all the males are taken out of the group and each replaced alone in turn with the females, each male is sexually active and shows a

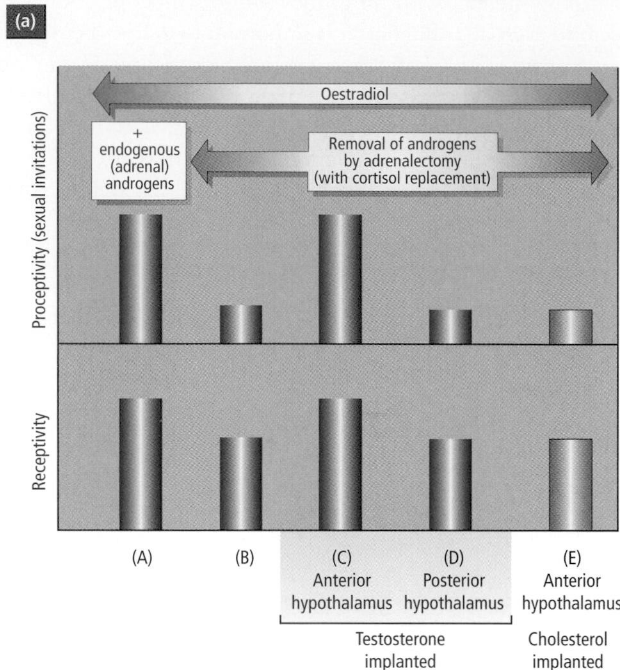

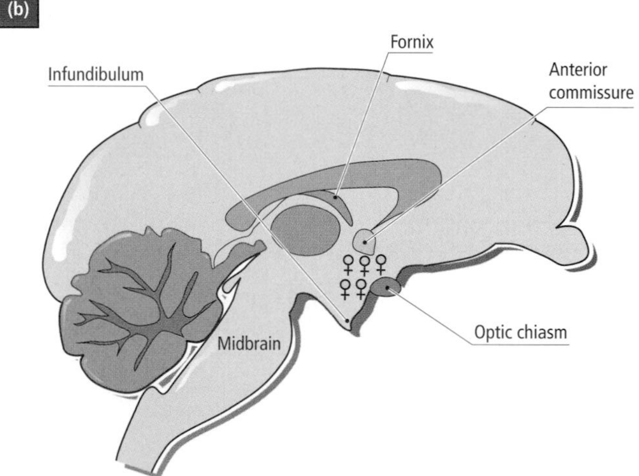

Figure 6.11 (a) A block diagram summarizing the results of experiments in which testosterone propionate was implanted in the CNS of androgen-deprived female rhesus monkeys. All females were ovariectomized and received injections of oestradiol benzoate throughout the experiment, so that they remained attractive to males. Adrenalectomy was followed, as shown in Figure 6.7, by decreased levels of proceptive and receptive behaviour (compare A with B). These changes in sexual activity were reversed by placing testosterone in the anterior hypothalamus (C) but not in the posterior hypothalamus (D). Cholesterol in the anterior hypothalamus was without effect (E). (b) Diagram of a sagittal section of the rhesus monkey's brain to show sites in the anterior hypothalamic area (♀) where testosterone implants reversed the behavioural effects of adrenalectomy in females (C).

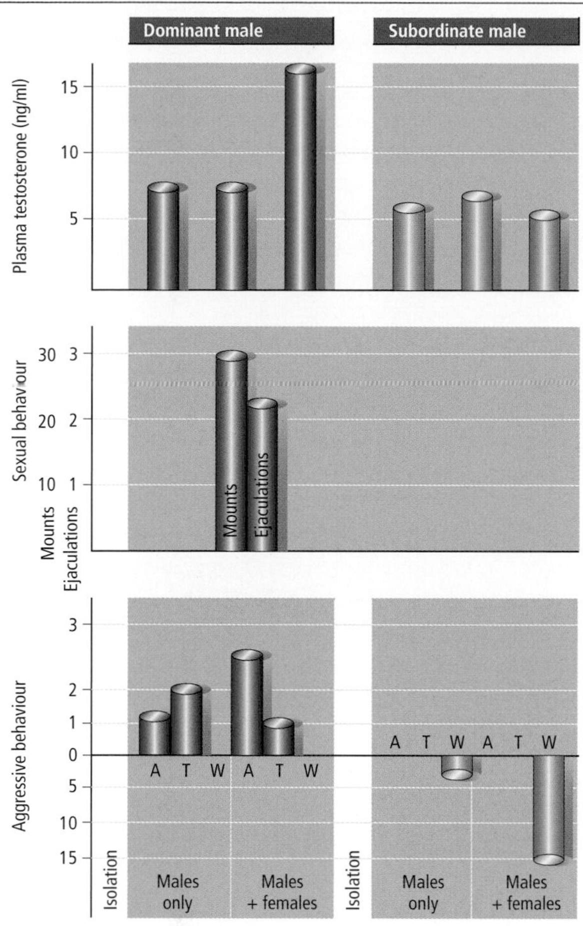

Figure 6.12 Changes in sexual and aggressive behaviour and plasma testosterone levels in both a dominant and a subordinate male talapoin monkey following a 12-week period in isolation, a 6-week period in an all-male group, and then a 7-week period in a social group with oestradiol-treated females. Note that only the dominant male's plasma testosterone increases in the social group, and that only he is sexually active and being aggressive towards, but not receiving aggression from, other males. The subordinate male receives, but does not give, aggression and withdraws more frequently, especially when females are in the group. A, attacks; T, threats (both measures of aggressive behaviour); W, withdrawals (a measure of submissive behaviour). (Mounts and ejaculations are both measures of sexual behaviour.) (Source: Keverne E.B. (1979) *Sex, Hormones and Behaviour*. Ciba Foundation Symposium 62. Excerpta Medica, Amsterdam. Reproduced with permission of John Wiley & Sons.)

significant rise in plasma testosterone. The addition of the dominant male to this group results in the subordinate male's plasma testosterone declining. These data suggest that subordination, especially being on the receiving end of aggression, causes the decrease in testosterone level. Conversely, being aggressive and/or displaying sexual behaviour, as in the dominant male, is associated with increased plasma testosterone. The significance of the elevated testosterone level in the dominant male, and the mechanism by which it is achieved, are not entirely clear. It has been suggested that testosterone may enhance the attractiveness of the dominant male to females by behavioural (e.g. posture) and/or non-behavioural (e.g. coat quality and odour) means.

A similar consequence of the dominance hierarchy is seen in the females of the group. The dominant female receives more sexual attention from the dominant male than do the subordinate females, and hardly any aggressive behaviour from other members of the group (Figure 6.13). Moreover, whilst dominant females show normal menstrual cycles, subordinate female have anovulatory and irregular cycles. When blood levels of the hormone prolactin are measured in these subordinate females, they are found to be high – a mark of stress (Box 6.3). These two examples emphasize the important and far-reaching consequences of social interaction in determining levels of sexual activity as well as endocrine and reproductive status.

Do these findings have relevance for humans? Since the evidence suggests that testosterone may be a key influence on sexual behaviour in both men and women, studies have investigated the relations between social interactions and testosterone blood levels. A key role for partnering patterns has been proposed. Thus, androgen levels were reportedly higher in single than in stably partnered men and women. However, among stably partnered women, testosterone levels were directly influenced by the frequency of partnered sexual activity, whilst in men sex with different partners led to the same response, an observation consistent with the acute rise in androgens elicited by potential sexual partners described above. These observations lead us into a consideration of how sexual partners are selected.

Selecting sexual partners

Thus far in this chapter, we have concentrated our discussion on the types of stimuli eliciting sexual responses, the factors modulating those responses and the possible mechanisms underlying them. In this last section, the focus returns to the reproductive role of sex in humans, and the extent to which 'sexual selection', in a Darwinian sense, operates. In the opening paragraphs of this book, we established that it is through reproduction that we pass our genes to a new generation. Each new generation in turn reproduces or dies out. The survivors are 'selected' by disease and by competition for resources for their fitness to live and to reproduce. In this way, the gene pool of surviving species is constantly adapting to the prevailing environment to provide the best available 'fit'. However, Darwin himself realized that it was not just fitness to survive and reproduce that determines whether our genes are passed on, but also the 'fitness' of those with whom we mate. By selecting the fittest available member of the opposite sex, we increase further our chances of our own genes surviving to be passed on in the next generation. Thus, **sexual selection as well as natural selection shapes our evolving gene pool**.

In most mammals, **females actively choose their mates**. The potential male partners compete against each other to be chosen and are less choosey. Mate choice benefits females by

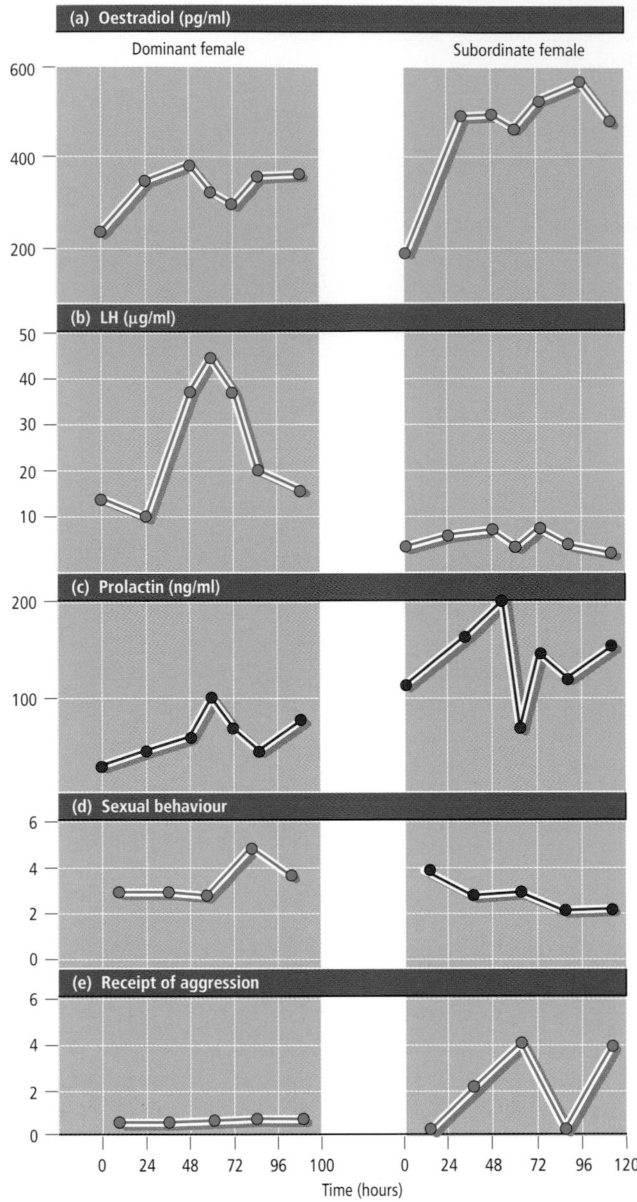

Figure 6.13 Sexual and aggressive behaviour received by dominant and subordinate female talapoin monkeys in a social group, and changes in plasma LH and prolactin levels when challenged with an oestradiol surge. (a) Increase in plasma oestradiol concentrations following the oestradiol surge. (b) Changes in plasma LH concentrations, which follow the oestradiol surge. (c) Plasma prolactin concentrations in the females. (d) Sexual interaction with males indicating mounts without ejaculation (red line) and mounts with ejaculations (green line). (e) Level of aggressive behaviour received by the females. Note that the dominant female receives high levels of sexual behaviour and low levels of aggression, has low plasma levels of prolactin and shows a surge of LH in response to an oestradiol surge. The converse is true of the subordinate female. (Source: Keverne E.B. (1979) *Sex, Hormones and Behaviour*. Ciba Foundation Symposium 62. Excerpta Medica, Amsterdam. Reproduced with permission of John Wiley & Sons.)

Box 6.3 Stress and libido

Markers of stress include prolactin and glucocorticosteroids, and both are implicated in reproduction. For example, the subordinate females with anovulatory and irregular cycles described in this chapter, also have high blood levels of prolactin. A similar association in women between infertility and elevated prolactin (**hyperprolactinaemia**) is observed, and treating the hyperprolactinaemia in these women can often restore menstrual cycles and fertility (see pages 347–8), implying causality. Quite how prolactin levels become elevated is less clear.

Glucocorticoid levels also rise in stressful situations and are chronically elevated in subordinate males compared with dominant males – unless an on-heat female is introduced when levels become transitorally elevated in the dominant male. Chronic elevation of glucocorticoids is known to divert energy from activities not required for immediate survival, such as immunity and reproduction, as well as to be neurotoxic. So, submissive males have lower levels of androgens and a generally higher risk of reduced 'fitness'.

Further reading

Roney JR, Simmons ZL, Lukaszewski AW (2010) Androgen receptor gene sequence and basal cortisol concentrations predict men's hormonal responses to potential mates. *Proceedings of the Royal Society: Biological Sciences* **277**, 57–63.

allowing them to select males capable of bestowing 'good genes' on their offspring and thereby increasing the quality of their offspring. Females thereby enhance their own fitness. The choosy nature of females and the indiscriminate approach of males reflect the different reproductive investments by each parent. It is crucial for females, with their more limited reproductive fecundity, to mate 'well'. Male fecundity is not compromised by pregnancy or lactation and so the more offspring they can father the better. Part of the observable sexual dimorphism observed in many mammals is related to these different 'selection' strategies. For example, larger aggression-displaying males can be more effective physically in competition with other males and in subduing females.

To what extent do these patterns apply to humans? Clearly, humans do select the person with whom they reproduce, but much of this selection operates at a social rather than an individual level. Thus, social or legal restrictions on, for example, the class, race or religion of the pool of eligible partners available for choice are common. In many societies, family members or 'elders' may select (or veto) partners. Dynastic or financial considerations can trump all others in some families in which no choice at all is offered to the potential partners involved.

However, often choice is available from within a limited pool of 'approved' candidates, whilst extramarital 'choices' are a constant and tempting possibility. So, the question asked here is: is there any evidence that sexual selection mechanisms seen in animals also operate in humans under such circumstances?

Sexual selection involves selection for 'fitness', and fitness is a genetic quality, so it must be reflected in the phenotypic traits of the males if females are to detect it reliably. In humans, sexual selection is likely to be complex, given that men also invest heavily in their offspring's upbringing, both financially and emotionally, and human females also show within-gender competition for men. Several potential phenotypic traits have been identified from animal studies and the three most pertinent to humans are described here.

High male social and economic status

Such men control resources and wield influence. They tend to produce more children who live longer and are healthier. These findings apply to virtually all agricultural or hunter societies studied, and status is shown to influence women's mate choice much more than men's. However, status alone is insufficient, as high status is only useful if the men show evidence of long-term commitment to family. Thus, **high status carries risks as well as benefits**, and there is thus a trade-off to be made. Indeed, some studies show that in selecting long-term partners, women prefer feminized or baby-faced men, perhaps because they are perceived as being more controllable or emotionally literate.

Symmetry as evidence of health

Evidence from anthropological and social studies indicates that male handsomeness, as manifested in height, regular strong facial features and a symmetrical muscled body shape plus a deeper voice, is also considered a selectable trait. Strong and symmetrical features are well documented in birds and animals to be associated with good health, both developmentally and as adults. Conversely, men with asymmetric features tend to be less confident, less physically active and more prone to mental and physical illness, and a causal association is implied, although clearly positive reinforcement may be at work too. What might the genetic basis for this selection trait be? One answer is that such men are more fertile, although studies that examined seminal quality against voice pitch or facial symmetry did not find a consistent association. An alternative much-touted answer is that such men show greater resistance to disease through their immuno-competence.

Immunocompetence as signalled through MHC loci and male odour

The **multi-histocompatibility complex (MHC)**, in humans called the **human leukocyte antigen (HLA)** system, is a highly polymorphic gene family involved in immune recognition of pathogens and parasites. Selection of sexual partners differing in MHC type (so-called MHC-disassortative mating) is advantageous because offspring enjoy enhanced resistance to infectious diseases. This increased resistance arises from the fact that MHC genes are expressed co-dominantly, such that heterozygotes can recognize a wider range of foreign antigens. It is now clearly established from studies in mice that females prefer to mate with MHC-dissimilar males. The MHC status of male mice is recognized phenotypically in the form of MHC-related peptides secreted by the extraorbital lachrymal gland and deposited on the facial hair. These are recognized by the olfactory system of the female.

Several studies on human populations have attempted to demonstrate MHC-disassortative pairing in marriages. In the few studies that claim a positive effect, it is weak and/or contested. If there is such an effect, is it also mediated by male odour? Again, some studies do suggest that women prefer the odour of fit men who are MHC dissimilar.

Summary

Clearly, women are exercising choice in their sexual partners but within highly constrained social circumstances, whether of a highly-gendered male-dominant society or a more female-liberated society in which peer and commercial pressures on women operate more subtly. The basis of women's choice also differs depending on whether sexual partners are being selected for social or for reproductive purposes. Indeed, even in birds there is now abundant evidence from genetic studies that females can chose to pair with males who will be good carers for the nestlings, but then mate with other males who are fitter and more attractive. So, in humans the relative roles played by status, symmetry or MHC in a marriage decision may differ from those when a choice is driven by desire alone. Indeed, there is evidence that the male features found attractive by a woman can change through the menstrual cycle, those for status and caring dominating outside the fertile period whilst classic 'fitness' traits were more emphasized at the time of maximum fertility. It is also the case that men, whilst in general less choosy than women, are not as passively acquiescing as the simple model above may suggest when it comes to commitment to reproductive sex. The complexity of humans is not reflected in the simpler animal models. However, it is not just choice of a partner that promotes reproduction, but also the commitment to that process, which in mammals can be long term. We will return to pair bonding in the context of parental care in Chapter 20.

Conclusions

In this first section of the book, we examined the developmental foundations of sexual reproduction in mammals and the many and wide-ranging aspects of the social life of mammals that flow from them. In subsequent chapters, we will look at the physiological processes regulating fertility and sexual behaviour that result in conception, pregnancy, parturition, lactation and maternal care.

Key learning points

- Sexuality involves the erotic and may be classified by the stimulus of erotic arousal. This system of classification is unsatisfactorily rigid. Many people find a range of stimuli arousing and the range may change with time.
- A sexual stereotype is the constellation of attributes and behaviours associated with people whose erotic arousal is classified according to a particular type of stimulus.
- The sexual identity of a person describes their inner state of feeling as a sexual being.
- Asexual people are not aroused erotically.
- Genes, brain structure, hormones and social learning have all been implicated in the development of sexuality, but there is no clear evidence directly linking any one element causally to a particular sexual identity.
- Hormones can act to influence sexual behaviour.
- Testosterone critically controls the sexual behaviour of male non-primate mammals.
- The effects of testosterone on masculine sexual behaviour are mediated by actions on both the brain and the genitalia.
- In some species (e.g. rodents), oestradiol is the active metabolite of testosterone, mediating its actions on sexual behaviour in the brain. Dihydrotestosterone is the active metabolite maintaining the integrity of the genital periphery.
- In male monkeys, testosterone clearly greatly influences sexual behaviour. However, sexual activity can persist after castration.
- In men, testosterone clearly affects sexual motivation, behaviour and fantasies, but only in hypogonadal males is chronic androgen administration of (variable) benefit.
- Sexual arousal in men is associated with a transient rise in androgen, and injection of androgen acutely can effect male arousal. There may be a reinforcing effect at work transiently.
- The tight control exerted by testosterone over sexual responses in non-primate males is not seen in male monkeys and men, where social and other factors are also important influences on sexual activity.
- Oestradiol and progesterone are critical determinants of sexual receptivity and proceptivity in female non-primates. Ovariectomized females are sexually inactive; appropriate sequential treatment with oestradiol and progesterone reinstates oestrous behaviour.
- In female monkeys, oestradiol and progesterone may affect sexual behaviour, but ovariectomized animals remain sexually receptive and proceptive.
- Ovarian steroids may profoundly affect sexual interaction between males and females by effects on the odour of female vaginal secretions.
- Androgens, of adrenal and ovarian origin, affect sexual proceptivity and receptivity in female monkeys.
- In women, ovariectomy does not generally affect libido.
- Androgen treatment in women improves sexual function, but with inter-individual variation.
- There is variation in coital activity through the menstrual cycle, but what causes it remains uncertain.
- The medial preoptic area is a critical site for the effects of testosterone, or its metabolites, on sexual behaviour in males.
- The ventromedial hypothalamus is a critical site for the effects of oestradiol and progesterone on sexual behaviour in female non-primates.
- The anterior hypothalamus is an important site mediating the effects of androgens on sexual behaviour in female monkeys.
- Other neural systems influence the expression of sexual behaviour in males and females.
- Social interactions can affect hormone levels in male and female monkeys, and possibly also in humans.
- In some societies, attributes of sexuality and gender have been conflated, implying that heterosexual arousal and a strong sense of gender identity are linked.
- Anthropological and social studies, as well as studies on trans people, show that it is possible to separate sexuality and gender identity.
- Selection of sexual partners by humans is subject to social as well as personal constraints.
- Factors shown to influence women's sexual partner selection include male socio-economic status, health and symmetry, but a role for MHC genetic make-up remains unproven.

Clinical vignette

Concerns about sexuality

A 32-year-old man, Mr. C, consults his general practitioner highly concerned about his sexual desires. He explains that he comes from a conservative family in whose cultural background it is usual to marry a partner of the opposite sex. He is considered by his extended family and community to have delayed marrying for too long, and is under a considerable amount of pressure to commit to a female wife, who has already been selected by his parents. Mr. C, however, explains that from a young age he has been sexually attracted exclusively to other males. While he likes the potential wife who has been selected for him by his family, he feels no sexual desire towards her. He feels trapped between wanting to please his parents and live a 'conventional' life, and denial of his sexuality. He asks the GP if there are any treatments available that would convert his sexual desire for men into a similar sexual desire for women, as he feels that this is the only way that he could commit to the proposed marriage and prevent being ostracized by his community. His GP listens sympathetically, but explains that there are no medical treatments that can change sexual desire. He offers counseling and support to Mr. C in helping him to explain the dilemma to his family. Mr. C attends the counselling sessions and joins a peer-support group. He is able to refuse the marriage with the support of this group. Eventually he feels able to explain his sexuality to selected members of his family, some of whom are able to accept his male partner.

There are often powerful social determinants of how sexual identities are expressed, which vary widely between societies. In many cultures, there is tight social regulation that limits sexual expression to a set of 'acceptable' normative behaviours. This regulation may be reinforced formally via legal systems, or informally by social approval, which can be an equally powerful restricting force on individuals who fear being shunned by a community. Risking social isolation by openly expressing a sexual identity that does not conform to societal expectation is a step that many individuals feel unwilling or unable to take. It is important that medical practitioners can listen without judgement to an individual's expression of their sexual identity, particularly if the consultation may be the first time that the individual has attempted to explain their sexual identity to a third party. The determinants of sexual identity are complex and incompletely understood. Furthermore, sexual identities can involve a variety of stimuli and can be fluid in time. Medical professionals can be helpful in enabling access to resources that will support individuals who are troubled by their sexual identities and by helping to put in place appropriate interventions to limit harm from the expression of sexual identity, for example, protection against sexually transmitted diseases, contraception, etc.

FURTHER READING

General reading

Auger AP, Auger CJ (2011) Epigenetic turn ons and turn offs: chromatin reorganization and brain differentiation. *Endocrinology* **152**, 349–353.

Berenbaum SA, Beltz AM (2011) Sexual differentiation of human behavior: effects of prenatal and pubertal organizational hormones. *Frontiers in Neuroendocrinology* **32**, 183–200.

Geary DC, Vigil J, Byrd-Craven J (2004) Evolution of human mate choice. *The Journal of Sex Research* **41**, 27–42.

Gooren L (2006) The biology of human psychosexual differentiation. *Hormones and Behaviour* **50**, 589–601.

Hines M (2010) Sex-related variation in human behavior and the brain. *Trends in Cognitive Sciences* **14**, 448–456.

Hines M (2011) Prenatal endocrine influences on sexual orientation and on sexually differentiated childhood behavior. *Frontiers in Neuroendocrinology* **32**, 170–182.

Wilcox AJ, Baird DD, Dunson DB *et al.* (2004) On the frequency of intercourse around ovulation: evidence for biological influences. *Human Reproduction* **19**, 1539–1543.

More advanced reading (see also Boxes 6.1 and 6.3)

Bancroft J (1989) *Human Sexuality and its Problems.* Churchill Livingstone, London (despite its age, a rich repository in which to dip).

Baum MJ, Kelliher KR (2009) Complementary roles of the main and accessory olfactory systems in mammalian mate recognition. *Annual Reviews in Physiology* **71**, 141–160.

Brunetti M, Babiloni C, Ferretti A *et al.* (2008) Hypothalamus, sexual arousal and psychosexual identity in human males: a functional magnetic resonance imaging study. *European Journal of Neuroscience* **27**, 2922–2927.

Byne W, Tobet S, Mattiace LA *et al.* (2001) The interstitial nuclei of the human anterior hypothalamus: an investigation of variation with sex, sexual orientation, and HIV status. *Hormones and Behavior* **40**, 86–92.

Connolly JM (2004) The development of preferences for specific body shapes. *The Journal of Sex Research* **41**, 5–15.

Crews D, Gore AC, Hsu TS *et al.* (2007) Transgenerational epigenetic imprints on mate preference. *Proceedings of the National Academy of Science of the United States of America* **104**, 5942–5946.

Davis S (1998) The clinical use of androgens in female sexual disorders. *Journal of Sex and Marital Therapy* **24**, 153–163.

Dixson AF (1998) *Primate Sexuality: Comparative Studies of the Prosimians, Monkeys, Apes and Human Beings*. Oxford University Press, Oxford.

Johnson MH (2004) A biological perspective on sexuality. In: *Sexuality Repositioned: Diversity and the Law* (ed. B Brooks-Gordon *et al.*), pp. 155–186. Hart Publishing, Oxford.

Keller M, Pillon D, Bakker J (2010) Olfactory systems in mate recognition and sexual behavior. *Vitamins and Hormones* **83**, 331–350.

Miller SL, Maner JK (2010) Scent of a woman: men's testosterone responses to olfactory ovulation cues. *Psychological Science* **21**, 276–283.

Mustanski BS, Dupree MG, Nievergelt CM *et al.* (2005) A genomewide scan of male sexual orientation. *Human Genetics* **116**, 272–278.

Nappi RE, Albani F, Santamaria V *et al.* (2010) Menopause and sexual desire: the role of testosterone. *Menopause International* **16**, 162–168.

Savic I, Berglund H, Lindstrom P (2005) Brain response to putative pheromones in homosexual men. *Proceedings of the National Academy of Science of the United States of America* **102**, 7356–7361.

Setchell JM, Smith T, Wickings EJ, Knapp LA (2010) Stress, social behaviour, and secondary sexual traits in a male primate. *Hormones and Behaviour* **58**, 720–728.

Simmons LW, Peters M, Rhodes G (2011) Low pitched voices are perceived as masculine and attractive, but do they predict semen quality in men? *PLOS One* **6**, e29271. doi:10.1371.

van Anders SM, Goldey KL (2010) Testosterone and partnering are linked via relationship status for women and 'relationship orientation' for men. *Hormones and Behaviour* **58**, 820–826.

van Wingen G, Mattern C, Verkes RJ *et al.* (2010) Testosterone reduces amygdala-orbitofrontal cortex coupling. *Psychoneuroendocrinology* **35**, 105–113.

Part 3
Preparing
for pregnancy

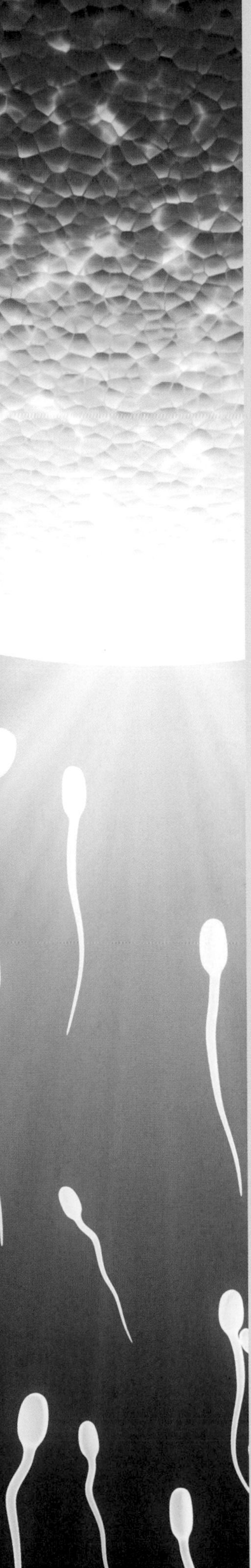

CHAPTER 7

Making sperm

Essential Reproduction, Eighth Edition. Martin H. Johnson.

Companion website: www.essentialreproduction.com

In the second part of this book, we examined how two mature fecund sexes develop to the point at which reproduction becomes a possibility. Part 3 examines in more detail, the structural and functional organization underlying this reproductive potential. We start with a consideration of the mature male.

Testicular organization

The adult testis has two major products: (1) spermatozoa, which transmit the male's genes to the embryo, and (2) hormones, mostly androgens, required for the maintenance of male reproductive functions in adulthood. Each of these is produced in **two anatomically discrete compartments** within the testis. **Spermatozoa develop within the tubules** in close association with **Sertoli cells**, while **androgens are synthesized between the tubules** in the **Leydig cells**. These two compartments are not only structurally distinct but are also separated physiologically by cellular barriers, which develop during puberty and limit the free exchange of water-soluble materials. The precise cellular location of these barriers can be visualized by injection of dyes or electron-opaque materials into the blood and following their distribution within the testis. Fairly free equilibration occurs between **blood** and the **interstitial and lymphatic tissues** (marked I in Figure 7.1), but penetration through the peritubular wall into the **basal compartment** of the tubule (marked B in Figure 7.1) is slower. Surprisingly, however, the major barrier to diffusion lies not at the peritubular boundary itself but between the basal compartment of the tubule and an **adluminal compartment** (marked A in Figure 7.1).

The physical basis for this barrier comprises multiple layers of adherens (inter-Sertoli cell anchoring), gap (inter-Sertoli cell communicating) and tight (para-Sertoli cell occluding) junctional complexes completely encircling each Sertoli cell, and linking it firmly to its neighbours (Figure 7.1, upper insert). Marker molecules are rarely seen penetrating through these junctional barriers between the adjacent Sertoli cells. The barrier constitutes the main element of the so-called **blood–testis barrier**. It is absent prepubertally, but develops before the initiation of spermatogenesis.

The presence of this barrier has three major functional consequences. First, its development during puberty coincides with the first progression of spermatocytes beyond preleptotene into meiosis, and if its development is delayed, or later reversed, so is the meiotic progression. Thus the barrier is essential for spermatogenesis. Second, and probably related to the first point, the composition of **intratubular fluid** differs markedly from that of the **intertubular fluids**: blood, interstitial fluid and lymph (Figure 7.2). If radioactive 'marker molecules' are injected into the blood and their equilibration between blood, lymph and interstitial fluid is measured over time, it is found that ions, proteins and charged sugars enter the interstitial fluid and lymph rapidly, confirming the absence of a significant barrier at the capillary level. In contrast, these molecules do not gain free diffusional access to the tubular lumen. The intratubular fluid has a quite distinct composition that results from a mix of actively transported and secreted molecules and in both cases the key player is the Sertoli cell. The presence of a blood–testis barrier means that the later stages of spermatogenesis occur in a quite distinct and controlled chemical microenvironment. Third, it prevents intratubular spermatozoa from leaking out into the systemic and lymphatic circulations. This function is important because the body's immune system is not tolerant of spermatozoal antigens, which are capable of eliciting an immune response. Such a response can lead to **antispermatozoal antibodies** and even to an autoimmune inflammation of the testis (**autoallergic orchitis**), both of which are associated with subfertility.

Testicular fluid secretion can occur against considerable hydrostatic and diffusional gradients. Thus, if the outflow of fluid from the tubules towards the epididymis is blocked pathologically or by ligation of the *vasa efferentia*, secretion continues nonetheless, and the seminiferous tubules dilate with fluid and spermatozoa, generating a hydrostatic pressure that, if unrelieved, will lead eventually to pressure necrosis and atrophy of intratubular cells.

In summary, the testis may be divided into two major compartments, each of which is further subdivided. The **extratubular compartment** consists of an **intravascular component** in free communication with an **interstitial component**, including the lymphatics, and in which androgen synthesis occurs in Leydig cells. The **intratubular compartment** consists of a **basal compartment** in restricted communication with the interstitium. A unique feature of the seminiferous tubule is that it also has a distinct **adluminal compartment**, which is effectively isolated from the other three compartments, and in which most of the spermatogenic events occur.

Spermatogenesis has three main phases

The mature spermatozoon is an elaborate, highly specialized cell. **Spermatogenesis**, the process by which spermatozoa are formed, has three sequential elements.

1. **Mitotic proliferation** produces large numbers of cells.
2. **Meiotic division** generates genetic diversity and halves the chromosome number.
3. **Cytodifferentiation** packages the chromosomes for effective delivery to the oocyte.

Large numbers of these complex cells are produced, about 1000 sperm per second in man! This prodigious production level amounts to a total output of 90 million sperm every day in men, or to 1.6 billion sperm per day in boars! How is this remarkable output achieved?

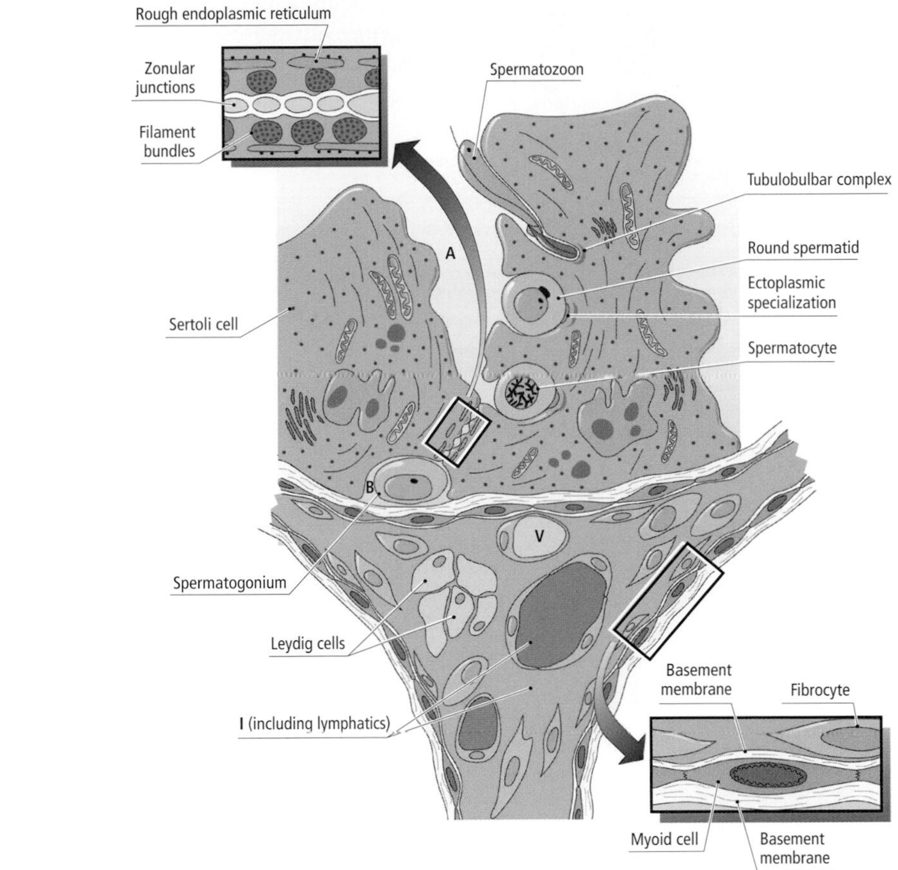

Figure 7.1 Cross-section through part of an adult testis to show the four compartments: vascular (V); interstitial (I), including the lymphatic vessels and containing the Leydig cells; basal (B); and adluminal (A). The latter two compartments lie within the seminiferous tubules. The interstitial and basal compartments are separated by an acellular basement membrane surrounded externally by myoid cells and invested with a loose coat of interstitial fibrocytes (see lower insert box). Myoid cells are linked to each other by punctate junctions. No blood vessels, lymphatic vessels or nerves traverse this boundary into the seminiferous tubule. Within the tubule, the basal and adluminal compartments are separated by rows of zonular tight, adherens and gap junctional complexes (see upper insert box), linking together adjacent Sertoli cells round their complete circumference and forming the **blood–testis barrier**. Intracellular to these junctional complexes are bundles of actin filaments running parallel to the surface around the 'waist' of the Sertoli cells, and internal to the filaments are cisternae of rough endoplasmic reticulum. Within the basal compartment are the spermatogonia, whilst spermatocytes, round and elongating spermatids and spermatozoa are in the adluminal compartment, in intimate contact with the Sertoli cells with which they form special junctions.

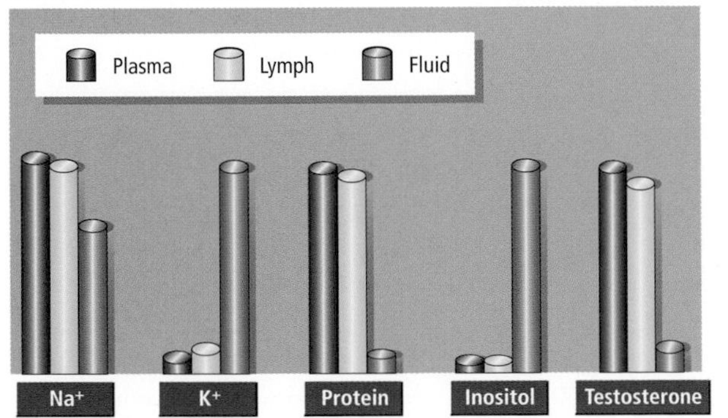

Figure 7.2 Relative concentrations of substances in the venous plasma (vascular compartment), lymph (interstitial compartment) and testicular fluid leaving the seminiferous tubules (adluminal compartment). Note that seminiferous tubule fluid differs markedly from plasma and lymph.

Mitotic proliferation increases cell number

The quiescent interphase prospermatogonial germ cells of the immature testis (Figure 7.3) are reactivated postnatally (in the mouse within days of birth) to enter rounds of mitosis in the basal compartment of the tubule. Henceforth they are known as **spermatogonial stem cells (SSCs)**. From within this small reservoir of self-regenerating SSCs there develop almost immediately a population of **transient amplifying progenitor cells (TAPCs)** or progenitor spermatogonia, up to 16 resulting by mitotic division from one SSC. Their emergence marks the **beginning of spermatogenesis**. These then transform into morphologically distinct type A1 spermatogonia, each of which undergoes a further five mitotic divisions in mice or rat as, sequentially, types A, intermediate and type B, with the type A being subdivided into type A1–4 (Figure 7.4; only three

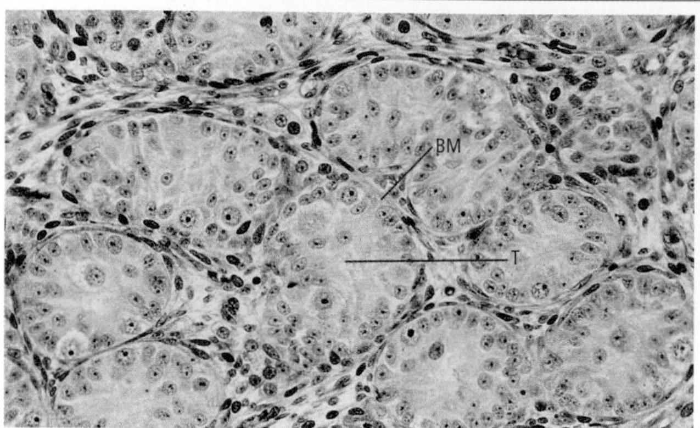

Figure 7.3 Section through immature testis. Note that each tubule is surrounded by a basement membrane (BM), and within the tubule there is no lumen (T) and a relatively homogeneous-looking set of cells comprising a very few prospermatogonial stem cells and mostly Sertoli cells.

divisions in man, who consequently produces proportionately fewer spermatozoa per day). The number of mitotic divisions from SSC to type B determines the total number of cells in the clone (a maximum of 512 in the rat), although cell death during mitosis reduces this number by about 60%. All the spermatogonia type B of the clone then divide to form **resting primary spermatocytes** (see Box 7.1 for further discussion of pluripotent SSCs).

A remarkable feature of this mitotic phase of spermatogenesis is that although nuclear division (**karyokinesis**) is completed successfully, cytoplasmic division (**cytokinesis**) is incomplete. Thus, all the primary spermatocytes derived from one SSC are linked together by thin cytoplasmic bridges, constituting effectively a **large syncytium**. Even more remarkable is the fact that this syncytial organization persists throughout the further meiotic divisions, and most individual cells are only released during the last stages of spermatogenesis as mature spermatozoa.

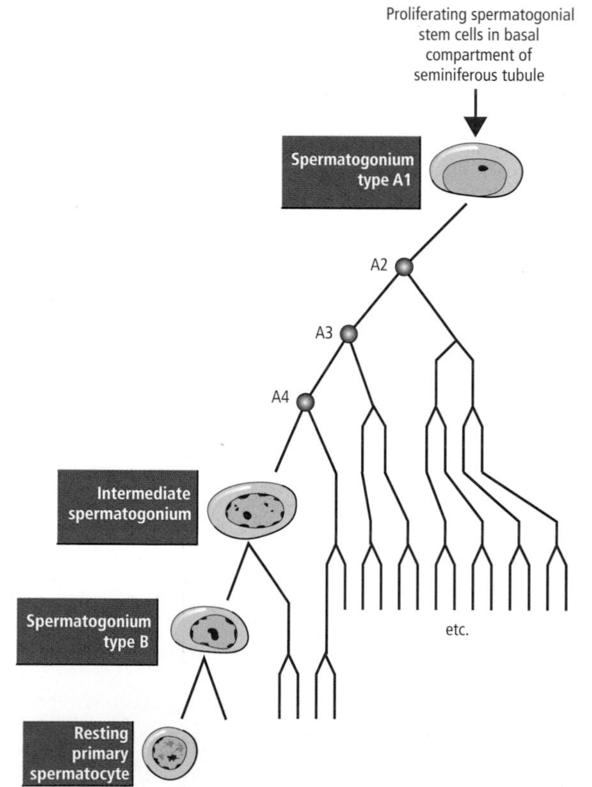

Figure 7.4 Cells in the mitotic proliferative phase of spermatogenesis in the rat (present in the basal intratubular compartment). From the population of proliferating spermatogonial stem cells, the so-called transient amplifying progenitor cells or TAPCs, arise type A1 spermatogonia, which have large, ovoid, pale nuclei with a dusty, homogeneous chromatin. The intermediate spermatogonia have a crusty or scalloped chromatin pattern on their nuclear membranes, and this feature is heavily emphasized in type B spermatogonia, in which the nuclei are also smaller and rounded.

Meiosis halves the chromosome number and generates genetic diversity

The proliferative phase of spermatogenesis takes place in the basal intratubular compartment of the testis. **Each resting preleptotene primary spermatocyte** so formed duplicates its DNA content and then **passes into the adluminal intratubular compartment** through a transient disruption of the zonular tight junctions between adjacent Sertoli cells. This brief breach of the blood–testis barrier seems to involve changes to both the actin and tubulin cytoskeletons. The spermatocytes then enter the first meiotic prophase, which is very prolonged (Figure 7.5; for details of meiosis see Figure 1.1). During prophase, the sister chromatid strands on the paired homologous chromosomes come together to form **synaptonemal contacts at pachytene**, during which the chromatids break, exchange segments of genetic material and then rejoin, thereby shuffling their genetic information, before pulling apart (Figures 1.1 and 7.5). Primary spermatocytes at different steps in this sequence can be identified by the characteristic morphologies of their nuclei, reflecting the state of their chromatin (Figure 7.5). During this prolonged meiotic prophase, and particularly during pachytene, the spermatocytes are especially sensitive to damage, and widespread degeneration can occur at this stage.

The first meiotic division ends with the separation of homologous chromosomes to opposite ends of the cell on the meiotic spindle, after which cytokinesis yields, from each primary spermatocyte, **two secondary spermatocytes** containing a single set of chromosomes. Each chromosome consists of two chromatids joined at the centromere. The chromatids then separate and move to opposite ends of the second meiotic spindle, and the short-lived secondary spermatocytes divide to yield **haploid early round spermatids** (Figure 7.5). Thus, from the maximum of 64 primary spermatocytes that entered meiosis (in the rat), 256 early spermatids could result. Again,

Box 7.1 Potency and spermatogonial stem cells (SSCs)

SSCs form a small self-renewing population

This small (0.02–0.03% of germ cells) population is located within the basal compartment of the tubule and from it some cells periodically divert to form the spermatogenic lineage. The SSCs, mostly studied in mice but increasingly in large farm animals and humans, can be isolated from the testis, purified using surface markers (including α6-integrin [CD49f], β1-integrin [CD29], Thy1 [CD90] and CD9) and studied *in vitro* or transplanted to other sites. When placed into a host 'SSC-depleted' seminiferous tubule, the transplanted SSCs can populate it and produce fertile spermatozoa – definitive proof of their stem cell potential. There is also evidence that the SSCs migrate along tubules *in vivo* to repopulate areas denuded of SSCs pathologically, suggesting a major role in the restoration of fertility.

How do the SSCs 'decide' to proliferate or differentiate?

SSC pluripotency is thought to be promoted by a variety of **growth factors** (GFs) and by the physical microstructure in the adluminal compartment, where Sertoli cells play a key role, although lesser roles for adjacent myoid and Leydig cells cannot be excluded. Some growth factors act juxtacrinologically and so their spatial organization on the Sertoli cell surface may be important in determining which GFs are active locally. The Sertoli cell also oscillates its activities with the spermatogenic cycle, so there may be temporal regulation of GF expression patterns too. These findings have given rise to the idea that the '**microenvironmental niche**' experienced by stem cells and their descendants determines the decision whether to remain a stem cell or to differentiate into spermatozoa. The peptide GF, **glial derived neurotrophic growth factor** (**GDNF**), secreted from both the Sertoli and myoid cells under testosterone control is identified as a key player in promoting SSC self-renewal, and genetic deletion of it or its receptor (expressed on SSCs) results in arrest of spermatogenesis. *In vitro* culture studies of SSC self-renewal suggest that the action of two further Sertoli cell GFs called: **fibroblast growth factors 2 and 8** (**FGF 2&8**) are also required, and that **colony stimulating factor 1** (**CSF-1**) from Leydig and myoid cells assists. Several genes whose expression is affected by GDNF stimulation have now been identified from microarray analyses, thereby facilitating identification of the regulatory pathways involved in maintaining pluripotency. The nature of the trigger to differentiate into transient amplifying progenitor cells is unclear, but may consist in part of the switching off or attenuating of those signals maintaining the SSCs, in which retinoic acid may play a role (see Box 7.4).

How pluripotent is the spermatogonial stem cell?

If instead of being placed into seminiferous tubules, the stem cells are held *in vitro* or transplanted to other sites in the body, they are found to have a broader stem cell potential than simply the spermatogenic lineage, being able to give rise to multiple cell types. In this regard, they resemble embryonic stem (ES) cells derived *in vitro* from the blastocyst (see Chapters 12 and 13). Conversely, both pluripotential embryonic stem cells and induced stem cells have now been shown to be capable under the influence of various GFs of giving rise to primordial germ cells, which can form SSCs and even after transplantation to testes, spermatozoa. Thus, spermatogonial and embryonic stem cells seem to be closely related, and so the potential use of SSCs therapeutically to treat degenerative disorders is being explored as an alternative to the use of ES cells, which is more controversial ethically (see Chapters 22 and 23).

Further reading

De Felici M, Farini D, Dolci S (2009) In or out stemness: comparing growth factor signalling in mouse embryonic stem cells and primordial germ cells. *Current Stem Cell Research and Therapy* **4**, 87–97.

Hayashi K, Ohta H, Kurimoto K, Aramaki S, Saitou M (2011) Reconstitution of the mouse germ cell specification pathway in culture by pluripotent stem cells. *Cell* **146**, 519–532.

Komeya M, Ogawa T (2015) Spermatogonial stem cells: progress and prospects. *Asian Journal of Andrology* **17**, 771–775.

Kostereva N, Hofmann MC (2008) Regulation of the spermatogonial stem cell niche. *Reproduction of Domestic Animals* **43**, 386–392.

Oatley JM, Brinster RL (2008) Regulation of spermatogonial stem cell self-renewal in mammals. *Annual Review of Cell Development and Biology* **24**, 263–286.

Ohta H, Yomogida K, Dohmae K *et al*. (2000) Regulation of proliferation and differentiation in spermatogonial stem cells: the role of c-kit and its ligand SCF. *Development* **127**, 2125–2131.

Walker MR, Patel KK, Stappenbeck TS (2009) The stem cell niche. *Journal of Pathology* **217**, 169–180.

Zhao G-Q, Garbers DL (2002) Male germ cell specification and differentiation throughout the phylogeny from invertebrates to mammals. *Developmental Cell* **2**, 537–547.

Zheng Y, Zhang Y, Qu R, *et al*. (2014) Spermatogonial stem cells from domestic animals: progress and prospects. *Reproduction* **147**, R65–R74.

Zhou G-B, Meng Q-G, Li N (2010) In vitro derivation of germ cells from embryonic stem cells in mammals. *Molecular Reproduction & Development* **77**, 586–594.

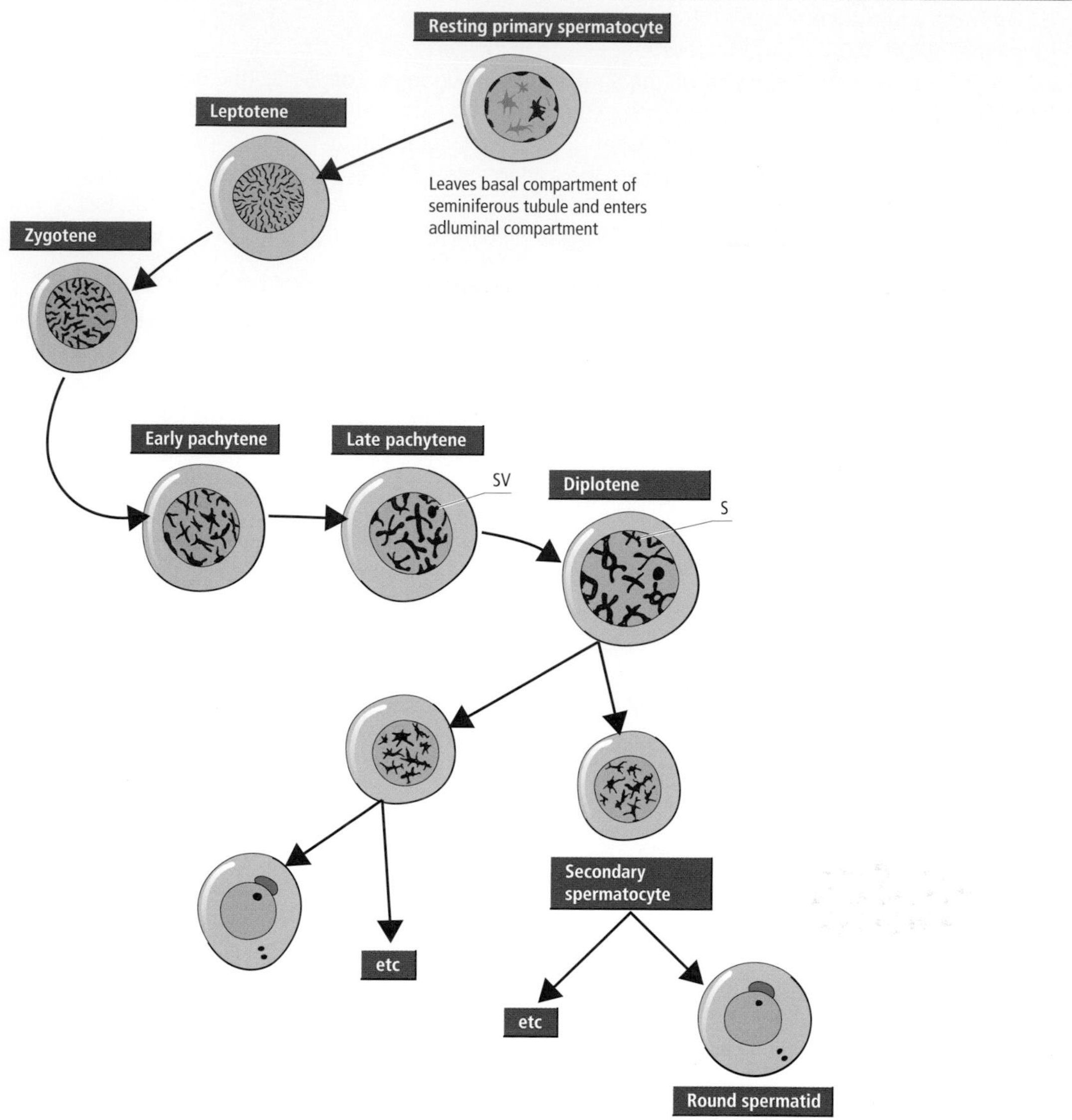

Figure 7.5 Progress of a rat primary spermatocyte through meiosis in the adluminal intratubular compartment. DNA synthesis is completed in the resting primary spermatocyte, although limited 'repair DNA' associated with crossing over occurs in late zygotene and early pachytene. In leptotene, the chromatin becomes filamentous as it condenses. In zygotene, homologous chromosomes thicken and come together in pairs (synapsis) attached to the nuclear membrane at their extremities, thus forming loops or 'bouquets'. In pachytene, the pairs of chromosomes (bivalents) shorten and condense, and nuclear and cytoplasmic volume increases. It is at this stage that autosomal crossing over takes place (the two sex chromosomes are paired in the **sex vesicle, SV**). The synapses (S) can be seen at light-microscopic level as chiasmata during diplotene and diakinesis, as the chromosomes start to pull apart and condense further. The nuclear membrane then breaks down, followed by spindle formation, and the first meiotic division is completed to yield two secondary spermatocytes each containing a single set of chromosomes. These are very short-lived and rapidly enter the second meiotic division, the chromatids separating at the centromere to yield four haploid round spermatids.

the actual number is much less than this as, in addition to any losses at earlier mitotic stages, the complexities of the meiotic process result in the further loss of cells. Yet again, the whole cluster of spermatids is linked syncytially via thin cytoplasmic bridges. With the formation of the early round spermatids, the important chromosomal reduction events of spermatogenesis are completed.

Cytodifferentiation packages the chromosomes for delivery

The most visible and major changes during spermatogenesis occur during a remarkable cytoplasmic remodelling of the spermatid that is called **spermiogenesis** (Figure 7.6). During this process, spermatids change shape from round to **elongated**.

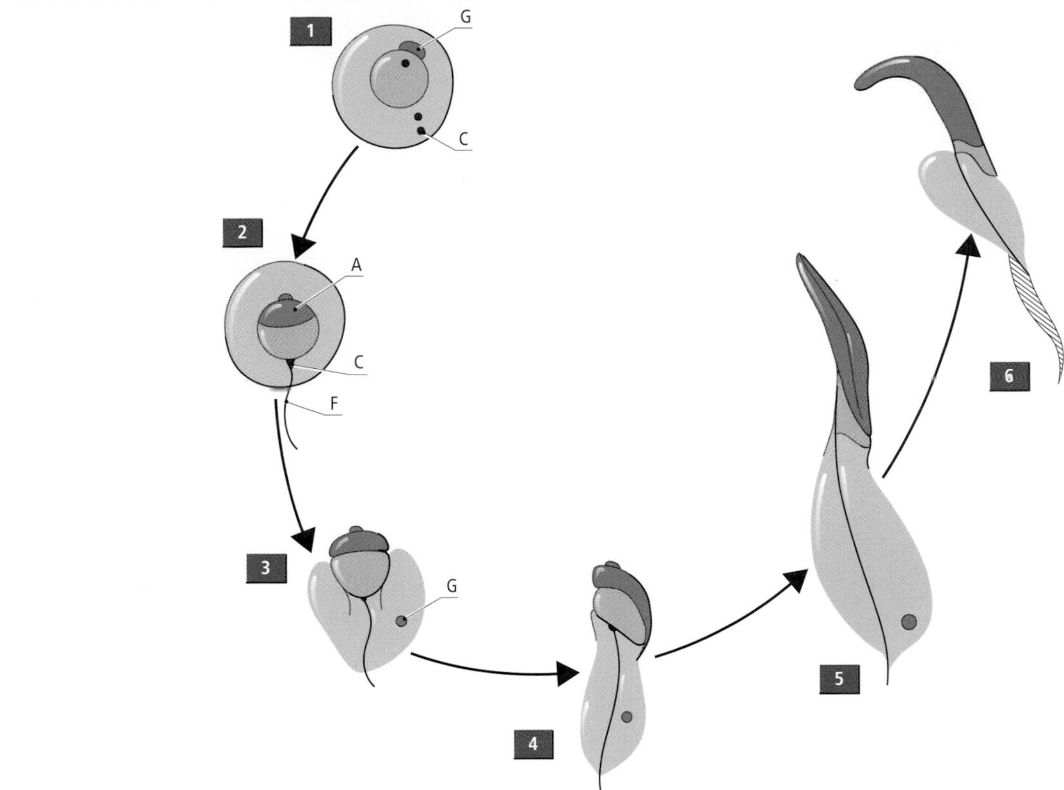

Figure 7.6 Progress of a rat round spermatid through the packaging phase of the spermatogenic lineage. The Golgi apparatus (G) of the newly formed round spermatid (1) gives rise to glycoprotein-rich lysosomal-like granules, which coalesce to a single acrosomal vesicle (A) that grows over the nuclear surface to form a cap-like structure (2). Between the acrosome and the nucleus, a subacrosomal cytoskeletal element, the perforatorium, forms in many species. The nuclear membrane at this site loses its nuclear pores. The two centrioles (C) lie against the opposite pole of the nuclear membrane, and a typical flagellum (F) (9+2 microtubules) grows outwards from the more distal centriole (2), while from the proximal centriole the neck or connecting piece forms, linking the tail to the nucleus. The nucleus moves with its attached acrosomal cap towards the cytoplasmic membrane and elongation begins (3 and 4). Chromatin condensation commences (Box 7.3) beneath the acrosomal cap, generating a nuclear shape, which is characteristic for the species (3–6), and superfluous nuclear membrane and nucleoplasm are lost. The Golgi apparatus detaches from the now completed acrosomal cap and moves posteriorly as the acrosome starts to change its shape. Nine coarse fibres form along the axis of the developing tail, each aligned with an outer microtubule doublet of the flagellum (see Figure 7.7 for details). In the final phase, the mitochondria migrate to the anterior part of the flagellum, and condense around it as a series of rods forming a spiral (see Figure 7.7). The superfluous cytoplasm appears 'squeezed' down the spermatid and is shed as the spermatozoa are released (hatched) (6). The mature spermatozoon has remarkably little cytoplasm left.

A **tail** is generated for forward propulsion; the **mid-piece** forms, containing the mitochondria (energy generators for the cell); the **equatorial and postacrosomal cap** region forms, and is important for sperm–oocyte fusion; the **acrosome** (a modified lysosomal structure) develops and functions like an 'enzymatic knife' when penetrating towards the oocyte; the nucleus contains the compact packaged haploid chromosomes; and the **residual body** acts as a dustbin for the residue of superfluous cytoplasm, and is phagocytosed by the Sertoli cell after the spermatozoon departs. The **spermatid centrioles** are of particular interest. They reduce to a central core structure linking the mid-piece to the sperm head. All or most of their pericentriolar material, which normally nucleates microtubules, is lost. The opposite happens in the oocyte (see Chapter 9), in which pericentriolar material is retained but centrioles are lost. This reciprocal pattern of reduction means that at fertilization there is centriolar complementarity of gametes (see Chapter 12).

Spermiogenesis is completed with the formation of a **spermatozoon** (Figure 7.7). With the appearance of the spermatozoa, the thin cytoplasmic bridges that make up the syncytium rupture, and the cells are released into the lumen of the tubule in a process called **spermiation**. They are washed along the seminiferous tubules in the testicular fluid secreted by Sertoli cells.

Genetic activity during spermatogenesis is special

Spermatogenesis is a complex and specialized process and, not surprisingly, requires a large number of genes for its successful completion. The processes of mRNA production and translation

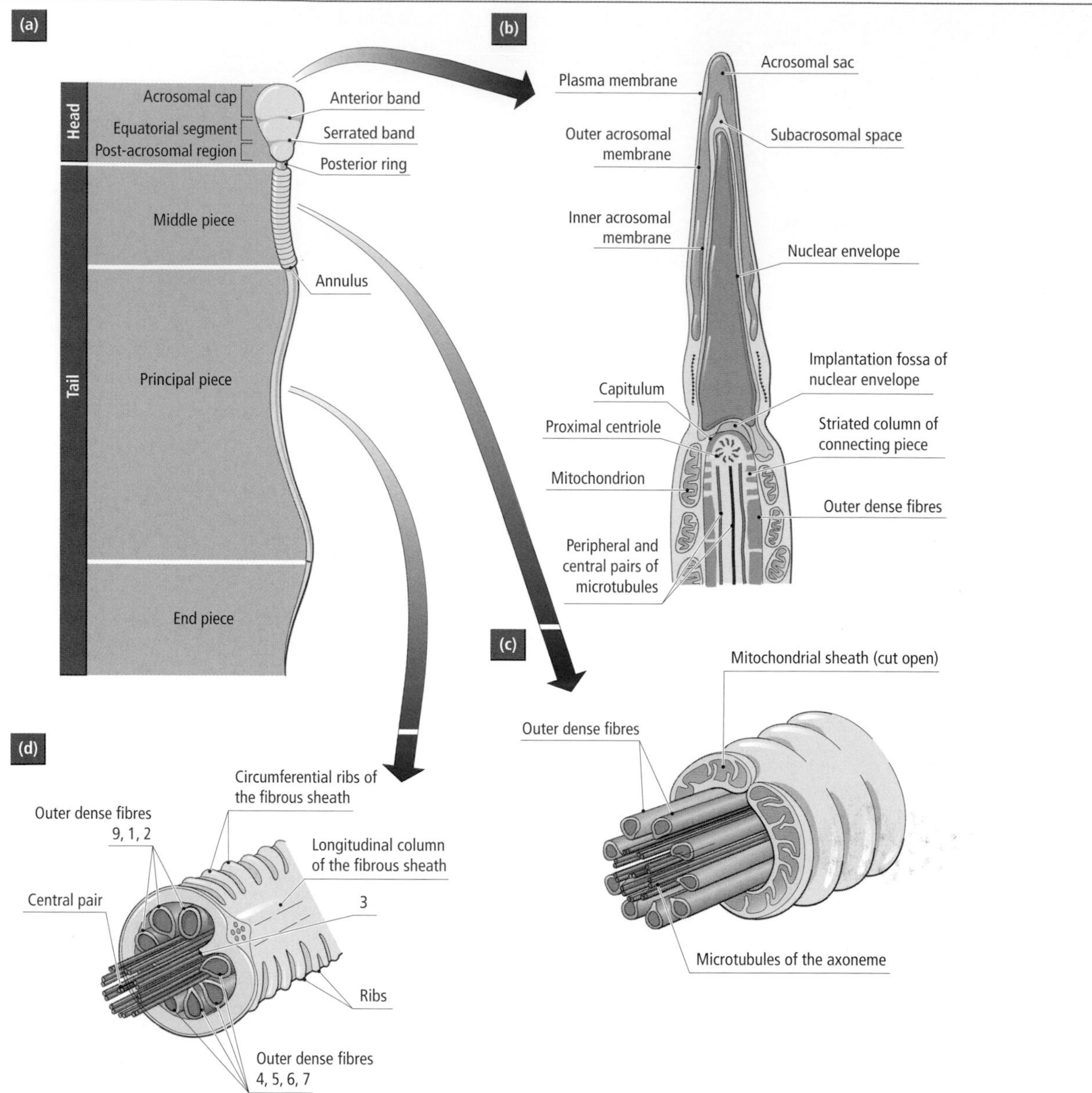

Figure 7.7 (a) Diagram of a primate spermatozoon (50 mm long) showing on the left, the main structural regions; and on the right, the boundaries between them. The surface membrane structure within each region is highly characteristic, having a unique lipid, sugar, surface charge and protein composition that differs from those in adjacent regions. These differences are probably maintained both by the inter-regional boundary structures and by underlying molecular attachments to cytoskeletal elements. The differences are important functionally (see Chapters 11 and 12). (b) Sagittal section of the head, neck and top of the mid-piece. Note the elongated (green) nucleus with highly compact chromatin and the acrosomal sac. The posterior end of the nuclear membrane is the only part to retain nuclear pores and forms the implantation fossa, which is connected to the capitulum by fine filaments. The capitulum, in turn, is connected to the outer dense fibres by two major and five minor segmented columns and is also the site of termination of the two central microtubules of the flagellum. The more distal of the two centrioles degenerates late in spermiogenesis. In a few species both centrioles are lost; they are not essential for sperm motility, only for the initial formation of the axoneme during spermiogenesis. (c) Sketch of the mid-piece (surface membrane removed). Note the sheath of spiral mitochondria, and the axoneme of the tail comprising nine circumferential doublets of microtubules and two central microtubules; peripheral to each outer doublet is a dense fibre. (d) Section and sketch of the principal piece (surface membrane removed). The mitochondria are replaced by a fibrous sheath, comprising two longitudinal columns interconnected by ribs. The two fibrous sheath columns connect to underlying outer dense fibres 3 and 8. The outer dense fibres terminate towards the end of the principal piece, and the fibrous sheath then attaches directly to outer microtubules 3 and 8 before itself fading away in the end piece.

continue throughout spermatogonial mitosis and meiosis (except on the sex chromosomes, which cease transcription from meiosis onwards, except for the activation of a few genes on the X and Y chromosomes that are essential for spermiogenesis); indeed, after the completion of meiosis there is a large transcriptional surge. **Autosomal transcription ceases during the transition from round to elongated spermatids.** The **burst of transcription immediately after meiotic completion** is characterized by two features not observed in somatic cells: the use of specialized transcriptional machinery and the expression of large numbers of spermatogenic-specific genes (Box 7.2). Because this postmeiotic burst occurs from the **haploid genomes**, it raises the possibility that the spermatozoa might differ from each other phenotypically in ways that reflect their unique haploid genetic composition. Were this so, it might then be possible to separate later spermatids and spermatozoa into subpopulations based on their carriage of distinctive genetic alleles. Such a separation might occur in the female genital tract, thereby exerting 'natural selection' on a population of spermatozoa that is genetically and, via haploid expression, phenotypically heterogeneous. Selection might also be made in the laboratory if spermatozoal enrichment for particular characteristics were wanted. For example, the separation of X- and Y-bearing spermatozoa might allow **prefertilization sex 'selection'**. However, successful evidence of haploid spermatozoal selection has been hard to come by. This is not entirely surprising as spermatids exist in a syncytial mass of cytoplasm, giving opportunity for mRNAs and proteins to diffuse into all spermatids regardless of their genotype. Additionally, the premeiotic inactivation of most genes on the X and Y chromosomes (see above) makes selection for sex by this approach very unlikely. More recently, the separation of X- and Y-bearing spermatozoa has been claimed not on the basis of the differential expression of the sex chromosomes, but as a result of their **different total DNA contents.** Thus, the percentage difference in DNA content of X- and Y-bearing human spermatozoa is 2.9% (boar 3%; bull 3.8%; stallion 4.1%; ram 4.2%). The separation of fertile spermatozoa by flow cytometry can achieve enrichment rates of over 75% in large farm animals. However, the prospects for a 100% successful separation of human spermatozoa for therapeutic use by this approach are more controversial.

Box 7.2 Gene expression during spermatogenesis

Highly distinctive processes occur during spermatogenesis

These include pluripotency retention in SSCs; meiosis; genetic recombination; haploid gene expression; chromatin remodelling and condensation; and acrosome and flagellum formation. These processes involve unique gene products, many of which have been identified recently through expression profiling of different spermatogenic cells isolated and purified by centrifugation and gravity sedimentation, combined with tests of function/expression using transgenic and knockout mouse models. Here we focus on haploid expression.

The transcription machinery of the haploid spermatogenic lineage shows several unique features

For example, **TATA-binding protein** (**TBP**), **transcription factor IIB** (**TFIIB**) and **RNA polymerase II** all accumulate in much higher amounts (approximately 100-fold) in postmeiotic cells than in somatic cells. In addition, testis-specific isoforms exist of many transcriptional complex proteins, including TBP, TAFII (TBF associated factors, such as TFIIAt and TAFIIQ) and TBP-related protein (TRF2). This highly characteristic complex of molecular machinery then facilitates expression of an equally characteristic set of spermatogenic genes.

Many genes expressed in somatic cells are also expressed in male germ cells

However, they use alternative promoters and/or splice isoforms, or are usurped by expression of homologous genes specific for the spermatogenic lineage. Other genes are uniquely expressed in that lineage. Many of the genes activated after meiosis have promoters containing **cAMP-responsive elements** (**CREs**), DNA sequences which recruit members of the **CRE-binding** (**CREB**) family of transcription factors. In somatic cells, CREBs bind to CREs, are phosphorylated and recruit a co-activator (**CREB-binding protein** or **CBP**). The CBP has a dual function as it acetylates histones, thereby contributing to the chromatin decondensation that precedes transcription and it also provides a link to the transcription machinery complex. However, there is little CREB expression in the testis (and most of it is in Sertoli cells). Instead, a protein called **cAMP response element modulator** (**CREM**; mostly present as its **CREMt isoform**) is highly expressed and interacts specifically with testis-specific TFIIAt and TRF2 in the spermatogenic transcription complex. The haploid expressing genes are also marked by a distinctive histone lysine modification, namely, crotonylation, which may act to maintain access of transcription factors in an otherwise repressive environment.

Mice null for CREM and TRF2 (but not CREB) block in early spermiogenesis

This observation indicates the critical role of these proteins in postmeiotic transcriptional control. Moreover, both proteins are located in the nuclei of round spermatids (genes expressing) but become cytoplasmic in elongating spermatids (genes shut down). CREMt levels increase under the influence of FSH, but interestingly the hormone acts by stabilizing the mRNA encoding the protein. CREM-controlled genes include many that are essential for mature sperm function, such as those encoding protamines 1 and 2 (essential for sperm DNA packing; see Box 7.3) and proacrosin (a sperm-specific protease precursor important in zona penetration; see Chapter 12).

Further reading

Grimes SR (2004) Testis-specific transcriptional control. *Gene* **343**, 11–22.

Monaco L, Kotaja N, Fienga G *et al.* (2004) Specialized rules of gene transcription in male germ cells: the CREM paradigm. *International Journal of Andrology* **27**, 322–327.

Montellier E, Rousseaux S, Zhao Y, Khochbin S (2012) Histone crotonylation specifically marks the haploid male germ cell gene expression program: post-meiotic male-specific gene expression. *Bioessays* **34**, 187–193.

Sassone-Corsi P (2002) Unique chromatin remodeling and transcriptional regulation in spermatogenesis. *Science* **296**, 2176–2178.

Schlecht U, Demougin P, Koch R *et al.* (2004) Expression profiling of mammalian male meiosis and gametogenesis identifies novel candidate genes for roles in the regulation of fertility. *Molecular Biology of the Cell* **15**, 1031–1043.

Venables JP (2002) Alternative splicing in the testes. *Current Opinion in Genetics and Development* **12**, 615–619.

Spermatozoal chromatin is modified during spermatogenesis

The cessation of transcriptional activity during spermiogenesis is due to a massive repackaging of the spermatogenic DNA, such that the chromatin becomes highly condensed (to about 5% of the volume of a somatic cell nucleus). This form of DNA is described as being **heterochromatic**. Condensation is achieved by the replacement of the histones that characterize somatic cell chromatin by **protamines** (Box 7.3). In this way, spermatozoa develop tightly compressed chromatin in which genetic expression is completely absent. As pointed out above, the sex chromosomes go through this process earlier than the autosomes and end up in a special nuclear compartment, the **sex vesicle**, which lacks RNA polymerase II. This repackaging and silencing of chromatin prevent or reduce the potential damage to the male genome that comes from exposure to free radicals during transit to the female tract (see Chapters 12 and 13).

Spermatogenesis is highly organized both temporally and spatially

Each mature spermatozoon is one sibling in a large family, derived from one parental SSC. The family is large because of the number of premeiotic mitoses, and the spermatozoa are only siblings and not 'identical twins' because meiotic chiasmata formation ensures that each is genetically unique despite having a common ancestral parent. Within each testis tubule, hundreds of such families develop side by side. There are 30 or so tubules within each rat testis. How is the development of these families organized temporally and spatially?

Spermatogenesis proceeds at a constant and characteristic rate for each species

One way to measure the length of time it takes to complete parts of the spermatogenic process is to 'mark' cells at different points during the process, and then to measure the rate of progress of the labelled cells through their completion. For example, if radioactive thymidine is supplied to the resting primary spermatocytes as they engage in the final round of DNA synthesis before they enter into meiosis, the cell nuclei will be labelled and their progress through meiosis, spermiogenesis and spermiation can be followed. In this way, the amount of time required for each spermatogenic step can be measured. In Figure 7.8 the times required for each step in the rat are represented visually in blocks, the length of each being a measure of relative time. The absolute time for the whole process, from entry into first mitosis to release of spermatozoa, is recorded for several species in Table 7.1. It is a lengthy process, taking several weeks.

Rounds of spermatogenesis are initiated at time intervals that are constant and characteristic for each species

So far we have considered the process of spermatogenesis from the viewpoint of a single SSC generating a family of descendant spermatozoa at a constant and characteristic rate. Once this process has been commenced at a particular point in any tubule, new SSCs at the same point do not commence the generation of their own clones until several days have elapsed. Remarkably, it has been found that this interval between successive entries into spermatogenesis is also constant and characteristic for each species (Table 7.1). Somehow, the stem cell

Box 7.3 Spermatogenic chromatin remodelling

In somatic cells, the nucleosome is the basic unit of chromatin

This consists of 146 base pairs of DNA wrapped round an octamer of core histones: two molecules of H2A, H2B, H3 and H4. A fifth histone, H1, protects the DNA fragments that link adjacent nucleosomes. The chromatin structure undergoes structural alterations fit for different local or global functions (gene expression/suppression, DNA repair/reproduction, chromosomal nuclear localization). These alterations include use of different non-allelic histone variants and various post-translational histone modifications (acetylation, methylation, phosphorylation, etc.), some of which may even be retained heritably across cell division cycles as **chromatin imprints** (see page 10 and Figure 1.7). The rich combinatorial possibility of multiple histone variants, each susceptible to multiple post-translational modifications, provides highly complex nucleosomal microenvironments – an informational treasure chest of 'epigenetic' information to rival the genetic code itself.

Histone changes occur during spermatogenesis

First, there is incorporation, mainly during meiosis, of histone variants, including testis-specific variants of H1 (H1t and HILS1), H2A (TH2A) and H3 (H3t). Their incorporation is also accompanied by a range of histone modifications. These changes seem to be associated with the events of meiosis (especially the H2A and H3 variants), the segregation of the sex chromosome to the sex vesicle (especially H2A variants), and the preparation for the next phase of chromatin remodelling during condensation (especially H1 variants).

The condensation of chromatin during the transition from round to elongating spermatid

This is accompanied by hyperacetylation and ubiquitination of histones, which is linked strongly to the removal of over 90% of them and their replacement with **transition nuclear proteins** (**TNPs 1 and 2**), small basic proteins unique to the testis. These appear to facilitate the subsequent change, during which they in turn are replaced to the extent of 85–95% during spermatid elongation with **protamines 1 and 2** (small basic proteins comprising 50% arginine). The protamine molecules, which like the histones they replace are extensively post-translationally modified by phosphorylation, acetylation, etc., on adjacent regions of DNA are cross-linked to each other, either directly via disulphide bonds on their constituent cysteines, or indirectly via zinc bridges between adjacent protamines. The end result is inactive condensed heterochromatin. This complex process of chromatin condensation has evolved in order to protect sperm DNA from oxidative damage during the passage through atmospheric oxygen. The potential disadvantage that the male's genes might have compared with those of the female through the 'unpacking' required in the egg, may be offset to some extent by the fact that those male genes critical for early development seem to be selectively excluded from protamine wrapping and remain enveloped in histones. Recently, it has been suggested that aberrant reprogramming of the sperm epigenome may be the basis of some forms of male subfertility.

Further reading

Aitken RJ, De Iuliis GN (2007) Origins and consequences of DNA damage in male germ cells. *Reproductive BioMedicine Online* **14**, 727–733.

Bittman EL (2016) Timing in the testis. *Journal of Biological Rhythms* **31**, 12–36.

Bjorndahl L, Kvist U (2010) Human sperm chromatin stabilization: a proposed model including zinc bridges *Molecular Human Reproduction* **16**, 23–29.

Carrell DT, Hammoud SS (2010) The human sperm epigenome and its potential role in embryonic development. *Molecular Human Reproduction* **16**, 37–47.

Castillo J, Estanyol JM, Ballesc JL, Oliva R (2015) Human sperm chromatin epigenetic potential: genomics, proteomics, and male infertility. *Asian Journal of Andrology* **17**, 601–609.

Govin J, Caron C, Lestrat C *et al*. (2004) The role of histones in chromatin remodelling during mammalian spermiogenesis. *European Journal of Biochemistry* **271**, 3459–3469.

Hammoud SS, Nix DA, Hammoud AO *et al*.(2011) Genome-wide analysis identifies changes in histone retention and epigenetic modifications at developmental and imprinted gene loci in the sperm of infertile men. *Human Reproduction* **26**, 2558–2569.

Jenkins TG, Carrell DT (2011) The paternal epigenome and embryogenesis: poising mechanisms for development. *Asian Journal of Andrology* **13**, 76–80.

Tanaka H, Baba T (2005) Gene expression in spermiogenesis. *Cellular and Molecular Life Sciences* **62**, 344–354.

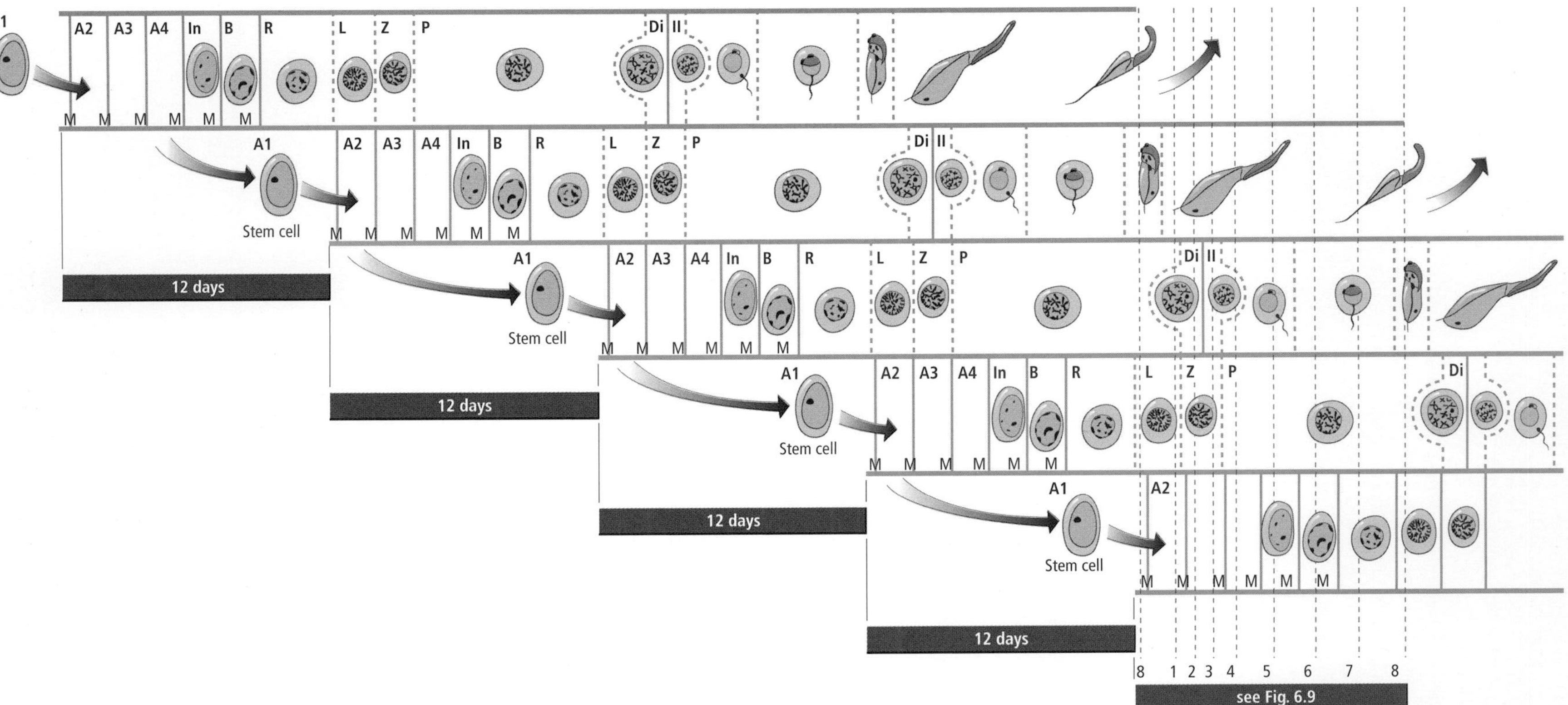

Figure 7.8 The top panel illustrates the passage of one rat spermatogonium through the spermatogenic process. The length of the block illustrating each cellular stage is proportional to the amount of time spent in that stage. When a new cell type arises by division, vertical solid grey lines separate adjacent blocks (M, mitosis). During meiotic prophase and spermiogenesis, however, cells change morphology by progressive differentiation, not quantal jumps. This continuum of change is indicated by the use of vertical broken lines to delineate blocks. A, In and B, spermatogonia; R, L, Z, P and Di, resting, leptotene, zygotene, pachytene and diplotene primary spermatocytes respectively; II, secondary spermatocytes. Each of the lower panels shows the history of other spermatozoa, which commenced development by cyclic generation of new type A spermatogonia from the stem cell population at progressively later time intervals. The interval between each of these events is about 12 days. Note that four such events occur before the upper spermatozoon (and its siblings in the family) have completed development and been released. As cells progress through spermatogenesis, they move progressively from the basal tubule to the luminal centre. Thus, several different cell types will be present in one cross-section of a tubule at the same time, although at different points on the radial axis through the tubule. This feature is illustrated more compactly in Figure 7.9, in which the sections indicated by the dashed lines numbered 1–8 are summarized.

population measures, or is told, the length of this time interval. This cyclic initiation of spermatogenesis is called the **spermatogenic cycle**.

In the case of the rat, the spermatogenic cycle is about 12 days long. This period is one-quarter of the 48–49 days required for completion of mature spermatozoal production, so it follows that four successive spermatogenic processes must be occurring at the same time (Figure 7.8). The advanced cells, in those spermatogenic families that were initiated earliest, are displaced progressively from the periphery towards the lumen of the tubule by subsequent rounds of developing spermatogenic cells. Thus, a transverse section through the tubule will reveal spermatogenic cells at four distinctive stages in the progression towards spermatozoa, each cell type representing a stage in separate, successive cycles (Figure 7.8).

Table 7.1 Kinetics of spermatogenesis

Species	Time for completion of spermatogenesis (days)	Duration of cycle of the seminiferous epithelium (days)
Man	64	16
Bull	54	13.5
Ram	49	12.25
Boar	34	8.5
Rat	48	12
Mice	35	8.6

As both the spermatogenic cycle and the spermatogenic process are of constant length, it follows that the cells in successive cycles will always develop in parallel. Therefore, the sets of cell associations in any radial cross-section through a segment of tubule taken at different times will always be characteristic (Figures 7.8 and 7.9). For example, as the cycle interval is 12 days and it also takes 12 days for the six mitotic divisions, entry into meiosis will always be occurring just as a new cycle is initiated by the first division of a spermatogonium (Figure 7.9, column 8). Similarly, it takes 24 days (i.e. two cycles) for the premeiotic spermatocyte to complete meiosis and the early phase of spermatid modelling. So, not only will entry into mitosis and entry into meiosis coincide, but so will the beginning of spermatid elongation (Figure 7.9, column 8). These events will also coincide with release of spermatozoa at spermiation, as it takes a further 12 days for the completion of spermatid elongation.

Up to this point, we have considered the organization of the SSCs and their descendants at one point in the tubule. However, there are thousands of spermatogonia throughout the tubules in any one testis at any one time: how are they organized in relation to each other?

The seminiferous epithelium cycles

Imagine all of the SSCs throughout the testis entering mitotic activity at exactly the same time. As the time to complete spermatogenesis is constant, there would be a simultaneous release of all the resulting spermatozoa. Moreover, as the spermatogenic cycle is constant for all stem cells, periodic pulses

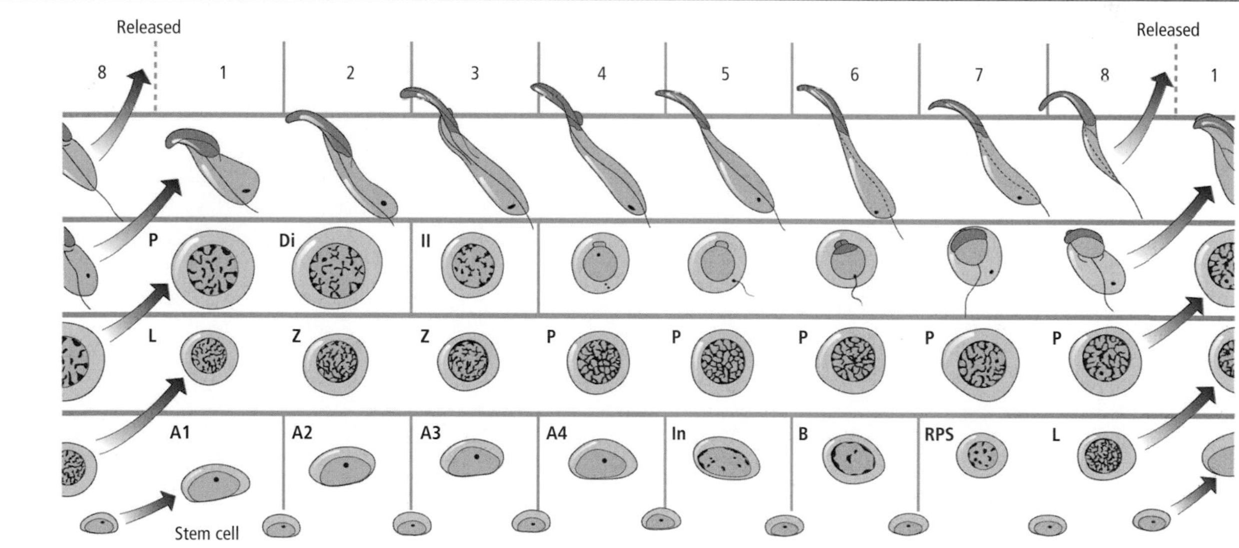

Figure 7.9 The sections indicated in Figure 7.8 are summarized here. Read from left to right. A vertical grey line between two cells in the sequence indicates a cell division; otherwise the changes are not quantal but occur by progressive differentiation. Abbreviations as for Figure 7.8. RPS, resting primary spermatocyte. (*Note*: because the spermatogenic process is continuous, it can be sub-divided more or less finely. The sub-division pattern used here is a basic one. A commonly used, more finely divided and thus more complex pattern uses 16 rather than eight sub-divisions numbered I to XVI. A rough equivalence is stage 1: IX–XI, stage 2: XII–XIII, stage 3: XIV–XVI, stage 4: I, stage 5: II–IV, stage 6: V–VI, stage 7: VII, and stage 8; VIII.)

of spermatozoal release would occur (e.g. every 12 days in the rat). This could result in an episodic pattern of male fertility. This problem could be circumvented if the SSCs throughout the testis initiated mitotic activity not synchronously but randomly. Then, their relative times of entry into spermatogenesis would be staggered, thus eliminating the pulsatile release of spermatozoa and smoothing it into a continuous flow. In fact, the testis functions in a manner somewhat between these two extremes, although the end result is continuous spermatozoal production.

Examination of cross-sections through the testes of most mammals shows that within a tubule the same set of cell associations is observed, regardless of the point on the circumference studied (Figure 7.10a–d). This means that all the stem cells in that section of tubule must be synchronized in absolute time. It is almost as though a message passes circumferentially around the segment of tubule, activating the stem cell population in that segment to initiate spermatogonial type A production together. This spatial coordination of adjacent spermatogenic cycles gives rise to the **cycle of the seminiferous epithelium**, as the whole epithelial cross-section goes through cyclic changes in patterns of cell association (see Figure 7.9).

The human testis (together with the testes of New World monkeys and great apes) is somewhat atypical, as a cross-section through an individual tubule reveals a degree of spatial organization that is more limited to 'wedges'. It is as if the putative activator message does not get all the way round a cross-section of tubule, and so the coordinated development of different SSCs is initiated over a smaller area. This does not mean, of course, that the control of either the spermatogenic cycle or the rate of spermatogenesis in man differs fundamentally from control mechanisms in other species. It means merely that the spatial coordination between adjacent individual stem cells is not so great.

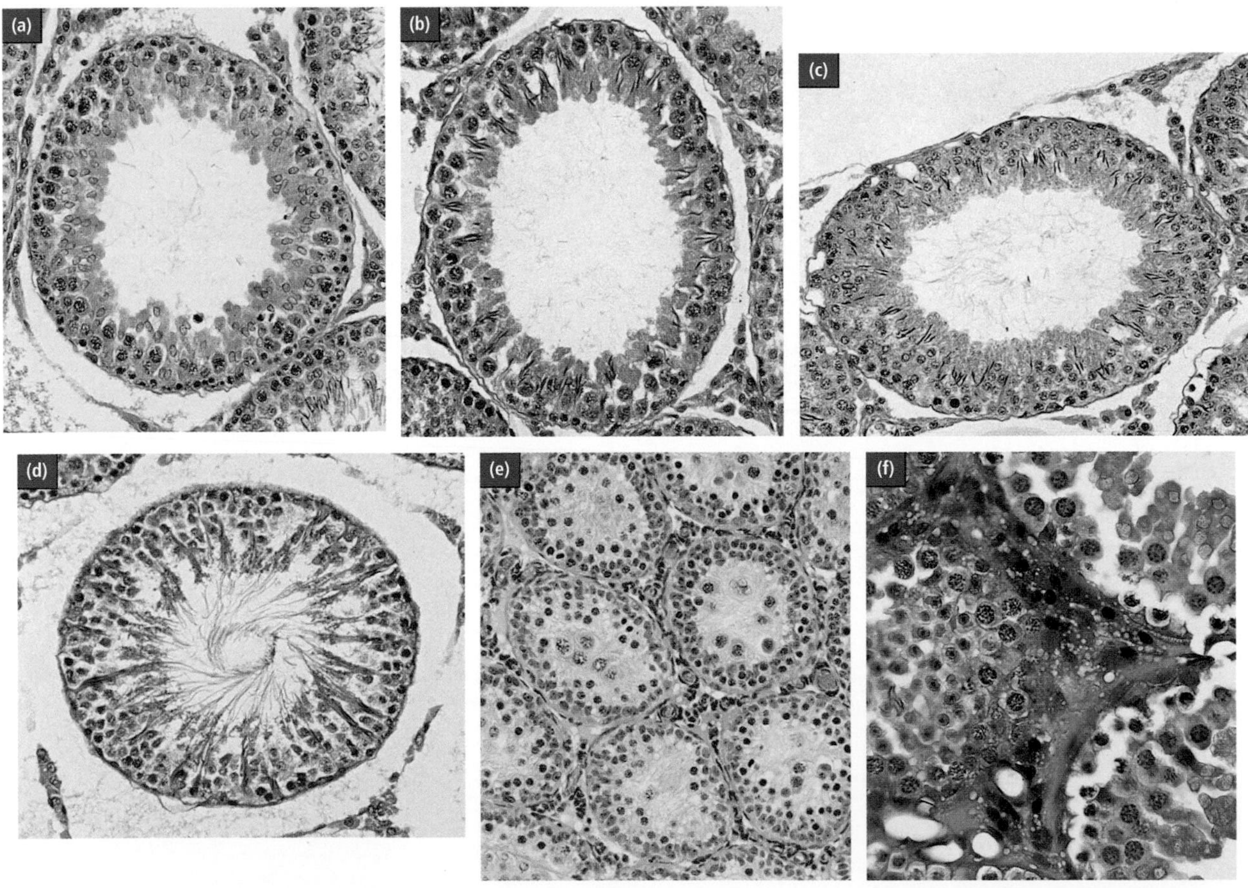

Figure 7.10 Cross-sections through rat seminiferous tubules. (a–d) Four adjacent tubules from the same adult testis. Note that, within each tubule, the sets of cell associations along all radial axes are the same. However, each tubule has a different set of cell associations from its neighbour. Thus, tubule (a) is at stage 8/1 in Figures 7.8 and 7.9, tubule (b) is at stage 2, tubule (c) is at stage 5, and tubule (d) is at stage 7. Tubule (e) is from an adult rat testis 4 weeks after hypophysectomy. Note that spermatogenesis fails during the early meiotic stage with no cells more mature than a primary spermatocyte present. Note also the lack of a tubular lumen, indicating cessation of fluid secretion. Panel (f) shows staining of the intertubular region of an adult intact testis for the Leydig cells. Note their reddish foamy cytoplasm indicative of steroidogenesis.

Spermatogenesis in adjacent regions along a seminiferous tubule appears to be phase advanced or retarded

If an individual seminiferous tubule is dissected and laid out longitudinally, and cross-sections are taken at intervals along it and classified according to the set of cell associations in it, a pattern similar to that in Figure 7.11 will then often result. Adjacent tubule segments, each containing synchronized populations of SSCs, seem to have entered the spermatogenic process slightly out of phase with each other. For example, in Figure 7.11 the most advanced segment (7) is at the centre; moving along the tubule in either direction leads to sets of cell associations characteristic of progressively earlier stages of the cycle of the seminiferous epithelium. It is as though the central segment has been activated first, and then another hypothetical 'activator message' has spread along the tubule in both directions, progressively initiating mitosis and, thereby, spermatogenic cycles. The resulting appearance in the adult testis, as shown in Figure 7.11, is sometimes called the **spermatogenic wave**.

It is important not to confuse the wave with the cycle of the seminiferous epithelium, although both phenomena appear very similar. Imagine that, whereas the sequence of cell associations forming the wave could be recorded by travelling along the tubule with a movie camera running, the same sequence of cell associations would only be captured in the cycle by setting up the movie camera on time-lapse at a fixed point in the tubule. Thus, **the spermatogenic wave occurs in space, while the cycle occurs in time.**

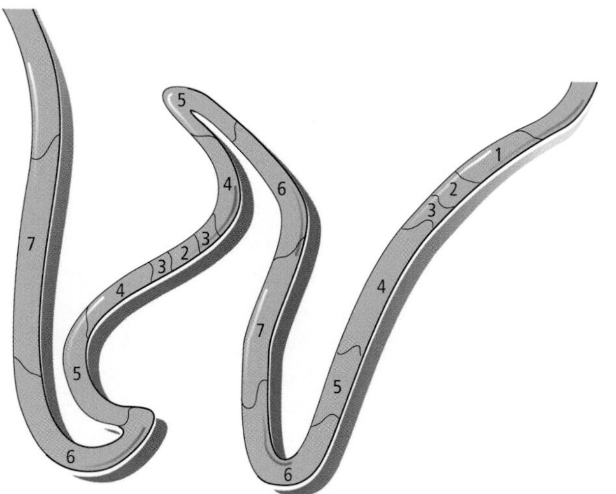

Figure 7.11 Dissected seminiferous tubule from a rat testis. Note that whole segments of the tubule are at the same stage (numbered) of the cycle of the seminiferous epithelium, and that adjacent segments tend to be either just advanced or just retarded.

The Sertoli cell coordinates the temporal and spatial organization of spermatogenesis

These observations on spermatogenesis imply a remarkable degree of temporal and spatial organization among the spermatogenic cells. The Sertoli cell is the probable organizer. Thus, the cytoplasms of adjacent Sertoli cells are in continuity with one another via extensive gap junctional contacts. These effectively provide a continuous cytoplasmic network along and around the tubule through which communication and synchronization might occur (see Figure 7.1). In addition, each Sertoli cell spans the tubule radially from peritubular basement membrane to lumen, thereby providing a potential radial conduit for communication, through which all its associated spermatogenic cells could be locked into the same rate of developmental progression. This latter possibility is made more attractive by the observation that the Sertoli cell engages in intimate associations with all the cells of the spermatogenic lineage. These associations are of three types:

- Pachytene spermatocytes communicate with, and receive material from, the Sertoli cell via gap junctional complexes.
- Most spermatocytes and spermatids form unique **ectoplasmic specializations** with Sertoli cells (ECs; adherens-like junctions which replace desmosomal junctional complexes present at spermatogonial stages; Figure 7.1), and these are largely thought to be concerned with anchoring and then releasing the spermatogenic cells and perhaps remodelling them during spermiogenesis. They are lost at spermiation.
- Elongating spermatids and Sertoli cells also form heavily indented **tubulobulbar complexes** (also adherens-like junctions) through which the Sertoli cell is thought to remove material during cytoplasmic condensation (Figure 7.1).

Finally, the Sertoli cell itself shows characteristic changes in morphology and biochemistry in concert with the cycle of the seminiferous epithelium. For example, the volume, lipid content, nuclear morphology, and number and distribution of secondary lysosomes vary cyclically, as do the synthesis and output of a number of testicular proteins, such as **androgen binding protein** (ABP), **transferrin** and **plasminogen activator**. Interestingly, the output of the latter protein is high at around the time of spermiation and the passage of preleptotene spermatocytes into the adluminal compartment, suggesting a potential function for its proteolytic activities. Indeed, it has been proposed that apically-generated **laminin peptides** released through proteolytic action at spermiation regulate the structural reorganization basally to permit passage of spermatocytes into the adluminal compartment. Sertoli cell cycling is initiated at puberty ahead of spermatogenesis, further suggesting that it leads and the spermatogenic process follows. However, the process is interactive and two way, since rat SSCs transplanted into mouse SSC-depleted seminiferous tubules impose a rat timing on spermatogenesis. Thus, the **spermatogenic cells may set the timer but the Sertoli cell transmits the time.**

Recent evidence has shed light on the molecular basis of this interactive role for Sertoli cells. These involve **Vitamin A** in its **retinoic acid form (RA)**, which has long been known to be essential for spermatogenesis, in its absence all spermatogenic stages from spermatogonia onwards being absent. Not only has RA been shown to be responsible for the transition of the pool of transient amplifying progenitor cells (TAPCs) to spermatogonia A, as well as the temporally associated changes to the blood-testis barrier and the release of mature spermatids, but the way that it does so been elucidated. First, it does not appear to work via the RA receptor, because, despite the receptor being present in Sertoli cells, when it is genetically knocked out spermatogenesis is unaffected. However, a pulse of RA can be detected at exactly the correct time and place, when TAPCs are transitioning to spermatogonia type A, and this pulse appears to reflect local changes in the balance between activities of enzymes synthesizing and destroying the RA (see Box 7.4). Two further observations are consistent with the idea that RA is the key organizing molecule of both the spermatogenic cycle and wave: (1) it has been shown that RA acts primarily in Sertoli cells at the initiation of spermatogenesis during puberty but that thereafter it also acts in the spermatogenic cells interactively with the Sertoli cells (see Box 7.4 for details), thereby providing a mechanism for the spermatogenic cells to time the spermatogenic cycle, and (2) if a bolus of RA is administered to either prepubertal mice, or to adult rats or

Box 7.4 Synthesis and degradation of retinoic acid in the testis

Vitamin A in the form of retinol originates in the diet or from stored reserves in the liver and is transported to the testis (bound to serum retinoid binding protein 4 and the transthyretin complex) where it is converted to RA in a process controlled by sequential oxidative steps catalyzed enzymatically by retinol dehydrogenases (Rdh) and retinaldehyde dehydrogenases (Aldh1a). Importantly, it has been shown that RA acts primarily in the Sertoli cells at the initiation of spermatogenesis during puberty but that thereafter it also acts in the spermatogenic cells interactively with the Sertoli cells, and a corresponding shift with time in the cell types, in which knock-down of these two enzymes affects spermatogenesis, has been observed. Thus, knock-down of *Rdh* or of *Aldh1a* in Sertoli cells has severe effects on the initiation of spermatogenesis, but there is little effect on spermatogenic initiation of knock-down of the same enzymes in germ cells. However, once spermatogenesis is initiated, knock-down of either enzyme in Sertoli cells does not have the same effect, for this, knock-down of the enzymes in the Sertoli cells and in both pachytene spermatocytes plus round spermatids is required to suppress spermatogenesis. Moreover, at the time when TAPCs are entering meiosis, a gene (*Stra8*) that is responsive to RA stimulation is expressed in both spermatogonia and preleptotene spermatocytes, coincidentally with the measured peak of a pulse of RA. Which leads us to consider how does RA exert its effects on target cells?

The RA produced interacts with heterodimers of the RA receptors (Rars) and the retinoid X receptors (Rxrs), and the complex then binds RA response elements (RAREs) in target genes, thereby recruiting repressors or activators of transcription. Both Rars and Rxrs are present in testicular germ cells and somatic cells, Rarα having been localized primarily to the Sertoli cells and some germ cells in the adult mouse testis, and knockout mice are sterile. Moreover, the Sertoli cell-specific Rarα gene knockout mouse produces a phenotype resembling that seen in the whole animal knockout, suggesting that Rarα is a key signalling molecule with a major function in the Sertoli cells. Rarγ has been detected in spermatogonia A, but Rarγ-knockout male mice can be fertile, displaying an altered spermatogonial differentiation. Of the RXRs, only the product of *Rxrβ* is shown to be essential for spermatogenesis, as deletion of Rxrβ globally or just in the Sertoli cells results in a delay in spermatid release, and eventually leads to testis degeneration. Thus, our knowledge of receptor roles is too fragmentary at the moment. As is our understanding of the balance between RA synthesis and its degradation, which is achieved by three cytochrome P-450 enzymes, Cyp26A1, Cyp26B1, and Cyp26C1. Cyp26B1 is expressed in immature Sertoli cells, degrading any RA, thereby preventing germ cells from prematurely entering meiosis, as evidenced by this occurring in a Cyp26B1 knockout mouse. However, the situation concerning the expression and effects of inhibition of Cyp26 enzymes in the postnatal mouse testis is unclear at the moment. What is clear is that both the germ cells and the Sertoli cells have the ability to generate, use and possibly also degrade RA. Finally, at least some of the actions of RA may be exerted post-transcriptionally, as, within the population of TAPCs, high levels of Nanos2, a mRNA sequestering protein, have been found. The treatment of TAPCs with RA reduces Nonos2 levels and allows pre-existing mRNAs hitherto unexpressed to make proteins known to be associated with the transition to spermatogonia A1.

These observations have led to a working model for the role of RA in spermatogenesis. Thus, in this model, RA is synthesized for the first round of spermatogenesis at puberty by the Sertoli cells and acts in a paracrine manner on germ cells. It does so in a patchy, non-uniform manner throughout the testis, such that the initiation of spermatogenesis occurs asynchronously. Subsequently, RA synthesized in spermatocytes or spermatids then stimulates the Sertoli cells through cycles of spermatogenesis in a paracrine manner, thereby setting the timing of spermatogenesis. So the unanswered questions then become: what is the cause of the first RA pulse and why is it initiated patchily?

mice, in which Vitamin A deficiency has led to complete arrest of spermatogenesis, then there is simultaneous initiation of spermatogenesis across the whole testis, resulting in the episodic production of spermatozoa and the loss the spermatogenic wave. These results are exactly what one would expect, strongly supporting a key role for RA in the control of both the spermatogenic cycle and wave.

Conclusions

In this chapter we have considered the organization of the testis and the production of spermatozoa by it – a complex and highly organized process, which is now well-described, even if its coordination remains imperfectly understood. How the Sertoli cell plays its role in the control of spermatogenic rates, cycles and waves remains to be fully established. In contrast, it is well established that the Sertoli cell plays a critical role in mediating the actions of hormones and other agents on spermatogenesis as will be described in the next chapter.

The testis is functionally and anatomically compartmentalized. In particular, its main endocrine secretion of testosterone by the Leydig cells occurs outside the seminiferous tubules, whereas the elaborately choreographed production of spermatozoa occurs inside the tubules in close association with the Sertoli cells. In the next chapter, the functional relationships between androgens and spermatogenesis are considered.

Key learning points

- The adult testis has two main products: spermatozoa and hormones.
- The testis has two main physiological compartments (inside and outside seminiferous tubules) each of which is subdivided: the extratubular compartment into intravascular and interstitial; and the intratubular compartment into basal and adluminal.
- A major physiological barrier (the blood–testis barrier) separates the basal and adluminal intratubular compartments.
- Spermatozoa are made within the tubules.
- Spermatogonial stem cells (SSCs) provide a pluripotent self-replenishing pool of spermatogenic precursor cells, and can also give rise to embryonic stem cells.
- Spermatogenesis has three main phases: proliferative, meiotic and cytodifferentiative.
- The proliferative mitotic phase occurs in the basal compartment and comprises SSCs, TAPCs and spermatogonia A, intermediate and B.
- The meiotic phase occurs in the adluminal compartment and comprises primary and secondary spermatocytes.
- The cytodifferentiative phase occurs in the adluminal compartment and comprises round and elongating spermatids. It is called spermiogenesis.
- As the spermatozoa from a single spermatogonium develop, they remain in contact with each other through cytoplasmic bridges.
- As spermatids elongate, they form an acrosome, condensed heterochromatin, a mid-piece containing mitochondria and centrioles, and a main-piece tail.
- The chromosomal condensation involves replacement of most histones first with basic transition nuclear proteins (TNPs) and then with protamines.
- Heterochromatization results in transcriptional inactivity in elongated spermatids and spermatozoa.
- Distinctive species of mRNA are made after meiosis (haploid expression) and survive in mature spermatozoa.
- It does not appear to be possible to separate haploid spermatozoa according to their genetic expression patterns.
- Surplus cytoplasm is discarded in the residual body.
- Not all the developing spermatozoa survive.
- When spermatogenesis is completed, the cytoplasmic bridges are broken and the spermatozoa are released into the tubular lumen in a process of spermiation.
- They are washed along the seminiferous tubules in testicular fluid secreted by the Sertoli cells.
- The process of spermatogenesis takes a time characteristic for the species and is remarkably constant.
- A new round of spermatogenesis is initiated at a time interval characteristic for the species and is remarkably constant. It is usually about 25% of the time taken to undergo spermatogenesis.
- This regular periodicity of spermatogenic initiation is called the spermatogenic cycle.
- Spermatogenesis is initiated at different times in different regions of the testis, thereby giving a smooth, nonpulsatile flow of spermatozoa.
- Spermatogenesis is initiated at the same time in closely adjacent regions of the same seminiferous tubule. Thus, cross-sections of tubule seem to 'cycle' together: the cycle of the seminiferous epithelium.
- Nearby sections of seminiferous tubule initiate spermatogenesis with a slight phase advance or retardation, giving the appearance of a spermatogenic wave.
- The Sertoli cells also cycle with their adjacent spermatogenic cells.
- The spermatogenic cells seem to set the timing of spermatogenesis.
- Sertoli cells coordinate the temporal and spatial organization of spermatogenesis.
- Retinoic acid has a key role in initiating and maintaining this temporal and spatial organization of spermatogenesis.

Clinical vignette

Oligospermia

A 35-year-old woman (Ms A) and her 40-year-old male partner (Mr B) had been trying to conceive without success for the past 2 years. Ms A had a 3-year-old daughter from a previous relationship, but Mr B had no children. Ms A's menstrual cycles were 30 days in length, and occurred on a regular basis. She was otherwise fit and well, with a BMI in the normal range. Mr B was extremely physically fit, and was a competitive road-cyclist. The couple attended the infertility clinic for investigation. Mr B's sperm count was 10 x 10^6 per ml of ejaculate, with 2% normal morphology and 20% normal motility. It was explained to the couple that these results were significantly below the normal range, and a diagnosis of oligozoospermia was made. The couple discussed the possible reasons for this diagnosis with their fertility expert. Mr B explained that he spent several hours each day riding his racing bicycle in tight shorts, and that he often had a long, hot bath afterwards to help with his muscle recovery. The fertility consultant advised that Mr B reduce his hours on the bicycle and explore other ways of maintaining his very high level of fitness during the period of trying to conceive. He also recommended that Mr B take shorter, cooler showers instead of baths, and wear looser fitting underwear whenever possible. The couple re-attended the clinic for review after 3 months. Although they had not yet conceived, further testing showed that Mr B's sperm count had improved to the normal range. The couple was encouraged to continue to try to conceive naturally. Ms A subsequently had a positive pregnancy test two months later. Their daughter was born healthy at full-term by vaginal delivery.

There is a number of lifestyle and environmental factors that exert important influences on rates of spermatogenesis in adulthood. These range from obesity to exposure to environmental toxins. Humans may be more vulnerable than other species to external influences impacting on spermatozoal production because of the low percentage of spermatozoa within the human ejaculate that can be classified as normal in terms of morphology and motility. In the case presented above, the main barrier to normal spermatogenesis is temperature. The human testes descends into the scrotum before birth, which allows them to be an average of 3–4 °C lower than normal body temperature. If testicular descent fails and the testes are retained in the abdominal cavity, then normal spermatogenesis does not occur and moreover there is a risk of malignancy in the undescended testes. The scrotum not only allows the testes to be held at a distance from the warmer core temperature of the abdominal cavity, but also provides a large surface area and a rich vascular bed (the pampiniform plexus) for heat exchange. Prolonged heat exposure causes hypoxia and oxidative stress in the germ line cells, and experimental evidence suggests that this is associated with both lower sperm counts in the ejaculate and longer times to conception in humans. In clinical practice, the impact of moderate scrotal heating on fertility is probably a minor effect in most men but in the case described, where the daily heat exposure is prolonged and the oligozoospermia mild, lifestyle changes may be enough to make a significant impact on spermatozoal production.

FURTHER READING

General reading

Busada JT, Geyer CB (2016) The role of retinoic acid (RA) in spermatogonial differentiation. *Biology of Reproduction* **94**, 10.

Cheng CY, Mruk DD (2010) A local autocrine axis in the testes that regulates spermatogenesis. *Nature Reviews: Endocrinology* **6**, 380–395.

Cheng CY, Wong EWP, Yan HHN, Mruk DD (2010) Regulation of spermatogenesis in the microenvironment of the seminiferous epithelium: new insights and advances. *Molecular and Cellular Endocrinology* **315**, 49–56.

De Kretser DM, Buzzard JJ, Okuma Y *et al.* (2004) The role of activin, follistatin and inhibin in testicular physiology. *Molecular and Cellular Endocrinology* **225**, 57–64.

Griswold MD (2016) Spermatogenesis: the commitment to meiosis. *Physiological Reviews* **96**, 1–17.

Hermo L, Pelletier RM, Cyr DG, Smith CE (2010) Surfing the wave, cycle, life history, and genes/proteins expressed by testicular germ cells. *Microscopic Research and Technology* **73**, 320–363.

Holdcraft RW, Braun RE (2004) Hormonal regulation of spermatogenesis. *International Journal of Andrology* **27**, 335–342.

Li N, Cheng EI, Wong CY (2016) Regulation of blood-testis barrier by actin binding proteins and protein kinases, *Reproduction* **151**, R29–R41.

Loveland KL, Hogarth C, Mendis S *et al.* (2005) Drivers of germ cell maturation. *Annals of the New York Academy of Sciences* **1061**, 173–182.

Sharpe R (1994) Regulation of spermatogenesis. In: *The Physiology of Reproduction* (ed. E Knobil, JD Neill), 1, pp. 1363–1434. Raven, New York.

Siu MKY, Cheng CY (2004) Dynamic cross-talk between cells and the extracellular matrix in the testis. *BioEssays* **26**, 978–992.

More advanced reading (see also Boxes 7.1–7.3)

Bedall MA, Zama AM (2004) Genetic analysis of Kit ligand functions during mouse spermatogenesis. *Journal of Andrology* **25**, 188–199.

Cheng CY, Dolores D, Mruk DD (2009) An intracellular trafficking pathway in the seminiferous epithelium regulating spermatogenesis: a biochemical and molecular perspective. *Critical Reviews of Biochemistry and Molecular Biology* **44**, 245–263.

Fawcett DW, Bedford JM (1979) *The Spermatozoon*. Urban & Schwarzenberg, Munich.

Gorczynska-Fjälling E (2004) The role of calcium in signal transduction processes in Sertoli cells. *Reproductive Biology* **4**, 219–241.

Krawetz SA (2005) Paternal contribution: new insights and future challenges. *Nature Reviews in Genetics* **6**, 633–642.

Luetjens CM, Weinbauer GF, Wistuba J (2005) Primate spermatogenesis: new insights into comparative testicular organisation, spermatogenic efficiency and endocrine control. *Biological Reviews of the Cambridge Philosophical Society* **80**, 475–488.

Miller D, Brinkworth M, Iles D (2010) Paternal DNA packaging in spermatozoa: more than the sum of its parts? DNA, histones, protamines and epigenetics. *Reproduction* **139**, 287–301.

Mruk DD, Cheng CY (2004) Cell–cell interactions at the ectoplasmic specialization in the testis. *Trends in Endocrinology and Metabolism* **15**, 439–447.

Roosen-Runge EC (1962) The process of spermatogenesis in mammals. *Biological Review* **37**, 343–377.

Sharpe RM, McKinnell C, Kivlin C, Fisher JS (2003) Proliferation and functional maturation of Sertoli cells, and their relevance to disorders of testis function in adulthood. *Reproduction* **125**, 769–784.

Wolgemuth DJ (2008) Function of cyclins in regulating the mitotic and meiotic cell cycles in male germ cells. *Cell Cycle* **7**, 3509–3513.

Xia W, Mruk DD, Lee WM, Cheng CY (2005) Cytokines and junction restructuring during spermatogenesis – a lesson to learn from the testis. *Cytokine & Growth Factor Reviews* **16**, 469–493.

CHAPTER 8
Men

Essential Reproduction, Eighth Edition. Martin H. Johnson.

Companion website: www.essentialreproduction.com

In Chapter 7, we examined the cell choreography underlying the organized progression of cells through spermatogenesis, which has a remarkable constancy, suggesting a high level of intrinsic organization. Indeed, hormones, or other externally applied agents, do not seem to speed up or slow down the spermatogenic process. However, whilst hormones do not seem to affect the rate at which spermatogenesis occurs, they do affect whether or not the process occurs at all.

Testicular hormones

The most important hormones produced by the testis are the androgens, which play a major and essential role during the fetal, postnatal and pubertal phases of male development (see Chapters 3 and 4). The adult testis also produces some oestrogens as well as the peptide hormone, oxytocin. Two other important peptides produced by the adult testis are the cytokines **inhibin** and **activin**. First, we ask: where in the testis is each of these hormones made?, how is its production controlled? and what does each do locally and systemically to make a fully fecund male? Then we explore androgen actions in the adult male further, for, in addition to their developmental role, androgens also play a critical role in the functioning of the mature male. We have already explored their possible influences on male sexual behaviour (see Chapters 5 and 6), and in this chapter we look at their other functions in adults.

Steroids of the testis

As discussed in Chapter 2 (see page 21) the androgens comprise a class of steroid with distinct structural and functional features. The principal testicular androgen is testosterone. It is synthesized from acetate and cholesterol (see Figure 2.8) in the smooth endoplasmic reticulum of the interstitial Leydig cells (Figures 7.1 and 7.10f). As expected, changes in Leydig cell biosynthetic activity correlate with testosterone output during puberty, in seasonally breeding animals, or pathologically.

In man, 4–10 mg of testosterone are secreted daily, and leave the testis by three routes. The hormone rapidly and freely enters both the blood and lymph, but most is carried away in the former due to its greater flow rate: 17 ml of blood/min versus 0.2 ml of lymph/min. However, the lymphatic flow is important as the lymph drainage carries testosterone to the adjacent testicular excurrent ducts and the male accessory sex glands, which are stimulated by it. The third export route of testosterone from the testis is exocrine. Being relatively lipid soluble, it passes freely into the tubule lumen where it binds to the androgen-binding protein (ABP) secreted by the Sertoli cells. The ABP carries it in the testicular fluid flowing into the excurrent ducts, which are stimulated by it (see pages 185–6).

However, **androgens also act within the testis**, where their levels are some 25–125-fold higher than in blood. The cells of the spermatogenic lineage do not express androgen receptors (ARs), but four testicular cell types in adults are potential targets, each possessing ARs. First, androgens act autocrinologically on the Leydig cells themselves in a short negative feedback loop. ARs are also present on vascular smooth muscle and endothelial cells, but their selective knock-out there does not impair spermatogenesis. In contrast, myoid cells express ARs, and the selective knock-down of this gene results in reduced tubular functional integrity and a reduction in the numbers of cells entering spermatogenesis. Finally, androgen within the seminiferous tubules enters Sertoli cells, where much of it is converted by 5α-reductase activity to the more active dihydrotestosterone (Figures 2.8 and 8.1). Both DHT and testosterone then bind to ARs within the Sertoli cell itself, levels of which rise during the juvenile prepubertal phase. Interestingly, whereas both adult Leydig and myoid cells show constitutive levels of AR, Sertoli cells show oscillations in AR levels in phase with the cycle of the seminiferous epithelium, peaking as the time of spermatozoal release is approached. These spermatogenic stages are also the ones most sensitive to androgenic deficit and it is during these stages that the production of most androgen-dependent proteins occurs. It is known that the promoter of the AR gene is sensitive to the androgen:AR complex, and so androgens themselves appear able to regulate AR expression and cyclical Sertoli cell activity. **Testosterone has four key functions in the Sertoli cells** as assessed by selective knock-down of AR therein. First, it **maintains the integrity of the blood–testis barrier** between the basal and adluminal compartments. Second, it is required for **the completion of meiosis**. Third, it is essential for **spermatid elongation** and the accompanying **Sertoli–spermatid adhesion**. Fourth, it is essential for the release of mature spermatozoa at **spermiation** and the **production of testicular fluid** to carry them away.

Testosterone can also be converted to oestrogen in the Sertoli cells, although recent evidence suggests that this pathway operates mainly in fetal life in humans. As the testis matures, most testicular oestrogen is derived directly from Leydig cell activity (Figure 8.1). In adult humans, the output is relatively small, but in the boar and stallion it matches the androgen output. Oestrogen's main function is not in the testis, but in the epididymis (see page 184 for details).

Cytokines and peptides of the testis

Oxytocin (see page 27; Figure 2.12) is produced in the Leydig cells and has been shown to **stimulate seminiferous tubule motility** via an action on the peritubular myoid cells, possibly aiding the flow of fluid to the excurrent ducts. Its passage from

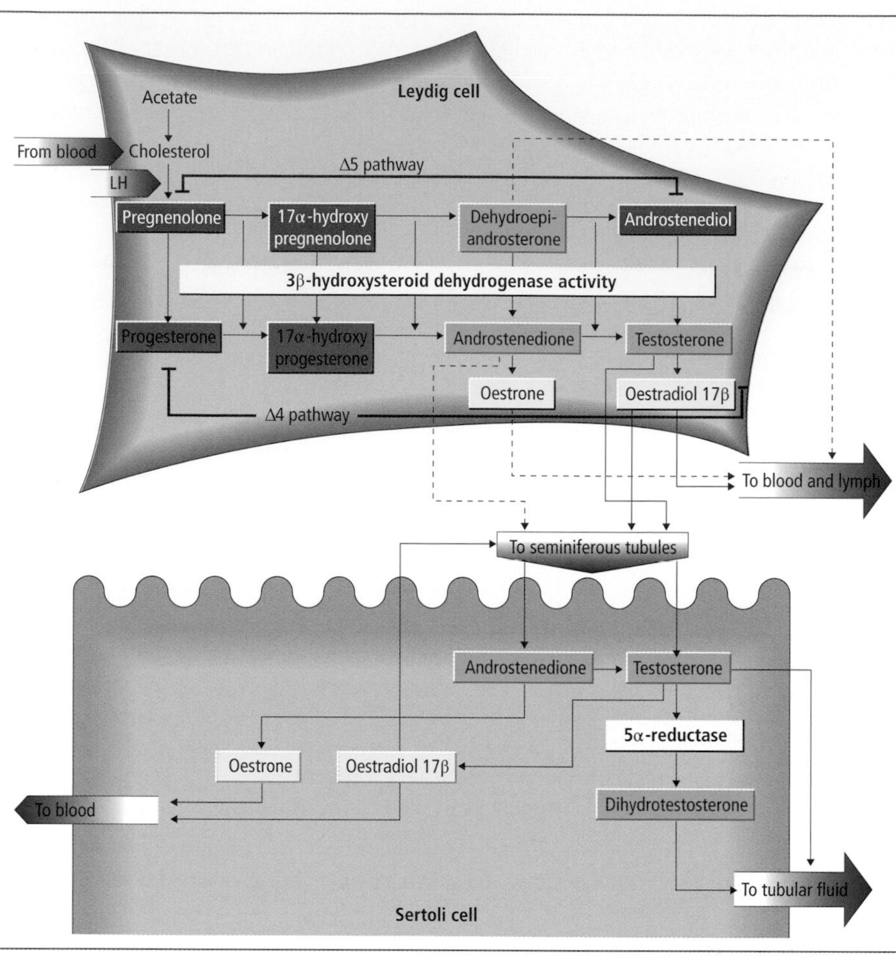

Figure 8.1 Summary of the steroidogenic pathways in the human testis. The principal (Δ5) path for testosterone synthesis in the Leydig cells of the human is indicated by heavy lines, but the Δ4 pathway is also used and may be more important in other species. In addition to testosterone, some of the intermediates in the pathway are released into the blood: androstenedione at 10% and dehydroepiandrosterone at about 6% of testosterone levels in man. Some testosterone and androstenedione enter Sertoli cells. Here they may bind to androgen receptors directly or after conversion to the more potent dihydrotestosterone. Androgens may also be converted to oestrogens. In humans prepubertally, this occurs predominantly in the Sertoli cells, but postpubertally in the Leydig cells. Oestrogen also enters the seminiferous tubule fluid (see Chapter 11).

the testis in the lymph may give it a further paracrine function: the stimulation of epididymal motility.

Two peptide growth factors called **inhibin B** (in humans, monkeys and rats/mice; but also inhibin A in rams, bulls and boars) and **activin A** are related dipeptide hormones (Figure 8.2) and both are produced mainly by the Sertoli cells. About 25% of their output leaves the testis via the lymphatic flow, with most of the remainder passing into the fluid of the seminiferous tubule, from which they are absorbed during passage through the epididymis. The blood level, and thus output, of inhibin reflect the number of functional Sertoli cells and are related to the successful completion of spermiogenesis. Thus, failure to complete spermatogenesis in man correlates with depressed inhibin levels. Conversely, stimulation of spermatogenesis in hypospermatogenic men is accompanied by a rising output of spermatozoa and also of inhibin. As Sertoli cells are implicated in the support of spermatogenesis, as well as being the site of inhibin production, it appears that **inhibin serves as the measure of spermatogenic effectiveness**.

Both peptides may exert undefined paracrine and autocrine roles within the testis, as receptors for inhibin are located on Leydig cells, and those for activin on both Sertoli and spermatogenic cells. A major endocrine role for the systemically transmitted peptides in men is via binding to target cells in the pituitary (see page 133).

Spermatogenesis is dependent on endocrine support

The production of androgens and spermatozoa is interrelated functionally. Thus, at puberty, androgen levels rise and spermatogenesis commences. In some adult mammals, androgen and spermatozoal production do not occur throughout the year (e.g. the roe deer, ram, vole, marine mammals and possibly some primates). Rather, the behaviour and morphology of the males show seasonal variations, which reflect the changing levels of androgen output. In these **seasonal breeders**, the spermatogenic output also varies in parallel with the changing endocrine pattern. If androgens are neutralized, spermatogenesis proceeds only as far as the very early preleptotene stages of meiosis. The male becomes **aspermatogenic**. Restoration of androgens restores spermatozoal output. As we saw earlier, the Sertoli cells are the primary target for androgen action, emphasizing their central role in spermatogenesis.

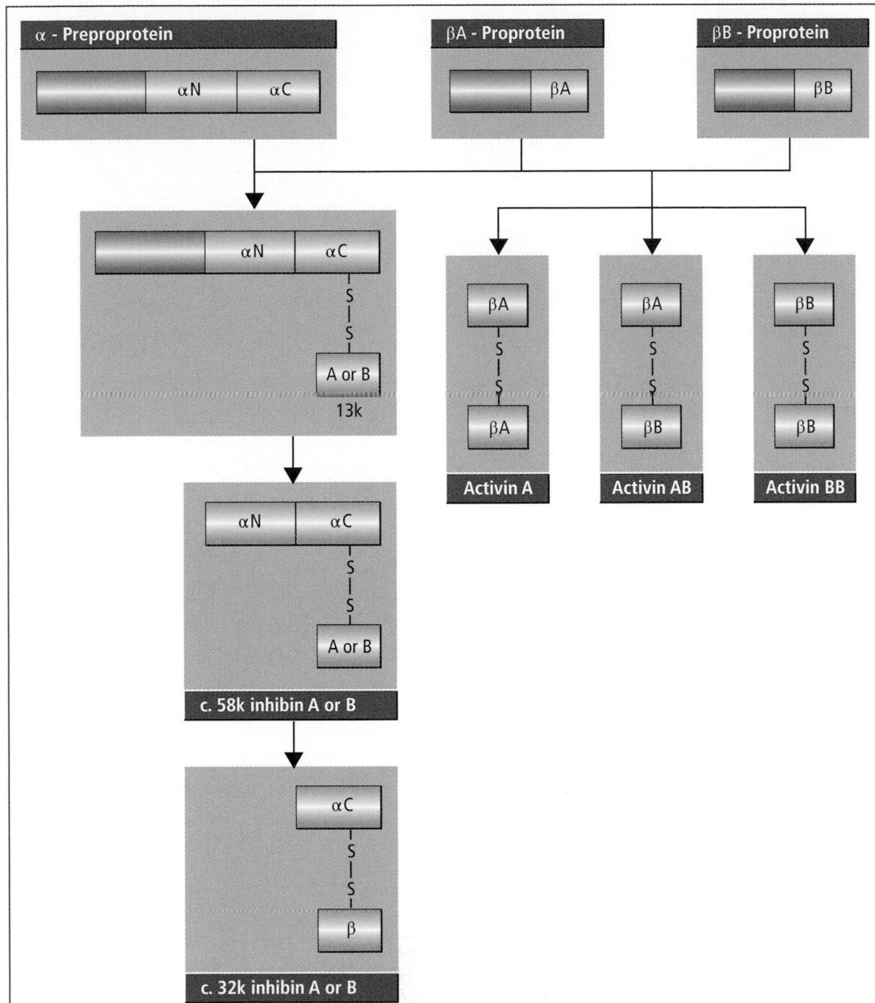

Figure 8.2 The biosynthesis of inhibins and activins occurs from three genes producing one **α-preproprotein** (specific for inhibin) and **two β-proproteins** (which can form part of either activin or inhibin). In each case, the N-terminus (dark blue) is cleaved off and the subunit peptides are then linked in different combinations. Activins take three forms depending on the β chain composition: **activin A** (AA β homodimer), **activin B** (BB β homodimer) and **activin AB** (AB β heterodimer). Two forms of **inhibin (A and B)** exist, depending on whether pairing is with an activin A or B β-subunit. The presence of glycosylation sites on the peptides means that the molecular weights of the hormones can vary considerably. Activin activity can be modulated by the inhibin chain sequestration of β chains, and the activin/inhibin hormones seem to exist in a functional equilibrium. Additionally, **follistatin**, a binding protein for activin, can sequester the cytokine in high-affinity complexes.

The causal association between the presence of androgens and the process of spermatogenesis ensures that mature spermatozoa are always delivered into an extragonadal environment suitably androgenized for their efficient transfer to the female genital tract. Given the importance of androgens for spermatogenesis and male function, how is their output regulated?

Luteinizing hormone acts on Leydig cells

It has been known for many years that removal of the pituitary gland (**hypophysectomy**) causes the testes to shrink, spermatozoal output to decline and spermatogenesis to arrest at the primary spermatocyte stage (see Figure 7.10e). The Leydig cells become involuted, testosterone output falls and the testosterone-dependent male genitalia hypotrophy. If testosterone is given at the time of hypophysectomy, spermatogenesis continues, albeit at a slightly reduced level, and the secondary sex characteristics show little sign of regression, although the Leydig cells do still involute. These experiments establish that the **secretion of testosterone by the Leydig cells is under the control of the pituitary**.

In experiments on the rat, it was shown that if, after hypophysectomy, **luteinizing hormone** (**LH**; see Table 2.4 for details) is administered instead of testosterone, then not only are secondary sex characteristics and spermatogenesis maintained, but the Leydig cells do not involute and testosterone output is maintained. Further confirmation of the role of LH comes from the effect of administration to an intact adult male of an antiserum to bind free LH. The level of plasma testosterone falls and regression of the androgen-dependent secondary sex characteristics follows. The results of these two experiments strongly suggest that pituitary-derived LH stimulates the Leydig cells to produce testosterone. LH has been shown to bind specifically to high-affinity **LH receptors on the surface of the Leydig cells**, the only testicular cells to express them. As a result, intracellular cAMP levels rise within 60 seconds, and testosterone output rises within 20–30 minutes. However, LH does not act alone on the Leydig cells: two other hormones also influence its activity. Both **prolactin**, a second anterior pituitary hormone (see Table 2.5), and

inhibin bind to **receptors on Leydig cells** and facilitate the stimulatory action of LH. Neither hormone alone, however, stimulates testosterone production.

It seems clear, then, that LH stimulates Leydig cells to produce testosterone, which passes into the tubules, binds to androgen receptors within the Sertoli cells and thereby supports spermatogenesis. However, it is important to stress that, although testosterone or LH administered immediately after hypophysectomy can prevent aspermatogenesis in rats, some reduction in testis size and a 20% reduction in spermatozoal output occur. In many other species, including primates, the decline in spermatozoal output after hypophysectomy is also observed, despite administration of high doses of testosterone or LH. Moreover, the first appearance of spermatogenesis at puberty is even more adversely affected in the absence of LH. Taken together, this evidence suggests a second pituitary hormone is required for full testicular function and recovery. This hormone is **follicle-stimulating hormone** (**FSH**; see Table 2.4).

Follicle-stimulating hormone acts on Sertoli cells

FSH binds to **receptors** (**FSH-Rs**) that are located **on the basolateral surface of Sertoli cells**, and thus freely accessible to the large blood-borne glycoprotein hormone. There FSH has two effects. First, at puberty it stimulates mitotic activity in Sertoli cells, thereby affecting testicular size. Second, in the mature adult, it acts as an anti-apoptotic survival factor for spermatogonia (and possibly also spermatocytes), thereby enhancing the number of spermatozoa that are formed. It does this in conjunction with androgens. Thus, FSH stimulates the production of intracellular ARs, as well as inhibiting its own FSH-R production. Androgens in turn stimulate the appearance of FSH-Rs. Thus, FSH-R levels vary with the cycle of the seminiferous tubule, but do so in opposite phase to the ARs, being lowest on the days when ARs are highest (stages 6–8). In this way, FSH and testosterone activities oscillate to act synergistically on the Sertoli cell to facilitate spermatogenic completion. The main second messenger system stimulated by FSH binding is adenyl cyclase, and so cAMP generation also cycles antiphase to ARs. Both hormones also stimulate calcium signalling in Sertoli cells. Thus, both hormones are intimately entwined functionally, but testosterone is the dominant partner, FSH activity not being an absolute prerequisite for fertility.

Paracrine actions by growth factors work locally with FSH and androgens

FSH and androgens interact to stimulate and regulate Sertoli cell function and thereby the proteins and growth factors (GFs) that it produces. The GFs then influence spermatogenesis (see Box 7.1). Over 300 FSH-inducible proteins have been described, including ABP (carries androgen in testicular fluid), transferrin (transports iron into germ cells), Gdnf (regulates proliferation of spermatogonial stem cells; see Box 7.1), aromatase (converts androgens to oestrogens), and CREMt (critically regulates postmeiotic gene expression; see Box 7.2). However, given the complex interplay between FSH, androgen and the cytokines they also influence, it is not clear whether these various genes are induced directly or indirectly by FSH and to what extent other hormones are also involved.

Testicular hormones modulate the output of pituitary hormones

We saw in the previous section that FSH and LH have key roles in regulating testicular output of both spermatozoa and androgens. However, we must now add a further layer of complexity, because it is also clear that the hormonal output of the testis can in turn act to influence the output of gonadotrophins via **feedback loops**. It is this feedback that we explore now. First, a reminder that in Chapter 2 (see page 27) and Chapter 4 (see pages 64–66) we introduced the peptide hormone **gonadotrophin-releasing hormone** (**GnRH**) (see Figure 2.11), which is released in pulses from the hypothalamus, passes via the portal system to the pituitary and there acts on GnRH receptors to release the gonadotrophins that in turn act on the testis.

Testosterone feedback acts to reduce LH and FSH output

LH stimulates testosterone secretion by the Leydig cells, and testosterone in turn acts to modulate LH secretion by **negative feedback**. Thus, male castration or neutralization of testosterone results in increased circulating levels of LH compared with those in intact males (Figure 8.3). In castrate males of all species including humans, the administration of exogenous testosterone causes an abrupt decline in LH levels. Even in intact men, synthetic androgens will depress LH output, as will synthetic progestagens, the basis for a possible male contraceptive (see page 358). How is this feedback effect achieved?

Figure 8.4 shows that LH and FSH output is pulsatile, reflecting the pulsed output of GnRH. Thus, alterations in the output of LH and FSH could be achieved (1) by increasing or decreasing either the **amplitude or the frequency of these pulses** of GnRH, or (2) by modulating the **response of the gonadotrophs to the pulses**. In fact, both mechanisms are employed.

The negative feedback effect of androgen on LH and FSH output is achieved by decreasing both the frequency and amplitude of the pulses. **Androgen receptors are found in abundance in both the hypothalamus and pituitary**, and implantation of testosterone in the mediobasal part of the periarcuate region of the hypothalamus of castrate male rats

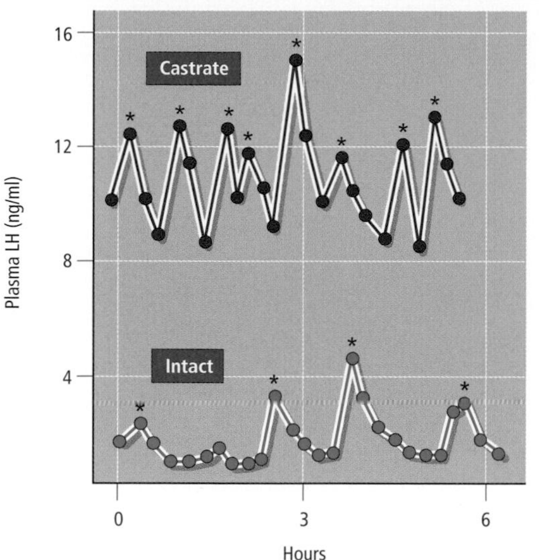

Figure 8.3 The negative feedback effect of testosterone on plasma LH levels in male red deer. Upper part: levels in the castrate male. Lower part: levels in the intact male. Note the increased frequency of pulses (*) after castration. (Source: Lincoln G.A. (1979) *British Medical Bulletin* (London), **35**(2), 167–172. Reproduced with permission from Oxford University Press.)

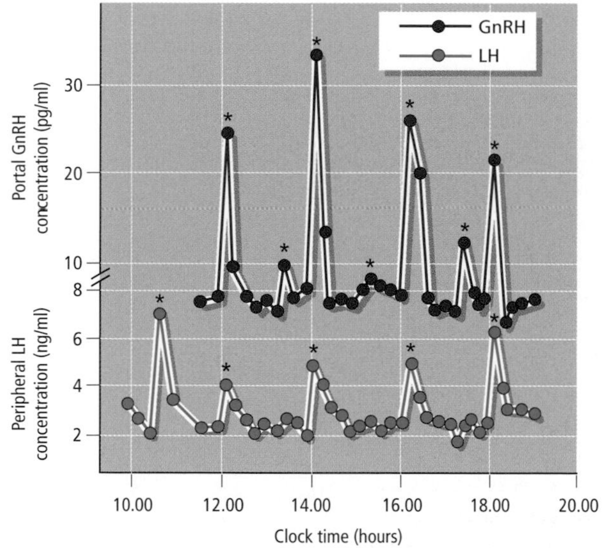

Figure 8.4 Concentrations of GnRH in portal plasma (red line) and LH in jugular venous plasma (green line) in this example of four castrated sheep. Asterisks indicate secretory pulses of GnRH and LH.

causes a significant fall in circulating LH levels. At least in rodents, 5α-dihydrotestosterone also has an inhibitory effect on LH secretion, whether given systemically or implanted in the hypothalamus.

Inhibin feedback acts to reduce mainly FSH output

As we have seen, testosterone does suppress FSH secretion, but its effects are less than those on LH. A more complete suppression of FSH comes from the combined action of androgens and the second testicular hormone, **inhibin, which acts entirely at the level of the pituitary to reduce sensitivity to GnRH**. Inhibin is produced by the Sertoli cells, and is a marker for the successful completion of spermiogenesis, declining when sperm production is defective. Since FSH stimulates Sertoli cells directly, a negative feedback loop seems a logical way of regulating Sertoli cell function. Accordingly, failure to complete spermatogenesis in man is correlated with depressed inhibin levels and elevated serum FSH levels. Conversely, stimulation of spermatogenesis in hypospermatogenic men is accompanied by a rising output of spermatozoa and inhibin, and by declining serum FSH levels.

Androgens play an essential role in the fecund male

In the foregoing chapters we have established the pivotal role of the testis and its secretions for male reproduction. We have been concerned mainly with hormone action in three areas: (1) in the generation and maintenance of sexual differentiation during fetal, neonatal and pubertal life; (2) in the divergent behaviour patterns of men and women; and (3) in the regulation of spermatozoal production. Now, we examine those remaining actions of the steroids in the adult male that ensure the attainment of full reproductive capacity.

The effects of sex steroids may conveniently be thought of as falling into two broad categories: some steroid actions are **determinative**, others are **regulatory**. Determinative actions involve essentially qualitative changes, which are irreversible or only partially reversible. Examples of this type of action are provided by the effect of androgens on the development of the Wolffian ducts and the generation of male external genitalia, the mild enhancement of these sexually distinct features by the low prepubertal androgen levels in males, and the changes in hair pattern, depth of voice, penile and scrotal size and bone growth that occur at puberty, as well as those in brain structure and function (especially in non-primates). These actions constitute part of a progressive androgenization, which establishes a clear and distinctive male phenotype. It represents the completion of a process initiated with the expression of the *SRY* (sex-determining region on the Y chromosome) gene.

In contrast, the regulatory actions of steroids are reversible, and can involve both quantitative and qualitative changes to established accessory sex organs and tissues. These actions are **not concerned with establishing the individual's sex** developmentally as a male or a female. They have evolved to **optimize functioning of the reproductive tracts and genitalia** in the reproductive process. In males, the regulatory actions of testicular steroids influence the activity of the **accessory sex glands**, **metabolism**, **erectile capacity** and, in some species, the more exotic secondary sexual characteristics, such as antlers in deer. These actions may be continuous or show seasonal variation.

Androgens and the male reproductive organs

Above (see page 129), we saw that testosterone was essential for the maintenance of spermatogenesis. Neutralization of testosterone by an antibody, or by synthetic antiandrogens, blocks or reduces spermatozoal production. Testosterone deprivation also has profound and immediate effects on the accessory sex glands of the male's genital tract (Figure 8.5). After castration, the **prostate**, **seminal vesicles** and **epididymides** (or their equivalents in various species; Table 8.1), involute, their epithelia shrink and secretory activity ceases (Figure 8.6). Direct measurement of their metabolic and synthetic activity shows a dramatic fall, and seminal plasma is no longer produced. If castrated animals are provided with exogenous testosterone, the involuted organs are fully restored, in both size and secretory activity. This reversible regression of accessory sex gland activity occurs naturally in seasonally active males, such as sheep and deer. The seasonal appearance of secondary sexual characteristics, such as antlers, and the behavioural interactions during which they are used, are similarly dependent on the actions of androgens. This androgen reaches its targets in blood, lymph and, in the case of the epididymis, in the fluids carrying the spermatozoa (see Chapter 11).

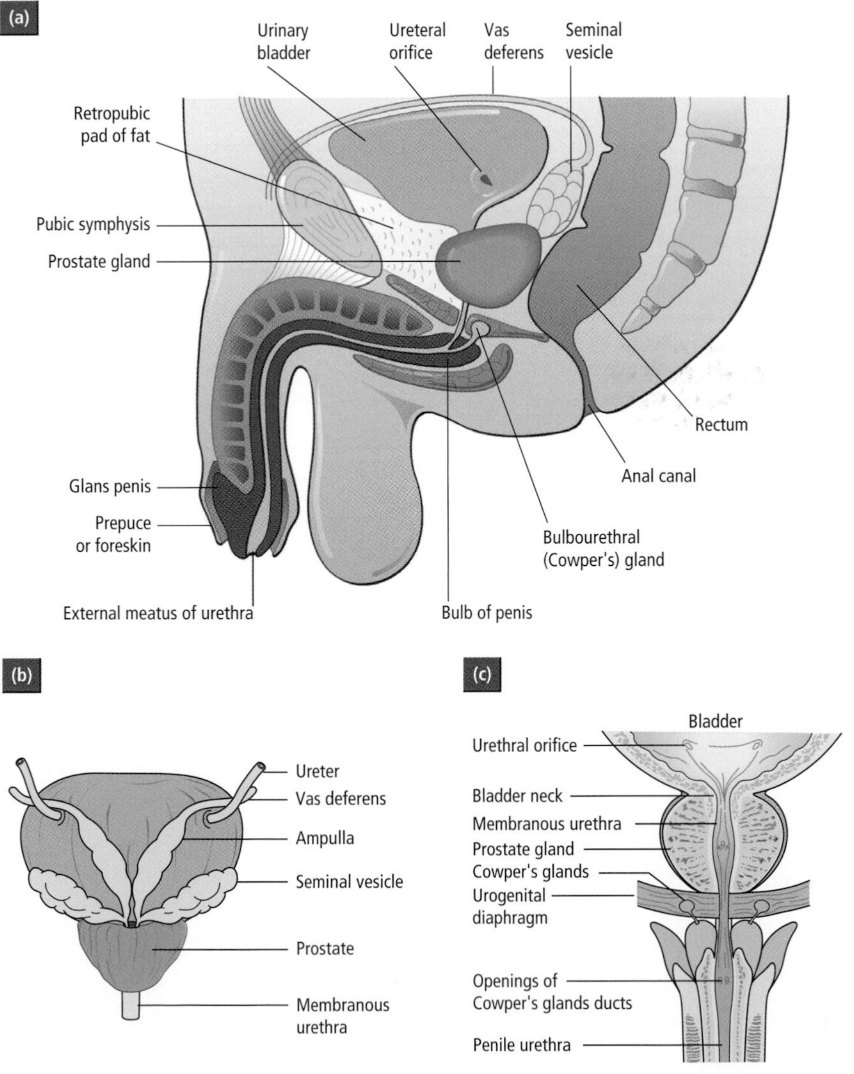

Figure 8.5 View of human male accessory sex glands: (a) midsagittal section through pelvis; (b) posterior view of dissected pelvic contents; (c) coronal section through (b) viewed anteriorly. Source: part (a) Ian Peate and Muralitharan Nair. *Anatomy and Physiology for Nurses at a Glance*, First Edition. Published 2015 by John Wiley & Sons, Ltd. Reproduced with permission of John Wiley & Sons.

Table 8.1 Relative size of principal male accessory sex glands

Species	Prostate	Seminal vesicle	Ampulla	Bulbourethral glands
Human	+++	++	+	±
Bull	+	+++	++	±
Dog	+++	–	–	–
Boar	++	+++	–	++
Stallion	++	+	++	±
Ram	++	+	++	++

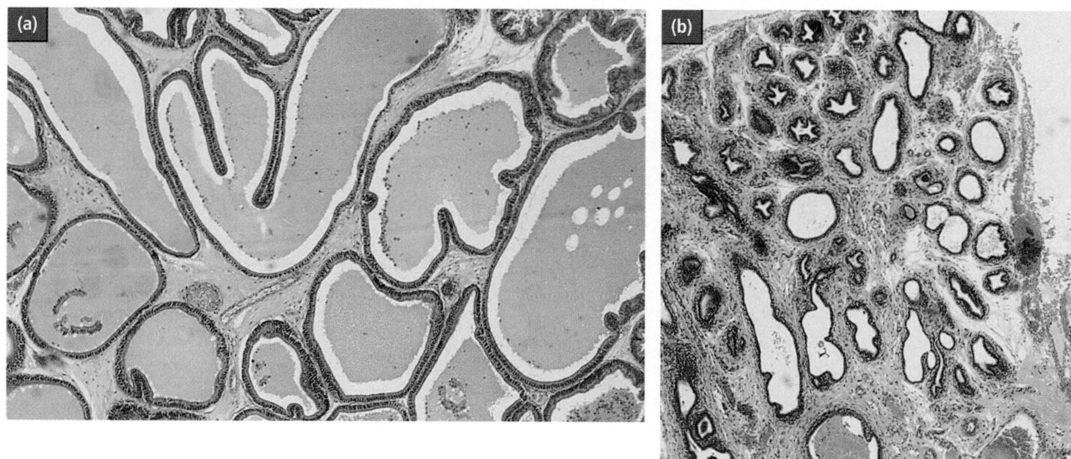

Figure 8.6 Sections through prostate from: (a) intact rat; and (b) rat 18 days after castration (both same magnification). Note that after castration there is a reduction in luminal size and secretory content, a reduced height of the luminal secretory epithelium and a relative expansion of the proportion of connective tissue.

Not surprisingly, these **target organs for androgen** activity are found to possess both **androgen receptors** and the enzyme **5α-reductase**, so androgenic stimulation is provided by both testosterone and its more potent derivative 5α-dihydrotestosterone. Earlier, we mentioned that the local activity of androgens is enhanced by **prolactin**. This relationship is reciprocated, as androgen-dependent prolactin receptors are present in both prostate and seminal vesicles, and prolactin itself is detectable in seminal plasma at higher levels than in blood.

Androgens exert anabolic effects

In addition to their effects on secondary sex organs, androgens also have distinctive **anabolic or myotrophic effects**, a reflection of which is the characteristically more muscular appearance of males that becomes progressively established after puberty. Androgens also **increase kidney and liver weight**, **depress thymus weight**, affect **fat metabolism and distribution** and **stimulate erythropoiesis**, however, the ways in which androgens exert their effects in these tissues remains poorly understood. Pharmacologists have devised a range of synthetic androgens with different relative androgenic and anabolic potencies for use medically and abuse socially (Box 8.1).

Conclusions

In this chapter we have considered the testicular hormones, their actions and controls. The main endocrine secretion of testosterone by the Leydig cells is dependent primarily on pituitary LH. Some of the testosterone enters the seminiferous tubules, where it acts on the Sertoli cells, together with FSH, in order to help maintain the cellular product of the testis: the spermatozoa. Without testosterone, spermatogenesis ceases. Although both FSH and testosterone are essential for full spermatogenesis, their actions appear to be mainly permissive.

Box 8.1 Anabolic effects of steroids: uses and abuses of anabolic-androgenic steroids (AASs)

What are AASs?

A range of synthetic testosterone derivatives has been developed (see Box 2.1) with different relative anabolic and androgenic potencies, solubilities (and thus routes of administration) and metabolic stabilities (variable hepatic degradation or tissue conversion to oestrogens and 5α-dihydrotestosterone).

The anabolic actions increase lean body mass

Muscle size, strength, repair and exercise tolerance are also increased, arising from increased muscle protein synthesis and possibly reduced degradation. Muscle fibre hypertrophy, increased myonuclear number and muscle pennation (to give improved high force, low velocity contractions) are observed. In addition, **anabolic enhancement of collagen synthesis and of bone density** (by osteoclast suppression) improves the mechanical effectiveness of the musculoskeletal system. AASs achieve this effect in part through direct binding to androgen receptors, but also perhaps by increasing the numbers of the receptors. Growth hormone and IGF1 stimulation may also assist anabolic impact.

AASs have been used medically

They are used to treat muscle repair or wasting (age-, immobility- or HIV-related), as well as male hypogonadism, and, more controversially, for treatment of the 'male menopause' and erectile dysfunction (testosterone replacement therapy; see Chapter 21). They have also been used to improve repair in elderly women suffering hip fractures.

Epidemiology of 'recreational use'

Some 3% of young American adults (roughly equal numbers of men and women), around 20% of weight trainers, and, worryingly, 3–12% of adolescents (1–3% in Europe) have taken an AAS at least once in their lives for competitive or cosmetic muscle-building purposes. However, it is lifetime prevalence usage data that are critical as sustained use is undesirable (see below). In general, supra-physiological doses of the more anabolic analogues such as stanozolol (30-fold more anabolic than testosterone) are used, but often at such high levels that their androgenic effects are also marked. Also, in an attempt to enhance anabolic effects, self-administration of multiple analogues occurs in combination (so-called 'stacking').

Health side effects of AAS use

These are rarely serious in adults and probably are all reversible without long-term damage, but may be compounded by physical damage from excessive exercising, exacerbation of any psycho-sociological damage which may in part underlie AAS use in the first place, and, perhaps related to this, use of other performance-enhancing and self-esteem-boosting drugs at supra-physiological doses (adrenaline, amphetamines, growth hormone, diuretics, thryroxine, etc.). Minor and reversible adverse effects common to most users include acne, infection, hepatotoxicity, elevated blood pressure, temperamental instability, and in men testicular atrophy, impotence, gynaecomastia, and in women hirsutism, voice deepening, clitoral enlargement and menstrual irregularities. For pubescent/immature users, longer-term problems associated with premature puberty and epiphyseal closure arise. For women, the long-term effects on fecundity are uncertain. Long-term administration to animals is associated with cardiovascular disease, hepatic tumours and infertility.

Ethico-legal aspects of AAS use

For the clinician these include counselling about self-harm and psychological support and acute awareness of the dangers for adolescents. The professional consequences of detection for athletes are severe. Legally, AASs are prescription-only drugs under the Medicines Act (UK), and thus only a doctor can legally prescribe them. They are classed as Schedule III Controlled Substances (USA) and Class C (UK, Misuse of Drugs Act; unlawful to possess or supply, although in practice possession purely for personal use is rarely pursued by the police).

Further reading

Bahrke MS, Yesalis CE (2004) Abuse of anabolic androgenic steroids and related substances in sport and exercise. *Current Opinion in Pharmacology* **4**, 614–620.

Cafri G, Thompson JK, Ricciardelli L *et al*. (2005) Pursuit of the muscular ideal: physical and psychological consequences and putative risk factors. *Clinical Psychology Review* **25**, 215–239.

Carson JA, Lee WJ, McClung J, Hand GA (2002) Steroid receptor concentration in aged rat hind-limb muscle: effect of anabolic steroid administration. *Journal of Applied Physiology* **93**, 242–250.

Evans NA (2004) Current concepts in anabolic-androgenic steroids. *American Journal of Sports Medicine* **32**, 534–542.

Livermore CT, Balzer DG (eds) (2004) *Testosterone and Aging: Clinical Research Directions*. Institute of Medicine Committee on Assessing the Need for Clinical Trials of Testosterone Replacement Therapy. National Academies Press, Washington, DC.

Thiblina I, Petersson A (2004) Pharmacoepidemiology of anabolic androgenic steroids: a review. *Fundamental & Clinical Pharmacology* **19**, 27–44.

Vermeulen A (2001) Androgen replacement therapy in the aging male: a critical evaluation. *Journal of Clinical Endocrinology and Metabolism* **86**, 2380–2390.

Key learning points

- Leydig cells make androgens, oestrogens and oxytocin, and are located in the interstitial compartment.
- Sertoli cells make inhibin and activin, and are located in the intratubular compartment.
- Spermatogenesis fails in the absence of the gonadotrophins: luteinizing hormone (LH) and follicle-stimulating hormone (FSH).
- In the absence of LH, androgens fall and spermatogenesis ceases.
- Prepubertal or quiescent testes cannot initiate spermatozoal production without both LH and FSH.
- Once spermatogenesis is underway, removal of FSH results in a significant fall in spermatozoal production and a reduction in testis size.
- LH binds to receptors on the Leydig cells and is required for synthesis of androgens.
- Prolactin and inhibin enhance the stimulation of Leydig cells by LH.
- Androgens enter Sertoli cells where testosterone is converted to dihydrotestosterone.
- Androgens are required for the integrity of the blood–testis barrier, Sertoli-spermatid adhesion and the release of mature spermatozoa at spermiation.
- Spermatogenesis fails during meiosis in the absence of androgens.
- FSH binds to receptors on the Sertoli cells and is required for synthesis of inhibin and activin.
- FSH also stimulates androgen receptors in the Sertoli cell.
- Androgens also stimulate receptors for FSH on Sertoli cells.
- FSH and androgens act synergistically on Sertoli cells to promote Sertoli cell function and the support of spermatogenesis.
- Androgens pass out of the testis in blood, lymph and testicular fluid (bound to ABP) and act systemically and locally in the male genetic tract.
- The anterior lobe of the pituitary makes and secretes LH and FSH.
- The preoptic/anterior hypothalamic areas and arcuate nuclei synthesize GnRH, which is secreted into the portal vessels in the median eminence and is carried in the blood to the anterior pituitary where it controls LH and FSH secretion.
- Hypothalamic GnRH secretion is pulsatile and thereby controls the pulsatile secretion of gonadotrophins.
- Testosterone exerts negative feedback control over LH secretion.
- These effects are mediated through both the hypothalamus and the anterior pituitary.
- Testicular inhibin exerts negative feedback control over FSH secretion by actions in the anterior pituitary.
- Steroids have determinative and regulatory actions.
- Testosterone and its reduced metabolite 5α-dihydrotestosterone maintain the functional integrity of the accessory sex glands.
- Androgens also have anabolic effects, and some synthetic androgens emphasize this property for use clinically and abuse recreationally.

Clinical vignette

Fertility after pituitary failure

A 53-year-old man, Mr E., attended the reproductive endocrinology clinic for advice. He had developed idiopathic hypogonadotrophic hypogonadism (pituitary gland failure of uncertain cause) five years previously. His pituitary gland failure had been diagnosed after he experienced symptoms reflecting a lack of androgens. These symptoms included hot flushes, mood changes, loss of muscle mass, and weight gain. Since diagnosis, he had taken daily topical testosterone replacement therapy, which was highly effective in reversing his symptoms. However, Mr E. subsequently began a new relationship with a 35-year-old female partner, with whom he wished to conceive. Despite regular penetrative vaginal intercourse function, the couple found it puzzling that they were unable to conceive. After a fertility consultation with a reproductive endocrinologist, Mr E's treatment for his hypogonadotrophic hypogonadism was changed to injections of human chorionic gonadotrophin (hCG). Despite a brief period during which Mr E's erectile function was compromised, the couple resumed regular sexual intercourse after the change in treatment. Nine months later Mr E's partner had a positive pregnancy test, and the couple welcomed the arrival of a healthy male baby the following year.

Testosterone is necessary to stimulate normal functioning of the Sertoli cells, which are required to support spermatogenesis. However, exogenous administration of androgens counter-intuitively suppresses spermatogenesis in humans and other primates. Evidence suggests that approximately 90% of men may be rendered azoospermic by exogenous testosterone therapy. The reason for this is that the exogenous administration of testosterone suppresses the normal production of testosterone from the testes. Therefore, while serum levels of testosterone may be normalized, the intra-testicular levels (which are normally approximately 20-fold higher than serum levels) remain low. Adequate intra-testicular testosterone is required to stimulate spermatogenesis. Although direct application of testosterone is an effective treatment to reverse the hypo-androgenic symptoms of pituitary failure (hot flushes, mood swings, weight gain, and gynaecomastia) and to stimulate normal male sexual function, it will not promote fertility. Therefore, if fertility is desired, treatment for hypogonadotrophic hypogonadism must focus on replicating the function of the hypothalamo-pituitary-gonadal axis at the level of the pituitary gland, rather than the gonad. Such stimulation is provided by injections of hCG, which are analogous to providing luteinizing hormone and are successful in many men in restoring spermatogenesis. At least one study has estimated that the live birth rate following HCG treatment of male partners with hypogonadotrophic hypogonadism may be as high as 50%.

FURTHER READING

General reading

Busada JT, Geyer CB (2016) The role of retinoic acid (ra) in spermatogonial differentiation. *Biology of Reproduction* **94**, 10.

De Gendt K, Verhoeven G (2012) Tissue- and cell-specific functions of the androgen receptor revealed through conditional knockout models in mice. *Molecular and Cellular Endocrinology* **352**, 13–25.

De Kretser DM, Buzzard JJ, Okuma Y *et al.* (2004) The role of activin, follistatin and inhibin in testicular physiology. *Molecular and Cellular Endocrinology* **225**, 57–64.

Griswold MD (2016) Spermatogenesis: the commitment to meiosis. *Physiological Reviews* **96**, 1–17.

Holdcraft RW, Braun RE (2004) Hormonal regulation of spermatogenesis. *International Journal of Andrology* **27**, 335–342.

Huhtaniemi I (2015) A short evolutionary history of FSH-stimulated spermatogenesis. *Hormones* **14**, 468–478.

O'Hara L, Smith LB (2015) Androgen receptor roles in spermatogenesis and infertility. *Best Practice & Research Clinical Endocrinology & Metabolism* **29**, 595e605.

Smith LB, Walker WH (2014) The regulation of spermatogenesis by androgens. *Seminars in Cell & Developmental Biology* **30**, 2–13.

More advanced reading (see also Box 8.1)

Bremner WJ, Millar MR, Sharpe RM, Saunders PT (1994) Immunohistochemical localization of androgen receptors in the rat testis: evidence for stage-dependent expression and regulation by androgens. *Endocrinology* **135**, 1227–1234.

Luetjens CM, Weinbauer GF, Wistuba J (2005) Primate spermatogenesis: new insights into comparative testicular organisation, spermatogenic efficiency and endocrine control. *Biological Reviews of the Cambridge Philosophical Society* **80**, 475–488.

Makanji Y, Zhu J, Mishra R, Holmquist C, Wong WPS, Schwartz NB, Mayo KE, Woodruff TK (2014) Inhibin at 90: from discovery to clinical application, a historical review. *Endocrine Reviews* **35**, 747–794.

McBride JA, Coward RM (2016) Recovery of spermatogenesis following testosterone replacement therapy or anabolic-androgenic steroid use. *Asian Journal of Andrology* **18**, 373–380.

Ramaswamy S, Weinbauer GF (2014) Endocrine control of spermatogenesis: role of FSH and LH/ testosterone. *Spermatogenesis* **4** (2), e996025.

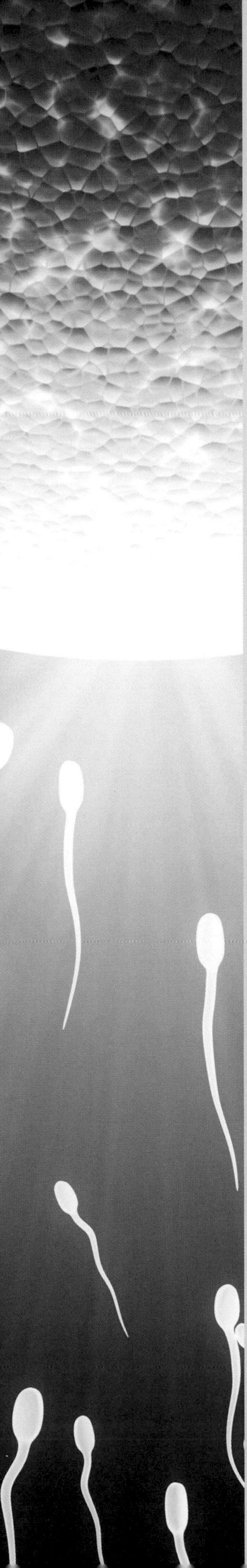

CHAPTER 9

Making eggs

Essential Reproduction, Eighth Edition. Martin H. Johnson.

Companion website: www.essentialreproduction.com

In Chapter 3 (see pages 40–3), we described how the fetal ovary formed from an indifferent genital ridge. It differentiates later than the testis and its endocrine activity is not required during fetal and neonatal life for development of the female phenotype. Significant ovarian endocrine activity occurs first during full sexual maturation at puberty (see pages 56–8) with two main steroid secretions: oestrogens and progestagens. From puberty onwards, the ovary also produces its haploid gamete: the oocyte. As in the adult testis, the endocrine activity of the ovary is coordinated with the production of its gametes. However, adult ovarian function differs fundamentally from testicular function in two ways. First, **relatively few oocytes are released**: up to 400 per lifetime compared with the millions of sperm daily. Second, **their release is not in a continuous stream but occurs episodically at ovulation** once each month. Moreover, the pattern of release of the oestrogens and progestagens reflects this episodic release of oocytes. Thus, the **period before ovulation is characterized by oestrogen dominance**, and the **period after ovulation is characterized by progestagen dominance**. Once this sequence of oestrogen–ovulation–progestagen has been completed, it is repeated. Therefore, we speak of a **cycle of ovarian activity**. The cyclic release of steroids imposes a corresponding cyclicity on the whole body and, in most species, on the behaviour of the adult female. These cycles are called the **oestrous cycle** in most animals and the **menstrual cycle** in higher primates.

Why is female reproductive activity cyclic? The genital tract of the female mammal, unlike that of the male, serves **two distinctive reproductive functions**, each with distinctive physiological requirements. It must act to **transport gametes** to the site of fertilization, and it also provides the site of **implantation of the conceptus** and its subsequent development. The **distinct phases of the female cycle reflect these two roles**. During the first, oestrogenic part of the cycle, the ovary prepares the female for receipt of the spermatozoa and fertilization of the oocyte. During the second, progestagenic part of the cycle, the ovary prepares the female to receive and nurture the conceptus should successful fertilization have occurred. Sandwiched between these two endocrine activities of the ovary, the oocyte is released at **ovulation**.

In this chapter, we will consider the **sequence of changes within the ovary** itself by which a coordinated and cyclic pattern of production of the oocyte and ovarian steroids is achieved. In the next chapter, we will consider how this ovarian cyclicity leads to the oestrous and menstrual cycles.

The adult ovary consists of follicles and interstitial tissue

The adult ovary is organized on a pattern comparable to the testis (Figure 9.1b) with an interstitial tissue made up of glandular tissue, the so-called **interstitial glands** (homologous to Leydig cells), set in **stroma**. The interstitial tissue surrounds **follicles** (homologous to tubules). In contrast to the testis, in which the contents of the tubules show cyclic cytological changes as the spermatozoa form (the cycle of the seminiferous epithelium, see pages 121–2), the **whole ovary cycles**. However, each ovarian cycle is made up of an interconnecting series of events in a subset of follicles as they mature. In the first part of this chapter, we will follow the maturation of a single follicle in some detail. In the second part of the chapter, we will relate the activity in a single follicle to the activity of the ovary as a whole, thereby explaining the ovarian cycle itself.

The pattern of gamete production in the female, like that in the male, shows processes of proliferation by mitosis, genetic reshuffling and reduction at meiosis, and cytodifferentiation during oocyte maturation. In the female, however, the need for mitotic proliferation is not as great because only one or a few oocytes are shed during each cycle, unlike the massive and continuous sperm output of the testis. Thus, as we saw in Chapter 1 (see pages 6 and 8), the female completes the proliferative phase during fetal and/or neonatal life, when the primordial germ cells or **oogonia** (equivalent to spermatogonia in the male) all cease dividing and enter meiosis where they arrest in the dictyate **germinal vesicle (GV)** stage (first meiotic prophase) to become **primary oocytes** (equivalent to primary spermatocytes). They do this within the **primordial follicle**, which consists of flattened mesenchymal cells, **granulosa cells**, condensed around what is first the oogonium and then the primary oocyte (Figures 9.2a and 9.3a), all enclosed within a basement membrane, the **membrana propria**. The **primordial follicle constitutes the fundamental functional unit** of the ovary.

The follicle is the fundamental reproductive element of the ovary

The primordial follicles can stay in this arrested state for up to 50 years in women, with the oocyte metabolically ticking over and waiting for a signal to resume development. Why oocytes have evolved to be stored in this extraordinarily protracted meiotic prophase is unknown. Although a few follicles may resume development sporadically and incompletely during fetal and neonatal life, **regular recruitment of primordial follicles into a pool of growing follicles occurs first at puberty**. Thereafter, a few follicles recommence growth every day, so that a continuous trickle of developing follicles is formed. When a primordial follicle commences growth, it passes through three stages of development *en route* to ovulation: first it becomes a **primary** or **preantral follicle**; then a

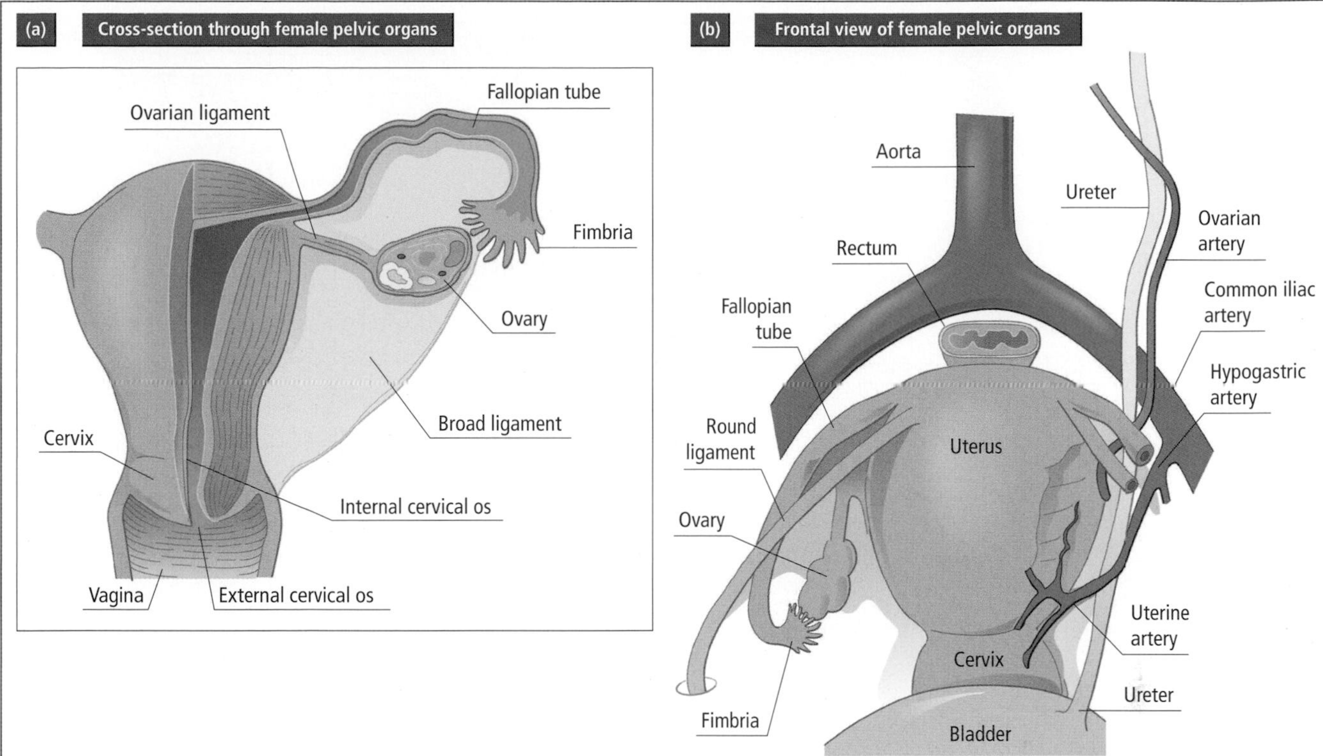

Figure 9.1 Posterior views of the human uterus and one oviduct and ovary: (a) intact; (b) sectioned. The ovaries have been pulled upwards and laterally, and would normally have their long axes almost vertical. Note that all structures are covered in peritoneum except the surface of the ovary and oviducal ostium. The ovary has a stromal matrix of cells, smooth muscle fibres and connective tissue containing follicles, corpora lutea and corpora albicans, and interstitial glands. Anteriorly at the hilus the ovarian vessels and nerves enter the medullary stroma via the mesovarium. (Source: Heffner L.J. and Schust D.J. *The Reproductive System at a Glance*, Fourth Edition. 2014, John Wiley & Sons, Ltd. Reproduced with permission of John Wiley & Sons.)

secondary or **antral follicle** (also called a **vesicular** or **Graafian** follicle); and finally a **tertiary** or **preovulatory follicle** in the run-up to ovulation itself. The times taken to traverse each of these stages differ (Table 9.1), the preantral phase being the longest and the preovulatory phase the shortest. We will now trace the development of one such primordial follicle.

Follicles grow and mature: primordial to preantral transition

The earliest preantral phase of follicular growth is characterized by an increase in the diameter of the primordial follicle from 20 μm to between 200 and 400 μm, depending on the species (Figure 9.3a,b). A major part of this growth occurs in the primary oocyte, which increases its diameter to its final size of 60–120 μm. During this period of oocyte growth, the oocyte remains arrested in meiosis. The dictyate chromosomes consist of decondensed dispersed DNA on which active synthesis of large amounts of ribosomal and messenger RNA occurs, the latter being used to build organelles and to generate stores of proteins, all of which are essential for later stages of oocyte maturation and for the first few days of development after fertilization (see pages 220–1). Towards the end of this phase of follicle development, the GV moves from a central position to an eccentric position in the egg by a mechanism that involves actin microfilaments. This period thus represents part of the cytodifferentiative phase, occurring in tandem with the arrested meiosis, not after meiotic completion as occurs in the male.

Early during growth of the oocyte, it secretes glycoproteins, which condense around it to form a translucent acellular layer called the ***zona pellucida***. The *zona* separates the oocyte from the surrounding granulosa cells, which divide to become several layers thick (Figure 9.3b). However, contact between granulosa cells and oocyte is maintained via granulosa cytoplasmic processes, which penetrate the *zona* and form gap junctions at the oocyte surface. Gap junctions also form in increasing numbers between adjacent granulosa cells, thus providing the basis for an extensive network of intercellular communication. Through this network, biosynthetic substrates of low molecular weight, such as amino acids and nucleotides, are passed to the growing oocyte for incorporation into macromolecules. This nutritional network is important, because the **granulosa layer**

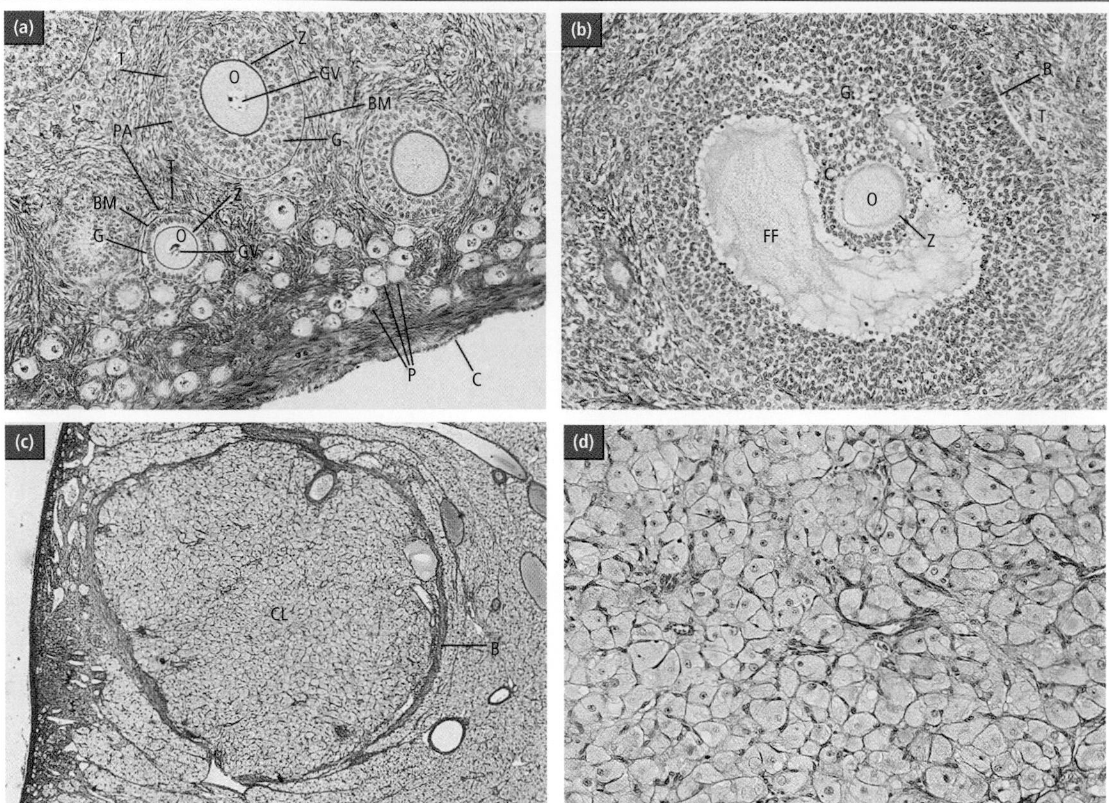

Figure 9.2 (a) Primordial follicles (P) in the cortical stroma of the ovary, lying adjacent to the coelomic epithelium (C), grow in size to give preantral follicles (PA) which contain an enlarged oocyte (O) containing a 'nucleus' or germinal vesicle (GV). Each oocyte is surrounded by a *zona pellucida* (staining blue; Z) and proliferated granulosa cells (G), which in the lower follicle are only one layer thick but in the upper one are several layers thick. Outside the granulosa cells is a basement membrane (BM) on which the theca (T) forms. (b) Further follicular growth leads to antral follicles in which the granulosa cells (G) have proliferated further and the fully grown oocyte (O) is surrounded by a *zona* (Z) and cumulus cells (C), set in the follicular antrum containing follicular fluid (FF). The thecal tissues (T) differentiate and remain separated from the granulosa cells by the basement membrane (B), which is not penetrated by blood vessels leaving the granulosa cells avascular. (c) After a preovulatory growth spurt by the follicle and the process of ovulation, those granulosa cells not ovulated with the oocyte increase in size and fill the emptied follicular antrum as they luteinize within a fibrous thecal capsule (B) to form a corpus luteum (CL). (d) A higher power view of the luteal cells. Note the relatively large size of the luteal cells and the vascularization of the previously avascular granulosa area, now luteinized. (a, b, d, same magnification; c, magnification 25% that of others.)

is completely avascular, with no blood vessels penetrating the membrana propria. The granulosa–oocyte complex thereby resembles the Sertoli cell–spermatogenic complex, in which a similar avascular close relationship occurs (see pages 110–3), perhaps reflecting their shared developmental origins (see pages 39–42).

In addition to oocyte growth and granulosa cell proliferation, the preantral follicle also increases in size and complexity through the condensation of ovarian stromal cells on the outer membrana propria. This loose matrix of spindle-shaped cells is called the **theca** of the follicle (Figure 9.3b). With further development and proliferation, the thecal cells can be distinguished as two distinct layers (Figure 9.3c): an **inner glandular, highly vascular theca interna**, surrounded by a **fibrous capsule, the theca externa**.

Preantral to antral transition

During this next phase, ooctye growth ceases and the numbers of trans-zonal processes fall under the influence of FSH, but the granulosa cells continue to proliferate, resulting in a further increase in follicular size. As they do so, a viscous fluid starts to appear between them (Figure 9.3c). This **follicular fluid** is composed partly of granulosa cell secretions, including mucopolysaccharides, and partly of serum transudate. The drops of fluid coalesce to form a **single follicular antrum** (Figures 9.2b and 9.3d). The appearance of the follicular antrum marks the beginning of the **antral phase** of development. Increase in follicular size from now on depends mainly on an increase in the size of the follicular antrum and the volume of follicular fluid, although granulosa cells do continue to proliferate.

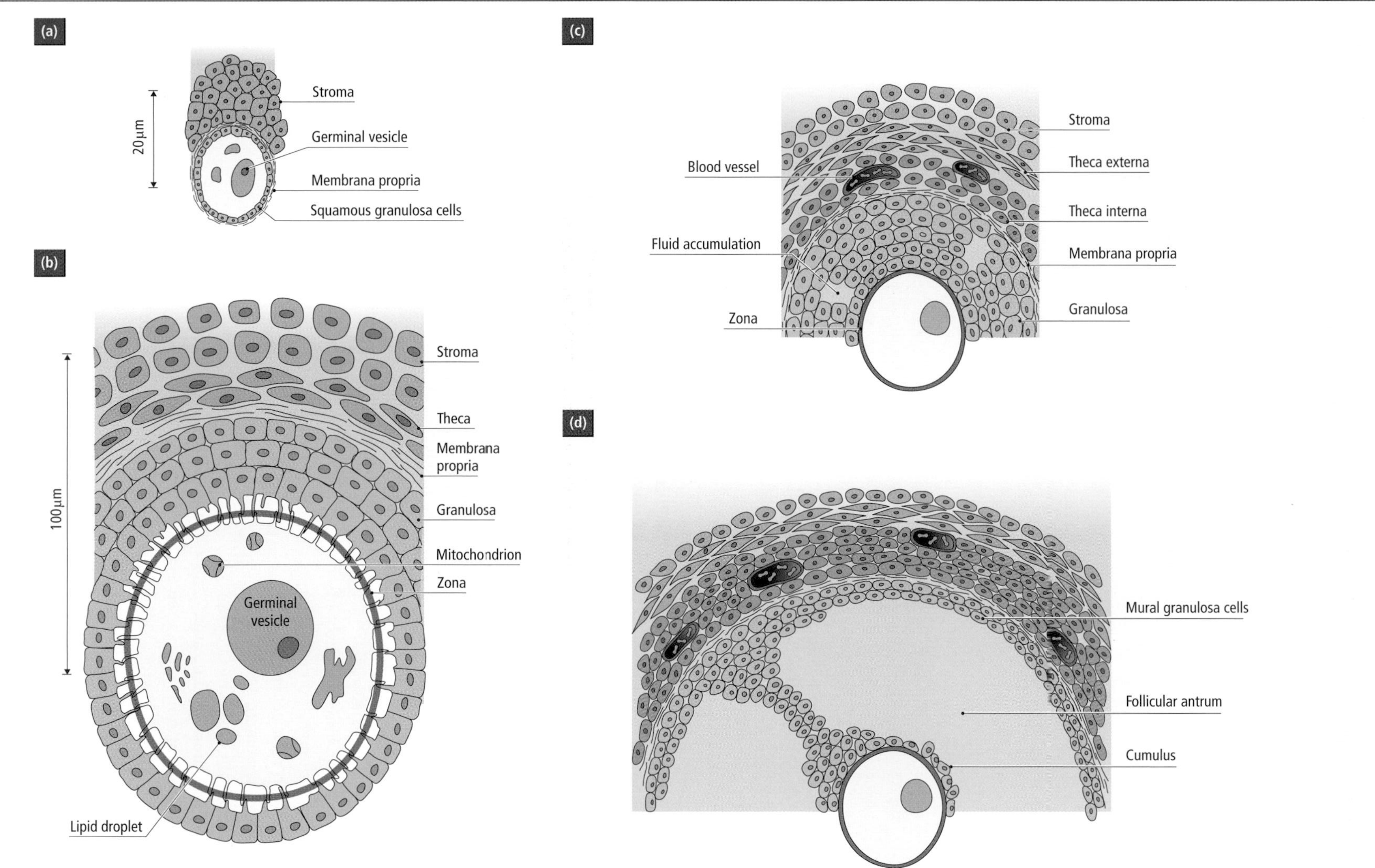

Figure 9.3 Stylized morphology of follicular development. (a) Primordial follicle embedded in stromal cells: note outer membrana propria, within which squamous granulosa cells enclose a small primordial oocyte (yellow) with its chromosomes at the dictyate stage within a nuclear membrane (the germinal vesicle: GV). (b) Preantral follicle: stromal cells condense on the membrana propria to form a thecal layer; the granulosa cells divide and become cuboidal; the oocyte grows and secretes the *zona pellucida*, through which cytoplasmic processes maintain contact between granulosa cells and the primary oocyte (not shown). Within the oocyte: mitochondria increase in number and are small and spherical with columnar cristae; the smooth endoplasmic reticulum breaks up into numerous small vesicles; the Golgi complex breaks into small vesicular units often associated with lipid droplets. The genes are transcriptionally active. (c) Early antral follicle in which the granulosa and thecal cells have proliferated. The thecal cells now comprise two layers, an outer fibrous theca externa and an inner theca interna, rich in blood vessels and large, foamy cells with abundant smooth endoplasmic reticulum. Within the membrana propria is the avascular granulosa layer, among which coalescing fluid drops are evident. Note that the GV has moved to an eccentric position within the oocyte. (d) Expanded antral follicle, with a fully developed follicular antrum, leaving the oocyte surrounded by a distinct and denser layer of granulosa cells, the cumulus oophorus, and the walls lined by mural granulosa cells. The oocyte itself shows only a small further increase in size. Note that throughout growth, the nuclear membrane of the GV is intact and its chromosomes remain arrested in meiotic prophase.

Table 9.1 Duration of phases of follicular development in the non-pregnant animal

Species	Preantral phase (days)	Antral phase (days)	Preovulatory phase (hours)	Luteal phase (days)
Mouse	6–10	3–4	11	2
Human	c.120*	8–12	30–36	12–15
Sheep	NK	4–5	22	14–15
Cow	NK	c.10	40	18–19
Pig	NK	c.10	41	15–17
Horse	NK	c.10	40	15–16

* Also includes very early antral development (see Table 9.2).
NK, not known.

Although the oocyte does not increase in size over the antral period, it initially continues to actively synthesize RNA and to turnover protein, but progressively its chromatin becomes more heterochromatic, less expressive and its histones increasingly deacetylated, and this change in chromatin configuration coincides with the capacity of the oocyte to resume meiosis if released from the follicle. As the follicular antrum grows, the oocyte is left suspended in fluid surrounded by a dense mass of granulosa cells called the **cumulus oophorus**. It is connected to the rim of peripheral, or **mural, granulosa cells** only by a thin 'stalk' of cells (Figure 9.3d). This maturing antral follicle is now ready to enter its preovulatory phase and to approach ovulation. However, before this remarkable process is described, the underlying control of growth from primordial to expanded antral follicle will be examined.

Early follicular growth occurs independent of external regulation

We do not understand fully how each day a few primordial follicles start to develop as preantral follicles, or how those that do are selected. **Intraovarian growth factors or cytokines** are certainly involved in the earliest stages of growth in a bewilderingly complex array (Boxes 9.1 and 9.2). It is also clear that preantral follicular development can occur **independently of any direct extraovarian controls**, as, although FSH receptors are detectable in the primary follicles, in their genetic absence or in hypophysectomized rats, mice or women, follicles still develop to the antral stage. However, in the absence of FSH or its receptors, fewer follicles develop and more become atretic, suggesting a facilitatory role. **Growth hormone** (GH; see Table 2.5) is generally facilitatory for early follicle growth and survival, probably mediated by **insulin-like growth factor 1** (**IGF1**) production from granulosa cells. What is clear is that once the primordial follicle is activated, the rate at which it develops is set by the oocyte: the '**folliculogenesis clock**' starts ticking.

There comes a point, however, when further **follicular development requires external support**, and, as was found for spermatogenesis in the male (see pages 130–2), this external support is provided by the **pituitary**. Thus, removal of the pituitary (hypophysectomy) prevents the completion of antral follicle development. The precise stage at which follicles arrest in the absence of a pituitary depends on the species: in the rat, arrest occurs at late preantral to early antral stages, but in the human, it occurs when antral follicles are slightly more advanced, with a diameter of 2 mm. The granulosa cells in arrested follicles show reduced protein synthetic activity, accumulation of lipid droplets and pyknotic nuclei. Apoptotic death of the oocyte and granulosa cells follows. Leukocytes and macrophages invade and fibrous scar tissue forms. This process is called **atresia**.

A similar follicular phenotype is also observed in mice genetically lacking gonadotrophins, indicating their obligatory involvement. Accordingly, atresia is prevented by follicle-stimulating hormone (FSH) and luteinizing hormone (LH). These hormones bind to follicular FSH and LH receptors that first appear on cells in the late preantral and early antral follicles. **FSH alone is sufficient for initial follicular growth**, but **LH assists further antral expansion**. Thus, FSH-knockout mice arrest follicular development preantrally, whereas LH-knockout mice block at the antral stage. So, **the gonadotrophins pick up preantral follicles and stimulate antral growth**. What do the gonadotrophins do and where in the follicle do they do it?

FSH and LH stimulate steroid synthesis during the antral stage

When the distribution of receptors in early antral follicles is analysed, it is found that **only the cells of the theca interna bind LH**, whereas **only the granulosa cells bind FSH**. Moreover, the effects of hormone binding at each of these sites produce very different consequences.

The antral follicles produce and release increasing amounts of steroids as they grow under the influence of the gonadotrophins. The main oestrogens produced are **oestradiol 17β and oestrone**. Antral follicles also account for 30–70% of the circulating androgens found in women, mainly **androstenedione and testosterone** (the remainder coming from the adrenal).

Box 9.1 Cytokines and primordial follicle recruitment

The activation of primordial follicles appears to be initiated by the activation of **mammalian target of rapamycin complex 1** (**mTORC1**), a complex kinase, in the flattened granulosa cells causing them to change shape and become cuboidal. This change then leads to the direct activation of **stem cell factor** (**SCF**; also known as **Kit**) on the granulosa cells which then combines with and activates its **receptor c-kit** on the dormant oocyte to stimulate its growth via a **phosphatidylinositol 3-kinase** (PI3k) signalling pathway. As part of this oocyte response, two key GFs are produced by the oocyte: namely **bone morphogenetic protein 15** (**BMP**) and **growth differentiation factor 9** (**GDF-9**) (see Box 9.2). The evidence for this sequence comes from the observation that activating mutations in *mTORC1*, *Scf/Kit* and PI3K are associated with premature wholesale maturation of primordial follicles and thus early egg loss. Conversely, the same outcome results from null mutations in **phosphatase and tensin homolog deleted on chromosome 10** (**Pten**), an intracellular negative regulator of PI3k, and in the TF, ***Forkhead Box O3*** (***FoxoO3***), a downstream effector of Scf/Kit which acts as a PI3K-dependent switch controlling initiation of oocyte growth. However, a full understanding of the pathways involved requires further research.

The features of the follicular microenvironment responsible for mTORC1 activation are unknown, but it seems likely that a balance of inhibitory and facilitatory factors operates to regulate the rate at which primordial follicles are recruited, thereby avoiding premature exhaustion of the follicular pool. **Anti-Müllerian hormone (Amh**; Box 3.2**)** from granulosa cells of growing preantral and small antral follicles appears to exercise a restraining negative feedback on the recruitment of primordial follicles, which is higher in mice null for Amh. Positive recruitment of primordial follicles seems to involve expression of **fibroblast growth factor-7** (**Fgf-7**) and **Fgf-2** (also known as basic fibroblast growth factor), both implicated as up-regulating Scf/Kit expression in granulosa cells. Finally, **insulin** has been shown to promote the primordial-to-primary follicle transition additively with Scf. However, these positive factors act upstream of the mTORC1 activator, the identity of which is currently unknown.

Box 9.2 Cytokines and follicle growth

Growth of follicles seems to involve expression of members of the **transforming growth factor β** (**Tgfβ**) super family, including **bone morphogenetic protein 15** (**Bmp15**) and **growth differentiation factor 9** (**Gdf-9**) from oocytes. The importance of Spf/Kit activation of the oocyte for its production of Bmp15 and Gdf-9 was described in Box 9.1. These GFs then play a major role in the development of follicular cells, as is shown by their knock-out. Thus, Gdf-9 null oocytes continue to grow and resume meiosis but granulosa cells do not proliferate or recruit thecal cells, and the oocyte eventually dies, the follicle failing at the preantral stage. The oocytic expression of Gdf-9 continues during the antral phase, driving the changing pattern of mural granulosa cell activity. Thus, the granulosa cells lining the follicle wall become increasingly different from the cumulus granulosa cells surrounding the oocyte. Whereas cumulus cells produce hyaluronic acid to provide the matrix that carries the egg out of the follicle at ovulation, mural cells express higher levels of LH receptor the further from the oocyte they are located. This granulosa heterogeneity is regulated by the oocyte through locally acting cytokines such as Gdf-9 that antagonize FSH stimulation locally. The depression of the LH receptor has an important role in the prevention of premature luteinization of the follicle by delaying progesterone synthesis in granulosa cells until the oocyte has left the follicle.

Likewise, Bmp15 null oocytes lead to preantral follicular arrest in sheep, but not in mice in which Bmp15 expression does not rise markedly until the late antral phase. However, even in mice, Bmp15 is thought to exert some co-stimulatory effects with Gdf-9 on granulosa cell proliferation in preantral follicles. In addition, if gap junctional contact is blocked via connexins knock-out in either oocyte or granulosa cells, then oocyte growth and differentiation cease, *zona* deposition is incomplete, and follicular development is arrested at the preantral phase, a finding that indicates how important gap junctional exchange of smaller molecules is for follicle development.

Remarkably, if oocytes from expanding antral follicles are transplanted surgically into primordial follicles, the time for the primordial follicles to mature is halved, leading to the suggestion that the oocyte orchestrates the rate of development. This intimate and interactive partnership between oocyte and follicular cells resembles that seen for spermatogenic cells and Sertoli cells in Chapter 7. In both cases, the germ cells set the timing, but the somatic cells provide the context.

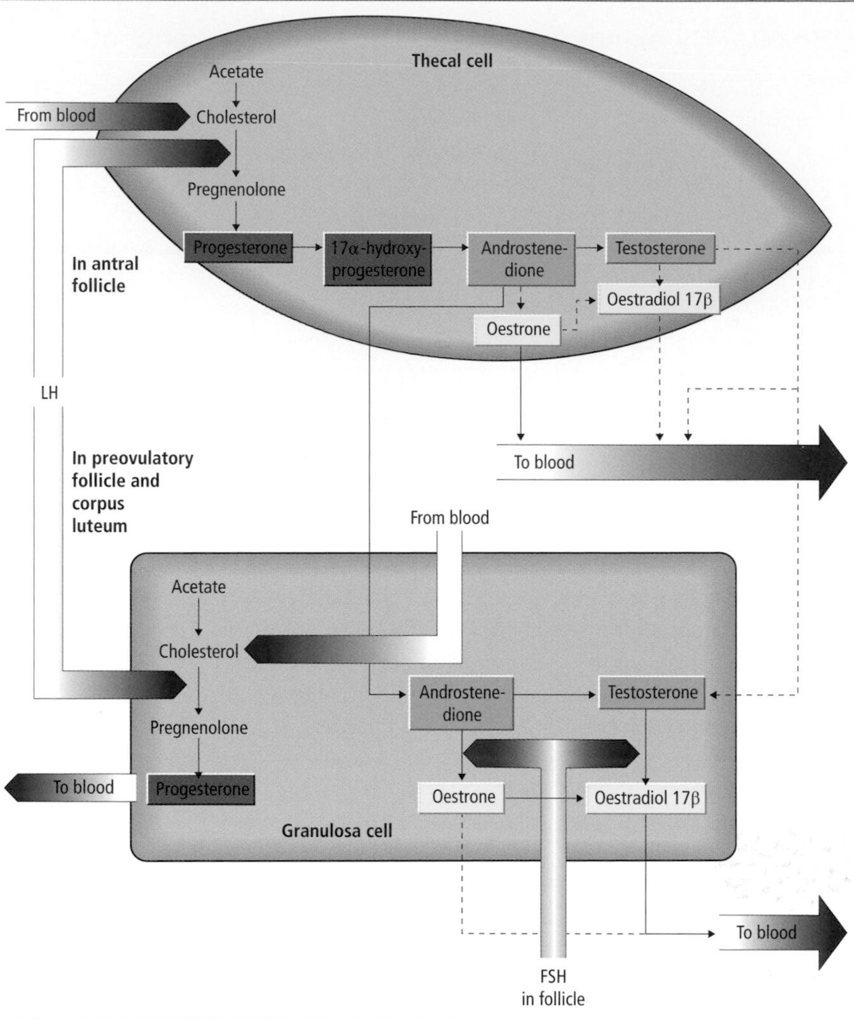

Figure 9.4 Scheme outlining the principal steroidogenic pathways and gonadotrophic stimuli in cells of the antral follicle – theca above and granulosa below – as well as the LH-stimulated pathway in the preovulatory follicle and corpus luteum. In the corpora lutea, the oestrogenic synthetic capacity of thecal cells only persists in those species in which the thecal cells become incorporated into the corpus luteum. Where alternative pathways exist, the minor pathway is indicated by a dashed line.

These various sex steroids are produced in different parts of the follicle. The antral follicle can be dissected surgically, so that its granulosa cells become separated from the cells of the theca interna. When grown separately *in vitro*, the **thecal cells are found to synthesize androgens** from acetate and cholesterol. This conversion is greatly **stimulated by LH** (Figure 9.4; see Figure 2.8 for overall biosynthetic pathway for steroids). Only very limited oestrogen synthesis by thecal cells is possible, particularly in the early stages of antral growth. The granulosa cells, in contrast, are incapable of forming androgens. If, however, the granulosa cells are supplied with exogenous androgens, they possess enzymes that will readily convert them to oestrogens. This **granulosa-mediated aromatization of androgens to oestrogens is stimulated by FSH** (Figure 9.4).

Thus, in the developing follicles the **androgens are derived exclusively from thecal cells**, whereas the **oestrogens can arise via two routes**. One involves **cell cooperation** in which **thecal androgens are aromatized by the granulosa cells**. The other route is by ***de novo* synthesis from acetate in thecal cells**. The balance between these two potential sources of oestrogen varies with different species, but it seems probable that all of the oestrogen within the follicular fluid and most of the oestrogen released from the follicle to the blood results from cell cooperation.

The production of steroids and the increase in size of antral follicles are intimately interlinked. It has become clear recently that the steroids, in addition to their release systemically via secretion into the blood, also have important **local intrafollicular roles**. The androgens play a key role here. First, as we saw above, **androgens serve as substrates for conversion to oestrogens**. Second, acting with FSH, **androgens stimulate the proliferation of granulosa cells and thereby follicular growth**. Third, **androgens stimulate aromatase activity**, thereby promoting oestrogen synthesis. Thus, the rising thecal output of androgens fuels a massive increase in oestrogen biosynthetic potential. This is further enhanced by the ability of the **oestrogens themselves to stimulate granulosa cell proliferation**. Thus, a powerful **positive feedback** system is operating that results in **a surge of circulating oestrogens** from the most advanced follicle(s) towards the end of antral

Table 9.2 Human follicular development

Morphology	Day of menstrual cycle	Diameter (mm)*	FSH/LH receptors present?	Oestrogen in peripheral blood (pmol/l)
Preantral	Throughout	<0.5	–	NA
Very early antral	Throughout	<2	–	NA
Early antral	1–6	2–7	+	<20
Expanding antral[†]	6–10	7–10	+	100–200
Expanded antral	10–12	10–20	+	200–400
Preovulatory	13–14	20–25	+	800+[§]

* Using ultrasound technology, it is possible to monitor these final stages of follicular growth in the conscious subject, and thereby to ascertain how near the follicles are to ovulation.
[†] In naturally cycling women a single dominant follicle emerges at this point and only it grows thereafter.
[§] 10^3–10^4 higher oestrogen concentrations within the follicular fluid itself.
NA, not applicable.

expansion. Daily monitoring of the rise in urinary oestrogen levels therefore provides a good guide to the state of maturity of the most mature follicles (Table 9.2). Not surprisingly, in females carrying mutations that prevent oestrogen synthesis or binding to receptors, follicular development arrests in the antral stages.

Thus, androgens and oestrogens are particularly important for follicular growth and maturation. Progesterone also assumes major significance, but only as ovulation approaches, and its role will be considered shortly.

FSH and LH also stimulate and interact with intrafollicular cytokines

Steroids are not the only paracrine actors within the antral follicle. The production and activity of several cytokines are also stimulated by the gonadotrophins during the antral phase, and then mediate and/or modulate the actions of steroids and gonadotrophins. The full details of the different cytokines, when and where they act and interact, and what controls them, are still being unravelled. However, it already seems clear that there is a **balance** between cytokines (such as IGFs, inhibins, Scf) that in general **support follicular progression**, acting mostly in cooperation with FSH and androgens, and cytokines that **depress or restrain follicular development or promote atresia** (such as Amh, TNFα, leptin and IGFBPs). Some of the putative actions of these cytokines are summarized in Boxes 9.1–9.3.

We should note in particular the secretion patterns of the two inhibins, the significance of which will become clear in Chapter 10. Both are produced by granulosa cells. However, whereas inhibin B production is stimulated by FSH, inhibin A is stimulated by *both* FSH and LH. Thus, the ratio of inhibin A:B rises as the follicle expands to peak at ovulation. The ratio thus acts as a marker for follicular expansion, in addition to the rising oestrogen output noted above (see Chapter 10 for significance).

Summary

Follicular growth and maturation are partly regulated endogenously and partly by exogenous gonadotrophins. The latter pick up preantral follicles and stimulate their antral development through actions involving both cytokines and sex steroids. However, some of the latter (notably sex steroids and inhibins) act not only paracrinologically within the follicle, but also have more widespread endocrinological actions, as we will see in Chapter 10. Before moving to endocrine actions, one further and critical paracrine action of oestrogens, in conjunction with FSH, is exerted within the expanded antral follicle. Together **FSH and oestrogen stimulate the appearance of LH-binding sites on the outer layers of granulosa cells**, which hitherto lacked them. These LH-binding sites are critical for the successful entry of the expanded antral follicle into the **preovulatory phase of follicular growth**.

Ovulation

Just as late preantral and early antral follicles trickling through the first, hormone-independent, phase of follicular growth will become atretic unless exposed to tonic levels of FSH and LH, so the expanding **antral follicles will also die unless a brief surge of high levels of LH coincides with the appearance of LH receptors on the outer granulosa cells**. If an LH surge occurs when both the granulosa and thecal cells can bind LH, then entry into the preovulatory phase of growth occurs. If it does not, the expanded antral follicle dies. How a subset of follicles is selected for ovulation is described in Box 9.3. The surge of LH affects these advanced follicles in two ways. First, it causes major changes in both the oocyte and the follicle cells, which result in the oocyte's expulsion from the follicle at **ovulation**. Second, it changes the whole endocrinology of the follicle, which becomes a **corpus luteum** at ovulation. Thus, although this

Box 9.3 Cytokines and ovulation: how is a follicle selected to be ovulated?

Of the 15–20 follicles recruited in the human, usually **only one emerges as dominant** during the later stages of follicular growth. It is only this one dominant follicle that subsequently undergoes preovulatory growth, develops LH receptors on its granulosa cells and is competent to ovulate in response to a surge in LH. This follicle is also the dominant source of oestrogen as ovulation approaches, as can be shown by selectively removing it and observing an abrupt decline in the surge levels of oestrogen. We do not know why one follicle emerges, but it is possible that cytokines are involved. FSH stimulates the production of **insulin-like growth factor 2** (**Igf2** in human, cow, sheep; Igf1 in pig and rat). Igf appears to mediate both the stimulation by LH of androgen output by the thecal cells and its FSH-dependent aromatization to oestrogen by granulosa cells. If Igf activity is neutralized, the effectiveness of FSH in stimulating oestrogen output is reduced. FSH is also involved in the regulation of ovarian production of the endogenous **Igf-binding proteins** (**IgfBPs**). These are also produced by the follicle, but FSH suppresses their synthesis. It also promotes production by granulosa cells of a protease, called **pregnancy-associated plasma protein A** (**Papp-A**), which cleaves the Igf from its IgfBP, thereby releasing it for action. Papp-A expression is markedly elevated in the dominant follicle. Overall, therefore, FSH stimulates the availability of Igf to promote follicular development. It is possible therefore that among growing follicles, IgfBPs are produced preferentially by those follicles with the least good supply of FSH and/or the fewest FSH receptors. The capture of Igf by its binding protein, especially in the absence of its cleaving protease, would further impair the effectiveness of FSH and so promote a downward spiral of the follicle to atresia. In contrast, those follicles best able to respond to FSH would spiral upwards to expansion and ovulation. Interestingly, it has been shown that **Bmp15**, produced in the oocyte, **inhibits Papp-A production by FSH**, suggesting that the oocyte itself has an input into which follicle becomes dominant and indeed how many dominant oocytes there are.

Once a dominant follicle has been selected, inhibin may also be involved in its maintenance. Inhibin A is produced in the late antral phase in dominant follicles under the influence of LH. Inhibin A then stimulates both thecal androgen production and its granulosa aromatization and so sets up a positive feedback loop leading to the oestrogen surge from the dominant follicle. In contrast, retarded follicles show low levels of inhibin but higher levels of activin, which attenuate androgen output by the theca and so suppress both inhibin synthesis and oestrogen output. Thus, dominant follicles are characterized by high ratios of both inhibin:activin and Igf:IgfBP. Retarded follicles show the reverse.

preovulatory phase of follicular growth is the shortest (see Table 9.1), it is also the most dramatic.

The oocyte undergoes major preovulatory changes

Within 3–12 hours of the beginning of a surge of LH (depending on the species), dramatic changes occur in the oocyte. The nuclear membrane surrounding the dictyate chromosomes breaks down (GVBD – germinal vesicle breakdown), and the **arrested meiotic prophase is at last ended**, many years after its initiation. The chromosomes progress through the remainder of the first meiotic division (see Figure 1.1 for details of meiosis), culminating in an extraordinary cell division in which **half the chromosomes, but almost all the cytoplasm including all the mitochondria, go to one cell: the secondary oocyte** (Figure 9.5). The remaining chromosomes are discarded in a small bag of cytoplasm called the **first polar body**, which dies subsequently. This **unequal division of cytoplasm conserves for the oocyte the bulk of the materials synthesized during earlier growth phases**.

The chromosomes in the secondary oocyte immediately **enter the second meiotic division** and come to lie on the second metaphase spindle. Then, suddenly, meiosis arrests yet again. The oocyte is ovulated in **this arrested metaphase state**. We do not know the biological significance of this second meiotic arrest, but we do know that it is caused by the presence of a protein complex called **cytostatic factor**. How it does so is considered in more detail in the context of fertilization (see pages 205–6).

The termination of the dictyate stage and the progress of **meiotic maturation through to second metaphase arrest and ovulation** are accompanied by the terminal **cytoplasmic maturation** of the oocyte (Figure 9.5). Both are stimulated by the surge of LH, yet it is clear that LH cannot, and does not, bind to the oocyte itself, only to the thecal and mural granulosa cells Therefore, its effect must be mediated via second messengers acting via the cumulus cells of the follicle, where it is thought to effect changes by loss of the intimate contact between the oocyte and the cells of the cumulus, due to withdrawal of the cytoplasmic processes, an event long considered to be causal in the reactivation of meiosis. Thus, simply removing the oocyte from the antral follicle at any stage up to this one, reactivates meiosis, and does so by depressing the cAMP levels in the oocyte. It was thought that cAMP was provided to the egg via the trans-zonal processes, but recent evidence from the mouse has shown that the egg can make large quantities of cAMP. The high levels of cAMP control the arrest of meiosis by regulating a protein complex called M-phase promoting factor (MPF), formed from the kinase CDK1 and cyclin B. High

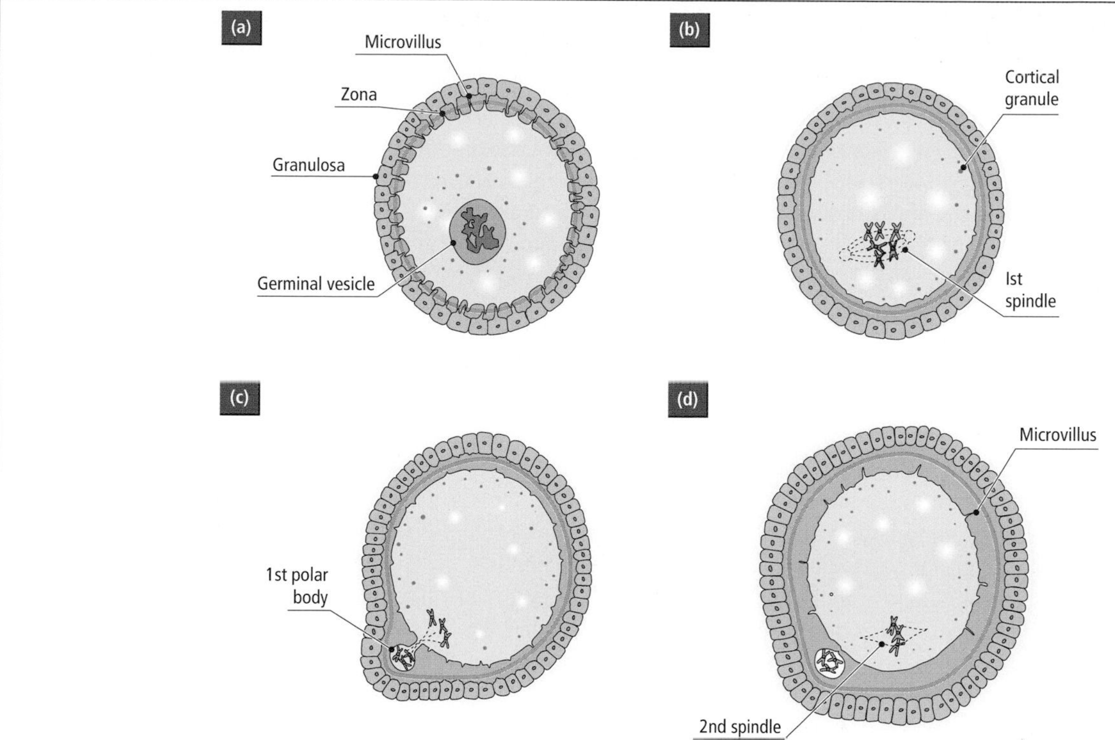

Figure 9.5 Reactivation of arrested prophase of the first meiotic division in the preovulatory oocyte. A few hours after LH stimulation (a to b), the germinal vesicle breaks down, and the chromosomes complete prophase and arrange themselves on the first meiotic spindle. Meanwhile, cytoplasmic contact between oocyte and granulosa cells ceases through withdrawal of microvilli and the appearance of a distinct sub-zonal space. Cortical granules made by the Golgi apparatus migrate to the surface. Subsequently, the first meiotic division is completed with expulsion of the first polar body (b to c). The oocyte chromosomes immediately enter the second meiotic division but arrest at second metaphase. The cytoplasm around the eccentrically placed spindle is devoid of cortical granules and the overlying membrane lacks microvilli in some species (c to d). The oocyte is ovulated in this arrested state (except in dogs and foxes in which the oocyte is ovulated at metaphase 1, and the first polar body is extruded after ovulation).

intra-oocyte concentrations of cAMP activate protein kinase A (PKA), which in turn phosphorylates and inactivates a phosphatase (Cdc25) and phosphorylates and activates two kinases (Wee1B and Myt1). As a result, tyrosine and threonine residues on CDK1 are phosphorylated, MPF activity is repressed and GVBD is inhibited. A decrease in intra-oocyte cAMP levels reverses this sequence, ultimately allowing activation of MPF and, as a direct consequence of this, meiotic resumption. So why do cAMP levels fall in response to the LH surge? It appears that cGMP, which is largely synthesized in the cumulus cells and supplied to the egg via the trans-zonal processes, is implicated. Thus, cGMP levels in the egg fall very early after the LH surge, due to the closure of tight junctional communication with the oocyte, and, since cGMP is known to inhibit an oocyte-specific phosphodiesterase that destroys cAMP, its fall leads to a secondary fall in cAMP (see Box 9.4).

Additional cytoplasmic changes that occur involve the Golgi apparatus of the oocyte, which synthesizes lysosomal-like granules that migrate towards the surface of the oocyte to assume a subcortical position **as cortical granules**. Remarkably, **centrioles are lost** at pachytene, and thereafter only **pericentriolar material** (**PCM**) organizes microtubules (contrast spermiogenesis, where PCM is lost, and reduced centrioles remain; see page 115). Protein synthetic activity continues at the same rate, but new and distinctive proteins are synthesized. This activity prepares the oocyte for fertilization (see Chapter 12). If an oocyte is shed from its follicle prematurely, or removed surgically before the completion of these cytoplasmic maturational events, then its fertilizability is much reduced. For this reason, in clinical *in vitro* fertilization programmes, human oocytes aspirated from preovulatory follicles may be cultured for a few hours before addition of spermatozoa. This procedure improves the chances of a complete maturation by the oocyte.

Preovulatory growth of follicle cells is associated with a switch to progesterone output

In addition to acting on follicle cells in order to generate signals to the oocyte, LH also affects directly the growth and endocrinological activity of the follicle cells themselves. A final and considerable **increase in follicular size occurs** (reaching 25 mm or more in diameter in humans; Table 9.2), almost exclusively due

Box 9.4 cGMP and the oocyte

Recently, cGMP has emerged as the critical player linking the LH surge to the events leading up to ovulation in the mouse, cow and pig. Thus, cumulus cells express preferentially the mRNA encoding the membrane-bound guanyl cyclase natriuretic peptide receptor 2 (*Npr2*), while mural granulosa cells express the mRNA encoding its ligand, natriuretic peptide precursor type C (*Nppc*) more intensely. Nppc is produced primarily by granulosa cells, then diffuses through the follicle, binds to its receptor and stimulates cumulus cells to produce cGMP, which then is transferred via gap junctions to the oocyte. Moreover, in the mouse, gonadotrophin stimulation of follicular development induces expression of Nppc by granulosa cell and of Npr2 by cumulus cells. However, for maintained expression of both Npr2 and Nppc, estradiol must be present, co-operating with FSH. The oocyte also plays an active role via the actions of its GFs, Gdf-9 and Bmp15, which also stimulate expression of Npr2 in cumulus cells.

The midcycle LH surge then neutralizes within 1 hour the inhibitory effects of cGMP by decreasing cGMP levels throughout the follicle through the suppression of Nppc synthesis and the inhibition of Npr2 activity in cumulus cells, as well as by suppressing gap junction communication by the activation of a mitogen-activated protein kinase (MAPK)-dependent pathway that leads to phosphorylation of connexin-43. LH's stimulation of GVBD is also influenced by two members of the EGF family, namely, amphiregulin and epiregulin. Thus, following LH stimulation of rat follicles, both are expressed at the mRNA level in the eggs before GVBD, and subsequently as membrane-bound precursors, from which the active forms are cleaved proteolytically. Intact follicles exposed to either show GVBD, and mice with an inactivating mutation of their receptor (EgfR) do not show GVBD after the LH surge, whilst inhibition of the phosphorylation-mediated activation of EgfR, that occurs normally in mural granulosa cells after LH treatment, blocks meiotic resumption. Finally, the ability of either GF to promote GVBD is accompanied by a large decrease in cGMP in follicle cells. Each of these observations suggests that the two GFs play a key role in the stimulation of GVBD, but precisely how in relation to the decrease in cGMP levels is unclear.

Although only shown in detail for rats and mice, there is evidence that the same mechanism applies more widely among species, including humans, in which ovulatory doses of hCG cause a reduction in levels of Nppc in follicular fluid. Likewise, amphiregulin levels in human follicular fluid, plus the mRNA encoding it in mural granulosa and cumulus cells are only detectable after an LH surge or the administration of hCG, as also is the expression in cumulus cells of mRNA encoding its receptor, suggesting that the latter cells are a target of amphiregulin. Moreover, low levels of this GF are associated with the recovery of immature oocytes, in which GVBD has not occurred.

to a rapid expansion of the volume of follicular fluid. This expansion is accompanied by a loosening of the intercellular matrix between the more cortical layers of granulosa cells and an increase in total blood flow to the follicle.

The preovulatory growth in follicular size is accompanied by changes in the pattern of steroid secretion. Within 2 hours or so of the beginning of the LH surge, there is a **transient rise in the output of follicular oestrogens and androgens, followed by a decline.** This rise coincides with distinctive changes in the thecal layer, which appears transiently stimulated and hyperaemic. The outer cells of the granulosa layer also show a marked change in their properties. First, they cease dividing and no longer convert androgen to oestrogen, but instead synthesize progesterone. **LH stimulates this granulosa synthesis of progesterone via the newly acquired LH receptors.** Second, the **granulosa cells lose their capacity to bind oestrogen and FSH, but gain the capacity to bind progestins.** This acquisition of the capacity to respond to LH by synthesizing progesterone (and then to be self-stimulated by it – **positive feedback**) results in an **exponential release of progesterone from the follicle**, which becomes significant in the human several hours before ovulation, although in most species only just before or immediately after ovulation.

This rising progestin output has three important functional consequences. First, it **depresses growth in the less mature developing follicles**. Second, **it is essential for ovulation** itself. Thus, progesterone inhibitors such as mifepristone (RU486) suppress ovulation, and females genetically null for progesterone receptors do not ovulate. Third, it promotes **the transition to the progestagenic phase of the ovarian cycle**.

The process of ovulation involves protease activity and prostaglandins

By the end of the preovulatory phase of follicular growth, the rapid expansion of follicular fluid has resulted in a relatively thin peripheral rim of mural granulosa cells, basement membrane and thecal cells, to which the oocyte, with its associated expanding cumulus cells, is attached only by a tenuous and thinning stalk of granulosa cells. The increasing size of the follicle and its position in the ovarian cortex cause it to bulge from the ovarian surface. At one point, the **stigma**, this bulging wall becomes even thinner and avascular, the connective tissue breaks down and the follicle ruptures. **The follicular fluid flows out carrying with it the oocyte and**

its surrounding mass of cumulus cells. The biochemistry of ovulation involves proteolytic enzymes, notably members of a large family of **matrix metalloproteinases** (MMPs) and their **natural tissue inhibitors** (TIMPs), and also serine proteases such as **plasmin** and **plasminogen activator**. Under LH influence, directly and/or indirectly via progesterone and/or prostaglandins, **collagenase** and **gelatinase** activities (both MMPs) increase, especially in the stigma, as do those of plasminogen activator (which cleaves procollagenase to generate active **collagenase**), while those of its inhibitor reach a nadir. The experimental inhibition of collagenase prevents ovulation. Proteases are also much involved in other follicular remodelling activities during conversion to a corpus luteum (see later).

Prostaglandin E2 (PGE2; Figure 9.7 on page 154) is a key paracrine mediator of the LH surge's effects on the process of ovulation, there being a rise in intrafollicular PGE2 in response to it. Thus, mice that lack expression of prostaglandin synthesis enzymes or in particular the PGE2 receptor, experience ovulatory failure. Moreover prostaglandin synthesis inhibitors prevent ovulation in rabbits, cows and primates, unless PGE2 was co-injected into the dominant follicle with the inhibitor, in which case, ovulation was restored. LH acts to stimulate PGE2 synthesis via an elevation of phospholipase A2 [G4A] and cyclo-oxygenase 2, the activities of both of which rise in mural granulosa cells following the LH surge, as do the levels of a transporter of PGE2 across the cell membrane, thereby facilitating its exit from the granulosa cells. However, a rise in the intracellular enzyme that destroys PGE2 early after the LH surge that then falls as ovulation approaches, suggests that a fine balance between PGE2 synthesis/export and its catabolism is crucial to its role in promoting the ovulatory events. PGE2 is thought to act on at least two sites: on the cumulus cells to enhance their expansion and on thecal cells to promote the formation and/or rupture of the stigma and the subsequent ingrowth of blood vessels, PGE2 achieves this by enhancing a local inflammatory reaction, thereby activating thecal fibroblasts, releasing proteolytic enzymes that promote angiogenesis and hyperemia, and activating collagenases and other proteolytic enzymes to degrade the follicle's connective tissue and cause ovulation.

In many species, including humans, the ovarian surface is directly exposed to the peritoneal cavity, but in some (e.g. the sheep, horse and rat) a peritoneal capsule or bursa encloses the ovary to varying degrees and acts to retain the oocyte cumulus mass(es) close to the ovary. There they are collected by cilia on **the fimbria of the oviduct**, which sweep the cumulus mass into the **oviducal ostium** (see Figure 9.1 and Chapters 11 and 12). The residual parts of the follicle within the ovary collapse into the space left by the fluid, the oocyte and the cumulus cells, and within this cavity a clot forms. Thus, the postovulatory follicle is composed of a fibrin core, surrounded by several collapsed layers of granulosa cells, enclosed within a fibrous outer thecal capsule. This structure becomes the **corpus luteum**.

The corpus luteum is the postovulatory 'follicle'

The fate of the oocyte will be described later in Chapters 11 and 12. For now we continue to focus on the remains of the postovulatory follicle.

The corpus luteum produces progestagens

The collapsed follicle is now transformed into a corpus luteum (Figure 9.2c). Within the follicular antrum, the fibrin core undergoes fibrosis over a period of several days; the membrana propria between the granulosa and thecal layers breaks down and blood vessels invade. Both granulosa cells and cells from the theca interna contribute to the corpus luteum, although many thecal cells also disperse to the stromal tissue. The granulosa cells have now ceased dividing and hypertrophy to form **large lutein cells**, rich in mitochondria, smooth endoplasmic reticulum, lipid droplets, Golgi bodies and, in many species, a carotenoid pigment, lutein, which may give the corpora lutea a yellowish or orange tinge. This transformation is referred to as **luteinization** and is associated with a steadily increasing **secretion of progestagens**. The thecal cells form **smaller lutein cells**, produce progesterone and androgens, and seem to be richer in LH receptors. After LH stimulation, these smaller cells can serve as a stem cell population for the more endocrinologically active, large lutein cells in some species.

In most species, the main progestagen secreted from the large lutein cells is progesterone, but secretion of significant quantities of 17α-hydroxyprogesterone in primates and of 20α-hydroxyprogesterone in the rat and hamster also occurs. In a few species, notably the great apes and humans, and to a lesser extent the pig, the corpus luteum also secretes oestrogens, particularly oestradiol 17β. Its source also seems to be the large luteal cells, using as substrate androgens derived from the small luteal cells (perhaps a hangover from granulosa function in preovulatory follicles?). In most species (e.g. the monkey, sheep, cow, rabbit, rat and horse), however, the corpus luteum secretes only trivial amounts of oestrogen.

The corpus luteum also secretes two other hormones. Inhibin A is secreted in large amounts in higher primates (see Figure 8.2). It acts to promote production of progesterone (see also pages 164–5). The second hormone is oxytocin (see Figure 2.12), which comes from the large lutein cells and the importance of which will become evident soon (see below).

A luteotrophic complex supports the corpus luteum

The endocrine support of the corpus luteum, like its cellular composition and secretory pattern, shows considerable variation among species. The conversion of a follicle to a corpus luteum requires that high surge levels of LH both provoke ovulation and initiate luteal conversion. This gonadotrophin is

Table 9.3 Luteotrophic/antiluteolytic hormones for the non-pregnant corpus luteum of different species

Species	LH*	Prolactin	Oestrogen	FSH†	Progesterone
Human	++	+?	Luteolytic?	–	+++
Cow	++	+?	Luteolytic?	–	+
Sheep	++	+	–	–	?+
Pig	+	+?	+	–	+
Rabbit	+	?	+++	+	First 3 days only
Rat/mouse	+	+++	–	–	–
Dog	+	+	?	?	±
Hamster	+	+	?	+	?

* LH is essential to stimulate the initiation of luteinization in all species. Here its role subsequently is assessed. It may be directly luteotrophic, but in species in which follicles are growing through the luteal phase, the LH may also stimulate some oestrogen synthesis locally.
† FSH may only be luteotrophic indirectly by stimulating oestrogen.

then also required, albeit at lower levels, for the maintenance of the corpus luteum. However, in some species, prolactin and/or progesterone also form an important component of the so-called **luteotrophic complex**. In those species (Table 9.3), appropriate receptors can be detected on granulosa cells from the preovulatory stage onwards.

Luteolysis: death of the corpus luteum may be active or passive depending on the species

The life of the corpus luteum in the non-pregnant female varies among species from 2 to 14 days (Table 9.3; see Chapter 13). Luteal regression or **luteolysis** involves a collapse of the lutein cells, ischaemia and progressive cell death with a consequent fall in the output of progestagens. The whitish scar tissue remaining, the **corpus albicans**, is absorbed into the stromal tissue of the ovary over a period that varies from weeks to months, depending on the species. Luteolysis can be caused by withdrawal or inadequacy of the luteotrophic complex. However, in many species, it is not primarily a **failure of luteotrophic support**, but active production of a **luteolytic factor** that brings about normal luteal regression. Thus, in these species, both luteotrophic support and antiluteolytic suppression are required to sustain the corpus luteum.

Most mammals studied, **with the notable exception of higher primates**, come into this latter category, and the source of the luteolysin has been identified as **the uterus**. Thus, luteal life can be prolonged by hysterectomy (removal of the uterus), but if the endometrium of the excised uterus is homogenized and injected, then luteolysis occurs (Figure 9.6e). Luteal prolongation can also be achieved by ligating the tissues, including the blood vessels, between the uterus and the ovary (Figure 9.6d). The results of these classical experiments led to the suggestion that a humoral factor passes from the endometrium to the ovary, and causes luteolysis. Two observations on the ewe suggested that this humoral factor was highly labile. First, if instead of removing the whole uterus, which in the ewe has two separate horns, only one horn was removed, then only the corpus luteum on the opposite side regressed (Figure 9.6b,c). Second, if the whole uterus was transplanted elsewhere in the body, luteolysis was prevented unless the ovaries had also been transplanted with the uterus as a unit.

The identity of the endometrial substance responsible has now been established as **prostaglandin $F_{2\alpha}$ ($PGF_{2\alpha}$**; Figure 9.7). $PGF_2\alpha$ is secreted in a series of pulses at 6–12-hour intervals from epithelial and glandular cells in the mid-to-late luteal phase. It passes from the endometrium into the uterine vein. From the vein, it passes by a local countercurrent transfer to the ovarian artery, and thence to the ipsilateral ovary. Corpus luteum regression follows shortly thereafter. Antibody neutralization of $PGF_2\alpha$, or inhibition of its synthesis, prevents luteolysis. Conversely, injections of exogenous $PGF_2\alpha$ 1 or 2 days in advance of its endogenous output lead to premature luteolysis; indeed, $PGF_2\alpha$ will cause regression of luteinized cells *in vitro* (see also Box 9.5).

The control of **luteolysis in higher primates, including humans**, unlike that for other species, does not involve uterine prostaglandins. The levels of prostaglandins secreted do rise in the late luteal phase, but neither hysterectomy nor antibodies to prostaglandins prolong luteal life. Moreover, injections of prostaglandins are without effect on the corpus luteum, unless very high doses are used, when a transient drop in progesterone output occurs. What then causes luteal decline in humans? It seems that it is **loss of adequate luteotrophic support** rather than the genesis of an active luteolytic agent that occurs in higher primates. Thus, the more extended preovulatory LH surge seen in higher primates followed by the relatively low sustained levels of LH during the luteal phase of primates are sufficiently

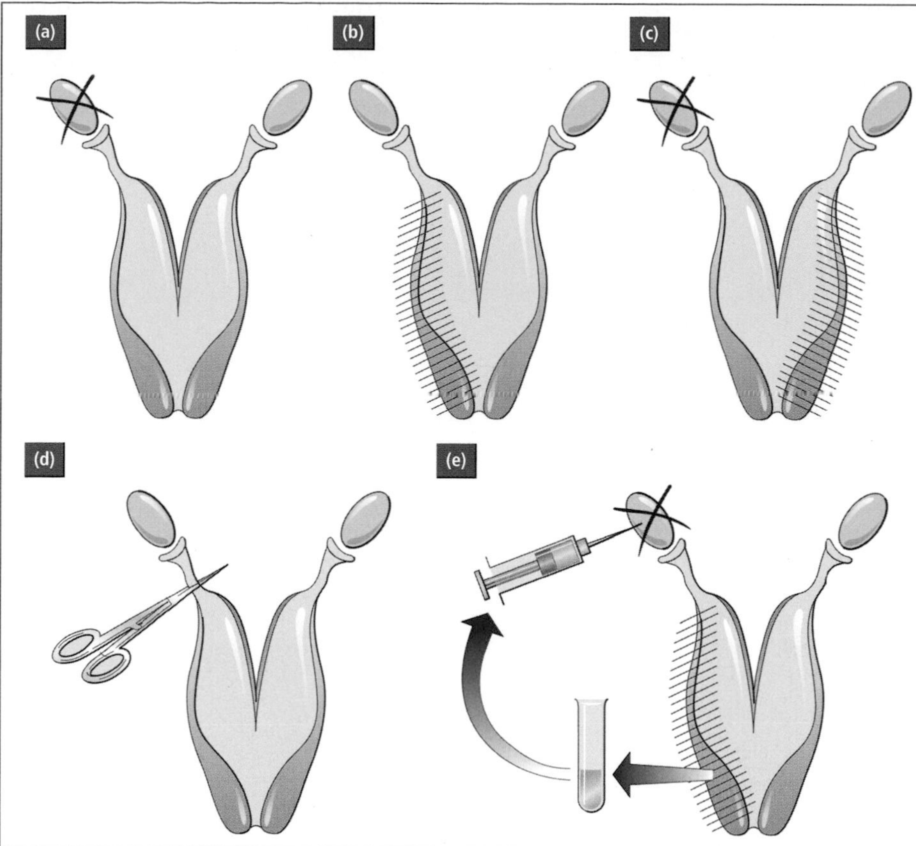

Figure 9.6 Non-pregnant sheep uterus and ovaries. (a) A single corpus luteum present in the left ovary regresses, indicated by a crossed through ovary. (b) Removal of the ipsilateral uterine horn (hatched) prevents regression. (c) Removal of the contralateral horn does not prevent regression. (d) Clamping the blood supply between the horn and ovary prevents regression. (e) If the endometrium of the removed ipsilateral horn is homogenized and re-injected into the ovarian artery, the corpus luteum regresses (compare b with e).

luteotrophic for a normal length luteal phase in the non-pregnant cycle.

We return to luteolysis in Chapter 13 when we consider the transition from the non-pregnant cycle to pregnancy.

Conclusions

Follicle maturation is complex and its control not fully understood. Moreover, there are species differences, making simple extrapolation from animal models to humans difficult. For example, the number of follicles ovulating in any one cycle is characteristic for each species and ranges from one to several hundred. However, there appear to be two points during follicular maturation at which this number can be regulated, and in both cases it is the **balance between survival and atresia** that is critical.

Follicular growth and differentiation from a primordial follicle through to a functional corpus luteum is a complex process that is only completed successfully by less than 0.1% of follicles. Most become atretic at some point during the process. The resumption of development by primordial follicles is hormone independent. During this preantral phase the oocyte undergoes its major growth, and the follicle acquires receptors for FSH and oestrogen (granulosa cells) and for LH (thecal cells).

In humans, about 7–10 early antral follicles in each ovary are rescued from atresia and recruited for development at the start of each menstrual cycle. FSH is the crucial hormone for follicular rescue, although of course LH is necessary for fully functional follicles. Thus, the number of follicles recruited can be increased if endogenous FSH levels are augmented by exogenous FSH or can be reduced if FSH levels are sufficiently diminished. This observation must mean that a delicate interplay between the levels of circulating FSH and the follicular levels of FSH receptors determines the number of follicles recruited from the early antral pool. If appropriate hormone levels and acquisition of sufficient receptors coincide, then follicular development continues.

Then, from amongst these 14–20 antral follicles, there emerges (usually) a single dominant follicle to enter and complete the pre-ovulatory phase through to ovulation. Again, it is the emergence of LH receptors within the granulosa cells of this follicle that must coincide with the LH surge for ovulation to proceed to completion. If appropriate hormone levels and acquisition of sufficient receptors do not coincide, then ovulation fails. **Thus, if hormone levels are inappropriate when the receptors develop, then follicular atresia ensues**. In the next chapter, we turn our attention to how the appropriate levels of LH and FSH are achieved.

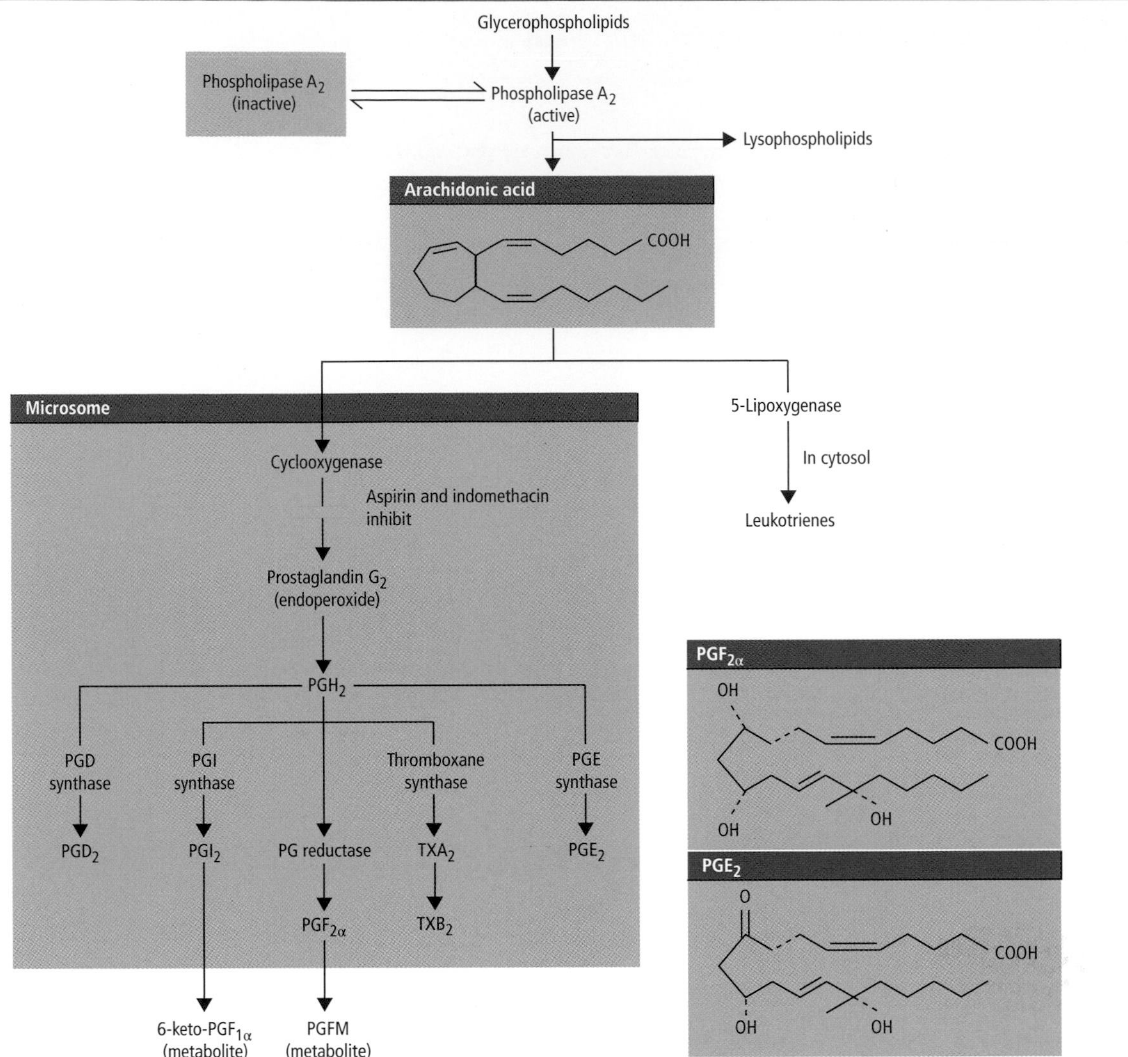

Figure 9.7 Biosynthetic pathway of **prostaglandins** (PGD_2, $PGF_{2\alpha}$ and PGE_2) and **leukotrienes** – both **eicosanoids:** lipidic reproductive hormones. PGs are synthesized in most tissues of the body, and have half-lives of the order of 3–10 minutes, and so act mainly as local hormones, either paracrinologically or after a short passage through the local bloodstream. They are inactivated totally by a single passage through the systemic circulation, and particularly through the lungs. The type of prostaglandin synthesized varies in different tissues at different times, as a result of variation in the relative activities of the downstream enzymes. The rate-limiting step in PG synthesis is the availability of the essential polyunsaturated fatty acid, arachidonic acid, which is derived from glycerophospholipids in the cell membranes by the actions of two enzymes: phospholipase A_2 (PLA_2, acting primarily on phosphatidylcholine and phosphatidyl-ethanolamine) and phospholipase C (PLC, acting on phosphatidyl-inositol). PLA_2 is present in an inactive membrane-bound form in lysosomes, and its release and activation depend primarily on reduction of the stability of the lysosomal membranes, which is affected by steroids, and which are therefore critical determinants of the rate of PG synthesis. Arachidonic acid is then converted to the leukotrienes by the action of cytosolic 5-lipoxygenase, and to the prostaglandins by the microsomal cyclooxygenases, which come in two isoforms: cox1 is constitutive and cox2 is inducible.

Leukotrienes stimulate myometrial contractility; PGE_2 stimulates myometrial activity, cervical ripening and autocrine output of PGE_2 from the amnion; $PGF_{2\alpha}$ stimulates myometrial activity and cervical ripening; PGI_2 may inhibit myometrial activity and is essential for decidualization. Each PG can act either through intracellular nuclear peroxisome proliferator-activated receptors (PPARs) or through G-protein–coupled cell surface receptors specific for each PG (called EP1–4 for PGE_2, FP for $PGF_{2\alpha}$, DP for PGD_2, IP for PGI_2 and TP for thromboxanes), PGI_2 = prostacyclin; TXA_2 and TXB_2 = thromboxane A_2 and B_2. The lower right box shows the PG structure.

Box 9.5 Molecular mechanisms underlying luteolysis by $PGF_{2\alpha}$

Luteal oxytocin regulates the production of $PGF_{2\alpha}$ via a receptor-regulatory mechanism

Why does the endometrium of most species release $PGF_{2\alpha}$ during the latter part of the luteal phase? It turns out that the timing mechanism is complex and hinges on oxytocin. Oxytocin secreted from the corpus luteum (and possibly also the pituitary), can be shown to be an essential player, as its neutralization delays luteolysis. Moreover, $PGF_{2\alpha}$ itself stimulates oxytocin release from the corpus luteum, thereby effecting a **positive feedback loop** – in this case leading to luteal destruction, a neat symmetry with the positive feedback loop involving progesterone that drives luteal formation after ovulation!

However, we still have a problem, because **oxytocin is released from the corpus luteum well before the later part of the luteal phase, so why does it not stimulate $PGF_{2\alpha}$ secretion earlier?** The answer seems to be that **oxytocin receptors in the endometrium are absent until late in the luteal phase**, and so the tissue is blind to oxytocin's presence. Why? We do not know, but it may depend on two distinct and paradoxical actions of progesterone on the uterine epithelium. Thus, progesterone stimulates $PGF_{2\alpha}$ production but depresses oxytocin receptor production (receptor regulation). However, if the elevation of progesterone is prolonged, then progesterone receptor production is depressed (auto receptor regulation), thereby allowing expression of the oxytocin receptor gene as the luteal phase progresses. $PGF_{2\alpha}$ release then occurs, followed by the precipitous decline of the corpus luteum. Epithelial oestrogen receptors also rise as progesterone receptors fall, and the oxytocin receptor promoter has a binding motif for activated E2-receptor; so, oestrogen has also been implicated in facilitating oxytocin-mediated $PGF_{2\alpha}$ release.

A role for endothelins in luteolysis?

How $PGF_{2\alpha}$ causes luteal regression is uncertain. One likely contributory factor is the adverse impact that it has on blood flow. The corpus luteum is highly vascularized and an early sign of luteolysis is reduced blood flow. Levels of **endothelin 1**, a potent vasoconstrictor peptide that is synthesized both by capillary endothelial cells and large and small luteal cells, rise during the mid–late luteal phase, in parallel with rising $PGF_{2\alpha}$. Moreover, premature $PGF_{2\alpha}$ injections lead to earlier and higher levels of endothelin 1 synthesis and premature luteolysis. This peptide reduces luteal blood flow *in vivo* and depresses progesterone production by bovine luteal cells *in vitro*. It also stimulates the production of $PGF_{2\alpha}$ by luteal cells, thereby setting up a **second luteolytic positive feedback loop**. Endothelin receptors are detectable on luteal cells, and receptor blockers specific for endothelin 1 reduce the antiprogesterone effects of $PGF_{2\alpha}$ *in vivo* and *in vitro*. Taken together, these results suggest that at least part of the luteolytic action of $PGF_{2\alpha}$ involves endothelin 1.

Key learning points

- Fertility in females is episodic: ovulation is preceded by a period of follicular/oestrogen dominance and followed by a period of luteal/progestagen dominance.
- The adult ovary produces oocytes and endocrine hormones: oestrogens, androgens, inhibins and progestagens.
- The follicle is the source of all of these and is the functional unit of the ovary.
- The follicle is made up of an interactive partnership between oocyte and follicular cells.
- Primordial follicles mature to preantral follicles in a constant trickle throughout adult reproductive life: this maturation does not require external hormonal support but is enhanced by growth hormone and does involve paracrine action by intra-ovarian cytokines.
- This maturation involves oocyte growth, secretion of the ZP glycoproteins and their assembly into the *zona pellucida* around the oocyte, granulosa cell proliferation and the deposition of interstitial cells on the outer membrana propria to form the theca interna and externa.
- The late preantral and early antral follicles develop FSH receptors on their granulosa cells and LH receptors on their theca interna cells.
- LH stimulates thecal cells to produce androgens.
- Androgens are aromatized by granulosa cells to produce oestrogens in an FSH-dependent process.
- FSH also stimulates the production of follicular fluid, expansion of the follicle and the appearance of the follicular antrum.
- Production of growth factor(s) by the follicle is stimulated by FSH, and mediates some of FSH's intrafollicular effects.
- Production of inhibin by the follicle is stimulated by FSH and LH, and the ratio of inhibin A to inhibin B rises as ovulation approaches.
- Under the influence of FSH and LH, the dominant antral follicle produces rising oestrogen levels, culminating in a surge of oestrogen.

- Under the influence of LH, the dominant follicle is induced to enter the preovulatory phase.
- The preovulatory phase is characterized by rising progesterone output from mural granulosa cells and the appearance of prolactin and progesterone receptors.
- During the preovulatory phase, the follicular antrum expands rapidly to protrude from the surface of the ovary, the granulosa cells form a thin rim lining the follicle with a stalk of cells connecting it to the oocyte surrounded by cumulus cells, the cytoplasmic links between oocyte and cumulus cells are severed, and the oocyte resumes meiosis.
- The oocyte extrudes a first polar body in an unequal cytoplasmic division and then arrests in second meiotic metaphase in which state it is ovulated.
- The oocyte is ovulated through the ruptured apex of the follicular protrusion at its stigma.
- The rupture is due to the action of proteases on the follicle wall.
- The ovulated oocyte is released onto the surface of the ovary.
- The empty follicle collapses, becomes fully vascularized and undergoes the process of luteinization to form the corpus luteum.
- Granulosa cells contribute large luteal cells and thecal cells contribute small luteal cells.
- Large luteal cells synthesize progestagens.
- Hormonal support of the corpus luteum is species variable: LH is required for luteinization; prolactin, oestrogen and progesterone may be involved in luteal maintenance.
- Luteolysis in most species is caused by prostaglandin $F_{2\alpha}$.
- Release of $PGF_{2\alpha}$ from the endometrium into the local circulation is stimulated by oxytocin produced by the corpus luteum and acting on endometrial oxytocin receptors that develop under the influence of sustained progesterone.
- Luteolysis in higher primates is caused by inadequate luteal LH support.

Clinical vignette

Turner's syndrome and fertility

A 28-year-old woman, Ms A, and her 30-year-old male partner had been trying to conceive for 18 months. She and her partner were both fit and well. Neither of the couple had any previous pregnancies with other partners. However, Ms A's menstrual cycles were irregular, and occurred on average every 35–40 days. The couple attended the fertility clinic to initiate investigations. The doctor in the clinic noted that Ms A was of unusually short stature (1.5m). In addition to the usual blood tests and imaging to investigate subfertility, a karyotype for Ms A was requested. The karyotype result was 45,X-/46,XX, which indicates a condition known as mosaic Turner's syndrome. In light of this result, Ms A was referred to a cardiologist to investigate her cardiac and aortic function further, which was found to be normal with no dilation of the aortic root. The couple then underwent pre-conception counselling with a maternal medicine specialist. These appointments were useful to the couple to understand the risks of proceeding with fertility treatment and pregnancy. As they had already tried to conceive spontaneously for two years without success, it was felt that using artificial reproductive technology would provide the couple with the best chance of pregnancy success. The couple opted to undergo IVF treatment using their own gametes. Two cycles of ovarian stimulation were unsuccessful in stimulating Ms A's ovaries to produce mature oocytes capable of fertilization. On the third cycle, however, with an enhanced stimulation protocol, a number of mature oocytes were retrieved and three fertilized successfully *in vitro*. The couple's first attempt at single blastocyst transfer resulted in a negative pregnancy and the second ended in an early miscarriage at 5 weeks. On the third attempt, however, a scan at 7 weeks revealed a viable fetus with a visible heart beat. Ms A had close monitoring from her maternal medicine team throughout pregnancy and subsequently gave birth to a healthy female baby at 36 weeks via Caesarean section.

Turner's syndrome provides a particular challenge for reproductive health services, as it impacts not only on gamete production, but also the risk of pregnancy to the mother and the chance of a successful full-term live birth. Most women with Turner's syndrome have 'streak ovaries', which do not contain sufficient primordial follicles to stimulate the production of a mature oocyte. Although normal numbers of germ line cells are present in the gonad during early fetal development, the number of oocytes declines during the second half of fetal life. This is likely to be a direct effect of the lack of a second X chromosome, which prevents successful pairing during meiosis. Hence women who have a mosaic karyotype of cells with one or two X chromosomes have higher numbers of oocytes within primordial follicles at birth than women with classical 45,X Turner's syndrome. Only 10% of women with a 45,X karyotype undergo menarche, but the percentage is higher in mosaic Turner's syndrome. The relatively small number of primordial follicles present in the ovaries of women with Turner's syndrome means that premature ovarian failure is highly likely, and many women consider cryopreservation of oocytes for future use. IVF using donor oocytes is a successful strategy for many women with Turner's syndrome who have a very low or non-existent ovarian reserve. It is important that such pregnancies are carefully monitored by a specialist maternal medicine service, as women with Turner's syndrome often have cardiac and other manifestations that make pregnancies high risk from a maternal point of view. Single-embryo transfer should always be considered in Turner's syndrome to minimize maternal complications of pregnancy.

FURTHER READING

General reading

Choi Y, Rajkovic A (2006) Genetics of early mammalian folliculogenesis. *Cellular and Molecular Life Sciences* **63**, 579–590.

Coticchio G, Dal Canto M, Renzini MM *et al.* (2015) Oocyte maturation: gamete-somatic cells interactions, meiotic resumption, cytoskeletal dynamics and cytoplasmic reorganization. *Human Reproduction Update* **21**, 427–454.

Drummond A (2006) The role of steroids in follicular growth. *Reproductive Biology and Endocrinology* **4**, 16.

Duffy DM (2015)Novel contraceptive targets to inhibit ovulation: the prostaglandin E2 pathway. *Human Reproduction Update* **21**, 652–670.

Guigon CJ, Magre S (2006) Contribution of germ cells to the differentiation and maturation of the ovary: insights from models of germ cell depletion. *Biology of Reproduction* **74**, 450–458.

Hsueh AJW, Kawamura K, Cheng Y, Fauser BCJM (2015) Intraovarian control of early folliculogenesis. *Endocrine Reviews* **36**, 1–24.

McLaughlin EA, McIver SC (2009) Awakening the oocyte: controlling primordial follicle development. *Reproduction* **137**, 1–11.

Monniaux D (2016) Driving folliculogenesis by the oocyte-somatic cell dialog: lessons from genetic models. *Theriogenology* **86**, 41–53.

Muttukrishna S, Tannetta D, Groome N, Sargent I (2004) Activin and follistatin in female reproduction. *Molecular and Cellular Endocrinology* **225**, 45–56.

Oktem O, Bulent U (2010) Understanding follicle growth in vivo. *Human Reproduction* **25**, 2944–2954.

Russell DL, Gilchrist RB, Brown HM, Thompson JG (2016) Bidirectional communication between cumulus cells and the oocyte: old hands and new players? *Theriogenology* **86**, 62–68.

Skinner MK (2005) Regulation of primordial follicle assembly and development. *Human Reproduction Update* **11**, 461–471.

Stouffer RL (2003) Progesterone as a mediator of gonadotrophin action in the corpus luteum: beyond steroidogenesis. *Human Reproduction Update* **9**, 99–117.

More advanced reading

Chang H-M, Qiao J, Leung PCK (2017) Oocyte–somatic cell interactions in the human ovary: novel role of bone morphogenetic proteins and growth differentiation factors. *Human Reproduction Update* **23**, 1–18,

Fortune JE, Rivera GM, Yang MY (2004) Follicular development: the role of the follicular microenvironment in selection of the dominant follicle. *Animal Reproduction Science* **82–83**, 109–126.

Makanji Y, Zhu J, Mishra R, Holmquist C *et al.* (2014) Inhibin at 90: from discovery to clinical application, a historical review. *Endocrine Reviews* **35**, 747–794.

Pangas SA, Matzuk MM (2005) The art and artifact of GDF9 activity: cumulus expansion and the cumulus expansion-enabling factor. *Biology of Reproduction* **73**, 582–585.

Peluso JJ (2006) Multiplicity of progesterone's actions and receptors in the mammalian ovary. *Biology of Reproduction* **75**, 2–8.

Rajareddy S, Reddy P, Du C *et al.* (2007) p27kip1 (cyclin-dependent kinase inhibitor 1B) controls ovarian development by suppressing follicle endowment and activation and promoting follicle atresia in mice. *Molecular Endocrinology* **9**, 2189–2202.

Reddy P, Liu L, Adhikari D *et al.* (2008) Oocyte-specific deletion of Pten causes premature activation of the primordial follicle pool. *Science* **5863**, 611–613.

Shirasuna K, Shimizu T, Hayashi K-G *et al.* (2007) Positive association, in local release, of luteal oxytocin with endothelin 1 and prostaglandin F2alpha during spontaneous luteolysis in the cow: a possible intermediatory role for luteolytic cascade within the corpus luteum. *Biology of Reproduction* **76**, 965–970.

Spencer TE, Bazer FW (2004) Conceptus signals for establishment and maintenance of pregnancy. *Reproductive Biology and Endocrinology* **2**, 49.

Uhlenhaut NH, Treier M (2011) Forkhead transcription factors in ovarian function. *Reproduction* **142**, 489–495.

Wang Y, Newton H, Spaliviero JA *et al.* (2005) Gonadotropin control of inhibin secretion and the relationship to follicle type and number in the *hpg* mouse. *Biology of Reproduction* **73**, 610–618.

CHAPTER 10
Women

Essential Reproduction, Eighth Edition. Martin H. Johnson.

Companion website: www.essentialreproduction.com

In the previous chapter, we described the development of an individual follicle through to either ovulation and luteinization, or (for most of them) atresia. But each of the two ovaries has many primordial follicles. How are they coordinated with each other and with the rest of the mature female's body? These topics form the subject of this chapter.

Follicular development and the ovarian cycle

First, we consider the relationships between the various follicles developing at different times, and in both ovaries, in order to obtain an overall picture of ovarian function and cyclicity.

The ovarian cycle

One complete **ovarian cycle** is the interval between successive ovulations, where each ovulation is preceded by a period of oestrogen dominance and followed by a period of progesterone dominance. As the oestrogens are derived from the follicles, the period preceding ovulation is often called the **follicular phase of the ovarian cycle**. Correspondingly, the period after ovulation is often called the **luteal phase**, because progesterone is derived from the corpus luteum. The duration of the ovarian cycle and its constituent follicular and luteal phases in various species is summarized in Table 10.1. It is immediately clear that there are major differences between species in both the absolute length of the ovarian cycle and in the relative duration of its follicular and luteal components. These apparent major differences mask a fundamentally similar organization, and result only from minor but significant modifications to a basic pattern. First, we will discuss the human ovarian cycle, as it is the easiest to understand. We will then relate that pattern to those of other species, because these are often used as models for the human.

Table 10.1 Ovarian cyclicity in different species

Species	Length of cycle (days)	Follicular phase (days)	Luteal phase (days)
Human	24–32	10–14	12–15
Cow	20–21	2–3	18–19
Pig	19–21	5–6	15–17
Sheep	16–17	1–2	14–15
Horse	20–22	5–6	15–16
Mouse/rat* (+ infertile male)	13–14	2	11–12
Rabbit* (+ infertile male)	14–15	1–2	13
Mouse/rat	4–5	2	2–3
Rabbit	1–2	1–2	0

* See text for discussion.

The ovarian cycle of the human

Figure 10.1a shows a basic outline of two sequential human ovarian cycles. The pattern of measured blood steroids and gonadotrophins is indicated below, and the activities of the four stages of follicular maturation are indicated above. A continuous trickle of developing primordial, preantral and early antral follicles occurs throughout the cycle; growth of these follicles does not require gonadotrophic support, and the follicles do not secrete significant levels of steroids, and so do not affect blood steroid levels. As we saw in Chapter 9, the early antral follicles so formed are doomed to atresia unless rescued by FSH and LH, which take them through full antral expansion. One (or two) of these rescued follicles will survive to become the dominant antral follicle, secreting high levels of oestrogens at maturity: witness the rising blood oestrogen levels. This advanced follicle is then converted to a preovulatory follicle by the transient high levels of LH that are measured in the blood at this time. The other antral follicles become atretic. The successful ovulatory follicle forms a corpus luteum, which secretes progesterone and oestrogen, until luteolysis 14 or so days later. During the luteal phase, LH and FSH levels are comparatively low and are insufficient to maintain antral development, so follicular atresia occurs. After luteolysis, a new cycle then begins as tonic gonadotrophin levels are elevated.

An important feature emerging from this description of the human ovarian cycle is that the terminal antral expansion phase, ovulation and the complete luteal phase occupy one complete ovarian cycle. This feature distinguishes humans from most other non-primate mammals. It is useful to explore what underlies this distinction for a full understanding of the human cycle.

The ovarian cycles of the cow, pig, sheep and horse have a shorter follicular phase

In these species, as indicated for the pig in Figure 10.1b, significant antral expansion occurs during the luteal phase of the previous crop of follicles to have ovulated. This growth is possible **because FSH and LH levels in these species do not fall to such low levels during the luteal phase**. Thus, in these species the complete antral expansion phase, ovulation and the complete luteal phase **occupy a longer period of time than does one complete ovarian cycle** (follow and compare the black lines in Figure 10.1a,b). It is almost as though in these

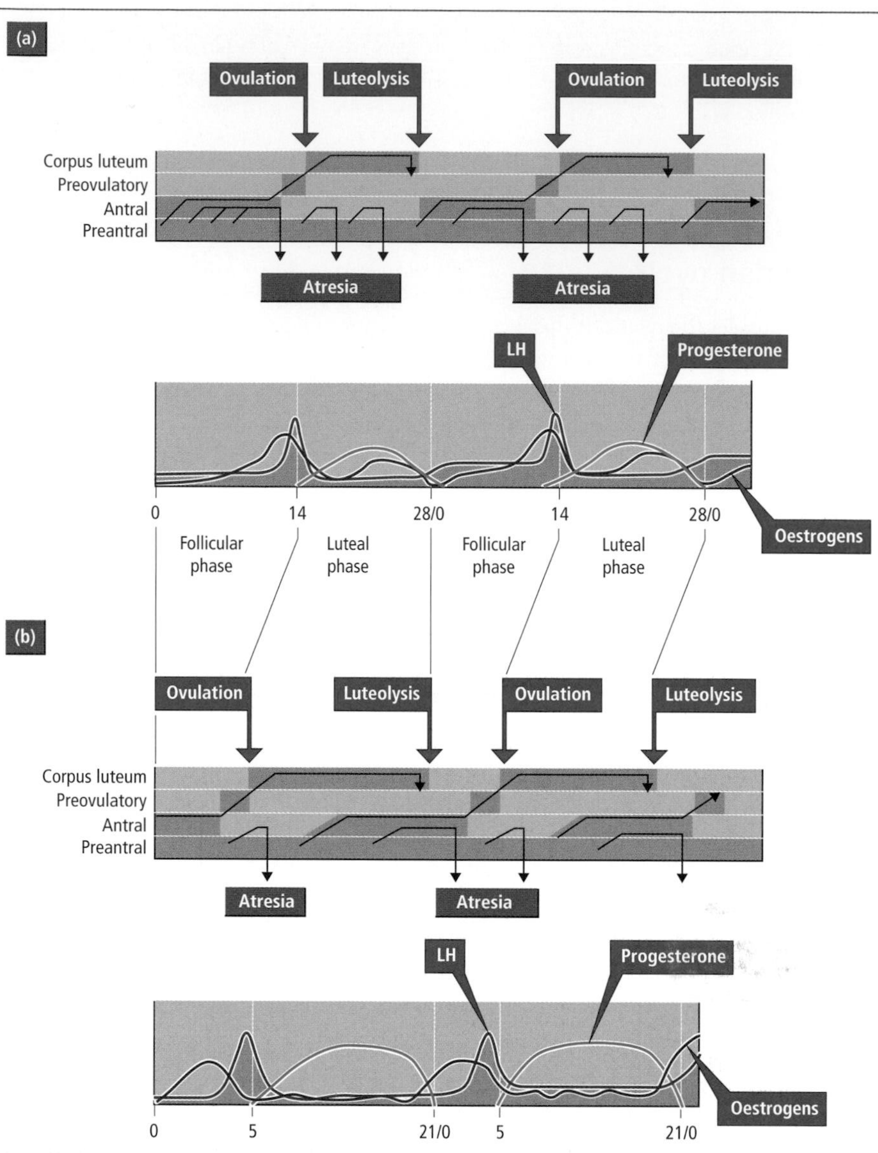

Figure 10.1 Schematic summary of follicular activity in two sequential ovarian cycles in: (a) the human and (b) the pig. The presence of the different follicular stages is indicated by darker shading within each horizontal bar (upper part of each panel). Relative blood hormone levels are also indicated below. The history of an ovulatory follicle from the preantral stage to luteolysis or atresia is indicated by the black lines marked with closed arrows. Note that the pig cycle is shorter than that of the human and this is due to a shorter follicular phase as indicated by the oblique arrows linking the human (a) and pig (b) panels: significant antral follicular growth occurs during the preceding luteal phase in pigs but not humans (see the difference in darker shading in the antral bars of the human and the pig).

species, the human cycle has been 'telescoped' by 'pushing' the follicular part of one cycle backwards into the luteal half of the previous cycle.

The ovarian cycles of most other species can be understood quite simply in terms of the above discussion. We will examine those of the rat, mouse and rabbit because they exhibit a certain distinctive feature of great importance in understanding the ovarian cycles of humans and the large farm animals. Our discussion on different ovarian cycles is summarized in Figure 10.2.

The ovarian cycles of the rat and mouse have abbreviated follicular *and* luteal phases

The ovarian cycles of the rat and mouse are basically of the 'telescoped' sort seen in large farm animals. They show, however, a curious and distinctive feature: the cycle differs in length, depending on whether or not the female mates. If the female has an infertile mating at the time of ovulation, e.g. with a vasectomized male, her luteal phase is 11–12 days in duration (often called a **pseudopregnancy**, and shorter than an actual pregnancy of 20–21 days), and the ovarian cycle resembles that of the pig. However, if she fails to mate at the time of ovulation the luteal phase is only 2–3 days long. In the latter case the corpora lutea become only transiently functional in producing progestagens, secreting a small amount of progesterone, but mainly 20α-hydroxyprogesterone.

The explanation for this curious phenomenon lies in the **mechanical stimulus to the cervix** provided naturally by the penis at coitus. The presence of such stimulation is relayed via sensory nerves from the cervix to the central nervous system (CNS), and activates the release of **prolactin** from the pituitary in twice-daily surges that last for about 10 days. This hormone,

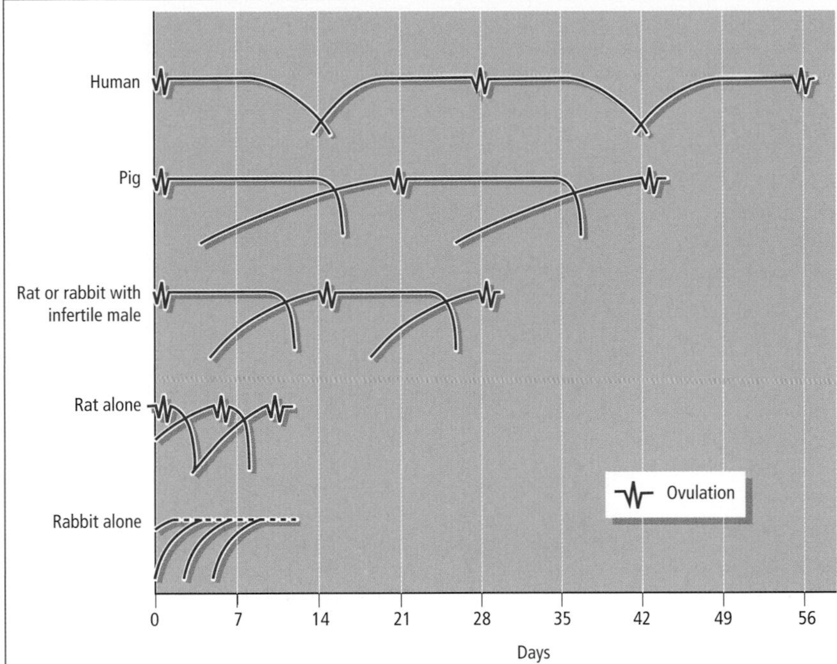

Figure 10.2 Schematic comparison of ovarian cycles of the human, pig, rat or rabbit caged with an infertile male, and the rat or rabbit caged alone. Day 0 indicates the day of the first ovulation. Each continuous line represents a complete and successful sequence of growth of one crop of antral follicle(s) through preovulatory follicle(s) to corpora lutea and ultimately luteolysis. Note cycle lengths and the proportions within each cycle vary (see text for explanation).

as pointed out earlier (see Table 9.3), is an essential part of the luteotrophic complex in the rat and mouse, and without it the luteal phase is abbreviated from the 'normal' extended pattern characteristic of the large farm animals. Not surprisingly, therefore, if the prolactin surges are blocked, pseudopregnancy does not occur.

Why such a complex cycle? The rat and mouse derive increased reproductive efficiency from this evolutionary modification, as without the abbreviating device they would only be fertile every 13 or 14 days instead of every 4 or 5 days. As their pregnancy only lasts 20–21 days, this is a highly significant economy. Evolutionary pressures for such a truncation of the luteal phase would not apply to the cow, pig, sheep and horse, where pregnancy is a very extended event lasting months compared with the luteal phase. This neat neural device, which shortens the luteal phase, emphasizes how the CNS can influence ovarian function – another example of which is provided by the rabbit.

The ovarian cycle of the rabbit is reduced to an extended follicular phase

A female rabbit caged alone shows little evidence of a cycle: blood levels of oestrogens are high; progestagens are low; she is always on heat and ready to mate, but ovulation cannot be detected. Yet her ovary contains waves of expanding antral follicles. It is as though she is in a **continuous follicular phase**. If she mates with a vasectomized buck, or if her cervix is stimulated mechanically, she ovulates 10–12 hours later and has a luteal phase (or pseudopregnancy) of 12 or so days. If the vasectomized buck is left in with her, she will show a 14-day cycle with a 2 + 12-day follicular + luteal pattern, similar to the porcine pattern. As in the rat, cervical stimulation in the rabbit is the source of a sensory input to the CNS, which, in this case, induces **a surge of LH**, high levels of which rescue any expanded antral follicles from atresia and ovulate them (see page 147). In essence, the rabbit has abbreviated her cycle even more than the rat or mouse by eliminating the luteal phase completely. This phenomenon is known as **induced ovulation** (see Box 10.1).

Summary

We have seen so far that the levels of gonadotrophins critically affect the successful completion of the ovarian cycle, and that the circulating levels of gonadotrophins vary naturally during the cycle. Moreover, we have observed that the CNS can affect the levels of circulating gonadotrophins. Next we examine how gonadotrophin levels are regulated, and how the CNS exerts its effects.

Ovarian hormones regulate gonadotrophin secretion

During the menstrual cycle, the dynamics of follicular maturation and ovulation are mainly orchestrated by the output of gonadotrophins. Moreover, in women, as in men (see page 132), the output of gonadotrophins is regulated mainly by the secretory products of the gonad. However, unlike in the male, where only negative feedback by androgens and inhibin B was observed, the situation in the female is more complex and involves two types of feedback: (1) **a negative feedback**, resembling that in males, in which **oestrogens, progestagens and inhibin depress gonadotrophin output**;

Box 10.1 Induced or reflex ovulation: coitus affects fertility in some species

In humans and other primates, sheep, rats and most other mammals, ovulation is said to occur 'spontaneously'. Thus, it depends on an endogenous event, which as we will see, is timed by the ovarian oestradiol surge, which results in an ovulatory discharge of LH. In induced or reflex ovulators, such as cats, rabbits and ferrets, the females remain in behavioural oestrus for long periods of time without ovulating until they copulate with a male (see Figure 10.2). Stimulation of the cervix and vagina during coitus evokes the reflex release of an ovulatory surge of LH via afferent, sensory pathways, which gain access to the GnRH release mechanism. Even in these species it seems that the hypothalamic–pituitary axis must be primed with high levels of oestrogen for the neural input to be effective.

Although the data supporting it are less than convincing, there are several reports of reflex ovulation in women. Ovulation has variously been reported to follow coitus very early in the follicular phase (even during menstruation), while termination of abnormally long follicular phases has sometimes been ascribed to coitus-induced, acute LH release. However, the subject has not been studied systematically, and these reports should therefore be viewed sceptically.

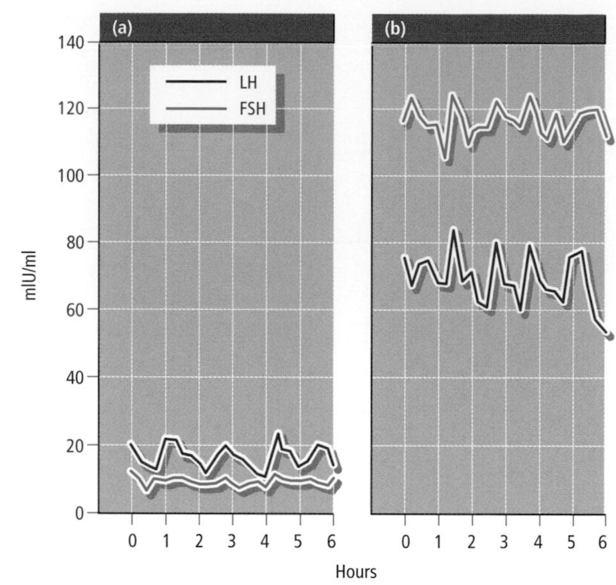

Figure 10.3 Circulating FSH (green) and LH (red) levels in: (a) a woman at day 7 in the (early) follicular phase; and (b) a postmenopausal woman. The difference between the two represents the negative feedback effect of oestradiol. Note: (1) the reversal of LH and FSH levels between (a) and (b), which may reflect preferential inhibition of FSH secretion in (a) due to inhibin (see text); and (2) the changes in pulsatility of FSH and LH between (a) and (b). (Source: Yen S.S.C. & Jaffe R. (eds) (1978) *Reproductive Endocrinology*. WB Saunders & Co, Philadelphia. Reproduced with permission from Elsevier.)

and (2) **a positive feedback** effect in which an increase, or **surge, of LH and FSH secretion is induced principally by oestrogen**. Throughout our discussion we will use data from the primate as a model, referring to other species where corroborative evidence or species differences exist.

Oestradiol regulates FSH and LH secretion both positively and negatively

Plasma concentrations of circulating FSH and LH increase markedly after ovariectomy or the menopause (Figure 10.3a,b). This rise is largely attributable to removal of oestradiol, as infusion of this hormone results in the rapid decline of FSH and LH levels. The important characteristics of this oestrogen action are: (1) only relatively small rises in circulating levels of oestradiol are required to exert a marked effect; and (2) the effect is very rapid in onset, detectable within 1 hour and maximal by 4–6 hours. As oestradiol is acting to suppress the gonadotrophins that stimulated its production by the follicle, the process is an example of **negative feedback**.

In contrast, if plasma concentrations of oestradiol increase greatly, e.g. 200–400% above those seen in the early follicular phase of the cycle, and remain at this high level for 48 hours or so, then LH and FSH secretion is enhanced, not suppressed (Figure 10.4). Under these conditions we speak of a **surge** of LH and FSH. The term **positive feedback** is often used to describe the relationship whereby high levels of oestradiol increase the secretion of gonadotrophins, and thereby (at least initially) of oestradiol.

Thus, oestradiol has a dual function in regulating gonadotrophin secretion. At low circulating levels, it exerts rapidly expressed, negative feedback control over FSH and LH. At higher, maintained circulating levels, positive feedback becomes the dominant force and a (relatively) delayed LH and FSH surge is induced.

Progesterone regulates FSH and LH secretion negatively

Progesterone also has **two effects** – but **both are negative** in outcome. First, the high plasma concentration of progesterone, such as is seen in the luteal phase (4–8 ng/ml in humans) **enhances the negative feedback effects of oestradiol**. FSH and LH secretion is held at a very low level. Second, the **positive feedback effect of oestradiol is blocked**. Thus, injection of oestradiol into women during the progesterone-dominated luteal phase of the menstrual cycle is *not* followed by an LH surge.

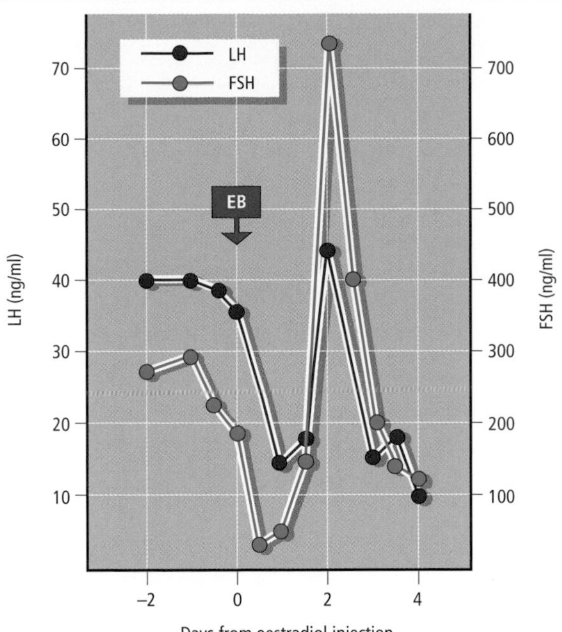

Figure 10.4 The negative and positive feedback effects of a large injection of oestradiol benzoate (EB) in a female rhesus monkey. Note the early action of oestrogen is to decrease FSH and LH levels (negative feedback, maximal at 6 hours or so). But if oestradiol levels are of sufficient magnitude and duration, a surge of the gonadotrophins occurs after 36–48 hours (positive feedback). Note that one effect naturally follows the other and each is critically dependent on different time–dose effects of the steroid. (Source: Yen S.S.C. & Jaffe R. (eds) (1978) *Reproductive Endocrinology.* WB Saunders & Co, Philadelphia. Reproduced with permission from Elsevier.)

The inhibins regulate FSH secretion negatively

The inhibins A and B (see Figure 8.2) also act to influence FSH secretion selectively and negatively, as in the male. Thus, intravenous administration of inhibin into ewes prevents the rise in FSH, which usually follows ovariectomy and blocks the FSH, but not the LH, response to a GnRH agonist (Figure 10.5). Infusion of antibodies to inhibin during the late antral phase of the rat causes an increase in FSH (but not LH) concentrations in plasma (Figure 10.6). In primates too, inhibin administration causes FSH to decline selectively. Clearly, these experimental data indicate a role for **inhibin in the negative feedback regulation of FSH secretion.**

Feedback by steroid hormones and the inhibins regulates the menstrual cycle

Let us now re-examine the blood levels of steroids, inhibins and gonadotrophins during a normal human menstrual cycle, and interpret them in the light of positive and negative feedback.

The follicular phase of the cycle

By convention, the menstrual cycle begins with the first day of menstruation (M in Figure 10.7). Menstruation is initiated by the preceding luteolysis, which results in falling levels of luteal oestrogen, progesterone and inhibin A. In consequence, **negative feedback inhibition is relaxed**, and both FSH and LH levels rise. These rises permit antral growth to proceed, resulting in first the rising output of inhibin B followed by androgens and oestrogens. These rises elicit negative feedback and so FSH levels fall and LH levels plateau. However, the selection

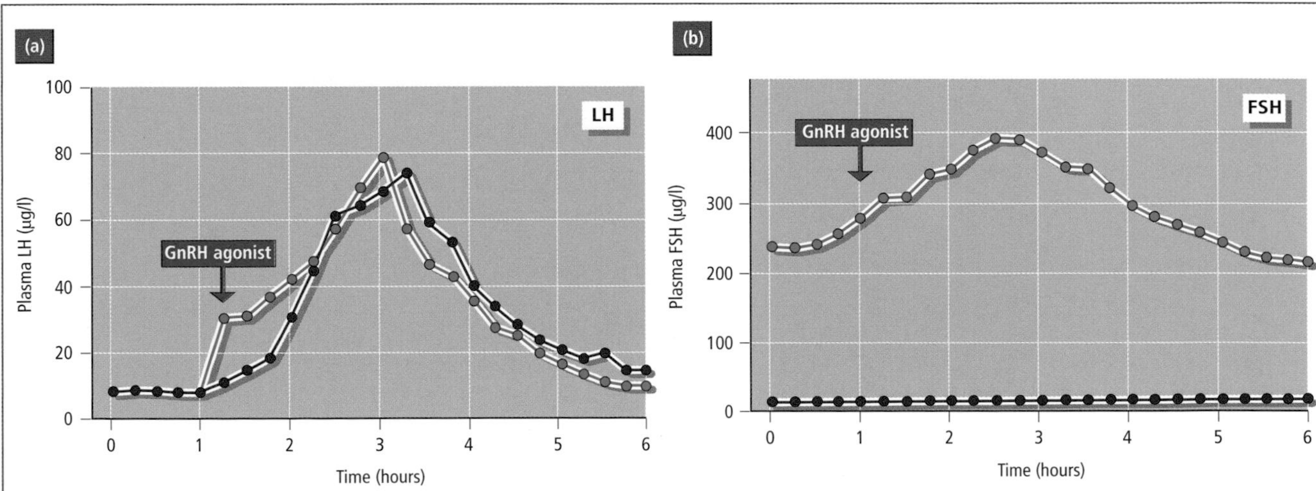

Figure 10.5 Changes in plasma (a) LH and (b) FSH concentration following an injection of a GnRH agonist, in control ovariectomized ewes (green line) and ovariectomized ewes treated with inhibin (red line). Note the complete absence of an FSH response to GnRH in the inhibin-treated ewes in (b). (Source: Martin G.B. *et al.* (1986) *Journal of Endocrinology* (London), **111**, 287–296. Reproduced with permission from Bioscientifica Ltd.)

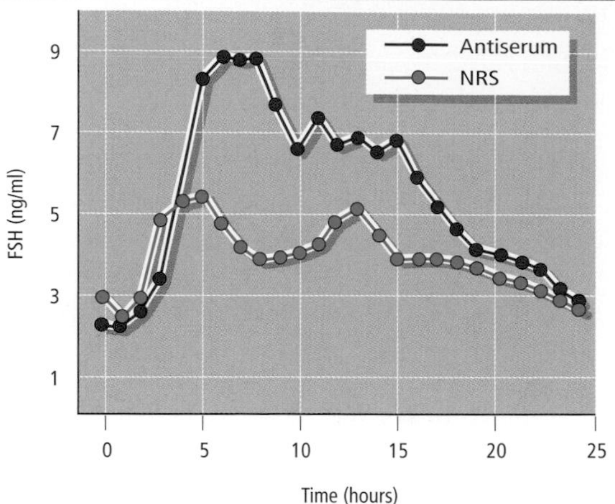

Figure 10.6 The effect of intravenous infusion of normal rabbit serum (NRS) (green line) or inhibin antiserum (red line) on plasma FSH levels during the late antral phase. Note the marked rise in plasma FSH concentrations in the female rats immunized passively against inhibin.

of a dominant follicle leads to a further rise in oestrogen (together with its biosynthetically associated androgens), culminating in a surge of oestrogen (and androgen) and a switch from inhibin B to inhibin A output under the combined stimulation of LH and FSH. This output of oestrogen and inhibin A reflects the development of only the **most advanced** follicle(s), and is thus a measure of nearness to ovulation. During the preovulatory phase, the **surge in oestrogen triggers a rapid rise (or surge) in LH and FSH levels via a positive feedback effect**, and ovulation follows. As a result of follicular collapse, androgen and thus oestrogen outputs fall, and progesterone levels rise as luteinization occurs. The LH and FSH levels now fall equally precipitously because, at least in part, they lack a continuing positive feedback stimulus. The first half of the cycle is complete.

The crucial role of the oestrogen surge in triggering the LH surge has been shown in female rhesus monkeys by actively immunizing them against oestradiol, when neither an LH surge nor ovulation occurs. However, if a replacement synthetic oestrogen, not neutralized by the antiserum, is given, the neutralization of endogenous oestradiol is bypassed and the synthetic oestrogen reinstates an LH surge. Thus, the pattern of FSH, LH, inhibin and steroid secretion during the first half of the cycle is explicable largely in terms of the positive and negative feedback effect of oestrogens (primarily oestradiol) and inhibin B on gonadotrophin secretion.

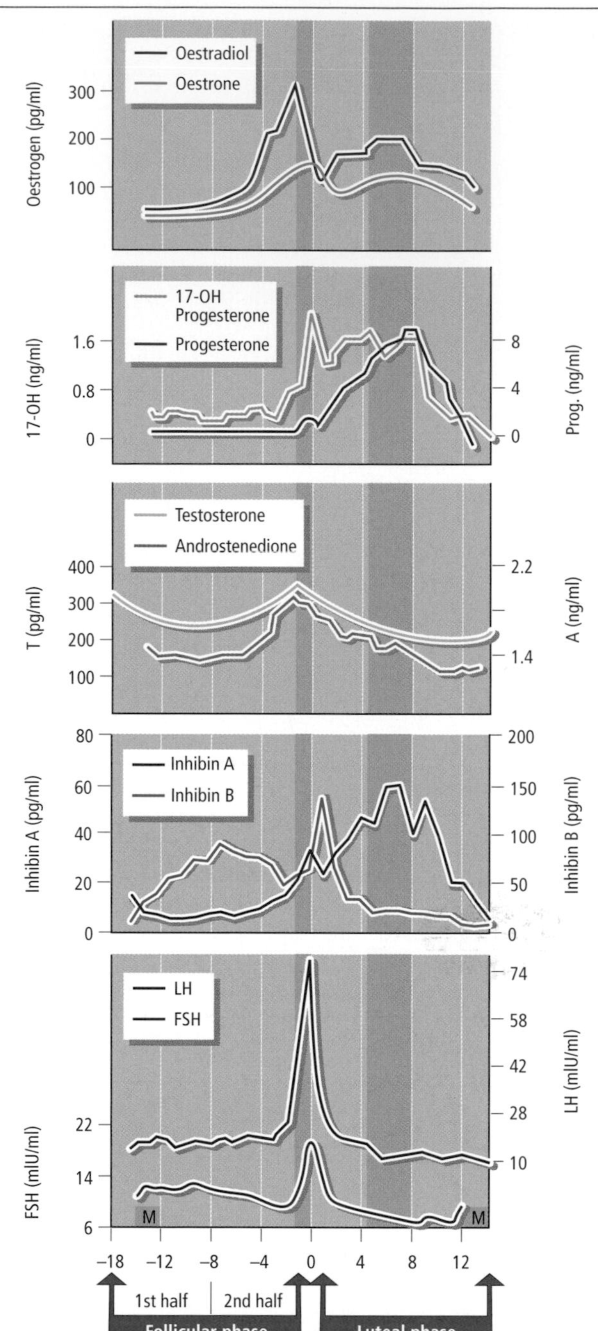

Figure 10.7 Serum hormone levels during the human menstrual cycle. Note the units of measurements; FSH and LH are expressed in milli-international units/ml; testosterone (T), oestradiol and the inhibins in pg/ml; 17α-hydroxyprogesterone (17-OH), progesterone and androstenedione (A) in ng/ml. M, menstruation. (Source: Yen S.S.C., Jaffe R. (eds) (1978) *Reproductive Endocrinology*. WB Saunders & Co, Philadelphia. Reproduced with permission from Elsevier.)

The luteal phase of the cycle

The luteal phase of the cycle is characterized by rising concentrations of plasma progesterone and 17α-hydroxyprogesterone, which peak around 8 days after the LH surge. In higher primates (but not in most other species), the luteinized cells of the corpus luteum also make large amounts of oestrogen and inhibin A (Figure 10.7). In all species, progesterone depresses levels of FSH and LH (negative feedback), a depression particularly marked in higher primates in which inhibin A is also active in suppressing FSH (compare the lower levels of gonadotrophins in the human in Figure 10.1a with those in the pig in Figure 10.1b). Growth of antral follicles is therefore suppressed and so androgens are also at a low level. Although in primates, oestrogens from the corpus luteum can rise to levels that previously induced positive feedback, they fail to induce an LH surge. This failure reflects the effects of the uniquely high levels of progesterone found during the luteal phase of the cycle that **inhibit positive feedback by oestradiol**. At the end of the luteal phase, if conception has not occurred, oestrogens, progesterone and inhibin A decline at luteolysis, the negative feedback effect of these hormones is relaxed, and LH and FSH levels start to rise, thereby permitting the rescue of preantral follicles and the initiation of another cycle.

Summary

The foregoing account reveals that the length of the follicular phase is determined by the rate at which the principal preantral follicle matures, since it is the main source of oestrogen and thereby a major arbiter of cycle length. Similarly, the length of the luteal phase is determined by the life of the corpus luteum. These findings have led to the concept of the **ovarian or pelvic clock** as the major temporal determinant of the menstrual cycle.

Positive and negative feedback are mediated at the level of both the hypothalamus and pituitary

Having described how the ovarian hormones influence gonadotrophin output during the menstrual cycle, we now consider the sites at which these hormones exert their regulatory influences. As for the male, the two obvious sites are the anterior pituitary and the hypothalamus: both are used.

The anterior pituitary mediates some of the feedback by steroids and all of that by inhibins

A convincing demonstration that the pituitary can respond to both the negative and positive feedback effects of oestradiol comes from experiments on ovariectomized rhesus monkeys. Large lesions of the mediobasal hypothalamus result in abolition of GnRH output, and a decrease in serum FSH and LH to undetectable levels. Hourly pulses of exogenous GnRH delivered by a programmable intravenous infusion pump restore pulsatile LH and FSH secretion. The subsequent injection of oestradiol, so as to reach surge levels, results first in a fall and then in a dramatic rise (surge) in serum FSH and LH levels. In the presence, therefore, of GnRH pulses of invariant amplitude and frequency, oestradiol can exert both its negative and positive feedback effects on gonadotrophin secretion. Clearly, this action can only have been mediated via the anterior pituitary. Clinical studies on postmenopausal or hypogonadal women point to the same conclusion. The demonstration that inhibin is able to reduce the FSH secretory response to GnRH (see Figure 10.5) strongly indicates that its effects are also mediated by actions on the anterior pituitary. Studies on the pituitary sensitivity of intact women undergoing normal cycles provide further support for this conclusion. These experiments were performed by measuring the change in plasma levels of LH and FSH induced by pulses of exogenous GnRH administered on different days of the menstrual cycle (similar results have been obtained during the oestrous cycles in rats). Figure 10.8 shows that secretion of FSH and LH by the pituitary, in response to constant GnRH pulses, increases during the follicular phase, reflecting an increased sensitivity of the secretory response with rising levels of plasma oestradiol.

However, note that the responsiveness of the pituitary to GnRH remains very high in the luteal phase of the cycle, a time when FSH and LH levels are normally at their lowest and when it is impossible to induce an LH surge with an oestrogen surge (see above). This must mean that oestradiol and/or progesterone, which are both high at this time, exert an important component of their negative feedback action on FSH and LH secretion somewhere other than the pituitary. Thus, taken together, these findings imply that **modulation of pituitary sensitivity to GnRH pulses can and does occur**, but is not in itself sufficient to explain all the changes observed in a normal cycle. Steroid-dependent alterations in the amplitude and frequency of the GnRH signal also play an important role. The mechanisms by which these feedback effects might be achieved is considered in Box 10.2.

The hypothalamus also mediates steroid hormone feedback

The pattern of LH and FSH pulses varies during the menstrual cycle. Thus, during the follicular phase, LH is secreted in a series of high-frequency, low-amplitude pulses occurring approximately once every hour. By contrast, the luteal phase of the cycle is characterized by a pattern of high-amplitude, low-frequency, irregular LH pulses, often with long intervals between them of up to 6 hours (Figure 10.10). As these gonadotrophin pulses reflect underlying GnRH secretory episodes, it is likely that the different steroid hormone environments in the follicular and luteal phases are influencing the GnRH output. The experimental manipulation of the steroid environment confirms this suspicion. Thus, **progesterone acts primarily to reduce pulse frequency while oestrogen acts to reduce pulse amplitude**.

More direct information on the modulation of GnRH secretory activity by steroids requires measurement of the GnRH peptide in portal blood or by microperfusion studies within the median eminence itself. Obviously this is impossible in humans and, even in experimental animals, presents considerable methodological difficulties. However, it has been achieved in experiments on rats, sheep and rhesus monkeys, and the data show clearly that **GnRH secretory activity**

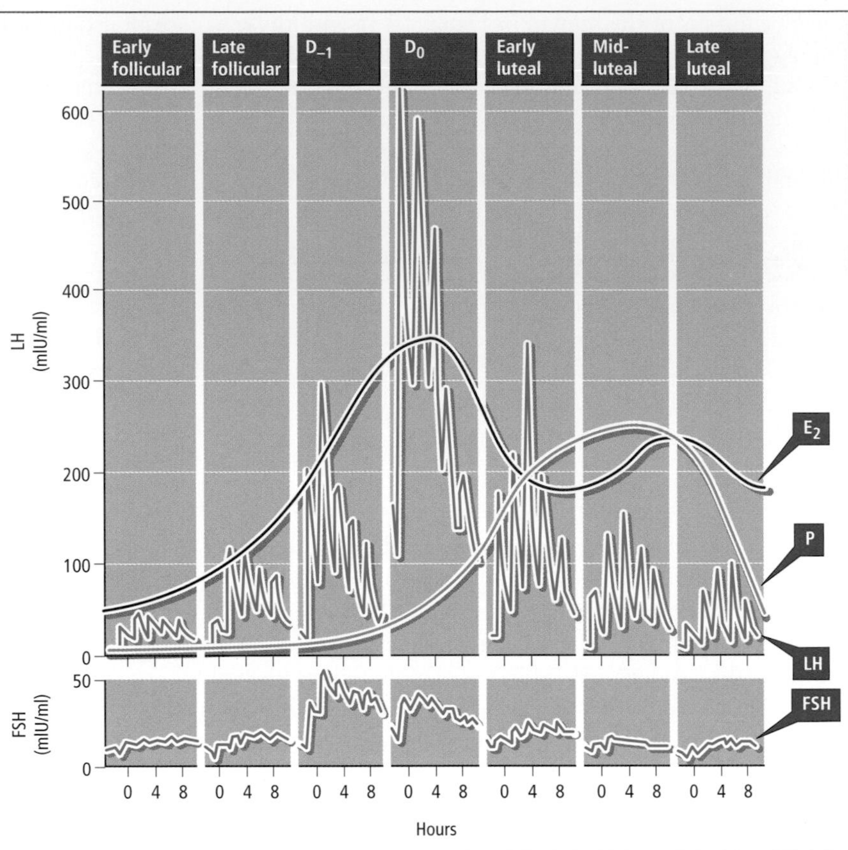

Figure 10.8 The release of LH and FSH (green) in response to 'pulse injections' of GnRH (10 mg at 5 x 2-hour intervals) during various phases of the menstrual cycle in women. The magnitude of the LH and FSH responses can be measured against the naturally changing circulating levels of oestradiol (purple) and progesterone (orange). Note the correlation between oestrogen levels and increasing pituitary responsiveness to the injected GnRH during the follicular phase. Note also that this responsiveness remains in the early and mid-luteal phases, at times when further LH/FSH surges cannot be elicited by oestrogen injections (see text). D^{-}_{1} and D_0, the day before and the day of the spontaneous LH surge. (Source: Yen S.S.C., Jaffe R. (eds) (1978) *Reproductive Endocrinology*. WB Saunders & Co, Philadelphia. Reproduced with permission from Elsevier.)

is subject to modulation by steroid feedback during the menstrual and oestrous cycles, so confirming the indirect evidence described above. Thus, in rats, GnRH secretion is increased on the afternoon immediately before the LH surge. Similarly, the LH surge induced by exogenous oestradiol administration in rhesus monkeys and ewes is associated with elevated concentrations of GnRH in portal blood (Figure 10.11), indicating a hypothalamic site of positive feedback. Indeed, in induced ovulators such as the rabbit (see Box 10.1) the coital stimulus is followed by a rise in GnRH output. In contrast, direct evidence for a hypothalamic site of negative feedback by oestradiol is less clear, as it is difficult to measure with confidence reductions in the already low levels of GnRH. However, experiments in which steroids are implanted in the hypothalamus suggest that negative effects occur there also – as we will now see.

Where in the hypothalamus do steroids exert their feedback effects?

Experiments in which oestradiol has been placed into the hypothalamus have consistently implicated the arcuate nucleus as a site of negative feedback influence on GnRH secretion. Thus, in rats, **oestradiol infused into the arcuate nucleus suppressed gonadotrophin secretion** without detectable amounts of the steroid reaching the anterior pituitary. Similar experiments in female rhesus monkeys showed that oestradiol, in amounts 1000-fold less than those administered systemically, exerted marked negative feedback actions when infused into the arcuate nuclei (Figure 10.12). In addition, the **negative feedback actions of progesterone during the luteal phase of the cycle also appear to operate in the arcuate region**, where oestradiol-induced progesterone receptors are found in considerable number.

What about positive feedback? In rats, **implantation of oestradiol in the anterior hypothalamic–preoptic area continuum has been shown to induce an LH surge** without evidence of diffusion of the steroid into the anterior pituitary. Moreover, if GnRH levels are monitored mid-cycle, GnRH levels rise in the anterior hypothalamic area. Whether existing neurons make more GnRH in response to high oestrogen or a new subset of neurons is recruited to become active in making GnRH (or indeed both) is unclear. However, in higher primates and humans, the anterior hypothalamic region is only sparsely populated by GnRH-containing neurons and the locus of positive feedback GnRH neurons seems to be the posterior arcuate nucleus. Indeed, evidence that a locus of hypothalamic feedback is necessary for the monkey is weak and is non-existent for the human, which may achieve the midcycle surge entirely at the level of pituitary sensitivity (Box 10.4).

These steroid feedback actions do not seem to involve GnRH-containing neurons directly, but to be mediated indirectly

Box 10.2 The mechanisms by which feedback control is achieved in the pituitary

What is the nature of the cellular mechanisms by which negative and positive feedback is exercised? What do the sex steroids and inhibin do to gonadotrophs to increase or decrease their sensitivity to GnRH?

Oestradiol appears to exert its positive feedback effects on the pituitary by inducing and maintaining GnRH receptors and by sensitizing a **self-priming process** whereby GnRH induces its own receptors. Thus, small-amplitude GnRH pulses, which do not by themselves cause an LH pulse, may prime a full LH response to the next adequate GnRH pulse. Figure 10.9 shows that the presence of oestradiol enhances this interaction between GnRH and its receptor, perhaps thereby contributing to the magnitude of the oestradiol-induced LH surge. There is less information on the ways in which oestradiol might cause a decrease in gonadotrophin secretion to mediate its negative feedback. Nor is there detailed information on the way that inhibin exerts its selective depressant effect on FSH secretion. A possible clue comes from the observation that the GnRH receptor has two *N*-linked glycosylation sites. Both oestrogen and inhibin increase glycosylation at one site, while prolonged oestrogen induces glycosylation at both. Since the stability of the intramembranous receptor is reduced by glycosylation, a potential negative feedback control mechanism is evident, but this possibility remains speculative.

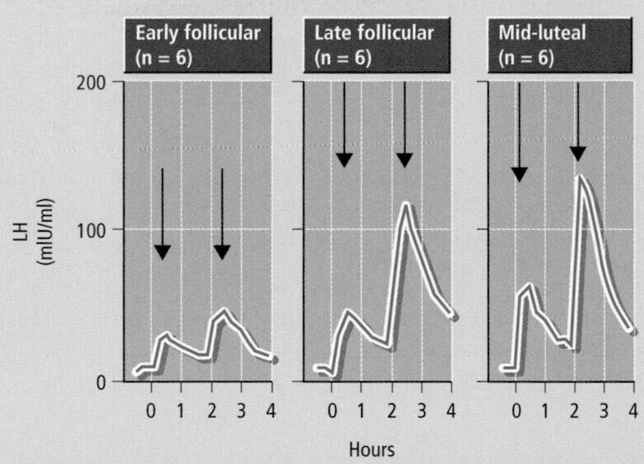

Figure 10.9 The 'self-priming' effect of GnRH and its dependence on steroid hormone levels. The figure shows the LH-enhanced response to a second injection of GnRH during three stages of the menstrual cycle (arrows indicate GnRH injection). Note the large increase between the early and late follicular phases, which correlates well with the rise in oestradiol secretion and, furthermore, how this effect remains pronounced in the mid-luteal phase when oestradiol, as well as progesterone, secretion is high. (Source: Yen S.S.C., Jaffe R. (eds) (1978) *Reproductive Endocrinology*. WB Saunders & Co, Philadelphia. Reproduced with permission from Elsevier.)

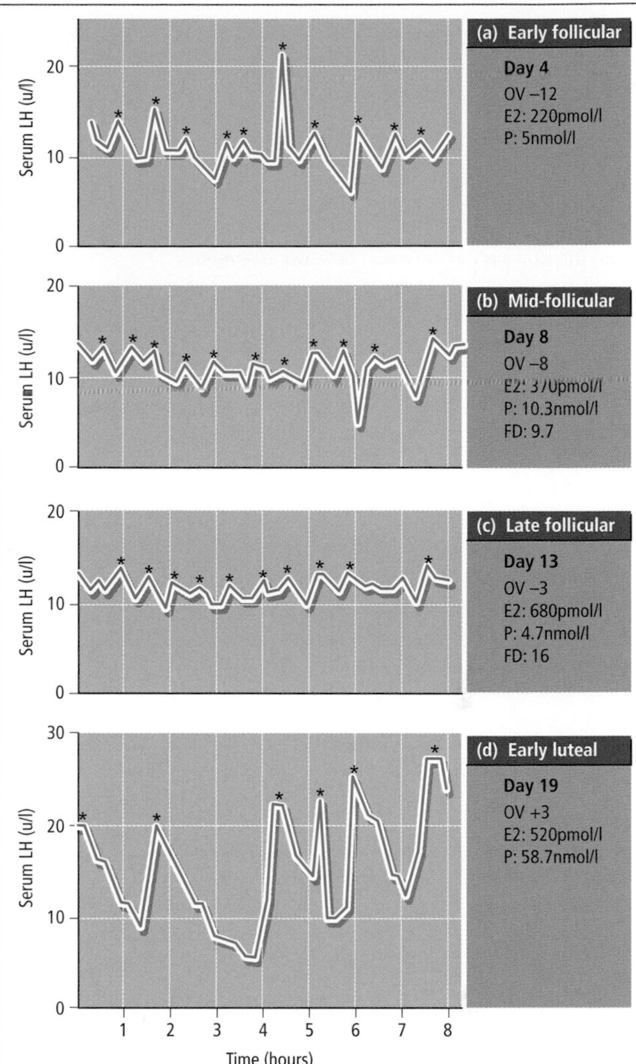

Figure 10.10 LH pulsatility at different points during a human ovulatory cycle (ovulation occurred on day 16) in 10-minute serum samples: (a) early follicular phase; (b) mid-follicular phase; (c) late follicular phase; (d) early luteal phase. Note the unique high-amplitude, low-frequency pulses characteristic of the luteal phase when progesterone plasma concentrations are high. *, LH peaks; OV, ± number of days after/before ovulation; E2, 17β-oestradiol; P, progesterone; FD, follicular diameter (mm). (Source: Clayton R.N. (1986) *Neuroendocrinology* (eds S.L. Lightman & B.J. Everitt). Blackwell Scientific Publications, Oxford. Reproduced with permission of John Wiley & Sons.)

through other neural systems with convergent effects on GnRH neurosecretion (see Box 10.3).

Summary

Taken together, these data indicate that while oestradiol and progesterone can act directly on the anterior pituitary to exert feedback effects on gonadotrophin secretion, the hypothalamus is also an important site for these effects via the modulation of

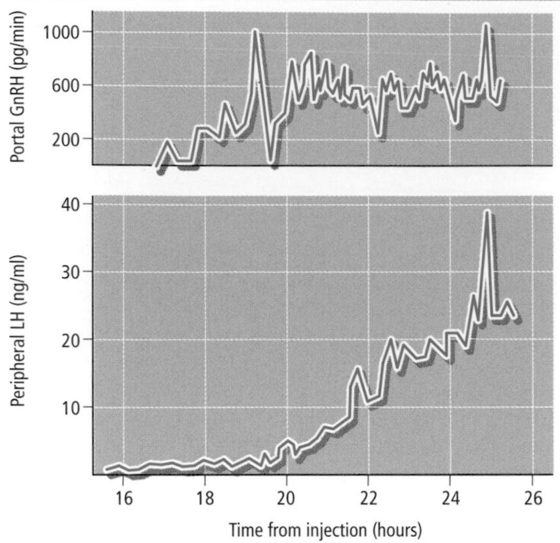

Figure 10.11 GnRH concentration in portal blood and peripheral plasma LH levels in an ovariectomized ewe given an injection of 50 mg of oestradiol monobenzoate to induce an LH surge. Note the increased frequency of both GnRH and LH pulses during the LH surge.

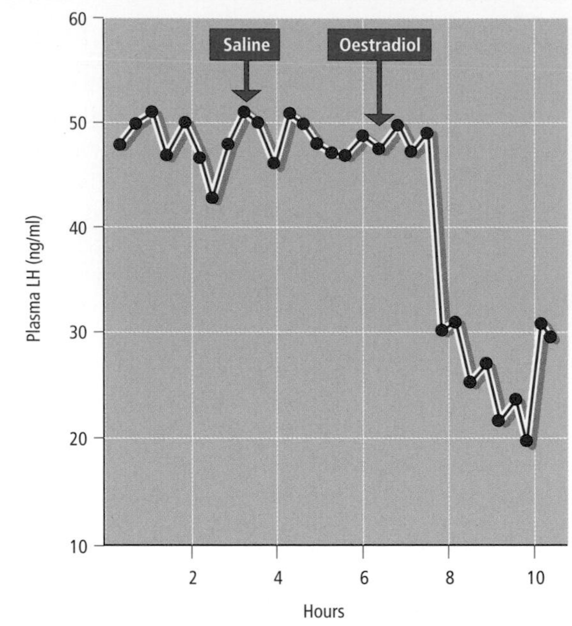

Figure 10.12 The negative feedback effect of oestradiol (800 pg) on plasma LH when infused into the hypothalamic arcuate nucleus of ovariectomized rhesus monkeys. Note the lack of response to earlier, control injections of saline and the decline in LH concentrations within a short time after oestradiol injection.

GnRH secretion. Inhibin appears to exert its effects on FSH secretion in the anterior pituitary. Thus, the female system of feedback resembles that in the male but is far more complex. Does this represent a fundamental difference between males and females or can males also show a gonadotrophin surge; and, if so, are the hypothalami fundamentally sexed (see Chapters 3 and 5)? The answer depends on the species studied, but, as we saw earlier, the answer for humans seems to be No (see Box 10.4).

The ovarian cycle in relation to the oestrous and menstrual cycles

Thus far, we have considered the cyclicity of reproduction in the female from the viewpoint of the ovary. However, the cyclical output of ovarian steroids imparts cyclicity to the anatomy and physiology of the whole female. As we saw in Chapter 6 (see pages 97–110) this external manifestation of the internal ovarian cycle can be recognized in many mammals by the behavioural characteristic of being only receptive to males or **on heat**, and therefore willing to mate, around the time of ovulation. This period of heat is conditioned by the hormonal milieu of the preovulatory phase, and can be accompanied by marked changes in behaviour patterns. The frisky state observed in some species, particularly in horses, was given the name **oestrus** (a noun describing 'a state as though attacked by gadflies'), and thus the internal ovarian cycle is manifested externally in the so-called **oestrous cycle** (Figure 10.14a). **Day 1 of each oestrous cycle is generally considered to be the day of first appearance of oestrous behaviour**.

Higher primates show little evidence of oestrus and thus it is inappropriate to speak of their oestrous cycle. However, in these primates, another external manifestation of ovarian cyclicity is observed: the shedding of bloody endometrial tissue via the vagina at the end of the luteal phase. This hormonally conditioned event is called **menstruation** (i.e. monthly event), and is used as the external basis for the measurement of the **menstrual cycle** in which **the first day of menstruation is considered day 1 of the cycle**, as may be observed in Figure 10.14b. Although both the oestrous and menstrual cycles reflect ovarian cyclicity, **the starting days for each cycle occur at different points in the underlying ovarian cycle**.

This brings us to a consideration of exactly what cyclical functional changes in the bodies of women are brought about by the ovarian cycle. The effects of steroids on the various tissues and organs may be studied by correlating changes in the cycle with changing steroid levels or experimentally by observing the consequences of injection of exogenous steroids into intact or castrate females.

The ovarian cycle and the oviduct (Fallopian tube)

The oviduct is a thin muscular tube covered externally with serosal tissue and peritoneum. A ciliated, secretory, high columnar epithelium overlies the stromal tissue internally. The oviduct is the site of fertilization and therefore the oocyte passes along it from the **fimbriated ostium** towards the spermatozoa, which

Box 10.3 Neuropharmacological systems and the steroid regulation of GnRH output

The arcuate nuclei and preoptic/anterior hypothalamic areas in rodents are rich in oestradiol receptors, and the GnRH content of neurons in this area changes in response to oestradiol. However, use of double-staining techniques has shown that GnRH + ve neurons do not express progesterone or oestradiol-α (ERα) receptors, and express ERβ only weakly. Thus, steroid effects on GnRH secretion must be mediated indirectly via other steroid-sensitive neural systems, which then converge onto GnRH neuronal cell bodies or terminals. The problem is that the GnRH + ve neurons receive, directly or indirectly, hundreds if not thousands of synaptic contacts from elsewhere in the brain – a complex network of potential neural influences. The main neurotransmitter systems implicated in the regulation of GnRH output during puberty were described in Box 4.2 and Figure 4.13. Negative regulators include γ-aminobutyric acid (GABA) and the opioid β-endorphin, whilst positive regulators include glutamate, noradrenergic and neurokinin B neurons. The same regulatory systems operate in the adult – but how? The task is to identify those key controlling pathways that seem to be steroid-sensitive and to establish some sort of hierarchy of importance among them.

Some 70–80% of rodent neurons containing GABA in the preoptic and mediobasal hypothalamic areas (yellow in Figure 4.13) bind oestradiol. Using *in vivo* microdialysis to measure extracellular transmitter levels in small regions of the brain, fluctuations in hypothalamic GABA release are found to correlate inversely with LH pulses in the peripheral circulation. Additionally, GABA receptor antagonists such as biciculline promote LH release. Conversely, progesterone enhances GABA release, which may mediate its negative feedback effects during the luteal phase. Moreover, systemic surge levels of oestradiol are associated with decreased GABA release in the mediobasal hypothalamus, implicating the lifting of GABA-mediated inhibition of GnRH output in the LH surge. Thus, **GABAergic neurons do seem to play a key inhibitory role in both negative and positive feedback in mice and rats**. β-Endorphin is found in a subset of arcuate nucleus neurons that richly innervate the medial preoptic area GnRH-containing neurons (green in Figure 4.13). Hypothalamic β-endorphin concentrations fluctuate during the cycle, with luteal highs and follicular lows. When given intraventricularly, β-endorphin suppresses pulsatile LH release and its preovulatory surge, whereas its antagonist naloxone accelerates LH pulses and elevates serum LH. Thus, these **β-endorphin–containing neurons may mediate part of the negative feedback effects of gonadal steroids, particularly progesterone**.

Glutamate levels in the vicinity of the glutamate receptor expressing GnRH cell bodies (but not terminals) increase around the time of the LH surge and this increase is oestrogen-dependent. Moreover, glutamate receptor antagonists interrupt GnRH pulsing and block the LH surge. In female rats or mice with poor LH surges, experimentally shifting the **GABA:glutamate balance** in favour of the latter restores full surges in the presence of oestrogen. Noradrenergic neurons in the brainstem medulla oblongata (black in Figure 4.13) richly innervate the hypothalamus and synapse directly on both preoptic and median eminence GnRH + ve cell bodies, which express both α- and β-adrenergic receptors. Noradrenergic neurons express oestrogen α-receptors, and an increase in noradrenergic transmission is observed mid-cycle as the ovulatory surge of GnRH occurs. Overall, this system may facilitate oestradiol positive feedback.

Finally, in Chapter 4 (see pages 64–5) the involvement of the neuropeptide transmitter, **kisspeptin 1** and its receptor **Kiss1R**, in the regulation of GnRH release at puberty was described. Kiss-1 + ve neurons, which are present in the arcuate nuclei of both mice and humans, but in the anteroventral periventricular nuclei (AVPV) of mice only, are also direct targets of oestrogens, nearly all expressing oestrogen receptor α (ERα), and some 30% expressing ERβ. Levels of *Kiss-1* mRNA in the arcuate nucleus of both mice and humans increase after gonadectomy and decrease with sex steroid replacement, an effect blocked by targeted deletion of the ERα. Thus, **changes in arcuate kisspeptin synthesis follow a pattern to be expected for negative feedback**, and the neurons of the arcuate nucleus thus appear well placed to mediate the hypothalamic negative feedback effects of steroids on GnRH secretion. Conversely, in the anteroventral periventricular nuclei (AVPV) of mice, castration depresses expression of *Kiss-1* mRNA, but oestrogen replacement increases it. As the AVPV nuclei project to the preoptic area implicated in generating the preovulatory GnRH/LH surge, **kisspeptin neurons seem to mediate the positive feedback loop** in mice. Moreover, the ventral posteromedial nucleus (VPMN) is sexually dimorphic in mice, being larger and having more Kiss-1 + ve neurons in females – and mouse brains are known to show functional dimorphism with only females capable of eliciting a positive feedback response (see Box 10.4). The issue of whether, and, if so, where in the human hypothalamus, kisspeptin is involved in positive feedback is unclear at present (see Box 10.4).

Box 10.4 Is the adult hypothalamus sexed for feedback?

If adult male rats are castrated and receive ovarian transplants, they fail to show any cyclical changes. In contrast, if ovaries are transplanted into recipient male rats castrated at birth, they undergo cyclical ovulation. Clearly, the presence of the testis at birth has prevented subsequent support of ovarian cyclicity. Testicular androgens are responsible for this effect. Thus, female rats injected with testosterone during the first few days after birth do not show oestrous cycles in adulthood. Their ovaries contain follicles that secrete oestrogens (they are said to be in constant oestrus), but as ovulation does not occur, there are no corpora lutea. Neonatal androgen causes acyclicity by suppression or modification of the oestradiol positive feedback mechanism. Thus, if male or female rats are castrated in adulthood and are subsequently injected with oestradiol, only the females show a surge of gonadotrophins (Figure 10.13a). If the same experiment is undertaken with female rats given testosterone during the first few days of life, no surge is observed. The neonatal testosterone acts on the rat hypothalamus to suppress the positive feedback response. Thus, the ovaries or the pituitaries of neonatally androgenized females are quite capable of secreting surge levels of oestrogen or LH if transplanted into normal females, so their functional capacity does not seem to be grossly impaired. Thus, **in the rat, the 'masculinizing' effect of neonatal testosterone is exerted on the hypothalamus**. Moreover, fewer kisspeptin containing neurons are found in the anteroventricular nucleus in males or in females androgenized early in life compared with females and early castrated males, suggesting a basis for the differential existence of positive feedback in the rodent (see Box 10.3). This 'masculinization' of the brain occurs in most species

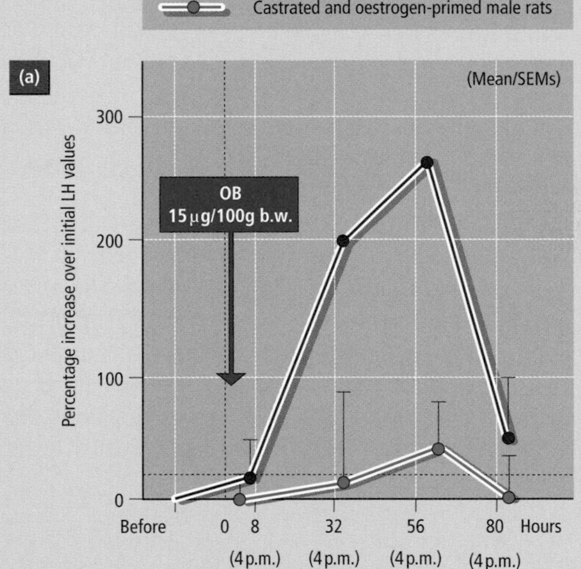

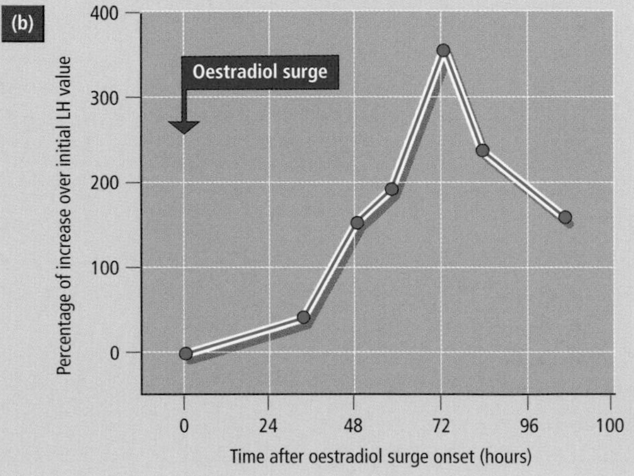

Figure 10.13 (a) The positive feedback effect of oestradiol in castrated male and female adult rats. The LH response, measured as a percentage change from resting values, to a single large injection of oestradiol benzoate (OB) is shown to be present only in the female (red) and not in the male (green). This is taken as evidence of sexual differentiation of the hypothalamus, neonatal androgens in the male preventing the ability to respond to an oestrogen surge with an LH surge in the adult. (b) Positive feedback effects of oestradiol in a castrate male talapoin monkey. Unlike the results in (a), the normal male monkey is able to respond to an oestradiol surge with an LH surge similar to that seen in the female. This suggests that exposure of the primate's brain to testosterone *in utero* does not 'masculinize' the hypothalamus as has been demonstrated in non-primate mammals. (Source: Dorner G. (1979) *Sex, Hormones and Behaviour*. Ciba Foundation Symposium 62. Excerpta Medica, Amsterdam. Reproduced with permission of John Wiley & Sons.)

(rodents, sheep and some carnivores), although the critical period of sensitivity to the effects of androgens varies considerably. For example, guinea pigs have a gestation period of 68 days, compared to 21 days in the rat, and are born in a state of relative maturity. The critical period during which androgens exert their effects on the brain is pre- and not post-natal in the guinea-pig. Similar considerations apply to large domestic animals and carnivores.

In contrast, evidence that a locus of hypothalamic positive feedback is necessary for the monkey is weak and is non-existent for the human, which may achieve the midcycle surge entirely at the level of the pituitary. Thus, the fact that the injection of exogenous kisspeptin in the preovulatory phase of the menstrual cycle, when oestrogen is highest, gives a much greater increase in LH secretion compared to injection at other time points, may simply reflect the greater sensitivity to GnRH of the pituitary. Indeed, experiments on primates indicate that 'masculinization' of the hypothalamus does not occur in the same way, and that the capacity for positive feedback exists in normal male monkeys and men, and this difference from the mouse may be a reflection of the absence of kisspeptin + ve neurons in the anterior hypothalamic region of the primate brain. Thus, in castrated male monkeys (as well as in hypogonadal and castrated men), an administered oestrogen surge induces a gonadotrophin surge (Figure 10.13b). Indeed, ovaries transplanted into castrated male monkeys undergo apparently normal monthly ovulatory cycles, in marked contrast to the results of similar experiments in rats. However, the positive feedback action of oestradiol cannot be elicited in intact male monkeys for reasons that are not clear, but presumably results from a suppression of positive feedback by testosterone akin that seen for progesterone. Female rhesus monkeys exposed to high levels of testosterone during fetal life are also able to show menstrual cycles as adults, although puberty occurs slightly later than usual. Indeed, in some of these monkeys the external genitalia are so 'masculinized' that menstruation occurs through a penis-like phallus. Similarly, human females exposed to high levels of androgens during gestation, e.g. as the result of the adrenogenital syndrome, also have menstrual cycles as adults, often after a somewhat delayed puberty. Clearly, **'masculinization' of the hypothalamic mechanism underlying positive feedback does not occur in male primates, including men**.

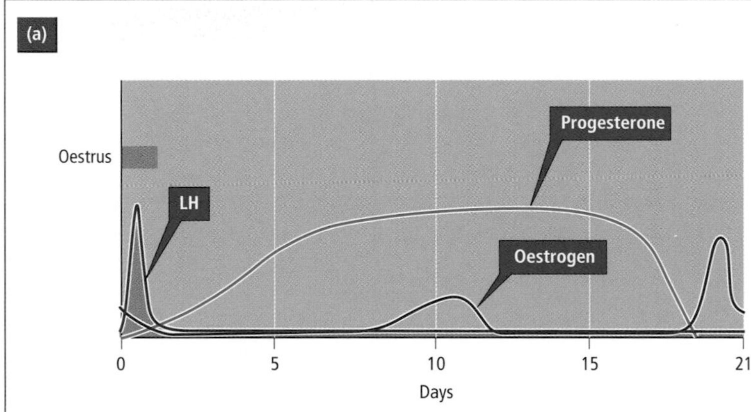

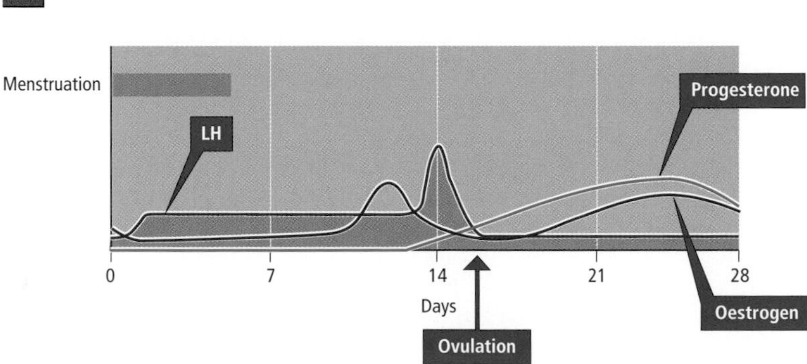

Figure 10.14 Schematic views of the ovarian cycles in (a) the cow and (b) the human, expressed as oestrous and menstrual cycles. The first day of the cycle coincides with the beginning of either oestrous or menstruation (plasma concentrations not to scale).

pass in the opposite direction from the isthmic junction with the uterus (see Figure 11.6). A few days later, the fertilized zygote reverses the path taken by the spermatozoa to enter the uterus (see Chapters 11 and 13). Clearly then, the **oviduct has a role in gamete transport**. It must also provide a **suitable environment for fertilization** and the earliest development of the conceptus.

After ovariectomy, oviducal cilia are lost, secretion ceases and muscular activity declines, indicating an important steroidal influence. Subsequent injections of oestrogen restore both the ciliated, high columnar epithelium and secretory activity, and increase spontaneous muscle contractions. When progesterone is imposed on this oestrogen background, the numbers of cilia decline and the quantity of the oviducal secretion also declines, the small volumes of fluid that are produced having a lower sugar and protein content. Raising the progesterone:oestrogen ratio may also exert a mildly depressant effect on oviducal musculature, particularly relaxing the sphincter-like muscle at the uterotubal junction (although data on this are somewhat conflicting; see Box 10.5 for more discussion).

The ovarian cycle and the uterus

The uterus shows even more prominent steroid-dependent cyclical changes in structure and function than does the oviduct (Figure 10.15). During each cycle, the uterus first prepares to

Box 10.5 The oviduct, hormones, emergency contraception, tobacco and cannabis

Historically, abrupt changes in the oestrogen:progesterone ratio caused disturbances in gamete and conceptus transport. Thus, when high doses of the synthetic oestrogen stilboestrol were taken within 72 hours of fertilization, the pregnancy rate was reduced. This reduction was thought to be due to either **defects in sperm or oocyte transport**, and thus reduced fertilization, or **premature expulsion** of the conceptus from the oviduct. The drug was also associated with an increased incidence of **ectopic oviducal pregnancy**, leading to the suggestion that in these cases the embryo had become '**tube-locked**' – perhaps due to spasm of the oviducal musculature *preventing* passage of the conceptus to the uterus.

Modern hormonal emergency contraception (the 'morning-after pill')

Two different pills are available, neither of which work after ovulation has occurred. One contains high-dose progestagen only (1.5 mg levonorgestrel for up to 72 hrs after unprotected sexual intercourse; prevents c. 69% of pregnancies), whilst the other contains a progesterone receptor modulator (30 mg tablet of ulipristal acetate; prevents c.85% of expected pregnancies if taken within 120 hrs of unprotected intercourse).In both cases, the efficacy declines with time since coitus. Both preparations have few side effects in around 10% of users, such as headache, nausea and dysmenorrheoa, and if vomiting is present, this may adversely affect efficacy by drug expulsion.

How do oral preparations act?

The answer to this question has ethical implications for those who place a higher moral value on a fertilized than an unfertilized oocyte. The balance of evidence suggests that both work primarily by ovulatory suppression (see page 185), but the progestagen is ineffective once the surge has got underway. Thus, these actions reduce fertilization rates, although this has not been measured directly in women. Thus, the cases of contraceptive failure are largely due to the fact that the women already had fertilized eggs. The most effective pre- and **post-fertilization emergency contraception** (within 5 days of fertilization) is achieved by inserting a copper intrauterine device (see pages 358–9), which acts by killing spermatozoa but also will kill any fertilized eggs that do form.

Tobacco, cannabis and tubal function?

Recent evidence has suggested that cannabinoids can induce abnormalities of gamete/conceptus transport through an action on noradrenergic release, raising the question: might use of cannabis influence ectopic oviducal pregnancy rates? Similarly, tobacco smoking is associated with effects on tubal motility and higher ectopic rates.

Further reading

Faculty of Sexual and Reproductive Health (2017) Emergency contraception. *FSRH Guideline.* March 2017.

Gemzell-Danielsson K, Marions L (2004) Mechanisms of action of mifepristone and levonorgestrel when used for emergency contraception *Human Reproduction Update* **10**, 341–348.

Talbot P, Riveles K (2005) Smoking and reproduction: the oviduct as a target of cigarette smoke. *Reproductive Biology and Endocrinology* **28**, 3:52.

Wang H, Guo Y, Wang D *et al.* (2004) Aberrant cannabinoid signaling impairs oviducal transport of embryos. *Nature Medicine* **10**, 1074–1080.

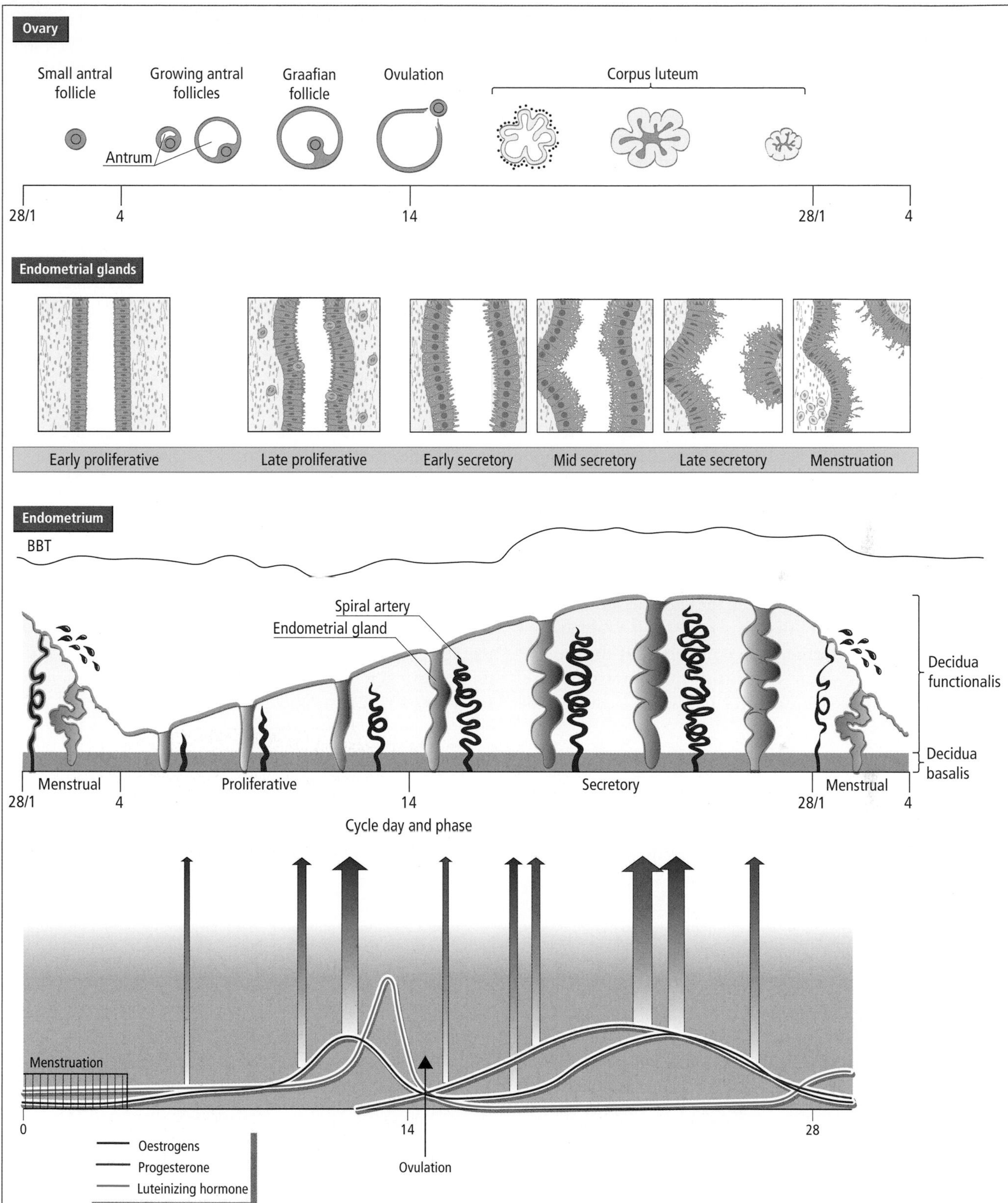

Figure 10.15 Changes in human endometrium during the menstrual cycle. Underlying steroid changes are indicated below and basal body temperature (BBT) is indicated above. Thickness of arrows (oestrogens, red; progestagens, blue) indicates strength of action on the uterus. Source: Heffner L.J., Schust, D.J. *The Reproductive System at a Glance*, Fourth Edition. 2014, John Wiley & Sons, Ltd. Reproduced with permission of John Wiley & Sons.

receive and transport the spermatozoa from the cervix to the oviduct (see Chapter 11), and subsequently prepares to receive the conceptus from the oviduct and nourish it through to term (Chapters 13 and 14). The uterus consists of an outer peritoneal and serosal investiture, over a thick **myometrium** of smooth muscle arranged in distinctly orientated layers, which show spontaneous **fundo-cervical, cervico-fundal and isthmo-fundic peristaltic activities** throughout the cycle, but the balance of each varies (see below). Internally, the **endometrium** consists of a **stromal matrix** over which lies a simple low columnar **luminal epithelium** with **glandular epithelial** extensions penetrating into the stroma (Figure 10.16).

After ovariectomy, all uterine tissues hypotrophy and the blood supply is reduced. Administration of oestrogen reverses this effect, with massive increases in mRNA and protein synthesis, cellular division and growth. In the normal cycle, the period of rapid oestrogen rise coincides with the latter half of the follicular phase of the menstrual cycle or, in non-primates, with the whole of the abbreviated follicular phase. During this **follicular phase, a similar uterotrophic effect of oestrogen** is observed. The **oestrogen-dominated myometrium increases both its contractility and its excitability**, and cervico-fundal peristalsis comes to dominate as ovulation approaches. Meanwhile in the endometrium, stromal thickening occurs, partly due to stromal cell proliferation. Reflecting this, the **uterine cycle equivalent of the ovarian follicular phase** is often called the **proliferative phase.** Stromal oedema also contributes to the thickening. The surface epithelium increases in surface area and metabolic activity. In primates and large farm animals, this also involves an increase in the number, and size, of the glandular invaginations of the stroma; this is less marked, however, in rabbits and rats. The oestrogen-primed epithelial cells secrete a fluid of a characteristically watery constitution, which contains a range of proteins, including proteolytic enzymes. These changes reach a maximum at the time of the oestrogen surge.

The oestrogens act by binding to oestrogen receptors (mainly the α receptors) present in abundance in uterine tissue. One of the most **critical actions of oestrogens** over this period is to **induce the synthesis of progesterone receptors.** At the beginning of the oestrogenic phase of the cycle, progesterone-binding receptors are at a low level, and progesterone therefore has little effect on the uterus. However, by the time of ovulation, the uterus is primed to bind progesterone, and so begins the progestagenic phase. In the rabbit and mouse, there is a rapid extension of the epithelial proliferation into glandular regions at this time (Figure 10.16b).

Progesterone stimulates the glands to synthesize a thick secretion rich in glycoprotein, sugars and amino acids. For this reason, the **luteal phase of the ovarian cycle** corresponds to the **secretory phase of the uterine cycle.** In many species, the release of this glandular secretion into the lumen requires, or is facilitated by, the secondary peak of luteal-phase oestrogen imposed upon the progesterone background (see pages 151–2). Stromal proliferation also increases under

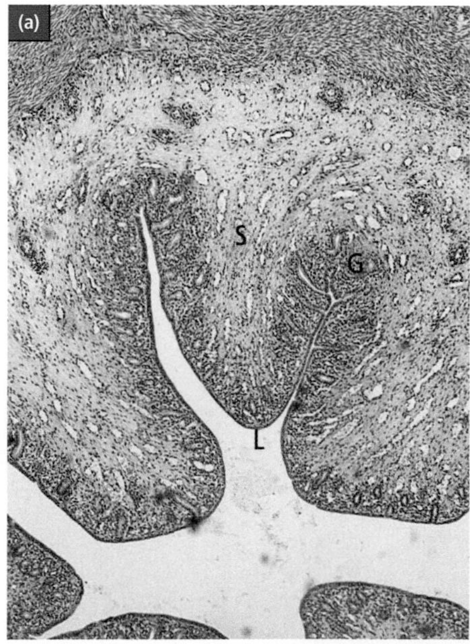

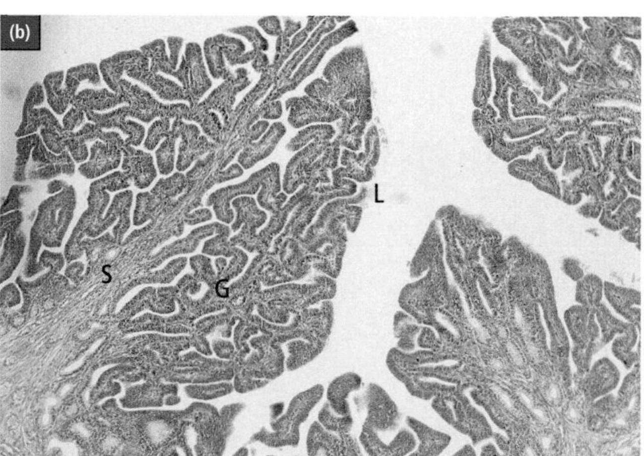

Figure 10.16 Sections through rabbit endometrium during: (a) the late follicular phase (oestrus) and (b) the secretory phase. Note the dramatic increase in invaginations of the glandular epithelium (G) from the luminal surface epithelium (L) into stromal tissue (S).

the influence of progesterone, the stromal cells becoming larger and plumper, particularly in rodents and primates. Within the stromal tissues of primates, characteristic **spiral arteries** become fully developed (see Figure 10.15). Curiously, but importantly, prolonged progesterone exposure of the uterine epithelia (but not of the stroma) causes down-regulation of progesterone receptors and a corresponding rise in oestrogen receptors, indicating that **luteal oestrogen may act primarily through epithelial cells**. Progesterone also acts on the myometrium causing further enlargement of cells but, in contrast to oestrogens, **progesterone depresses the overall excitability** of the uterine musculature such that the fundus is quiescent, and only limited range peristalsis occurs from cervical and isthmic foci. It is important to re-emphasize that these actions of progesterone will only occur in an oestrogen-primed uterus, another example of receptor regulation.

With the withdrawal of steroid support at the end of the luteal phase of the cycle, the elaborate secretory epithelium collapses, with evidence of apoptotic cell death. In most mammals, the endometrium is resorbed and a thin stromal layer overlain with epithelium replaces it, ready for entry into a new uterine cycle as oestrogens rise. In humans, apes and Old World monkeys, the endometrial tissue is shed via the cervix and vagina, together with blood from the ruptured arteries, as the **menses**. The spiral arteries contract to reduce bleeding.

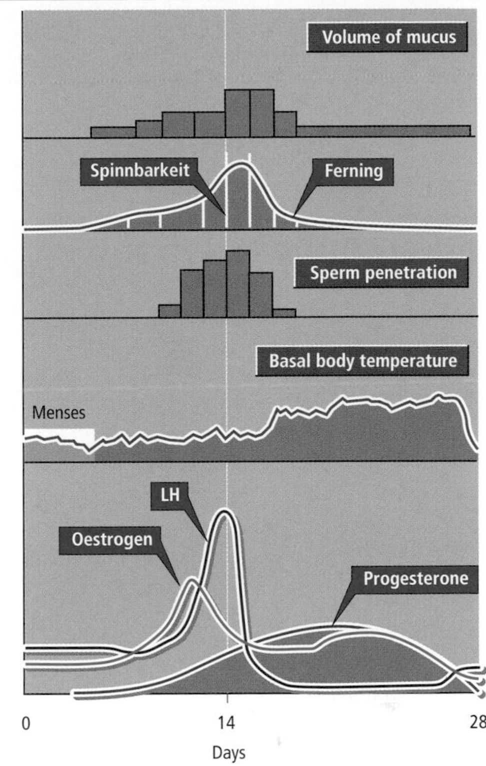

Figure 10.17 Schematic view of changes in properties of cervical mucus at various days of the human cycle (blood hormone levels and basal body temperature shown). Parameters changing under the influence of high oestrogen and low progesterone are: volume of mucus; **spinnbarkeit** of mucus (bar height illustrates relative length to which mucus thread can be stretched before snapping); **ferning** (curve illustrates proportion of crystallized mucus that shows a ferning pattern when dried on a slide); and *in vitro* tests of the ability of spermatozoa to penetrate mucus. Note luteal oestrogen does not induce changes due to elevated progesterone at the same time. Progestagenic contraceptives, including the emergency 'morning-after' pill, prevent or reduce normal periovulatory changes (see page 357). LH, luteinizing hormone.

The ovarian cycle and the cervix

The cervix is traversed by the spermatozoa at coitus and the neonate at parturition (see Chapters 11 and 18). In many species, including humans, the spermatozoa must actively swim through the cervix, which may also act as a storage reservoir for them (see Chapter 11). The properties of the cervix show marked steroid-dependent changes during the cycle and these can be crucial for fertility. During exposure to oestrogen **in the follicular phase, the muscles of the cervix relax and the epithelium becomes secretory**. However, during the luteal phase, **when progesterone levels are elevated, secretion is reduced and the cervix is firmer**.

Cervical mucus may be collected for examination during the human cycle and tested in a variety of ways for its steroid-dependent properties. The test of greatest functional significance is that of **sperm penetration**, in which the capacity of spermatozoa to swim into and through a smear of mucus on a slide is assessed. Characteristically, sperm penetration is low in the early follicular and the luteal phases of the cycle, and reaches a maximum around the time of ovulation (Figure 10.17). These effects can be mimicked by administration of exogenous steroids: **oestrogens enhance sperm penetration** while addition of **progesterone depresses penetration**, even in the presence of oestrogens. Thus, continuous administration of progestagens throughout the cycle, or the local release of progestagens from capsules placed in the uterus, suppresses sperm penetration even at the time of ovulation and the oestrogen surge. Use is made of this property in low-dose progestagenic contraceptives (see Table 21.4).

The steroids act via an effect on the amount and nature of the glycoproteins secreted by the cervical epithelium. Rising oestrogens stimulate the production of **Muc5B** and **Muc4**, both hydrophilic mucins that may form an aqueous matrix through which spermatozoa can swim, and which also may trap pathogens. The mucinous gel matrix may also hold the cervical canal patent. If mucus from oestrogenic cervices is allowed to dry on a slide, its distinctive molecular composition results in a characteristic pattern known as

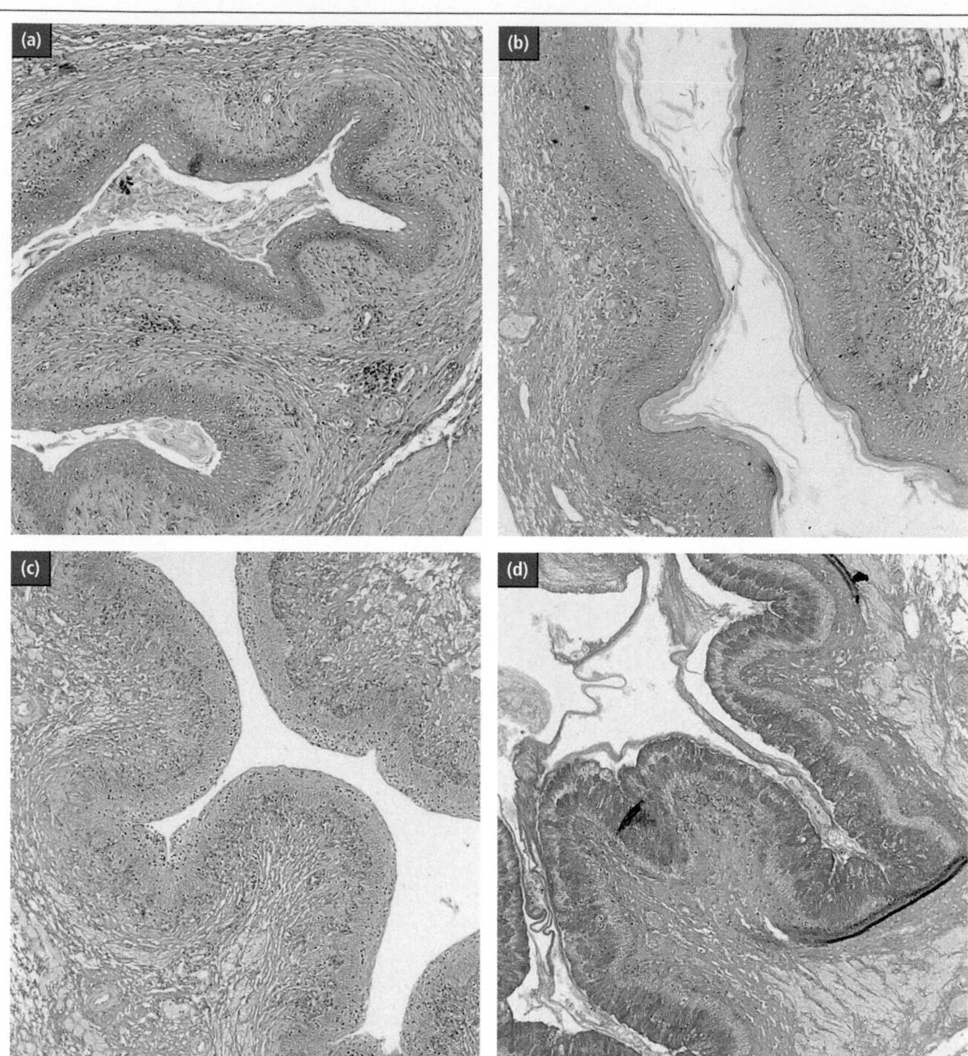

Figure 10.18 Sections through the rat vagina to illustrate changes in the vaginal epithelium in response to changing hormones during the reproductive cycle. (a) Rising oestrogen during the follicular phase leading up to oestrus (**pro-oestrus**) causes nucleated cells (small dark spots) to be shed into the lumen and a keratin layer starts to develop on the surface epithelium. (b) This process is completed at oestrus with a heavily keratinized layer (yellow), some of which desquamates into the lumen. (c) During the short transition from oestrogen to progesterone dominance immediately after ovulation (the progestational phase of **dioestrus**), the epithelial cells are nucleated and non-keratinized, and leucocytes are visible in the epithelium and pass into the lumen. (d) Under progesterone dominance during pregnancy, the surface epithelium is glandular and secretory (blue).

ferning (Figure 10.17). Under progestagen dominance, secreted mucin levels decline precipitously, and strands of mucus can only be stretched a short length before the threads snap: a low **spinnbarkeit** (Figure 10.17). These simple tests of cervical mucus are important, because a hostile, impenetrable cervix will reduce the progress of the sperm towards the oviduct and thus fertility.

The ovarian cycle and the vagina

The vagina is the initial site of sperm deposition and is vulnerable to infection (Chapter 11). It must distend sufficiently for parturition (Chapter 18). It shows marked structural changes during the cycle in some species. In the guinea pig, for example, a membrane completely closes the **vaginal os** throughout most of the cycle, and only breaks down under the influence of rising plasma oestrogen concentrations at oestrus. In many mammals, including humans, oestrogens induce an increased mitotic activity in the columnar epithelium of the vagina, with a tendency to keratinize. This change is particularly marked in rodents, in which the stages of the oestrous cycle can be assessed reliably by examination of the different cell types present in daily smears from the vaginal epithelium (Figure 10.18). The fluids within the vagina also change during the cycle, and one effect of this is to vary the metabolic substrates available to the bacterial flora there. Cyclical changes in the vagina, induced by oestrogen and progesterone, result in the generation by bacteria of differing proportions of volatile aliphatic acids. These give distinctive odours to vaginal secretions and may have marked behavioural consequences, as was described in Chapter 6. The leukocyte population of the vagina is also steroid-sensitive, an influx of polymorphonuclear neutrophilic leukocytes (PNLs) occurring under progesterone domination (Figure 10.18).

The ovarian cycle and immune function in the female genital tract

The female tract has evolved to receive and transmit spermatozoa. This function exposes it to pathogens, which is particularly dangerous given the patent continuity between the vagina and the peritoneal cavities. However, spermatozoa and the conceptus itself differ antigenically from the mother. There is thus a potential immune conflict for the uterus, the resolution of which is not yet fully understood. In addition, immune cells play roles in tissue remodelling, which is substantial during the menstrual cycle. It is clear that a dynamic, steroid-dependent population of immune cells and molecules characterizes the endometrium, some 7% of its cells being leukocytes. Chemokines that attract immune cells are synthesized in the uterus in patterns that vary with the steroid balance. Macrophages peak in the late secretory phase, and are thought to be involved in tissue breakdown associated with menses. Neutrophils are present in small numbers through most of the cycle, increasing massively as menstruation approaches because of progesterone withdrawal. During the secretory phase, **uterine-specific natural killer cells** (uNK cells) predominate in close contact with glands and blood vessels. Progesterone may stimulate uNK cells indirectly via stromal cells, which possess progesterone receptors and secrete interleukin 15 (IL-15) and prolactin (which in turn stimulates IL-2 production). We return to the vexed question of maternal immune function in Chapter 16 (see pages 274–5).

The ovarian cycle and other tissues

Many other features of a female's anatomy and physiology may change with the ovarian cycle. Oestrogens have general effects on the cardiovascular system and metabolism that may be revealed cyclically or in women taking oral contraceptives. Thus, **oestrogens depress appetite, are mildly anabolic and maintain bone structure**. They also appear to be involved in the reduced capillary fragility and higher levels of low- and high-density lipoproteins, and thus in the ability to bind cholesterol, which may account for the reduced incidence of thrombosis seen in premenopausal women when compared to men. Oestrogens may also increase the ability of the cardiovascular system to withstand high blood pressures. The mechanisms by which oestrogens act in this way are unclear. However, maintained elevated levels of oestrogens, such as occur in pregnancy or during treatment with some contraceptive pills, may in some women actually cause hypertension and, via effects on lipid metabolism, an increased blood-clotting rate (see Chapter 21).

Progesterone, in contrast to oestrogens, is mildly catabolic in humans. It has two major sorts of systemic action that may become manifest during the menstrual cycle. The first is on the CNS: so-called neuroactive effects. **Progesterone elevates basal body temperature** (Figure 10.17) by a direct action on hypothalamic areas concerned with thermoregulation. The rise in temperature occurs only after ovulation and thus is only useful as a basis for establishing the regularity of a woman's cycle if the rhythm method of contraception is contemplated (see page 353). Progestogenic steroids, such as progesterone, pregnenolone, dehydroepiandrosterone (DHEA) and 5α-reduced pregnenolone (allopregnenolone) also have **anxiolytic effects**, and the fall in progestagens towards the end of the menstrual cycle can result in the loss of this property and the development of anxiety, excitability and disturbances of mood (**premenstrual tension**). Progesterone shows some affinity for aldosterone receptors in the kidney, presumably because of similarities of stereochemical structure. However, after binding by progesterone, the receptor is not activated and, in consequence, progesterone acts as an inhibitor such that natriuresis ensues. A compensatory rise in aldosterone output occurs to restore sodium (Na^+) retention. Retention of Na^+ in women may also be enhanced in the luteal phase by a direct stimulatory effect of luteal oestrogen on angiotensinogen production. The consequence of these events may be a **net retention of Na^+ and water towards the end of the luteal phase**, which contributes to some symptoms characteristic of the premenstrual period, e.g. heavy, tender breasts.

These examples of the widespread cyclical changes in structure and function of many of the somatic tissues of a female, and not just her reproductive organs, show how the ovary and its secretions play such a dominant part in day-to-day physiology.

Conclusions

The basic principles underlying male and female reproductive function in the adult are similar, but considerable sex differences have evolved to yield the reproductive strategies described in Chapter 1 (see pages 4–8). Thus, the unique female role resulting from internal fertilization and viviparity imposes very different temporal fecundity patterns on her. Nonetheless, these patterns do vary in detail among different mammals, tailored through natural selection to their own reproductive strategies.

Thus far, we have considered reproductive development and function in the male and the non-pregnant female. We have seen how the gonadal hormones play a coordinating role in the production of mature gametes and the conditioning of reproductive function and sexual behaviour to maximize the chance of fertilization. Should it occur, a dramatic change must ensue, particularly for the female. Her whole anatomy and physiology must change from cyclical fertility patterns to nidatory and pregnancy patterns. It is this critical transition that constitutes the fourth part of this book.

Key learning points

- The ovarian cycle and the menstrual cycle are interlinked: the follicular and luteal phases of the ovarian cycle corresponding to the proliferative and secretory changes of the menstrual cycle.
- Oestradiol and progesterone secreted by the ovary exert negative feedback control over GnRH, FSH and LH secretion by actions within both the hypothalamus and the anterior pituitary.
- Oestradiol, in addition to negative feedback effects on LH and FSH secretion, has a unique ability to induce the ovulatory surge of LH secretion by a positive feedback action at both sites in the rodent but not in humans, for which only a pituitary site of action seems likely.
- Ovarian inhibin controls FSH secretion by negative feedback actions in the anterior pituitary.
- The pattern of hormone secretion during the menstrual cycle is determined by these feedback interactions between ovarian hormones and GnRH, LH and FSH secretion.
- Ovarian steroid feedback regulation of GnRH secretion is mediated indirectly by a variety of neural mechanisms within the hypothalamus that seem to converge on kisspeptin 1.
- Coitus itself may induce ovulatory discharges of LH by a neural mechanism in species, therefore, known as reflex ovulators.
- Oestradiol and progesterone regulate cyclical changes in the oviduct, uterus, cervix and vagina that are critical for gamete transport, fertilization and implantation.
- Oestrogens and progesterone have other effects on, for example, appetite, bone metabolism, vascular function, body temperature regulation, mood and water/mineral balance.

Clinical vignette

Polycystic ovarian syndrome

A 19-year-old woman, Ms R., was distressed by long-standing acne on her face and upper back, which she perceived as unattractive. She felt that her acne was preventing her from being confident in social situations and from progressing in her career as a journalist. She sought help from her general practitioner (GP) about possible treatments. Her GP established that she had a long-standing history of irregular periods, which occurred approximately 2–3 times a year. She suffered from increased hair growth on her upper lip and chin, for which she underwent regular waxing. Her body mass index was 32 (classified as obese). Ms. R's GP requested an ultrasound scan of her pelvis. The pelvic ultrasound scan revealed bulky ovaries with multiple small follicles arranged around the outside of the ovaries bilaterally. The GP made a diagnosis of polycystic ovarian syndrome (PCOS). Ms R was given health advice regarding her diet and exercise, and advised about the long-term health implications of her diagnosis. She was advised that weight loss is a helpful way to reduce the symptoms that she experienced and to reduce the risks of cardiovascular disease, endometrial cancer, and type 2 diabetes that are associated with PCOS. Although Ms R had no desire for pregnancy immediately, she was also informed that being in the normal range for weight would improve her chances of conceiving if she wished to have children in the future. Ms R was highly motivated to follow this advice and managed to lose 8 kg in weight over a 3-month period. She began taking the oral contraceptive pill, which improved her acne and regulated her menstrual cycle. With adherence to a calorie-controlled diet and regular exercise, Ms R managed to continue her weight loss and lessen the lifetime health risks posed by her PCOS.

Polycystic ovarian syndrome is a relatively common endocrine condition, with an estimated prevalence of 5–15% of women worldwide. PCOS is characterized by clinical symptoms of hyper-androgenism (acne, hirsuitism, male-pattern alopecia) and oligomenorrhoea, in the absence of other adrenal or pituitary pathology. Despite the name of the syndrome, the presence of multiple ovarian cysts on ultrasound scan is not particularly sensitive or specific for PCOS. The pathogenesis of PCOS remains only partially understood, however, the clinical manifestations are mainly related to increased ovarian androgen levels. It is often assumed that unregulated steroid hormone production by the ovary is the primary pathogenic mechanism in PCOS. The ovaries in PCOS have large numbers of follicles arrested in early developmental stages (primary, secondary, and early antral phase). Follicular arrest may be a direct result of increased androgen levels or may be connected to hyperinsulinaemia. Insulin resistance is a prominent feature of PCOS, which has led to speculation that the primary disorder may be endocrine, with reproductive system implications, rather than vice versa. PCOS increases the likelihood of developing type 2 diabetes, endometrial cancer, dyslipidaemia, and obesity. If the endometrium is then not shed on a regular basis, the chance of endometrial cancer increases, particularly in older women. Excess adipose tissue is common in women with PCOS, and the common symptoms (particularly acne and hirsuitism) are exacerbated when body mass is increased. Weight control is therefore an important and effective symptom reduction measure, in particular even small reductions in adipose mass can be highly effective in improving insulin sensitivity and restoring ovulation.

FURTHER READING

General reading

Clarkson J,_Herbison, AE (2009) Oestrogen, kisspeptin, GPR54 and the pre-ovulatory luteinising hormone surge. *Journal of Neuroendocrinology* **21**, 305–311.

Cortes ME, Carrera B, Rioseco H et al. (2015) The role of kisspeptin in the onset of puberty and in the ovulatory mechanism: a mini-review. *Journal of Pediatric and Adolescent Gynecology* **28**, 286–291.

Jayasena CN, Dhillo WS, Bloom SR (2009) Kisspeptins and the control of gonadotropin secretion in humans. *Peptides* **30**, 76–82.

Jayasena CN, Nijher GM, Comninos AN *et al.* (2011) The effects of kisspeptin-10 on reproductive hormone release show sexual dimorphism in humans. *Journal of Clinical Endocrinology & Metabolism* **96**, E1963–E1972.

Plant TM (2013) A comparison of the neuroendocrine mechanisms underlying the initiation of the preovulatory LH surge in the human, Old World monkey and rodent. *Frontiers in Neuroendocrinology* **33**, 160–168.

Plant TM (2016) Neuroendocrine control of the onset of puberty. *Frontiers in Neuroendocrinology* **38**, 73–88.

Popa SM, Clifton DK, Steiner RA (2008) The role of kisspeptins and GPR54 in the neuroendocrine regulation of reproduction. *Annual Reviews in Physiology* **70**, 213–238.

Tena-Sempere M (2010) Kisspeptin signaling in the brain: recent developments and future challenges. *Molecular and Cellular Endocrinology* **314**, 164–169.

Thackray VG, Mellon PL, Djurdjica Coss D (2010) Hormones in synergy: regulation of the pituitary gonadotropin genes. *Molecular and Cellular Endocrinology* **314**, 192–203.

van Gestel I, Ijland MM, Hoogland HJ, Evers JL (2003) Endometrial wave-like activity in the non-pregnant uterus. *Human Reproduction Update* **9**, 131–138.

More advanced reading

Gipson IK (2001) Mucins of the human endocervix. *Frontiers in Bioscience* **6**, D1245–D1255.

Kayisli UA, Guzeloglu-Kayisli O, Arici A (2004) Endocrine-immune interactions in human endometrium. *Annals of the New York Academy of Sciences* **1034**, 50–63.

Kunz G, Leyendecker G (2001) Uterine peristaltic activity during the menstrual cycle: characterization, regulation, function and dysfunction. *Reproductive BioMedicine* **4** (Suppl. 3), 5–9.

Neal-Perry GS, Zeevalk GD, Shu J, Etgen AM (2008) Restoration of the luteinizing hormone surge in middle-aged female rats by altering the balance of GAGA and glutamate transmission in the medial preoptic area. *Biology of Reproduction* **79**, 878–888.

Tomikawa J, Homma T, Tajima S *et al.* (2010) Molecular characterization and estrogen regulation of hypothalamic *kiss1* gene in the pig. *Biology of Reproduction* **82**, 313–319.

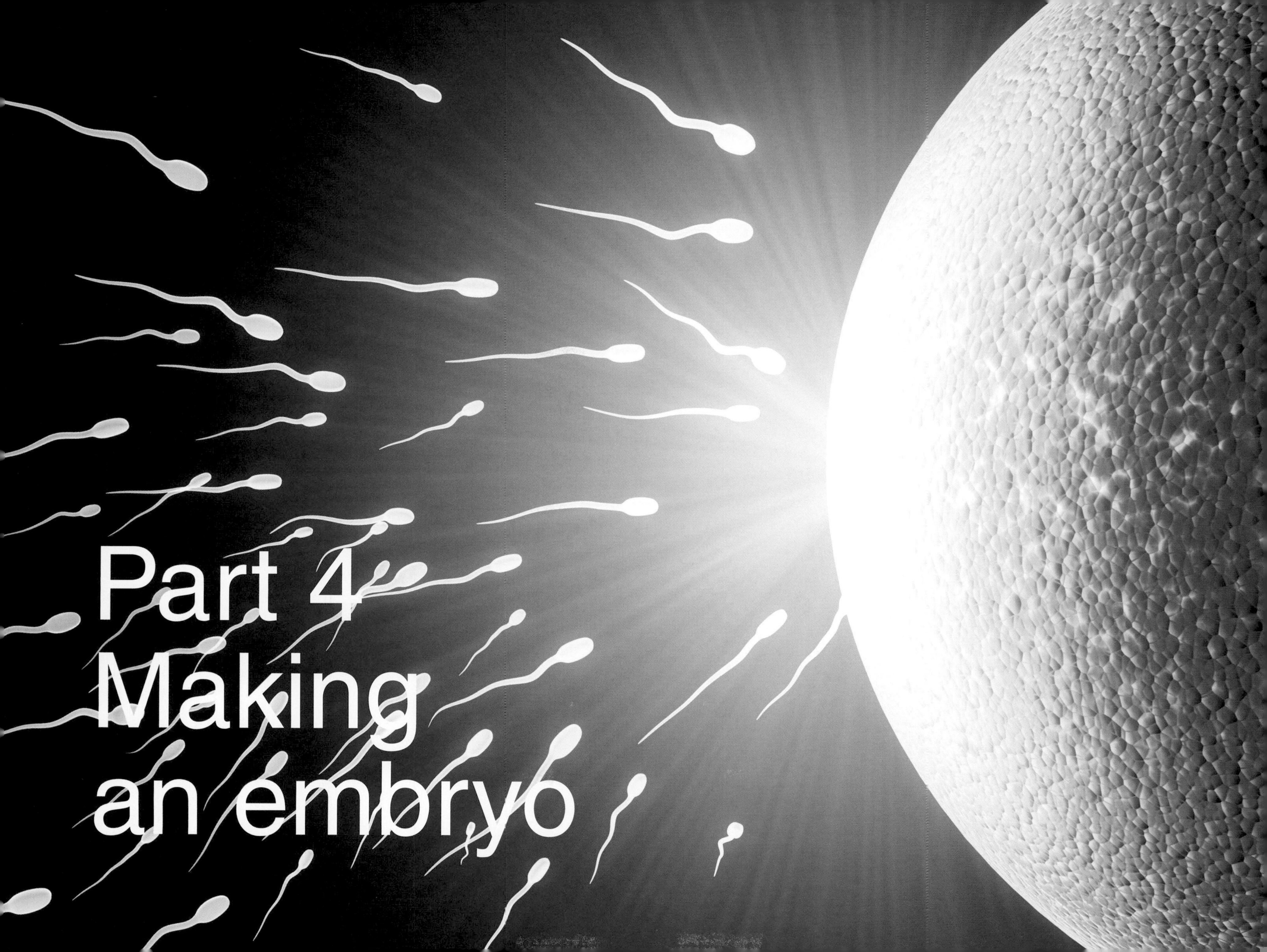
Part 4
Making
an embryo

CHAPTER 11
Sperm and eggs

Essential Reproduction, Eighth Edition. Martin H. Johnson.

Companion website: www.essentialreproduction.com

In the foregoing chapters, we have considered reproductive function in the male and the non-pregnant female. A central concept underlying all of our discussion is the coordinating role played by the gonadal hormones in the production of mature gametes and the conditioning of reproductive function and sexual behaviour. In this way, the chance of fertilization is maximized. Reproduction requires fertilization to be successful. If it is, then a dramatic change must ensue for both reproductive partners, but particularly for the female. Her whole anatomy and physiology must change from a cyclic fertility pattern to a nidatory and pregnancy pattern. It is this critical transition that constitutes the subject of the fourth part of this book.

In this chapter, we start the process with a consideration of the hazardous journeys that the spermatozoa undertake during their travel from their sites of production to the site of fertilization in the oviduct. In Chapter 7, we left the spermatozoa in the lumina of the seminiferous tubules. Human spermatozoa, a few microns in length, must travel through some 30–40 cm of the male and female reproductive tract, or more than 100 000 times their own length, to reach the oviduct. During this long and hazardous journey, several major obstacles must be overcome, including transport between individuals at coitus. **Fewer than one in a million of the spermatozoa produced ever complete the journey**. It is not just that the journey itself is difficult, but also that the spermatozoa must successfully undergo a series of changes in both the male and female genital tracts before they gain full fertilizing capacity. These changes are termed **maturation** in the male tract, and **capacitation, activation** and **the acrosome reaction** in the female tract. The spermatozoon achieves all of these complex changes while expressing none of its own genes, being transcriptionally inactive and heterochromatic.

Spermatozoa require a period of epididymal maturation

Spermatozoa are released from their close association with the Sertoli cells into a fluid secreted by these cells such that a continuous flow rich in spermatozoa washes towards the ***rete testis*** (Figure 11.1). As the fluid passes through the *rete testis*, the composition of its ions and small molecules changes, probably mainly by diffusional equilibration through the tubule walls, since the absence of inter-Sertoli cell junctions renders the blood–testis barrier much less complete. The spermatozoa are then carried through the short and delicate ***vasa efferentia***, which come together within the **initial segment of the epididymis** (Figure 11.2). The *vasa efferentia* **absorb over 90% of the fluid** carrying the spermatozoa, and, if they are ligated, the seminiferous tubules literally 'blow up' with accumulating fluid, and spermatogenesis ceases as a result of pressure atrophy. The **absorption is dependent on oestrogen**, which is carried in the fluid at high concentrations, having been synthesized in Leydig cells. The dependence on oestrogen of the absorptive epithelium of the *vasa efferentia* is seen dramatically in mice genetically lacking in the oestrogen α receptor. At puberty, fluid secretion begins, but it is not absorbed, and so backpressure builds and eventually pressure aspermatogenesis and infertility results – a sort of 'genetic ligation'. Even before this, however, sperm recovered from the epididymis are subfertile, due to an alkalization of the fluid and altered cAMP production within the sperm. The fluid absorption continues in the epididymis to concentrate the spermatozoa some 100-fold, and their **further onward transport then becomes dependent on the activity of epididymal musculature.**

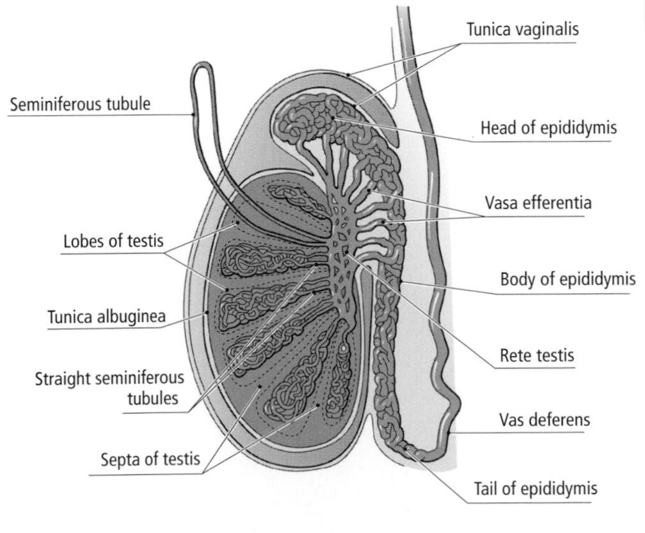

Figure 11.1 Section through an adult human testis to show the general structure of the excurrent ducts.

In addition, the **epididymis adds secretory products** (both exocrine and apocrine), including carnitine, glycerophosphorylcholine, fructose and various proteins and glycoproteins, that interact with the surface of the spermatozoa and many of which are implicated in sperm–oocyte interactions. This interaction is associated temporally with the presence of two unique macromolecular structures in the epididymal lumen: membrane-bound, prostasome-like, so-called **epididymosomes** and dense bodies made up of **molecular chaperones.** These structures juxtapose to the sperm membrane to facilitate the transfer of proteins into and onto the sperm surface. Several of these proteins

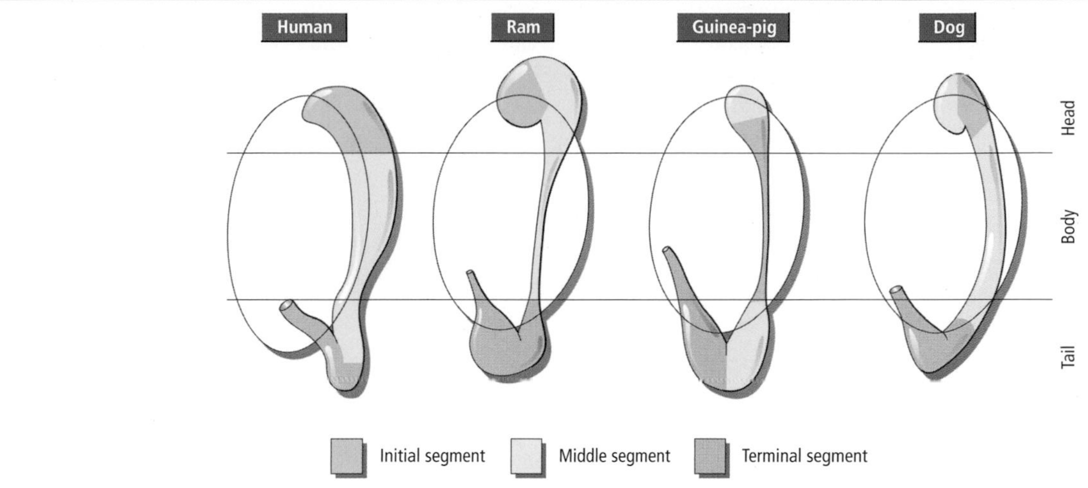

Figure 11.2 Schematic of the excurrent ducts of the human, ram, guinea-pig and dog to illustrate anatomical variation. *Vasa efferentia* (varying in length and from 10 to 20 in number between the species) connect the *rete testis* to a long, highly convoluted tube: the epididymis. Anatomically, the coils of the tube form a head (***caput***, which also contains the termini of the *vasa efferentia*), body (***corpus***) and tail (***cauda***). Three histological segments, which usually do not correspond with the anatomical divisions, are present: an **initial segment** of high, ciliated epithelium and smallish lumen; a **middle segment** with wider lumen and shorter cilia; and a wide **terminal segment** of low cuboidal, relatively poorly ciliated epithelium with more smooth muscle in the underlying *stroma*. Micropinocytosis and fluid absorption occur mainly in the *vasa efferentia* and initial segment, with secretion in the middle and terminal segments, which also store spermatozoa at a high density in the lumen (see also Figure 11.1).

are located in the membranes via glycosyl phosphatidylinositol (GPI)-linkage, and are implicated in sperm–egg binding (see page 202). In addition, **EPPIN**, for epididymal protease inhibitor, is secreted by both Sertoli and epididymal cells and coats sperm. It has been shown to protect sperm from proteolytic attack, thereby preventing premature capacitation (see Chapter 12) and to inhibit premature motility expression.

Passage through the *vasa efferentia* takes about a day and through the epididymis a further 5–11 days depending on the species. This period in the epididymis has a profound effect on spermatozoal behaviour. Thus, spermatozoa entering the *vasa efferentia* are quite incapable of movement (beyond an infrequent twitch) and, when inseminated into females, cannot attach to or fertilize an oocyte. However, **by the time they arrive in the *cauda epididymidis*, spermatozoa have acquired the potential to fertilize oocytes and to swim progressively** (although they do not swim actively *in vivo*, but only after their release from the male tract). These maturational changes in functional capability are accompanied by changes in the biochemistry and morphology of the spermatozoa (Table 11.1). This whole process is called **maturation** and is crucially dependent on adequate **stimulation of the epididymis by androgens**.

If the androgens are removed by castration, the epididymis hypotrophies. Injection of testosterone restores activity. Most of the androgens that stimulate epididymal function are derived not from the circulation but from the lymph and the fluid entering from the *vasa efferentia*. Thus, ligation of the *vasa efferentia* to impair these flows results in considerable functional and structural regression of the epididymis. Within

Table 11.1 Maturational changes to spermatozoa in the epididymis

Property	Details of changes
Concentration	100-fold; 50×10^6/ml entering suspended in fluid; densely packed 50×10^8/ml on leaving
Completion of sperm modelling	Nuclear condensation and acrosomal shaping completed; cytoplasmic drop 'squeezed' down tail and shed
Metabolism	Cholesterol and phospholipids selectively metabolized, shifting lipid balance towards diacylglycerol, unsaturated fatty acids and desmosterol Increased dependence on external fructose for glycolytic energy production; little oxidative metabolism; pH rises
Mobility	Increase in disulphide linkages between proteins in outer dense fibres of tail, yielding a more rigid flagellum with a stronger potential beat; cAMP content of tail rises; acquires capacity for forward motion
Membrane	Coated with glycoproteins. Rise in surface charge (due to sialic acid increase) and change in profile of surface proteins; membrane fluidity increases

this fluid, testosterone is bound to androgen-binding protein and reaches concentrations approaching those of testicular venous blood (between 30 and 60 ng/ml; dihydrotestosterone is also present at about half this level). Within the epididymis, intracellular receptors take up the androgens, and 5α-reductase converts testosterone to dihydrotestosterone to yield very high tissue levels of this more active androgen.

In some species, spermatozoa may be stored for several weeks in the *cauda*, but in humans storage seems to occur for a few days only. After leaving the tail of the epididymis, spermatozoa enter the ***vas deferens*** as a very densely packed mass. Ligation of the *vas deferens* (**vasectomy**) does not lead to accumulation of masses of fluid behind the ligature, as occurs with ligation of the *vasa efferentia*, and so there is no pressure atrophy within the seminiferous tubule. However, spermatozoa do build up behind the vasectomy ligature and these must be removed either by phagocytosis within the epididymis or by leakage through the epididymal wall. The normal non-ligated *vas deferens* serves as a storage reservoir for spermatozoa. In the absence of ejaculation, spermatozoa dribble through the **terminal ampulla** of the *vas deferens* into the urethra and are washed away in the urine.

Semen is made up of spermatozoa and seminal plasma

Ejaculated spermatozoa are carried to the female tract in a fluid called **seminal plasma**; the two together are called **semen**. Seminal plasma is derived largely from the major accessory sex glands (Figure 11.3), with only a small contribution from the epididymis. Different species have bewilderingly different patterns of accessory sex gland structure (see Table 8.1) and

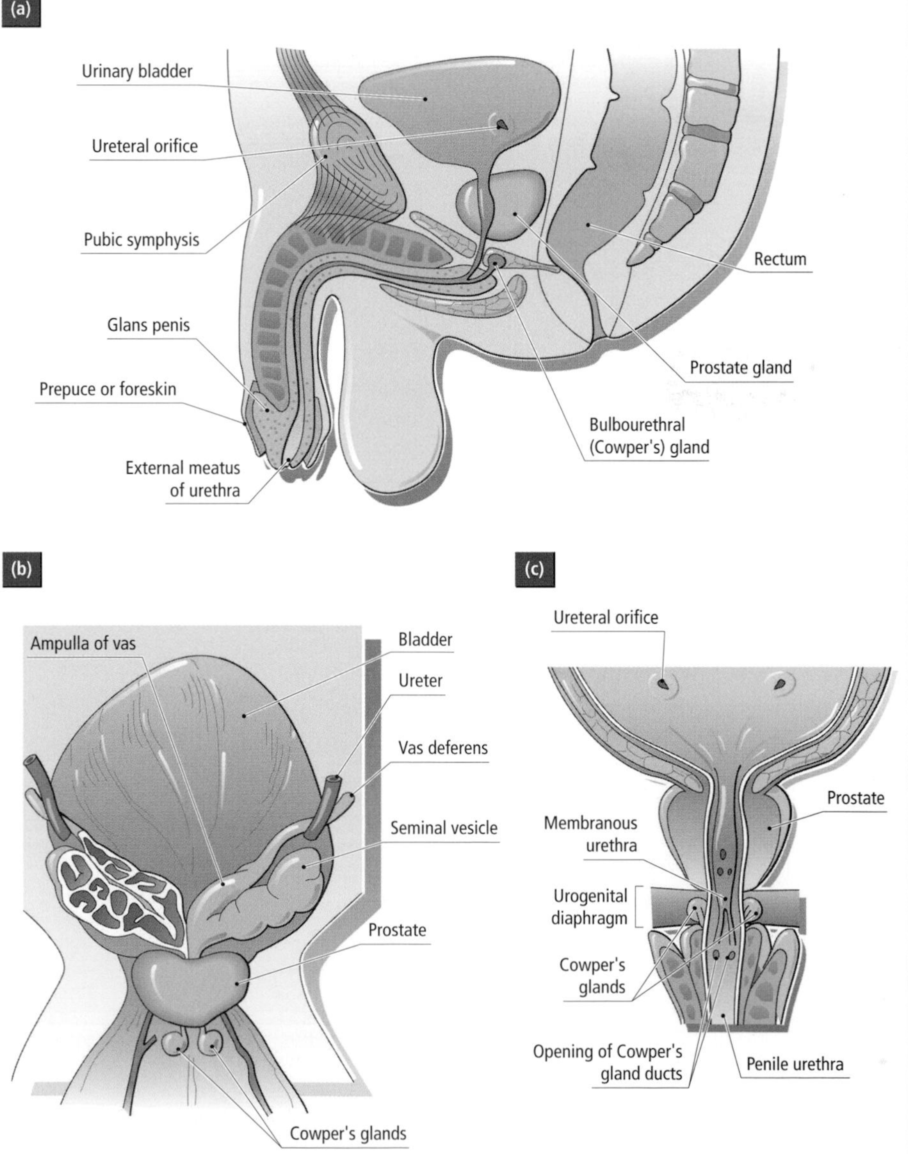

Figure 11.3 View of human male accessory sex glands: (a) mid-sagittal section through pelvis; (b) posterior view of dissected pelvic contents; (c) coronal section through (b) viewed anteriorly.

function (Table 11.2). Knowledge of the origin of some of the main constituents of seminal fluid, as shown in Table 11.2, can help to diagnose deficiencies of function, in particular of accessory sex glands.

Seminal plasma is not essential for effective sperm function, as spermatozoa taken directly from the vas deferens can fertilize oocytes in the test-tube. However, *in vivo*, spermatozoa require a **fluid vehicle** for their normal transport and the seminal plasma supplies this, either exuberantly in half-litre volumes in the boar, or with a more conservative 3 ml or so in the human. In addition to providing a transport medium, the **seminal plasma also provides nutritional factors** such as fructose or sorbitol, buffering capacity to counteract the acid pH of vaginal fluids, and reducing agents such as ascorbic acid, hypotaurine and ergothioneine to protect against potential oxidation following exposure of spermatozoa to atmospheric oxygen. Seminal plasma also supplies the sperm surface with further materials that stabilize it and prevent premature capacitation (see Chapter 12), so-called **decapacitation factors**. It has been proposed that prostaglandins in semen might stimulate muscular activity in the female tract.

Seminal plasma does not only carry spermatozoa and contain substances to assist the maintenance of spermatozoal fertility. Large numbers of leucocytes may also be present in

Table 11.2 Composition of ejaculate* of man and domestic animals

Constituent	Species	Concentration range no./nl	Principal source	Function
Spermatozoa (ejaculate volume in ml in brackets)				
	Boar (150–500)	20–300	Testis	
	Bull (2–10)	300–2000		
	Dog (2–15)	60–300		
	Man (2–5)	50–150		
	Ram (1–2)	2000–5000		
	Stallion (30–300)	30–800		
Fructose		mM		
	Man, ram, bull	8–37	Seminal vesicle and ampulla	Anaerobic fructolysis (sorbitol also used in ram)
	Boar, dog, stallion	<0.5		
Inositol				
	Man, bull, ram, stallion	1–3	Testis and epididymis (seminal vesicle in bull, boar)	Preserves seminal osmolarity
	Boar	28		
Citric acid				
	Bull, ram	15–45	Seminal vesicle/prostate (stallion, ram, boar, bull); prostate (man, dog)	Ca^{2}+ chelator (limits rate of Ca^{2+}-dependent coagulation to prevent seminal 'stones'?)
	Man	5–73		
	Stallion	0.5–2.5		
	Boar	2.5–10		
Glycerylphosphorylcholine				
	Man, stallion	2–3	Epididymis	for use in phospholipid metabolism
	Ram	58–73		
	Bull	4–18		
	Dog, boar	5		
Acid phosphatase (expressed in activity units/ml)				
	Man	2470	Prostate	
	Bull	6		
	Boar	2		

*pH = 7.2–7.8; also contains significant amounts of various prostaglandins (PGs), especially 19-hydroxylated PGE_1 and PGE_2 (humans, monkeys); PGE_1 and PGE_2, and 19-OH PGFs, and $PGF_{1\alpha}$ and $PGF_{2\alpha}$ (humans).

seminal plasma, as well as potentially infectious agents. **Sexual interaction between individuals provides one occasion when these genitourinary infectious agents can be transmitted**. Of particular concern is the presence of hepatitis B or C virus, human immunodeficiency virus (HIV, the cause of AIDS) and genital human papillomavirus (HPV, associated with genital warts and cancers) in the semen of infected men, as these agents cause seriously debilitating and/or potentially fatal diseases. The chance of transmitting these viral infections during ejaculatory vaginal, oral or anal intercourse depends on: the virulence of the viral strain; the viral load within the seminal plasma of the infected individual; concurrent infections or inflammatory conditions in the genital tracts of either sexual partner; and whether or not safer sex with condoms is practised. In general, rates of transmission of hepatitis B virus are about 10 times those of HIV, and the transmission risk increases with oral, vaginal to anal intercourse. Potential transmission is, of course, not unidirectional, although infected females seem to transmit to males during vaginal intercourse at about half the reciprocal rate – and particularly so where the male is not circumcised. Effective prophylactic vaccination is available against hepatitis B and HPV and pharmacological prophylaxis against HIV. Proper use of the right type of condom, lubricant and viricide (see Table 21.3) can provide effective protection against viral transmission – if, of course, a pregnancy is not desired.

Coition

In mammals, fertilization is internal and the male gametes must be deposited in the female tract at coitus. Coitus itself is of variable duration (minutes in man, hours in the camel) and is accompanied by extensive physiological changes not just in the genitalia but also in the body as a whole.

It is only since the mid-1960s that research into human sexual physiology has become an accepted part of the study of reproduction. Masters and Johnson, from their studies on human heterosexual interaction and masturbation, proposed a widely accepted model for sexual responses in men and women. Their so-called ‘EPOR’ model describes: (1) an initial **excitement phase** (E) during which psychogenic or somatogenic stimuli raise sexual arousal; (2) the **plateau phase** (P) during which arousal becomes intensified; (3) the **orgasmic phase** (O), which is reached if the level of stimulation is adequate and entails the few seconds of involuntary **climax** in which sexual tension is relieved, usually in an explosive wave of intense pleasure; and (4) the **resolution phase** (R) during which sexual arousal is dissipated and pelvic haemodynamics resolve to the unstimulated state. The specific physiological changes occurring during these phases will be discussed below. In men, an **absolute refractory period** occurs after orgasm during which time sexual re-arousal and orgasm are impossible. Its duration depends somewhat on age, being abbreviated in boys, and also on a variety of situational factors, such as novelty of partner or context. Women may not generally experience an absolute refractory period, although relevant data are relatively sparse.

The male

Penile erection

Penile erection can be elicited by **psychogenic stimuli**, such as visual cues and erotic imagery or memory. These are integrated in the brain, presumably involving mechanisms within the **limbic system**, which is then able, via descending projections to the spinal cord, to influence somatic and autonomic efferents to the genitalia. These same efferents can be **activated reflexly by tactile stimulation** of the penis and adjacent perineum, this being a most effective means of inducing erection. Data from animals and from men with spinal cord transection reveal that the afferent limb of the reflex is carried by the **internal pudendal nerves**.

Three efferent outflows influence erection: (1) the **pelvic nerve** (parasympathetic outflow) promotes erection; (2) the **hypogastric nerve** (sympathetic outflow) carries fibres that depress erection and possibly some that promote it; and (3) the **pudendal nerve** (somatic) promotes erection. In man, erection is achieved entirely by **haemodynamic changes**, the two ***corpora cavernosa*** (trabeculated sinus spaces surrounded by a tough fibrous capsule: Figure 11.4b,c) providing the main erectile tissue. The haemodynamics of erection are described in Box 11.1.

Erectile dysfunction

Erectile dysfunction (**ED** or **impotence**) is defined as the consistent or recurrent inability of a man to attain and/or maintain a penile erection sufficient for sexual activity. ED may be caused by tears in the fibrous capsule of the *corpora cavernosa*, obstructions to the vessels supplying the penis, diabetes which impairs release of nNOS and eNOS (Box 11.1), use or abuse of drugs (including alcohol) which antagonize neurotransmitters mediating tumescence, physical nerve damage (a potential side effect of prostatectomy), psychogenic factors, or a mixture of these. Smoking is a strong predictor of ED, associated as it is with several of the causative factors above. Some types of ED can be treated therapeutically by the **pharmacological stimulation of erection** by the intracavernosal injection of the synthetic prostanoid prostaglandin E_1 (also called alprostadil; registered trade names Caverject, Viridal) or its injection intraurethrally (trade name MUSE). More usual treatment now, as less physically invasive, is with orally active agents such as sildenafil (Viagra), tadafil (Cialis) and vardenafil (Levitra). These are phosphodiesterase-5 inhibitors that enhance the action of NO by stabilizing cGMP. Side effects may include more widespread consequences of vascular relaxation, such as headaches, vasocongestion (especially rhinitis) and skin flushes, as well as risk of ischaemic priapism due to a prolonged erection.

The importance of testosterone in erectile function is not entirely clear. Nocturnal erections, which occur during each episode of **rapid eye movement sleep** (REM, dreaming or paradoxical), are known to be testosterone-dependent, whereas erections in response to visual erotic stimuli, for example, are much less dependent on testosterone and occur readily in

Box 11.1 The haemodynamics of erection

The sequence of events underlying the change from **flaccidity** through **tumescence** to **erection** and then **detumescence** of the penis and clitoris is as follows. In the flaccid state, **myogenic tone** within the smooth muscle fibres of the cavernous trabeculae and of the arteries supplying the penis is maintained by the sympathetic outflow of the hypogastric nerve (**adrenergic tone**). During a natural erection, this sympathetic effect is countered by stimulation of the parasympathetic outflow to reduce myogenic tone in the arterial smooth muscle (causing arterial dilatation and an increased blood flow into the *corpora cavernosa*) and in the cavernous trabecular muscle (decreasing intracavernous resistance and expanding cavernous volume). Additionally, arteriovenous shunts, which bypass the sinuses of the *corpora cavernosa* when the penis is flaccid, now direct blood into them. Finally, the venous outflow from the *corpora cavernosa* is reduced by compression of the sub-tunical venous plexus resulting from the rapidly developing turgor. This sequence of events changes the intracavernous space from a **low-volume, low-pressure system** into a **large-volume, high-pressure** one. In man, all this is achieved by **increasing the inflow of arterial blood** but **reducing its through-flow**, such that in a fully rigid state in-flow and outflow are almost absent. Prolonged erection (**priapism**) thus endangers the oxygenated blood supply to the penis (**ischaemic priapism**), and may need to be relieved pharmacologically. The ***corpus spongiosum*** (Figure 11.4) also increases in turgor, but not as much as the *corpora cavernosa*, thereby avoiding compression of the urethra.

Several neurotransmitters have been implicated in erection, both facilitatory (dopamine, acetycholine, oxytocin, VIP, prostanoids, nitric oxide) and inhibitory (noradrenaline, encephalins, angiotensin II), but the most important locally is **nitric oxide** (NO), production of which by **nitric oxide synthase** (NOS) leads to vasodilatation through the production of cGMP within the vascular smooth muscle cells. The **initiation of erection** seems to depend on release of **neuronal NOS (nNOS)**, but **maintenance of erection** requires further amplificatory release of NOS from the local **vascular endothelium** itself (**eNOS**), perhaps as a result of increased sheer pressure from blood flowing through the vessels.

In other species, the haemodynamic changes may be complemented by the parasympathetic-mediated **relaxation of a retractor muscle**, which, in the flaccid state, pulls the penis back into the prepuce (e.g. the bull and macaque) and/or by a penile bone or ***os penis*** attached to the capsule of the corpora cavernosa (e.g. the dog, macaque, mink and sea lion).

hypogonadal men. It has been suggested, therefore, that measuring nocturnal penile tumescence (NPT) provides a method of assessing the capacity for sexual arousal independently of the complicating cognitive factors that may compromise sexual function.

Emission and ejaculation

As ejaculation approaches, the turgor in the penis increases further and the penile circumference at the **coronal ridge of the glans penis** increases, largely achieved by the action of the pudendal somatic outflow to the perineal striated muscle fibres (the **ischiocavernous and bulbocavernous muscles**) surrounding the **corporeal *crura*** (see Figure 11.4). Coincidentally, the testes are drawn reflexly towards the perineum and may increase their volume by as much as 50% as a result of vasocongestion. The scrotal skin thickens and contracts due to activity in the *dartos* muscle. With further stimulation, a sequence of contractions of the muscles of the prostate, *vas deferens* and seminal vesicle is induced, and the components of the seminal plasma, together with the spermatozoa, are expelled into the urethra. This process of **emission** is mediated largely by noradrenergic sympathetic fibres via the hypogastric plexus, and administration of drugs that interfere with the α-adrenergic system (as in treatment for hypertension) leads to **dry orgasms**: erection is not impaired (and may be assisted: see earlier), but emission is. The efficient emission of sperm from the *vas deferens* also seems to require the co-release of ATP with noradrenaline acting via purinergic receptors on the smooth muscle of the *vas*.

Ejaculation, whereby semen is expelled from the posterior urethra, is achieved by contraction of the smooth muscles of the urethra and striated muscles of the *bulbocavernosus* and *ischiocavernosus*. It is usually associated with contractions of the pelvic floor musculature innervated by the pudendal nerves. The passage of semen back into the bladder is normally prevented by contraction of the **vesicular urethral sphincter**; failure of this can lead to **retrograde ejaculation** into the bladder. The composition of the early and late fractions of the human ejaculate reflects the sequential nature of the contractions and the relative lack of mixing of the various seminal components within the urethra. The early fraction is rich in acid phosphatase (prostate), the mid-fraction is rich in spermatozoa (vas deferens) and the late fraction is rich in fructose (seminal vesicle).

Concomitant with penile erection in men, erection of the nipples and increases in heart rate and blood pressure occur as sexual excitement increases. Immediately before ejaculation, skin rashes may develop over the epigastrium, chest, face and neck together with involuntary muscle spasms. At ejaculation, the cardiovascular changes, skin rash and muscle spasms intensify and are often accompanied by hyperventilation, contractions

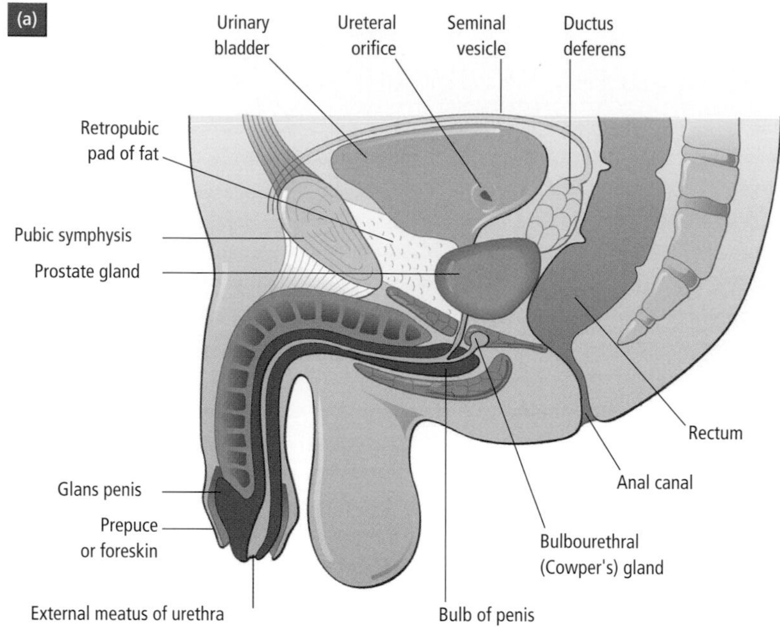

Source: Heffner L & Schust D The Reproductive System at a Glance, 4e (2014)

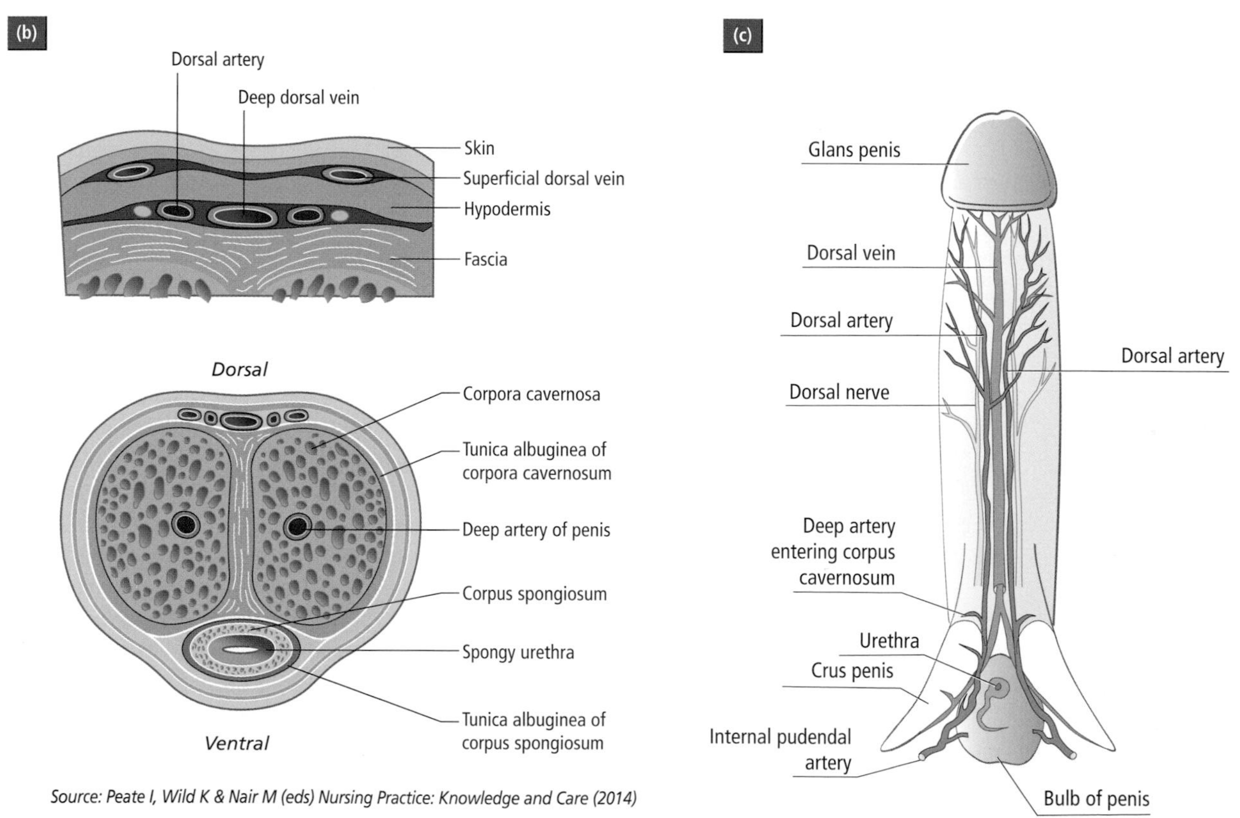

Source: Peate I, Wild K & Nair M (eds) Nursing Practice: Knowledge and Care (2014)

Figure 11.4 Structure of the human adult penis: (a) mid-sagittal section through the pelvis; (b) transverse section through shaft of the penis: note fibrous sheaths enclosing the *corpora cavernosa*, which allow the generation of hydrostatic pressure required for erection; tears in the fibrous capsule lead to failure of erection; and (c) coronal section showing entry of prostatic, ejaculatory (*vasa deferentia* and seminal vesicles) and bulbourethral gland ducts. Penile secretory glands of Littré open into the roof of the penile urethra. The internal pudendal arteries supply both the dorsal arteries to the glans and the pudendal arteries to the *corpora cavernosa*. There is considerable species variation in glans structure: mushroom-shaped in the human, corkscrew-shaped in the boar and bull, spiny in the cat and rat, and having thin urethral vermiform processes in the goat and ram. Source: (a) and (b) Peate I., Nair M. *Anatomy and Physiology for Nurses at a Glance*, First Edition. 2015 John Wiley & Sons, Ltd. Reproduced with permission of John Wiley & Sons; (c) Heffner L.J., Schust D.J. *The Reproductive System at a Glance*, Fourth Edition. 2014 John Wiley & Sons, Ltd. Reproduced with permission of John Wiley & Sons.

of the rectal sphincter and vocalizations. Associated with ejaculation is **orgasm** whereby sexual tension and arousal are released and an intense sensation of pleasure occurs. Penile detumescence occurs through the activity of the pelvic nerve sympathetic outflow restoring smooth muscle tone.

The female

At coitus, frictional stimulation of the glans penis is provided by movements against the external female genitalia and vaginal walls (Figure 11.5). The human female undergoes a remarkably similar sequence of reflex responses to that observed in the

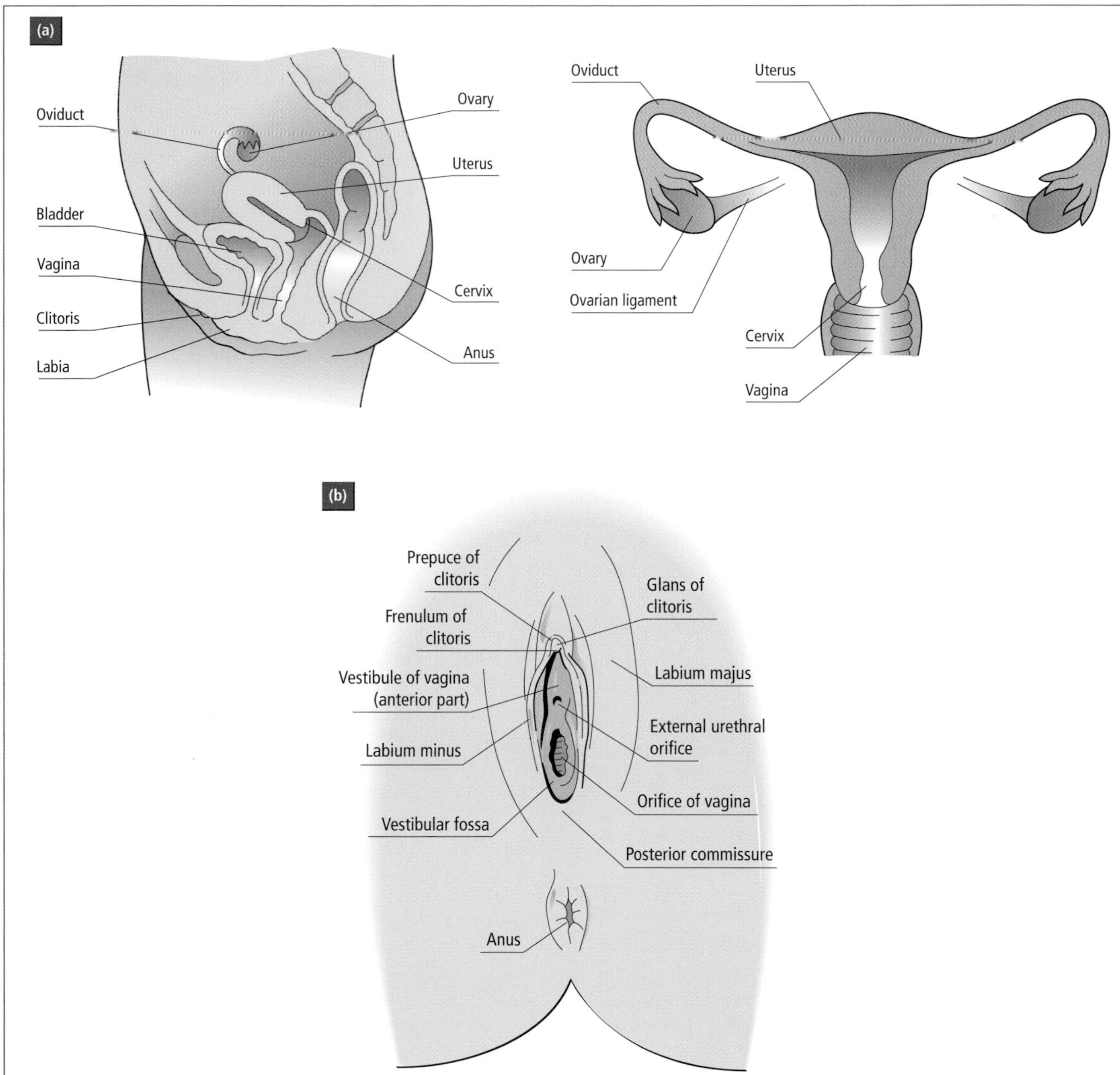

Figure 11.5 Structure of human adult female genitalia. (a) Mid-sagittal section through the pelvis: note ventroflexure of the uterus, which tends to lift upwards during sexual stimulation, moving the cervical *os* anteriorly in the tenting effect. The smooth muscular wall of the vagina is overlain by extensive highly vascularized *stroma* covered in a stratified squamous epithelium. It is not a secretory epithelium; vaginal fluids are derived either by transudation, or from secretions of the cervix and Bartholin's glands, as well as minor vestibular glands at the vaginal vestibule. (b) View of genitalia from the exterior. In some species, the clitoris contains a small bony structure, the os clitoris, homologous to the *os* penis in the male. (Source: (a) *Physiology at a Glance*, Fourth Edition. (Eds Ward J.P.T., Linden R.W.A.). (2017) John Wiley & Sons, Ltd. Reproduced with permission of John Wiley & Sons. (b) Heffner L.J., Schust D.J. *The Reproductive System at a Glance*, Fourth Edition. 2014 John Wiley & Sons, Ltd. Reproduced with permission of John Wiley & Sons.)

male, and its elicitation is also dependent on tactile and psychogenic stimulation. Tactile stimulation in the perineal region, and in some women particularly of the **glans clitoris**, provides the primary afferent input reinforced by vaginal stimulation after penile penetration. A vascular response may cause engorgement of the *corpora* of the clitoris, with consequent **clitoral erection**, although the extent of clitoral involvement varies considerably. The pharmacological agents used to treat erectile dysfunction in men may also be of use in facilitating genital sexual responsiveness in women. Vaginal lubrication occurs by **transudation** of fluid through the vaginal wall, which vasocongests and becomes purplish red. The vagina expands and the *labia majora* become engorged with blood. With increased stimulation, the width and length of the vagina increase further and the uterus elevates upwards into the false pelvis, lifting the cervical os to produce the so-called **tenting effect** in the mid-vaginal plane. At orgasm, frequent vaginal contractions occur and uterine contractions beginning in the fundus spread towards the lower uterine segment.

Systemically, the female may experience increased heart rate and blood pressure and manifest skin flushes, vocalizations and muscle spasms (notably rhythmic contractions of the pelvic striated musculature), and intense sensations of pleasure. Post-orgasmically, clitoral erection is lost, the labia and the vaginal *os* detumesce, and the uterine and vaginal walls relax to their original positions. It has not proved possible to differentiate between the orgasms that follow clitoral or vaginal stimulation in terms of their physiological manifestations.

The general descriptions of orgasm recorded by men and women are so similar as to be indistinguishable. The time course of sexual responses in the female is generally longer than in the male. Indeed, ejaculation and orgasm are the end result of coitus within minutes in men (an average of 4 minutes in the Western world according to one study). Proportionately fewer women appear to achieve orgasm from coitus, various surveys giving figures of 30–50%. Larger numbers achieve orgasm by clitoral stimulation, but even then a significant number, 10–20%, do not achieve orgasm despite being highly sexually aroused. It has been suggested that orgasm in women is not a reflex action, as it appears to be in men, but is learned. Cross-cultural studies suggest strong psychosocial influences on reflex-mediated events; e.g., women are more likely to enjoy sex and have orgasms in societies where this is expected than in societies where it is not. Phosphodiesterase-5 inhibitors such as sildenafil have been reported to improve women's arousal, orgasm and enjoyment, but it is unclear pharmacologically how this improvement is achieved.

The **ability to communicate sexually** and to give and receive sexual pleasure within a relationship is very often a cornerstone of its success in modern Western society. Social and physical factors affect the capacity to respond sexually to a partner by experiencing orgasm and, when communication and the relationship itself are becoming poor, efforts to define at least some of these factors can have considerable value.

Semen is deposited in the vagina, cervix or uterus depending on the species

At coition, the semen is ejaculated within the vagina and onto the cervical *os* in the human, sheep or cow. In some other species, e.g. the pig, dog, horse, mouse and rat, there is a direct deposition into the cervix and/or uterus. In many species studied, the semen **coagulates** rapidly during or immediately after deposition. The coagulation may become gelatinous (e.g. in the human, pig and horse), fibrous (e.g. in the guinea pig) or calcareous (e.g. in the mouse) and results from an enzyme:substrate interaction. In humans, coagulating enzymes derived from the prostate interact with a fibrinogen-like substrate derived from the seminal vesicle. The coagulum may act to retain spermatozoa in the vagina, reducing their physical loss (although back-leakage occurs in most coituses studied) or perhaps buffering them against the hostile acidity of the vaginal fluids (pH 5.7). In some cases of ejaculatory disorder, coagulation takes place within the urethra or (in retrograde ejaculation) within the bladder, and obstruction to urinary flow can result. In normal circumstances in the human female tract, the coagulum is dissolved within 20–60 minutes by progressive activation of a proenzyme derived from the prostatic secretion of the ejaculate.

Gamete transport through the female genital tract

After their deposition in the female tract, the spermatozoa must reach the oviduct, and must do so at the same time as the ovulated oocyte. So next we consider how each gamete achieves this.

Spermatozoa are transported largely by their own activity

Spermatozoa deposited in the vagina face a journey of some 160–200 mm if they are to reach the site of fertilization. Most do not make it. Within a minute or so of mating in all species studied, some spermatozoa can be detected in the cervix or the uterus. It is not at present clear how human spermatozoa deposited in the vagina actually enter the cervix. Perhaps the ciliated surface of the cervical *os* wafts them towards the cervical canal. However, in the human, over 99% of spermatozoa do not enter the cervix and are lost by post-coital leakage from the vagina. The few successful spermatozoa may survive for many hours deep in the cervical crypts of the mucous membrane (Table 11.3), where they are nourished by mucoid secretions. Their further progress to the uterus then depends on the mucus consistency. Only in the absence of progesterone domination does the mucus permit

Table 11.3 Estimates of the survival of viable fertile gametes in the female genital tract

Species	Spermatozoa (d)	Oocytes (h)
Human	3–4	6–24
Cow	3–4	8–12
Sheep	3–4	15–25
Horse	4–6	6–8
Pig	3–4	8–10
Mouse	0.5–1	6–15
Rabbit	1.5–2	6–8

sperm penetration (see pages 175–6), and even then morphologically abnormal spermatozoa are prevented from passing further up the female tract.

Studies on the transport of spermatozoa through the uterus to the oviduct are difficult to undertake. It seems clear that the vaginal, cervical and uterine movements often present in the pre-orgasmic and orgasmic phases are not required for effective sperm transport but do assist it. Nor is it likely that the prostaglandins present in the semen are required as a stimulant to the female tract. In humans, the **spermatozoa probably move through the uterus to the uterotubal junction aided and directed by peristaltic contractions**, but they then **traverse the junction and enter into the ampulla of the oviduct by their own propulsion**. This conclusion fits with the timescale of spermatozoal migration, since the earliest that living spermatozoa are recovered from the oviduct is 2–7 hours after coition. However, the capacity to swim is not in itself sufficient to ensure that spermatozoa arrive in the oviduct, as shown by the failure of spermatozoa from male mice with various genetic deletions that affect expression of the spermatozoal surface glycoprotein ADAM1 (see page 204), despite being able to swim, to traverse the uterotubal junction, which is very narrow and full of mucus. These results argue for active spermatozoal engagement with oviducal cells.

The number of living oviducal spermatozoa detected during the early stages of fertilization can be measured in tens or hundreds, and even subsequently the total number present at any one time rarely exceeds several hundred. It is not at all clear how the flow of sperm to the oviduct is regulated. It has been proposed but not demonstrated that the cervical crypts act as a reservoir, slowly releasing sperm into the uterus. The **uterotubal junction** does regulate entry to the oviduct by its action as an intermittent sphincter. Having reached the **isthmus of the oviduct**, spermatozoa linger, binding temporarily to oviducal epithelial cells in a **spermatozoal reservoir**. As a result of this binding, the oviducal cells are stimulated to secrete factors that serve to maintain the stored sperm in a viable state. Only at ovulation do spermatozoa detach from the oviducal cells, re-acquire motility and swim to the ampullary–isthmic junction (Figure 11.6) and the site of potential fertilization. *In vitro* studies suggest that this process may depend on **the release of chemo-attractants** by both the oocyte and the cumulus mass. The nature of the chemo-attractant(s) is unclear, although the examination of spermatozoa for potential chemo-attractant-sensing receptors has revealed the presence of **odorant receptors** that resemble those found in the olfactory epithelium. They are located in the sperm mid-piece and flagellum base and are activated by small aldehyde molecules, the molecular nature of which is species-specific. Receptor stimulation results in G-protein activation of an adenylate kinase, leading to cAMP-mediated stimulation of a rise in intracellular Ca^{2+}. This rise is associated with **sperm chemotaxis** and **chemokinesis** (enhanced swimming speed) and **hyperactive flagellar beating**. It remains to be proved that such a system of chemo-attraction actually functions *in vivo* and, if so, what the molecular identity of the *in vivo* chemo-attractant molecule(s) is. Two reports suggest that progesterone acts *in vitro* as a chemo-attractant from the cumulus cells in humans, but likewise their relevance for the *in vivo* situation is unclear.

Oocyte transport depends on the activity of the oviduct

While the spermatozoa are moving towards the oviducal ampulla, the ovulated oocyte(s), with enclosing cumulus cells, is picked up from the surface of the ovary in the peritoneal cavity by the fimbriated *ostium* and passes into the oviducal ampulla (see Figure 11.6) and adheres initially to its epithelium via the cumulus cells. Then, as the cumulus cells start to disperse, the egg mass is swept by a combination of propulsive contractions and oviducal cilia beats along the ampulla towards the junction with the isthmus. Oocyte transport is affected adversely if cumulus cells are lacking, and if oviducal cilia and/or smooth muscle are malfunctional. Pacemaker activity in the **interstitial cells of Cajal** (ICC) has been reported as responsible for regulating oviducal motility and egg transport. If inhibited, then tube locking of the embryo may occur (see Box 10.5). The oocytes and the spermatozoa come together at the ampullary–isthmic junction and it is here that the final events leading to fertilization occur.

Conclusions

The transport of gametes within and between men and women is the necessary prelude to internal fertilization. For each of the gametes the journey is perilous. In both male and female reproductive tracts the spermatozoa face a difficult journey in which there is progressive attrition and also progression through a series of maturational steps undertaken by the dwindling survivors in order to prepare them for fertilization. In the next chapter we turn to examine the process of fertilization itself.

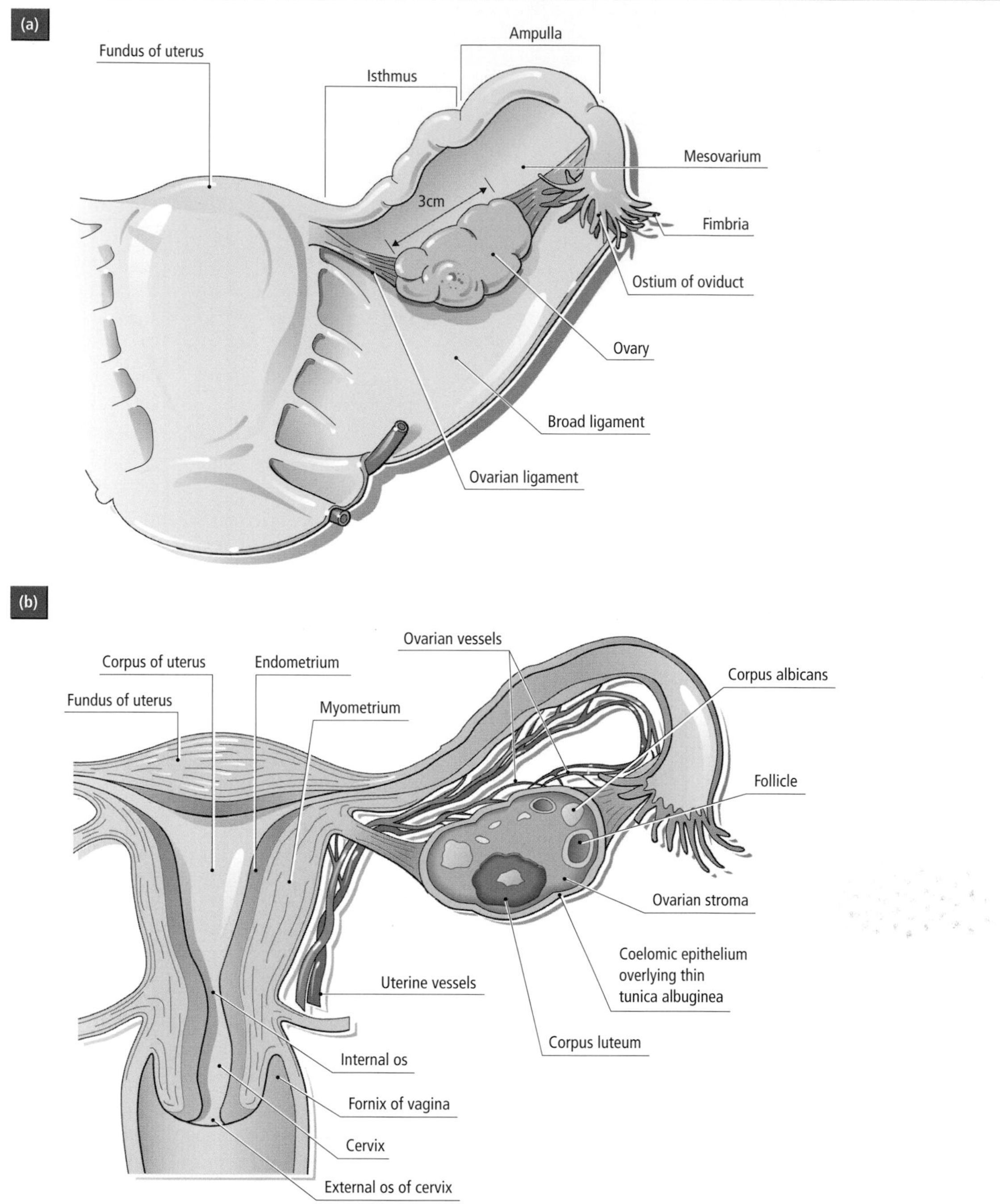

Figure 11.6 Posterior views of the human uterus and one oviduct and ovary: (a) intact; (b) sectioned. The ovaries have been pulled upwards and laterally, and would normally have their long axes almost vertical. Note that all structures are covered in peritoneum except the surface of the ovary and oviducal ostium through which there is continuity between the female tract and the peritoneal cavity. At ovulation the fimbria pick up the ovulated egg(s) from the surface of the ovary and waft them in their cumulus mass into the oviduct as can be seen in the video viewable at www.talbotcentral.ucr.edu

Key learning points

- Spermatozoa leave the testis through the *rete testis* and *vasa efferentia* carried in fluid secreted by the Sertoli cells.
- This fluid is absorbed by the *vasa efferentia* in an oestrogen-dependent process.
- Thereafter, spermatozoa are moved through the male genital tract by the activity of its musculature.
- Spermatozoa mature during transit through the epididymis.
- Epididymal maturation is androgen-dependent and most androgen arrives via lymph and testicular fluid.
- Spermatozoa may be stored in the cauda epididymis and/or *vas deferens* depending on the species.
- Seminal plasma is produced in the male accessory sex glands, especially the prostate and seminal vesicle.
- Seminal plasma forms a fluid vehicle for spermatozoa and may provide buffering, antioxidant and metabolic support for them.
- Seminal plasma is also a vehicle for transfer of infection between individuals, as is vaginal fluid.
- Seminal plasma combines with spermatozoa at emission and ejaculation to form semen.
- Coition involves sexual responses and genital reflexes.
- The phases of sexual responsiveness are called excitement, plateau, orgasmic and resolution (EPOR).
- These phases are accompanied by a range of systemic and emotional manifestations.
- Erection involves haemodynamic changes to the *corpora cavernosa* (of the penis and clitoris) converting each to a high-volume/high-pressure system.
- In some species, relaxation of penile retraction and a penile bone aid the erectile process.
- The haemodynamic changes depend on increased blood inflow due to vasodilatation mediated by nitric oxide which reduces sympathetic myogenic tone; vascular shunts may also be closed and turgor-induced pressure may reduce blood outflow.
- Erectile dysfunction (or impotence) may be treated with smooth muscle relaxants such as prostaglandin E_1 (alprostadil) or phosphodiesterase-5 inhibitors (sildenafil, tadafil and vardenafil).
- Semen is deposited in the vagina, cervix or uterus depending on the species.
- After vaginal deposition, most spermatozoa fail to enter the cervix and those that do may be stored there for up 2 days.
- Spermatozoa reach the oviduct by swimming, perhaps helped by uterine movements.
- Spermatozoa wait in the oviducal isthmus until ovulation when they move into the ampulla, possibly by chemotaxis involving odorant receptors on spermatozoa.
- The ovulated oocyte is picked up actively by the oviducal *ostia* and transported to the ampullary–isthmic junction.

Clinical vignette

Erectile dysfunction

A 45-year-old man, Mr C, attended his general practitioner (GP) complaining of erectile dysfunction. He described being able to intermittently become aroused during sexual activity with his female partner but quickly losing his erection and being unable to reach orgasm. He continued to wake with early morning erections and was able to reach orgasm by masturbation. Mr. C had been in an exclusive relationship with his female partner for the past 6 years. He worked as an accountant for a busy firm in London, and had been particularly busy and stressed in the last 6 months. Mr. C's GP carefully examined him, and concluded that the testes and genitalia were of normal size and appearance. There were no central nervous system signs suggestive of a space-occupying lesion intra-cranially. He performed blood tests to check his pituitary function, gonadotrophin levels and free testosterone. An MRI of the brain was also performed. To Mr C's surprise and relief, these tests all came back normal. A diagnosis of psychogenic erectile dysfunction was reached. MrC and his partner attended psychosexual therapy, and reviewed his work/life balance. Several months later Mr C was again seen in clinic, and reported his work had become much less stressful, his relationship with his partner was improved and that he was now normally sexually active again.

Erectile dysfunction is an important health issue that can have a seriously detrimental impact on a man's quality of life and his relationship with his partner. It is a sensitive area that many patients find it difficult to bring up with medical practitioners. The important fact in the history of erectile dysfunction given by Mr C is that his early morning erections were unaffected during the period of sexual problems. Morning erections occur during REM sleep but the precise mechanism by which they come about remains unknown. It is thought that neurotransmitter release that is normally inhibitory to penile blood flow may cease during REM sleep and allow testosterone-mediated tumescence. Morning erections are hence not influenced by psychological considerations, such as stress or relationship difficulties, and are thus considered to be a sensitive diagnostic tool in determining the aetiology of erectile dysfunction. If Mr C had reported that he was no longer experiencing morning erections, then the likelihood of tests revealing an underlying problem with the hypothalamus, pituitary or testes is increased. If erectile dysfunction has a physical basis, then the likelihood is that the problem lies at the level of the pituitary gland. Such pathologies can include micro or macro prolactinomas (benign tumours of the pituitary gland) or vascular problems with the supply to the gland.

FURTHER READING

General reading

Bancroft J (1989) *Human Sexuality and Its Problems.* Churchill Livingstone, Edinburgh.

Cirino G, Fusco F, Imbimbo C, Mirone V (2006) Pharmacology of erectile dysfunction in man. *Pharmacology & Therapeutics* **111**, 400–423.

De Jonge C (2017) Biological basis for human capacitation revisited. *Human Reproduction Update* **23**, 289–299.

Gadella BM, Bourke A (2016) An update on post-ejaculatory remodeling of the sperm surface before mammalian fertilization. *Theriogenology* **85**, 113–124.

Hunter RHF (2008) Sperm release from oviduct epithelial binding is controlled hormonally by peri-ovulatory Graafian follicles. *Molecular Reproduction and Development* **75**, 167–174.

Kuo Y-W, Li S-H, Maeda K-I, *et al.* (2016). Roles of the reproductive tract in modifications of the sperm membrane surface. *The Journal of Reproduction and Development* **62**, 337–343.

Masters WH, Johnson VE (1966) *Human Sexual Response.* Churchill, London.

Masters WH, Johnson VE (1970) *Human Sexual Inadequacy.* Little, Brown, Boston.

Moore HDM (1996) The influence of the epididymis on human and animal sperm maturation and storage. *Human Reproduction* **11**, 103–110.

Sakkas D, Ramalingam M, Garrido N, Barratt CLR (2015) Sperm selection in natural conception: what can we learn from Mother Nature to improve assisted reproduction outcomes? *Human Reproduction Update* **21**, 711–726.

Suarez SS (2016) Mammalian sperm interactions with the female reproductive tract. *Cell and Tissue Research* **363**, 185–194.

Talbot P, Riveles K (2005) Smoking and reproduction: the oviduct as a target of cigarette smoke. *Reproductive Biology and Endocrinology* **28**, 3: 52 (also details oviducal transport features, with web video links).

More advanced reading

Burnett LA, Xiang X, Bieber AL, Chandler DE (2008) Crisp proteins and sperm chemotaxis: discovery in amphibians and explorations in mammals. *International Journal of Developmental Biology* **52**, 489–501.

Dixon RE, Hwang SJ, Hennig GW *et al.* (2009) Chlamydia infection causes loss of pacemaker cells and inhibits oocyte transport in the mouse oviduct. *Biology of Reproduction* **80**, 665–673.

Domeniconi RF, Ferreira Souza AC, Xu B, *et al.* (2016) Is the epididymis a series of organs placed side by side? *Biology of Reproduction* **95**, 1–8.

Dunn PM (2000) Purinergic receptors and the male contraceptive pill. *Current Biology* **10**, R305–R307 (emission control).

Eisenbach M (1999) Sperm chemotaxis. *Reviews of Reproduction* **4**, 56–66.

Gervasi MG, Rapanelli M, Ribeiro ML *et al.* (2009) The endocannabinoid system in bull sperm and bovine oviductal epithelium: role of anandamide in sperm–oviduct interaction. *Reproduction* **137**, 403–414.

Griffiths GS, Galileo DS, Aravindan RG, Martin-DeLeon PA (2009) Clusterin facilitates exchange of glycosyl phosphatidylinositol-linked spam1 between reproductive luminal fluids and mouse and human sperm membranes. *Biology of Reproduction* **81**, 562–570.

Hess RA (2000) Oestrogen in fluid transport in efferent ducts of the male reproductive tract. *Reviews of Reproduction* **5**, 84–92.

Hess RA (2003) Estrogen in the adult male reproductive tract: a review. *Reproductive Biology and Endocrinology* **1**, 52.

Holt WV, Van Look KJW (2004) Concepts in sperm heterogeneity, sperm selection and sperm competition as biological foundations for laboratory tests of semen quality. *Reproduction* **127**, 527–535.

Hurt KJ, Musicki B, Palese MA *et al.* (2002) Aktdependent phosphorylation of endothelial nitric-oxide synthase mediates penile erection. *Proceedings of the National Academy of Sciences of the United States of America* **99**, 4061–4066.

Joseph A, Hess RA, Schaeffer DJ *et al.* (2010) Absence of estrogen receptor alpha leads to physiological alterations in the mouse epididymis and consequent defects in sperm function. *Biology of Reproduction* **82**, 948–957.

Kolle S, Dubielzig S, Reese S, *et al.* (2009) Ciliary transport, gamete interaction, and effects of the early embryo in the oviduct: ex vivo analyses using a new digital videomicroscopic system in the cow. *Biology of Reproduction* **81**, 267–274.

O'Rand MG, Widgren EE, Hamil KG, *et al.* (2011) Functional studies of eppin. *Biochemical Society Transactions* **39**, 1447–1449.

Oren-Benaroya R, Orvieto R, Gakamsky A *et al.* (2008) The sperm chemoattractant secreted from human cumulus cells is progesterone. *Human Reproduction* **23**, 2339–2345.

Partridge JM, Koutsky LA (2006) Genital human papillomavirus infection in men. *Lancet Infectious Diseases* **6**, 21–31.

Spehr M, Schwane K, Riffell JA *et al.*(2006) Odorant receptors and olfactory-like signaling mechanisms in mammalian sperm. *Molecular and Cellular Endocrinology* **250**, 128–136.

Sullivan R, Mieusset R (2016) The human epididymis: its function in sperm maturation. *Human Reproduction Update* **22**, 574–587.

Teves ME, Barbano F, Guidobaldi HA *et al.* (2006) Progesterone at the picomolar range is a chemoattractant for mammalian spermatozoa. *Fertility and Sterility* **86**, 745–749.

Toda N, Ayajiki K, Okamura T (2005) Nitric oxide and penile erection function. *Pharmacology and Therapeutics* **106**, 233–266.

CHAPTER 12

Fertilization

Essential Reproduction, Eighth Edition. Martin H. Johnson.

Companion website: www.essentialreproduction.com

In Chapter 11, the gametes were brought into proximity in the oviduct – a necessary prelude to fertilization. In this chapter we will consider the critical events of fertilization itself. Fertilization has recently assumed a highly significant status. It has assumed this status biologically, because it sets the genetic constitution of an individual – including their sex (see Chapter 3). This undue emphasis on the genetics of fertilization is questionable, as will become clear. Fertilization has also assumed ethical significance with the success of **in-vitro fertilization** (**IVF**), a process that generates more human conceptuses than can be used to produce babies. The ethically problematic *in vitro* conceptus has in turn generated a tangle of legal definitions of fertilization. So overall, the process of fertilization has moved from the realms of biology onto a wider stage. We will examine its place on that stage later (see Chapter 23). Here we describe some of the key ideas and knowledge about the biology of fertilization. A key point to grasp is that **fertilization is not an event but a protracted process** taking many hours from initiation to completion.

Spermatozoa gain their full fertilizing capacity in the female tract

If mature spermatozoa recovered at ejaculation are placed with oocytes *in vitro*, fertilization either does not occur or does so only after a delay of several hours. In contrast, spermatozoa recovered from the uterus or oviduct a few hours after coitus are capable of immediate fertilization. The process by which this attainment of a full fertilizing capacity is achieved within the female tract is called **capacitation**. A fully capacitated spermatozoon is distinguished by two characteristics: (1) a change in its movement pattern to a **hyperactivated motility** state in which the regular undulating wave-like flagellar beats are replaced by stronger, wide-amplitude or '**whiplash' beats** of the tail that push the spermatozoon forwards in vigorous lurches (Figure 12.1b); and (2) a change in the surface membrane properties that renders the spermatozoon responsive to signals encountered in the immediate vicinity of the oocyte, which themselves then induce a further change in the spermatozoa called the **acrosome reaction** (see later). However, it should be stressed that this full capacitatory state is the culmination of a number of more incremental changes.

An oestrogen-primed uterus or oviducal isthmus is optimal for capacitation, critical features of their secretions being the proteolytic enzymes, cholesterol binding 'sinks', and higher ionic strength compared with seminal plasma. In addition, there is evidence that, at ejaculation and exposure to atmospheric oxygen, reactive oxygen species form and stimulate production of hydrogen peroxide (H_2O_2), a potent capacitating agent. For most species, including humans, it is possible to mimic capacitating conditions *in vitro* by preparation of suitable culture media in atmospheric oxygen.

The process of capacitation involves a complex series of physiological and biochemical events, the nature of which is still being unravelled. Some key physiological changes include a stripping from, and/or modification of, many of the glycoproteins that coated the spermatozoal surface during passage through the epididymis and in the seminal plasma. This stripping is associated with changes to the surface charge and to the macromolecular organization and lipid structure of the spermatozoal membrane, especially the loss of cholesterol. These collectively act to reduce membrane stability and enhance its fluidity, leading to the creation within the membrane of aggregates of lipids and proteins in multimeric micro-domains or 'rafts' that are important – immediately, in the acrosome reaction (see pages 200 and 202), and ultimately, in fusion with the oocyte (see pages 203–4). The capacitation

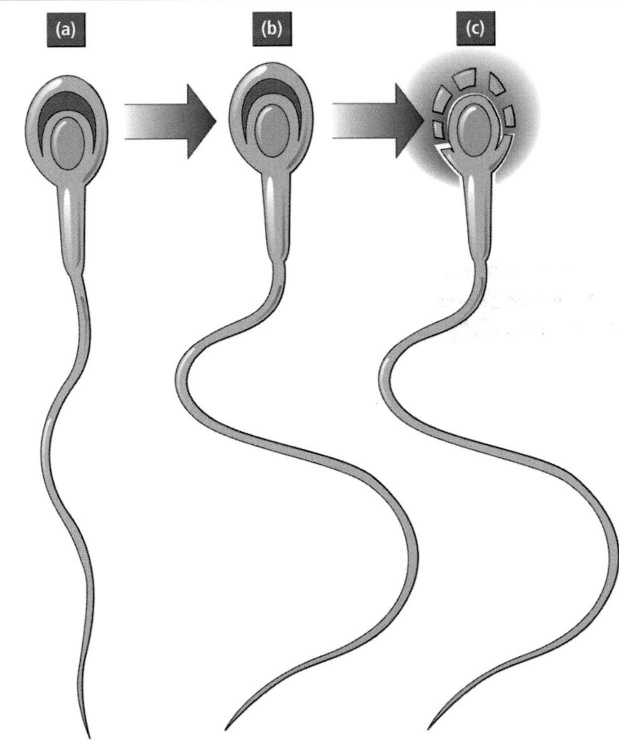

Figure 12.1 (a) Schematized spermatozoon prior to capacitation; a consequence of capacitation is (b) hyperactivated tail movements, and development of the capacity subsequently to undergo (c) the acrosome reaction, in which multiple sites of fusion between the plasma membrane and the outer acrosomal membrane occur, first at the tip of the acrosome and then at the equatorial region. As a result of the acrosome reaction, the plasma membrane remaining in the equatorial and post-acrosomal regions acquires the potential to fuse with the plasma membrane of the oocyte.

Box 12.1 Difficulties in studying fertilization

It will be evident from our account that our understanding of the events of fertilization leading up to spermatozoal–oocyte fusion is provisional and hedged with qualifications. There are good reasons for this state of knowledge.

- These events occur at very inaccessible sites *in vivo*, and the use of *in vitro* models may introduce all manner of artefacts – not least that the egg mass *in vitro* is surrounded by many more spermatozoa than it ever would experience *in vivo*, where a ratio of approximately 1 egg:1 spermatozoa is encountered at any one time.
- The human spermatozoa for *in vitro* use or study are collected only after their exposure to semen for some 30 minutes or more in atmospheric oxygen, while liquefaction occurs, a very different scenario to the spermatozoa *in vivo*, which escape rapidly from the seminal plasma into the cervix. Thus, spermatozoa *in vitro* spend much longer exposed to decapacitation substances, and so may take correspondingly longer to capacitate.
- Of the millions of spermatozoa produced, very few make it through to fuse with the oocyte, yet it is not possible to identify these until most of the prolonged transport and capacitation process is complete: for most of the time, we may be studying the failures!
- Because the process is both prolonged and complex, many studies are undertaken on a heterogeneous mixture of spermatozoa at different points along the same pathway.
- Some molecular players clearly play more than one role.
- Spermatozoa are a key locus of sexual selection. Thus, when data on the protein (proteomic) profiles of spermatozoa of different species are examined, evidence is found of accelerated evolution for some 22% of spermatozoal cell membrane genes (including genes encoding acrosomal surface proteins) relative to other genes expressed in spermatozoa. Thus, considerable variation between species, and complexity within species, from duplication and modification of genes, are possible among these proteins. It therefore makes it imperative to undertake more studies on human spermatozoa, rather than relying on animal models. These sorts of comparative proteomic study may explain why there seems to be considerable gene redundancy when studying the effects of genetic knock-outs on fecundity.

process is experimentally reversible, at least in its earlier stages; thus, if capacitating spermatozoa are re-incubated in epididymal fluid or seminal plasma, they become **decapacitated**.

The biochemical events occurring during capacitation are the subject of ongoing research and as yet there is no universal agreement on the full sequence. Moreover, most experiments are conducted *in vitro* and few of them use human spermatozoa. Recent experiments on mouse spermatozoa recovered and examined freshly at various stages in the fertilization process from *in vivo* have cast doubt on some of the results obtained exclusively *in vitro*. Box 12.1 features some of the difficulties arising from the exclusive use of *in vitro* approaches to the question of capacitation.

The first steps in capacitation seem to be: removal of coating glycoproteins (such as EPPIN; see page 185) and cholesterol efflux, which may commence in the cervix, and which are followed by increased pH and Ca^{2+} entry, generation of cAMP, activation of protein kinase A (PKA), and the phosphorylation of tyrosines on a number of proteins (see Box 12.2). Once capacitated, a spermatozoon, and in particular its membrane system, have traditionally been regarded as being in a metastable state. Indeed, the isthmic region of the oviduct helps to stabilize the spermatozoa by mechanisms that are unclear, but, on the ovulatory release of an egg mass and its putative chemo-attractants (see page 193), the spermatozoa move towards the ampullary–isthmic junction and complete the capacitatory process. Finding the oocyte is not straightforward for the spermatozoa as each oocyte is ovulated surrounded by a *zona pellucida* and its associated outer cumulus of granulosa cells (Figure 12.2a). Both of these investments must be penetrated before contact with the oocyte can occur.

Penetrating the egg investments

The cumulus cells

The cumulus cells form a dense and sticky mass around the freshly ovulated oocyte and may aid its postovulatory pick-up and transfer into and along the oviduct. The cells are held together by an extracellular matrix that is rich in hyaluronan, an unsulphated glycosaminoglycan. The spermatozoal acrosome is a rich source of **hyaluronidase**, the enzyme that digests the matrix and causes the cumulus cells to fall apart, exposing the *zona pellucida* that surrounds the egg. So how does the hyaluronidase escape from the acrosome? One idea is that the earliest capacitated spermatozoa to arrive in the vicinity of the egg are ‘sacrificial’, in the sense that they are destined not to fertilize it but, because of their metastable state, to undergo a premature release of hyaluronidase, so clearing the path to the *zona pellucida* for the later spermatozoa. Indeed, there is evidence that the high concentrations of progesterone in the cumulus mass may induce this release via an acrosome reaction (see also next section).

Box 12.2 The biochemistry of capacitation

- Alkalization of the spermatozoal cytoplasm occurs during transit through the female tract. Exactly how this is achieved is not agreed, but several mechanisms have been described, including active proton export, Na^+/H^+ exchange and bicarbonate entry, of which the latter seems to be of greatest importance.
- The elevated pH leads to elevated calcium permeability of the spermatozoal membrane and modest rises in internal calcium levels. This rise is coupled with a loss from the spermatozoal surface of calmodulin-binding proteins, which may make the spermatozoa more responsive to the effects of calcium. The development of hyperactivated motility is particularly sensitive to calcium levels.
- Probably as a result of elevated calcium and bicarbonate, adenyl cyclase activity within the spermatozoa increases, leading to elevation of cAMP levels and increased cAMP-dependent phosphorylation of spermatozoal proteins. Spermatozoal motility is enhanced by addition of either exogenous cAMP or inhibitors of phosphodiesterase (such as pentoxyfylline, thereby preventing destruction of cAMP).
- The cAMP activates spermatozoal protein kinase A (PKA; a serine-threonine kinase), the activity of which then leads to the activation of phosphotyrosinases that phosphorylate tyrosine residues on PKA and other intracellular proteins, including elements within the proteosome complex, inhibition of which prevents completion of capacitation. The identities of the key phosphotyrosinating kinases are not yet determined, although it has been suggested that members of the Src family may be involved.
- If the phosphorylation of the PKA is prevented, the ability of spermatozoa to undergo a subsequent acrosome reaction is impaired, and thus the event seems to be a critical component of the capacitatory process.
- In addition, other changes, both up and down, in levels of serine and threonine phosphorylation occur, notably the dephosphorylation, and thereby deactivation, of protein kinase C (PKC).
- Actin polymerization occurs in the cytosol between the surface and acrosomal membranes, and is stimulated by phosphatidylinositol-4 kinase (PI4K). This F-actin intervenes to prevent a premature fusion of the now destabilized surface membrane with the underlying outer acrosomal membrane.
- However, recent results on murine epididymal spermatozoa have described successful acrosome reactions, hyperactivation and fertilization by exposing spermatozoa transiently to calcium ionophore despite inhibition of cAMP generation, suggesting that high calcium can circumvent the requirement for cAMP and events downstream thereof, at least in spermatozoa not exposed to seminal plasma.

The acrosome reaction

According to conventional ideas, those capacitated spermatozoa that gain access to the *zona pellucida*, bind to it via receptors on the spermatozoal anterior head region, thereby inducing a dramatic morphological transformation in the spermatozoa called the **acrosome reaction** (Figure 12.1c; Box 12.3). During this process, the acrosome swells, its membrane fuses with the overlying plasma membrane, a vesiculated appearance is created and the contents of the acrosomal vesicle and the inner acrosomal membrane both become exteriorized in a process of **exocytosis**. The acrosome reaction, although visibly distinctive and thus conceptually separable from capacitation, is best seen as **the terminal phase of the capacitation process**.

It has been generally accepted that the agent responsible for inducing the acrosome reaction is a constituent of the *zona pellucida*. The *zona* is made up of three sulphated glycoproteins called ZP1, 2 and 3 in the mouse, the human *zona* having four ZPs, of which 1–3 are homologues of those in mice. ZP1 is a minor component that cross-links filaments made up of ZP2–3 dimers. ZP1 is not essential for binding of capacitated spermatozoa. The evidence suggests a **dominant primary binding role for ZP3**, but only fully effectively if organized structurally in conjunction with ZP2 binding in mice. In humans, **ZP3 functions in conjunction with ZP4** in binding and inducing the acrosome reaction. This **three-dimensional ZP molecular framework is species-specific** – mouse spermatozoa bind mouse ZP2/3 and human spermatozoa bind human ZP3/4 – and constitutes the major natural **block to cross-species fertilization**.

What about the ZP binding site on spermatozoa and what is the nature of the molecular interaction? Despite decades of research, the answers remain controversial. The primary interaction seems to involve ZP3 carbohydrates and a sugar-binding protein on the spermatozoal surface. Thus, sugars block the interaction, as does deglycosylation of the ZPs – whether fully or of O-linked residues only. However, alternative or additional binding pathways involving the protein variable region of ZP3 may also operate, given the importance of the three-dimensional matrix structure of the *zona* for generation of the acrosome reaction (see Box 12.4 for more details).

Most of the spermatozoal receptor for ZP2/3 is shed on the vesiculating membrane during the acrosome reaction, and so spermatozoal **binding to ZP3 is short-lived**. In addition, the acrosome releases **β-hexosaminidase B**, which digests any local ZP3 receptor, so preventing further binding. The acrosome-reacted spermatozoa now present an exposed inner acrosomal membrane, on which are located several proteolytic enzymes,

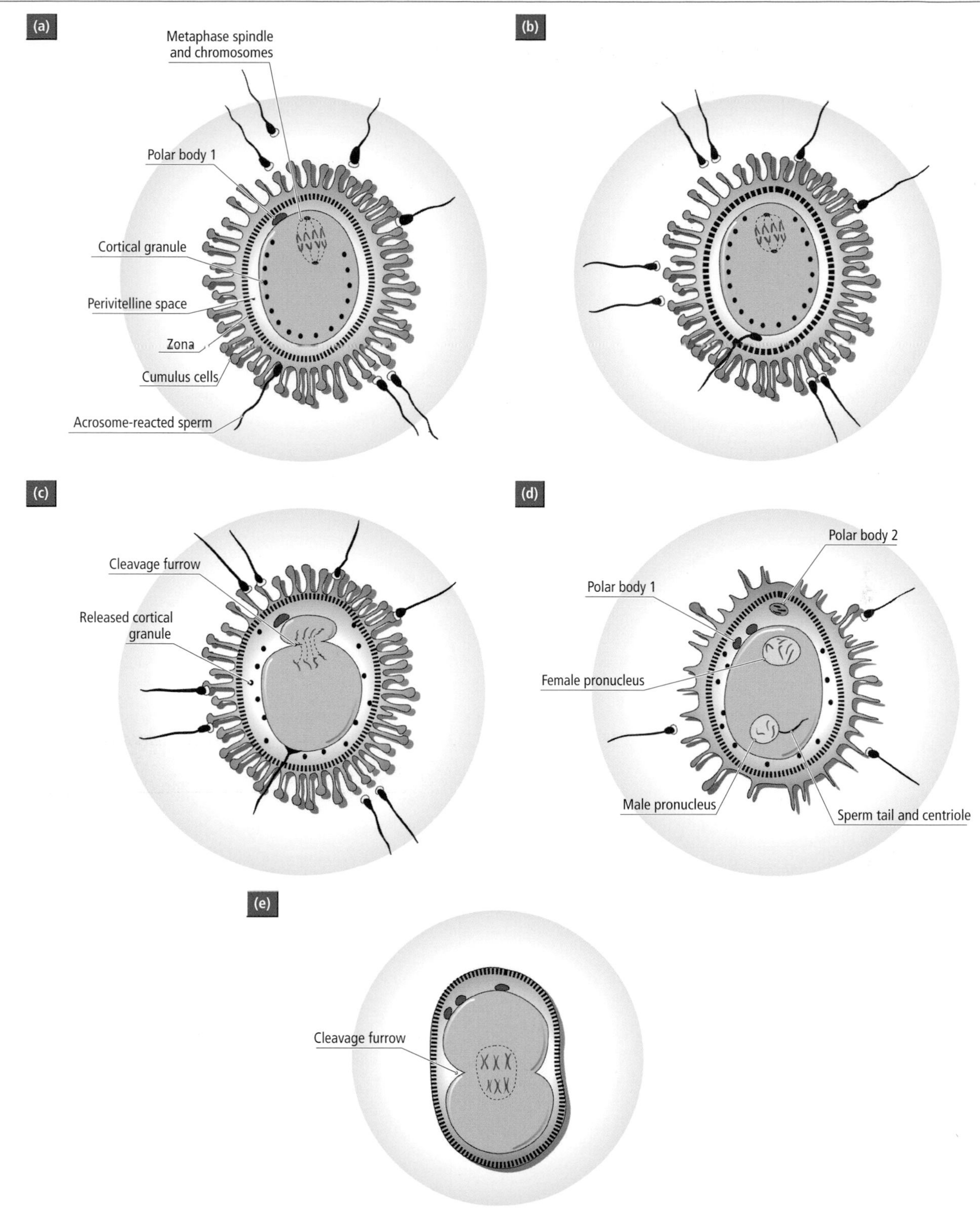

Figure 12.2 (a) Spermatozoa approach an ovulated oocyte with its first polar body, chromosomes on the second metaphase spindle, cortical granules and perivitelline space between the oolemma of the oocyte and the *zona pellucida*. One capacitated spermatozoon binds to the ZP2/3 components of the *zona pellucida* inducing the acrosome reaction and then (b) penetrates the *zona* to lie in the perivitelline space, where it (c) binds to and then fuses with the plasma membrane via its equatorial and post-acrosomal membrane, thus activating a series of calcium waves (see Figure 12.3). The calcium induces a release of cortical granules and resumption of meiosis with formation of an asymmetrically located cleavage furrow. (d) Meiotic division is completed, yielding the second polar body and female pronucleus. Meanwhile the spermatozoal nucleus decondenses and the male pronucleus forms, the spermatozoal centriole and tail being adjacent. (e) Chromosomes duplicate their DNA, pronuclei migrate together and their membranes break down as the first mitotic spindle forms. After its formation, chromatids separate and move apart, and a cleavage furrow develops. By this stage the cumulus cells have dispersed.

Box 12.3 The biochemistry of the acrosome reaction

The acrosome reaction represents a distinctive terminal phase of the capacitation process, and accordingly is associated with the following:

- Further large increases in intracellular calcium and cAMP. This calcium elevation seems central to the acrosome reaction, as is shown by its failure in the absence of calcium and in its premature induction with use of calcium ionophores that enhance calcium entry.
- A key function of the elevated calcium is the depolymerization of the recently formed F-actin in the cytosol separating the surface and acrosomal membranes, which is achieved by calcium-activated actin-severing proteins. So, what is it about the ZP binding by spermatozoa that triggers the increased calcium levels?
- During the terminal stages of capacitation and the start of the acrosome reaction, the activation of phosphatidylinositol-3 kinase (PI3K) by cAMP-PKA is observed. This activation can be countered by PKC activity. Thus, the up- and down-regulation of PKA and C respectively during capacitation may set the stage for the final acrosome reaction.
- PI3K then phosphorylates phosphatidylinositol 4,5-bisphosphate (PIP2) to form PIP3, which in turn activates several proteins implicated in the acrosome reaction, including phosphokinase Cζ and Akt kinase.
- The identity of the biochemical trigger for the acrosome reaction mediated by ZP binding remains, however, unclear. One suggestion is that it stimulates production of PIP3 by promoting PI3K activation.

Thus, the biochemistry of the acrosome reaction builds on that of capacitation.

Box 12.4 The molecular basis of spermatozoal–zona interactions

Several molecular candidates have emerged for spermatozoal binding and induction of the acrosome reaction. Indeed, more than one may be involved, given the multimeric micro-domains or 'rafts' in the capacitated spermatozoal membrane and the likely three-dimensional structure of the ZP binding site. Species differences complicate the issue. Here we summarize some candidates.

- **Lectins** are a class of proteins, each having a high affinity for a specific carbohydrate. Several such proteins have been implicated, of which one called ZP3R is of particular interest. ZP3R (also called SP56) binds galactose, and most of it actually resides within the acrosomal vesicle. However, as capacitation proceeds, small amounts translocate to the apical membrane – possibly by micro-exocytosis – further supporting the view that the acrosome reaction should be viewed as a terminal phase of capacitation.
- The enzyme β1,4-**galactosyl transferase** 1 (GalT 1) is an N-acetyl glucosamine-binding enzyme that is present on the anterior spermatozoal membrane overlying the acrosome. Site-directed mutation or knock-out of the *GalT 1* gene greatly reduces ZP binding. However, the residual spermatozoal binding suggests involvement of one or more additional spermatozoal membrane players.
- **Selectin** on spermatozoa is also implicated via binding to a *zona* sequence of sugar molecules called sialyl-Lewisx (SLeX) that is made up of branching chains consisting of four sugars – galactose, fucose, **N**-acetylglucosamine and sialic acid terminally. SLeX can reduce spermatozoal binding to human *zona* extracts.
- **Zonadhesin**, like ZP3R, is acrosomal until exposed during terminal capacitation. Its neutralization by antibody blocks *zona* binding, which is highly species-specific. However, mice spermatozoa null for zonadhesin remain fertile and bind to the *zona* – but not only of mice; they bind promiscuously to *zonae* of other species. Thus, it has been suggested that zonadhesin provides the species specificity observed in spermatozoal–zona binding.

Indeed, immunolocalization analyses have shown that several other intra-acrosomal proteins with sugar binding properties behave like ZP3R and relocate to the surface as capacitation proceeds. There they aggregate into multimeric micro-domains in rafts with a distinctive lipid composition in the apical membrane at the site of binding with the ZP. These multiprotein–lipid complexes seem to be the real sites of interactions with the *zona*, and unravelling their exact molecular architecture should advance understanding of the various contributors to the specificity of the interactions.

such as **acrosin**. These digest a path through the *zona*, along which the spermatozoon passes aided by the whiplash forward propulsion of the hyperactivated tail. Penetration of the *zona* takes between 5 and 20 minutes.

However, although it has been generally accepted that spermatozoa do not undergo the acrosome reaction until they encounter the *zona pellucida*, which then induces the reaction via an interaction with ZP3, recent experiments on mouse eggs and spermatozoa recovered from the oviduct at various stages of fertilization reveal that virtually all spermatozoa appear to be acrosome-reacted **before** they reach the cumulus cells, let alone the *zona pellucida*, suggesting that some other acrosome reaction inducer is active *in vivo* – possibly progesterone? Whether the fertilizing spermatozoon is included among these is, however, unclear – might they be the 'sacrificial spermatozoa' described above?

It was also thought to be important that the acrosome reaction occurs close to the oocyte, because acrosome-reacted spermatozoa had a very short lifespan. Herein lay the explanation for the rather complex series of changes that spermatozoa undergo in the female tract. Hours (or even days, in the human) may intervene between deposition of the ejaculate into the female tract and the ovulation of an oocyte. Higher fecundity may be achieved by establishing, first, **cervical and then isthmic reservoirs of spermatozoa**, releasing a gradual trickle through the capacitating uterus and oviduct to the ampullary isthmic junction where their **final fertilizing capacity can only be realized in the presence of an oocyte**. However, recent experiments have shown that even spermatozoa that have reached the peri-vitelline space can be recovered and used to fertilize a second egg, rather overturning the idea of short lifespan!

Spermatozoal–oocyte interactions

After penetrating the *zona pellucida*, the spermatozoon eventually comes to lie tangential to the oocyte surface between the *zona pellucida* and the oocyte membrane (**oolemma**) in the **perivitelline space**, where microvilli on the surface of the oocyte envelop the spermatozoal head (Figure 12.2b). Two sequential events then follow, each involving distinct molecular partnerships: first **spermatozoal–oocyte binding** and second **spermatozoal–oocyte fusion**. It is important to note that **only capacitated spermatozoa that have undergone an acrosome reaction are capable of binding and fusion**. This requirement seems to arise because it is only after the acrosome reaction that surface molecules essential for subsequent events become exposed on the spermatozoal surface.

Binding occurs between the surface membrane of the oolemma and the spermatozoal surface membrane overlying the middle and the posterior half of the spermatozoal head (the equatorial and post-acrosomal region). Recent research has complicated rather than resolved much of our understanding about the molecular basis of spermatozoal–oocyte binding and fusion, there being evidence to implicate several distinct types of interaction between egg and spermatozoa (see Box 12.5).

It is probable that a number of other molecules can influence the binding and fusion processes through interactions with these central molecular players (see Box 12.5). Calcium is again required for the fusion process, and its action may be mediated by calmodulin, as antagonists to this molecule can block spermatozoal–oocyte fusion. Once fusion has occurred, the spermatozoon dramatically ceases to move and its nucleus (together with variable parts of the mid-piece and tail contents, depending on the species) passes into the ooplasm.

Activation of the oocyte

Within 1–5 minutes after fusion of the oocyte with the spermatozoon, there is a **dramatic increase in the level of free intracellular calcium in the egg**, due largely to the release of calcium from internal stores. The rise lasts about 2–3 minutes and does not occur synchronously over the whole oocyte, but rather sweeps in a wave across the oocyte starting from the point of spermatozoa entry. This first rise is followed by a series of calcium spikes, each spike being of 1–2-minute duration, becoming more synchronous over the whole oocyte and occurring every 3–15 minutes depending on the individual oocyte (Figure 12.3). This spiking activity can last for several hours. This **calcium activity is critical for all the subsequent events of fertilization**, and if it is blocked, so are fertilization and development.

The fusion of the spermatozoon with an oocyte triggers the calcium release through the action of a spermatozoal enzyme that passes into the oocyte after fusion. This enzyme is identified as a spermatozoal-specific **phospholipase Cζ** (zeta). Thus, injection of either mRNA encoding PLCζ or recombinant PLCζ protein itself into oocytes is sufficient alone to activate calcium pulsing and development, whereas RNAi-mediated neutralization of endogenous PLCζ truncates calcium pulsing and blocks development. Moreover, some types of male infertility have been associated with defects in PLCζ expression or activity. PLCζ exerts its effects by stimulating the release of the second messengers, **inositol triphosphate** (**IP_3**) and **diacylglycerol** (**DAG**). The former then activates the calcium release, while the latter activates **protein kinase C** (**PKC**) to stimulate the phosphorylation of proteins essential for the further development of the conceptus (Figure 12.4).

This first phase of fertilization, from entry into the cumulus mass to fusion, is completed. The oocyte is now called a **zygote**. The ensuing events of fertilization last some 20 hours or so and are concerned with distinctive activities by both chromosomal and cytoplasmic elements. The calcium and the phosphoproteins stimulated by the activity of PLCζ play critical roles in both activities.

Fertilization completion

The penetrated oocyte must undergo four transitions for the successful completion of fertilization. First, diploidy is established; second, the cytoplasmic constitution of the oocyte is secured; third, syngamy occurs; and fourth the developmental programme is initiated. The first three of these transitions are considered now and the fourth is the subject of Chapter 13.

Box 12.5 Specificity of interaction between the oolemma and the spermatozoa

Several plausible types of interaction have been described

An interaction between an **integrin-like molecule in the oolemma** with the glycoprotein **disintegrin ADAM (A Distintegrin And Metalloproteinase domain-containing protein) on the spermatozoon is implicated**. ADAM is implicated in intraoviducal aspects of spermatozoal transport, as well as in binding and fusion. The integrin family has many members, and several of them are expressed on the surface of the oocyte. Use of antibodies to neutralize individual integrins and of specific genetic knock-outs of some integrin genes in the oocyte has not yet resolved which, if any, of the integrin $\alpha\beta$ dimers mediate binding of the spermatozoa. In contrast, a similar approach on spermatozoa has been more successful. There are three families of *ADAM* genes (1, 2 and 3; also called ***fertilin*** α, ***fertilin*** β and ***cyritestin***), each containing multiple variants, and each represented on the spermatozoal surface and involved in spermatozoa–oocyte binding. Expression of all three families appears to be co-regulated. Thus, knock-out of one reduces expression of the others and reduces spermatozoal binding to the oocyte. However, mice with genetic knock-outs for ADAM and integrins are not completely infertile, suggesting that other molecular players are involved.

A second spermatozoal membrane protein **Izumo1** that becomes exposed on the equatorial region only after the acrosome reaction is a possible candidate, as an antibody to it inhibits spermatozoal–oocyte binding and/or fusion. Izumo1-male mice are completely sterile, despite spermatozoa penetrating the *zona* and contacting the oolemma. If, however, a mutant spermatozoon is injected into the ooplasm, activation occurs and the embryos develop to term. Other mutations that affect Izumo1 expression or distribution also interfere with aspects of fertilization. Recently, an oocyte protein called **Juno** has been identified as a possible partner for Izumo1. Likewise, female mice knock-outs for Juno are infertile. However, there is a problem with Juno as an egg receptor, as it is expressed over the whole surface of the mouse egg, including an area to which spermatozoa do not bind. This failure of spermatozoa to bind to that region of the oolemma overlying the metaphase spindle has been known for some time, but recently a detailed study has explained why. Thus, this region of the mouse egg lacks **ganglioside-enriched microdomains**, and, in the absence of these domains, spermatozoa do not bind to the egg. Thus, the Juno explanation is at best incomplete. In addition to the uncertainty about the primary binding mechanism, fusion must also be addressed.

Two molecular partners in the oolemmal complex have been identified as important for successful fusion to follow binding. **Tetraspanins** are large membrane-spanning proteins involved in membrane fusion in several situations (viral infection, myotube fusion). Three tetraspanins (**CD9, CD81 and CD151**) have been identified on the oolemma, and antibodies to CD9 block the fusion of spermatozoa with normal oocytes. Oocytes from CD9-null mice have reduced fertility, do bind spermatozoa, but show reduced fusion. Mice doubly null for CD9 and 81 are completely infertile, unless rescued by injection of mRNA encoding CD9 or of a spermatozoon directly into the oocyte cytoplasm. A second molecular suspect is a class of proteins called **glycosyl-phosphatidylinositol-anchored proteins** (**GPIAPs**), of which two members (**CD55 and CD59**) have been detected on the oolemma. These are deficient in oocytes, lacking a critical biosynthetic enzyme subunit of N-acetyl glucosaminyl transferase, and oocyte-specific GPIAP-knock-out female mice show severely reduced fertility.

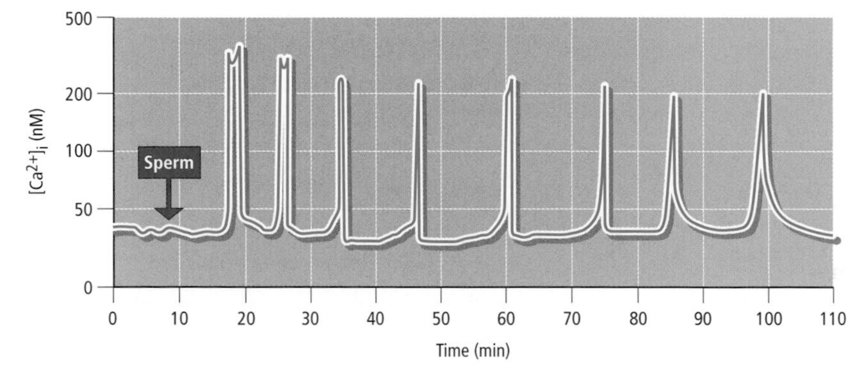

Figure 12.3 The calcium spiking pattern of a fertilized zona-free oocyte inseminated with spermatozoa at the time indicated by the arrow. After 12 minutes, a spermatozoon fused with the oocyte, generating a transient 5–10-fold rise in internal calcium. Thereafter, calcium pulses followed at 3–15-minute intervals (characteristic for each oocyte) and lasted for several hours. (Source: Courtesy of Dr Orla McGuinness).

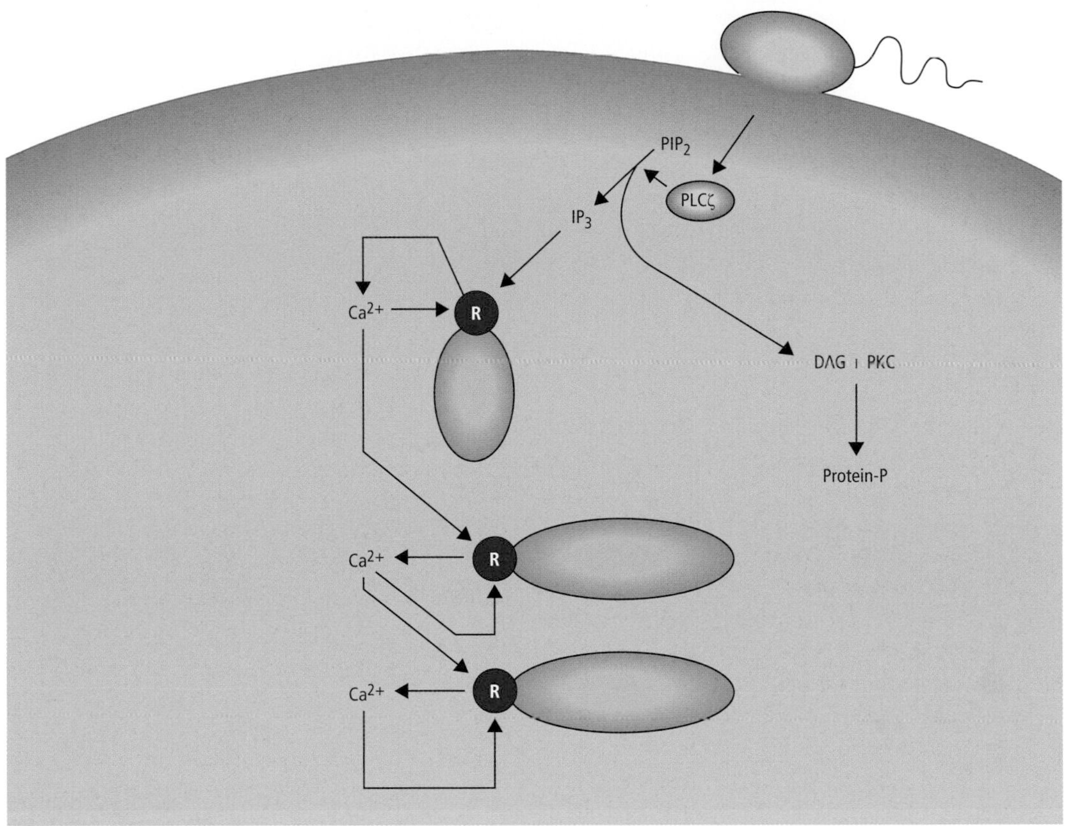

Figure 12.4 Model for activation of the oocyte calcium waves by the fertilizing spermatozoon. After fusion, the spermatozoon introduces **phospholipase Cζ (zeta)** into the oocyte. The PLCζ stimulates the release of the second messengers **inositol triphosphate** (IP_3) and **diacylglycerol** (DAG). The IP_3 activates the calcium release, while the DAG activates **protein kinase C** (PKC) to stimulate the phosphorylation of proteins essential for the further development of the conceptus. R = IP_3 receptor. (Source: Dr Orla McGuinness.)

Diploidy established

Diploidy requires that only one haploid spermatozoon fertilizes it – and no more (the **block to polyspermy**), as this would cause **triploidy** (for one extra spermatozoa) or **polyploidy** (for several). Polyploidy of this sort is described as **androgenetic**, as the extra sets of haploid chromosomes are derived from the male. Second, the ovulated oocyte was arrested in its second meiotic division and therefore has two haploid sets of chromosomes. If it is to transmit only one set to the next generation, it must complete its second meiotic division, dispatching one set of chromosomes to the **second polar body** (Figure 12.2c), and enter the interphase of the first cell cycle. The failure to complete the second meiotic division and jettison a set of female chromosomes would lead to **gynogenetic triploidy**.

The calcium pulsations resulting from gamete fusion play a key role in achieving diploidy, as can be demonstrated clearly by blocking them experimentally. The pulses probably work at least in part by activating calcium-binding proteins, which in turn directly or indirectly phosphorylate and/or dephosphorylate several target proteins critical for the next steps in the fertilization process. So how exactly is diploidy achieved?

First, the elevated calcium results in mobilization and then fusion of the zygotic subcortical **cortical granules** with the overlying oolemma (Figure 12.2c). The calcium acts via at least two calcium-binding proteins. **Calmodulin kinase II** (**CaMKII**) is important for mobilizing the granules into close proximity to the oolemma, and **synaptotagmin** mediates their docking and fusion. Fusion leads to the release of their contents into the perivitelline space (**the cortical reaction**). Among the contents are enzymes that act on the *zona pellucida* to prevent, or impair, further binding and penetration by spermatozoa (**the *zona* reaction**). These enzymes have at least three actions. A protease cleaves glycoprotein ZP2, and β-hexosaminidase B digests the oligosaccharide receptor on glycoprotein ZP3. In this way, the ZP spermatozoal-binding properties are removed, and spermatozoal binding to the *zona* ceases. In addition, tyrosine residues on adjacent ZPs are cross-linked, rendering the *zona* indissoluble to proteolytic cleavage and so impenetrable by spermatozoa. A final consequence of spermatozoal–oocyte fusion, the mechanism for which is not yet understood, is a reduction in the spermatozoal-binding properties of the oolemma itself. All these events occur rapidly as a result of the calcium pulses to reduce the chances of androgenetic polyploidy.

Second, the avoidance of gynogenetic triploidy also depends on the calcium rise. At ovulation, the oocytes are arrested in the second meiotic metaphase (M phase). This arrest has long been known to depend on the balance between two types of cytoplasmic activity called the **maturation or M-phase promoting factor** (**MPF**) and the **cytostatic factor** (**CSF**) (Figure 12.5; biochemistry summarized in Box 12.6). MPF stabilizes meiotic M-phase and is in turn itself stabilized by CSF. Calcium pulses lead to the inhibition of CSF and so to the destabilization of MPF. As a result, the oocyte exits from the M phase and progresses through the second meiotic division. In most species, the metaphase spindle lies just under and perpendicular to the surface (Figure 12.2a). This means that as meiosis is completed, the cleavage furrow is eccentrically placed, generating a large zygote and a small, second polar body (Figure 12.2c). The whole process takes about 30–45 minutes. It is important to note that spermatozoa do not bind to the oocyte membrane immediately overlying the second metaphase spindle; indeed, this area is devoid of ganglioside-enriched microdomains, integrins, CD9 and GPI-anchored proteins (Figure 12.2b,c and Box 12.5). This makes sense functionally, as the coincidence of a spermatozoon entering the oocyte and a second polar body being ejected could lead to mutual interference and **aneuploidy** (deviation from the normal euploid chromosomal number). With the successful attainment of one maternal and one paternal set of haploid chromosomes, **a euploid zygote is formed**.

Box 12.6 The biochemistry of meiosis resumption

- MPF has been shown to consist of a complex of at least two proteins: a kinase (phosphorylating) enzyme called **cdk1** and a smaller protein called **cyclin B**, and is active stabilizing the spindle until oocyte activation.
- A key component of **CSF** is **Emi2**, which stabilizes and inhibits another enzyme complex called **anaphase-promoting complex/cyclosome (APC/C)**, preventing it from disrupting MPF by **ubiquitinating cyclin B**, and so leading to MPF's degradation.
- With the fertilization-induced rise in calcium, **calmodulin kinase II** (**CaMk2**) is activated, which in turn phosphorylates Emi2 to create a docking site for the **kinase Plx1**, which further phosphorylates Emi2, marking it for degradation.
- With the loss of Emi2, the APC/C complex is released from inhibition, and degrades cyclin B. MPF activity is lost and meiosis resumes.
- Experimental evidence for the above comes from use of inhibitors of calcium pulsing or CaMKII. In contrast, constitutively active CaMKII leads to premature meiotic progression, as does destruction of Emi2.

Gametes provide more than chromosomes

The establishment of diploidy is accompanied by other contributions to the zygote by each gamete. We have already learnt that the fertilizing spermatozoon contributes phospholipase Cζ and, in most species, also contributes another essential zygotic component in the form of a **centriole.** The centrioles together with **pericentriolar material** make up the **centrosome.** Although the oocyte contributes pericentriolar material, only the spermatozoon contributes a centriole (in all mammals studied except rodents, see Chapter 7). The centrosome is a key player in the regulation of karyo- and cyto-kinesis at mitosis. Without it, cellular division during early development is compromised and eventually fails (see page 208).

The oocyte provides the cell membrane, cytoplasm, cell organelles and macromolecular matrix in which the two sets of chromosomes and the centrosome operate: the so-called **maternal cytoplasmic inheritance**. It thus is critical for successful development, and defects in oocyte maturation lead to developmental defects or failure. Especially significant are the **oocyte's mitochondria**, for, whilst those of the spermatozoon may enter the oocyte at fertilization, they do not survive. **All of the mitochondria in the adult are maternally derived**. Since mitochondria contain a mini-chromosome, which encodes some of the mitochondrial proteins, there is thus a small part

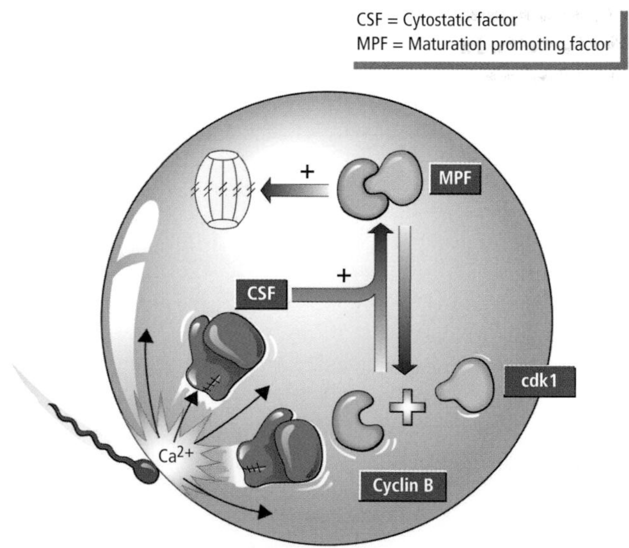

Figure 12.5 Schematic view of the stabilizing effect (+) of maturation promoting factor (MPF = cyclin B + cdk1) on the second meiotic spindle (shown in yellow). MPF is in turn stabilized (+) by cytostatic factor (CSF). Raised calcium, as occurs when the spermatozoon fuses with the oocyte, blocks CSF activity and leads to the destruction of cyclin B (see Box 12.6). The MPF in equilibrium with cyclin B unravels and so the spindle is no longer stabilized and meiosis resumes.

of the total genetic inheritance that is exclusively maternally derived. Genetically based defects of mitochondrial function in a mother will therefore be transmitted to her offspring (see Figure 22.12).

Syngamy

During oocyte–spermatozoal fusion and second polar body expulsion, the cytoplasmic contents of the spermatozoal cell membrane (now fused with the oocyte membrane) pass into the oocyte cytoplasm. The spermatozoa nuclear membrane breaks down and the highly condensed chromatin starts to swell, releasing filamentous strands of chromatin into the cytoplasm. The protamines that created such compressed chromatin are released and replaced by normal histones. This chromatin decondensation is actively induced by factors in the oocyte cytoplasm that develop in the terminal phases of intra-follicular maturation, such as high levels of the reducing agent **glutathione**.

Between 4 and 7 hours after fusion, the two sets of haploid chromosomes each become surrounded by distinct membranes and are now known as **pronuclei** (see Figure 12.2d). The male is usually the larger of the two. Both pronuclei contain several nucleoli. During the next few hours, each pronucleus gradually moves from its subcortical position to a more central and adjacent cytoplasmic position. During this period, the haploid chromosomes synthesize DNA in preparation for the first mitotic division, which occurs about 18–24 hours after gamete fusion. The pronuclear membranes around the reduplicated sets of parental chromosomes break down (Figure 12.2e), the mitotic metaphase spindle forms and the chromosomes assume their positions at its equator. The final phase of fertilization has been achieved: **syngamy** (or coming together of the gametic chromosomes) has occurred. Immediately the first mitotic anaphase and telophase are completed, the **cleavage furrow** forms, and the **one-cell zygote becomes a two-cell conceptus**.

Anomalous fertilization

Fertilization is a process that should result in a euploid two-cell conceptus equipped with all the materials required to take it through early development – the subject of Chapter 13. Before we follow it on this journey, however, we will explore what happens when fertilization goes wrong or when novel methods of manipulating a quasi-fertilization state are applied *in vitro*.

Aneuploidy

The euploid zygote contains one maternal and one paternal set of haploid chromosomes. However, frequently in humans the process goes wrong, and an aneuploid conceptus is formed. Aneuploidy causes embryonic abnormality and loss in most cases. Other aneuploid conceptuses may develop into tumours, such as **hydatidiform mole** (a benign trophoblastic tumour) or **choriocarcinoma** (its malignant derivative). Thus, aneuploidy, apart from being of potential danger to the mother, is reproductively wasteful and can be distressing. What then is the scale of aneuploidy in humans, what types of aneuploidy occur, how do they arise and with what consequences?

After fertilization *in vitro* and culture of fertilized oocytes in the IVF clinic, only 30–50% result in blastocysts. Interestingly, this proportion is higher than that found *in vivo* after lavage of blastocyst from uteri of women engaging in coitus at mid-cycle. However, on their transfer to women, these IVF-generated blastocysts give early pregnancy rates of between 20% and 50%. This massive **early developmental failure of the human conceptus is associated with a high incidence of abnormalities of chromosomal distribution or number**. Thus, many of the conceptuses examined have whole sets of chromosomes missing or duplicated (**errors of ploidy**), or individual chromosomes have gone astray or are present in more than two copies (**errors of somy**), or structural rearrangements of chromosomes (**translocations**), or a mixture of cells (**genetic mosaics**) – some lacking nuclei or having multiple nuclei, others being of varied chromosomal constitution. A high level of genetic abnormality is also detected in recognized spontaneous non-IVF clinical pregnancies (in 40–60% of spontaneous miscarriages, especially those occurring in the first trimester, 5% of stillbirths, and 0.5% of all live births). The type of each genetic abnormality observed at birth, and its approximate incidence in spontaneous miscarriages, are recorded in Table 12.1.

It is clear from Table 12.1 that some types of abnormality are more common in miscarriage and others are compatible

Table 12.1 Incidence of chromosomal abnormalities in spontaneously aborted human conceptions

	Number/100 aborted conceptions	Surviving to birth (%)
Triploidy (three sets of chromosomes)	12–15	<0.01
Tetraploidy (four sets of chromosomes)	3–5	<0.01
Sex chromosome trisomies (three sex chromosomes)	<1	>99
Sex chromosome monosomies (one sex chromosome)	10	<1
Trisomy for one or two autosomes	20–40	3
Monosomy for one or two autosomes	1	None
Structural rearrangement of chromosomes	2–3	35

with survival to birth. It is also clear that some types of abnormality are missing altogether (e.g. haploids) or are underrepresented, e.g. autosomal and sex chromosomal monosomies, which might be expected to occur with the same frequency as trisomies, since when one nucleus gains a chromosome at division, the other will lose one. The analysis of early mouse development tells us that these types of abnormality are lethal very early in development (lack of genetic material being deleterious earlier than excess), whereas trisomic, triploid and tetraploid conceptuses survive for longer. It has been calculated from clinical data, and by extrapolation from data derived from comparative studies, that 50% or more of all human conceptions may result in genetically abnormal embryos and fetuses. It is clear that humans are genetically a mess when it comes to reproduction, especially when compared with other mammals. Why?

This massive genetic burden cannot be due entirely to constitutional genetic defects in all the germ cells of one or both parents, and many, if not most, genetic abnormalities in fetuses have indeed been shown to arise prior to, at, or shortly after fertilization. For example, **disorders of ploidy** are likely to result from failure of polar body formation (10–20% of triploids), polyspermy (80–90% of triploids) or failure of one early cleavage division (tetraploidy). Many of the **mono- and tri-somic conceptuses arise from abnormalities of oocyte meiotic divisions**. Moreover, these failures of the normal mechanisms for establishing euploidy increase in certain conditions. Thus, the fertilization of prematurely ovulated (and thus not fully matured?) oocytes is associated with defective cortical granule release and poor spermatozoal head incorporation. The fertilization of oocytes that have been ovulated and passed into the oviduct 24 hours or more before spermatozoal arrival can also be problematic. In most mammalian species, in which ovulation and behavioural oestrus are closely synchronized, the problem of ageing oocytes will not be acute. In humans, however, where coitus and ovulation are not necessarily, or even usually, synchronized, the fertilization of old oocytes is much commoner. There is evidence that the high incidence of genetic abnormality and early pregnancy loss in humans in part reflects this dissociation of mating from fertility. In addition, anomalies have been associated with exposure of a female to alcohol or anaesthesia around the time of ovulation.

Other chromosomal abnormalities in the conceptus arise from events in the ovary or testis that affect gametes directly, well in advance of the acute events of fertilization and early cleavage. For example, abnormalities may be induced by exposure to X-irradiation or mutagenic chemicals, and such exposure correlates with a subsequently increased natural miscarriage rate. Moreover, an increased incidence of fetal chromosomal abnormality is observed with increasing maternal age, which may reflect the fact that oocytes are maintained in a prolonged meiotic dictyate stage (see Chapter 3) during which they are susceptible to such damage. Oocytes seem to have less effective surveillance systems than somatic cells for detecting chromosome distribution abnormalities, meaning that genetically flawed oocytes survive to be fertilized. Similarly, cells of the very early conceptus inherit a relaxed system of cell-cycle mitotic check points, leading to greater possibilities for chromosomal mis-segregation. Overall, it appears that many of the chromosomal anomalies originate in women from environmental or social causes.

Spermatogenic cells, in contrast, are highly sensitive to chromosomal anomalies and so few survive to pass them on. However, spermatozoa do lack the effective DNA repair systems of somatic cells, and consequently show three-fold higher mutation rates than eggs – indeed, this high mutation rate has been described as the '**engine of evolution**'. At first sight, it seems curious that the germline is less well genetically policed than somatic cells. Is it possible that we have evolved a system that encourages genetic diversity generation and relies on effective screening out of the failures during pregnancy – truly evolution in action with each throw of the fertilization dice?

Parthenogenesis

Although the spermatozoon induces many remarkable changes in the oocyte, it does not appear to be essential for many of them. An oocyte may be activated **parthenogenetically** by a variety of bizarre stimuli, such as electric shock, or exposure to various enzymes or to alcohol. Activation by these stimuli is especially easy in 'aged' oocytes that have been ovulated several hours previously. These stimuli have the common feature that they induce a calcium rise, thereby mimicking the act of spermatozoal fusion. As a result, cortical granules exocytose, meiotic metaphase is resumed and the oocyte's developmental programme is activated. The **parthenote** so formed may undergo some cell divisions but eventually fails in most species, including humans and the large farm animals. In a few species, such as the mouse, rat and hamster, cell division is unimpaired and a blastocyst may form, implant and develop to a stage where a beating heart, somites and forelimbs are present. However, even in these species, most parthenotes die fairly early on in this sequence and none survives to term. Two observations explain this failure to survive.

- Because, in most mammals, parthenogetically activated oocytes lack a centriole and thus the fully functional centrosome that a spermatozoon would provide, an early cessation of cell division occurs. In the mouse, in contrast, the cells of the conceptus appear to be able to generate their own centrosome *de novo* such that cell division is relatively unimpaired, hence the further development of parthenotes in these animals. However, even in the mouse, parthenogenetic development fails eventually and the second observation explains why.
- We described in Chapter 1 (see page 11), the process of **parental imprinting**, in which a sex-specific epigenetic imprint marked a subset of around 100 genes such that each 'remembered' its origins as paternally or maternally derived. These **maternal and paternal imprinting processes** mean that, although the oocyte and the spermatozoon each

contribute one complete set of chromosomes and genes to the conceptus, each set is not on its own fully competent to direct a complete programme of development. Only when a set of genes from an oocyte is combined with a set of genes from a spermatozoon is a fully functional genetic blueprint achieved. A parthenogenetically-activated oocyte lacks access to some crucial genetic information, which, although present in its chromosomes, cannot be accessed because of the maternal imprinting to which it was subjected during oogenesis. In the normal zygote, this information would be provided by genes on the paternally-derived set of chromosomes. These ideas are supported by the observation that the experimental erasure of all parental imprints genetically allows parthenote mice to develop to term. Thus, it is parental imprinting that **compels us to reproduce sexually** and means that parthenogenesis is not possible in mammals.

There are costs to obligatory sexual reproduction, as discussed in Chapter 1. The time and energy taken up with seeking a sexual partner, courting and mating are costly. Males cannot be dispensed with, but consume resources. Asexual reproduction is advantageous for rapid and efficient propagation of the species. Sexual reproduction may be more advantageous at times of stress when maximum genetic flexibility is required for the species to survive – the more extreme the threat, the greater the chance of the species surviving if its genetic variability is maximized. Many organisms retain an option on asexual reproduction, even where sexual reproduction can occur, but mammals have lost that option during evolution. One of the great puzzles of mammalian biology is why, and the answer may lie in the origins of viviparity – thus parental imprinting is observed in all mammals except egg-laying monotremes. It is in mammals that the close and prolonged association of a genetically alien fetus survives in the uterus. The basis for this survival will be discussed in Chapter 16, but here we can note that many or most parentally imprinted genes are expressed in placental tissues. Moreover, those genes that are expressed from paternal chromosomes tend to promote fetal growth at maternal expense, whilst those maternally active genes tend to constrain fetal development and protect the mother. These observations have led to the suggestion that parental imprinting evolved in parallel with viviparity as part of a **battle between the male and female genomes for control over the transmission of each to the next generation** – the '**parental conflict hypothesis**'. In addition, the discovery of parental imprinting has aided our understanding of the origins of some developmental pathologies (see Box 12.7).

Somatic cell nuclear transfer

Although asexual reproduction naturally by parthenogenesis does not occur in mammals, the ingenuity of scientists has circumvented this block experimentally. A diploid somatic nucleus, taken from an adult or fetal differentiated cell, can be placed into the cytoplasm of an oocyte from which the metaphase 2 spindle has been removed together with its chromosomes (Figure 12.6). The reconstituted oocyte can then be activated to develop and the developing diploid conceptus placed in a uterus. Some conceptuses can develop to term. The individual animal produced shares all its nuclear chromosomes with the donor of the adult or fetal nucleus and is a **reproductive clone**. The process leading to it is called **somatic cell nuclear transfer** or **SCNT**.

Despite examples of SCNT resulting in live young, the vast majority of cloned conceptuses die at early to mid-gestation stages of development. The small proportion of conceptuses that survives to late fetal stages or birth is characterized by a high mortality rate, frequently grossly increased placental and birth weights and respiratory distress, and curtailed healthy lifespan of any surviving offspring. Why? The answer to this question is complex and the subject of continuing research, but certain explanations already seem likely.

Box 12.7 Pathologies associated with parental imprinting

Parentally imprinted genes are effectively hemizygous as only one of the two inherited alleles is functional. Thus, if the only functional allele carries a mutation, then the individual is effectively null for that gene. Although there are relatively few parentally imprinted genes, their expression during placental and fetal development means that the absence of expression can have profound consequences clinically.

One region on the long arm of human chromosome 15 is a major site of parentally imprinted genes: some maternally imprinted and others paternally imprinted. If a paternal contribution to this region, called 15q11–q13, is lacking because of mutation or deletion, then the developing child suffers from **Prader–Willi syndrome**, a constellation of developmental disturbances involving neonatal hypotonia, lethargy and difficulty feeding. Adults are short in stature, obese and intellectually impaired. Conversely, the absence of a normal maternal copy of the same region causes **Angelman syndrome**, a neurodevelopmental disorder associated with a happy disposition, intellectual disturbance, seizures and generally jerky movements. Similarly, **Beckwith–Wiedemann syndrome**, which manifests as disproportionate growth and various other developmental defects, is associated with failure of normal biparental inheritance of chromosome 11p15, and loss of parental imprinting is associated with certain childhood cancers, such as **Wilms' tumour**. Whilst these conditions are fortunately rare, they do provide insights into the phenomenon of parental imprinting.

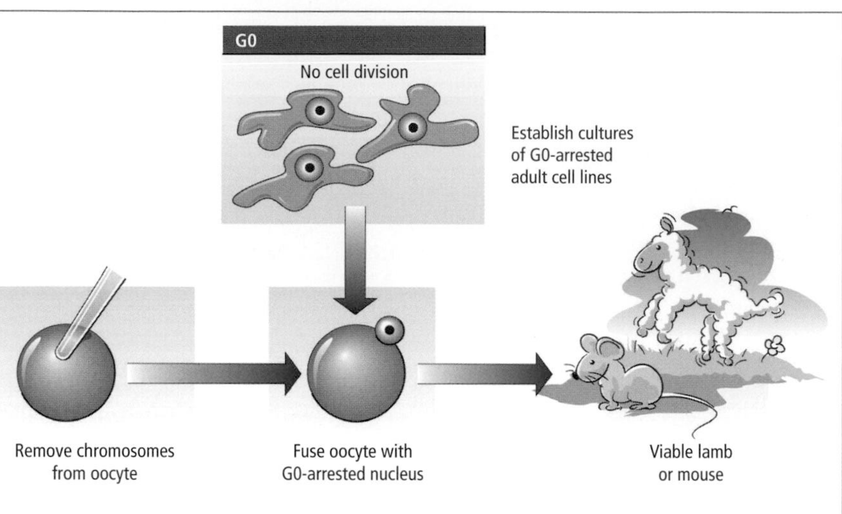

Figure 12.6 Schematic summary of the procedure for somatic cell nuclear transfer (SCNT) in sheep or mice. A differentiated somatic cell is cultured and its division cycle arrested by removal of nutrients (G0 stage). A **karyoplast** (the nucleus with a small amount of cytoplasm and cell membrane surrounding it) is then prepared from the quiescent cell. It is placed next to an M2-arrested oocyte from which the second meiotic spindle and chromosomes have been removed by suction. A fusogenic signal is then given. The nucleus and enucleated oocyte fuse and initiate cleavage. The cleaving conceptus is placed into the uterus of a ewe (or mouse) and a viable offspring can result. (Source: Stewart C. (1997) *Nature*, **385**, 769. Adapted with permission from *Nature*.)

First, although the conceptuses created by SCNT are diploid, most if not all show abnormal gene expression patterns compared with those developing from fertilized oocytes. The reason for this difference is, in part if not entirely, due to **abnormal epigenetic modification patterns**. We saw in the last section the important role of parental imprinting in forcing sexual reproduction. However, epigenetic changes to the genome also occur naturally in non-parentally determined ways as cells divide and differentiate to generate all the multitude of cell types making up the body – these are called **developmental epigenetic imprints**. This process ensures that the genes needed for a particular tissue are available for expression in that tissue and that other unrequired genes are closed down to expression (see pages 10–11 and Figure 1.7). Thus, when a diploid nucleus from, say, the skin is used in SCNT, it carries with it whole sets of epigenetic modifications to its genes that must be erased if it is to be able to express all the genes required to make an entire organism. Although it seems that the oocyte cytoplasm is very effective at erasing these developmental epigenetic imprints, after SCNT it may not have sufficient time or capacity to erase them all. Evidence to support this explanation is shown by the greater ability of nuclei from very early developmental stages to support cloned conceptus development than nuclei from more differentiated cells. In addition, serial transfer of differentiated nuclei through one ooplasm into a second one improves the capacity of these nuclei to support development.

A second explanation for the failure of the cloned conceptuses to develop may arise from the fact that, although differentiated nuclei have properly parentally imprinted genes, these parental imprint patterns may also be disturbed during the process of SCNT, with further adverse developmental consequences.

Finally, it has been shown that the transferred somatic cell nuclei carry with them a coating of cytoplasm containing mitochondria and that these mitochondria, unlike those brought in by spermatozoa, survive. Thus, the cells of the cloned conceptus have a mixture of mitochondria from different sources and it is suggested that incompatibility between them and the nuclear genome may be developmentally problematic.

There is clearly a long way to go before we fully understand what happens after SCNT in animals. On safety grounds alone, therefore, the prospect of **human reproductive cloning is unethical**. However, ethical concerns go beyond safety, and these concerns have led to the legal prohibition of human reproductive cloning in many jurisdictions. However, in some countries, including the UK, **non-reproductive or therapeutic human SCNT** is permitted under strict licence for research purposes only. This research aims to understand the earliest post-fertilization molecular events in humans, such as how, where and when epigenetic imprints are erased or modified, and what problems mitochondrial heterogeneity may present. Such research is of particular importance if human embryonic stem cells are to be medically useful. But more of that in Chapter 22!

Conclusions

The interval between the departure of the gametes from the gonads and the successful formation of a two-cell conceptus encompasses an extraordinarily complex sequence of events involving biochemical, behavioural, endocrine, physiological and genetic components. Not surprisingly, aspects of this process often go wrong and infertility results (see Chapter 22). Fortunately, biomedical advances mean that several of these component events can now be carried out *in vitro*. Mature oocytes can be aspirated laparoscopically directly from the follicles; spermatozoa can be recovered from ejaculates or even microsurgically from the testis and epididymis, and can be capacitated and activated in defined media; fertilization and the formation of a zygote can be achieved outside the body. These techniques have provided help for many otherwise infertile couples (see Chapter 22), and the pioneering research that underlies these achievements led to Professor Sir Robert 'Bob' Edwards, the dedicatee of this book, being awarded the 2010 Nobel Prize for Physiology or Medicine.

Key learning points

- Spermatozoa are capacitated in the female genital tract to gain forward activated movement and the capacity to undergo the acrosome reaction.
- Capacitation involves removal of adsorbed macromolecules and cholesterol, destabilization of the spermatozoal surface, a rise in pH, entry of calcium, activation of adenyl cyclase, a rise in cAMP, phosphorylation of PKA and activation of tyrosine kinases.
- The acrosome reaction involves fusion of the acrosome with the overlying plasma membrane and release of acrosomal contents.
- It can be induced by progesterone and binding of spermatozoal proteins to *zona* glycoproteins 2 and 3.
- Acrosome-reacted spermatozoa digest a path proteolytically through the *zona*.
- A spermatozoon binds at its equatorial post-acrosomal segment with the oolemma, possibly involving an Izuma1–Juno interaction.
- Fusion of spermatozoon and oocyte involves the further involvement of CD9- and GPI-anchored proteins.
- Fusion results in the entry of spermatozoal phospholipase Cζ into the oocyte, where it generates a series of calcium spikes, which lasts for several hours.
- The fertilized oocyte is now called a zygote.
- The calcium activates release of cortical granules which prevent polyspermy by effects on the *zona* and the oolemma.
- The calcium also reactivates arrested meiosis and second polar body extrusion by destroying cytostatic factor and cyclin B, and thereby maturation promoting factor (MPF) activity.
- Spermatozoa also contribute their centrioles to the zygote in most species and this is essential for continuing cell division during development.
- Oocytes contribute most other cell organelles to the zygote including mitochondria, which are inherited exclusively via the maternal route.
- A male pronucleus forms and its heterochromatin decondenses, protamines being replaced by histones.
- A female pronucleus forms and is usually smaller than the male pronucleus.
- Over an 18–24-hour period, the pronuclei replicate their DNA, move together, lose their nuclear membranes and form a mitotic spindle, and the male and female chromosomes come together for the first time on the spindle at syngamy.
- Aneuploidy can arise at fertilization, especially errors of ploidy. Errors of somy occur more frequently when oocytes from older women are fertilized.
- Spermatozoal chromosomes show a high rate of mutation, because they lack an efficient DNA repair system, and have been called the 'engines of evolution'.
- Oocytes can be activated in the absence of a spermatozoon, but show only limited development because of the absence of a centriole (in most species) and the fact that some genes are differentially parentally imprinted (in all species) and are therefore not formally equivalent on homologous chromosomes.
- Reproductive cloning by transfer of the nucleus from an adult somatic cell to an enucleated oocyte (SCNT) can result in the development of a few live young which differ genetically from the donor of the nucleus mainly in their mitochondrial DNA.
- Most cloned conceptuses die early in development, and many of those surviving have enlarged placentae or other pathologies, probably mainly as a result of epigenetic abnormalities.

Clinical vignette

Meiotic non-dysjunction

A 42-year-old woman, Ms W, found herself unexpectedly pregnant. She had two children from a previous relationship, aged 10 and 12 respectively, who were fit and well. She had recently entered into a relationship with a new partner, when the conception occurred. The couple decided to continue with the pregnancy, and attended their local maternity unit for a dating scan at 12 weeks of gestation. At this time, Ms W and her partner accepted an offer of combined screening for trisomies 13, 18 and 21. Unfortunately the screening test revealed a high risk (1 in 5) of the pregnancy being affected by trisomy 21 (Down syndrome). Ms W was referred to a fetal medicine centre where both the limitations of screening and the risks versus benefits of a diagnostic test were explained in detail. Ms W and her partner decided that they wished to proceed with a diagnostic test for trisomy 21 (chorionic villus sampling), despite the associated risk of miscarriage of approximately 1%. A sample of the placental tissue was obtained using a needle inserted through Ms W's abdomen into the uterus, and sent to the lab for analysis. Fluorescence *in situ* hybridization was performed in order to determine the number of copies of chromosome 21 that were present in the placental cells. The test results were available 3 days later, and revealed that the fetus was affected by trisomy 21. Ms W was unsurprised to receive this news and had been researching the implications of Down syndrome while awaiting the result. After a number of long discussions with each other, other family, and with parents of children affected by Down syndrome, the couple decided to continue with the pregnancy. Baby W was born small but healthy at 38 weeks gestation by spontaneous vaginal delivery. He was able to leave hospital after a week, and was a source of great joy to his parents and older siblings.

The risk of a pregnancy being affected by trisomy 21 increases with maternal age: initially this is a linear increase until the age of 30, but increases exponentially thereafter. At the age of 42, the rate of Down syndrome affecting pregnancies in the first trimester is approximately 1 in 30, and at birth is approximately 1 in 50–60. The discrepancy between rates at different gestational ages is partly due to the higher rate of spontaneous pregnancy loss in fetuses with trisomy 21, and partly due to elective termination of pregnancy where the diagnosis is known. Most human aneuploidy arises from segregation errors during meiosis I in oocytes, and the frequency of such segregation problems is influenced by chromosome cohesion during the dictyate stage. In a mother aged 40, the oocytes have been held suspended in meiotic prophase I for twice as long as in a mother aged 20. It is thought that this lengthy period suspended in prophase makes oocytes vulnerable to decreased chromosomal cohesion and therefore to meiotic non-dysjunction. In contrast to most other trisomies, trisomy 21 is survivable and therefore most data is available on trisomy 21. The accelerated risk of trisomy 21 with increasing maternal age contrasts with the relationship between trisomy 21 risk and paternal age. Trisomy 21 risk increases gradually with paternal age, but this association is highly likely to be confounded by the strong association between maternal and paternal age rather than an independent effect. Biologically, this reflects the de novo synthesis of spermatozoa from germ line cells every 60 days, rather than being held in a state of suspended meiosis over a number of years.

FURTHER READING

General reading

Ambartsumyan G, Clark AT (2008) Aneuploidy and early human embryo development. *Human Molecular Genetics* **17**, R10–R15.

Armon L, Ben-Ami I, Ron-El R, Eisenbach M (2014) Human oocyte-derived spermatozoa chemo-attractant is a hydrophobic molecule associated with a carrier protein. *Fertility and Sterility* **102**, 885–890.

Barroso G, Valdespin C, Vega E *et al.* (2009) Developmental spermatozoa contributions: fertilization and beyond. *Fertility and Sterility* **92**, 835–848.

Bedford JM (2011) Site of the mammalian sperm physiological acrosome reaction. *Proceedings of the National Academy of Sciences of the United States of America* **108**, 4703–4704.

Bianchi E, Doe B, Goulding D, Wright JG (2014) Juno is the egg Izumo receptor and is essential for mammalian fertilization. *Nature* **508**, 483–487.

Blake D, Proctor M, Johnson N, Olive D (2007) Cleavage stage versus blastocyst stage embryo transfer in assisted conception. *The Cochrane Collaboration*. John Wiley & Sons, Ltd, Chichester.

Boerke A, Tsai PS, Garcia-Gil N *et al.* (2008) Capacitation-dependent reorganization of microdomains in the apical sperm head plasma membrane: functional relationship with zona binding and the zona-induced acrosome reaction. *Theriogenology* **70**, 1188–1196.

Breitbart H, Rotman T, Rubinstein S, Etkovitz N (2010) Role and regulation of PI3K in sperm capacitation and the acrosome reaction. *Molecular and Cellular Endocrinology* **314**, 234–238.

Clark GF (2011) Molecular models for mouse spermoocyte binding. *Glycobiology* **21**, 3–5.

De Jonge C (2017) Biological basis for human capacitation revisited. *Human Reproduction Update* **23**, 289–299.

Drobius EZ, Overstreet JW (1992) Natural history of mammalian spermatozoa in the female reproductive tract. *Oxford Reviews of Reproductive Biology* **14**, 1–46.

Gadella BM, Boerke A (2016) An update on post-ejaculatory remodeling of the sperm surface before mammalian fertilization. *Theriogenology* **85**, 113–124.

Jones KT (2005) Mammalian egg activation: from Ca^2+ spiking to cell cycle progression. *Reproduction* **130**, 813–823.

Johnson MH (2001) The developmental basis of identity. *Studies in History and Philosophy of Biological and Biomedical Sciences* **32**, 601–617.

Kashir J, Heindryckx B, Jones C *et al.* (2010) Oocyte activation, phospholipase C zeta and human infertility. *Human Reproduction Update* **16**, 690–703.

Kuo Y-W, Li S-H, Maeda K-I *et al.* (2016) Roles of the reproductive tract in modifications of the sperm membrane surface. *The Journal of Reproduction and Development* **62**, 337–343.

Malcuit C, Kurokawa M, Fissore RA (2006) Calcium oscillations and mammalian egg activation. *Journal of Cell Physiology* **206**, 565–573.

Reid AT, Redgrove K, Aitken RJ, Nixon B (2011) Cellular mechanisms regulating sperm–zona pellucida interaction. *Asian Journal of Andrology* **13**, 88–96.

Sakai RR, Tamashiro KL, Yamazaki Y, Yanagimachi R (2005) Cloning and assisted reproductive techniques: influence on early development and adult phenotype. *Birth Defects Research (Part C)* **75**, 151–162.

Sakkas D, Ramalingam M, Garrido N, Barratt CLR (2015) Sperm selection in natural conception: what can we learn from Mother Nature to improve assisted reproduction outcomes? *Human Reproduction Update*, **21**, 711–726.

Schatten H, Sun Q-Y (2011) New insights into the role of centrosomes in mammalian fertilization and implications for ART. *Reproduction* **142**, 793–801.

Stewart C. (1997) Nuclear transplantation. An udder way of making lambs. *Nature*, **385**, 769.

Suarez SS (2016) Mammalian sperm interactions with the female reproductive tract. *Cell and Tissue Research* **363**, 185–194.

Swann K, Yu Y (2008) The dynamics of calcium oscillations that activate mammalian eggs. *International Journal of Developmental Biology* **52**, 585–594.

Van Blerkom J, Zimmerman S (2016) Ganglioside-enriched microdomains define an oolemma that is functionally polarized with respect to fertilizability in the mouse. *Reproductive BioMedicine Online* **33**, 458–475

Wassarman PM (2011) The sperm's sweet tooth. *Science* **333**, 1708–1709.

Wassarman PM, Litscher ES (2010) Egg's ZP3 structure speaks volumes. *Cell* **143**, 337–338.

More advanced reading

Baibakov B, Gauthier L, Talbot P *et al.* (2007) Sperm binding to the zona pellucida is not sufficient to induce acrosome exocytosis. *Development* **134**, 933–943.

Breitbart H, Cohen G, Rubinstein S (2005) Role of actin cytoskeleton in mammalian sperm capacitation and the acrosome reaction. *Reproduction* **129**, 263–268.

Chiu PCN, Wong BS, Chung MK *et al.* (2008) Effects of native human zona pellucida glycoproteins 3 and 4 on acrosome reaction and zona pellucida binding of human spermatozoa. *Biology of Reproduction* **79**, 869–877.

Dorus S, Wasbrough ER, Busby J *et al.* (2010) Sperm proteomics reveals intensified selection on mouse sperm membrane and acrosome genes. *Molecular Biology and Evolution* **27**, 1235–1246.

Ducibella T, Rafael Fissore R (2008) The roles of Ca2+, downstream protein kinases, and oscillatory signaling in regulating fertilization and the activation of development. *Developmental Biology* **315**, 257–279.

Ensslin MA, Shur BD (2003) Identification of mouse sperm SED1, a bimotif EGF repeat and discoidindomain protein involved in sperm–egg binding. *Cell* **114**, 405–417.

Goto M, O'Brien DA, Eddy EM (2010) Speriolin is a novel human and mouse sperm centrosome protein. *Human Reproduction* **25**, 1884–1894.

Hino T, Muro Y, Tamura-Nakano M *et al.* (2016) The behavior and acrosomal status of mouse spermatozoa in vitro, and within the oviduct during fertilization after natural mating. *Biology of Reproduction* **95**, 1–11.

Holt WV, Van Look KJW (2004) Concepts in sperm heterogeneity, sperm selection and sperm competition as biological foundations for laboratory tests of semen quality. *Reproduction* **127**, 527–535.

Hunter RHF, Rodriguez-Martinez H (2004) Capacitation of mammalian spermatozoa in vivo, with a specific focus on events in the Fallopian tubes. *Molecular Reproduction and Development* **67**, 243–250.

Hurt KJ, Musicki B, Palese MA *et al.* (2002) Aktdependent phosphorylation of endothelial nitric-oxide synthase mediates penile erection. *Proceedings of the National Academy of Sciences of the United States of America* **99**, 4061–4066.

Ikawa M, Inoue N, Benham AM, Okabe M (2010) Fertilization: a sperm's journey to and interaction with the oocyte. *The Journal of Clinical Investigation* **120**, 1–10.

Jones R, Howes E, Dunne PD, *et al.* (2010) Tracking diffusion of GM1 gangliosides and zona pellucida binding molecules in sperm plasma membranes following cholesterol efflux. *Developmental Biology* **339**, 398–406.

Knott JG, Kurokawa M, Fissore RA, *et al.* (2005) Transgenic RNA interference reveals role for mouse sperm phospholipase Cz in triggering Ca^2+ oscillations during fertilization. *Biology of Reproduction* **72**, 992–996.

Kong M, Diaz ES, Morales P (2009) Participation of the human sperm proteasome in the capacitation process and its regulation by protein kinase A and tyrosine kinase. *Biology of Reproduction* **80**, 1026–1035.

Kono T, Obata Y, Wu Q *et al.* (2004) Birth of parthenogenetic mice that can develop to adulthood. *Nature* **428**, 860–864.

Lalande M (1997) Parental imprinting and human disease. *Annual Reviews in Genetics* **30**, 173–195.

Lefevre B, Wolf J-P, Ziyyat A (2010) Sperm-egg interaction: is there a link between tetraspanin(s) and GPI-anchored protein(s)? *BioEssays* **32**, 143–152.

Lefievre L, Conner SJ, Salpekar A *et al.* (2004) Four zona pellucida glycoproteins are expressed in the human. *Human Reproduction* **19**, 1580–1586.

Lishko PV, Botchkina IL, Fedorenko A, Kirichok Y (2010) Acid extrusion from human spermatozoa is mediated by flagellar voltage-gated proton channel. *Cell* **140**, 327–337.

Ono Y, Kono T (2006) Irreversible barrier to the reprogramming of donor cells in cloning with mouse embryos and embryonic stem cells. *Biology of Reproduction* **75**, 210–216.

Oren-Benaroya1 R, Orvieto R, Gakamsky A *et al.* (2008) The sperm chemoattractant secreted from human cumulus cells is progesterone. *Human Reproduction* **23**, 2339–2345.

Santos F, Dean W (2004) Epigenetic reprogramming during early development in mammals. *Reproduction* **127**, 643–651.

Spehr M, Schwane K, Riffell JA, *et al.* (2006) Odorant receptors and olfactory-like signaling mechanisms in mammalian sperm. *Molecular and Cellular Endocrinology* **250**, 128–136.

Swales AKE, Spears N (2005) Genomic imprinting and reproduction. *Reproduction* **130**, 389–399.

Tardif S, Brady HA, Breazeale KR *et al.* (2010) ZonadhesiD3-polypeptides vary among species but are similar in *Equus* species capable of interbreeding. *Biology of Reproduction* **82**, 413–421.

Tateno H, Krapf D, Hino T, Sanchez-Cardenas C, *et al.* (2013) Ca2+ ionophore A23187 can make mouse spermatozoa capable of fertilizing in vitro without activation of cAMP-dependent phosphorylation pathways. *Proceedings of the National Academy of Sciences of the United States of America* **110**, 18543–18548.

Van Blerkom J, Zimmermann S (2016) Ganglioside-enriched microdomains define an oolemma that is functionally polarized with respect to fertilizability in the mouse. *Reproductive BioMedicine Online* **33**, 458–475.

Vjugina U, Zhu X, Oh E *et al.* (2009) Reduction of mouse egg surface integrin alpha9 Subunit (ITGA9) reduces the egg's ability to support sperm-egg binding and fusion. *Biology of Reproduction* **80**, 833–841.

Vogt E, Kirsch-Volders M, Parry J, Eichenlaub-Ritter U (2008) Spindle formation, chromosome segregation and the spindle checkpoint in mammalian oocytes and susceptibility to meiotic error. *Mutation Research* **651**, 14–29.

CHAPTER 13

Initiating pregnancy

Essential Reproduction, Eighth Edition. Martin H. Johnson.

Companion website: www.essentialreproduction.com

The tiny two-cell conceptus sitting in the oviduct now has to perform a heroic task. It must somehow communicate its presence to the mother and convert the whole of her physiology and anatomy from a cyclic reproductive state to a pregnant one. How it does so is the subject of this chapter.

The preimplantation conceptus

The conceptus remains at the site of fertilization for a further few days (Table 13.1). It is then transferred through the isthmus of the oviduct and enters the uterus. Transfer is facilitated by the changing endocrine milieu of the early luteal phase with its rising ratio of progesterone to oestrogen, which affects the oviducal and uterine musculature and relaxes the isthmic sphincter (see page 172). There is also evidence that the conceptus itself actively influences its rate of transport through local effects on epithelial secretion and ciliary activity.

During its period in the oviduct, the two-cell conceptus proceeds through cellular divisions at a rate characteristic for each species (Table 13.1). Each cell or **blastomere** undergoes a series of divisions during which the total size of the conceptus remains much the same (Figure 13.1). In consequence, with each of these so-called **cleavage divisions**, the size of the individual blastomeres is reduced progressively, converting the very high cytoplasmic:nuclear ratio of the oocyte and zygote to adult levels. As it does so, cleavage parcels out the maternal ooplasmic inheritance progressively in smaller and smaller packages to each cell.

On reaching the uterus, the conceptus engages in an elaborate conversation with the mother in which several messages are transmitted in both directions. This interaction has two important and distinctive components: one short range and one long range. First, the conceptus establishes physical and nutritional contact with the maternal endometrium at **attachment** and **implantation**; failure to do so properly would deprive the conceptus of essential nutritional substrates and delay or arrest its growth. The conversation operating during implantation involves short-range messages. Second, the conceptus **signals its presence to the maternal pituitary–ovarian axis**; failure to do so would result in the normal mechanisms of luteal regression coming into operation, causing a fall in progesterone levels and loss of the conceptus. Somehow the conceptus must convert the whole of the female from a cyclic pattern with oscillating dominance of oestrogens and progestagens, to a non-cyclic pregnant pattern, in which progestagens dominate throughout. The conversation operating during this **maternal recognition of pregnancy** involves long-range signals. Thus, pregnancy is not initiated with fertilization but only when the conceptus has signalled its presence successfully to the mother.

Early development is 'embryogenic'

Up to now, we have referred to the sequence of fertilization stages initially as an oocyte, then a zygote, culminating with the two-cell conceptus. Strictly speaking, in the mammal an embryo does not exist at these early stages, despite the fact that many scientists use this term loosely to describe any stage after sperm–egg fusion. In fact, the first 14–16 days of human development are concerned mainly with the elaboration of two distinct types of tissue within the proliferating cells of the conceptus: the pluripotent **embryonic cells**, that will give rise later exclusively to a **fetus**, are set apart from, and enveloped within, various **extraembryonic cells** (meaning outside of the embryo), that will provide the support structures for viviparous pregnancy – mainly by contributing to the **placenta** (Figure 13.2). Indeed, within 3–4 days of the two-cell stage, over 95% of the cells present in the conceptus are extraembryonic! This approximately 14-day

Table 13.1 Times (in days) after ovulation at which various developmental and maternal events occur

Species	Cleavage to four cells	Major burst of transcription	Conceptus enters uterus	Formation of blastocyst	Time of attachment	Luteal regression time if mating infertile	Duration of pregnancy
Invasive							
Mouse	1.5–2	2-cell	3	3	4.5	10–12	19–20
Rat	2–3	2-cell	3	4.5	4.5–5.5	10–12	21–22
Rabbit	1–1.5	8–16-cell	3.5	3.5	7–8	12	28–31
Human	2	4–8-cell	3.5	4.5	7–9	12–14	270–290
Non-invasive							
Sheep	4	8–16-cell	2–3	6–7	15–16	16–18	144–152
Pig	1–3	4-cell	2	5–6	18	16–18	112–115
Cow	2–3	8–16-cell	3–4	7–8	30–45	18–20	277–290
Horse	1.5–2	?	5–6	6	30–40	20–21	330–345

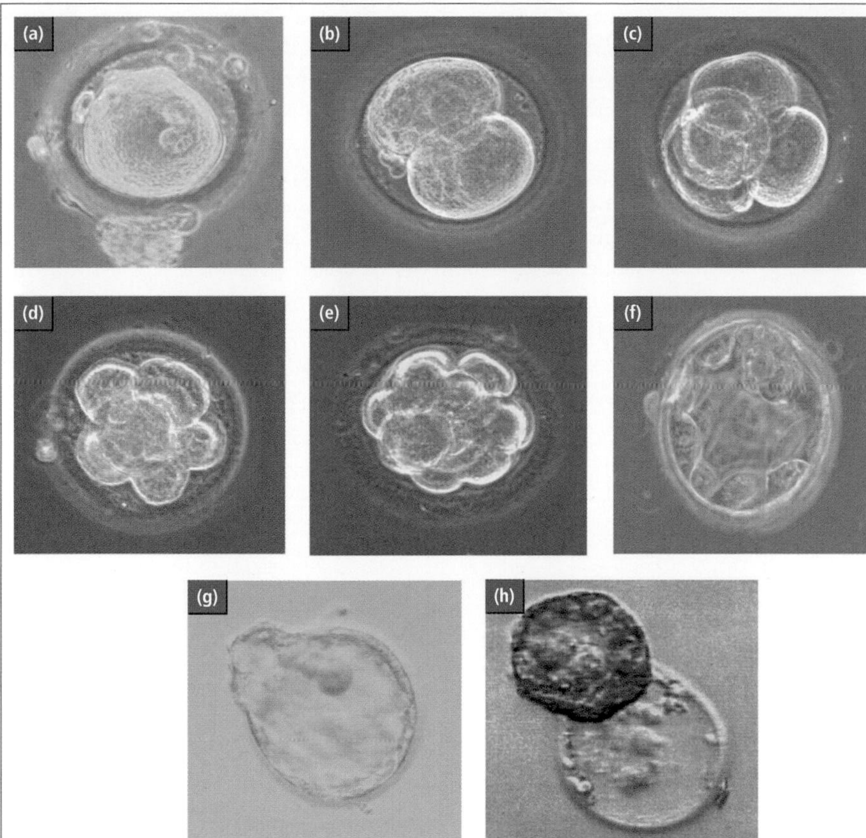

Figure 13.1 Photographs of various stages of human preimplantation development showing that during these cleavage divisions the cells get smaller but the embryo remains roughly the same total size. In each case the *zona pellucida* is visible. (a) Zygote or newly fertilized oocyte: note cumulus cells attached to outer surface of *zona*, a few non-fertilizing spermatozoa visible, two pronuclei internally and a clear second polar body to the left. (b) Two-cell stage conceptus: polar bodies clearly visible between blastomeres. (c) Four-cell stage. (d) Eight-cell stage. (e) Early morula stage, approximately 16 cells: the blastomeres are smaller and are flattened on each other due to the process of compaction. (f) Early blastocyst stage: note the blastocoelic cavity and the small cluster of cells at the top, which is the pluriblast or inner cell mass. (g) Blastocyst hatching through the *zona pellucida* at top left: note that the *zona* is much thinner. (h) Hatched blastocyst with the empty *zona* lying beneath it and partially covered by it. (Source: Photographs courtesy of Professor P.R. Braude.)

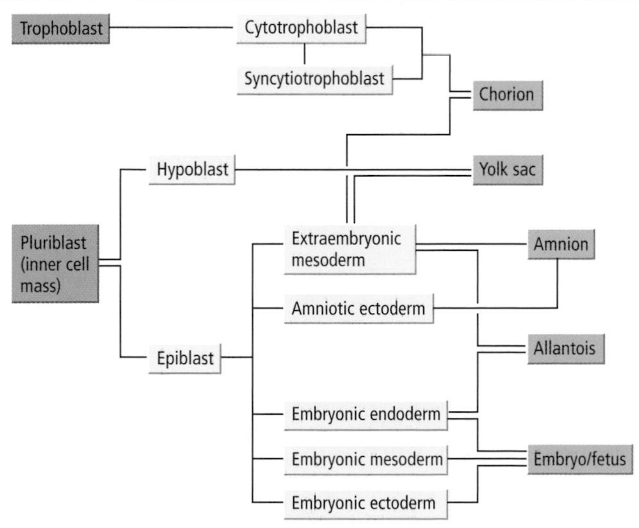

Figure 13.2 Derivatives of the blastocyst. The trophoblast lineage is characterized by expression of *Cdx2* and *Eomes* genes. The epiblast (also called primary embryonic ectoderm) is characterized by expression of *Nanog* and *Oct 3/4* genes. The hypoblast (also called primary endoderm) is characterized by expression of *GATA6* genes.

period is therefore called the **embryogenic phase** of development (generating an embryo). Before this stage the total product of fertilization is properly called the **conceptus** (also called **pre-embryo, pro-embryo or embryogen**).

Once an embryo is generated, the **embryonic phase** is under way. It lasts about 6 weeks and during this time the various embryonic cell and tissue types differentiate, and the basic body plan is laid down such that eventually a tiny fetus is formed, shorter than the length of a matchstick. The transition from embryo to fetus marks the end of the **first trimester** (=3 months or the first third of pregnancy). The **fetal phase** of development then proceeds through the second and third trimesters to term and production of a baby, whereupon the **neonatal phase** begins. A summary of the approximate lengths of each phase for selected species is given in Table 13.2, but these times are necessarily approximate as they are imposed artificially on what is a continuous developmental process.

The embryogenic conceptus

The embryonic and extraembryonic cells and tissues generated during the embryogenic phase derive from a complex series of cell interactions. At around the 8- to 16-cell stage (or **morula**

Table 13.2 Comparison of the durations (in days post-fertilization) of the main developmental phases in five mammals (modified from Johnson and Selwood, 1996)

Species	Embryogenic	Embryonic	Fetal
Mouse	6	5	9
Human	14	36	220
Cow	16	25	240
Opossum	7	4.5	1.5
Wallaby	14	8	5

Note: the process of development is continuous, not categorical, as this descriptive system implies, so the durations of each phase are necessarily approximate, as are their beginnings and ends!

stage) in most species studied, the cleaving conceptus changes its morphology by undergoing the process of **compaction** (Figures 13.1 and 13.3). This process involves maximizing intercellular contacts and also transforming the cell phenotype from radially symmetrical to highly polarized or epithelioid (Figure 13.3c). This polarized phenotype is important developmentally, as when each 8-cell blastomere divides to the 16-cell stage, one of its cellular offspring may inherit its basal domain while the other inherits its apical domain. Each of these offspring is located in different relative positions: those inheriting the basal domain tend to be internal and those inheriting the apical domain tend to be external. This differential inheritance and relative position then prompt these two cells to develop into different cell types in the **blastocyst**, which forms shortly thereafter at the 32- to 64-cell stage in most species (Figures 13.1 and 13.3d,e; Table 13.1). The precise form of the blastocyst

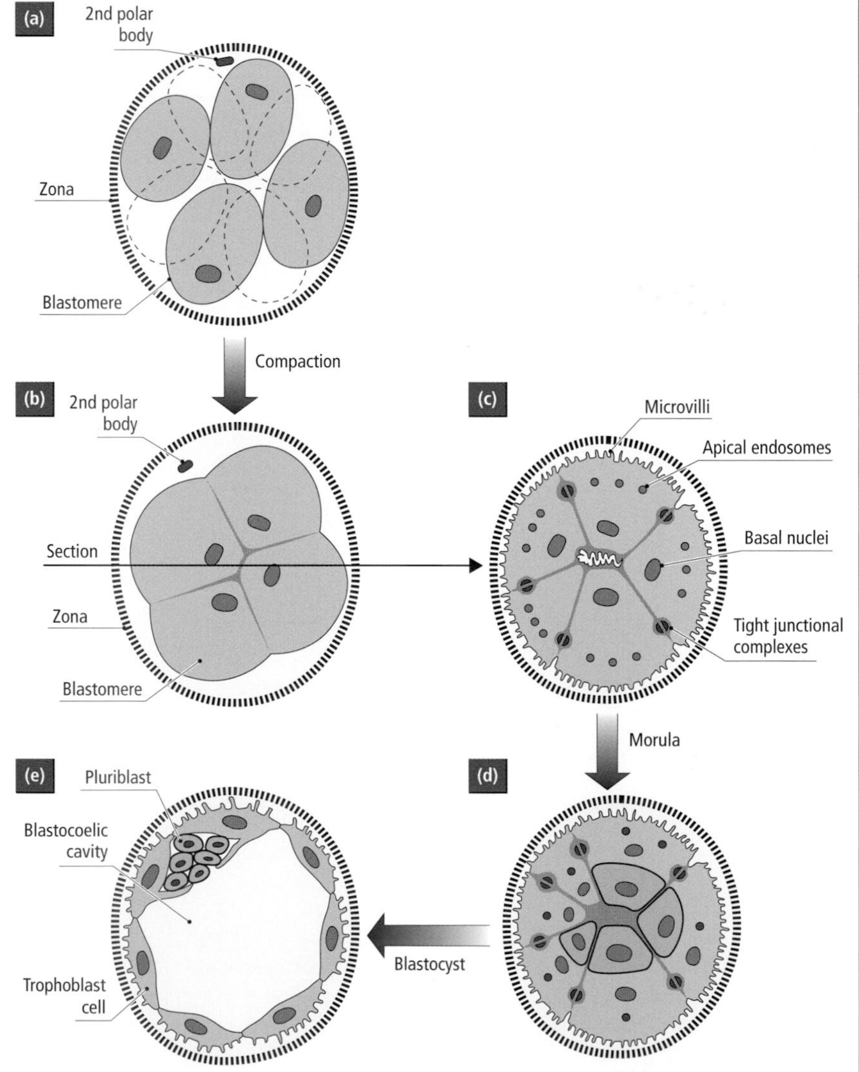

Figure 13.3 (a–c) Compaction of eight-cell conceptus. Spherical cells (a) become wedge shaped (b,c in which not all 8 cells are shown) and, by apposing adjacent surfaces, maximize cell contact. In cross-section (c), it can be seen that tight junctional complexes develop between the outer membranes of adjacent cells; these are punctate at first, but later become zonular, forming a barrier to intercellular diffusion between the inside and outside of the conceptus. Each cell also becomes polarized: the nucleus occupying a more basal position, endosomes and other organelles being apical and microvilli being restricted to the exposed surface and points of contact with other cells basally. (d) During cell division to the 16- and 32-cell stages (shown in section), two populations of cells form: the precursors of the outer trophoblast and inner pluriblast (blue) cells. The numbers of each cell type forming depend upon the orientation of the cleavage plane in each cell; in this example only the top left 8-cell has not divided to give an inner and outer cell, but to give two outer cells. (e) Section through a 64-cell blastocyst; fluid accumulation within the blastocoelic cavity becomes possible when the tight junctional complexes between adjacent trophoblast cells become zonular and prevent its escape. Note the eccentric position of the pluriblast or inner cell mass.

varies in different species, but in all or most eutherians, the blastocyst contains two distinctive types of cell: (1) an outer rim of **trophoblast** cells surrounding a **blastocoelic cavity** containing **blastocoelic fluid**; and (2) an inner group of pluripotent **pluriblast** or **inner cell mass** (**ICM**) cells, which is eccentrically placed within the blastocoelic cavity against, or embedded within, the wall of trophoblast.

The trophoblast cells constitute the first **extraembryonic tissue**, as they do not contribute to the embryo or fetus itself. Instead, they give rise to part of an accessory fetal membrane called the **chorion**, which is concerned with the nutrition and support of the embryo and fetus. Throughout development to the blastocyst, the conceptus remains enclosed within the *zona pellucida*. The *zona* has two functions. First, it prevents the blastomeres of the conceptus from falling apart during early cleavage, before compaction. If the conceptus becomes divided into two distinct groups of cells at this stage, **monozygotic twins** result (derived from one zygote and therefore genetically identical, compared with **dizygotic twins**, which are derived from the independent fertilization of two oocytes from the same ovulatory episode; Figure 13.4). Second, and conversely, the *zona* prevents two genetically distinct conceptuses from sticking together to make a single **chimaeric** conceptus composed of two sets of cells each of distinct genotype. It is during the transition from a morula to a blastocyst that the conceptus enters the uterus (Table 13.1).

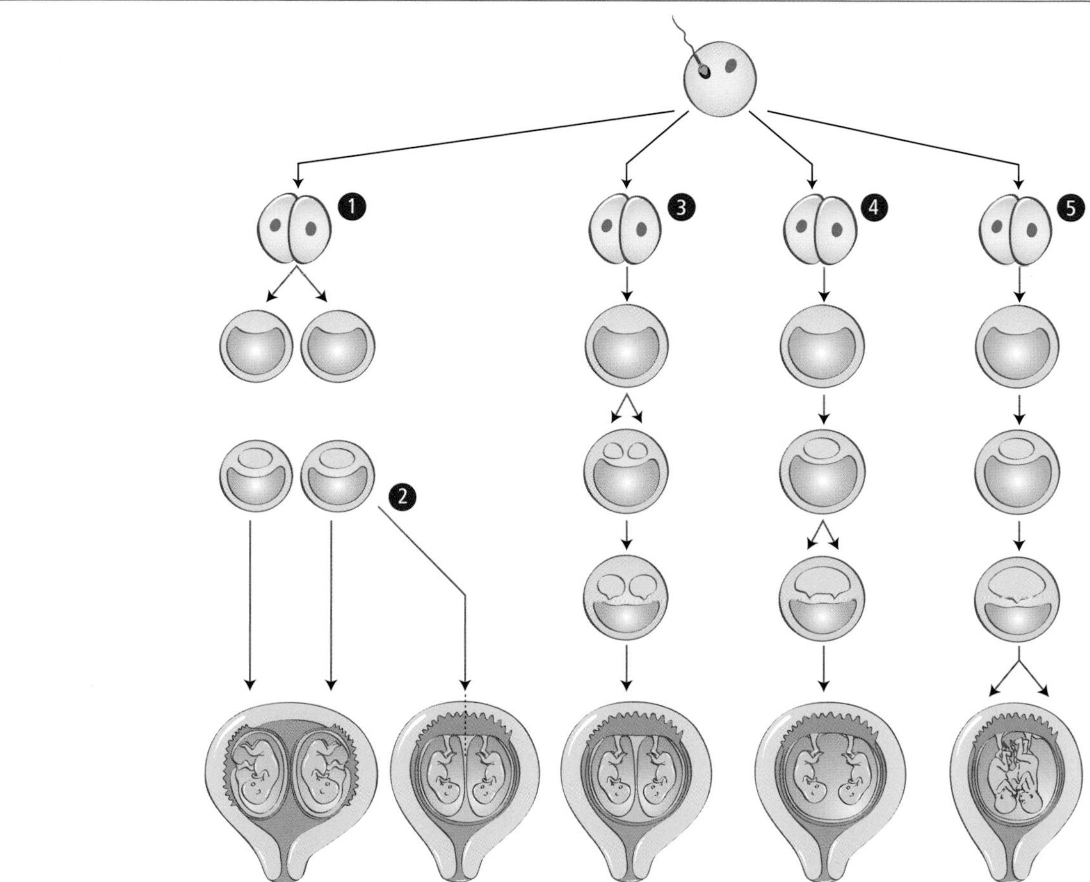

Figure 13.4 The mechanism underlying the formation of monozygotic twins depends on the developmental time at which they arise, which also then affects the disposition of placental membranes. All monozygotic twins arise from a single fertilized egg, in contrast to dizygotic twins in which two eggs are fertilized and carried through to term. Early embryo splitting results in approximately 40% of twins (Type 1), yielding two discrete embryos that can implant separately and form two monochorionic and monoamniotic fetoplacental units (20%), or which can secondarily fuse again (20%) to give the outcome shown for Type 2, in which each fetus has its own amniotic cavity but shares a fused dichorionic placenta. Dizygotic twins (about 40% of cases) can also undergo fusion as for Type 2. Where secondary fusion occurs, vascular anastomoses are extremely rare – hence the designation 'fused dichorionic'. When later splitting of the inner cell mass (ICM) occurs (60% of monozygotic twins), then a monochorionic diamniotic arrangement of membranes results (Type 3), in which vascular continuity between both halves of the placenta is often present. This can result in imbalance of blood flows and thus in **twin transfusion syndrome**, in which one twin is disadvantaged and may be smaller or even die. Even later splitting – at the embryonic disc stage (Type 4) – is very rare (<2%) and results in a monochorionic and monoamniotic set of membranes. Finally, such late splitting may be incomplete (Type 5) in which case **conjoined twins** are formed. For more details on placental membranes, see Figure 14.1.

The *in vitro* conceptus

Whilst some of the morphological observations described above can be observed in descriptive snapshots of conceptuses flushed freshly from the oviduct or uterus, most have been obtained after variable periods of culture *in vitro*. For dynamic biochemical studies, culture *in vitro* is obligatory. However, culture brings its own problems. Thus, *in vivo* the developing conceptus experiences a hormonally-dependent changing milieu that cannot be reproduced *in vitro*. In this environment, exposure to infectious agents and higher oxygen tension with the possibility of free radical damage is risked. Thus, the sterile use of differing media sequentially under reduced oxygen tension and/or in the presence of free oxygen scavengers is now increasingly in use in IVF clinics. However, culture *in vitro* is not successful for the conceptuses of most species, and even for the most studied murine, bovine and human conceptuses, the extended period of culture from one cell to blastocyst has only recently or problematically been achieved. Even then, it is not clear whether the development and survival of the conceptus *in vitro* are due to its adaptation to a suboptimal environment, and if so, whether this adaptive response might have adverse long-term consequences (see Box 13.1). However, the improving technology of culture *in vitro* has made possible the dynamic studies that are described next.

Developmental control transfers from mother to conceptus soon after fertilization

The large volume of oocyte cytoplasm, distributed to blastomeres during cleavage, contains most of the materials required for this cleavage process. These include: ribosomes and the full protein biosynthetic apparatus for making proteins, including a vast diversity of mRNA species; the maternal mitochondria and an ATP-generating system, which is based initially on the use of pyruvate and then on glucose as a metabolic substrate; a Golgi system for production and modification of glycoproteins; and a cytoskeletal system essential for cyto- and karyokinesis. The spermatozoon has little cytoplasm compared with the **maternal cytoplasmic inheritance of the oocyte**, and contributes, in most species, only its centriole in addition to its parentally imprinted genetic inheritance. It is also clear that the earliest stages of development are marked by very low levels of mRNA synthesis that are not essential for proximate development, because, if blocked by inhibitors, development can continue. Thus, these early stages are, in effect, controlled by the products of oogenesis and thus by the maternal genome. This almost total dependence of the early conceptus on its maternal cytoplasmic inheritance means that any deficiency in oocyte maturation will result in impaired, slowed or failed early development with an associated failure to establish pregnancy.

However, eventually this reservoir of maternally inherited cytoplasmic information ceases to be the sole controller of development. At a developmental stage that is characteristic for the species (Table 13.1), the numbers of conceptus genes being transcribed increases markedly and become essential for further development. At the same time, most of the maternally inherited mRNA is destroyed. However, despite the fact that most **newly synthesized proteins** switch from being **maternally encoded to embryonically encoded**, many proteins that were synthesized earlier on maternal mRNA templates persist until

Box 13.1 Culturing the early human conceptus *in vitro*

In vitro culture can only approximate the *in vivo* microenvironment experienced by the conceptus – an environment that changes temporally in response to the changing balance of maternal hormones, as well as in response to a series of interactions between the developing conceptus and the oviduct and uterus. Indeed, until very recently, media for the culture of conceptuses of most species, and even for those from most strains of mouse – the main model for the human – were inadequate. Thus, most, if not all, conceptuses 'blocked' during development at a stage characteristic for each species and usually correlating with the time of major gene activation (see Table 13.1). As culture media have improved, such that more conceptuses survive, a crucial question therefore is: are the survivors healthy, or just 'good enough', and, if the latter, are they in some way defective? The question then may be rephrased as: how do we know which among them are the 'best' of the batch? This question is of particular importance for human IVF culture, as it has raised several other issues.

- If some conceptuses are less 'capable' than others, should several be transferred to maximize the chance of pregnancy – given the problems that a multiple birth can bring (see page 358)?
- Is it dangerous to prolong culture in a suboptimal *in vitro* medium, or is it better to assume that the 'best' conceptuses will be the survivors and culture for longer – but risk there being no survivors to transfer?
- How can we develop better criteria for identifying the 'best' conceptuses?

Full answers to these questions are not yet to hand, but empirical evidence from clinical practice strongly supports transferring one or at most two embryos in most cases, and culture media that support development to the blastocyst now result in outcomes at least as good as culture and transfer at earlier stages. We return to these issues in Box 13.2 and in Chapter 23.

the blastocyst stage and beyond and continue to influence development. Thus, the **handover of genetic control from mother to conceptus is extended**.

Shortly after this major burst of gene activation, the conceptus also shows a marked quantitative increase in its biosynthetic capacity. Net synthesis of RNA and protein increases, transport of amino acids and nucleotides into the cells rises, and changes occur in the synthetic patterns of phospholipids and cholesterol. Maturational changes occur in mitochondria, Golgi and the endocytic and secretory systems of the conceptus.

From this time onwards, the development and metabolic activity of the preimplantation conceptus *in vitro* can be stimulated by a number of growth factors, which vary with species (see Box 13.2). The synthesis of many of these growth factors has been detected either in the conceptus itself or in maternal endometrial tissues, as a result of which they appear in uterine fluids. So, it is therefore reasonable to conclude that *in vivo* these factors act as **autocrine or paracrine** agents to promote early development. An at least adequate autocrine secretion may be indicated by the observation that oocytes fertilized *in vitro* can develop in defined media lacking the uterine growth factors.

It is during the transition from a morula to a blastocyst that the conceptus enters the uterus (Table 13.1), and it is therefore the blastocyst that will engage in the conversations with the mother that are vital to its survival and future development. So, next we will consider how the blastocyst and the uterine endometrium interact locally with one another to effect implantation.

Implantation

The free-living blastocyst is bathed in uterine secretions from which it draws the oxygen and metabolic substrates required for continued growth and survival. Through the trophoblast cells, it actively accumulates organic molecules and ions by specific transport mechanisms, while the exchange of oxygen and carbon dioxide is diffusional. There is a limit to the size that a free-living conceptus can attain before such exchanges become inadequate and a more intimate association with the mother's endometrial tissue is one feature of viviparity. This intimate association occurs through the process of implantation, the ultimate outcome of which is the placenta (see Chapter 14).

On entering the uterus, the conceptus is positioned to implant at a site (or sites) within the uterus that is characteristic for each species. In the human, where characteristically a single blastocyst implants, the posterior wall of the uterine fundus is typical. However, in polytocous species such as the mouse, where up to 20 blastocysts may implant, an appropriate location and spacing of implantation sites are important in minimizing physical and nutritional competition among conceptuses. Uterine muscular activity may be important in controlling the site(s) of implantation, which can be abnormal if inhibited.

Implantation involves an initial process of **attachment**. Attachment has two phases: close **apposition** is followed by **adherence** of the trophoblast cells of the blastocyst to the luminal epithelial cells of the endometrium. However, before

Box 13.2 Growth factors and early human development

Growth factors have not routinely been added to human conceptus cultures, despite abundant evidence from the culture of animal conceptuses of beneficial effects on cell number and developmental competence. A number of growth factors has been identified in the human female reproductive tract. These include some derived from the granulosa cells that are stimulated by hCG and thus are potentially responsive to the developing conceptus (e.g. activin A, IGF2, inhibin and VEGF). Others are expressed by oviduct cells during the luteal phase, including epidermal growth factor (EGF), granulocyte macrophage colony stimulating factor (GM-CSF), heparin-binding epidermal growth factor (HB-EGF), IGF1, LIF, TGFα and VEGF. The human luteal endometrium also expresses activin A, inhibin, colony stimulating factor (CSF), EGF, fibronectin, GM-CSF, HB-EGF, IGF1 and 2, LIF, platelet-derived growth factor (PDGF) and VEGF. Receptors must be present for growth factors to exert effects and those for some of the above are detectable on human preimplantation stages.

However, it is important to realize that growth factors may be present in *in vitro* cultures lacking their specific addition, either as a result of their production by the embryo itself or as contaminants in macromolecular additives, such as serum or albumen. Also, animal embryos can develop in the absence of added growth factors, and even results from rigorously controlled animal studies, which show benefits from the addition of exogenous growth factors, may not be applicable to the human situation. Hence, caution in applying the animal work to the clinical situation. However, a few clinical trials report interesting results. Thus, both GM-CSF and IGF1 reduced the number of apoptotic cells, and the addition of GM-CSF also increased cell numbers in cleaving embryos, and in inner cell masses and trophoblasts of human blastocysts. Blastocyst rates improved if GM-CSF, IGF1, HB-EGF and LIF were in the culture medium, and hatching from the *zona pellucida* also was higher with GM-CSF and HB-EGF. However, these *in vitro* outcome measures have not been supported by improved pregnancy outcomes, GM-CSF supplementation failing to increase clinical pregnancies. There is no current consensus of the value, or dangers, of adding growth factors to human IVF cultures, nor is there a clear understanding of how they might act. (Zhou W, Chu D, Sha W *et al*. (2016) *Journal of Assisted Reproduction and Genetics* **33**, 39–47. DOI:10.1007/s10815-015-0627-7.)

attachment can occur, the *zona pellucida* interposed between these cells must be removed. The proteolytic enzymes required can come from either the trophoblast cells themselves or the uterine secretions, the balance of each varying by species (Figure 13.5a–c). Attachment induces changes in the endometrial epithelium and the underlying endometrial stromal tissue, initiating its development as the maternal component of the placenta. How this is achieved varies with the species. In some,

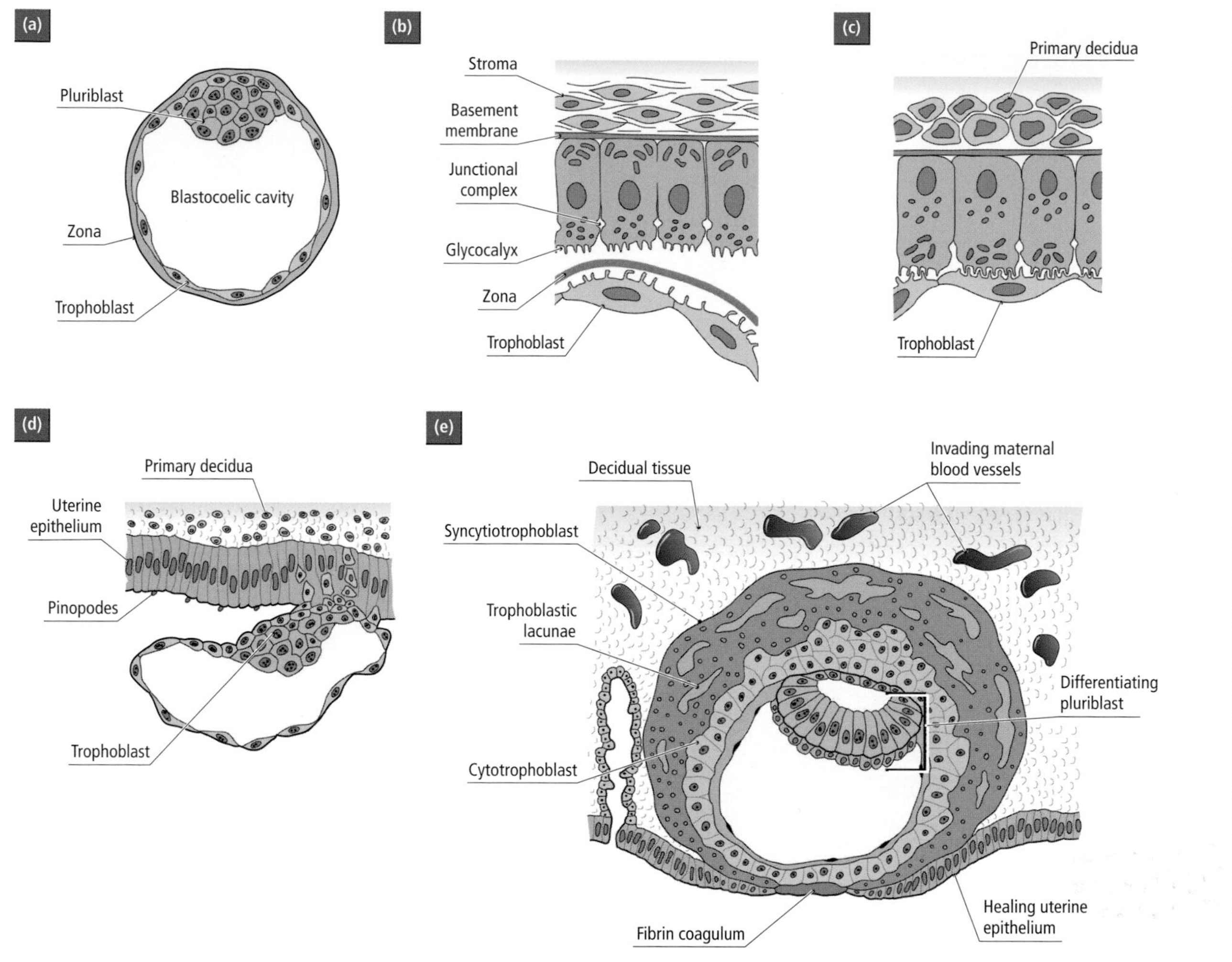

Figure 13.5 Schema of conceptus–uterine relations during invasive implantation in the human. (a) At 4–5 days: free-living *zona*-enclosed blastocyst. (b) Detail of blastocyst in (a) showing microvillous trophoblast and *zona* interposed between the blastocyst and the uterine luminal epithelial cells, which are linked to each other by zonular junctional complexes, with surface microvilli interdigitating with a thick electron-dense glycocalyx. Within epithelial cells are apical pinocytotic vesicles, central nuclei, basal lipid droplets and mitochondria. The epithelial cells rest on a basement membrane underlain by connective tissue containing spindle-shaped endometrial stromal cells in an extensive extracellular matrix abundant in collagen fibres. (c) At attachment, *zona* and glycocalyx have been lost, trophoblast and epithelial cells become very closely apposed and their microvilli become shorter, flatter and interdigitated. Attachment is seen to be inducing changes in the uterine epithelial and underlying stromal tissues. Epithelial nuclei assume a basal position and mitochondria are apical, with dispersed fat droplets in between. The endometrial stroma becomes oedematous due to increased vascular permeability, which is accompanied by loss of collagen fibres and swelling of stromal cells that subsequently develop extensive endoplasmic reticulum, polysomes, enlarged nucleoli, lysosomes, glycogen granules and lipid droplets, and become primary 'decidual cells'. Extensive intercellular gap junctions are also seen. Nuclei frequently become polyploid. Peripherally, sprouting and ingrowth of maternal blood vessels occur. *Note:* in the human endometrium, some decidual-like changes may occur in stromal cells in the absence of a conceptus during the late luteal phase. These changes are often called 'predecidualization'. (d) A slightly later post-attachment lower-power schematic view in which the trophoblast is starting to penetrate the epithelium. (e) Decidualization in underlying stromal tissue spreads out rapidly from the attachment site; the trophoblast rapidly erodes surface epithelium, invades and destroys adjacent primary decidual tissue, and becomes embedded. The vascularization response in the decidua is evident. (Source for Figures 13.5 b,c: Hamilton W.J., Boyd J.D., Mossman H.W. (1972) *Human Embryology*. Williams & Wilkins, Baltimore. Reproduced with permission from LWW.)

implantation is **invasive**, the conceptus breaking through the surface epithelium to invade the underlying stroma. In other species, implantation is **non-invasive**, epithelial integrity being retained (or at least only breached locally, transiently or much later in gestation; see later) and the epithelium becomes incorporated within the placenta.

Invasive implantation

Invasive implantation occurs in the human, primates (except lemurs and lorises), dogs, cats, mice and rabbits. The free-living phase of blastocysts of these species is relatively short-lived (Table 13.1). In consequence, invasive conceptuses tend to be smaller at attachment and only a few trophoblast cells are involved in making contact with the maternal epithelium. Yet, within a few hours, an increased vascular permeability in the area of stromal tissue underlying the conceptus is observed. This change is associated with an oedema, localized changes in the intercellular matrix composition and stromal cell morphology, and a progressive sprouting and ingrowth of capillaries (Figure 13.5c–e). This **stromal reaction** is particularly marked in primates and rodents, where it is called the **primary decidualization reaction**. After 2–3 days, the decidualization spreads to give a larger **secondary decidualization reaction**, as the major endometrial component of the placenta is prepared. In other invasively implanting species the decidualization reaction may be less marked but is functionally equivalent, and in humans, weak and delayed decidualization is seen even in the absence of a blastocyst. However, in general, it is fair to say that a **primary, highly localized signal** from the small conceptus has been effectively **transduced through the epithelium** to the stroma where it is then **amplified** rapidly.

Within a few hours of attachment, the surface epithelium underlying the conceptus becomes eroded (Figure 13.5d). Trophoblastic processes seem to 'flow' between adjacent epithelial cells, isolating and then dissolving and digesting them. Some trophoblast cells fuse together and form a syncytium (**syncytiotrophoblast**), while others retain their cellularity (**cytotrophoblast**) and serve as a proliferative source for generating more trophoblast cells (Figure 13.5e). The uterine glandular tissue and the decidual tissue immediately adjacent to the invading trophoblast of the conceptus are destroyed, releasing primary metabolic substrates (lipids, carbohydrates, nucleic acids and proteins), which are taken up by the growing conceptus.

The depth to which the conceptus invades maternal tissues varies with the species and also in certain pathological conditions. Those species in which the conceptus invades the stroma so deeply that the surface epithelium becomes restored over it are said to implant **interstitially** (e.g. the human, chimpanzee, mouse and guinea pig; Table 13.3 and Figure 13.5e). In other species, the stroma is only partially invaded and the conceptus continues to project to varying degrees into the uterine lumen: so-called **eccentric** implantation (e.g. the rhesus monkey, dog, cat and rat). Indeed, in these species, secondary contact and attachment by the conceptus to the uterine epithelium on the opposite side of the uterine lumen may result in two sites of placental development, a **bidiscoid** placenta (e.g. the rhesus monkey), or a complete 'belt' of placenta round the 'waist' of the conceptus, a **zonary** placenta (e.g. the dog and cat; see Table 13.3 and Figure 13.6a,b).

The **invasiveness of the conceptus is influenced by the extent of the decidual response**. Thus, where the decidual response is inadequate, for example, at the site of a caesarian scar in the uterus or after implantation at non-uterine **ectopic** sites, invasion is much more aggressive and penetrating. This observation has been taken by some to suggest a 'restraining' influence of the decidual tissue over the conceptus, but it could reflect an attempt by the invading trophoblast to overcome the inadequate nutritional support of a defective decidual response.

There are also species differences in the degree of proximity ultimately established between the circulations of the invasive conceptus and the mother. In some species, a relatively high

Table 13.3 Classification of implantation and placental forms in several species

Species	Depth of invasion	Extent and shape of attachment (Figure 13.6)	Maternal tissue in contact with conceptus (Figure 13.7)	No. of layers of chorionic trophoblast (Figure 13.7)	Histological type (Figure 13.7)
Invasive					
Human	Interstitial	Discoid	Blood	1	Haemomonochorial
Rabbit	Eccentric	Discoid	Blood	2	Haemodichorial
Rat/mouse	Eccentric	Discoid	Blood	3	Haemotrichorial
Rhesus monkey	Eccentric	Bidiscoid	Blood	1	Haemomonochorial
Dog/cat	Eccentric	Zonary	Capillary endothelium	1	Endotheliochorial
Non-invasive					
Pig/mare	Central	Diffuse	Epithelium	1	Epitheliochorial
Ewe/cow	Central	Cotyledonary	Syncytium of epithelium and binucleate cells	1	Synepitheliochorial

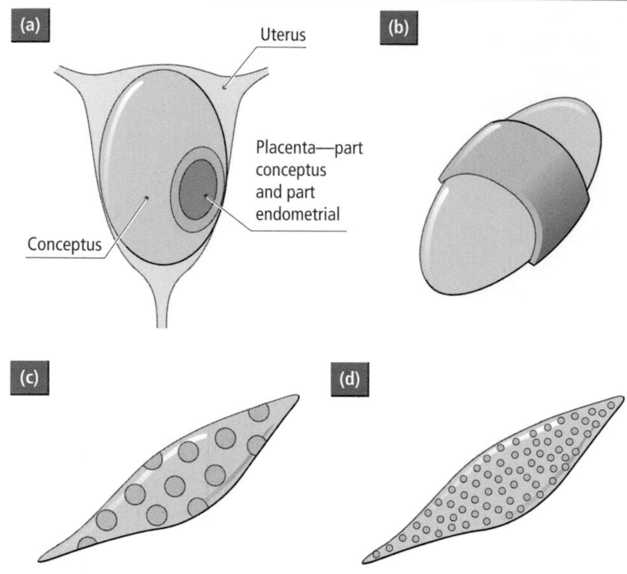

Figure 13.6 Highly schematic representation of placental morphologies. (a) A discoid human placenta. The egg-shaped sac is the conceptus; it has a specialized disc on its surface that forms the placental component. A corresponding area on the endometrium also contributes to the placenta. Discoid placentae are also found in the mouse, rat, rabbit, bat and insectivores. Other types of placentae (uterine part is not shown) are: (b) zonary (e.g. the cat, dog, bear, mink, seal and elephant); (c) cotyledonary (e.g. the cow, sheep, giraffe, deer and goat); and (d) diffuse (e.g. the horse, pig, camel, lemur, mole, whale, dolphin and kangaroo). (Source: Steven D.H. (ed.) (1975) *Comparative Placentation*. Academic Press, London. Reproduced with permission from Elsevier.)

degree of integrity of maternal tissue is retained, but in others, such as the human, the maternal blood vessels are eroded so that the trophoblast cells are eventually bathed in maternal blood (summarized in Table 13.3 and Figure 13.7a–d). There is little evidence that the number of layers of tissue that ultimately lie between the circulation of the mother and the conceptus can in any way be related to the **efficiency of placental transfer**. However, it may well influence **the types of transport mechanism** used (see Chapter 14 for more details). With the formation and invasion of the decidua, implantation is completed, a physical hold and an immediate nutritional source are established.

Non-invasive implantation

Non-invasive implantation occurs in the pig, sheep, cow and horse, in which species attachment is, in general, initiated relatively later than that of invasive conceptuses (Table 13.1). During the prolonged interval before attachment, the free-living conceptuses draw on rich and copious secretions from uterine glands and grow to a much greater preimplantation size than the human or rodent invasive conceptus. The growth may be prodigious and rapid. The pig conceptus, for example, elongates from a spherical blastocyst 2 mm in diameter on day 6 of pregnancy to a membranous, highly convoluted thread some 1000 mm long by day 12. This growth is confined almost exclusively to the extraembryonic tissues of the conceptus, and establishes a vast surface area over which exchange of metabolites with the uterine milk can occur.

One consequence of this large surface area of extra-embryonic chorionic trophoblast is a more extensive eventual attachment to the uterine epithelium. In the horse and pig, attachment occurs at multiple sites over most of the external surface of the conceptus, but in ruminants attachment is limited to **uterine caruncles**, distinct areas of projecting aglandular uterine mucosa, which in sheep and goats may be up to 90 in number. As a result, the ultimate form of the placentae in these non-invasive species differs from that of invasive conceptuses, and is said to be **diffuse** (e.g. in the mare and sow) or **cotyledonary** (e.g. in the cow and ewe; see Figure 13.6c,d). Because penetration and erosion of the epithelium do not occur at implantation in these species, the conceptus is said to implant **centrally**. This means that more layers are interposed between the two circulations developing in the placenta (Figure 13.7e,f). However, junctional complexes are observed between the trophoblast and the adjacent uterine epithelial cells (in the sow and ewe), and some invasion of conceptus cells into the endometrium may occur either transitorily (in the mare) or much later in development (in the ewe). In these species, the epithelium acts to control invasion, and no typical decidual response is induced in the underlying stroma. However, stromal vascularity increases and distinct changes in cellular morphology occur, providing evidence of the recognition of the presence of the conceptus. These changes constitute the beginnings of the formation of the placenta and mark the end of the protracted preimplantation period seen in these species.

The molecular conversations at implantation

The blastocyst provokes a response of considerable magnitude in the uterus. However, embryo transfer experiments show it can only respond to the embryos signals over a very narrow window of time during a normal oestrous or menstrual cycle. During this time, the endometrial epithelial surface is said to be receptive to the conceptus – the so-called **window of implantation**.

The implantation window

Before this window (days 1–7 of the luteal phase), the endometrium is said to be **prereceptive** and is characterized by long apical microvilli, a high surface charge and a thick glycocalyx, which includes transmembrane **mucin glycoproteins** integral to the endometrial epithelial cells (see Figure 13.5b): all features likely to impair attachment. Accordingly, the

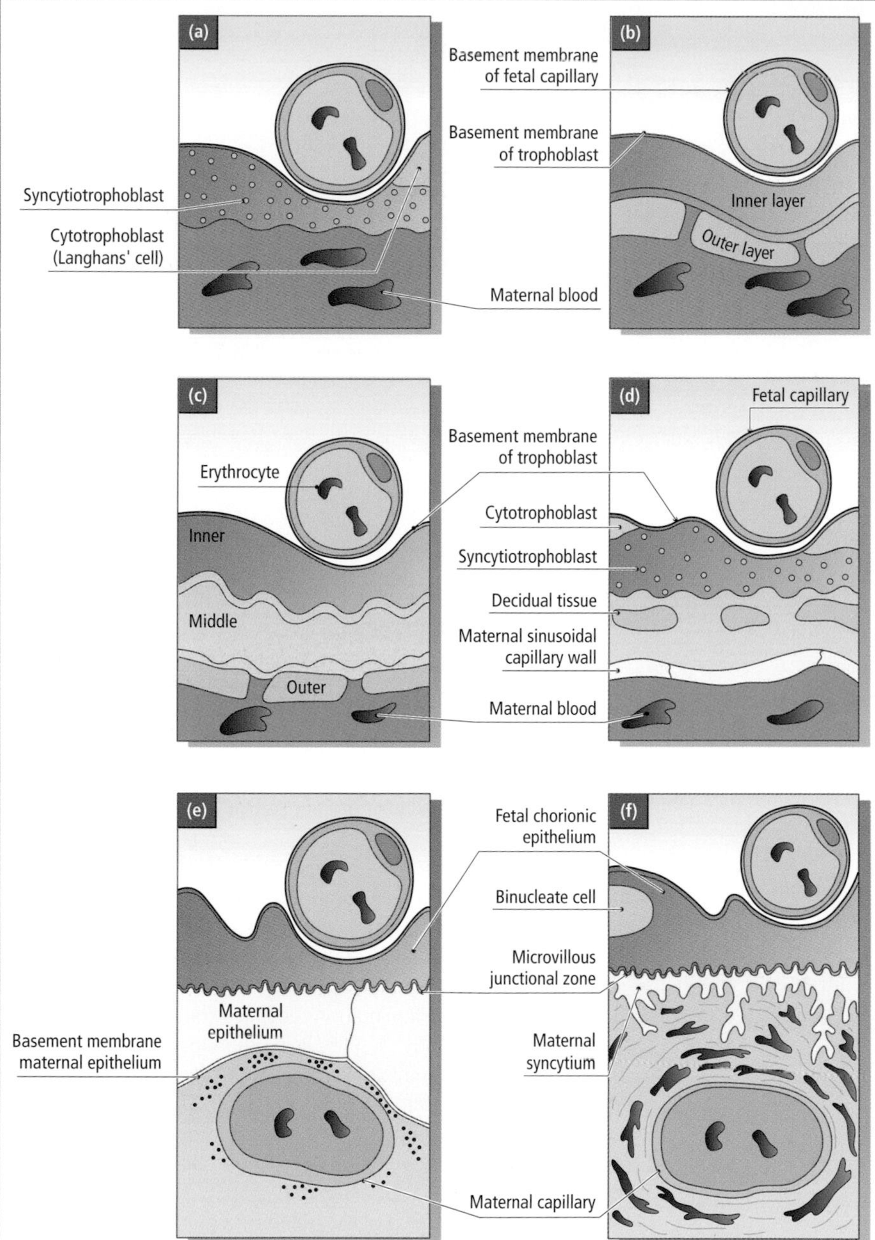

Figure 13.7 Schematic view of microstructure at the mature placental interface of various species. (a) Human: haemomonochorial (one layer of trophoblast); (b) rabbit: haemodichorial (two layers of trophoblast); (c) rat and mouse: haemotrichorial (three layers of trophoblast); (d) dog (also bears, cats and mink): endotheliochorial; (e) horse and pig (also whales, lemurs, dolphins, deer and giraffe): epitheliochorial; (f) sheep and cow: synepitheliochorial: the maternal uterine epithelial cells fuse with binucleate cells from the fetal side to form a syncytium; the binucleate cells may also invade maternal tissues later in pregnancy.

transition to receptivity is associated with two broad sorts of structural change. First, apical protrusions known as **pinopodes** appear over much of the epithelium, which may act to absorb uterine fluids, reduce the volume of the uterine cavity and so bring into close apposition the opposing epithelia (**occlusion**). Occlusion is assisted by a generalized oedema in the uterine stroma, which further compresses the lumen. Second, there is loss of surface negative charge, shortening of microvilli and **thinning of the uterine mucin coat** together with changes in its molecular composition, either globally or locally at the site of attachment (Figure 13.5c,d), changes which will facilitate close apposition of trophoblast and uterine epithelium.

If this receptive phase, which lasts approximately 3 days from day 7 to day 10, does not lead to attachment (no embryo), then it is followed by a **refractory** phase. Uteri in both the pre-receptive and refractory phases not only resist attachment by a blastocyst but in many species are hostile to any conceptus that leaves the oviduct prematurely or arrives too late (e.g. in rodents, sheep and rabbits, but not humans or monkeys). Thus, in the former species the uterus can be thought of as a primarily hostile environment able to carefully control a potentially dangerous invasive trophoblastic tissue. Clearly, for the conceptus to survive, its early development and transport must be coordinated precisely with the changing receptivity of the uterus. This coordination is achieved through the mediation of the steroid hormones.

The endocrine basis for the implantation window

Progestagenic domination is required if the uterus and implanting blastocyst are to engage effectively. We described in Chapter 10 the impact of the cyclic sequence of follicular oestrogen (E2) followed by luteal progesterone on the distinctive proliferative and secretory patterns of uterine secretions. Both sex steroids act via their respective receptors (ERα and β and PRα and β), and we saw that follicular E2 stimulates the appearance of PRs in the luminal, glandular and stromal cells of the endometrium. In this way, E2 plays a significant role in setting up the luteal phase of the cycle. However, the roles of both E2 and progesterone change as the implantation window approaches. First, the PRs in the uterine glandular and luminal epithelium are down-regulated. This occurs because whilst **progesterone stimulates glandular production, it inhibits glandular secretion** and so this receptor down-regulation lifts that inhibition. Second, in several species, including the mouse but not the human or primates, a secondary burst of **nidatory E2 is required to actively stimulate the release of the glandular secretion.** This secretion then stimulates activity within the blastocyst. The nidatory oestrogen also acts on the luminal epithelial cells to make them responsive or sensitive to a blastocyst signal, so that they can attach to the trophoblast and transmit evidence of the blastocyst's presence to the underlying stromal cells: the oestrogen opens the window of receptivity. Thus, in these species, **oestrogen must rise during the luteal phase** (Figure 13.8a). The oestrogen required for attachment and implantation comes from the ovary in many of these species, but can be synthesized **by the conceptus itself**. In the pig, for example, the luteal phase lacks a significant rise in oestrogen unless a conceptus is present, when a clear oestrogen peak is detectable.

This nidatory oestrogen requirement is most clearly demonstrated in female rodents, which can be deprived of ovarian oestrogen by ovariectomy during the first 2–3 days of pregnancy while progesterone levels are maintained by daily injections. Under such conditions the conceptus enters the uterus as normal, but instead of implanting, remains free in the uterine lumen. The blastocyst may spend many days in the uterus in this quiescent state, its metabolism 'ticking over'. If a single injection of oestrogen is then given, blastocyst metabolism increases, attachment occurs rapidly and decidualization is initiated. This facility of the rat and mouse to suspend the blastocyst *in utero* for several days is called **delayed implantation** (see Box 13.3). However in humans, although E2 increases during the luteal phase, there is no evidence that post-ovulatory oestrogen is required to open the implantation window (Figure 13.8b). During artificial cycles in which E2 is followed by progesterone alone, embryo transfer results in successful pregnancy. All species do, however, share the requirement for sequential exposure to E2 followed by progesterone in order to develop a receptive endometrium.

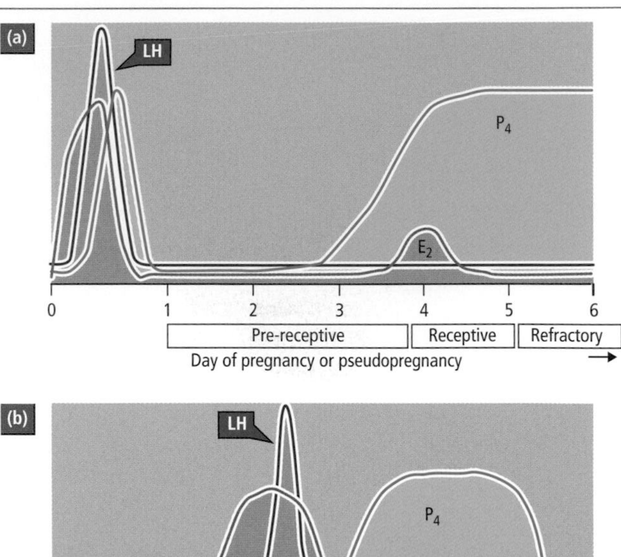

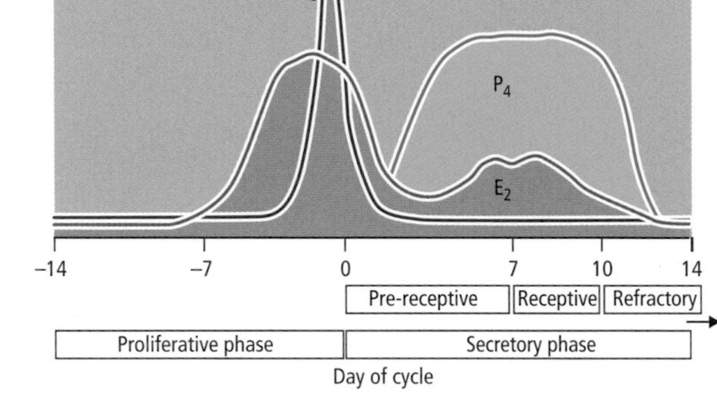

Figure 13.8 The window of uterine receptivity in mice and humans. The endometrium differentiates into a receptive state following sequential exposure to oestrogen (E2) followed by progesterone (P4). (a) In mice, an additional burst of 'nidatory' E2 early on day 4 of pregnancy is required to open the window of receptivity. Blastocyst transfer during this 'implantation window', which lasts for approximately 20 hours on day 4 of pregnancy or pseudopregnancy becoming non-receptive (refractory) by the afternoon of day 5, may then result in pregnancy. (b) In humans the uterus is classified histologically and functionally into proliferative (follicular) and secretory (luteal) phases during the average 28–30-day menstrual cycle. The uterus becomes receptive during the mid-secretory phase, spanning approximately days 7–10 after ovulation. Embryo transfer experiments show that in humans, the exact timing and duration of the receptive phase vary considerably between women compared to rodents where timings are more consistent, but that a burst of nidatory oestrogen is not essential for implantation. (Source: Wang H., Dey SK. (2006) *Nature Reviews Genetics* **7**, 185–199. Reproduced with permission of Nature Publishing Group.)

Identifying biomarkers of a receptive endometrium

Receptivity is defined as the ability to support embryo implantation but **how do we know when the endometrium is functionally receptive?** This is an important question when seeking to understand the causes of implantation failure in patients with infertility or when deciding the best time to transfer an embryo for those undergoing IVF. Microarray technologies and more recently RNA sequencing allow the simultaneous quantification of thousands of RNA transcripts

Box 13.3 Delayed implantation and diapause

Delayed implantation should not be thought of as purely an artificial or experimentally induced phenomenon. It occurs naturally in the rat, because of the suppression of endogenous oestrogen secretion in females that are suckling young of the previous litter (see Chapter 19 for details). This type of natural delay is often called **facultative delayed implantation**, as it only occurs under conditions of suckling. It is also shown by mice, gerbils, the bank vole and some marsupials, and is clearly useful to the mother since it delays the growth of her uterine litter for as long as she is suckling the previous litter. If this delay did not occur, the helpless newborn of the second litter would have to compete for milk with older and bigger siblings of the first litter.

In addition, there are many species in which an **obligatory delayed implantation** or **diapause** is an essential and normal part of their pregnancy (e.g. in the roe deer, badger, elephant seal, fruit bat, mink, stoat and brown bear). In these species, a prolonged period of weeks or months may be spent with blastocysts in diapause, an evolved trait that confers distinct biological advantages. The roe deer, for example, mates in July and August when the adults are well fed and in their prime, thereby ensuring effective competition for mates. The conceptus remains as a blastocyst until January when it reactivates for delivery of young in May and June when the nutritional conditions will have become optimal for the new offspring.

There is no evidence for delayed implantation or diapause in rabbits, pigs, guinea pigs, hamsters, primates or humans, which may explain in part the emancipation in humans/primates from the required nidatory oestrogen surge seen in those species showing a facultative or obligatory delay of implantation.

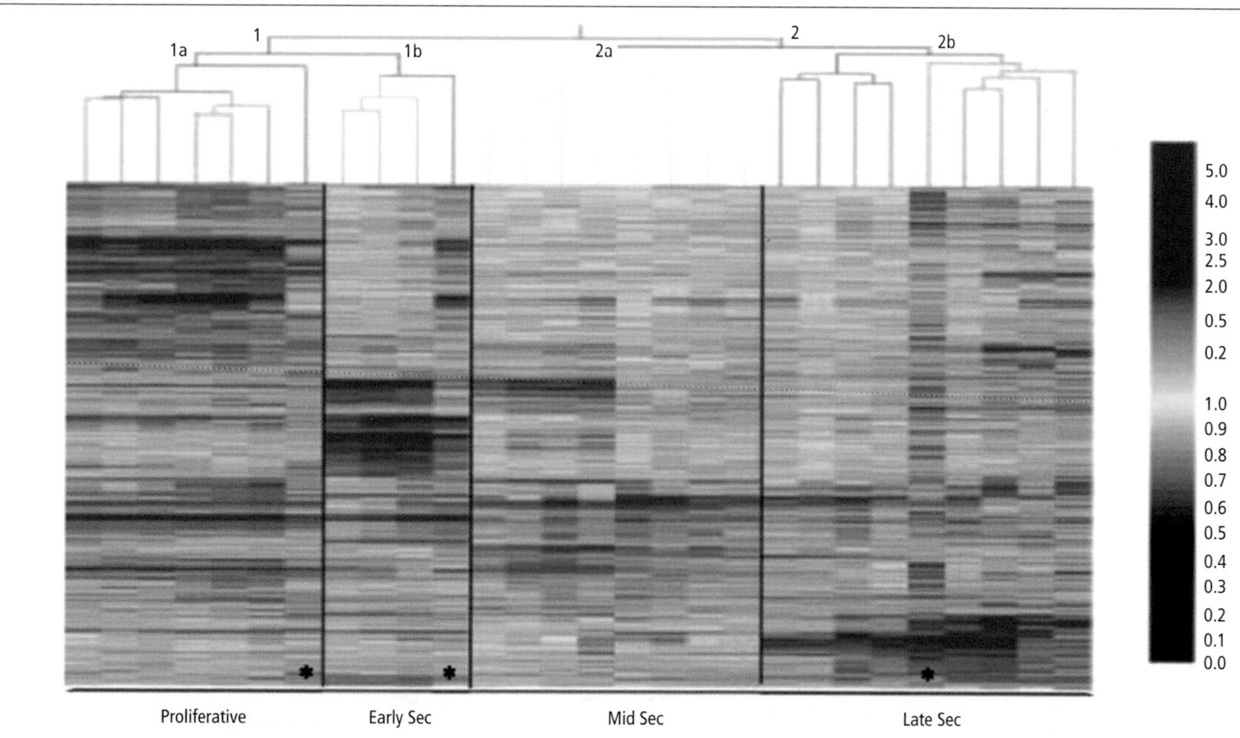

Figure 13.9 RNA expression profiling of human endometrial tissue across the menstrual cycle. Microarrays were used to measure expression levels of RNA transcripts in endometrium sampled across the menstrual cycle. The relative expression of each gene is colour-coded in this heat map as high (red) or low (blue) (colour legend on the right). Each horizontal line represents a single gene and each column represents a single patient sample. Using the technique of unsupervised hierarchical clustering, samples cluster according to their gene signature, corresponding to four main cycle phases as shown at bottom of the heat map: proliferative, early-secretory (Early Sec), mid-secretory (Mid Sec), late secretory (Late Sec). Some patient samples (marked *) show unusual gene expression patterns that depart to some degree from the samples they are clustered with, indicating considerable patient-to-patient variation in gene expression profiles even in samples taken at the same point in the menstrual cycle. How much this variation contributes to the low implantation rates in humans compared to other species is not clear. (Source: Talbi S. *et al*. (2006) *Endocrinology* **147**, 1097–1121. Reproduced with permission from Oxford University Press.)

in endometrial biopsies and thereby the response of the tissue as a whole integrated system to be determined. These techniques have been used in recent years to understand the complex cascade of changes brought about by steroid hormones that lead to a receptive endometrium. Coupled with sophisticated bioinformatics techniques, they show that the endometrium has a characteristic 'RNA expression profile' at each stage of the menstrual cycle as shown in Figure 13.9. This powerful approach allows identification of transcripts or proteins, the up- or down-regulation of which coincides with the transition from the pre-receptive to the receptive state. Many of these proteins have subsequently been shown to play important functional roles in mouse or primate models. Comparison of endometrial biopsies from women of known fertility and those with endometrial dysfunction using these gene-profiling techniques has greatly advanced our understanding of receptivity and infertility. However, it has proved difficult to develop a single biomarker that reliably defines the endometrium as receptive in humans. This is in part because of the wide variation in timing and expression levels of specific transcripts between endometria from different women, even when taken at the same stage in the cycle (see asterisked examples in Figure 13.9). An alternative approach in which the overall gene expression profile (rather than the expression of individual genes) is used to define the receptive state is now in a clinical trial to determine the optimal cycle day to replace embryos during IVF treatment for each specific patient to see whether this approach improves patient outcomes.

The molecular messages at attachment

Once a receptive endometrium is achieved, it can respond to signals from the conceptus to initiate implantation. What is the nature of these messages passing between conceptus and uterus? The answer to this question is not fully resolved. The main players are being identified and some aspects of their roles clarified. However, there are two general points to bear in mind. First, there are differences between species in some details, most notably whether a nidatory oestrogen surge is required (in those species in which obligatory or facultative delay can occur (e.g. rat and mouse) or not (human and primates). Second, within a species, there appears to be considerable redundancy between different molecular messages, that is, several types of messenger may be able to fulfil similar functions. In what follows, a general outline of current thinking is presented with these cautions in mind.

Two types of exchange occur during the transition of the endometrium from the pre-receptive to the receptive state. First, **'go away' messages** from the luminal epithelium **are switched off in order to remove barriers to attachment and adhesion**. Then **'come hither' messages** are sent to **promote active engagement of the conceptus**. A key 'go away' message seems to be provided by a complex glycoprotein called **Muc1**. Expression of Muc1 at the epithelial surface increases during the early pre-receptive progestagenic phase, but it then declines during the receptive phase, either globally (mouse) or locally in the vicinity of the blastocyst (rabbit, human). In the human, it is thought that the conceptus produces an enzyme that locally cleaves Muc1's carbohydrate side-chains, thereby removing its repulsive properties. A critical molecular 'come hither' message is **leukaemia inhibitory factor** (**LIF**). LIF is cytokine produced by the cells of the endometrial glands in response to nidatory oestrogen in those species, such as the mouse, which show this phenomenon (Figure 13.10), but are simply responsive to increasing levels of progesterone in humans. The uteri of mice genetically deficient in LIF cannot support implantation, although the blastocysts from these uteri are fully competent to attach if transferred to non-affected recipients. Receptors for LIF are up-regulated in both the epithelium and stroma of the endometrium around the time of implantation. LIF acts to

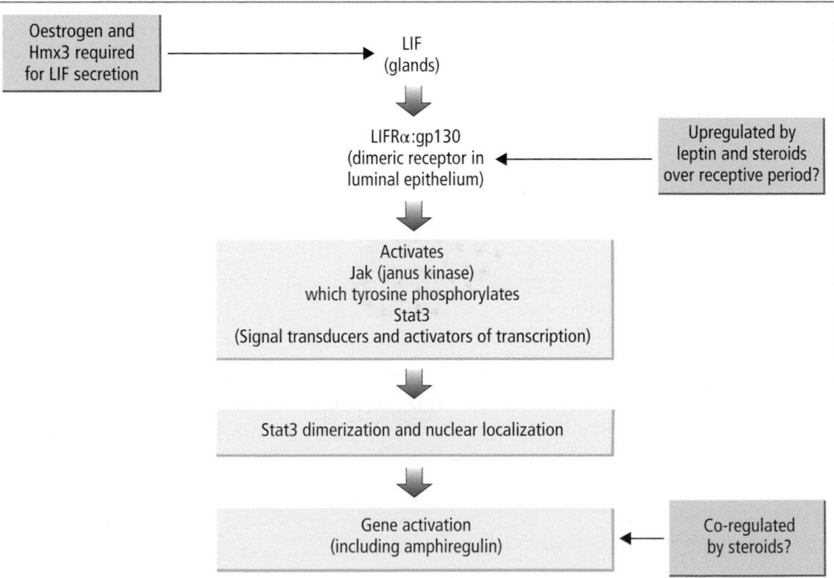

Figure 13.10 Schematic view of the mechanism of action of LIF in the mouse. LIF is produced in uterine glands under the influence of oestrogen (in the mouse, but not in the human). Gene knock-out studies suggest that expression of the homeobox gene family *Hmx* (especially *Hmx3*) is required, possibly in the underlying stroma. LIF acts on luminal epithelial (and stromal) cells via a dimeric receptor, which is up-regulated in the epithelium over the receptive period, probably by leptin and sex steroids. Receptor binding activates the Jak–Stat3 intracellular signalling pathway, which results in activation directly or indirectly of several genes involved in attachment and invasion.

permit the cells of the luminal epithelium to respond to local signals from activated blastocysts. The earliest **localized uterine responses to a blastocyst** in the mouse appear shortly after LIF acts on this epithelium, namely, the appearance in the endometrial luminal epithelium of several members of the epidermal growth factor (EGF) family. One of the first to appear is **heparin-binding EGF-like growth factor (HB-EGF)**. This LIF-dependent appearance occurs only in epithelial cells adjacent to the mouse blastocyst, and precedes dissolution of the *zona pellucida*, attachment and any increase in stromal vascularity, strongly suggesting that signals of an unknown nature from the blastocyst elicit its expression (Figure 13.11). However, its expression in several other species, including baboons and humans, may not be so localized. Both membrane-bound and soluble forms of HB-EGF are detected, raising the possibility that it might be able to act on the conceptus after passing through the *zona pellucida*. Significantly, trophoblast cells do express both **EGF receptors** and **heparan sulphate proteoglycans (HSPG)**, thereby providing double binding sites for HB-EGF. Moreover, when HB-EGF binds to EGF receptors on the blastocyst, it induces receptor phosphorylation and activation of the intracellular second messenger cascade (see Figure 2.9). This is followed by local dissolution of the *zona pellucida*, and attachment and invasion by trophoblast. Exposure of mouse blastocysts to inhibitors of HSPG renders them incompetent to attach.

These observations enable us to start to build a picture of the molecular language used in the conversations occurring between uterus and conceptus. In the presence of oestrogen (mice) or rising progesterone (humans), the endometrium produces LIF, which sensitizes the epithelium to the blastocyst (Figure 13.11). One response to this sensitization is HB-EGF production locally or globally, which binds to the conceptus, which in turn responds by shedding its *zona* and attaching. Other EGF family members in both the epithelium (and underlying stroma) may then assist the attachment process, as they too are up-regulated following *zona* loss and blastocyst attachment, and include **beta-cellulin, epiregulin and neuroregulin-1**. In addition, another family member, the LIF-responsive globally elevated **amphiregulin** (see above) becomes restricted to the region of attachment at this time. Receptors for these family members (the ERB receptor family, which includes the EGF receptor) are indeed expressed on the trophoblast. Although this experimental work has been undertaken largely on mice, humans also up-regulate LIF and HB-EGF expression over the receptive period, and HB-EGF binds to and stimulates human blastocysts. Overall, the EGF family seems to contribute the major molecular players in attachment, but several other adhesion molecules have been implicated in attachment and adhesion, including certain **integrins**, extracellular matrix molecules such as **laminin**,

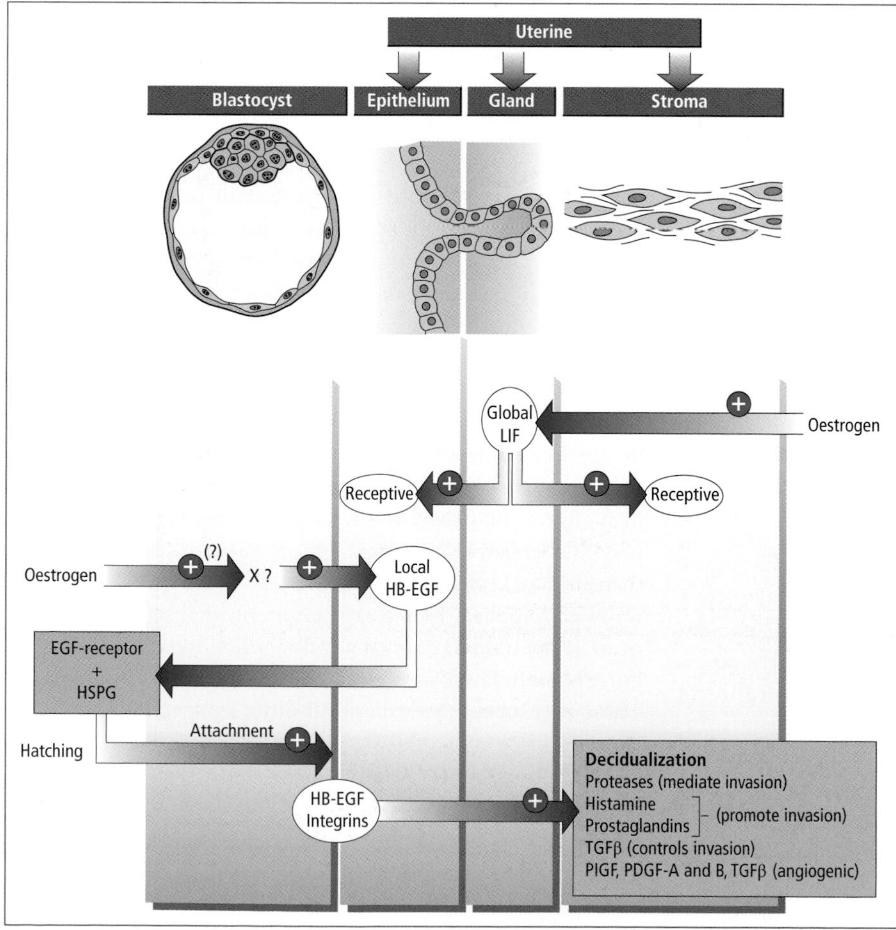

Figure 13.11 The language of implantation between blastocyst, uterine epithelium (luminal and glandular) and stroma. + indicates a stimulatory interaction; ? indicates uncertainty; X indicates a putative blastocyst signal to the endometrium. LIF, leukaemia inhibitory factor; HB-EGF, heparin-binding epithelial growth factor; HSPG, heparan sulphate proteoglycan; TGF, transforming growth factor; PIGF, placental growth factor; PDGF, platelet-derived growth factor. *Note* that oestrogen is only required in those species in which a nidatory surge of oestrogen is essential. The molecular nature of X is unclear but may consist of several factors. For example, CG has been shown to induce up-regulation of a number of genes identified as critical for attachment and invasion, and so may be one element in a complex of molecular signals from the conceptus.

fibronectin and **collagen family** members, **selectins** and the **trophinin–bystatin–tastin complex**. However, it is not yet clear whether or how they might act *in vivo*.

The molecular messages at invasion

During invasion, two types of message exchange are also in play. On the one hand, there is the language of containment and control, keeping the trophoblastic invasion in check. On the other, there is the language of incitement and encouragement, promoting the invasive events. For example, **metalloproteinases (MMPs)** digest stromal components, but the action of **tissue inhibitors of MMPs (TIMPs)** holds them in check.

The precise sequence of molecular events that follows the initial attachment of the blastocyst to the luminal epithelium is unclear, but somehow the epithelial cells signal to the underlying stromal tissue that invasion is imminent. The resulting decidualization response by stromal cells is often described as resembling a '**proinflammatory endometrial reaction**', because **prostaglandins (PGs)**, key molecular players in inflammatory responses, such as **hyperaemia, oedema and angiogenesis**, are also involved in implantation. Although decidualization requires the presence of progesterone, the hormone is not the primary trigger of the process. Thus, **cyclooxygenase-2 (Cox2)**, an inducible enzyme with a key role in PG synthesis by converting arachidonic acid to PGH_2 (see Figure 9.7), becomes elevated in both the luminal epithelium and the stroma over the receptive period. However, **this elevation occurs only at the site of the implanting blastocyst**, which seems to induce it locally (Figure 13.11). Moreover, inhibitors of Cox2 or of PGs themselves (especially of PGI_2) reduce decidualization, and implantation fails in mice genetically null for Cox2, a defect at least partially overcome by the local injection of PG stabilizers. In addition, null mutants for phospholipase A_2 (PLA_2), a major supplier of arachidonic acid for PG synthesis, show delayed implantation, and PLA_2 itself rises in the luminal epithelium over the receptive period. Thus, PGs are strong candidates as molecular cheerleaders for invasion. PGs can exert their actions through both conventional cell surface PG receptors and nuclear **peroxisome proliferator-activated receptors (PPARs)**. Endometrial localization and pharmacological studies suggest that it is **PPARδ** that is the most important mediator of the decidual response to locally produced PGs.

An important feature of the decidual response is the increased vascular permeability and the growth of new capillaries that occurs (**angiogenesis**). In addition to the role of PGs in these processes, several families of placental growth factors (and receptors to them) are also involved, namely, **vascular endothelial growth factors (VEGFs)**, the **angiopoietins** and prolactin-related proteins, such as **proliferin** and **proliferin-related protein (PRP)**. The cellular events of decidual remodelling and trophoblastic invasion are complex, and a number of other cytokines, extracellular matrix molecules and cell surface adhesion molecules have also been implicated in the process. Amongst these, a key player in both decidualization and also possibly invasion containment is **interleukin 11**. This cytokine is maximally expressed in the decidua and its receptor is expressed on both decidual cells and the invading trophoblast. Null mutations in the cytokine or its receptor in mice are associated with defective decidualization and result in pregnancy failure.

Human endometrial stromal cells can be decidualized *in vitro* and emerging data suggests that in some women with recurrent miscarriage, decidualization may be abnormal. However, the causes of this abnormal decidualization and its clinical importance are not yet clear. Thus, although in humans IL-11 has been shown *in vitro* to promote decidualization, the evidence that it is dysregulated in the endometrium of women with infertility is very weak. The contributions of all these various factors, where they act in the sequence of interactions between blastocyst, uterine epithelium, stroma and decidua, and precisely how the endocrine milieu might influence their roles, remain to be elucidated and are the subjects of continuing clinical research. It is clear, however, that the conversation occurring between the conceptus and the uterine tissues is complex and employs a rich molecular language.

Summary

Whilst the conceptus is securing an anchorage in the endometrium, the luteal phase of the cycle is progressing and lasts only 2 weeks – much shorter than pregnancy. The conceptus must somehow also prevent corpus luteum regression. How does the conceptus achieve this task? How does it signal its presence to the maternal ovary to maintain progesterone output?

The prolongation of luteal life

In a few species, such as the dog, the luteal phase is of the same duration as pregnancy (Table 13.1). In others, such as the rat, rabbit and mouse, coitus is equated with possible pregnancy, and so the coital stimulus neurogenically activates a pituitary secretion of luteotrophic prolactin that prolongs the normal brief luteal phase of the oestrous cycle into a pseudopregnancy, which is about half the length of pregnancy (Table 13.1; see Figure 10.2). In primates and the large farm animals, however, pregnancy greatly exceeds the normal life of the corpus luteum (Table 13.1), and even in the rat, rabbit and mouse, the prolonged luteal phase is not as long as pregnancy itself.

We saw in Chapter 9 (see pages 151–3) that luteal survival depends on the balance between a positive luteotrophic complex and negative luteolytic factors, which, in species other than primates, seem to be uterine prostaglandins. If luteal life is to be prolonged, then clearly either normal luteolytic factors must be neutralized or normally dwindling luteotrophic factors must be stimulated or supplemented. In practice, both mechanisms are probably at work – but differ by species.

Chorionic gonadotrophin prolongs luteal life in primates

The primate blastocyst prolongs the life of the corpus luteum by production of a luteotrophic factor of its own, which takes over from the inadequate support provided by the pituitary luteinizing hormone (LH) present during the luteal phase (see pages 151–2). If blood samples from pregnant women 8–12 days after fertilization are compared with those taken from women in the comparable period of a non-pregnant cycle, they are found to contain rising levels of the glycoprotein **human chorionic gonadotrophin** (**hCG**; see Table 2.4). hCG is synthesized in the trophoblast of the implanting blastocyst from as early as 6–7 days after fertilization and is released to pass into the maternal circulation (Figure 13.12b). It is carried to the ovary where it binds to LH receptors on the luteal cells to exert a luteotrophic action by sustaining **progesterone output that itself actively promotes luteal survival by autocrine stimulation of luteal cells** – yet another example of positive feedback.

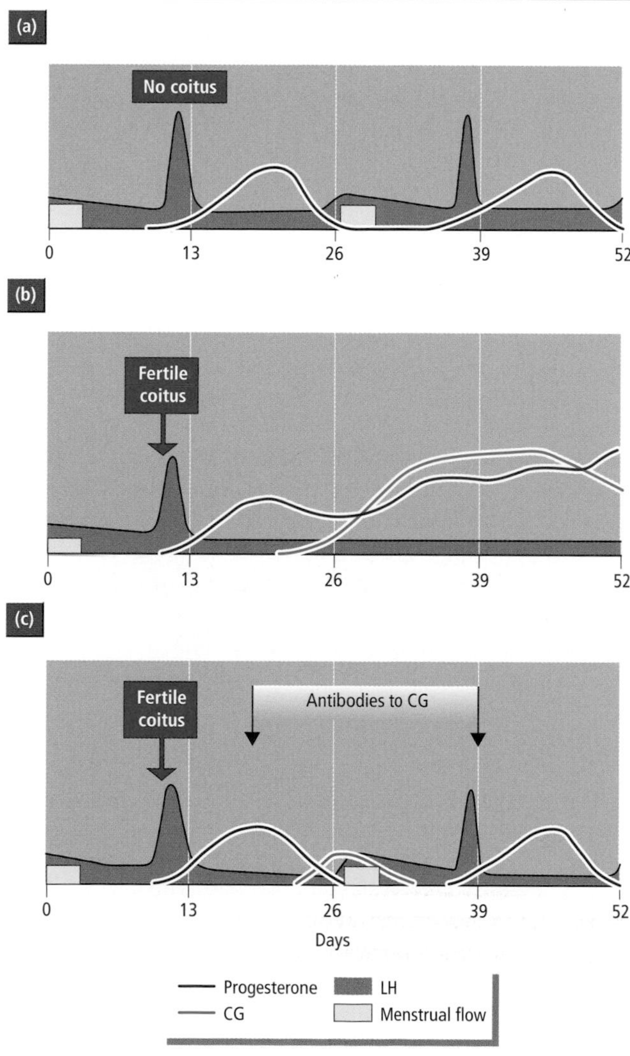

Figure 13.12 Levels of hormones in the blood during two menstrual cycles of a higher primate. (a) Non-pregnant cycles. (b) Cycles in which fertile mating occurs (arrow). (c) Similar to (b) but passive administration of a highly specific antichorionic gonadotrophin (CG) antibody is given (arrowheads); depression of CG and loss of pregnancy occur but there is no effect on LH levels or cyclicity.

In addition, production of a polypeptide hormone called **luteal relaxin** also rises in response to hCG and supports the transition to pregnancy. This combined luteotrophic stimulus overcomes any tendency to luteolysis.

The evidence for this critical effect of hCG comes from two types of experiment. If non-pregnant women are given daily injections of hCG, starting in the mid-luteal phase, regression of their corpora lutea is prevented or delayed and progestagen levels remain elevated. Conversely, antibodies specific to the terminal amino acid sequence unique to the hCG β-chain (or indeed to CG derived from baboons, rhesus monkeys and marmosets, which make bCG, rhCG and mCG, respectively) will bind to CG but not to LH. Injection of this antiserum throughout the luteal phase of a pregnant human or monkey cycle neutralizes CG production and luteal regression occurs on time (Figure 13.12c). Female monkeys actively immunized to their own CG have normal LH levels and menstrual cycles, and fertile matings, but no pregnancies. These lines of evidence clearly implicate CG as necessary and sufficient for the prevention of luteal regression during early pregnancy in the primate. So, detection of hCG in luteal phase blood or urine is taken as evidence of a **biochemical pregnancy**.

Suppressed luteolysis prolongs luteal life in large domestic animals

In the large farm animals, luteolysis is normally induced by secretion of uterine prostaglandins. We saw that in the absence of these prostaglandins, e.g. after a hysterectomy, luteal life is prolonged or lasts indefinitely. So, in these species, the luteotrophic support by the pituitary is presumably quite adequate. Not surprisingly, therefore, CGs have not been detected in the blood of pigs, cows and sheep bearing pre-implantation conceptuses. The pig conceptus produces oestrogens from about day 12 onwards, and oestrogen is luteotrophic in this species (see Table 9.3). However, in the large farm animals the primary role of the conceptuses is **to suppress luteolytic activity by neutralizing the action of prostaglandin $F_{2\alpha}$** ($PGF_{2\alpha}$).

In pregnant sheep and cattle, the levels of $PGF_{2\alpha}$ and its metabolites in the blood draining the uterus are reduced compared with those in non-pregnant animals. As we saw in Chapter 9 (see pages 152–3), luteal oxytocin provides the stimulus for $PGF_{2\alpha}$ release, but the oxytocin receptors only appear in the epithelial and glandular cells of the uterus towards the end of the luteal phase. In the presence of a conceptus, however, oxytocin receptor expression is blocked, thereby preventing oxytocin stimulation of PG release (but not its synthesis, which persists). The oxytocin receptor depression is caused by the paracrine action of **trophoblast interferon** (**IFN-τ**; also called **trophoblastin** and **trophoblast protein 1** or **TP-1**), a peptide belonging to the type-1 interferon family of cytokines. IFN-τ secretion from mononuclear trophoblast cells is restricted to the conceptus over precisely the period during which suppression of $PGF_{2\alpha}$ activity is required. Infusion of IFN-τ into non-pregnant uteri has anti-luteolytic effects resembling the presence of a conceptus. IFN-τ acts via its receptors **IFNAR1 and 2** to promote expression of the gene ***IFN regulatory factor 2*** (***IRF-2***), which is a powerful repressor of gene transcription. However, in parallel with its **repressor role** in the uterine epithelia, **IFN-τ also acts as a promoter** in the endometrial stroma to activate expression of several genes, which seem to be important for sustaining pregnancy.

The detailed situation in the pig differs from that in ruminants, although the general principles are similar. Thus, in the non-pregnant cycle, oxytocin is involved in the exocrine release of $PGF_{2\alpha}$. However, the conceptus switches $PGF_{2\alpha}$ release from an endocrine to an exocrine route, so most of it no longer passes in the blood to the corpus luteum, but is secreted into the uterine lumen, where it is metabolized. This diversionary secretion is mediated by two periods of oestrogen secretion from the pig conceptuses at around days 11–13 and 15–30 of pregnancy. In addition, the oestrogen promotes PGE production (which antagonizes the luteolytic action of $PGF_{2\alpha}$), and also promotes LH and prolactin receptors in the corpus luteum, thereby boosting the effective luteotrophic support. Its actions can be mimicked in non-pregnant pigs by oestrogen injections. Interestingly, the pig conceptus does not produce IFN-τ but does secrete IFN-γ and -δ. Neither is anti-luteolytic, but they do, like IFN-τ in ruminants, seem to have a key role in sustaining early pregnancy.

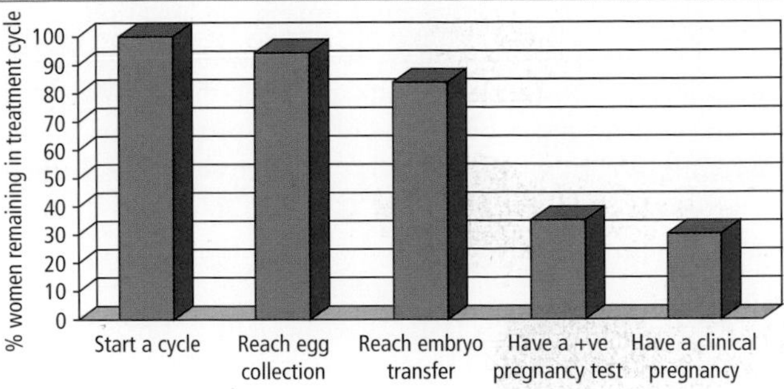

Figure 13.13 Plot of data of where in an IVF treatment cycle failure occurs. A treatment cycle is taken to begin when hormonal administration to the women commences. Note the big increase in treatment failure over the implantation period.

Conclusions

The early conceptus faces a hazardous few days of early life. Not only must it pass from the oviduct to the uterus at exactly the right time, but it must also establish adequate nutritional support for its growth and development. It is perhaps not surprising that in human *in vitro* fertilization (IVF) programmes, the major treatment failures occur over the implantation period (Figure 13.13). Thus, a sound understanding of the dynamics of implantation, the molecular language used and the roles of steroids in influencing these processes is of profound practical importance to infertile couples and those treating them. The days leading up to implantation are marked by very little conceptus growth, but the rising progesterone stimulates the secretory phase of the uterus to provide a nutritious glandular fluid support. In those species in which facultative or obligatory delay of implantation occurs, the superimposition of an oestrogenic stimulus allows attachment and implantation to occur, but in other species rising progesterone is sufficient to allow uterine receptivity. The conceptus can then secure a physical hold on the mother and so begin the process of building a placenta – of which more in Chapter 14. However, the initiation of a successful pregnancy requires more than this. The uterus will only provide a congenial environment as long as appropriate endocrine conditions persist. While the conceptus is establishing a secure physical and nutritional anchorage within the uterus, the luteal phase of the cycle is progressing and lasts only two weeks – much shorter than pregnancy. The conceptus also prevents corpus luteum regression by secreting luteotrophic CG in primates or by inhibiting the luteolytic prostaglandin in most animals. Once these dual tasks are achieved, pregnancy is underway. We next explore how it is maintained.

Key learning points

- Development may be divided into embryogenic, embryonic, fetal and neonatal phases.
- The embryogenic phase involves the separation of a small embryonic stem cell population from the primitive extraembryonic tissues.
- Early development involves a switch from almost exclusive use of maternally inherited molecular information to increasing use of mRNA and proteins made by the genes of the conceptus.
- Early cell divisions are cleavage divisions, which reduce the cytoplasmic:nuclear ratio, and involve no net growth.
- Compaction, blastomere polarization, blastocoele formation and the differentiation of pluriblast and trophoblast cells evident in the blastocyst characterize preimplantation development.
- The preimplantation period ends with attachment of the trophoblast to the uterine epithelium.
- Attachment marks the initiation of implantation.
- Total or focal digestion of the *zona pellucida* is required before this can occur.
- The length of the preimplantation period depends on the species, varying from 4 to longer than 30 days.
- The endometrium requires sequential exposure to oestrogen then progesterone to become receptive.
- Attachment in some species (e.g. mice but not humans) additionally requires the imposition of oestrogen on a progestagenic background (so-called 'nidatory oestrogen').

- This oestrogen can come from the ovary or conceptus, depending on the species.
- Some species always delay this oestrogen production to control the time of year at which implantation occurs through a process called obligate delay of implantation or diapause.
- In other species this oestrogen production can be delayed by the suckling of young from a previous litter through a process called facultative delay of implantation.
- In species with nidatory oestrogen, the oestrogen renders the endometrium receptive.
- In these species, oestrogen causes loss of surface negative charge, shortening of microvilli and thinning of the mucin coat, together with changes in its molecular composition such as loss of Muc-1 expression to facilitate close apposition of trophoblast and uterine epithelium.
- In these species, oestrogen stimulates secretion from the endometrial glandular epithelium – but only if the inhibitory effect of progesterone on secretion is relaxed by the disappearance of progesterone receptors from the epithelial cells.
- Leukaemia inhibitory factor (LIF) production and secretion are a functionally important component of the glandular secretion.
- LIF acts on uterine luminal epithelium and stroma via LIF receptors; without it implantation does not occur.
- In species with nidatory oestrogen, the oestrogen stimulates heparin-binding EGF-like growth factor (HB-EGF) production by the uterine luminal epithelium adjacent to the blastocyst.
- HB-EGF binds to receptors on the blastocyst and stimulates *zona* hatching and attachment.
- HB-EGF, together with amphiregulin, epiregulin and other EGF family members, may also be involved in binding blastocysts to the epithelial cell surface.
- In species not requiring nidatory oestrogen (e.g. human), a similar set of actions is elicited by rising progesterone,
- Attachment results in immediate changes to the underlying stromal tissues, which includes primary decidua formation in invasive implanters.
- Conceptuses that implant early also tend to be invasive and form compact placentae with few layers interposed between maternal and fetal circulations.
- Conceptuses that implant late tend to be non-invasive and form diffuse placentae with multiple layers between maternal and fetal circulations.
- Invasive implantation involves prostaglandins, metalloproteinases and their inhibitors, and angiogenic growth factors.
- The implanting conceptus must signal its presence to the mother and prevent the withdrawal of progestagenic support by luteolysis.
- The primate blastocyst prolongs the life of the corpus luteum by production of the luteotrophic factor, chorionic gonadotrophin (CG).
- CG is synthesized in the syncytiotrophoblast of the implanting primate blastocyst from as early as 6–7 days after fertilization and is carried to the ovary where it binds to LH receptors on the luteal cells to exert a luteotrophic action.
- Progesterone itself is luteotrophic, an example of positive feedback.
- In the large farm ruminants the conceptus suppresses luteolytic activity by neutralizing the oxytocin-dependent release of $PGF_{2\alpha}$ from uterine epithelial cells.
- It does so by producing trophoblast interferon, which suppresses the development of oxytocin receptors in the uterine epithelial cells.
- In pigs, the conceptus produces oestrogen, which has both antiluteolytic actions (diverting $PGF_{2\alpha}$ from endocrine to exocrine secretion) and luteotrophic ones (promoting LH and prolactin receptors).

Clinical vignette

Monozygotic twins

A 25-year-old woman, Ms N, attended for an ultrasound scan after experiencing some light vaginal bleeding at 12 weeks into her second pregnancy. The scan revealed two fetuses, both with visible heartbeats. There was a single placental mass, but the sonographer was unable to visualize any membrane between the two fetuses. Ms N was referred to a specialist fetal medicine service, where the consultant confirmed the diagnosis of monochorionic, monoamniotic twins. The consultant explained that such pregnancies are very unusual and carry high risks of poor fetal outcome. Careful monitoring of the pregnancy was initiated, with measurements of the fetal growth and well-being every 1–2 weeks. As the pregnancy progressed, the twins showed concordant growth and good placental blood-supply. A delivery by Caesarean section was planned at 32 weeks gestation, in order to avoid the risks of cord entanglement, which are increased at later gestational ages in monochorionic pregnancies. Unfortunately, at 30 weeks, Ms N attended the emergency assessment unit of her local hospital with decreased fetal movements. Monitoring of the heartbeats of both twins was undertaken. The monitoring of twin 1 was extremely concerning, with deep and frequent decelerations. The obstetricians on duty undertook an emergency Caesarean section to deliver the twins as rapidly as possible. Both babies were born alive, but in poor condition. The neonatal team commenced resuscitation, and eventually both twins were stabilized sufficiently for transfer to the intensive care unit. After a prolonged stay in hospital, both babies were discharged home. In childhood it became apparent that both twins had some residual neuro-cognitive impairment, and that twin 1 also had mild cerebral palsy.

Twin pregnancies can arise from two distinct fertilizations (dizygotic twins) or from the splitting of a single conceptus (monozygotic twins). Splitting that occurs earlier post-fertilization results in fewer extra-embryonic parts of the pregnancy being shared between the twins. If the conceptus divides within the first 3 days post-fertilization, then this will result in dichorionic (two placentas), diamniotic (two gestational sacs) twins. Between days 4 to 8 post-fertilization, the twins will share a placenta, but develop within their own separate amniotic sacs (monochorionic, diamniotic). After day 8, splitting of the conceptus becomes much more rare, and these twins will share not only a placenta but also a single amniotic sac (monochorionic, monoamniotic). Splitting that occurs later than day 13 can result in conjoined twins. Approximately 1:330 conceptions are monozygotic, but only 1–2% of these are monochorionic and monoamniotic. The survival rate for these twin pairs is around 50%. One of the major causes of death is cord entanglement, and for this reason it is usually recommended that these twins are delivered by elective preterm Caesarean section. However, the increasing risks of cord entanglement are balanced against the risks of prematurity and this can be a difficult balance to strike. In the case presented here, the fetal distress that led to emergency delivery was likely the result of cord entanglement. Elective delivery earlier than 30 weeks would have been possible and would have avoided the acute event, but could ultimately have led to increased disability because of the problems posed by being born preterm. Any form of monochorionic twin pair will share at least some vascular placental connections, and hence the circulatory disruption caused by death or severe compromise of one of a monochorionic pair will very often lead to compromise in the co-twin, as occurred here.

FURTHER READING

General reading

Dimitriadis E, Nie G, Hannan NJ *et al.* (2010) Local regulation of implantation at the human fetal-maternal interface. *International Journal of Developmental Biology* **54**, 313–322.

Herington JL, Guo Y, Reese J, Paria BC (2016) Gene profiling the window of implantation: Microarry analyses from human and rodent models. *Journal of Reproductive Health and Medicine* **(Suppl 2)**, S19–S25. doi: 10.1016/j.jrhm.2016.11.006.

Johnson MH (2009) From mouse egg to mouse embryo: polarities, axes, and tissues. *Annual Reviews in Cell and Developmental Biology* **25**, 483–512.

Johnson MH, Selwood L (1996) The nomenclature of early development in mammals. *Reproduction, Fertility and Development* **8**, 759–764.

Li L, Zheng P, Dean J (2010) Maternal control of early mouse development. *Development* **137**, 859–870.

Lim HJ, Dey SK (2009) HB-EGF: a unique mediator of embryo-uterine interactions during implantation. *Experimental Cell Research* **315**, 619–626.

Lopes FL, Desmarais JA, Murphy BD (2004) Embryonic diapause and its regulation. *Reproduction* **128**, 669–678.

Paiva P, Menkhorst E, Salamonsen L, Dimitriadis E (2009) Leukemia inhibitory factor and interleukin-11: critical regulators in the establishment of pregnancy. *Cytokine & Growth Factor Reviews* **20**, 319–328.

Richter KS (2008) The importance of growth factors for preimplantation embryo development and in-vitro culture. *Current Opinion in Obstetrics and Gynecology* **20**, 292–304.

Rosario GX, Stewart CL (2016) The multifaceted actions of leukaemia inhibitory factor in mediating uterine receptivity and embryo implantation. *American Journal of Reproductive Immunology* **75**, 246–255. DOI: 10.1111/aji.12474.

Selwood L, Johnson MH (2006) Trophoblast and hypoblast in the monotreme, marsupial and eutherian mammal: evolution and origins. *BioEssays* **28**, 128–145.

Sharkey AM, Smith SK (2003) The endometrium as a cause of implantation failure. *Best Practice & Research Clinical Obstetrics & Gynaecology* **17**, 289–307.

Smith A (2006) The battlefield of pluripotency. *Cell* **123**, 757–760.

Steven DH (1975) *Comparative Placentation*. Academic Press, London.

Talbi S, Hamilton AE, Vo KC *et al.* (2006) Molecular phenotyping of human endometrium distinguishes menstrual cycle phases and underlying biological processes in normo-ovulatory women. *Endocrinology* **147**, 1097–1121.

Wang H, Dey SK. (2006) Roadmap to embryo implantation: clues from mouse models. *Nature Reviews Genetics* **7**, 185–199.

Zhou W, Chu D, Sha W *et al.* (2016) Effects of granulocyte-macrophage colony-stimulating factor supplementation in culture medium on embryo quality and pregnancy outcome of women aged over 35 years. *Journal of Assisted Reproduction and Genetics* **33**, 39–47. DOI:10.1007/s10815-015-0627-7.

More advanced reading

Campbell EA, O'Hara L, Catalano RD *et al.* (2006) Temporal expression profiling of the uterine luminal epithelium of the pseudo-pregnant mouse suggests receptivity to the fertilised egg is associated with complex transcriptional changes. *Human Reproduction* **21**, 2495–2513.

Caniggia I, Mostachfi H, Winter J *et al.* (2000) Hypoxiainducible factor-1 mediates the biological effects of oxygen on human trophoblast differentiation through TGFβ3. *Journal of Clinical Investigation* **105**, 577–587.

Das SK, Wang XN, Paria BC *et al.* (1994) Heparin-binding EGF-like growth factor gene is induced in the mouse uterus temporally by the blastocyst solely at the site of its apposition: a possible ligand for interaction with blastocyst EGF-receptor in implantation. *Development* **120**, 1071–1083.

Evans J, Catalano RD, Brown P *et al.* (2009) Prokineticin 1 mediates fetal–maternal dialogue regulating endometrial leukemia inhibitory factor. *FASEB Journal* **23**, 2165–2175.

Gipson IK, Blalock T, Tisdale A *et al.* (2008) MUC16 is lost from the uterodome (pinopode) surface of the receptive human endometrium: in vitro evidence that muc16 is a barrier to trophoblast adherence. *Biology of Reproduction* **78**, 134–142.

Hamatani T, Carter MG, Sharov AA, Ko MS (2004) Dynamics of global gene expression changes during mouse preimplantation development. *Developmental Cell* **6**, 117–131.

Kolle S, Dubielzig S, Reese S *et al.* (2009) Ciliary transport, gamete interaction, and effects of the early embryo in the oviduct: ex vivo analyses using a new digital videomicroscopic system in the cow. *Biology of Reproduction* **81**, 267–274.

Lee KY, DeMayo FJ (2004) Animal models of implantation. *Reproduction* **128**, 679–695.

Malik NM, Carter ND, Murray JF *et al.* (2001) Leptin requirement for conception, implantation, and gestation. *Endocrinology* **142**, 5198–5202.

Sherwin JRA, Sharkey AM, Cameo P *et al.* (2007) Identification of novel genes regulated by chorionic gonadotropin in baboon endometrium during the window of implantation. *Endocrinology* **148**, 618–626.

Spencer TE, Johnson GA, Bazer FW, Burghardt RC (2004) Implantation mechanisms: insights from the sheep. *Reproduction* **128**, 657–668.

Strumpf D, Mao CA, Yamanaka Y *et al.* (2005) Cdx2 is required for correct cell fate specification and differentiation of trophectoderm in the mouse blastocyst. *Development* **132**, 2093–2102.

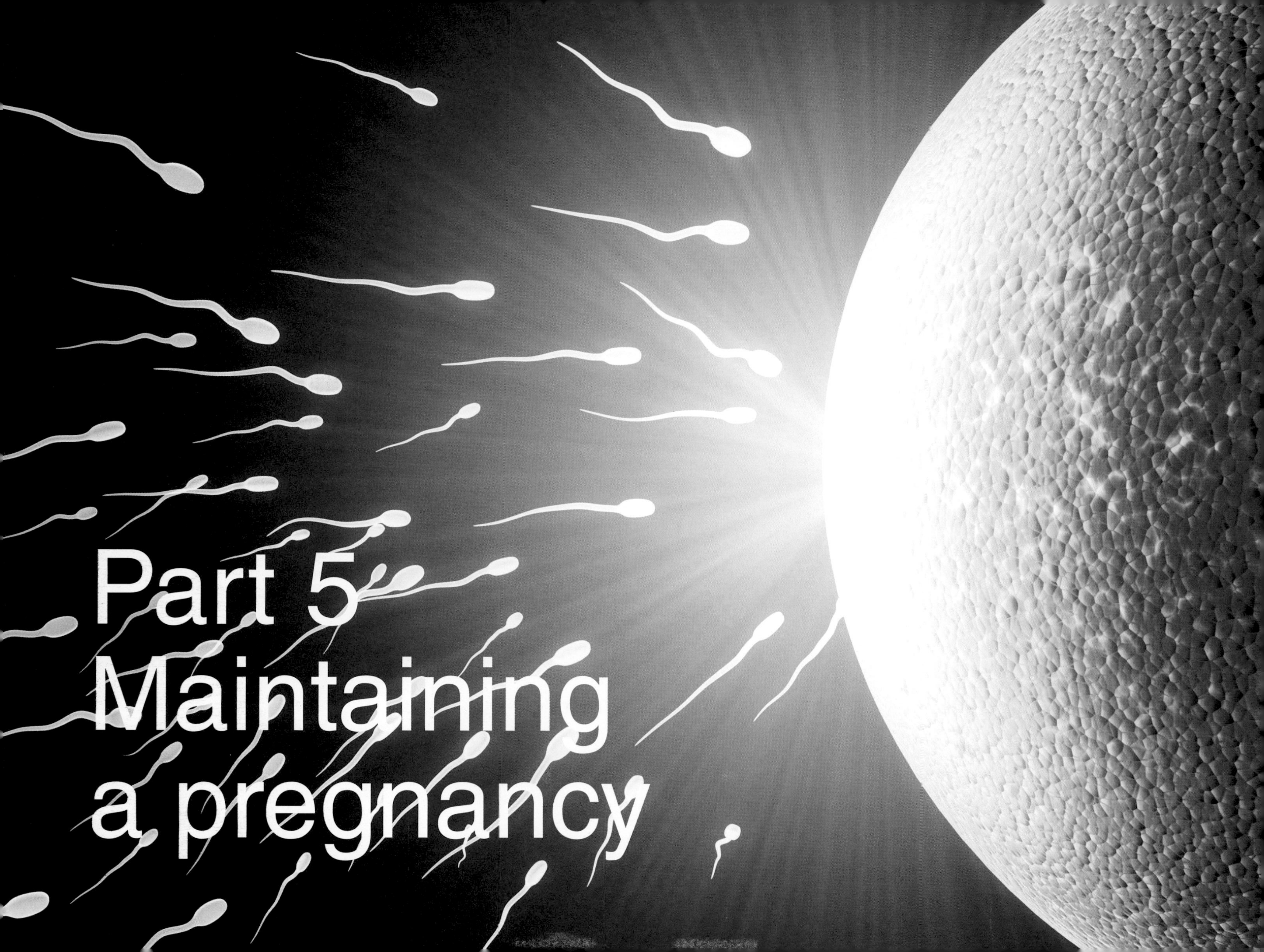
Part 5
Maintaining
a pregnancy

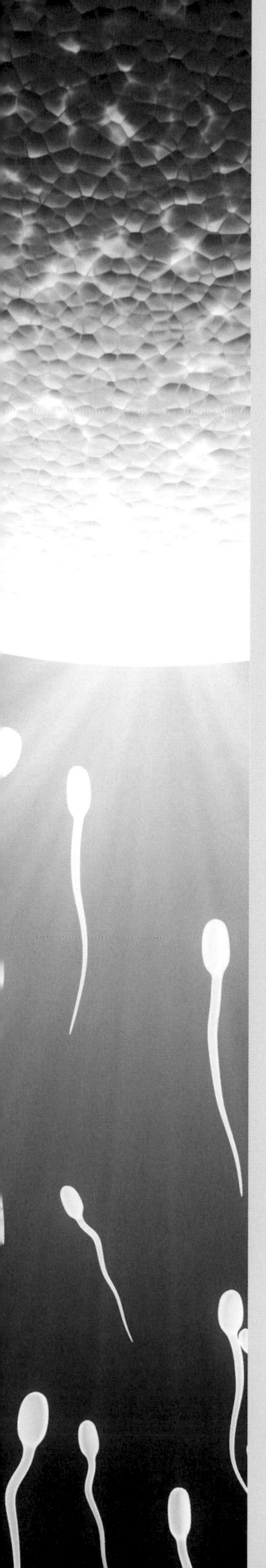

CHAPTER 14

Supporting the embryo and fetus

Essential Reproduction, Eighth Edition. Martin H. Johnson.

Companion website: www.essentialreproduction.com

The successful initiation of pregnancy creates a new and extraordinary parabiotic liaison between mother and conceptus, which may last for a period of months in some species. In humans, in whom pregnancy approximates 9 months, its course is often described as lasting three **trimesters** (each trimester being approximately 3 months). Conventionally, human pregnancy is timed from the last menstrual period. Given that this event marks the beginning of a menstrual cycle rather than ovulation, a pregnancy designated '12 weeks' is in fact only 10 weeks after ovulation and fertilization, and 8 weeks after corpus luteum salvage.

In Chapter 13, we saw how the conceptus made contact with the uterine endometrium at implantation (see pages 221–4) and signalled its presence to the mother so as to delay luteolysis (see page 230). If both actions are successful, pregnancy is initiated. However, the first few weeks of human pregnancy are precarious, and if first trimester loss is to be avoided, each of these initiating processes must be sustained. This chapter builds on the previous one to explore how the early fetus is supported. A consideration of **nutritional strategies** is followed by a description of the **endocrine strategies** that have evolved to support the early embryo.

Nutritional strategies

Implantation initiates the more intimate association with the mother's endometrial tissue, the ultimate outcome of which is the **placenta**, and is essential if viviparous embryonic growth is to occur. The **placenta comprises both maternal and conceptus components**, and so the process by which it is formed involves major structural and functional changes to both. Even as implantation is occurring, the growing conceptus is developing its own blood-producing and blood-distributing vascular system that can function to exchange essential metabolites at its extraembryonic surface, and it distributes them throughout its tissues (see page 241). Conceptuses also develop one or more anatomically distinct and highly vascularized regions of their extraembryonic surface through which the interchange of materials with maternal tissues is particularly facilitated. These vascularized regions draw on two major maternal sources of nutrition.

Histiotrophic support

During the initial **pre-, peri- and early post-implantation stages throughout the first trimester** of pregnancy, the conceptus utilizes almost exclusively maternal 'tissue juices' of various sorts. **This sort of nutrition is called histiotrophic**. Thus, the breakdown products released from the decidual tissue immediately adjacent to the invading trophoblast provide primary metabolic substrates (lipids, carbohydrates, nucleic acids and proteins), which are taken up by the growing conceptus. However, more sustained histiotrophic support comes from the uterine glands adjacent to the implantation site, the secretions from which bathe the conceptus. Their secretions continue copiously for at least the first 12 weeks of human pregnancy, and are rich in glycoproteins, growth factors and micronutrients that are taken up by the trophoblast (see Box 14.1). It is during these 10–12 weeks, and using this glandular secretion, that the embryo develops within the extraembryonic membranes, assuming in miniature the fetal form that is recognizably that of a vertebrate. The key developmental features of this relatively long period of histiotrophic support are the relatively **limited growth in size** of the embryo that contrasts with its **complex differentiation** into the multiple tissues that make up the basic body form. Only when the complexities of the embryonic phase are largely completed does the **switch from histiotrophic to haemotrophic** support

Box 14.1 Early pregnancy loss and histiotrophic support

The discovery that nutrition of the first trimester conceptus is largely histotrophic has led to the examination of whether deficient glandular activity, often called **luteal phase defect**, might result in early pregnancy failure. Ultrasonographic studies indicate that an **endometrial thickness of at least 8 mm** is required for successful implantation, and some biochemical markers of glandular activity in uterine flushings at days LH surge + 10 and LH surge + 12 are lower in many women who go on to miscarry.

Furthermore, the down-regulation of progesterone receptors in glandular epithelial cells that would normally herald the implantation window does not occur in women with luteal phase defect. Thus, the normal inhibition of expression of uterine milk proteins exerted by the progesterone receptor is not lifted, thereby compromising the glandular secretory activity. In addition, it has been suggested that prolactin secreted by decidual cells and placental lactogen by the syncytiotrophoblast during the first trimester may act in concert with human chorionic gonadotrophin to stimulate the secretory activity of glandular epithelial cells once it has been initiated. There is some evidence that mild deficiencies in the production of these hormones are associated with early pregnancy failure. However, many losses over this period come from aneuploidies in the conceptus itself.

occur as the second trimester begins, and this latter form of nutrition then sustains the **growth and maturation** of the fetus through the remainder of pregnancy.

Haemotrophic support is initiated with the full functional development in the maternal endometrial tissue of a corresponding specialized and vascularized region. This zone of adjacent and highly vascularized contact between mother and conceptus is called the **haemotrophic placenta**. In the placenta, the two discrete circulations lie sufficiently close that rapid and efficient transfer of materials between them can occur. The placenta, then, is the ultimate outcome of the primary interactions between mother and conceptus that occur at attachment and implantation.

The change from histiotrophic to haemotrophic support

In order to understand the functional changes involved in this transition, we need first to examine the structure of the developing conceptus and the principal routes of metabolic exchange that become available.

The extraembryonic membranes give rise to the fetal membranes

The development of the fetal membranes and the interspecies variety of their organization provide one of the most enduring confusions for students of embryology. A simplified scheme to explain the developmental route linking blastocyst, embryo and fetus with the extraembryonic and then feto-placental membranes is shown in Figure 14.1, and this figure and its legend should be studied carefully before you read further. During the early histiotrophic phase of growth, the blastocyst takes up materials from, and excretes waste products into, the surrounding endometrial fluids. These materials pass through the thin 'shell' of trophoblast and are distributed by simple diffusion through the cavities and tissues of the conceptus itself. However, as the **mesoderm of the conceptus** forms, blood vessels develop within it, and then link up to form an extensive vascular network. This **vasculogenesis** is stimulated by **vascular endothelial growth factors (VEGFs) and by placental growth factor (PlGF)** secreted by the trophoblast cells and by natural killer cells (see Chapter 15), both induced by the action of hypoxia-inducible factor HIF-1α (HIF1 α) in response to hypoxia (see Box 14.3), and leads to the rapid formation of a network of blood vessels. Progesterone and oestrogen also play a role in vasculogenesis by promoting differentiation of decidual cells into endothelial cells, and also being positively involved in the remodelling of the blood vessels by extravillous trophoblast. Conversely, androgens depress vasculogenesis, and, if present in excess, reduce placental size. **Blood formation occurs in the yolk sac mesoderm**, and a **primitive heart forms in the cardiac mesoderm** (Figure 14.1f) within the developing embryo itself. Blood can thus be pumped throughout the extensive mesodermal tissue of the whole conceptus in both its embryonic and its extraembryonic parts. This blood passes throughout the chorionic mesoderm where equilibration with maternal glandular secretory and decidual fluids occurs. By this stage, it has become possible to image an implanting conceptus for the first time using **ultrasound imaging** and thereby to diagnose **a clinical pregnancy with its evidence of a beating heart.**

With further development, the vascularity in the conceptus becomes particularly marked in the **yolk sac mesoderm and where the yolk sac and chorionic mesoderm fuse together** (arrowed in Figure 14.1e), and a corresponding vascularity develops within the endometrium adjacent to this site of fusion. Together, the two adjacent, highly vascular sites form **the yolk sac (or choriovitelline) placenta**. The yolk sac, and the yolk sac placenta to which it contributes, are a structure homologous to the yolk-containing sac found in the eggs of reptiles, birds and monotreme mammals such as the platypus. This comparatively primitive origin is reflected in the transitory existence and function of the yolk sac placenta in most mammals. However, in some mammals (e.g. marsupials), the yolk sac placenta functions alone throughout pregnancy and in others (e.g. the rabbit, rat and mouse), it persists and remains functional, serving a specific subset of transport functions during pregnancy. However, in most mammals a second exchange site develops called the **chorioallantoic placenta** (arrowed in Figure 14.1f).

The chorioallantoic placenta forms as a result of the outgrowth of an **endodermal diverticulum** (the **allantois**) from the hindgut region of the developing embryo (Figure 14.1f). Together with its ensheathing mesoderm, containing the pro-umbilical blood vessels, it fuses with the chorionic mesoderm and thereby determines the site (or sites) of chorioallantoic placentation. As we saw in Chapter 13, the consequence of allantoic outgrowth in species with early invading conceptuses is a restricted discoidal, bidiscoidal or zonary placenta, whereas late-attaching conceptuses acquire a more extensive cotyledonary or diffuse chorioallantoic placenta. Throughout these developments, the embryo and later the fetus itself are suspended in **amniotic fluid** produced by and contained within the **amnion**, itself an extraembryonic membrane system composed of extraembryonic ectoderm and mesoderm (Figures 14.1 and 14.2). As pregnancy progresses, the amnion becomes progressively pushed against the outer chorion until it obliterates the extraembryonic coelom (Figure 14.2e).

It is important to re-emphasize that the functional placenta incorporates a contribution not only from the conceptus, as shown developing in Figure 14.1, but also from endometrial tissue. This is shown schematically in Figure 14.2 for the developing human conceptus, which has a discoid, chorioallantoic placenta. Having considered the general disposition of fetal membranes, fetus and placenta, we will now take a closer look at the organization of the human chorioallantoic placental interface itself. Often used as a human surrogate in functional studies, the sheep placenta is described in Box 14.2.

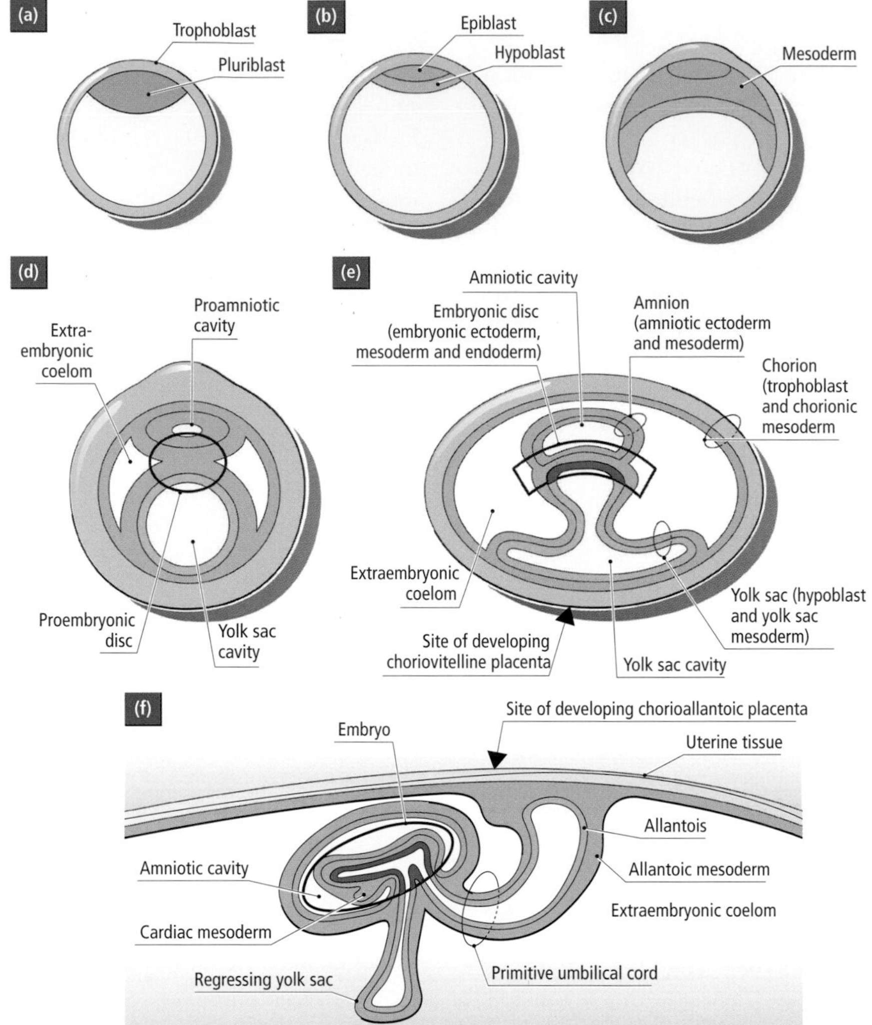

Figure 14.1 Schematic view of the development of the extraembryonic and then fetal membranes. (a) Blastocyst with trophoblast and pluriblast. (b) Development has occurred within the pluriblast: two layers of cells have formed, namely epiblast (salmon pink, also called primary ectoderm) and hypoblast (blue, also called primary endoderm). (c) A third mesodermal tissue develops (pink) between the epiblast and hypoblast. The hypoblast spreads out over the underside of the trophoblast. (d) The hypoblast edges meet to form a hollow spherical cavity: the yolk sac cavity. Mesoderm extends between the hypoblast and trophoblast and between the epiblast and trophoblast. Cavities develop within the mesodermal tissue (the extraembryonic coelom) and within the epiblast tissue (proamniotic cavity). The proembryonic disc forms. (e) The extraembryonic coelom, proamniotic cavity and yolk sac cavity enlarge and change shape. The epiblast is now divisible into amniotic and embryonic ectoderm, the hypoblast is now divisible into yolk sac and embryonic endoderm, and the mesoderm is divisible into extraembryonic and embryonic mesoderm. Between the amniotic and yolk sac cavities, a triple cell layer structure of embryonic ectoderm, mesoderm and endoderm is now evident as the trilaminar or embryonic disc of the definitive embryo, which will eventually give rise to the fetus. All the other tissues are extraembryonic and will form the fetal membranes. Thus, the amnion develops from a layer of extraembryonic ectoderm fronting the amniotic cavity and a layer of extraembryonic mesoderm outside it. The yolk sac develops from a layer of extraembryonic endoderm fronting the yolk sac cavity and a layer of extraembryonic mesoderm outside it. The chorion develops from a layer of trophoblast fronting the uterine tissue and a layer of extraembryonic mesoderm within it. The point at which the yolk sac mesoderm and chorionic mesoderm fuse is arrowed; this is the site of formation of the yolk sac (or choriovitelline) placenta. Blood vessels develop throughout the embryonic and extraembryonic mesoderm, and the embryonic blood starts to flow through them. (f) The trilaminar embryonic disc curls up with its outer ectoderm surrounded by amniotic fluid in the amniotic cavity; the derivatives of this ectoderm will include the outer 'skin' of the fetus, as well as most of the nervous system. Within this curled-up disc, there is an endoderm-lined cavity (continuous with the yolk sac cavity) that is the primitive gut. The interposed filling of embryonic mesoderm (which will form many of the fetal tissues between the gut and the skin and their derivatives) is highly vascular and includes, at the anterior or head end of the embryo, the primitive heart (cardiac mesoderm) that pumps blood through the network of embryonic and extraembryonic vessels. A diverticulum of the endoderm, the allantoic endoderm, develops at the posterior or tail end of the embryo and it grows out surrounded by mesoderm to form the allantois. The allantoic and chorionic mesoderm (both rich in blood vessels) fuse (arrowed) and mark the site of formation of the chorioallantoic placenta. The connection that the allantois makes between the embryo and the chorioallantoic placenta will become the umbilical cord (see Figure 14.2). The disposition of these membranes in twin pregnancies is summarized in Figure 13.4. (Source: Hamilton W.J., Boyd J.D., Mossman H.W. (1972) *Human Embryology*. Williams & Wilkins, Baltimore. Reproduced with permission from LWW.)

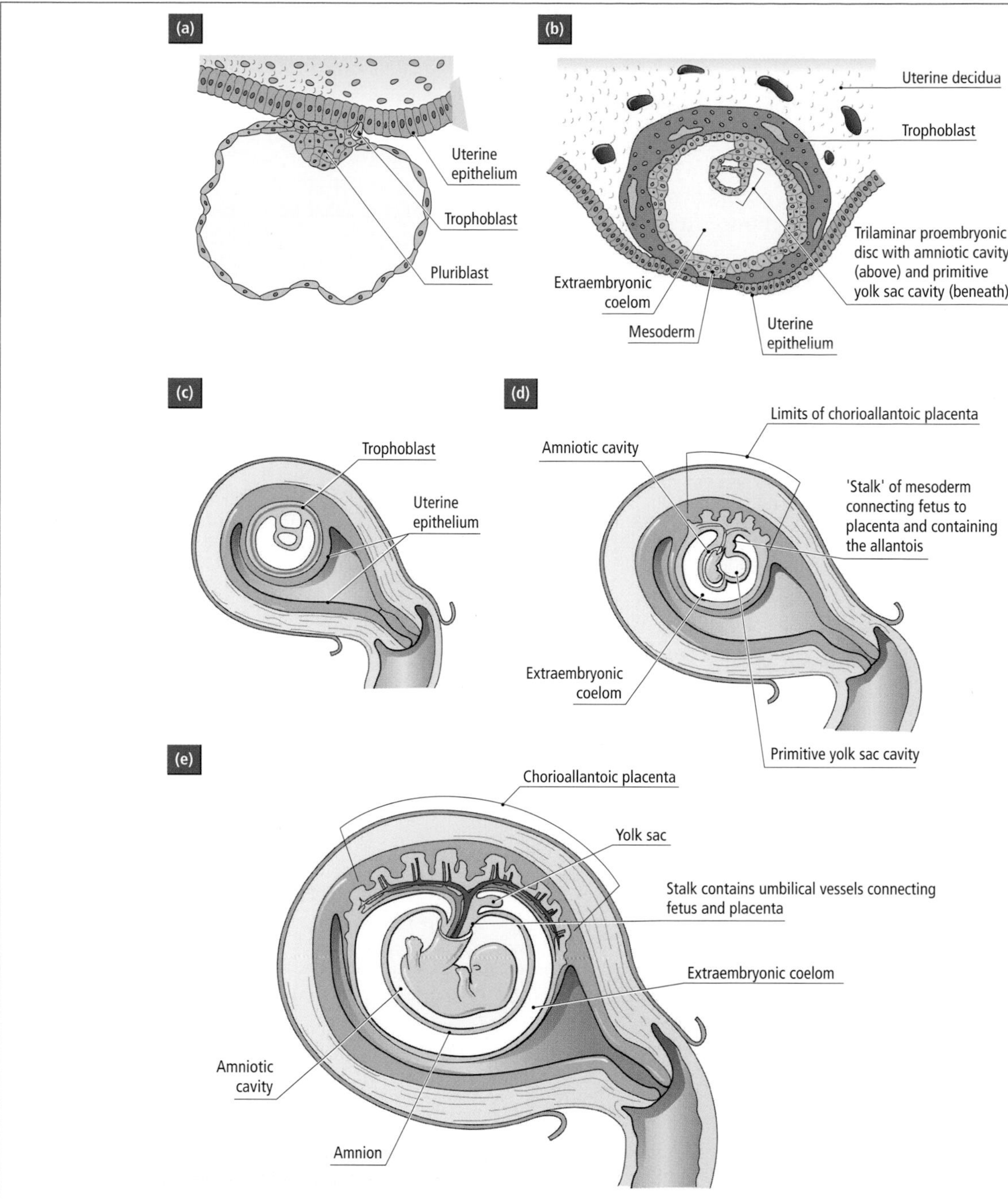

Figure 14.2 Highly schematic views (not to scale) of a human conceptus at: (a) attachment and beginnings of invasion; (b) well-advanced invasion; and (c–e) progressively later stages of pregnancy. The general relationship of fetus, fetal membranes, placenta and maternal vessels is shown. Note how the fetus itself ends up floating in amniotic fluid and linked to the placenta by an umbilical stalk. Ultimately the **extraembryonic coelom becomes obliterated** as the amnion presses up against the outer chorion.

The placental interface is organized to facilitate exchange between maternal and fetal circulations

The chorioallantoic placenta is characterized by:

- Extensive proliferation of the chorionic tissue to give a large surface area for exchange. The critical tissue for this process is the trophoblast, which comprises a proliferative group of cells called the **cytotrophoblast** that can differentiate into a number of more specialized trophoblast derivatives, the most important of which for our present discussion is the **syncytial trophoblast**. Syncytiotrophoblast, as its

Box 14.2 Organization of the sheep haemochorial placenta

Often used as a human surrogate in functional studies, **the sheep placenta is synepitheliochorial and cotyledonary**, having some 80–90 independent sites of close vascular proximity. Attachment occurs at each **uterine epithelial caruncle**, the **fetal chorion** at the point of attachment showing specialization as a **cotyledon**. The **fetal cotyledon and maternal caruncle together constitute the functional unit of the placenta, known as a placentome** (Figure 14.3). In the mature cotyledon, an ingrowth of chorionic villi indents and compresses the epithelium of the maternal caruncles. Within the villi, a core of vascularized fetal mesoderm develops. Each cotyledon receives one to three branches of the umbilical vessels. The vessels divide and ramify within the villi where they come to lie under the surface of the trophoblast at the tips of the villi (Figure 14.3). On the maternal side, the surface epithelium of the caruncle becomes syncytial, and the underlying stroma becomes acellular and vascular. Binucleate cells form in the fetal cotyledon, and some of these contribute to this maternal syncytium (an equivalent penetration into maternal tissues may occur in other large farm animals that are nonetheless appropriately classified as non-invasive at implantation itself). Tortuous, coiling maternal arteries supply each caruncle and split into capillaries between the penetrating terminal villi of the fetal cotyledon. The capillaries then run back along the long axis of the terminal villus towards the tip, where they drain into maternal veins (Figure 14.3). This organization of fetal and maternal vessels means that, in the capillary beds, the blood could flow in opposite directions, which would maximize opportunities for metabolic exchange.

name implies, forms by fusion of cytotrophoblast cells to form a multinuclear tissue.

- Highly developed vascularity of both fetal and maternal components that are organized into **intimately juxtaposed, but physically separate**, fetal and maternal blood flows. In the human haemochorial placenta, it is the syncytiotrophoblastic layer with its underlying cytotrophoblast and mesodermal tissues containing fetal blood vessels that is interposed between the fetal and maternal circulations, being bathed in maternal blood.

In the mature haemochorial human placenta, tongues or **villi** of chorionic syncytiotrophoblast, containing cores of mesodermal tissue in which fetal blood vessels run, penetrate deeply into the maternal tissue to form an extensive network (Figure 14.4b). Each villous blood vessel, by progressive branching of the main divisions of the umbilical vessels, is separated from the surface by a very thin syncytiotrophoblastic layer (Figure 14.4c). At the tips of the **terminal villi**, the capillaries are dilated and form tortuous loops (Figure 14.5). Thus, fetal blood flow through the tips will be slow, allowing for exchange of metabolites with maternal blood. The branches of the villi are arranged, with a somewhat variable degree of regularity, to form 'fenestrated bowls' (rather in the shape of the bowl of a brandy glass). Their terminal villi project inwards into the central space of the bowl, between the adjacent villi that form the fenestrated wall of the bowl, and outwards into the space peripheral to the bowl. Each 'bowl' unit is sometimes called a **fetal lobule**.

It is suggested that, in the human, the **maternal spiral artery** at the decidual base of the placenta ejects its blood into the space that forms the bowl of this lobule, filling the brandy glass, as it were. The pressure of the blood causes its circulation through the fenestrated wall of the lobule over the fine terminal villi. The blood is then thought to drain back via the basal venous openings into maternal veins (Figure 14.3b). Up to 200 such lobular units form in the mature human placenta, which is a 'pancake' 15–20 cm in diameter and 3 cm thick. Several lobular units are grouped together to form a **lobe**, the boundaries of which may be seen grossly on the placenta, defined by **fibrous septa**.

This anatomical description is probably somewhat idealized, and the consistency with which such a regular arrangement of maternal vessels and villi occurs in the human placenta is perhaps questionable. The important point to understand is that maternal blood circulates across the fine terminal villi containing the fetal capillaries.

Establishing mature patterns of blood flow in the human placenta

So much for placental structure, but what about the functional aspects of placental blood flow?

Extravillous trophoblast and placental function

The general anatomical organization of the placental interface is achieved in the human chorioallantoic placenta by 3–4 weeks of pregnancy, but is not fully functional until 10–12 weeks. This is because a **fully matured maternal blood flow does not develop until 10–12 weeks**. What underlies this functional delay? It is now clear that a second differentiation product of the cytotrophoblast, namely, the **extravillous trophoblast** cell is responsible Thus, in parallel with the development of villi, non- or extra-villous trophoblast cells detach in groups from the cytotrophoblast in the trabeculae and **migrate into the endometrium and myometrium**. Two migratory routes are observed. Some cells migrate through the maternal tissue and are called **interstitial extravillous trophoblast**, whilst others migrate into and along the maternal spiral arteries and are called **endovascular extravillous trophoblast** (Figure 14.6). It is these endovascular extravillous cells that regulate blood flow patterns in the placenta and they do so in two ways.

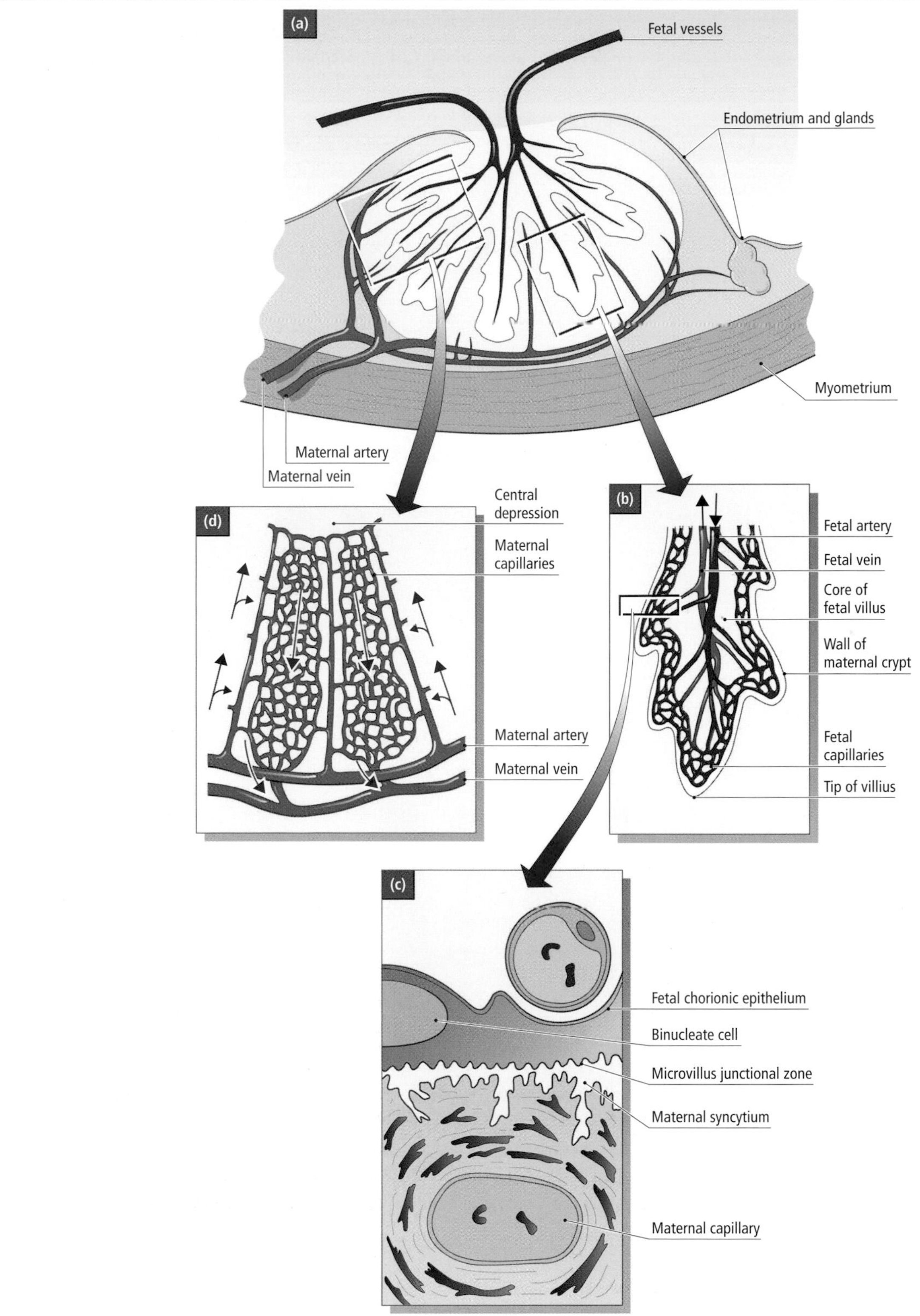

Figure 14.3 (a) General structure of a sheep placentome. Fetal villi shown in longitudinal section in (b) interdigitate with maternal tissue giving close apposition of circulations (c). Maternal blood flows around the fetal villus, as indicated in (d). A similar interdigitation of fetal villi with maternal tissue also occurs in the horse and the cow.

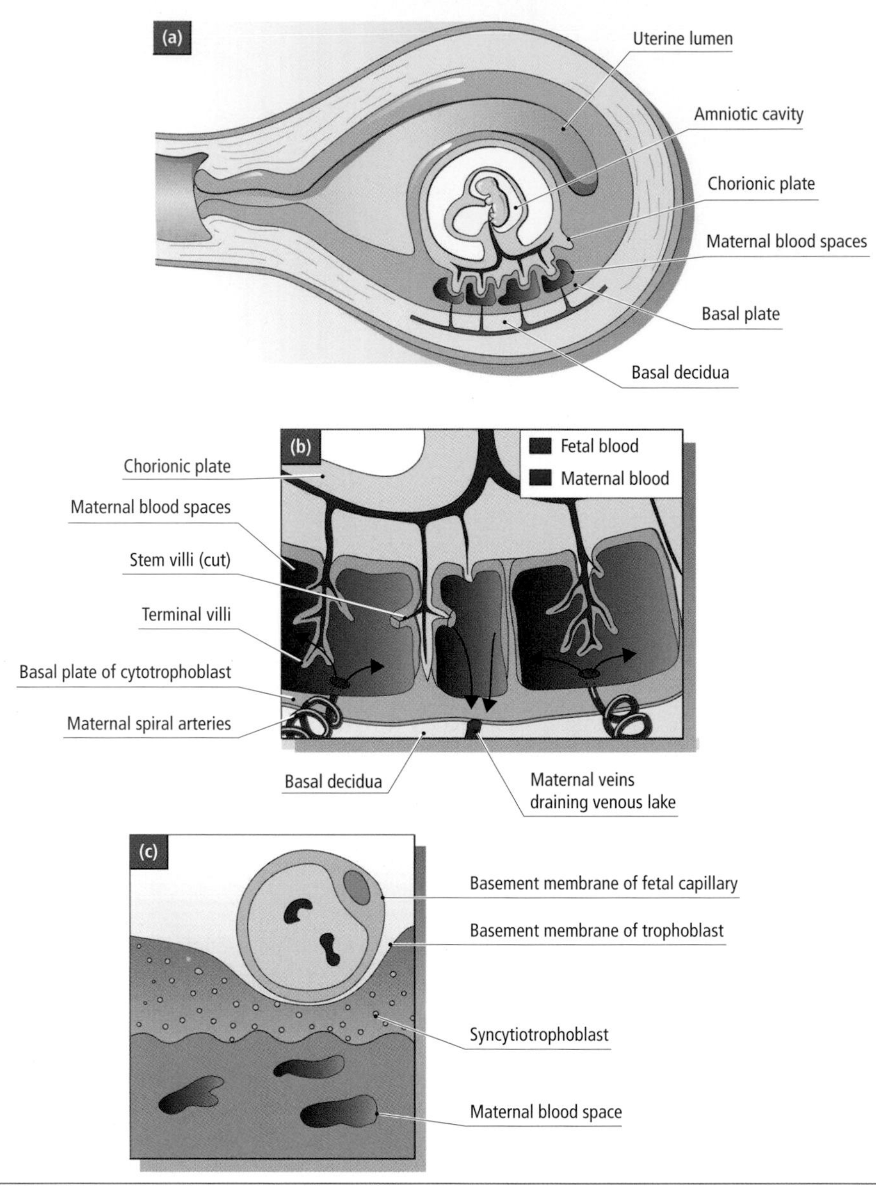

Figure 14.4 Schematic representation of structures of the human placental interface. (a) Stem villi connect the chorionic plate to the basal plate forming a labyrinthine series of spaces. (b) From the stem villi and the chorionic plate, smaller villi ramify into the intervillous space, forming a network of fine filamentous terminal villi, which are the principal sites of metabolic exchange. (c) At these sites, only a thin layer of chorionic syncytiotrophoblast separates the fetal blood vessels from maternal blood. Villi are classified as: **primary**, when composed of solid trophoblast; **secondary**, when mesoderm invades the villous core; and **tertiary**, when blood vessels penetrate the mesoderm. As a villus grows and extends, it goes through each of those stages.

First, they stimulate the **terminal dilatation of the spiral arteries** in a process called '**conversion**' (Figure 14.6). This dilatation lays the foundation for the blood flow patterns observed in the fully functional chorioallantoic placenta. Second, the endovascular trophoblast **partially occludes these terminal dilatations**, thereby impeding the maternal blood flow. Indeed, during the first trimester the intervillous space is filled with a clear fluid made up of maternal plasma percolating through these trophoblastic plugs and mixed with the secretions of the uterine glands. Because this fluid lacks an oxygen carrier, the oxygen supply within the placenta is as low as 2–3% oxygen. Why does the placenta, having created the possibility of good oxygenation, then go to such lengths to delay it from happening?

The answer seems to lie in the two-edged sword that is oxygen. **Oxygen is essential for efficient ATP production** via mitochondrial oxidative phosphorylation. However, **oxygen can also be very toxic** via the production of **reactive oxygen species (ROS)** such as the **superoxide anion**, which is a byproduct of the less than 100% efficient mitochondrial respiratory chain. Excess superoxide anions (and active derivatives of it) are mopped up by antioxidants such as **vitamins C and E** and by a series of protective enzymatic reactions involving **superoxide dismutase, catalase and glutathione peroxidase**. However, ineffective mopping results in the excess ROS inflicting oxidative damage on proteins, lipids and nucleic acids with severe consequences such as cell stress, death and even carcinogenesis. It seems that the risk of ROS production affecting adversely the critical events of embryonic differentiation and morphogenesis, with the potential for malformation and even embryonic death and pregnancy loss, is just too great.

Thus, the placenta has evolved a mechanism to hold oxygenation levels in check and to restrain growth until the crucial embryonic differentiative events are completed (see Box 14.3). Thus, **natural selection has sacrificed growth to protection**. Then, from 10 weeks onwards, the trophoblastic plugs start to disperse and the circulation of maternal blood into villous spaces gets underway. Thus, haemochorial placentation, and haemotrophic support, in the human do not truly occur functionally until the second trimester. Until then, histiotrophic support reigns.

Haemotrophic support is more efficient than histiotrophic support at establishing metabolic gradients to drive diffusional and carrier-mediated exchange of metabolic substrates and excretory products. So, the **second trimester onwards is characterized by growth** and maturation as the fetus rapidly increases in size, a counterpoint to the preceding embryonic

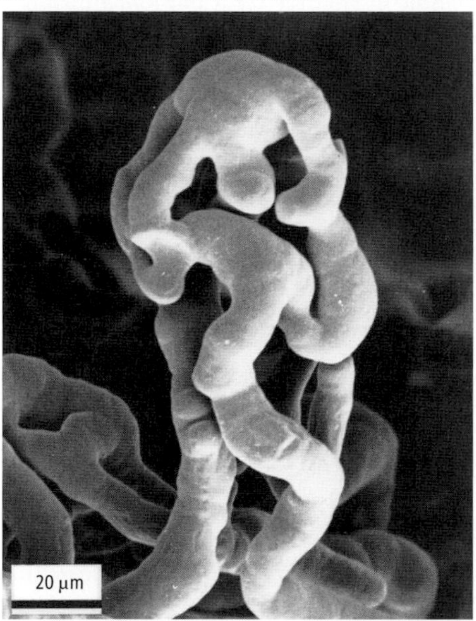

Figure 14.5 Low-power view of a cast made from the microvascular capillaries in a terminal villus on the fetal side of the placental circulation. Note how the terminal vessels form a convoluted knot supplied by straight capillaries. Note also the terminal dilatations of the vessels in which blood flow is slower and at which exchange of metabolites between fetal and maternal blood takes place. (Source: Courtesy of Dr Graham Burton.)

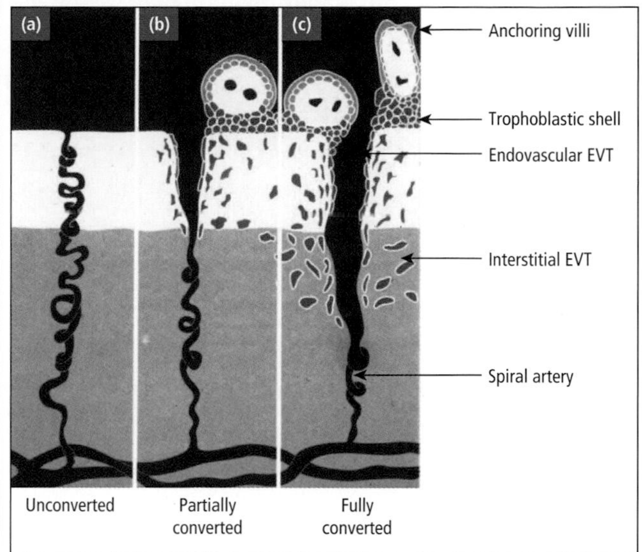

Figure 14.6 Extravillous trophoblast (EVT) migration and spiral artery conversion. (a) Unconverted spiral artery. (b) During conversion, some EVT migrates through the maternal tissue and is called interstitial EVT, whilst other EVT migrates (c) into and along the maternal spiral arteries and is called endovascular EVT. It is this latter EVT that regulates blood flow patterns in the placenta. In the first and second trimesters, endovascular EVT stimulates the terminal dilatation of the spiral arteries in a process called 'conversion', which converts the maternal vessel into a low pressure, high capacity, flaccid conduit. Conversion also involves loss of smooth muscle and endothelial lining and of vasoreactivity, and lays the foundation for the blood flow patterns observed in the fully functional chorioallantoic placenta (see later). However, in the first trimester, the EVT partially occludes these terminal dilatations, thereby impeding the maternal blood flow during embryogenesis. Failure of conversion is associated with complications such as pre-eclampsia and intrauterine growth restriction.

Box 14.3 The molecular basis of oxygen sensing

The first trimester occurs under conditions of low-oxygen levels of 2–3%. These low levels, whilst hypoxic for most tissues, maintain trophoblastic stem cells and inhibit their differentiation. Only when the cytotrophoblast cells migrate into the spiral arteries, and are exposed to oxygen levels of 10% or more do they differentiate into endovascular extravillous trophoblast. Low-oxygen responses are mediated by the transcriptional activation of **hypoxia-inducible factors** (**HIFs**). HIF1 is the main regulator of hypoxia responsiveness, having an oxygen-sensitive subunit, HIF1α, and a β subunit that is oxygen-insensitive. Both subunits are expressed constitutively and ubiquitously, but the **HIF1α protein has a short half-life of under 5 minutes in the presence of oxygen** and is rapidly degraded. However, under hypoxic conditions HIF is stabilized, binds to **hypoxic response elements** (**HREs**) in DNA, to activate genes that promote cell survival under low-oxygen conditions. Genetically *HIF1*-null mice show aberrant placental development and poor embryonic survival, and aberrant HIF induction is associated with pre-eclampsia. A second HIF, called HIF2, is also regulated by low oxygen levels to restrict tissue-specific gene expression.

period when the primary developmental process was the increase in complexity of form and tissues. The significance of this histiotrophic to haemotrophic transition occurring effectively and at the correct time is seen in the consequences arising from situations in which it **occurs too early** (leading to **pregnancy loss**) or **too late or incompletely** (resulting in **placental insufficiency, fetal retardation and eclampsia**; Figure 16.2b).

Haemotrophic blood flow at fetal stages

Once haemotrophic nutrition is established, the effectiveness with which gradients for exchange of metabolites are established depends largely on how effectively rates of blood flow through the two sides of the placental circulation are regulated, as well as on the diffusional barriers and special transport mechanisms that might exist between them. The latter are considered in detail in Chapter 15. Here we are concerned with rates of blood flow and the factors affecting them.

Cardiac output and total blood volume increase by up to 40% in pregnancy, in response to the additional peripheral load imposed by the feto-placental unit. Maternal blood reaches the placenta via uterine and ovarian vessels and constitutes about 25% of the mother's total cardiac output by the end of pregnancy, achieved through a three-fold increase in flow due to both vascular dilation and proliferation. In the human, the **uterine arteries** course along the lateral walls of the uterus giving off 9–14 branches, each of which penetrates the outer third of the myometrial tissue. At this level, anastomosis of these arteries with the ovarian arteries may occur, and **from the anastomosis a series of arcuate arteries** runs within the anterior and posterior myometrial walls of the uterus, thereby encircling it. From this enveloping vascular network, **radial arteries** penetrate through the remaining myometrium into the basal endometrial tissue. Here the so-called **basal arteries** distribute **spiral arteries** to supply the endometrial decidua. This convoluted or spiral nature of terminal endometrial arteries, together with the **terminal dilatations** that result from conversion, and their **lack of responsiveness to vasoconstrictor** transmitters and drugs, are a feature common to many species. Each of these features tends to **diminish the arrival velocity** of the maternal blood, which is further diminished to 0.1–10 ml/s in the primate by the extensive volume of the intervillous blood spaces.

This sluggish flow may protect the conceptus from being dislodged by 'spurts' of blood, and also gives ample time for the exchange of metabolites at the placental interface (estimated mean transit time of 15 seconds in the full-term monkey placenta). A similar slowing of flow occurs on the fetal side of the circulation, where the total cross-sectional area of blood vasculature increases as a result of both the profusion of vascular branching and the **terminal capillary dilatations** (Figure 14.5) at which exchange occurs. This massive expansion of the fetal blood vasculature means that the fetal circulation operates at low pressure. The protection of the fetal blood vessels from collapse is assured by a corresponding low maternal perfusion pressure (4–10 mmHg) within the intervillous spaces.

The maternal arteries have a sympathetic innervation restricted to their myometrial course, which enables them to constrict in response to sympathetic nerve stimulation or sympathomimetic drugs. **Reduced placental perfusion** can therefore result either from local vasoconstriction or from lowered systemic pressure. Transient reductions in perfusion pressure do not seem to have adverse effects on placental exchange or fetal growth, but chronic reductions do, particularly later in pregnancy. Thus, **chronic anxiety** or **heavy smoking** during pregnancy result in **smaller term babies**, probably caused in part by effects on placental perfusion. Similarly, administration of drugs to relieve maternal hypotension as, for example, in asthma, will cause an increase in visceral vasoconstriction, which will already be elevated reflexly, and thereby further reduce placental perfusion. During maternal exercise, some reduction in blood flow to the uterus occurs, but the conceptus itself seems to be relatively protected, unless exercise is severe and prolonged.

In addition to the adverse effects of impaired flow towards the placenta, occlusion or **impaired blood flow** at the placental interface will also **reduce the efficiency of metabolite exchange**. Such an effect occurs in conditions of increased maternal blood viscosity, e.g. **sickle cell anaemia**, or after the excessive expansion of placental villous structures, with a consequent reduction in the intervillous space available for circulation. This occurs in conditions of **increased umbilical vein pressure**, such as occurs in **erythroblastosis (haemolytic disease of the fetus)** or **vascular occlusion of the fetal liver**, both of which cause distension of the villi. **Smoking in pregnancy** also exerts direct effects on the fetal vasculature in the placenta, there being fewer, narrower and less convoluted capillaries in the terminal villi. Finally, occlusion of the maternal venous drainage, as can occur during compression of the inferior vena cava when lying supine or during uterine contractions at parturition, will reduce flow through the placental interface. Engorgement of the intervillous space can then, via pressure effects, reduce fetal blood circulation and the efficiency of placental exchange.

Summary

The extraembryonic membranes, and the placenta which develops in part from them, are evolved elements central to mammalian viviparity. In this chapter, we have focused on their role in providing, restricting and then releasing blood flow to the developing embryo/fetus. But this function is but one of many. Throughout pregnancy, the conceptus is totally dependent on the mother for its protection and nutrition. It subverts many of her metabolic and physiological activities to its own ends, so that there is adequate mobilization of oxygen, salts and organic precursors to supply its needs.

One way in which it achieves this is by inducing and taking part in the formation of the vascularized placenta where the bulk of these substances is exchanged for its own metabolic waste. During pregnancy, the mother is also preparing for the future requirements of the fetus and neonate. Thus, hypertrophy of the uterine musculature, which participates in fetal expulsion at parturition, and the development and maintenance of mammary glands for postpartum lactational nutrition, are both stimulated during pregnancy. All this support for the embryo, then fetus and then neonate is achieved by the **pregnancy hormones**. In the remainder of this chapter, we consider the nature and origin of these pregnancy hormones. It will become clear that the **placenta can be thought of as a major and versatile endocrine gland**.

Endocrine support strategies

The extended secretion of progesterone beyond the natural decline seen at luteolysis is critical for the initiation of pregnancy. This **absolute requirement for progesterone persists throughout pregnancy**. Indeed, in some species, progesterone levels in maternal blood rise continuously as pregnancy proceeds (Figure 14.7). Moreover, as pregnancy advances, the level of **oestrogens in the maternal blood also rises** (Figure 14.7). These steroid hormones may reach plasma levels many times greater than those seen in a normal luteal phase. In many species, the ovarian corpora lutea under the control of the maternal pituitary continue to secrete an essential proportion of these steroids throughout pregnancy. Removal of either the ovary or pituitary at any time therefore results in pregnancy loss (Table 14.1, group C). In other species (Table 14.1, groups A and B), the whole of pregnancy clearly does not depend totally on steroids secreted by pituitary–ovarian interactions, because, at varying periods during pregnancy, one or both glands may be removed without inducing pregnancy loss. Where does the endocrine support come from in these species? Some of the clearest answers to this question have come from studies on human pregnancy, which shows the least dependence on the maternal ovarian–pituitary axis of any species. We will therefore examine the endocrinology of human pregnancy in some detail and discuss the evidence from other species in relation to the human pattern.

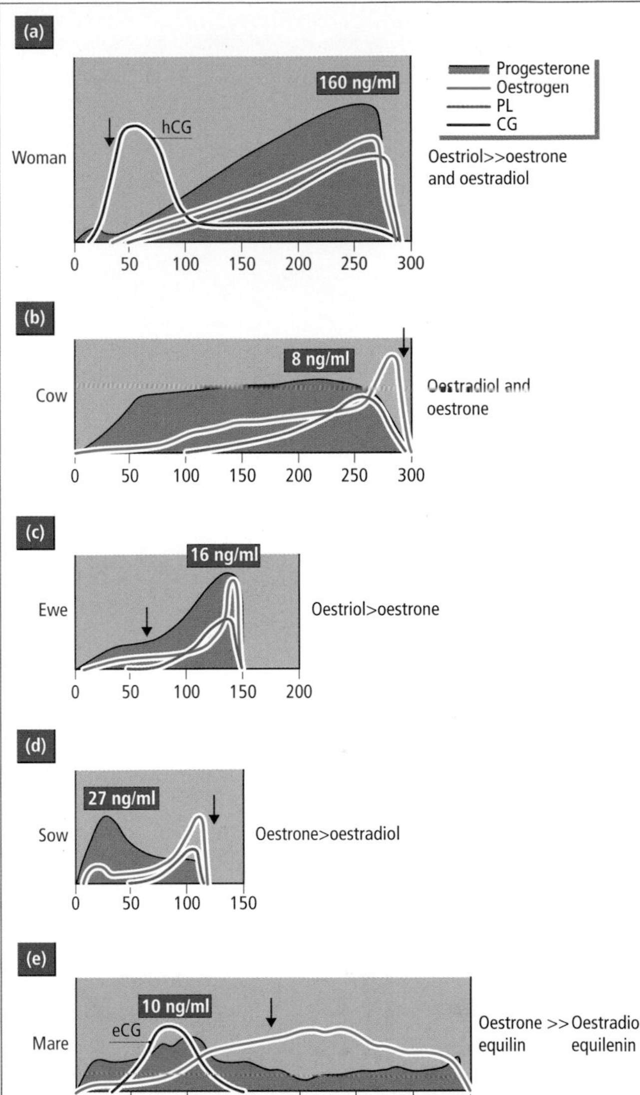

Figure 14.7 Patterns of plasma hormones during pregnancy (in days) in (a) woman, (b) cow, (c) ewe, (d) sow, and (e) mare. The maximal progesterone level is indicated in the blue box and the principal oestrogens are recorded to the right of each panel. PL, placental lactogen; hCG and eCG are human and equine chorionic gonadotrophins. The arrow indicates the time at which ovariectomy no longer terminates pregnancy.

The human conceptus synthesizes steroid hormones

We saw in Chapter 13 (see pages 230–1) that within 8 days of fertilization the trophoblast of the human conceptus synthesizes and releases the hormone hCG. This hormone then maintains the progestagenic activity of the corpus luteum. Within a further 2–3 weeks, the conceptus is also synthesizing all the steroidal hormones required for pregnancy, and although the maternal corpus luteum remains active for the whole of pregnancy, it can be dispensed with after only 4–5 weeks and plays only a trivial role in total progesterone output at later stages. **The human conceptus thus shows a remarkable endocrine emancipation**. It asserts its complete endocrine independence of the mother.

Our understanding of the role of the conceptus in steroidogenesis has come from several sources. For example, much has been learnt from observations on steroid output and interconversions in abnormal pregnancies, and from isolation of tissues from different parts of the normal conceptus and assessment of their steroid biosynthetic capacity *in vitro*. However, by far the most informative and important data have come from *in vivo* studies using the fetuses of induced

Table 14.1 Dependence of pregnancy on maternal ovarian and pituitary function in various species

Species	Duration of pregnancy (days)	Duration of non-pregnant luteal phase (days)	Day of pregnancy when hypophysectomy is without effect	Day of pregnancy when ovariectomy is without effect
Group A				
Human	260–270	12–14	?	40
Monkey (*M. mulatta*)	168	12–14	29	21
Sheep	147–150	16–18	50	55
Guinea pig	60	16	3	28
Group B				
Rat	22	10–12	12	Term
Mouse	20–21	10–12	11	Term
Cat	63	30–60	Term?	50
Horse	330–340	20–21	?	150–200
Group C*				
Cow	280–290	18–20	?	Term
Dog	61	61	Term?	Term?
Pig	115	16–18	Term	Term
Rabbit	31	12	Term	Term
Goat	150	16–18	Term	Term

* NB: These results do not mean that these species are solely dependent on pituitary and ovarian function. It is established that some are partially dependent on support from placental sources, which are inadequate when acting alone (see Table 14.3 and Box 14.5).

pregnancy terminations. Direct sampling of umbilical and maternal blood, and the infusions of minute quantities of radiolabelled steroid precursors into the maternal, fetal or placental circulations, with subsequent analysis of their interconversions, has provided information of great clinical value about the sites of steroid synthesis and interconversion in the conceptus (Figure 14.8).

Progesterone is secreted by the placental trophoblast

By late human pregnancy, the output of progesterone exceeds 200 mg/day. Blood absolute levels are high, partly due also to a three-fold increase in **transcortin** (cortisol-binding globulin), which increases the proportion of bound progesterone (see Table 2.7). From where in the conceptus does the progesterone come?

A clue comes from observations on pathological pregnancies, in which an embryo fails to develop but progesterone output is only marginally reduced. Thus, **choriocarcinomas** or **hydatidiform moles** (malignant and benign tumours of the chorion, respectively) secrete progesterone in the absence of any embryonic tissue, suggesting that the placental trophoblast itself is the principal progesterone source. *In vitro* studies on the biosynthesis of progesterone by cultures of syncytiotrophoblast confirm this conclusion, and also indicate that the trophoblast can only use cholesterol (not acetate) as a substrate. The cholesterol is usually derived from the maternal rather than the fetal circulation. The rising output of progesterone through pregnancy appears to be completely autonomous; no external controlling mechanism has yet been discovered.

Clinical observations on patients whose ovaries have been removed at various times in early pregnancy indicate that the placenta is capable of synthesizing an adequate, supportive level of progesterone by 5–6 weeks. In the normal pregnant woman, there is a plateau, or even a slight fall, in the concentration of blood progesterone between 6 and 9 weeks. The plateau coincides with a marked fall in 17α-hydroxyprogesterone (an ovarian progestagen), indicating that over this period the placenta takes over progestagenic support. Both pregnenolone and progesterone pass from the placenta into both the fetal and maternal circulations, infused radiolabelled progestagens being distributed widely within the fetal and maternal tissues. The principal excreted metabolite of progesterone (15% of total) is **pregnanediol**, but its urinary level is not a useful indicator of fetal well-being and only a crude indicator of placental function. The failure to correlate progesterone output with fetal well-being is not surprising, as the fetus itself plays no part in progesterone synthesis. The poor correlation with placental function is mainly a result of the wide interindividual variation in the level of progesterone's urinary metabolites in regular pregnancies.

The human fetus and placenta cooperate to produce oestrogens

The conceptus is also the source of the high level of oestrogens in human pregnancy (Figure 14.7). The principal oestrogen is not oestradiol 17β but the less potent **oestriol** (Tables 2.3 and 14.2). Oestrogen output differs from that of progesterone, however, in that it is severely reduced in cases of choriocarcinoma

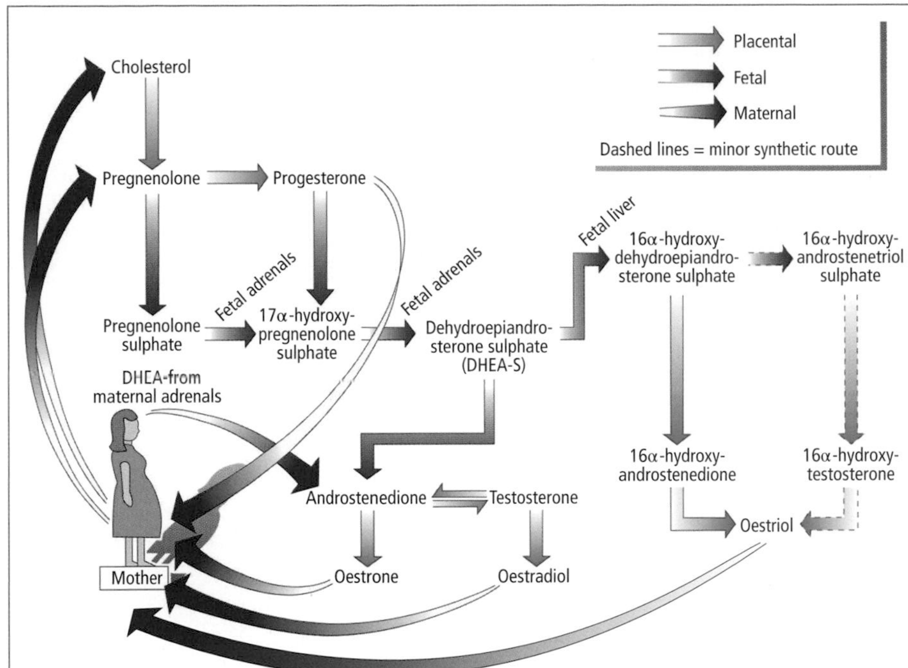

Figure 14.8 Summary of the principal routes by which the human maternofeto-placental unit synthesizes oestrogens. An X-linked placental sulphatase deficiency exists and is associated with deficiencies in the aromatization of C_{19} steroids to oestrogens, with a consequent maternal oestrogen deficiency.

Table 14.2 Plasma levels of various steroids in women during late pregnancy

Steroid	Values in pregnancy (ng/ml)		Values in luteal phase of cycle (ng/ml)
Progesterone	125–200		11
Oestriol	7	113	–
Oestriol conjugates	106		
Oestrone	7	46	0.2
Oestrone conjugates	46		
Oestradiol 17β	10	15	0.2
Oestradiol conjugates	5		

and hydatidiform mole, in which a fetus is lacking. Moreover, although the incubation of placental tissue with radiolabelled cholesterol or pregnenolone yields labelled progesterone, no labelled oestrogens are produced. Clearly, the placenta alone is inadequate for synthesis of oestrogens. Could the fetus itself produce them?

When radiolabelled pregnenolone was infused into the circulation of isolated perfused human fetuses, little or no labelled oestrogen was produced. Therefore, the fetus alone cannot be the site of oestrogen synthesis. However, when labelled pregnenolone was injected into an intact fetoplacental unit, complete synthesis of oestrogens occurred. These observations show that the human fetus and placenta cooperate to produce oestrogens. A series of observations on the result of infusing various radiolabelled steroid substrates has shown that, whilst the placenta is capable of synthesizing oestrogens from C_{19} androgens, it cannot synthesize its own androgens from progestagens. The **fetal adrenal**, in contrast, can synthesize the C_{19} androgens (DHEA and some androstenedione) from progestagens, but cannot aromatize them to oestrogens. These C_{19} steroids, which are made in a **special fetal zone of the adrenal**, must therefore pass to the placenta, which can then convert them to oestrogens. The principal pathways involved are summarized in Figure 14.8. This cooperative organization is reminiscent of that observed between luteal and granulosa cells in the production of ovarian oestrogens (see Chapter 9).

Two important points should be noted about this cooperative event. First, the initial step in the fetal handling of all steroids arriving from the placenta or mother is conjugation to sulphates. This conjugation occurs mainly in the fetal liver but also in the adrenals. The conjugated steroid is then converted biosynthetically by the fetal adrenal to other steroid derivatives via sulphated intermediates. Conversely, the placenta takes conjugated steroids arriving from the fetus and deconjugates them, releasing free steroids into the maternal (and fetal) circulation. Thus, the **fetus sulphates** and the **placenta desulphates**. This apparently bizarre manoeuvre is in fact extremely important, as conjugated steroids are both water-soluble and biologically inactive. The conjugation may protect the fetus against untoward steroidal penetration into, and activity within, fetal tissues (e.g. masculinization of females by weak androgens; see Chapter 3). Thus, this strategy permits the **high ambient levels of steroid required by the maternal organism** for the maintenance of pregnancy to co-exist with the **low level of active steroids essential for protection of the fetal organism** from unwanted side effects.

Second, the placental synthesis of oestrone and oestradiol 17β can occur from DHEA sulphate derived from either the fetal adrenal (40% but rising towards term) or the maternal adrenal. Oestriol synthesis also occurs in the placenta – but the 16α-hydroxylated substrate required comes *solely* from the **fetal liver**. It follows therefore that 16α-hydroxylated steroids, such as oestriol, should provide an indicator not only of placental function but also of **fetal well-being**. In pregnancies at risk, a consistent decline over a period of days of 16α-hydroxylated steroids correlates with fetal distress and may indicate that premature delivery should be induced. A parallel decline in oestradiol 17β is not seen because 50–60% of its synthesis by the placenta utilizes maternal not fetal DHEA sulphate. In anencephalic fetuses, in which fetal adrenal function is impaired, the ratio of oestradiol + oestrone: oestriol is greatly increased, as would be expected. Thus, **16α-hydroxylated steroids can have a diagnostic value**.

Corticosteroids rise during pregnancy under the influence of oestrogens

Maternal blood levels of cortisol rise in pregnancy, and this rise can be mimicked by oestrogen injection into non-pregnant women. The elevated levels are partly due to a **decrease in the metabolism** of free cortisol, but predominantly due to an **oestrogen-stimulated synthesis of transcortin** from 3.5 mg/100 ml to 10 mg/100 ml. The transcortin binds both cortisol and progesterone.

The major placental protein hormones

These protein hormones were introduced in Chapter 2 (Tables 2.4 and 2.5), where we noticed that pregnancy-specific variants of several existed (e.g. CG versus LH; GH-N and GH-V versus GH; the placental lactogens versus prolactin). The elaboration of these variants may reflect the fact that the needs of the mother and the fetus sometimes conflict, and so some competition for control of resources may be occurring.

Chorionic gonadotrophin is not required for luteal maintenance once placental steroid synthesis is established

We saw earlier that hCG from the syncytial trophoblast is critical for the initiation of pregnancy by extending luteal life from 2 to 6–7 weeks until placental steroids take over. The rising blood and urinary levels of hCG over this period can usefully be measured when testing for pregnancy, a positive test indicating a **biochemical pregnancy**. The action of hCG is limited to this short-term luteal support, as its blood levels fall after 8 weeks (Figure 14.7), correlating with the fall in 17α-hydroxyprogesterone (a luteal product) and the emancipation of the feto-placental unit from ovarian dependence. However, a **hyperglycosylated form of hCG** is produced by the extravillous trophoblast which acts as an autocrine factor to promote extravillous trophoblast migration and survival, and thereby spiral artery conversion. Low levels of hCG remain detectable for the remainder of pregnancy (although in the rhesus monkey rCG levels are trivial from 40 days onwards). How the variation in CG output is controlled is uncertain. The placenta produces its own pulses of GnRH and the pulsatile output of CG is sensitive to exogenous GnRH pulses, suggesting a causal relationship. Moreover, inhibin A and progesterone depress, while activin A and oestrogen enhance, CG production. However, although the basic building blocks of an intraplacental control system are in place, just how they might be coordinated to give the observed temporal pattern of CG secretion remains to be elucidated.

The chorionic somatomammotrophins are synthesized by the placenta towards the end of the first trimester

As hCG levels decline, syncytiotrophoblastic expansion is associated from about 6–8 weeks with **increased chorionic secretion** of **human placental lactogen** (hPL) and the placental variants of human **growth hormone (hGH-N and V**; see Table 2.5). Blood levels then increase further during the last trimester. The nature of the physiological control of this switch in hormone production is unclear. As plasma levels of hGH-V rise, those of pituitary hGH fall, suggesting that the placental variant is exerting a negative feedback control systemically. Indeed, pregnancy survives in the face of maternal GH deficiency, and placental GH does not itself enter the fetal circulation. Placental GH modulates maternal metabolism through stimulation of placental growth and the induction of placental IGF1 to induce nutrient repartitioning to the fetus. A deficiency of placental GH is associated with fetal growth retardation.

In addition, release of pituitary prolactin is stimulated by oestrogens (see Box 14.4), and maternal plasma concentrations reach over 200 ng/ml by the last trimester. Prolactin is also synthesized by the decidual tissue. Some synthesis is detectable by the spontaneous decidual cells that appear towards the end of each luteal phase (see Chapter 13), but much larger amounts are secreted between 10 and 30 weeks of pregnancy. Progesterone is implicated in the control of its secretion. Little decidual prolactin appears in either the fetal or maternal circulation. Most of it seems to pass into the amniotic fluid where it may have a role in maintaining the water and electrolyte balance of the fetus (see pages 262–3). Different strategies achieve the endocrine support of pregnancy in other species (see Box 14.5).

The richness of the endocrine placenta

The foregoing account captures the complexity of the placental endocrine gland. Nonetheless, it is a relatively impoverished account. The placenta resembles a 'mini-organism': in addition to steroid and protein hormones, it also synthesizes hypothalamic-releasing hormones (GnRH, CRH) and pituitary-like peptide hormones (ACTH, oxytocin, vasopressin) together with a large range of cytokines (including TGFβ, activin A, inhibins A and B, IGF1 and 2, FGFs, EGF), vasoactive peptides (VEGF, endothelin), neurohormones (monoamines, neuropeptide Y) and metabolic hormones (leptin, ghrelin). This list is far from

Box 14.4 Prolactin production in pregnancy

The rising prolactin levels as pregnancy progresses parallel rising oestrogens, which induce the **hyperprolactinaemia** by binding to ERα in the lactotrophs and stimulating prolactin synthesis. Increased spontaneous output of prolactin occurs. Chronic oestrogen exposure results in increased lactotroph numbers, which are therefore more numerous and larger even than in non-pregnant females, who have higher ambient prolactin levels than males. An oestrous rhythm of prolactin secretion is observed in some animals, such as the rat, with a mid-cycle prolactin surge coincident with LH. This appears to result from the preovulatory surge in oestrogen, and is associated with a decline in dopamine receptor level and thereby a diminished inhibitory influence of dopamine. Blocking the oestrogen surge blocks the prolactin surge. However, in women, there seems to be no clear menstrual rhythm in serum prolactin levels, and prolactin secretion does not alter significantly after the menopause.

exhaustive. Many of these hormones spill over into maternal and/or fetal blood. Others act locally within the placenta itself. It is not yet clear exactly what all these hormones do, as the complexity of their production is matched by the even greater complexity of their interactions. Some of these individual players we will pick up in later chapters, but for the moment grasp the extraordinary biosynthetic richness of this transient organ.

Conclusions

Three comments of general relevance need to be made when comparing the available data on pregnancy hormones. First, the tendency for the feto-placental unit to take over endocrine control from the mother in whole or in part is seen in most species. Second, while the levels of plasma oestrogens and progesterone rise, there is great species variation in the absolute level achieved (compare 160 ng of progesterone/ml in the human with 8 ng/ml in the cow). Third, the temporal patterns of plasma steroids recorded through pregnancy differ among various species. The large variation, both qualitative and quantitative, in plasma hormone levels is difficult to explain. This difficulty is compounded by the fact that we do not yet have a complete understanding of the functions of many of these hormones during pregnancy. These issues are addressed in the ensuing chapters.

Box 14.5 Different strategies achieve the endocrine support of pregnancy in other species

Elevated steroid levels are a feature of pregnancy in most mammals studied, although there is considerable variation in the patterns and absolute levels from species to species (Figure 14.7). As we saw in Table 14.1, the degree of independence from the pituitary–ovarian axis also varies, but even in species such as the cow and the pig that require both the pituitary and ovary to be present throughout pregnancy, there is a fetal steroid contribution (Table 14.3). Conceptuses of both the horse and the sheep become independent of the ovary by about one-third of the way through pregnancy. Both are comparable to the human, with a fetal source of DHEA sulphate being deconjugated and aromatized by the placenta. In the horse, the principal oestrogens formed are oestrone, together with two oestrogens found only in equids: **equilin and equilenin**. The fetal DHEA sulphate used for aromatization in the horse is derived

Table 14.3 Major sites of hormone synthesis during established pregnancy: a comparison of different species

Species	Progestagens	Oestrogens	Gonadotrophins
Human	Placenta	Fetal adrenal and placenta	Placenta (hCG and hPL)
Horse			
Early	Corpus luteum	Ovarian follicles	Placenta (eCG)
Mid–late	Placenta	Fetal gonad and placenta	–
Sheep			
Early	Ovary and placenta	Placenta (oestradiol and oestrone sulfate)	–
Mid–late	Placenta	Ovary (oestrone), placenta (oestrone sulfate)	Placenta (oPL)
Cow	Ovary (and placenta)	Placenta (and ovary)	Pituitary and placenta later (for bPL)
Pig			
Early	Ovary	Placenta	Pituitary
Late	Ovary (and placenta)	Placenta	Pituitary

not from the fetal adrenal but from the hypertrophied interstitial tissues of the fetal gonads, which show a spectacular increase in weight between 100 and 300 days of gestation, regressing by birth. The sheep fetus, like the human, synthesizes sulphated DHEA in the adrenal, and maternal sources of DHEA may also be available. In addition, the sheep placenta, unlike the human placenta, can undertake conversion of progesterone to oestrogens, and this property is of great importance at parturition (see Chapter 18).

From the time of attachment (around 16–30 days; see Table 13.1) biosynthesis of **placental lactogens** (PLs; see Table 2.5) by the binucleate trophoblast cells of **sheep, cow and goat** conceptuses can be detected. This synthesis is reflected subsequently in their rising blood levels as pregnancy progresses (Figure 14.7). However, in these species these hormones are much more closely related to prolactin than hPL is to human prolactin, and so have less GH-like activity than does hPL. Placental GH-V synthesis occurs in sheep and goats (but not cows, which use fetal and maternal GH), roughly paralleling PL production and coming from the same binucleate cells. Together these two placental hormones act to stimulate the uterine glandular secretions, which remain so important throughout the pregnancy of large farm animals with their non-invasive, central implantations (see Chapter 13). There is evidence that IFN-τ, directly or indirectly, may sensitize the glands to these hormones, independent of its actions in suppressing oxytocin receptor synthesis.

Between days 40 and 120 of pregnancy in the mare, a gonadotrophin with weak FSH- and strong LH-like activity, called originally **pregnant mares' serum gonadotrophin** (**PMSG**) but now **equine chorionic gonadotrophin** (**eCG**) on the basis of its structural similarity to other CGs, is secreted from the trophoblastic cells of the endometrial cups. How its secretion is controlled is not known, although the quantity of its secretion is determined by the genetic constitution of the mare. One effect of the eCG is to promote follicular growth and even secondary ovulation in the maternal ovaries between days 40 and 120 of pregnancy; as a result, secondary corpora lutea may develop. These remain an active source of progesterone until the decline of eCG and the take-over by placental progesterone at around 140–150 days.

In many species, but notably in the guinea pig and pig, a cytokine called **relaxin** (a member of the insulin family) has been detected in blood at low levels during pregnancy, rising just before parturition. This hormone appears to be produced by the corpus luteum of pregnancy, and has also been detected in humans. Its actions will be discussed in the context of parturition (see Chapter 18). Finally, and also in the pig, **placental oestrogens** continue to play a major role in the **support of histiotrophic secretion**, acting in part via induction of **uterine IGF1 and FGF7**.

Key learning points

- Extraembryonic tissues develop into fetal membranes (chorion, yolk sac, amnion, allantois).
- A primitive yolk sac (or choriovitelline) placenta is functional throughout pregnancy in marsupials, for part of pregnancy in the rabbit, rat and mouse, and marginally in higher primates.
- In most mammals, the chorioallantoic placenta assumes the sole or major role in fetal nutrition, but only becomes fully functional after the embryonic phase of development during the second and third trimesters.
- The embryonic stage of development is protected from damaging reactive oxygen species by a low oxygen tension.
- The placenta is a site where the maternal and fetal circulations pass close to each other but do not mingle, and where exchange of materials is facilitated.
- Impairment of maternal blood flow to or through the placenta is associated with fetal maldevelopment.
- The placenta is a rich source of many hormones.
- In humans, the conceptus synthesizes all the steroidal hormones required for pregnancy from about 5–6 weeks of pregnancy.
- Dependence of steroid synthesis on the pituitary–ovarian axis varies among species, as does the pattern of hormones during pregnancy.
- The placental trophoblast is the principal source of progesterone in human pregnancy.
- The human trophoblast also synthesizes oestrogens from C_{19} androgens, notably DHEA.
- The human trophoblast cannot synthesize its own androgens from progestagens.
- The human fetal adrenal synthesizes the C_{19} androgens in the fetal zone of the adrenal but cannot aromatize them. These C_{19} steroids pass to the placenta, which converts them to oestrogens.
- The fetus conjugates steroids and the placenta deconjugates them.
- Corticosteroids rise during pregnancy under the influence of oestrogens.
- CG falls after the first trimester and is no longer needed.
- Placental trophoblastic lactogen and growth hormone rise during the second and third trimesters.
- The decidua tissue is a source of rising prolactin, especially from 6–8 weeks of pregnancy.

Clinical vignette

Serum biomarkers of placental function

A 38-year-old woman, Ms G, was pregnant for the first time. At 12 weeks of gestation she opted to have combined screening for Down syndrome performed via the NHS screening programme. Although her risk of Down syndrome was low (1:5,826), the tests demonstrated that she had an unusually low level of pregnancy-associated protein A (papp-A; 0.3 multiples of the median) in her serum. Ms G's obstetrician explained that the low papp-A level indicated an increased risk of placental dysfunction in the pregnancy, which could lead to fetal growth restriction and a high risk of adverse outcomes, including stillbirth. It was recommended that Ms G should have additional monitoring of fetal growth and placental blood supply during the pregnancy. At her 20-week ultrasound scan, the fetus was growing on the 50th centile with normal anatomy. However, a subsequent scan at 23 weeks showed that the pulsatility index of both uterine arteries was elevated. A further growth scan was performed at 28 weeks gestation. At this scan, the fetal abdominal circumference measurement had reduced from the 50th centile to the 20th centile and the femur length had fallen to the 3rd centile. The pulsatility index in the umbilical artery was at the 95th centile, but the end diastolic flow in the vessel was positive. By 30 weeks, the picture had further deteriorated. The liquor volume had reduced and there was reversed end diastolic flow in the umbilical artery. There was evidence of cerebral redistribution of blood flow within the fetus and Ms G reported a decrease in the fetal movements. On this basis, her obstetrician admitted Ms G to hospital for close monitoring. The next day, the fetal heart rate variability had reduced and the obstetrician opted to deliver the baby by Caesarean section. Ms G's baby was born in good condition although small-for-gestational age. After a prolonged stay in the neonatal intensive care unit, the baby went home feeding and growing well.

The value of serum biomarkers in predicting placental function and hence fetal growth has been the subject of intense interest and research activity in obstetrics. Of all of the markers identified that can predict fetal well-being, papp-A has proven to be one of the most useful. Papp-A is a metalloproteinase that increases the bioavailability of insulin-like growth factor (IGF), by cleaving its high-affinity binding proteins into smaller, lower-affinity fragments. Hence, lack of papp-A in early pregnancy is associated with low IGF activity. IGF is a major factor that regulates placental development: lack of IGF in early gestation results in poor placentation and reduced fetal growth. In the case outlined above, ongoing checks on fetal well-being are performed using the pulsatility index (PI) in various blood vessels. The PI is calculated using the difference between maximal systolic velocity and the end diastolic velocity, divided by the average flow velocity in the vessel. The PI reflects the compliance of the vessel and the afterload generated by the downstream vascular bed. Hence, in the umbilical artery, a high PI reflects high resistance in the placental vasculature. Poor placental blood flow predicts fetal compromise and is a useful clinical marker of fetal well-being.

FURTHER READING

Burton GJ *et al.* (2006) Anatomy and genesis of the placenta. In: *The Physiology of Reproduction*, 3rd edn. (ed. JD Neill). Elsevier, St Louis.

Burton GJ, Jauniaux E (2004) Placental oxidative stress: from miscarriage to preeclampsia. *Journal of the Society for Gynecological Investigation* **11**, 342–352.

Burton GJ, Jauniaux E, Charnock-Jones DS (2007) Human early placental development: potential roles of the endometrial glands. *Placenta 28, Supplement A, Trophoblast Research* **21**, S64–S69.

Cartwright JE, Fraser R, Leslie K et al. (2010) Remodelling at the maternal–fetal interface: relevance to human pregnancy disorders. *Reproduction* **140**, 803–813.

Cole LA (2009) New discoveries on the biology and detection of human chorionic gonadotropin. *Reproductive Biology and Endocrinology* **7**, 8.

Duncan WC (2000) The human corpus luteum: remodelling during luteolysis and maternal recognition of pregnancy. *Reviews of Reproduction* **5**, 12–17.

Maliqueo M, Echiburú B, Crisosto N (2016) Sex steroids modulate uterine-placental vasculature: implications for obstetrics and neonatal outcomes. *Frontiers in Physiology* **7**, 152–160.

Spencer TE, Bazer FW (2004) Conceptus signals for establishment and maintenance of pregnancy. *Reproductive Biology and Endocrinology* **2**, 49.

Steven DH (ed.) (1977) *Comparative Placentation*. Academic Press, New York.

Stouffer RL (2003) Progesterone as a mediator of gonadotrophin action in the corpus luteum: beyond steroidogenesis. *Human Reproduction Update* **9**, 99–117.

Tuuli MG, Longtine MS, Nelson DM (2011) Oxygen and trophoblast biology – a source of controversy. *Placenta, Supplement B, Trophoblast Research* **25**, S109–S118.

More advanced reading

Ashworth CJ, Hoggard N, Thomas L *et al.* (2000) Placental leptin. *Reviews of Reproduction* **5**, 18–24.

Gootwine E (2004) Placental hormones and fetal–placental development. *Animal Reproduction Science* **82–83**, 551–566.

Gultice AD, Kulkarni-Datar K, Brown TL (2009) Hypoxia-inducible factor 1alpha (HIF1α) mediates distinct steps of rat trophoblast differentiation in gradient oxygen. *Biology of Reproduction* **80**, 184–193.

Hempstock J, Cindrova-Davies T, Jauniaux E, Burton GJ (2004) Endometrial glands as a source of nutrients, growth factors and cytokines during the first trimester of human pregnancy: a morphological and immunohistochemical study. *Reproductive Biology and Endocrinology* **2**, 58.

Kalinka J, Hanke W, Sobala W (2005) Impact of prenatal tobacco smoke exposure, as measured by mid-gestation serum cotinine levels, on fetal biometry and umbilical flow velocity waveforms. *American Journal of Perinatology* **22**, 41–47.

Poswillo D, Alberman E (1992) *Effects of Smoking on the Fetus, Neonate and Child*. Oxford University Press, Oxford.

CHAPTER 15
Growing the fetus

Essential Reproduction, Eighth Edition. Martin H. Johnson.

Companion website: www.essentialreproduction.com

The emergence of a fully functional chorioallantoic placenta as the first trimester draws to a close heralds the beginnings of fetal growth and maturation that are so critical for a successful birth outcome (Figure 15.1). This process presents challenges for the mother. Too large a fetus is associated with a difficult delivery, which is risky for both mother and neonate. Too small a fetus is associated with both immediate and long-term health problems for the neonate. An understanding of how growth is regulated, and how it is related to birth timing, is thus important for pregnancy management, the clinical care of small neonates and the general health of the population. Babies may be of **low birthweight** (<2500 g) because they are **born preterm** (formerly called **premature**), that is, prior to 37 weeks since the last menstrual period, or because they arise from a multiple pregnancy (twins, triplets, etc., usually also born preterm) or, if born at full term, because they suffered from **intrauterine growth retardation** (**IUGR**). As many as one-third of low-birthweight singleton babies come into the latter category and are said to be **small for gestational age** (**SGA**, or sometime **small for dates**; defined as being of a weight that is two standard deviations below the weight expected of a baby of a particular gestational age).

So, it is important to understand how growth rate is regulated – how the fetal growth trajectory balances the demands of the fetus with the capabilities of the mother. In considering how, we will see that the fetus cannot be regarded as a quiescent, passively growing product of conception tucked neatly away in its protected uterine environment. Although undoubtedly dependent on the mother for its growth and survival, the fetus nonetheless enjoys considerable independence in the regulation of its own development. Indeed, the fetus exerts profound effects on maternal physiology via hormones secreted into the maternal circulation by the fetal part of the placenta. These hormones in part determine the mother's ability to meet the metabolic requirements of pregnancy. Thus, mother, placenta and fetus are each implicated in the endocrine regulation of fetal growth.

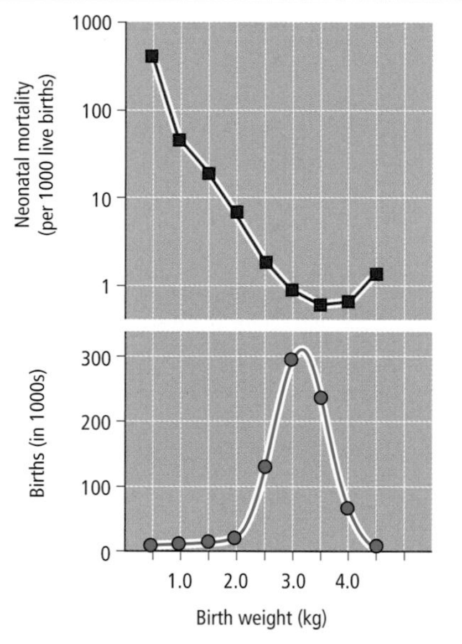

Figure 15.1 Impact of birthweight on fetal morbidity and mortality for the USA. Note that both high and low birthweights are associated with increased neonatal mortality.

Patterns of fetal growth

The rate of **fetal growth is relatively slow up to the 20th week** of pregnancy but accelerates to reach a **maximum around weeks 30–36**, declining thereafter until birth (Figure 15.2a). **Placental size**, and/or maternal uterine blood flow through it, may be limiting growth after week 36, because although the placenta increases in size slowly and steadily until birth, the relatively rapid growth of the fetus just prior to birth means that the **ratio of placental weight to fetal weight falls significantly** during the later stages of pregnancy (Figure 15.2b). A **postnatal peak in growth velocity occurs during week 8**.

Growth initially occurs through an increase in cell number, but from week 32 increase in cell size dominates. **Protein accumulation** occurs early in fetal development to reach its **maximum of about 300 g by week 35**, and precedes fat deposition, most of which is subcutaneous, and which only comes to exceed the weight of protein by week 38. By term, **some three times as much energy is stored as fat as is stored as protein**. The relative amount of water (95% in the young fetus) also decreases.

Maternal contributions to growth control

The pregnant woman has two conflicting demands placed on her physiology: she needs to **support fetal growth** but also to **constrain it** so that the fetus does not become too large to deliver (Figure 15.1). Her physiology adapts to the demands of the conceptus by **modifying her caloric intake and her metabolic activity**. Routine monitoring of maternal weight gain during pregnancy is generally unhelpful diagnostically, as it varies widely among women, averaging at 13 kg (distributed as: fetoplacental unit 5.8 kg; uterus 0.9 kg; breasts 0.4 kg; fat 3.5 kg; blood 1.2 kg; extracellular fluid 1.2 kg). The wide variation in weight changes makes advice about energy consumption difficult. Pregnant women have been shown to vary in both basal metabolic rate and total energy expenditure, and can be classed as either **energy-sparing or profligate**. The latter require an additional 80 000 kcal to meet the increased energy expenditure of pregnancy, whilst the former can show a net reduction in expenditure of around 13 000 kcal, mainly

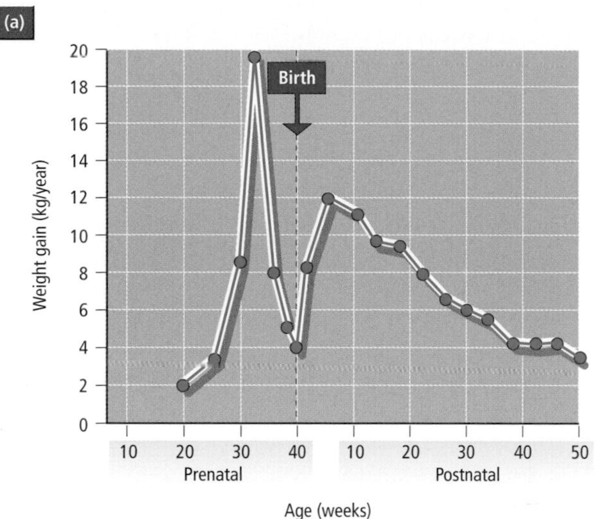

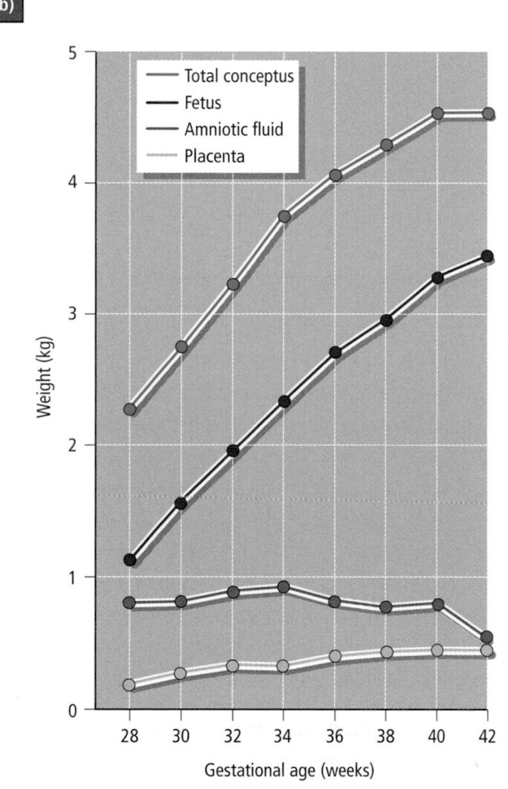

Figure 15.2 (a) Change in velocity of growth in weight of singleton fetuses and neonates. (b) Weight changes of conceptus, fetus, placenta and amniotic fluid during pregnancy. (Source: Biggers J.D. (1979) *Medical Physiology* (ed. V. Mountcastle). Mosby, St Louis. Reproduced with permission from Elsevier.)

due to their basal metabolic rate declining in the first two trimesters. Thus, the best advice for most women is '**eat to appetite**'. Dietary calorific supplements are only found to be of clear value in preventing SGA babies in poorer, undernourished women on less than 1600 kcal/day, suggesting that the **availability of nutrients is not limiting in well-nourished healthy women**. However, **healthy eating** is recommended, both to avoid known teratogens, such as high levels of vitamin A (liver), potential infection sources (toxoplasmosis in undercooked meat; listeria in unpasteurized milk) and vitamin deficiencies. Thus, vitamin D supplementation is advised for some classes of women who are not exposed to adequate levels of sunlight or whose diet lacks it. In addition, for some diets or women, additional iron, folate or iodine is recommended.

Additional maternal factors that affect birthweight include ethnicity, parity (**primiparous** mothers tend to have smaller babies than **multiparous** mothers), maternal size, multiple pregnancy (more than one fetus carried simultaneously reduces birthweight), maternal height (linked to uterine capacity and thus potential for growth), health problems such as diabetes (see page 264) and self-inflicted damage, e.g. smoking, drug and/or alcohol abuse (see page 248 for discussion).

Fetal contributions to growth control

Fetal sex affects growth, male fetuses being on average larger than females. The pattern of fetal growth is determined primarily through the **genome of the fetus**, but additional fetal and maternal factors modulate the effects of its expression. **Insulin-like growth factors 1 and 2** (**IGFs**; also called **somatomedins**) are produced by a range of fetal cell types, the mix of which varies with stage of development in species-specific ways. Synthesis of both IGFs rises with progress through pregnancy, IGF2 being present in fetal blood at 2–3-fold higher levels than IGF1. IGF1 provides a major direct endocrine stimulus to fetal growth through its anabolic effects; IGF2 may do so mainly indirectly by stimulating placental growth and transport mechanisms. **Genetic knock-out of either or both IGFs results in fetal growth retardation**. Fetal IGF1 production is responsive to nutrient levels, and thus matches fetal growth to nutrient supply. Thus, it declines when nutrients fall, and responds to nutrient-sensitive hormones such as insulin, thyroxine and glucocorticoids. IGF1 production is not stimulated by fetal GH, probably because of a deficit in fetal GH receptors. Fetal thyroid hormones also stimulate growth in the latter part of pregnancy. Leptin levels, produced not only by adipocytes but also the syncytiotrophoblast and possibly other fetal tissues, rise during pregnancy and correlate strongly with fetal and placental growth rates, suggesting but not proving a functional link with growth.

If normal growth and development are to occur, the fetus must be provided with the basic building materials and energy sources: essential amino acids, fatty acids, sugars, vitamins and minerals. About 50% of the calories needed for growth and metabolism come from **carbohydrates**, 35% from **fats** and the remainder from **amino acids**. All these must come from the mother, most via the placenta, but some via the amniotic membranes to enter the amniotic fluid in which the growing fetus is suspended (see Box 15.1). We now consider the placental transport mechanisms themselves, and the changes in maternal and fetal metabolism and physiology that determine the relative blood levels of metabolites, and thereby the gradients across the placental interface.

Placental transport

The discrete nature of the maternal and fetal circulations, separated by cellular and acellular layers, confers an important barrier property on the placenta. 'Bleeds' across the placenta are rare (except at parturition) and probably occur mainly in a fetomaternal direction. This circulatory separation means that **simple diffusional exchange** between the circulations will only be significant in the case of low-molecular-weight molecules, such as blood gases, sodium ions, water and urea, or in the case of non-polar molecules, such as fatty acids, ketone bodies and non-conjugated steroids, that can pass through cell membranes. In contrast, hexose sugars, conjugated steroids, amino acids, nucleotides, water-soluble vitamins, plasma proteins, maternal cells, potentially infective agents such as viruses and bacteria, and lipoprotein complexes, including cholesterol, gain significant access to the fetal circulation by **special transport mechanisms** (Figures 15.3 and 15.4).

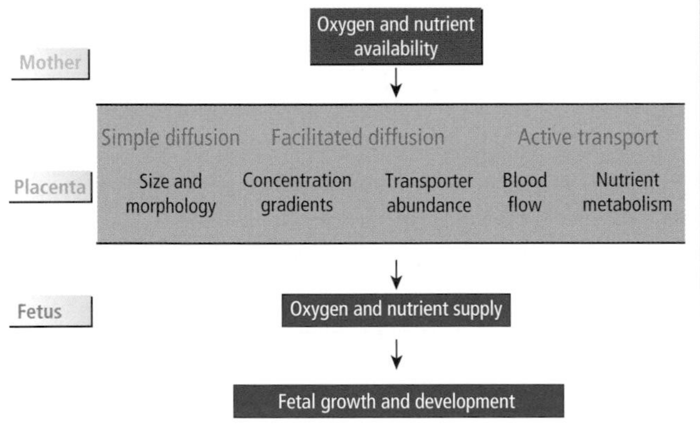

Figure 15.3 Schematic summary of some of main factors influencing transplacental transport.

The identification of transfer routes and rates between mother and fetus is not straightforward. Thus, both maternal and fetal components of the placenta are dynamic and their anatomy and physiology change during pregnancy. Several potential routes of transfer exist (chorioallantoic placenta, yolk sac placenta and amnion), each of which may also have some heterogeneity in its transport systems. Moreover, the placental components may each utilize or produce the substance under study and so complicate quantitative transfer measurements. Studies have used labelled markers and serial sampling in intact materno-feto-placental units, perfused fetoplacental units, isolated placentae, cultured placental fragments or trophoblastic cells *in vitro*. Each approach has its experimental and ethical complications.

Regardless of whether transport is diffusional or transport-mediated, the two factors that influence exchange of a substance between mother and fetus are (1) its relative fetal and maternal blood concentrations, and so the transplacental **gradient of the substances to be transported**, and (2) the organization and rates of **maternal and fetal blood flows through the placenta** (see pages 244–8). The two types of transport mechanism are distinguished by the **relative significance of the thickness, organization and surface areas** of the tissues interposed between the two circulations (summarized in Figure 15.3).

Diffusional transport

The interspecies differences in placental microstructure of the layers separating maternal and fetal circulations (see Table 13.3; Figure 13.6) do influence the relative efficiency of diffusional exchange. Thus, diffusion of Na^+ is faster across the human haemochorial than the sheep synepitheliochorial placenta. However, the more freely diffusible molecules such as O_2 are much more affected by blood flow than placental barrier thickness. The important point to appreciate about the microstructure of the placental interface is that it either permits adequate diffusional exchange (i.e. has a large safety factor) or employs special transport systems to promote selective transport of less freely diffusible, but essential, substances.

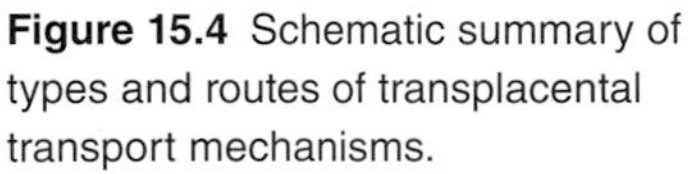
Figure 15.4 Schematic summary of types and routes of transplacental transport mechanisms.

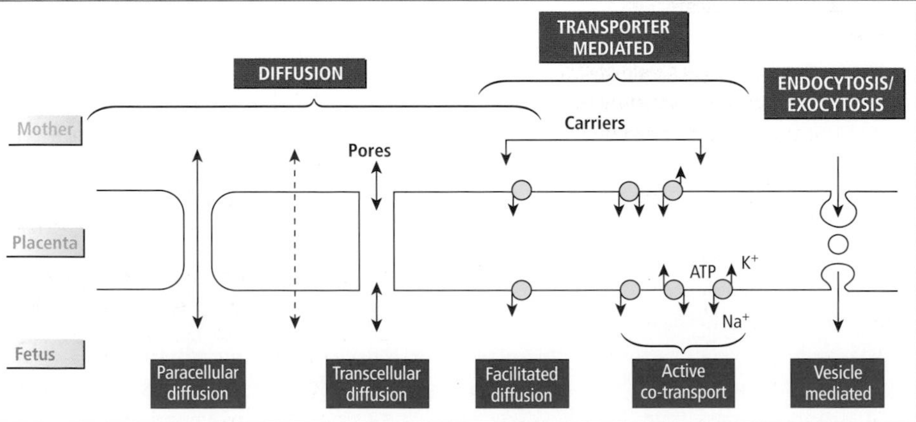

The facility with which diffusional exchange occurs varies during human pregnancy. For example, in early pregnancy, terminal villi in the human placenta are large (150–200 µm in diameter) and the fetal vessel is located centrally beneath 10 µm of syncytiotrophoblast, so metabolites must diffuse a considerable distance. Furthermore, the syncytiotrophoblast is itself metabolically active and intercepts and utilizes maternal metabolites. As pregnancy progresses, the **villi thin** to 40 µm in diameter and the **fetal vessel occupies a more eccentric position**, indenting the overlying syncytiotrophoblast to leave only a 1–2-µm layer separating it from the maternal blood (Figure 14.4c). This altered anatomical relationship increases the capacity for diffusional exchange and reduces consumption by the trophoblast of oxygen. A similar thinning of diffusional barriers also occurs as the synepitheliochorial placenta of sheep matures. Finally, diffusion may be '**facilitated**' by transport mechanisms that are not active, in the sense of energy-consuming, but do occupy a somewhat intermediate position between simple diffusion and active transport. We now consider the factors affecting the diffusional transport of particular molecules in more detail.

Table 15.1 O_2 and CO_2 composition of human maternal and fetal blood

	Maternal blood (arterial)	Fetal blood (venous)	Umbilical artery	Umbilical vein
Oxygen tension (Po_2) (mmHg)	90	35	15	30
Oxygen saturation (%)	95	70	25	65
Oxygen content (vol. %)	14	10	5	13
CO_2 tension (Pco_2) (mmHg)	30	35	53	40
pH	7.43	7.40	7.26	7.35

Oxygen and carbon dioxide

During pregnancy, **maternal O_2 consumption** at rest and during exercise is increased proportionate to the growing tissue mass of the conceptus compared with non-pregnant females. Physical working capacity and the efficiency with which work is performed are not affected significantly by pregnancy. **Cardiac output** (both stroke volume and rate) increases by about 30% during the first trimester, but little more thereafter. The increased cardiac output is accommodated by a reduction of peripheral resistance by as much as 30% owing to the demands of the conceptus, and only a slight blood pressure increase. **Blood volume** rises by up to 40% near term in humans, due partly to a 20–30% increase in erythrocytes and partly to increasing plasma volume (up 30–60%).

Maternal **pulmonary ventilation** increases by 40% during pregnancy, possibly because of a direct effect of progesterone on respiratory mechanisms in the brainstem. A decrease of about 25% in maternal Pco_2 results, with a corresponding fall in bicarbonate concentration, a slight increase in pH and greater changes in pH with exercise. The fetus requires O_2 in relatively continuous supply because fetal stores of the gas are very small: a 3-kg fetus near term requires 18 ml of O_2/min, but stores are only 36 ml or 2 minutes' worth. In addition, the placenta itself is very active metabolically and consumes a massive 40–60% of the total glucose and O_2 supplied to the feto-placental unit by the mother! Because it is a non-polar molecule, O_2 diffuses readily across the placental interface, as does the CO_2 generated by fetal metabolism, with its diffusion constant 20 times higher than that of O_2.

The gradients of the gases at the transplacental interface may be estimated from the figures shown in Table 15.1. Clearly the O_2 tension in the fetal blood is low and that of CO_2 is high relative to the tension of the gases in maternal blood. Gradients to drive diffusional exchange therefore exist. However, inspection of Table 15.1 reveals that although the Po_2 in the oxygenated fetal venous blood leaving the placenta in the umbilical vein is relatively low, the O_2 saturation and content are not much less than in maternal arterial blood. Thus, a much lower O_2 tension leads to a very similar O_2 content, indicating that the '**O_2 trapping**' properties of fetal blood must be more effective than those of maternal blood. This higher affinity of fetal blood is shown graphically in Figure 15.5. This is achieved by the programmed production of **developmental haemoglobins**.

The earliest site of **erythropoiesis** in the implanting mammalian conceptus is the **yolk sac mesoderm** (see Figure 14.1e). The primitive **embryonic erythrocytes** formed are nucleated, and are replaced later by **fetal erythrocytes** (also nucleated) made in the **liver**. During the last trimester, the **spleen and bone marrow** take over fetal erythropoiesis. Embryonic and fetal erythrocytes, like those of the adult, contain haemoglobin as the O_2-carrying molecule, comprising four globin chains coupled to a haem group. However, they differ from the adult in that the constituent globin chains are **not two α- and two β-chains**. Rather, **embryonic haemoglobin** contains **two φ- and two ε-chains**, and **fetal haemoglobin** contains **two α- and two γ-chains**. Each globin chain is coded for by a different gene, and during development a programme of gene switching occurs to yield the embryonic, the fetal and the adult sequences. It is this difference in globin chain composition that results in the different O_2-binding curves (Figure 15.5).

Additionally, in adults, the highly charged molecule **2,3-diphosphoglycerate** (**DPG**) binds to sites on the β-chain, which are exposed only on deoxygenation. This binding **stabilizes the deoxyhaemoglobin**, thereby reducing the opportunity

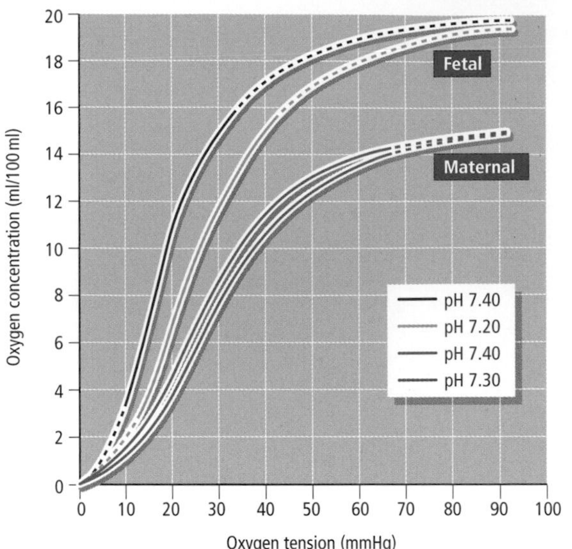

Figure 15.5 The effect of varying oxygen tension on the oxygen content of human maternal and fetal blood under conditions likely to be encountered in the placenta. Note the greater oxygen concentration in fetal blood at any given Po_2. Note also that the pH shifts occurring in the placenta (falling pH in maternal blood and rising pH in fetal blood) will further facilitate unloading of oxygen from maternal blood and its uptake by fetal blood.

for O_2 binding. Thus, a higher Po_2 is needed to load haemoglobin with O_2. The β-chain is lacking in embryonic and fetal haemoglobins, and its equivalents (the ε- and γ-chains) are relatively insensitive to DPG. Therefore, **at any given Po_2, fetal and embryonic haemoglobins will bind more O_2 than will maternal haemoglobin**. In some species, the fetal DPG levels may also be lower, which will further assist the O_2 loading process. Finally, as Figure 15.5 shows, a **double Bohr effect** occurs in the placenta to facilitate transfer of about 10% of the O_2 in a fetal direction. Thus, the fall in pH of maternal blood due to uptake of fetal CO_2 drives release of maternal O_2, and the rise in fetal pH due to the removal of its CO_2 facilitates uptake of O_2.

Under non-pathological conditions the maternal Po_2 and placental perfusion are unlikely to be limiting, and thus the rate of transfer of O_2 across the placenta will vary simply with the Po_2 of the fetal blood in the umbilical circulation. Thus, **the level of fetal oxygenation will be regulated simply by the fetal requirement for O_2.**

Water and electrolytes

During pregnancy, the elevated progesterone binds competitively to the maternal renal aldosterone receptor but does not activate it, leading initially to reduced sodium retention and **natriuresis**. However, a compensatory 10-fold **increase in aldosterone** results: an exaggerated version of the luteal aldosterone rise observed in the menstrual cycle. In addition, oestrogens **stimulate angiotensinogen output** by the liver four- to six-fold, further stimulating aldosterone output. **The net overall effect is an increase in maternal sodium and water retention.**

Exchange of water between mother and fetus occurs at two main sites: the placenta and the remainder of the non-placental chorion where it abuts the amnion internally (Figures 14.2e and legend; Box 15.1). The relative quantitative contributions of each route, particularly early in pregnancy, are unclear, although the placenta is suspected to be the main site of exchange. Most studies in humans and in animals indicate that both amnion and chorion are freely permeable to water molecules. There is no evidence for active transport or water secretion by the membranes themselves, so either diffusional or hydrostatic fluxes must account for water transfer. However, a large hydrostatic pressure difference between maternal and fetal circulations within the placenta cannot occur, because if it were large in the maternofetal direction, the fetal vessels would collapse and impair fetoplacental exchange; and if large in a fetomaternal direction, the fetus would dehydrate. Most likely, then, diffusion plus small or intermittent hydrostatic gradients are responsible for moving the large amounts of water necessary for the fetus.

There is a large measurable traffic of sodium and other electrolytes across the placenta in both directions. Much of this flux is in **association with energy-dependent co-transport** of other molecules via a **transcellular route**, but the evidence that this route is used primarily for ion accumulation, or even contributes significantly to net ion flux, is not secure, at least for humans. As ions can also diffuse, and so equilibrate, via a **paracellular route** (Figure 15.4), it is probable that the net flux is mainly driven by diffusion. Significant quantities of water, sodium and other electrolytes may also cross the amniotic membranes. There is some evidence that the high concentration of decidual prolactin in the amniotic fluid of late pregnancy may assist this exchange.

Glucose

The **fetus has little capacity for gluconeogenesis** largely because the necessary enzymes, although present, are inactive at low arterial Po_2, and it is **only at birth, when arterial Po_2 rises, that gluconeogenesis is initiated**. The fetus must therefore obtain its glucose from maternal blood, the glucose levels of which depend on maternal nutritional status and the integrated endocrine control mechanisms, which maintain free plasma glucose levels within narrow limits. Thus, the secretion of **insulin** from the maternal pancreas prevents glucose levels from rising too high (**hyperglycaemia**) by increasing glucose utilization for glycogen and fat synthesis and storage. Conversely, absorption of glucose from the gut and glycogenolysis (by the utilization of glycogen stores under the action of catecholamines, corticosteroids, glucagon and growth hormone) help prevent maternal glucose levels from falling (**hypoglycaemia**).

Box 15.1 Amniotic fluid

The volume of amniotic fluid increases during pregnancy from about 15 ml at 8 weeks post-conception to 450 ml at week 20, after which net production declines to reach zero by week 34. The composition of amniotic fluid (Table 15.2) suggests that it is a dialysate of maternal and/or fetal fluids, save that the concentration of protein is only 5% that of serum. During the last trimester, total solute concentration in amniotic fluid falls, while the concentrations of urea, uric acid and creatinine increase.

Amniotic fluid is in a dynamic state, complete exchange of its water component occurring every 3 hours or so. There are several routes by which water and solutes enter and leave the amniotic cavity. Exchange with the fetus occurs via its gastrointestinal, respiratory and urinary tracts, and until week 20 or so also with fetal extracellular fluid through the non-keratinized skin of the fetus. Thus, the fetus swallows from 7 ml of amniotic fluid/h at week 16 to around 120 ml/h at week 28. Radio-opaque dye injected into amniotic fluid becomes concentrated in the fetal stomach, presumably because after swallowing, most of the water is absorbed by the fetus. The fetal lungs produce a fluid that fills the alveoli and also contributes to amniotic fluid. Some 3–5 ml/h of hypotonic urine passes into the amniotic cavity at 25 weeks, rising to 26 ml/h (500–600 ml/day) by week 40, after which time it drops rapidly. The relative importance of these pathways for exchange at various times of gestation has not been determined, although the fetal kidney seems to be a principal source of amniotic fluid later in pregnancy. Renal agenesis causes **oligohydramnios** (insufficient amniotic fluid, **Potter's syndrome**), while excessive accumulation of amniotic fluid (**polyhydramnios**) is associated with impaired or no swallowing, e.g. with anencephaly or oesophageal atresia. In addition, the amniotic epithelium, which during the later stages of pregnancy obliterates the extraembryonic coelom and apposes both chorion and the umbilical cord (see Figure 14.2), is also a route of exchange. Indeed, the finding in rhesus monkeys that removal of the fetus does not prevent formation of amniotic fluid emphasizes this capability of the amnion.

The composition and turnover of amniotic fluid became a subject of particular interest with the advent of the sampling procedure, **amniocentesis**, in the diagnosis of fetal abnormalities. Thus, the glycoprotein **α-fetoprotein** is normally present in very low concentrations, but is elevated markedly if the fetus has a defect in neural tube formation, as in **spina bifida** or **anencephaly**. The recovery of fetal cells in amniotic fluid allows fetal karyotyping for gross chromosomal abnormalities (e.g. **trisomy 21** or **Down syndrome**) and assessment of fetal genetic sex, or for diagnosis of various genetic disorders (e.g. those associated with Duchenne muscular dystrophy, phenylketonuria or haemophilia) using gene amplification techniques and DNA sequence analysis of candidate mutant genes. The recovery of amniotic fluid for these diagnostic tests is discussed further in Box 21.2.

Table 15.2 Compositions of amniotic fluid in early and late pregnancy and of the full-term maternal and fetal serum

Fluid	Total osmotic pressure (mosmol)	Na (mM)	Cl (mM)	K (mM)	Non-protein nitrogen (mg/100 ml)	Urea (mg/100 ml)	Uric acid (mg/100 ml)	Creatinine (mg/100 ml)	Protein (mg/100 ml)	Water (%)
Amniotic fluid (first and second trimesters)	283	134	110	4.2	24	25	3.2	1.23	0.28	98.7
Amniotic fluid (third trimester)	262	126	105	4.0	27	34	5.6	2.17	0.26	98.8
Maternal serum (full term)	289	137	105	3.6	22	21	–	1.55	6.5	91.6
Fetal serum (full term)	290	140	106	4.5	23	25	3.6	1.02	5.5	–

Source: Biggers J.D. (1979) *Medical Physiology* (ed. V. Mountcastle). Mosby, St Louis. Reproduced with permission from Elsevier.

Early in gestation, progesterone increases maternal appetite and stimulates the deposition of glucose in maternal fat stores. Later in pregnancy, placental lactogen uses its growth hormone-like activity to mobilize fatty acids from these fat 'depots'. These fatty acids are important for maternal metabolism and essential for fetal growth, because later in pregnancy maternal tissues become progressively less sensitive to insulin. The consequences of this **insulin insensitivity** are two-fold. First, **latent diabetes mellitus** in women may appear for the first time from mid-pregnancy onwards. Second, the **maternal blood glucose is taken up less by maternal tissues and more by placental transfer to the fetus**. This transfer occurs by **facilitated diffusion**, in which specific carriers within both the maternal and the fetal surfaces of the trophoblast use concentration gradients to drive the transport of d-glucose, thereby enhancing diffusional rates to a maximum flux of 0.6 mmol/min/g of placenta. The levels of fetal glucose are related simply and directly to those in the mother, as the carrier system saturates only at supraphysiological maternal serum concentrations (approx. 20 mmol/L). The placenta also utilizes metabolically a considerable amount of the 'in transit' glucose and generates lactate, some of which is distributed to the fetal circulation at about one-third the equivalent flux of glucose, thereby contributing significantly to fetal metabolism. Under normal conditions there is little direct flux of lactate from mother to fetus.

The rate at which glucose is utilized by growing fetal tissues is probably determined largely by the actions of fetal insulin secreted by the fetus' own pancreas in response to glucose load. If the load is high, as can occur in maternal diabetes mellitus, fetal growth and fat storage are promoted and overweight neonates result. The storage of glucose as glycogen, particularly in the fetal liver, is important if the metabolic needs of the neonate are to be provided until feeding begins, and so a progressive increase in the deposition of liver glycogen occurs as parturition approaches (see Chapter 17).

Fetal cardiac muscle contains a high concentration of glycogen, which probably explains why the heart can maintain its contractile activity in the face of severe hypoxia. By contrast, the brain has no glycogen stores and therefore relies totally on a supply of circulating glucose. Thus, **fetal hypoglycaemia can have deleterious effects on the brain**, especially if accompanied by hypoxia, and the risk is higher in a fetus with low cardiac glycogen reserves, as occurs through the glucoprivation accompanying placental insufficiency. **Neonatal hypoglycaemia** may also be seen in babies born to a diabetic mother, as the reduction in the maternal supply of glucose at parturition is not immediately accompanied by a corresponding reduction in the fetal hypersecretion of insulin. A sharp drop in neonatal blood glucose results. Fortunately, neonatal hypo-glycaemia, although potentially a major cause of brain damage, is highly treatable.

The storage of glucose as fat in the fetus is regulated primarily by insulin. Thus, when glucose levels are maintained optimally, such as occurs when maternal nutrition is good, the glucose available after the requirements for growth are fully met is diverted by fetal insulin into **white fat stores**. In conditions where glucose supply to the fetus is reduced, the needs of growth have priority over storage, glucose is released from fetal stores, and the newborn consequently has an emaciated appearance. In addition to these storage depots of white fat, **brown adipose tissue** or **brown fat** is also found in the fetus, newborn and infant. It is deposited in five sites: (1) between the scapulae, in a thin diamond shape; (2) in small masses around blood vessels in the neck; (3) in the axillae; (4) in the mediastinum between the oesophagus and trachea, as well as around the internal mammary vessels; and (5) in a large mass around the kidneys and adrenal glands. It is different in structure to white fat, the lipid being distributed multilocularly and having large numbers of mitochondria-bearing prominent cristae. These deposits of brown fat are of immense importance in **temperature regulation neonatally** and can generate large quantities of heat, independent of increased muscular movement and shivering. Indeed, this form of heat production is termed **non-shivering thermogenesis**. As postnatal development proceeds, brown fat becomes less important thermogenically, and so regresses.

Fatty acids

Maternal lipid metabolism changes during pregnancy. Early in gestation, progesterone increases maternal appetite and stimulates the deposition of glucose in maternal fat stores. Later in pregnancy, placental lactogen mobilizes fatty acids from these fat 'depots' leading to lipidaemia. The fetal demand for lipids is met by both placental transfer and endogenous fetal synthesis. The main dietary **essential fatty acids (EFAs)**, **linoleic and α-linolenic acids**, are substrates for the supply of **long-chain polyunsaturated fatty acids (LCPUFAs**; especially **arachidonic and docosahexaenoic acids**), which are also ingested in meat and dairy foods. FAs are essential throughout pregnancy, but the greatest requirement for the LCPUFAs occurs during the third trimester of pregnancy, when they are critical for brain and retinal growth and function. Several studies have shown that low levels of neonatal infant docosahexanoic acid correlate with impaired cognitive, visual and behavioural functions. For example, if the normal early accumulation of lipids in maternal tissues and the later maternal hyperlipidaemia are limited or prevented by hypothyroidism or diabetes early in pregnancy, fetal growth is reduced and brain development late in gestation can be damaged irreversibly.

The placenta is almost impermeable to lipids other than free FAs and ketone bodies. Free FAs for transfer may be released locally, as **placental lipase** activity increases in the last trimester, and changes in lipase activity correlate with growth retardation (lower) and maternal diabetes and larger babies (higher). The released FAs are thought to diffuse into the trophoblast cells, although lipoprotein receptors have been detected on placental syncytiotrophoblast and are also involved in fatty acid uptake by the placenta. In the trophoblast much of the FAs is used in membrane biosynthesis, in intracellular signalling systems or as energy sources. Some, however, passes to the fetal circulation. Whether exit to the fetal circulation is by diffusion or involves transportation by binding proteins is unclear, but trophoblast is rich in a range of FA-binding proteins – some of which selectively bind certain FAs, particularly LCPUFAs. Thus, estimates of the proportion of FAs passing transplacentally purely by diffusion vary from 10% to 50%. Studies on isolated, perfused human placentae have revealed the order of selectivity in the placental transfer of free FAs to the fetal circulation to be: docosahexanoic > arachidonic > α-linolenic > linoleic, an order influenced presumably by selective use as well as selective transfer?

Blood comparisons in late pregnancy and at birth reveal that fetal/neonatal levels of LCPUFAs are 2–14 times those in the mother, whereas EFAs are lower. The steady-state concentration of each FA in the fetal circulation is presumably a function of its selective delivery by the placenta, removal from the fetal circulation and, in the case of LCPUFAs, fetal synthesis from their essential fatty acid precursors.

Urea

Fetal urea is excreted across the placenta by diffusion, but is discussed below in connection with its production from amino acids.

Special transport mechanisms

Active transport depends on energy input, in addition to relative maternofetal gradients and blood flow, and many of the key metabolic building blocks for fetal growth use these mechanisms.

Amino acids

In the adult, amino acids are derived directly from digestion of both dietary and endogenous protein, and by interconversion from other amino acids. Deamination of amino acids during catabolism results in release of ammonia, levels of which are kept low by hepatic conversion to **urea**, which constitutes the major source of urinary nitrogen excretion. Traditionally, maternal protein supplements have been considered a desirable feature of pregnancy diets, to cope with the increased protein synthetic demands of the growing conceptus. However, except in cases of extreme malnourishment, there is little or no evidence to support this position, and protein supplements appear largely to be used for conversion by deamination to energy sources. As the fetus clearly does grow and thereby increases the total protein content of the pregnant maternofetal unit, where do the required 'extra' amino acids come from? There is no evidence for improved maternal digestion of dietary protein (which exceeds 95% normally). However, the efficiency of the maternal intermediary metabolism of amino acids appears to increase. Thus, maternal **urea excretion falls markedly in pregnancy**, suggesting that the same intake of dietary amino acids is being utilized more efficiently. A reduced capacity of the maternal liver to deaminate amino acids occurs during pregnancy as a result of progesterone action. Thus, the human placental progesterone regulates maternal metabolism of amino acids such that no extra dietary protein intake is required to support fetal growth. The 'extra' amino acids retained in the mother are transported actively to the fetal circulation, and fetal urea produced by the limited catabolism of fetal amino acids diffuses passively into maternal blood with its already lowered endogenous urea levels.

The fetal blood levels of most amino acids are higher than those in the mother, evidence of their active transport across the placenta. Moreover, experimental infusion studies indicate that transport is driven simply by the concentration of amino acids in maternal blood. **Amino acid transporter proteins** exist in the trophoblast membranes fronting both maternal and fetal circulations. Amino acids are taken up actively from the maternal circulation, diffuse through the trophoblastic cytoplasm and then are transported out into the fetal circulation. Numerous different classes of amino acid transporter, with differing amino acid specificities and transport kinetics and characteristics, have been identified in the two surfaces of the trophoblast, and their patterns of expression change during development in species-specific ways. Quite how this dynamic complexity is regulated is uncertain. There is evidence that at least part of the growth-promoting effects of the IGFs (especially IGF2) is mediated through stimulation of placental amino acid transport. Indeed, when the maternal amino acid supply is limited, the fetus signals this to the placenta by synthesizing more transporter in an attempt to make up for the deficit. Thus, the fetus seems to control the amino acid transporter levels.

Iron

Iron is present in fetal and maternal blood both in a free form and bound to the protein **transferrin**. However, fetal blood contains iron at two to three times the concentration of maternal blood. Since the iron-binding capacities are similar, the higher fetal concentration is due to more unbound iron, accumulated by active placental transport. Trophoblast contains intracellular iron in a **ferritin complex** as part of the transport process. In pregnancy, additional iron is required to provide an average of 300 mg to the fetus, 50 mg to the placenta, 200 mg in blood loss after labour and about 500 mg to increase the maternal haemoglobin mass. **Iron absorption is enhanced** from 10% in the first trimester to 30% or more in the third. A food intake of 12 mg/day and 10% absorption provides an estimated 335 mg, sufficient for the majority of pregnant women, but a supplement in the second half of pregnancy is recommended for some 40% showing maternal serum iron deficiency, although **anaemia** is less frequent.

Calcium

Fetal levels of calcium, phosphorus and magnesium in late gestation are all higher than in maternal blood, reflecting their transport actively across the placenta and their requirement for bone formation. Calcium is transferred to the fetal circulation via the syncytio-trophoblast. There the Ca channel (**transient receptor potential cation channel, subfamily V, member 6: TRPV6**) opens in the maternal-facing basement membrane allowing Ca to enter the cells, where it binds to **calbindin-D9k**, and is transported to the fetal-facing basement membrane, across which it is transported actively by **Ca^{2+}-ATPase**. During the last few days of gestation, expression of all three transporter elements increases, and knock-down or inhibition of the first two in rodents results in a reduced transplacental Ca transport and in severe fetal hypocalcemia. Little is known about the transfer of phosphorous, although the **Na^+/Pi cotransporters, Npt2b, and Npt3** have been detected in mouse and human syncytio-trophoblast, increasing towards term, and are plausible transporter candidates. Even less is known about the mechanism of placental Mg transport.

Fetal **ossification** requires some 21 g of calcium, 80% of it accumulated in the last trimester at a daily rate of up to 150 mg/kg of fetus, placing a considerable demand for calcium on the mother. However, the average daily maternal intake of calcium greatly exceeds adequacy and during pregnancy maternal dietary calcium absorption increases by 30–35%. The absorption efficiency increases further during lactation when higher levels of **maternal parathormone** (**PTH**) stimulate renal conversion of **vitamin D** to the active derivative **1,25-dihydroxyvitamin D**. In addition, maternal mobilization of bone calcium increases, bone mineral density reducing by 4–8%. Overall, there is a positive calcium balance throughout pregnancy and lack of vitamin D is a more likely cause of **maternal osteomalacia** than is dietary deficiency of calcium. Maternal PTH, calcitonin and 1,25-dihydroxyvitamin D do not cross the placenta, but 25-hydroxyvitamin D does, where it is hydroxylated to its active form (see Chapter 17).

Vitamins

Folic acid and vitamin B_{12} are two essential dietary compounds with fundamental metabolic actions vital for normal fetal development. Folic acid and its coenzyme forms are involved in 1-carbon transfers, and thereby in nucleoprotein synthesis and amino acid metabolism. Vitamin B_{12} is a co-factor in folate metabolism and in the metabolism of some fatty acids and branched amino acids. Both are provided at the expense of maternal stores, making fetal deficiency unlikely. However, vitamin deficiencies in the mother may affect the fetus indirectly via the resultant maternal metabolic disorders.

Red cell folate activity accounts for 95% of the maternal blood folate, levels of which decrease progressively towards term. The incidence of **folate deficiency** is around 2% and that of **megaloblastic anaemia** is much lower. Folate blood levels in megaloblastic anaemias of pregnant women are generally less than in non-pregnant women. The precise consequence of folate deficiency for the fetus is uncertain, but appears to be associated with prematurity and abortion. Folic acid supplements are generally considered to be desirable in pregnancy as a means of preventing the development of maternal anaemia.

Vitamin B_{12} is absorbed relatively slowly across the mucosa of the terminal ileum to be transported in blood mainly by **transcobalamin II**. Maternal serum levels of vitamin B_{12} decrease during pregnancy, falling to a minimum at 16–20 weeks, but not low enough to be a true deficiency, which is rare in pregnancy. Indeed, vitamin B_{12}-deficient women are unlikely to become pregnant in the first place.

Bilirubin

Bilirubin is a lipid-soluble product of haemoglobin catabolism and is present in fetal, neonatal and adult plasma in both a free form and bound to serum albumin. In the mother, bilirubin passes in the blood to the liver where the enzyme **uridine diphosphate (UDP) glucuronyl-transferase** converts it to **bilirubin glucuronide**. This polar conjugate cannot cross the placenta but is excreted in the maternal bile (Figure 15.6). The fetal liver lacks the conjugating enzyme until late in pregnancy in preparation for neonatal bilirubin excretion. So, the non-conjugated bilirubin diffuses readily across the placenta down a concentration gradient, the mother's hepatic conjugation activity preventing its return to the fetus (Figure 15.6).

Bilirubin transport and metabolism are important clinically. **Hyperbilirubinaemia** is common in the newborn and in its mild form (so-called '**physiological hyperbilirubinaemia**') is known as **jaundice** because of the associated yellow coloration of the skin and mucous membranes. Its occurrence can be associated with accelerated red blood cell breakdown where an infant has a large blood volume, as can arise if the umbilical cord is clamped too late, when some 30% of babies may show jaundice. Alternatively, the effectiveness of bilirubin conjugation by the neonate may be depressed as a result of dehydration, low caloric intake or even the actions of steroid hormones, particularly progestagens, which actually inhibit conjugation. Severe or **pathological hyperbilirubinaemia** may result in **encephalopathy**, also known as **kernicterus**. It may arise as a result of an increased fetal red cell breakdown, such as occurs if the mother develops an immune response to fetal erythrocytes (see Chapter 16) or impairment of maternal bilirubin conjugation. Damage to the fetal liver as a result of injection of drugs, maternal diabetes, congenital defects and prematurity are also associated with poor conjugation of bilirubin. The hyperbilirubinaemia (or its consequences) may be prevented by a number of approaches: exposure to ultraviolet light (**phototherapy**) causes the breakdown of bilirubin to a non-toxic product; early feeding prevents hypoglycaemia and dehydration; early clamping of the umbilical cord decreases red cell volume and hence bilirubin plasma levels. If the above fail, or in severe cases, exchange transfusion is used.

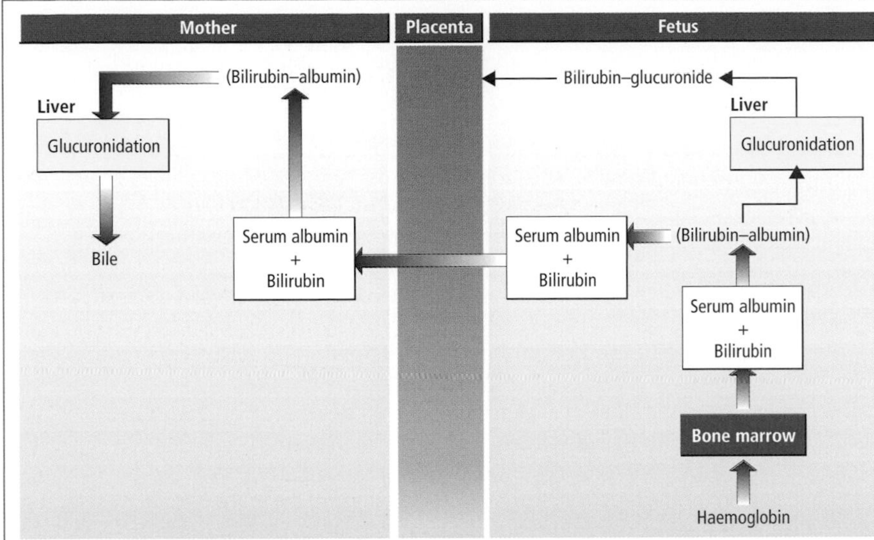

Figure 15.6 The metabolism of bilirubin. The red arrows represent the principal route of metabolism during pregnancy. Fetal glucuronidation does not occur significantly until after birth. (Source: Biggers J.D. (1979) *Medical Physiology* (ed. V. Mountcastle). Mosby, St Louis. Reproduced with permission from Elsevier.)

Macromolecules

Macromolecular transport across the human placenta is limited, very low rates of diffusion of dextran, albumin, horseradish peroxidase and heparin being observed using an *in vitro* perfusion system. The only **exceptional macromolecule is immunoglobulin G**. This molecule is transferred by an Fc-receptor-mediated transcellular mechanism involving syncytial trophoblast endocytotic transfer. It is through this route that the mother provides the fetus with **passive immunity**, to be supplemented postnatally by further antibody during suckling (see page 326). A smooth linear rise in fetal IgG is evident from as early as 13 weeks' gestation to term. In rats, mice and rabbits, the yolk sac placenta performs this function, while in farm animals, a similar protection is afforded solely by antibodies transmitted postnatally in the milk. Whilst clearly of adaptive value for neonatal survival, this transfer of antibody can also pose a threat to the fetus, as is discussed in Chapter 16 (see pages 375–6).

Conclusions

At the start of this chapter, we saw how the control of fetal growth was critical for a successful birth outcome (Figure 15.1). However, growth and maturation are both intrinsically complex processes in which much can and does go wrong. In addition, the embryo/fetus has to reside inside the mother for approximately 9 months. Although nurturant, that mother is a distinct individual with her own needs and problems, and her own legal and moral status. Conflicts between the developing embryo/fetus and mother can and do arise: biological, medical, ethical and legal conflicts. In Chapter 16, some of these various intrinsic problems and conflicts are examined.

Key learning points

- The rate of fetal growth is relatively slow up to week 20 of gestation, but peaks in weeks 30–36, then to decline thereafter until birth.
- Fetal growth and fetal and postnatal well-being are affected by fetal genetic make-up, insulin-like growth factors, maternal nutrition and maternal health and well-being.
- The transfer of nutrients, blood gases and other essential chemicals across the placenta from mother to fetus is influenced by the thickness and area of the placental interface, maternal and fetal blood flows, fetal and maternal concentrations of substances to be transported, and the types of transport mechanisms available.
- Maternal Po_2 and placental perfusion are not normally the rate-limiting factors affecting transfer of O_2 across the placenta; instead, the rate of transfer of O_2 across the placenta varies simply with the Po_2 of fetal blood.
- Fetal forms of globin allow effective oxygen trapping by the fetal blood.
- The fetus has little capacity for gluconeogenesis and must obtain glucose from maternal blood, the glucose levels of which depend on maternal nutritional status and endocrine control mechanisms.
- The rate at which glucose is utilized by fetal tissues is determined mainly by the actions of fetal insulin secreted by the pancreas in response to fetal glucose load.
- Storage of glucose as glycogen in the fetal liver is important to meet the metabolic needs of the neonate until feeding begins.

- Brown fat is critically important for thermogenesis in the newborn. It regresses during childhood.
- The largest fetal requirement for essential fatty acids, and therefore for the supply of long-chain polyunsaturated fatty acids, occurs during the third trimester of pregnancy, and is met by both placental transfer and endogenous synthesis.
- Fetal deposition of long-chain polyunsaturated fatty acids is rapid during this period of maximum brain growth, and a failure to accomplish a specific component of neural growth owing to inadequacy of critical membrane lipids may result in irreversible brain damage.
- Fetal amino acids and urea are derived directly from more efficient maternal digestion of dietary and endogenous protein; non-essential amino acids are derived by interconversion from other amino acids.
- Exchange of water between mother and fetus occurs at two main sites: the placenta and the remaining non-placental chorion where it abuts the amnion internally.
- There is a large traffic of Na^+ and other electrolytes across the placenta in both directions.
- Significant quantities of water, Na^+ and other electrolytes may also cross the amniotic membranes.
- Fetal blood contains iron at two to three times the concentration of maternal blood, as a result of an increased concentration of unbound iron, which accumulates through active transport across the placenta.
- Maternal iron deficiency and anaemia can occur, when additional iron intake in pregnancy is required.
- The fetus places a considerable demand for calcium on the mother, largely during ossification in the last trimester; calcium is transferred to the fetal circulation against a concentration gradient by a saturable active transport mechanism.
- Folic acid and vitamin B_{12} are vital for normal fetal development, and are provided for the fetus at the expense of maternal stores.
- Diffusion of bilirubin from fetus to mother, and maternal hepatic conjugation activities preventing its return to the fetus, ensure adequate elimination of fetal bilirubin. Hyperbilirubinaemia is common in the newborn and in its mild form is known as jaundice.
- The volume of amniotic fluid increases during pregnancy from about 15 ml at 8 weeks after conception to 450 ml at week 20, after which time net production declines to reach zero by week 34.
- Amniotic fluid is in a dynamic state, complete exchange of its water component occurring about every 3 hours.

Clinical vignette

Gestational diabetes

A 30-year-old woman of Egyptian origin, Ms J, presented for antenatal care in her first pregnancy. Her pre-pregnancy BMI was 32. Her ethnic origin and her obesity were both risk factors for developing gestational diabetes and hence a glucose tolerance test (GTT) was performed at 28 weeks. The GTT demonstrated that she had developed gestational diabetes. Ms J was referred to a specialist antenatal clinic for diabetes in pregnancy, where close monitoring of the pregnancy by obstetricians, endocrinologists, and dieticians was initiated. An initial trial of managing her diabetes with dietary and exercise modifications was commenced, but her blood sugars still remained high. An ultrasound scan demonstrated that Ms J's baby's abdominal circumference was measuring above the 95th centile for gestational age. Insulin therapy was commenced, and this was successful in controlling her blood sugars to levels within the normal range. An induction of labour was undertaken at 38 weeks, and Ms J had a vaginal delivery of a baby boy weighing 3.6 kg. After delivery, the baby required monitoring of his blood sugars, but was discharged well at day 3 of life. Ms J was informed that she should have a further glucose tolerance test 6 weeks after delivery, and annually thereafter.

This case illustrates the importance of stringent application of screening tests for gestational diabetes. All pregnant women experience a lowering of their baseline glucose tolerance during pregnancy. The lowered glucose tolerance is driven primarily by placental hormones (human placental lactogen, growth hormone), which act to antagonize the effects of endogenous insulin. This is a helpful adaptation to pregnancy because it allows glucose to remain available in the circulation for longer and hence become available to the fetus. It also promotes gluconeogenesis and glycogenolysis to ensure a ready availability of glucose for fetal growth. However, in women whose glucose tolerance is already low, this adaptive mechanism may become pathological. If blood glucose levels remain consistently high during the pregnancy, then fetal macrosomia (excessive fat deposition) and placental insufficiency (driven by glycosylation) become risks. These lead to higher incidences of birth trauma and of stillbirth. For this reason, delivery is usually recommended for women with gestational diabetes at around 38–40 weeks of gestation.

FURTHER READING

General reading

Desforges M, Sibley CP (2010) Placental nutrient supply and fetal growth. *International Journal of Developmental Biology* **54**, 377–390.

Duttaroy AK (2009) Transport of fatty acids across the human placenta: a review. *Progress in Lipid Research* **48**, 52–61.

Fowden AL (2003) The insulin-like growth factors and fetoplacental growth. *Placenta* **24**, 803–812.

Fowden AL, Giussani DA, Forhead AJ *et al.* (2006) Intrauterine programming of physiological systems: causes and consequences. *Physiology* **21**, 29–37.

Hay WW Jr (2006) Recent observations on the regulation of fetal metabolism by glucose. *Journal of Physiology* **572**, 17–24.

Jansson T (2001) Amino acid transporters in the human placenta. *Pediatric Research* **49**, 141–147.

Kane SV, Acquah LA (2009) Placental transport of immunoglobulins: a clinical review for gastroenterologists who prescribe therapeutic monoclonal antibodies to women during conception and pregnancy. *American Journal of Gastroenterology* **104**, 228–233.

Kudo Y, Boyd CA (2002) Human placental amino acid transporter genes: expression and function. *Reproduction* **124**, 593–600.

Murphy VE, Smith R, Giles WB, Clifton VL (2006) Endocrine regulation of human fetal growth: the role of the mother, placenta, and fetus. *Endocrine Reviews* **27**, 141–169.

Ohata Y, Ozono K, Michigami T (2016) Current concepts in perinatal mineral metabolism. *Clinical Pediatric Endocrinology* **25**, 9–17.

RCOG (2010) Nutrition in pregnancy. *Scientific Advisory Committee Opinion Paper* **18**, 1–7.

RCOG (2011) Diagnosis and treatment of gestational diabetes. *Scientific Advisory Committee Opinion Paper* **23**, 1–5.

Regnault TR, Friedman JE, Wilkening RB *et al.* (2005) Fetoplacental transport and utilization of amino acids in IUGR: a review. *Placenta* **26** (Suppl. A), S52–S62.

Sack J (2003) Thyroid function in pregnancy – maternal–fetal relationship in health and disease. *Pediatric Endocrinological Reviews* **1** (Suppl. 2), 170–176.

More advanced reading

Bass JK, Chan GM (2006) Calcium nutrition and metabolism during infancy. *Nutrition* **22**, 1057–1066.

Bloomfield FH, Oliver MH, Hawkins P *et al.* (2003) A periconceptional nutritional origin for noninfectious preterm birth. *Science* **300**, 606.

Gluckman PD (1986) The role of the pituitary hormones, growth factors and insulin in the regulation of fetal growth. *Oxford Reviews of Reproductive Biology* **8**, 1–60.

Gootwine E (2004) Placental hormones and fetal–placental development. *Animal Reproduction Science* **82–83**, 551–566.

Haggarty P, Page K, Abramovich DR *et al.* (1997) Long-chain polyunsaturated fatty acid transport across the perfused human placenta. *Placenta* **18**, 635–642.

Henson MC, Castracane VD (2006) Leptin in pregnancy: an update. *Biology of Reproduction* **74**, 218–229.

Herrera E, Munilla MA (1997) Maternal lipid metabolism and its implications for fetal growth. In: *Placental Function and Fetal Nutrition*, Nestlé Nutrition Workshop Series, Vol. **39** (ed. FC Battaglia), pp. 169–182. Nestlé Ltd, Vevy/Lippincott-Raven, Philadelphia.

Jumpsen J *et al.* (1997) Fetal lipid requirements: implications in fetal growth retardation. In: *Placental Function and Fetal Nutrition*, Nestlé Nutrition Workshop Series, Vol. **39** (ed. FC Battaglia), pp. 157–167. Nestlé Ltd, Vevy/Lippincott-Raven, Philadelphia.

Prentice AM, Spaaij CJ, Goldberg GR *et al.* (1996) Energy requirements of pregnant and lactating women. *European Journal of Clinical Nutrition* **50** (Suppl. 1), S82–S111.

Rodien P, Jordan N, Lefèvre A *et al.* (2004) Abnormal stimulation of the thyrotrophin receptor during gestation. *Human Reproduction Update* **10**, 95–105.

Tobias JH, Cooper C (2004) PTH/PTHrP activity and the programming of skeletal development in utero. *Journal of Bone and Mineral Research* **19**, 177–182.

CHAPTER 16
Fetal challenges

Essential Reproduction, Eighth Edition. Martin H. Johnson.

Companion website: www.essentialreproduction.com

Human pregnancy is associated with an unusually high rate of complications presenting serious risks for the mother and her fetus. Given the extraordinary complexity of embryonic and fetal development, it is perhaps not surprising that errors and problems occur. Some result in loss of the pregnancy, others in problems during pregnancy or neonatally, and yet others do not manifest until later in postnatal life. Some of these problems arise from errors intrinsic to the developing embryo/fetus, others arise from conflicts between its own developmental demands and the needs of the mother in whom it grows, and others arise from the programming by an adverse intrauterine environment of an increased susceptibility to stresses in later life. This chapter explores some of these various intrinsic problems and conflicts.

Pregnancy loss

Pregnancy is precarious, losses occurring at all stages and especially up to the end of the first trimester. We saw that the earliest sign that implantation is likely to have occurred comes from the detection in the blood, and later in the urine, of human chorionic gonadotropin (hCG) during the period 18–30 days after the initiation of the last menstrual flow (see page 221). This observation leads to the diagnosis of a **biochemical pregnancy**, although some tumours can also produce hCG. Definitive evidence of a **clinical pregnancy** is obtained by ultrasonographic investigation from as early as 5 weeks, at which time the presence and number of gestational sacs can be assessed (Figure 16.1). Using an ultrasound probe in the vagina (or trans-abdominally 7–10 days later), the fetal heartbeat should be detectable by 7 weeks after the last menstrual period – indeed, the **8-week ultrasound scan** has become an almost routine event in modern medicine. However, fertilization and the early development of the conceptus do not lead inevitably to a sustained pregnancy. Loss of the human conceptus is common and can occur at any stage (see Box 16.1), either because of its inherent deficiencies or as the result of an environmental insult or inadequate maternal support. Overall, the cumulative outcome of such losses makes it possible that over 60% of human conceptions do not survive to birth. Why are they lost?

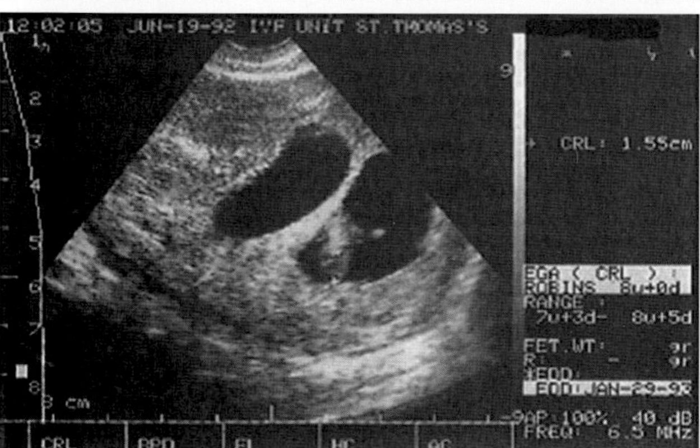

Figure 16.1 Transvaginal ultrasound scan of a twin gestation in a woman 8 weeks' pregnant. Both sacs can be seen on this longitudinal section, although only one embryo (in the lower right-hand sac) is visible in this plane. The gestational age of the embryos can be determined from measurement of the crown–rump length (between the white crosses). (Source: Courtesy of Professor Peter Braude.)

Box 16.1 Estimated losses during pregnancy

One-third of cycles in which frequent, unprotected intercourse over the fertile period occurs nonetheless fail to yield a pregnancy. Some of this failure can be accounted for by very early loss of the conceptus rather than failure of fertilization. Thus, when human conceptuses were recovered by **uterine lavage** (flushing fluid through the uterus) 4.5 days after the detection of ovulation and the insemination of spermatozoa, only 20% were blastocysts, the rest being retarded or abnormal. Moreover, of those conceptions surviving to the blastocyst stage and signalling their presence by the production of hCG, between 8% and 25% may fail, as this proportion of menstrual cycles is characterized by detectable but transient levels of hCG during the latter part of the (often slightly prolonged) luteal phase. This hCG is assumed to derive from lost peri-implantation conceptuses. Such a loss could arise from (1) production of abnormal conceptuses that develop to the blastocyst stage but then fail; (2) failure of the uterine–conceptus interaction at implantation; or (3) failure of the corpus luteum to respond adequately to the hCG stimulation. There is some evidence to suggest that the hCG rise is delayed slightly in those cycles destined to fail, which suggests that an embryonic deficiency in hCG production may be responsible. Of pregnancies that survive for longer and are detected clinically, some 10–15% miscarry subsequently, the vast majority during the first trimester, some unfortunate couples experiencing recurrent miscarriage (see Box 16.2). Overall, the cumulative outcome of such losses makes it possible that well over 60% of human conceptions do not survive to birth.

Causes of miscarriage

Miscarriages can occur in any pregnancy, the two main independent risk factors being maternal age and previous miscarriage. Indeed, some women experience several miscarriages serially. **Recurrent miscarriage** is defined as the loss of three or more consecutive pregnancies. In **primary recurrent miscarriage**, the first three pregnancies are affected, whereas **secondary recurrent miscarriage** follows one or more live births. In the UK, about 1% of pregnant women experience the condition. Some of the factors associated with recurrent miscarriage are summarized in Box 16.2.

For the more common non-recurrent pregnancy loss, a range of underlying causes has been identified, but most, especially the majority occurring in the first trimester, result from placental disorders. Thus, some two-thirds of miscarriages are associated with reduced extravillous trophoblast invasion into the placental bed. As we saw in Chapter 14 (see pages 244–8), during normal pregnancy this invasion plays a critical role in reducing the blood flow to the developing embryo during the first trimester, ensuring that embryonic development occurs in a low oxygen environment, and thereby reducing the risk of damage by free oxygen radicals. In cases of early miscarriage, incomplete plugging of the lumen at the tips of the spiral arteries occurs and their conversion is likewise incomplete (Figure 16.2). These defects lead to premature onset of the maternal circulation throughout the entire placenta, leading to direct mechanical effects due to the relatively unconstrained blood flow and, indirectly, to widespread O_2-mediated trophoblastic damage, including lipid peroxidation and apoptosis.

Box 16.2 Recurrent miscarriage

Although a relatively small proportion of pregnancies, recurrent miscarriages are amongst the most distressing for the couples involved, and for their families and carers.

- In 3–5% of such couples, one partner carries a **balanced chromosomal reciprocal (Robertsonian) translocation**, which gives a 40–50% chance of an affected pregnancy. Preimplantation genetic diagnosis is a high-tech, high-cost option (see page 377).
- **Uterine anatomical anomalies** are associated with 2–38% of recurrent miscarriages, the range reflecting different study thresholds for inclusion of anomalies. The miscarriages tend to be late rather than early.
- **Antibodies to phospholipids (αPLs).** Up to 20 antibodies directed against phospholipid-binding proteins can be detected, of which two are associated with recurrent miscarriage: anti-cardiolipin and lupus anticoagulant. These are associated with early (pre-10 week) recurrent miscarriage, as well as later loss of normal fetuses and severe pre-eclampsia. αPLs inhibit trophoblast function and differentiation and are also associated with thrombosis of the utero-placental vasculature later in pregnancy. Combined treatment with aspirin and low-dose heparin improves outcome to 70% live births, but also increases the risk of pregnancy complications and so needs careful monitoring.
- **Endocrine anomalies** do not seem to be a major cause of recurrent miscarriage, and there is no evidence that luteal progesterone or hCG therapy improves outcomes, other than following IVF in which induced ovulation can benefit from progesterone support.
- **Immunological incompatibility.** Haemolytic diseases of the fetus and neonate account for a diminishing proportion of fetal loss in Europe. Immunological incompatibility of the mother and father, and thereby of the fetus, at either the ABO or rhesus blood group loci poses the major problem (see page 275–6).
- **Unexplained recurrent miscarriage**. Where none of the above associated factors is present (c. 50% of total), supportive care can result in live births, but the older the mother and the more prior miscarriages, the lower the likelihood of this outcome. A number of studies has recently been published claiming an association between various genetic isotypes and recurrent miscarriage, see Table 23.1, but as yet none has been usefully translated to the clinic.

Further reading

Porter TF, Scott JR (2005) Evidence-based care of recurrent miscarriage: best practice and research. *Clinical Obstetrics and Gynaecology* **19**, 85–101.

Regan L, Braude PR, Trembath PL (1989) Influence of past reproductive performance on risk of spontaneous abortion. *British Medical Journal* **299**, 541–545.

Royal College of Obstetricians and Gynaecologists Guideline 17 (2003) *The Investigation and Treatment of Couples with Recurrent Miscarriage*. RCOG Press, London.

Stirrat GM (1990) Recurrent miscarriage. *Lancet* **336**, 673–675.

van der Linden M, Buckingham K, Farquhar C *et al.* (2011) Luteal phase support for assisted reproduction cycles. *Cochrane Database of Systematic Reviews* 2011, (10), CD009154.

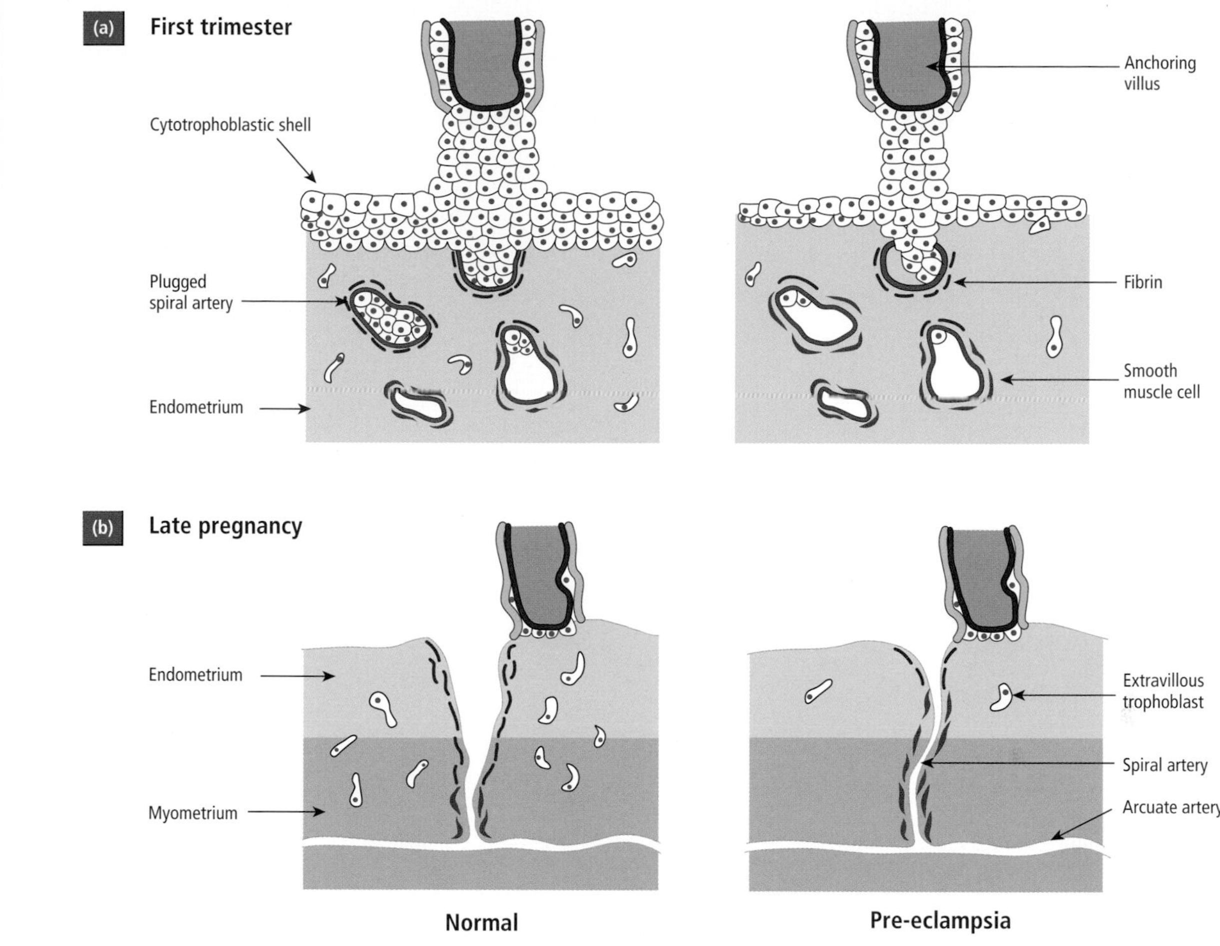

Figure 16.2 The utero-placental interface (a) in the first trimester and (b) later in pregnancy to show the reduced cytotrophoblastic plugging and incomplete transformation of the spiral arteries in pregnancies complicated by pre-eclampsia.

Detachment of the placenta from the uterine wall follows. These recent findings have defined a mechanism, but beg the question: why is extravillous trophoblast invasion defective in these cases?

Chromosomal anomalies characterize at least 50% of early miscarriages, and seem to underlie many of the trophoblastic defects. The nature and origins of genetic abnormalities in the embryo/fetus as causes of embryo loss were discussed in Chapter 12 (see pages 207–8). Genetic defects in oxygen metabolism have also been associated with pregnancy loss, as has smoking, itself a generator of oxygen-free radicals. **Anatomical problems**, such as cervical incompetence or implantation in eccentric uterine positions or ectopically, account for other losses. The potential role of immune problems in pregnancy loss is discussed in more detail below (see pages 275–9).

IVF and fetal morbidity

The advent of **in-vitro fertilization** (IVF) has brought with it concerns for the well-being of the offspring. A number of epidemiological studies has not indicated any marked increase in pregnancy loss, other than would be expected in older women or in women carrying more than one fetus. Several surveys have suggested a small increase in the incidence of congenital or neonatal disorders. A common assumption is that these disorders arise from some aspect of the IVF procedure itself. However, whilst difficult to exclude this possibility entirely, it is probable that most if not all of the increase is due to the patient profile. Thus, those undergoing IVF treatment are likely to be subfertile through age or other factors (see Chapter 22). Indeed, the vast majority of anatomical congenital disorders observed after IVF involves the reproductive tract. Moreover, the increased incidence is so small that concerns about the risk posed by them are outweighed for most patients by the benefits of having their own child.

Maternal death and morbidity

Worldwide, pregnancy and childbirth remain major causes of female mortality (Figure 16.3). The fifth of the Millennium Development Goals recognizes this fact by identifying the improvement in maternal health by reducing maternal mortality by three-quarters between 1990 and 2015 as an aim,

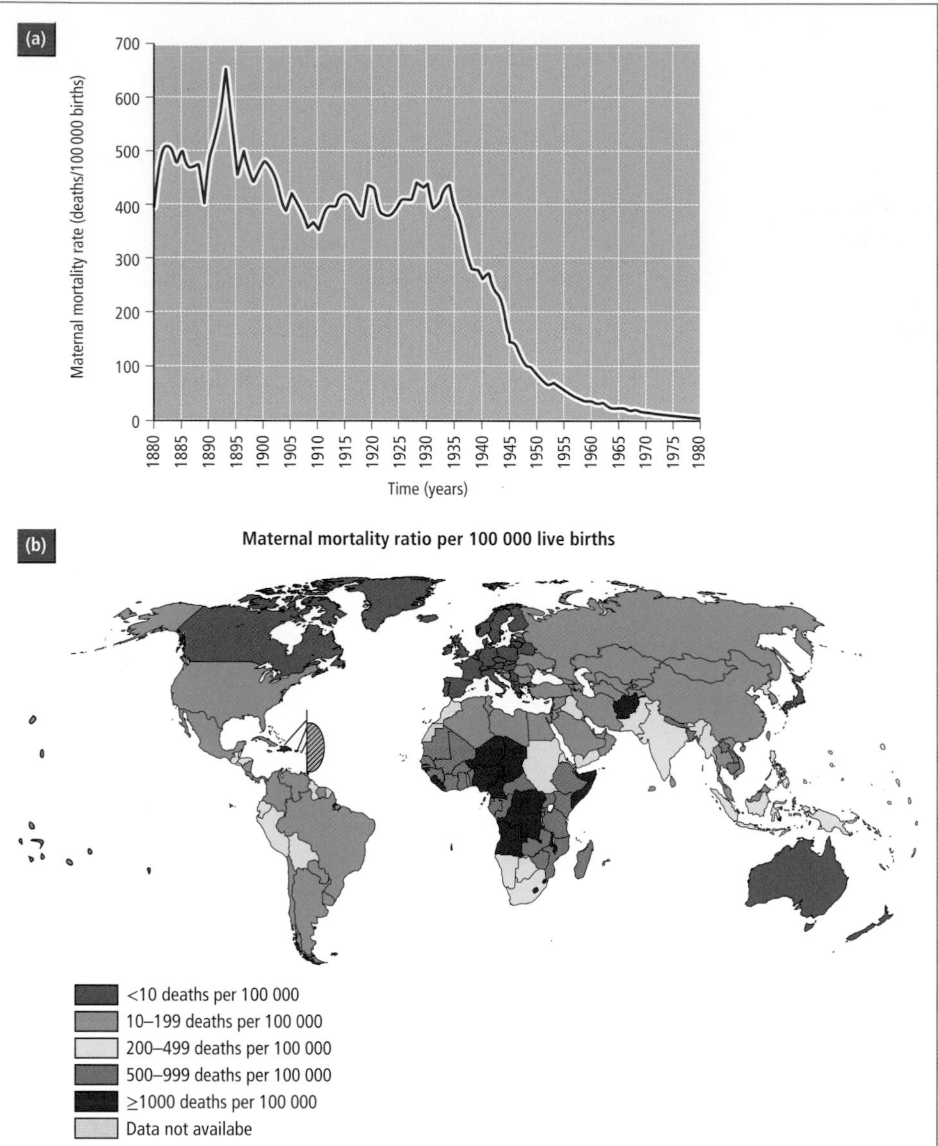

Figure 16.3 Annual maternal mortality expressed as deaths per 100 000 pregnancies (a) historically in the UK from 1880 to 1980 and (b) worldwide for the year 2000. Note the decline in mortality in the UK from around 500 to negligible from 1940 onwards. Note also that rates of 500 are still widespread internationally, notwithstanding the aim to reduce them.

one on which little progress has been made. In part, maternal deaths occur as the result of increased pressure on already stressed physiological systems. For example, pregnancy itself raises the relative risk of acute myocardial infarction three- to four-fold, the risk increasing by 30-fold for women over the age of 40 years compared with those under 20 years. Although still a small absolute risk – in the UK, maternal death from ischaemic heart disease was only 1 in 132 000 pregnancies – the data on relative risk illustrate the point. Similarly, the incidence of frank diabetes rises during pregnancy – for reasons discussed in Chapter 15 (see page 264), as does the possible exacerbation of existing health conditions such as HIV infection or autoimmune disease.

Of conditions specific to human pregnancy, a major cause of morbidity and death is **pre-eclampsia** (also called **toxaemia** or **gestosis**). This condition, which affects about 4% of all human pregnancies and more rarely those of chimpanzees and gorillas, is characteristically a disease of first pregnancies, and usually emerges in the last trimester. In pre-eclamptic women, diastolic blood pressure is periodically or continuously elevated, and oedema and in more severe cases albuminuria occur. Severe pre-eclampsia can lead to the convulsive state of **eclampsia.** Throughout, both mother and fetus are at risk, but with the improved survival and care of prematurely delivered neonates, fewer deaths from eclampsia are occurring in developed countries. Pre-eclampsia is more common in some ethnic groups such as Afro-Caribbeans and African Americans, and where parents differ ethnically. These observations suggest that whether or not pre-eclampsia develops seems to be affected by a genetically-based interaction between the female and male partners.

The aetiology of pre-eclampsia is unclear, but it seems to represent a less extreme form of the failure of extravillous trophoblast invasion seen in most early miscarriages. Thus, although invasion is sufficient to anchor the conceptus, it does not fully convert the spiral arteries into low-resistance channels. This failure means that some smooth muscle is retained within their walls and with it some vasoreactivity in at least parts of the placental vascular bed. In consequence, intermittent perfusion of the intervillous space occurs, leading to repeat transient bursts of oxygenation and hypoxia, which are known to generate free radicals. Maternal anti-oxidants such as ascorbic acid fall, and so lead to a vicious positive feedback circle in which the lipid peroxides formed then mediate maternal endothelial dysfunction, leucocyte activation, platelet adhesion and aggregation, and the release of vasoconstrictive agents, leading over time to the elevation of blood pressure. Thus, in pre-eclampsia placental damage is progressive but for a while can be compensated for by the intrinsic placental anti-oxidant capacity. Indeed, a clinical trial in women at high risk of pre-eclampsia showed that anti-oxidant vitamin C and E supplementation during the second trimester reduced oxidative stress and decreased the occurrence of clinical disease. Pre-eclampsia also seems to be more common when the mother and her fetus express certain combinations of one of the major histocompatibility antigens (HLA) loci: which brings us to a consideration of one of the most enduring puzzles of viviparity.

Fetomaternal immune relations

On the first page of this book, we discussed the particular value of sexual reproduction in generating genetic diversity. The whole edifice of sexual differentiation, reproductive cyclicity and pregnancy, with their social ramifications, is constructed on the basis of the biological advantages conferred by producing a genetically distinct individual. Yet we now come full circle, for the genetically distinct individual will also be phenotypically unique. A component of this unique phenotype is the array of cell-surface glycoproteins that constitute the system of histocompatibility antigens. Thus, the biological advantages of genetic variation appear to confront and conflict with those of viviparity. The problem may be illustrated dramatically. If the skin of a newborn child is grafted to its mother, she rejects it. Why then did she not reject the whole fetus, 'grafted' as it was into the maternal uterus? Indeed, the pregnant mother is competent to respond immunologically to the fetus and examination of maternal blood in late pregnancy indicates that a regular feature of pregnancy is an immunological reaction against the conceptus. Even more striking, the active pre-sensitization of the mother, by injecting or grafting paternal tissues, does not necessarily prejudice the establishment and maintenance of subsequent pregnancies by the same father. Thus, neither a generalized nor a specific depression of maternal immune responsiveness can adequately explain fetal survival. The survival of the fetus *in utero* has for many years fascinated and puzzled biomedical scientists. Recently, the story has become both clearer and more puzzling! Several types of explanation have been advanced.

The fetus does express antigens

The developing fetus does not lack target antigens. The **major histocompatibility antigens (MHCs) appear on embryonic cell surfaces** shortly after implantation and, although present in smaller amounts than in the adult, are detectable throughout pregnancy. Thus, grafting of fetal tissues to another individual (or to the mother) results in their rejection. They can elicit immune responses and be destroyed by them. However, the fetus is not normally in direct contact with the mother, but is separated from her by the placenta – admittedly part of which originates from the conceptus. How effective is this separation and could it explain the fetal survival?

The placenta intervenes

The fetus and its circulating blood are separated from the mother by the investments of fetal membranes (see Figure 14.2). The discrete nature of the fetal and maternal circulations prevents appreciable passage of maternal cells to the fetus. In many species, including the pig, sheep, cow and horse, antibodies are also excluded. However, in humans and monkeys towards the end of pregnancy, immunoglobulin G (IgG) antibodies normally pass across the chorioallantoic placenta into the fetal circulation via a special transport mechanism (see page 267). The maternal antibodies will include some directed against prevalent bacteria and viruses and will thus confer passive immunity on the fetus and afford the neonate temporary protection for a few weeks postnatally. However, in addition to antibodies reactive to bacteria, IgG antibodies directed against fetal antigens presumably could also be transferred. That such transfer does indeed occur is seen in cases of sensitization to the major blood group antigen called rhesus. If a woman lacks the **rhesus antigen** (rhesus negative) and carries a fetus possessing it (rhesus positive), she may mount an IgG immune response to the antigen, particularly at parturition when extensive fetal bleeding into the mother may occur. In subsequent pregnancies, the IgG antibody is transferred across the placenta and destroys the fetal erythrocytes in a rhesus-positive fetus – **haemolytic disease of the fetus/newborn**. Here, then, is a clear example of the mother rejecting her fetus immunologically. But the rhesus antigen is only one of many antigens by which mother and fetus may differ. Can we gain any clues from the rhesus example as to why fetuses are not normally rejected?

The rhesus antigen differs in two ways from most of the other cellular antigens expressed on fetal cells. First, it is present only on red blood cells. Most of the other important histocompatibility and blood group antigens are also present on several other types of fetal cell and are thus widely distributed among the tissues of the fetus. Second, the rhesus antigen exists only as a structural component of the cell membrane. Other antigens, such as ABO blood group and major HLAs, seem to be present not only as structural membrane components but also in solution in the fluids of the fetus, such as the blood and amniotic fluid. Thus, if an antibody, or indeed the odd lymphocyte, directed against these other antigens should enter the fetal circulation, it could first be '**mopped up**' harmlessly by free soluble antigen. Any remaining antibody will be distributed among a wide range of cell types and so will effectively be '**diluted out**'. Any given single cell will be unlikely to bind a large number of antibody molecules and, as only the binding of a large number of antibodies will debilitate the cell, gross tissue damage is avoided. In the case of the rhesus antigen, 'mopping up' and 'diluting out' cannot occur and so the chance of tissue damage increases considerably with obvious pathological consequences.

So the placenta seems to provide at least a **filter function**, if not a barrier function, between maternal immune responsiveness and fetal antigenicity. However, the membranes separating the two organisms are themselves part of the conceptus, and therefore genetically alien to the mother, so why is the placenta not rejected?

The trophoblast does express some fetal antigens

The outermost layer of the placenta is usually the chorionic syncytial trophoblast, which is in intimate contact with the mother; in haemochorial placentae remarkably so, as it is bathed in maternal blood (see Figure 13.7). In addition, the extravillous trophoblast breaks away to invade maternal spiral arteries during the first trimester. Fragments of syncytiotrophoblast or this invading trophoblast can become detached and be carried into the maternal circulation where it can be detected for several months. Both cytotoxic T cells and antibody responses are generated by the mother's systemic immune system against these shed fetal antigens. So, why does this strong immune rejection response not endanger the fetus? Is the chorionic trophoblast able to resist maternal rejection and, together with its filter function, thereby protect the fetus?

Studies on the **antigenicity of the syncytiotrophoblast** have revealed convincing evidence that its **surface lacks any HLA antigens**, the main ligands for immune receptors on the cells of the immune system. This means syncytiotrophoblast cannot be directly recognized by maternal T cells. Likewise, the **extravillous trophoblast** cells that invade the uterus and especially the termini of the spiral arteries also lack **conventional HLA class I and class II antigens**. However, in contrast to the syncytiotrophoblast, the extravillous trophoblast cells express an unusual array of HLA class I antigens: **HLA-C, -G and -E**. Of these, HLA-G is unique to the trophoblast, not apparently being expressed by any other fetal or maternal cells, and HLA-E and HLA-G are effectively monomorphic so will not vary from pregnancy to pregnancy. In contrast, HLA-C is highly polymorphic and both paternal and maternal allotypes of this class I MHC antigen are expressed by extravillous trophoblast cells. What might be the outcome when maternal immune cells encounter these paternal alloantigens on the trophoblast invading into the decidua?

Role of T cells and allorecognition in pregnancy

T cells comprise 10–20% of decidual leucocytes during the first trimester. CD8+ and CD4+ T cells recognize antigenic peptides presented by class I and class II HLA molecules respectively. The unusual HLA profile of trophoblast means only HLA-C restricted T cells can recognize trophoblast directly. However in HLA-C mismatched pregnancies (where the paternal HLA-C differs from that of the mother), there is an increase in the CD4+ regulatory T cells (Tregs) in the decidua. This was not seen in HLA-C matched pregnancies. The Tregs are a CD4+ subset that normally act to suppress immune responses against self antigens. Decidual Tregs may therefore contribute to a tolerogenic environment in the decidua. Since trophoblast does not express class II HLA, these CD4+ Tregs presumably recognize fetal HLA-C indirectly by presentation of fetal HLA-C peptides by class II HLA on maternal antigen-presenting cells in the decidua. There is no evidence for CD8+ T cells that directly recognize paternal alloantigens presented by HLA-C on trophoblast. Another mechanism proposed to regulate T cell responses in both mouse and human decidua includes reducing T cell entry into decidua by silencing local chemokine expression. There is circumstantial evidence that high local levels of progesterone, corticosteroids and/or chorionic gonadotrophin might modulate local responsiveness, and that a high local metabolism of tryptophan might also contribute to T cell tolerance, but decisive evidence is lacking. Thus, although it is possible that these qualitative changes contribute to the survival of the fetal 'graft', they alone cannot easily provide a complete explanation. Currently there is no evidence in humans that T cell-dependent responses to fetal antigens, either systemically or locally in the decidua, act to damage placental cells or cause pregnancy failure.

Interactions between uterine NK cells and the trophoblast may regulate trophoblast invasion

A stronger possibility is that a more unusual **local immune regulation** occurs in the close vicinity of the conceptus. Among the population of leucocytes in the uterus, **uterine natural killer cells** (**uNK cells**) predominate. Their number varies during the menstrual cycle, being sparse in the proliferative phase and high in the decidua during the early stages of gestation, particularly in the *decidua basalis* where trophoblast cells invade the uterus at implantation. HLA-G on extravillous trophoblast stimulates secretion of tolerogenic cytokines such as IL-6 and IL-10 by both uNK cells and decidual macrophages. This has been shown to reduce T cell proliferation to alloantigens. Thus HLA-G on invading trophoblast can deliver a pregnancy-specific signal to all the main decidual leucocyte populations. This signalling may contribute to the relative scarcity of T cells and their apparent lack of immune reactivity to the trophoblast.

These uNK cells also express a variety of receptors that are capable of recognizing HLA-C molecules on extravillous trophoblast, including the killer-cell immunoglobulin-like receptors (KIR). The KIR gene family comprises up to 14 different KIR genes present in a linear array (haplotype) on chromosome 19. Some KIRs can bind to HLA-C resulting in

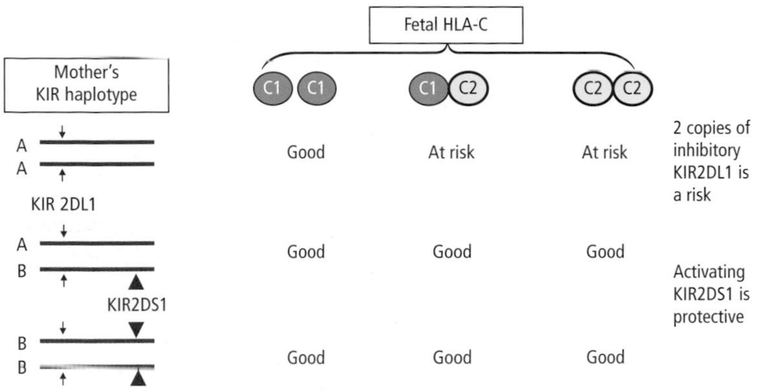

Figure 16.4 Certain combinations of maternal KIR and fetal HLA-C genotypes increase susceptibility to pre-eclampsia. If the mother has a *KIR AA* genotype, then KIR2DL1 binds strongly to fetal HLA-C allotypes on trophoblast bearing a C2 epitope resulting in strong inhibition of uNK cells. This is associated with defective spiral artery remodelling and increased risk of pre-eclampsia. In contrast, when the mother has a *KIR AB* or *BB* genotype, the *KIR2DL1* alleles on the B haplotype tend to inhibit uNK function more weakly when they bind C2. The *KIR B* haplotypes also contain the activating KIR2DS1 which can bind to the C2 epitope. In this situation, uNK cells are stimulated to produce increased levels of cytokines such as GM-CSF that can enhance placentation. Similar results are seen with patient and control cohorts for fetal growth restriction, or recurrent miscarriage. Presence of *KIR2DL1* on the *KIRA* haplotype or *KIR2DS1* on the *KIRB* haplotype is indicated by black or red arrow respectively. (Source: Moffett A, Hiby SE (2007) *Placenta* **28** Suppl. A, S51–S56. Reproduced with permission from Elsevier.)

uNK activation or inhibition depending on the specific KIR receptor. The gene family is highly polymorphic with variation in KIR gene number between individuals, allelic diversity at individual KIR loci and differences in copy number. Despite this extreme variation, a broad distinction into two basic haplotypes has been informative in epidemiological studies. The *KIR A* haplotype has fewer genes expressing mainly inhibitory KIRs, whereas *KIR B* haplotypes express additional activating KIRs. The polymorphic HLA-C ligands for KIRs on trophoblast cells also belong to two groups, C1 and C2, based on a dimorphism at position 80 of the α1 domain. All individuals therefore have C1 and/or C2 epitopes that bind KIR. Three KIRs are known to bind HLA-C; inhibitory KIR2DL1 and activating KIR2DS1 bind HLA-C allotypes bearing a C2 epitope, whilst inhibitory KIR2DL2/L3 bind C1. So each pregnancy is characterized by different combinations of maternal *KIR* and fetal *HLA-C* genes. Since both maternal and paternal HLA-C are expressed by trophoblast, this can result in variable uNK inhibition or activation for each pregnancy.

Genetic studies now strongly suggest that specific combinations of maternal *KIR* and fetal *HLA-C* genes influence reproductive success. Mothers with two *KIR A* haplotypes (*KIR AA* genotype) are at increased risk of disorders of placentation such as pre-eclampsia or recurrent miscarriage, only if the fetus carries a C2 epitope inherited from the father (Figure 16.4). Conversely, mothers with a *KIR B* haplotype (containing activating KIR2DS1 that can also bind C2) are at low risk of pre-eclampsia even if the fetus is C2, but instead these mothers have an increased risk of delivering a large baby. When the fetus is C1 homozygous, the mother's *KIR* genotype has no effect, so C2 is the crucial fetal ligand. Increased risk of pre-eclampsia, fetal growth restriction and recurrent miscarriage are all associated with KIR/HLA-C combinations that result in strong inhibition of uNK functions such as cytokine secretion. This could influence the differentiation or migration of the invading trophoblast via a paracrine network resulting in decreased trophoblast invasion and compromised fetal development. It has led to the suggestion that pre-eclampsia is fundamentally a disease of immune incompatibility between the parents, with secondary vascular and tertiary oxygen radical components that ultimately manifest as pathology. In contrast, activation of uNK cells through the binding of the activating KIR2DS1 to its C2 ligand results in secretion of cytokines such as GM-CSF, which has been shown to enhance trophoblast invasion *in vitro*. This may lead to improved spiral artery remodelling, and hence increased fetal growth (Figure 16.5 and Box 16.3). This model suggests that the specialized interaction between KIR on uNK cells and HLA-C on extravillous trophoblast can regulate placental development and hence influence birth weight to keep it mostly within the desired limits.

Summary

The protection of the fetus from the immune response of the mother appears to depend on:

- An antigenically unique trophoblast forming the front-line defence, possibly via locally mediated regulation of immune reactivity.

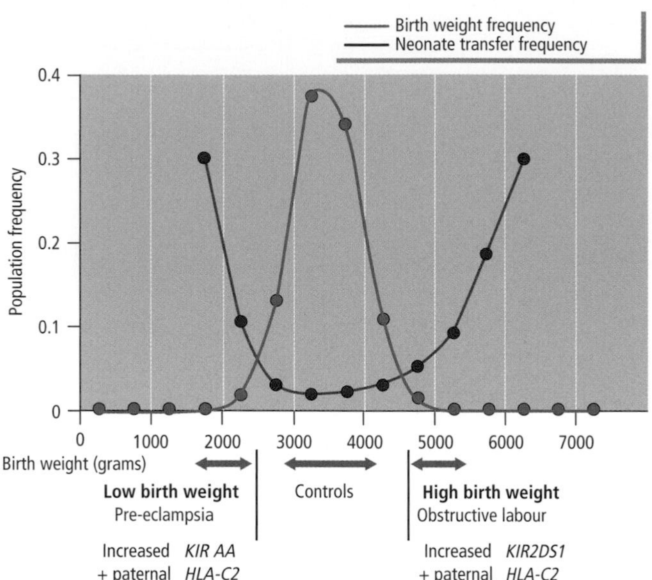

Figure 16.5 Presence of maternal *KIR2DS1* is associated with increased birth weight. Distribution of birth weights correlates with increased neonatal morbidity (frequency of transfer to special care baby unit) at the extremes of birth weight. The cohort was divided into low (<5th centile) normal (6–89th centile) and high (>90th centile) birth weight babies. The frequency of maternal *KIR AA* genotype + fetal *HLA-C2* is increased in the small babies. Presence of maternal *KIR2DS1* and fetal *HLA-C2* combination is associated with increased birth weight. (Source: Moffett A., Hiby S.E., Sharkey A.M. (2015) The role of the maternal immune system in the regulation of human birth-weight. *Philosophical Transactions of the Royal Society B* **370**:20140071 https://doi.org/10.1098/rstb.2014.0071.)

Box 16.3 Evolutionary aspects of HLA-C-KIR interactions

Comparative anatomical investigations in higher primates have shown that extravillous trophoblast migration in Old World monkeys and gibbons occurs only by the endovascular route and seldom reaches as deep as the myometrium, but that the deeply invasive human arrangement also occurs in the chimpanzee and gorilla. It has been suggested, but not demonstrated, that the deeper penetration observed in great apes and humans evolved to enable a greater blood supply in late pregnancy to allow for brain expansion and development. Interestingly, HLA-C is not present in Old World monkeys or gibbons, but appears in the orangutan, and becomes polymorphic in the lineage leading to the gorilla, bonobo, chimpanzee and human. These observations suggest that the co-evolution of the HLA-C/KIR polymorphic systems in great apes may have been one prerequisite for deep trophoblast invasion. However, its association with pre-eclampsia in a small but significant sub-population of women must mean that the reproductive losses are outweighed by other evolutionary advantages.

Recent research has indicated that after their migration out of Africa, modern humans mated with archaic humans (such as Neanderthals) who had been residing in Eurasia for over 200,000 years, and whose immune systems were better adapted to local pathogens. Successful interbreeding provided a way for rapid acquisition of HLA variants already adapted to local pathogens. The archaic HLA-A, -B and -C allotypes present in abundance within modern human populations indicate that positive selection for these genes has occurred. Since several of these archaic allotypes are strong ligands for KIRs, the adaptive introduction and expansion of these HLA alleles seem to have been driven by their role in controlling NK cells. Given the key reproductive role that KIRs play, it is tempting to conclude that the advantages conferred by these archaic genes provided one route to rapid evolution.

Further reading

Abi-Rached L, Jobin MJ, Kulkarni S *et al*. (2011) The shaping of modern human immune systems by multiregional admixture with archaic humans. *Science* **334**, 89–94.

Carter AM (2011) Comparative studies of placentation and immunology in non-human primates suggest a scenario for the evolution of deep trophoblast invasion and an explanation for human pregnancy disorders. *Reproduction* **141**, 391–396.

Moffett A, Colucci F (2015) Co-evolution of NK receptors and HLA ligands in humans is driven by reproduction. *Immunological Reviews* **267**, 283–297.

Wildman DE (2011) Toward an integrated evolutionary understanding of the mammalian placenta. *Placenta* **32** (Supplement B, Trophoblast Research 25), S142–S145.

- Special populations of uNK cells in the decidua that are able to recognize specific HLAs on invading trophoblast cells and via paracrine mechanisms regulate invasion and maternal immune reactivity.
- A complete (or, in humans, highly selective) barrier to the transmission of immune cells or antibodies from mother to fetus.
- Properties of fetal antigens such that any aggressive immune cells or antibodies that get across the placenta are mopped up and diluted out before they can cause extensive tissue damage.

Fetal hypoxia

One of the most common challenges that the fetus faces during pregnancy, in particular later in gestation, is an episode of reduced oxygenation or **acute fetal hypoxia**. Episodes of acute fetal hypoxia may occur as result of compression of the umbilical cord during late gestation or even during the actual process of birth. For instance, the maternal uterine contractions during labour may transiently reduce blood flow from the placental and umbilical circulations to the fetus, thereby reducing its oxygenation. In contrast to glucose or protein, there is no mechanism for the longer-term storage of oxygen in our bodies either before or after birth. Therefore, before and after birth, we have evolved well-defined compensatory responses to even short-term reductions in oxygenation. However, these responses differ dramatically when the fetal and postnatal periods are compared. Outside the womb, there is a vast supply of oxygen from the atmosphere. Therefore, after birth, a reduction in oxygen supply to the tissues is met with an increase in ventilation to increase the level of oxygenation in our pulmonary blood. This **luxurious supply** of atmospheric oxygen allows the cardiovascular system in the adult to increase blood flow – even to the peripheral circulation, maintaining its oxygenation, during periods of systemic hypoxia. Within the womb, the supply of fetal oxygenation is dependent on the placenta rather than on pulmonary ventilation. In contrast to pulmonary ventilator processes, mechanisms within the placenta to increase the input and output of oxygenated blood are limited. Therefore, during an acute period of fetal hypoxia, the fetus has to deal differently with this **finite oxygen supply** by reducing oxygen consumption and/or by making best use of the available oxygen supply. The fetus achieves these outcomes by calling into play cardiovascular defence responses to acute hypoxia. These responses have primarily been investigated in the sheep fetus, since the developmental time course of the ovine cardiovascular system resembles that in humans.

The fetal cardiovascular defence against acute hypoxia

One of the main consumers of oxygen is the heart. Therefore, during acute fetal hypoxia, there is a reduction in fetal heart rate or **bradycardia**, which reduces myocardial oxygen consumption. In addition, the fetal cardiac output is redistributed away from peripheral circulations, such as those perfusing the limbs, to more essential vascular beds, such as those taking blood to the **brain, heart and the adrenal glands**. This **redistribution of blood flow** is known as the **fetal brain sparing effect**. We now know that the increase in blood flow to the fetal brain during acute hypoxia occurs as a result of local increases in powerful vasodilators in the fetal cerebral vasculature, such as **adenosine, nitric oxide and prostanoids**. In contrast, the fetal bradycardia and the fetal peripheral vasoconstriction are triggered by a **carotid chemoreflex**, which is sensitive to the reduction in fetal PO_2. This chemoreflex increases **vagal outflow** to the heart, leading to bradycardia, and increases **sympathetic outflow** to the fetal peripheral circulations, triggering vasoconstriction. Experiments in late gestation fetal sheep show that section of the carotid sinus nerves (**carotid body denervation**) prevents the fetal bradycardia and the fall in fetal femoral blood flow during an episode of acute fetal hypoxia (Figure 16.6).

Like many neurally-triggered responses, which once triggered are then maintained by endocrine mechanisms, the chemoreflex-initiated peripheral vasoconstriction is then maintained by the release of constrictor hormones into the fetal circulation, such as catecholamines and vasopressin. More recently, it has been discovered that the fetal vasculature has an **intrinsic oxidant tone** that is determined by the ratio between the **superoxide anion** and **nitric oxide** in the fetal peripheral circulations, an increase promoting vasoconstriction and a decrease in vasodilatation. Thus, we now know that the magnitude of the fetal peripheral vasoconstrictor response to acute hypoxia, part of the fetal brain-sparing response, represents the result of the combined influences of carotid chemoreflexes, endocrine constrictor responses and a vascular oxidant tone acting at the level of the fetal circulation (Figure 16.7). Fetal hypoxia, if significant or prolonged, will be accompanied by fetal metabolic acidosis, leading to fetal asphyxia. The fetal cardiovascular defences against acute asphyxia triggered, for instance, by complete occlusion of the umbilical cord are broadly similar to those of acute fetal hypoxia but the challenge is substantially more severe, the cardiovascular responses are more intense and the end result is more injurious to the hypoxia-sensitive organs, such as the fetal brain, the heart and the adrenal glands. Cerebral injury as a result of unrelieved cardiovascular defence responses to intrapartum asphyxia may, if survived, lead to cerebral palsy.

Maturation of the fetal cardiovascular defences

It is now established that the pre-partum increase in fetal plasma cortisol promotes a number of maturational changes in the physiology of the fetus in preparation to sustain postnatal life. These include the stimulation of surfactant production by type II pneumocytes in the fetal lung and a switch from lung liquid secretion to lung liquid reabsorption at birth (see pages 292 and 294). In addition, there is accumulating evidence to suggest that the rise in fetal plasma cortisol towards term is

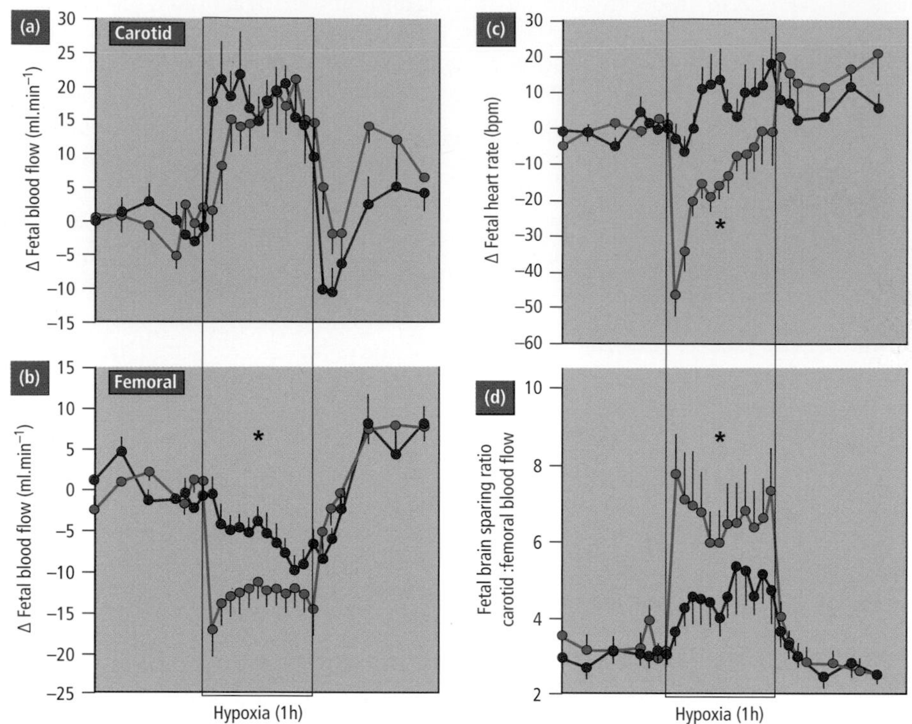

Figure 16.6 Fetal cardiovascular defence responses to acute hypoxia. The data (mean ± SEM) show the change from baseline in (a) fetal carotid blood flow, (b) fetal femoral blood flow, and (c) fetal heart rate in intact (blue symbols) and carotid body-denervated (red symbols) chronically-instrumented sheep fetuses in late gestation during a 1-hour episode of acute hypoxia. Calculation of the ratio between simultaneous measurements of fetal carotid and femoral blood obtained with implanted transonic flow probes yields the fetal brain-sparing index (d). Carotid body denervation prevents the fetal bradycardia and diminishes the fall in fetal femoral blood flow and the increase in the fetal brain-sparing index during acute hypoxia. *$P < 0.05$, intact *vs*. denervated. (Source: Giussani D.A. (2016) *Journal of Developmental Origins of Health and Disease* **4**, 328–337. Reproduced with permission of John Wiley and Sons.)

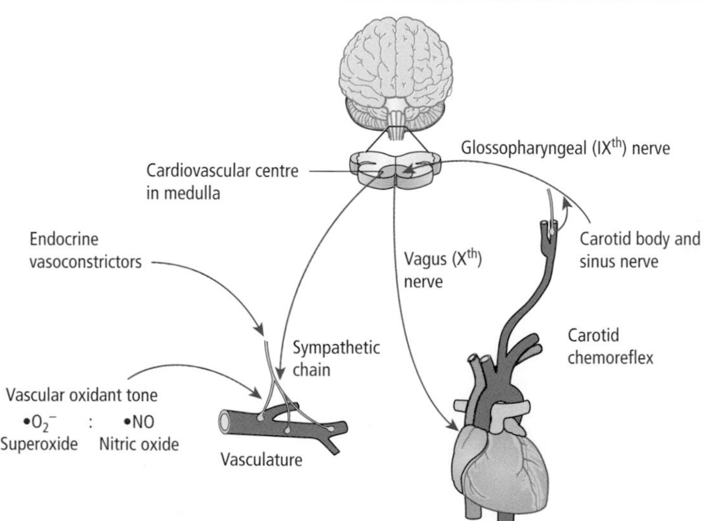

Figure 16.7 The fetal brain-sparing response to hypoxia. This fetal cardiovascular response to acute hypoxia is triggered by a carotid chemoreflex, which leads to bradycardia and an increase in peripheral vasoconstriction. The neurally triggered peripheral vasoconstriction is maintained by the release of constrictor hormones into the fetal circulation as well as a local vascular oxidant tone, determined by the interaction between nitric oxide (NO) and the superoxide anion (•O_2^-). (Source: Giussani D.A. (2016) *Journal of Developmental Origins of Health and Disease* **4**, 328–337. Reproduced with permission of John Wiley and Sons.)

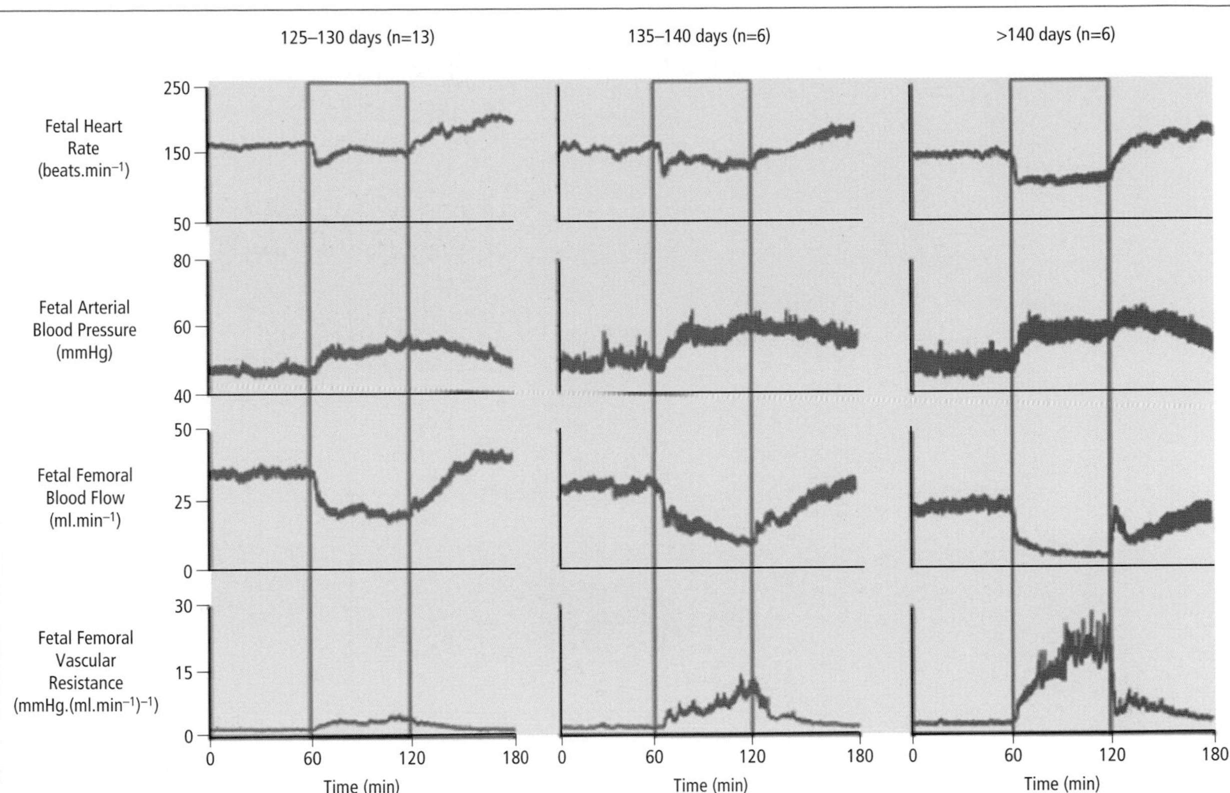

Figure 16.8 Ontogeny of the fetal cardiovascular responses to acute hypoxia. The data show from the top down the mean values of the fetal heart rate, fetal arterial blood pressure, fetal femoral blood flow and fetal femoral vascular resistance during a 1-hour episode of acute hypoxia (box) in 13 fetuses between 125–130 days of gestation, 6 fetuses between 135–140 days of gestation and 6 fetuses >140 days (term is *ca*. 145 days). Basal fetal heart rate and basal fetal femoral blood flow decrease with advancing gestation. In addition, during acute hypoxia, the bradycardia becomes enhanced and persistent and the femoral vasoconstriction is more intense as the fetus approaches term. (Source: Fletcher A.J. *et al*. (2006) *American Journal of Physiology. Heart and Circulatory Physiology* **291**, H3023–H3034. Reproduced with permission of the American Physiological Society.)

involved in the maturation of the fetal cardiovascular defence to acute challenges, such as fetal hypoxia. As the fetus advances through gestation towards term, there is not only an ontogenic fall in basal fetal heart rate and an ontogenic increase in basal fetal arterial blood pressure, but the pattern and magnitude of the fetal cardiovascular responses to acute hypoxia also change. With advancing gestation, there is a switch from a transient to a sustained bradycardia during acute hypoxia. Further, the magnitude of the fetal femoral vasoconstrictor response to acute hypoxia increases with advancing gestational age towards term (Figure 16.8).

The physiology underlying the sustained bradycardic response to acute hypoxia in the term compared to the preterm fetus involves maturation of the carotid chemoreflex sensitivity to acute hypoxia as well as reciprocal changes in the reactivity of the fetal heart rate to autonomic agonists. Comparison of the term and preterm fetus reveals there is an increase in the fetal heart rate sensitivity to muscarinic receptor stimulation and a fall in its responsiveness to β_1-adrenergic receptor stimulation. Similarly, the greater femoral vasoconstrictor response to acute hypoxia in the term relative to the preterm fetus is mediated by maturation of the carotid chemoreflex sensitivity to acute hypoxia and a greater plasma endocrine vasoconstrictor response to acute hypoxia. Plasma concentrations of catecholamines and of vasopressin increase to much greater levels during a similar degree of acute hypoxia in the term compared with the preterm fetus. The more sustained bradycardia in response to acute hypoxia ensures a greater fall in myocardial oxygen consumption in the term relative to the preterm fetus. The greater peripheral vasoconstrictor response to acute hypoxia in the term fetus represents a more mature fetal brain-sparing response to acute hypoxia.

In Chapter 18, we will see that antenatal glucocorticoid therapy of pregnant women threatened with preterm labour

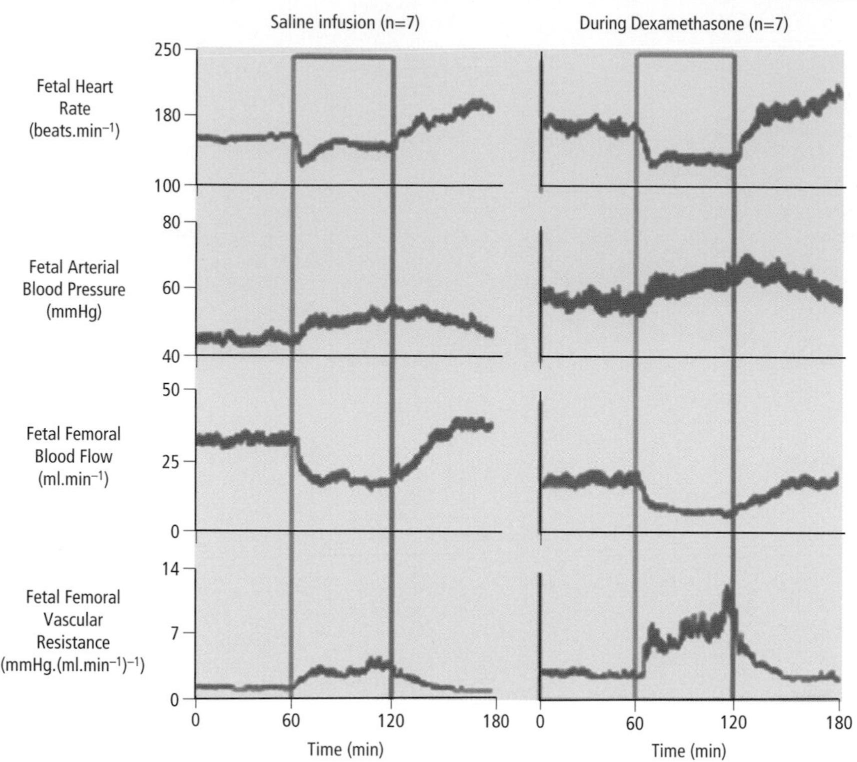

Figure 16.9 Antenatal glucocorticoid therapy and maturation of the fetal cardiovascular defence to acute hypoxia. The data show mean values for the fetal heart rate, fetal arterial blood pressure, fetal femoral blood flow and fetal femoral vascular resistance during a 1-hour episode of acute hypoxia (box) in 14 fetuses at 127 days of gestation (term is *ca.* 145 days) following 2 days of continuous fetal i.v. infusion with saline or with dexamethasone. Fetal treatment with dexamethasone switches the pattern and the magnitude of the fetal heart rate and the femoral vascular resistance responses to acute hypoxia towards those seen in fetuses close to term (see Figure 16.8). (Source: Fletcher *et al.* (2003) *Journal of Physiology* **549**(Pt 1), 271–287. Reproduced with permission from John Wiley and Sons, Ltd.)

mimics the endogenous fetal plasma cortisol surge observed towards term (Chapter 17). There is now evidence that exposure of fetal sheep to clinically relevant doses of synthetic glucocorticoids, such as dexamethasone, also accelerates maturation of the fetal cardiovascular defences to acute hypoxia. Thus, treatment of immature fetal sheep with dexamethasone switches the fetal bradycardic and peripheral vasoconstrictor response to acute hypoxia from the preterm to the term pattern and magnitude (Figure 16.9).

Developmental programming

We have thus far considered the more immediate hazards to mother and fetus, and to the conflicts between them during pregnancy. In this last section, we examine how immediate interactions between mother and fetus can have longer-term consequences. Thus, epidemiological studies of human populations have suggested that **adverse pregnancy conditions** can influence physiological function and susceptibility to disease of the offspring in adult life. Common adverse conditions during pregnancy include maternal under-nutrition, maternal obesity, maternal stress and chronic fetal hypoxia, which may result from an increase in placental vascular resistance, such as occurs in pre-eclampsia or placental insufficiency. Alterations in fetal oxygenation, nutrition or exposure to stress hormones have all been associated with a constellation of overlapping adult pathologies, increasing the risk of cardiovascular disease, which together are often given the name **metabolic syndrome** (Table 16.1). This syndrome is experienced regardless of the contemporary levels of oxygenation, nutrition, activity patterns, stress exposure or the presence of obesity of the adult concerned. Examination of the babies born to mothers experiencing these adverse pregnancy conditions shows that frequently they are smaller (**intrauterine growth retardation or IUGR**) and sometimes disproportionate too (**asymmetric IUGR**). Moreover, if postnatally they experience an early growth spurt, often called **catch-up growth**, then the chance of them having metabolic syndrome later in life is enhanced. In addition, they are much more likely to be obese as children, often with poor appetite control and irritable behaviour patterns. It will not escape notice that these features characterize one of the major contemporary health problems internationally, an epidemic of child and adult obesity and associated pathologies, such as heart disease and diabetes, sweeping through not only the more affluent modern worlds in all continents, but also in developing societies. Is it possible that there are **fetal origins for much of this adult risk of disease**? And if so, what exactly is happening?

Table 16.1 Adult diseases that have been associated with suboptimal intrauterine conditions

Cardiovascular disease	Hypertension, coronary heart disease, cardiac hypertrophy, dilated cardiomyopathy, stroke, atherosclerosis, peripheral vascular disease, coagulation disorders, pre-eclampsia
Metabolic disease	Impaired glucose tolerance, insulin resistance, dyslipidaemia, obesity, type 2 diabetes
Reproductive problems	Polycystic ovary, early adrenarche/menarche, early menopause
Respiratory disease	Chronic obstructive pulmonary disease, asthma
Endocrine disease	Hypercortisolism, hypothyroidism
Renal disease	Reduced nephron endowment, glomerulosclerosis
Nervous disease	Schizophrenia, dementia

Experimental studies: mechanisms, cause and effect?

Controlled experimental studies in pregnant animals, in which a single variable such as oxygenation or diet can be altered and its impact on both size of offspring and adult health patterns assessed, clearly suggest a causal link. This observation has led to the concept that **the fetus is developmentally programmed by the quality of the intrauterine environment**. Moreover, although the exact pattern of diseases generated may depend on the variable under study, it is more the balance between different pathologics that is influenced than the range of pathologies, which come as a constellation. Given the variety of precipitating conditions, it is tempting to think that their effects must be funnelled through a limited pathway to produce such a similar spectrum of pathologies. What might that pathway be? Again, controlled animal studies are most informative but have not yet provided an unambiguous answer.

One possible scenario comes back to the fetal brain-sparing response to acute hypoxia. In response to chronic fetal hypoxia, redistribution of blood flow away from peripheral circulations may persist. This may explain why some babies born to complicated pregnancies show asymmetric IUGR, as blood flow is constantly driven away from peripheral circulations. Hence, these babies may have a normal size head with a shorter body length or be thin for their length, thereby having a low ponderal index. The less well-described adverse side-effect of persistent redistribution of blood flow away from peripheral fetal circulations (peripheral vasoconstriction) during complicated pregnancy may be an increase in the load against which the fetal heart pumps, known as **the fetal cardiac afterload**. The fetal heart may thus respond to this load like any muscle, laying down more muscle – the basis of body building exercise. Hence, this may result in thicker walls in the fetal heart and major vessels, or **ventricular and aortic wall hypertrophy**, in essence, hallmarks of cardiovascular disease. This is how chronic fetal hypoxia may thus trigger a fetal origin of heart disease and/or programme an increased susceptibility to later cardiovascular dysfunction in the adult offspring. Sustained redistribution of blood flow away from peripheral circulations during pregnancy complicated by chronic fetal hypoxia may also explain the lower endowment of nephrons in the fetal kidney and of pancreatic β-cells in the fetal Islets of Langerhans. This is of interest because both conditions have also been associated with increased susceptibility to the metabolic syndrome in adult offspring of complicated pregnancies, particularly when asymmetric IUGR is present.

Another possibility is that maternal hormonal profiles are affected by many of the maternal stressors and that secondarily this affects both the endocrine exposure of, and the endocrine activity in, the fetus. Hormones are intimately involved in growth control and metabolic regulation and are thus plausible mediators of programming. **Glucocorticoids** are particularly implicated, since they are **sensitive barometers of maternal stress** and have widespread **growth inhibitory effects on the fetus**. Moreover, experimental over-exposure to glucocorticoids mimics many of the effects of maternal stressors, the negative programming effects of which are also ameliorated by depression of maternal glucocorticoid responses. Consistent **abnormalities in the glucocorticoid feedback loop** involving the hypothalamus, pituitary and adrenal have been described **in the programmed offspring of stressed mothers**.

Yet another possibility is that many adverse intrauterine conditions, such as chronic fetal hypoxia, maternal obesity or pregnancy exposed to excess glucocorticoids are all associated with increased levels of oxidative stress in the placental and fetal compartments. Oxidative stress during development has been linked with programming of increased susceptibility to heart disease and the metabolic syndrome in the adult offspring. Indeed, recent studies show that maternal supplementation with antioxidants can protect against the programming effects of adverse pregnancy on cardiometabolic dysfunction in the adult offspring.

However, developmental programming can occur preconceptually or when maternal stresses are experienced for limited periods during pregnancy, including periovulatory, embryogenic and embryonic periods as well as fetal periods. This observation places some constraints on the underlying mechanism(s) that might affect the subsequent adult physiology. For example, it is unlikely that these early developmental effects are exerted directly on the physiological systems regulating fetal physiology and metabolism, because these are not yet matured at the earliest of these stages. These sorts of observations make us wonder whether changes induced early in development leave some sort of memory trace, the impact of which is seen later in disturbed physiological patterns?

Memory traces?

One hypothetical memory trace might result from the impact of stressors on cell proliferation in the early conceptus to affect the relative and/or absolute numbers of cells allocated to and within the different parts of the placenta and embryo. This might easily lead to **disproportionality**, small babies and different organs (including the placenta) being affected differently, depending on whether exposure to the stressor occurred during a rapid growth period in that particular organ. A possible and plausible mechanism by which such growth and physiological differentials might be generated is through metabolic effects on the embryonic cells.

A second memory trace for which direct evidence exists is the epigenetic modification of gene promoters that affect the level of gene expressibility and thus activity in adults. We introduced the concept of epigenesis in Chapter 1 (see pages 9–10), from which you will recall that two sorts of epigenetic modification can leave an imprint: the post-translational modification of chromatin histones and the direct methylation of cytosines within the DNA sequence itself (see Figure 1.7). This latter methylation of selected cytosines is a post-replicative event mediated by a methyl transferase, and once initiated, it is then copied at each round of DNA replication as long as the maintenance methylase remains present. So, it is this type of epigenetic modification that is most clearly heritable and a possible candidate for the maternally programmed memory trace. Indeed, in animal studies, experimental manipulation of nutritional factors and glucocorticoids or induction of placental insufficiency have all been shown to alter patterns of cytosine methylation in individual gene promoters at all stages of development. Moreover, these changed patterns then seem to persist into adulthood, and in some cases have been associated with altered expression patterns at the level of functional proteins. So a **programming route via epigenetic modification seems highly likely**. Further analysis of individual genes and gene families is ongoing. Potentially interesting genes are the adipokines such as leptin, because of their association with obesity and their elevation during pregnancy in both maternal and fetal circulations. Resetting of the fetal leptin regulatory system, directly through epigenetic changes or indirectly downstream of such changes, to promote fat deposition might provide an explanation for child and adult obesity.

Finally, it is known that certain polymorphic variants of some genes involved in growth and metabolism predispose to metabolic syndrome, and environmental factors seem to interact with these to produce pathology. Analysis of the epigenetic modifications to these polymorphic variants under different maternal stressor conditions will be interesting. (There is further discussion of gene–environment interactions and epigenesis in Chapter 20.) For the moment, however, we do not yet have a definitive answer to the question: by what mechanism(s) do maternal stressors heritably influence adult health?

Why programming?

There is now convincing evidence that maternal programming of development occurs, even if we do not understand fully the mechanism(s) mediating it. What possible evolutionary reason could there be for programming adult ill health in this way? That is. of course. the wrong question. The process of evolution selects adaptively for those genotypes that result in phenotypes most able to survive and reproduce. There may well have been survival value in selecting for genes that reduced fetal growth in response to maternal nutritional stress as a way of preparing the offspring for optimized survival in a resource-poor world – **a predictive adaptive response**. Where the postnatal environment turns out to be nutritionally rich, the predictive adaptive response is inappropriate and catch-up growth predicates poorer middle-aged health despite survival to reproductive age. So, perhaps the question we should really ask is: what do these studies mean for optimized maternal and neonatal nutrition during pregnancy, and by optimization can we avoid disruptive programming? We hope so. However, the worrying scenario is that the mother's metabolism has already, through her own lifetime experiences including that *in utero*, been set in ways that prevent useful manipulations to her diet from being effective in ameliorating adverse impacts on her offspring. Indeed, there is already some evidence for **transgenerational effects of the maternal environment** on patterns of offspring morbidity. Exactly how these effects are mediated is also uncertain, but is clearly of concern if the wave of obesity is to be addressed not simply symptomatically but also preventively. In the case of obesogenic pregnancy, the health focus may have to be on regulating the postnatal diet to prevent the worst negative impact of maternal programming.

Conclusions

Viviparity seems to provide the developing embryos and fetuses with the optimal environment for growth. The uterus may be viewed as the ultimate 'nest': temperature-controlled, a continuous supply of food and protection from predators. This environment is created at the mother's expense and her metabolism is brought, to varying degrees, under the control of the fetus, which to a large extent functions autonomously within its protected environment. It should, however, be clear from this account that we still have much to learn about fetal and maternal function in pregnancy and its control. Central to the maintenance of pregnancy is the trophoblast of the placenta. This remarkable tissue elaborates and secretes the steroid and protein hormones; it is involved in the transplacental passage of a variety of essential substances and also the products of metabolism, both by diffusion and by active transport; it acts as a selective barrier between the two circulations, it presents an antigenically bland tissue to the mother's immune system – except for an extravillous subpopulation which seems to play a key role in balancing trophoblast invasion against effective fetal nutrition and oxygenation. The role of the trophoblast ends with delivery of the new infant, after which the mother still has a major and crucial role to play postnatally. How delivery is approached and achieved is the subject of Part 6 of this book.

Key learning points

- Human pregnancy is associated with an unusually high rate of complications, presenting serious risks for the mother and her fetus.
- Early pregnancy is diagnosed chemically (by hCG testing; 18–30 days) and/or clinically (by ultrasound; 7–8 weeks).
- Most pregnancies fail, the majority of them in the first trimester.
- About 1% of women have recurrent miscarriages.
- Most early miscarriages are associated with a failure of the migration of the extravillous trophoblast, leading to failure to convert the spiral arteries.
- IVF *per se* is not associated with significantly increased fetal loss or congenital abnormality.
- Maternal morbidity and mortality remain high through much of the world.
- Pregnant women die or become ill either as a result of increased pressure on already stressed physiological systems or through pregnancy-specific diseases such as pre-eclampsia.
- Pre-eclampsia appears to be fundamentally a disease of immune incompatibility between the parents, with secondary vascular and tertiary oxygen radical components.
- In most pregnancies, protection of the genetically and antigenically distinct fetus from the immune response of the mother appears to depend on: (1) an antigenically relatively inert trophoblast that lacks HLA-A and -B antigens forming the front-line defences, possibly in association with local, endocrinologically mediated depression of immune reactivity; (2) special populations of natural killer cells in the decidua that are able to recognize HLA-C antigens on invading extravillous trophoblast cells and, via paracrine mechanisms, regulate invasion and maternal resistance; (3) a placental filter or barrier to the transmission of immune cells or antibodies from mother to fetus; and (4) properties of fetal antigens such that aggressive immune cells or antibodies that do cross the placenta are mopped up and diluted out before they can cause extensive tissue damage.
- Adverse conditions during pregnancy, such as chronic fetal hypoxia, fetal under- or over-nutrition, and excessive fetal glucocorticoid exposure are all associated with programming of the metabolic syndrome in the adult offspring.
- Intrauterine programming of fetal development adjusts fetal metabolic and growth responses to predicted ambient conditions.
- Should the actual conditions experienced postnatally differ, the physiology of the neonate may be set inappropriately with consequences for the later development of cardiovascular and metabolic diseases.
- Developmental programming may work through several types of mechanisms, including epigenetic changes to certain genes, via effects on cell division and tissue growth and/or via effects on cell metabolism.

Clinical vignette

Stillbirth

A 37-year-old Caucasian woman, Ms H, presented at 37 weeks in her third pregnancy with reduced fetal movements. In her previous two pregnancies, she had experienced an early miscarriage at 8 weeks, and a spontaneous vaginal delivery of a normally grown daughter at full term. She had attended for all antenatal care and accepted all screening offered during for her pregnancy. Sadly, on arrival at the maternity unit, no fetal heartbeat could be detected. The obstetrician on duty performed an ultrasound scan that revealed an *in utero* fetal death. Condolences were offered to Ms H and her partner on the sad loss of their pregnancy. The next day, labour was induced and Ms H had an uncomplicated vaginal delivery of a 1.6 kg (<3rd centile for gestational age) stillborn baby girl. The couple were understandably devastated by the death of their baby and asked that everything be done to help understand the cause. Maternal blood was taken to search for causes including gestational diabetes, pre-eclampsia, infection, and clotting disorders. Swabs and cultures of the placenta, urine, and vaginal secretions were also sent for microbiology investigations. All of the serum and microbiological investigations showed normal results. Autopsy and DNA analysis demonstrated that, despite being small, Ms H's baby had no structural or chromosomal abnormality that could have explained the death. However, inspection of the placenta following delivery demonstrated extensive areas of calcification and infarction.

The likely cause of this stillbirth was placental insufficiency. All placentas will age and fail eventually if a pregnancy does not deliver, but some will fail at earlier stages. Currently, there are no reliable screening tests to pick up early signs of placental failure that are suitable for use in a low-risk unselected population. In the UK, screening for fetal growth restriction is limited to measurement of symphysal-fundal height (measurement of the maternal abdomen from the top of the uterus to the pubic symphysis). Symphysal-fundal height is measured by midwives throughout the third trimester, but has a poor detection rate for fetuses that are failing to achieve their growth potential. Urgent further research is required into promising strategies to detect placental insufficiency, which currently include routine third trimester scanning for fetal growth, measurement of proteins in the maternal blood (including pregnancy-associated plasma protein A) and more extensive use of Doppler blood flow measurements.

FURTHER READING

General reading

Aiken CE, Ozanne SE (2013) Transgenerational developmental programming. *Human Reproduction Update* **20**, 63–75

Allegrucci C, Thurston A, Lucas E, Young L (2005) Epigenetics and the germline. *Reproduction* **129**, 137–149.

Fernandez-Twinn DS, Ozanne SE (2010) Early life nutrition and metabolic programming. *Annals of the New York Academy of Science* **1212**, 78–96.

Fletcher AJ, Gardner DS, Edwards CM *et al.* (2003). Cardiovascular and endocrine responses to acute hypoxaemia during and following dexamethasone infusion in the ovine fetus. *Journal of Physiology* **549**(Pt 1), 271–287.

Fletcher AJ, Gardner DS, Edwards CM *et al.* (2006). Development of the ovine fetal cardiovascular defense to hypoxemia towards full term. *American Journal of Physiology. Heart and Circulatory Physiology* **291**, H3023–H3034.

Fowden AL, Forhead AJ (2004) Endocrine mechanisms of intrauterine programming. *Reproduction* **127**, 515–526.

Fowden AL, Giussani DA, Forhead AJ (2006) Intrauterine programming of physiological systems: causes and consequences. *Physiology* **21**, 29–37.

Giussani DA (2016) The fetal brain sparing response to hypoxia: physiological mechanisms. *The Journal of Physiology* **594**, 1215–1230.

Giussani DA, Davidge ST (2013) Developmental programming of cardiovascular disease by prenatal hypoxia. *Journal of Developmental Origins of Health and Disease* **4**, 328–337.

Gluckman PD, Hanson MA (2004) Living with the past: evolution, development, and patterns of disease. *Science* **305**, 1733–1736.

Hochberg Z, Feil R, Constancia M *et al.* (2011) Child health, developmental plasticity, and epigenetic programming. *Endocrine Reviews* **32**, 159–224.

Hunt JS (2006) Stranger in a strange land. *Immunological Reviews* **213**, 36–47.

Hunt JS, Morales PJ, Pace JL *et al.* (2007) A commentary on gestational programming and functions of HLA-G in pregnancy. *Placenta* **28** (Supplement A, Trophoblast Research 21), eS57–S63.

Jauniaux E, Poston L, Burton G (2006) Placental-related diseases of pregnancy: involvement of oxidative stress and implications in human evolution. *Human Reproduction Update* **12**, 747–755.

Jellyman JK, Gardner DS, Edwards CM *et al.* (2005) Fetal cardiovascular, metabolic and endocrine responses to acute hypoxaemia during and following maternal treatment with dexamethasone in sheep. *Journal of Physiology* **567**(Pt 2), 673–688.

Loudon I (2012) Maternal mortality in the past and its relevance to developing countries today. *American Journal of Clinical Nutrition* **72** (Suppl.), 241S–246S.

Moffett A, Chazara O, Colucci F, Johnson MH (2016) Variation of maternal KIR and fetal HLA-C genes in reproductive failure: too early for clinical intervention. *Reproductive BioMedicine Online* **33**, 763–769.

Moffett A, Hiby SE (2007) How does the maternal immune system contribute to the development of pre-eclampsia? *Placenta* **28** Suppl. A:S51–56. Epub 2007 Feb 8.

Moffett A, Hiby SE, Sharkey AM (2015) The role of the maternal immune system in the regulation of human birthweight. *Philosophical Transactions of the Royal Society B* **370**, 20140071. https://dx.doi.org/10.1098/rstb.2014.0071

Moffett A, Loke C (2006) Immunology of placentation in eutherian mammals. *Nature Reviews in Immunology* **6**, 584–594.

Owen D, Andrews MH, Matthews SG (2005) Maternal adversity, glucocorticoids and programming of neuroendocrine function and behaviour. *Neuroscience and Biobehavioral Reviews* **29**, 209–226.

More advanced reading

Barker DJP (1997) The fetal origins of coronary heart disease. *Acta Paediatrica* **422** (Suppl.), 78–82.

Bloomfield FH, Oliver MH, Hawkins P *et al.* (2003) A periconceptional nutritional origin for noninfectious preterm birth. *Science* **300**, 606.

Burton GJ, Jauniaux E (2004) Placental oxidative stress: from miscarriage to preeclampsia. *Journal of the Society for Gynecological Investigation* **11**, 342–352.

Carter AM (2011) Comparative studies of placentation and immunology in non-human primates suggest a scenario for the evolution of deep trophoblast invasion and an explanation for human pregnancy disorders. *Reproduction* **141**, 391–396.

Giussani DA, Camm EJ, Niu Y *et al.* (2012) Developmental programming of cardiovascular dysfunction by prenatal hypoxia and oxidative stress. *PLoS One* **7(2)**, e31017.

Henson MC, Castracane VD (2006) Leptin in pregnancy: an update. *Biology of Reproduction* **74**, 218–229.

Hiby SE, Apps R, Sharkey AM *et al.* (2010) Maternal activating KIRs protect against human reproductive failure mediated by fetal HLA-C2. *Journal of Clinical Investigation* **120**, 4102–4110.

Lash GE, Otun HA, Innes BA *et al.* (2010) Regulation of extravillous trophoblast invasion by uterine natural killer cells is dependent on gestational age. *Human Reproduction* **25**, 1137–1145.

McMillen IC, Edwards LJ, Duffield J, Muhlhausler BS (2006) Regulation of leptin synthesis and secretion before birth: implications for the early programming of adult obesity. *Reproduction* **131**, 415–427.

Royal College of Obstetricians and Gynaecologists (2011) *Cardiac Disease and Pregnancy*. RCOG Good Practice Notes No. 13. RCOG Press, London.

Waterland RA, Jirtle RL (2004) Early nutrition, epigenetic changes at transposons and imprinted genes and enhanced susceptibility to adult chronic disease. *Nutrition* **20**, 63–68.

Weinstock M (2005) The potential influence of maternal stress hormones on development and mental health of the offspring. *Brain, Behavior and Immunity* **19**, 296–308.

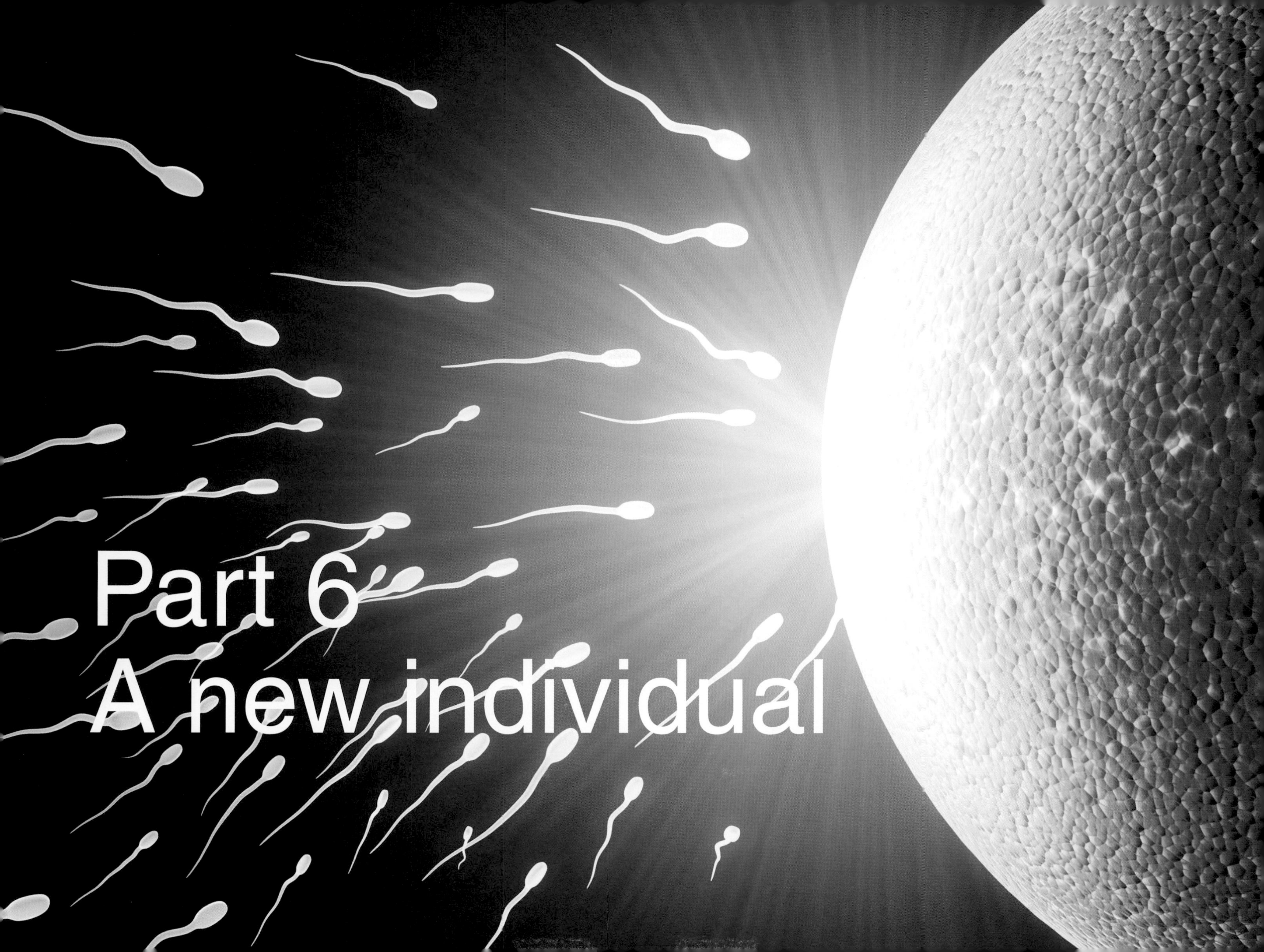
Part 6
A new individual

CHAPTER 17
Preparing for birth

Essential Reproduction, Eighth Edition. Martin H. Johnson.

Companion website: www.essentialreproduction.com

The foregoing chapters have charted the reproductive course that leads to birth. It should now be evident that we are the remarkable survivors of a hazardous journey. Birth itself also brings challenges as both the fetus and mother must develop physiological mechanisms that anticipate the transition from a uterine to an external, independent existence at parturition. This extraordinary change of circumstance must be achieved while maintaining the internal environment of the neonate.

Fetal systems develop and mature in preparation for postnatal life

The fetus must exist both for the present and the future. The physiological basis of life within the maternal environment is very different from that experienced after parturition. The fetus must therefore develop mechanisms that anticipate this change and gear its own metabolism to adapt more or less instantaneously. A **rising fetal output of corticosteroids** towards term plays an important part in orchestrating the maturation of fetal physiology as well as the timing of birth itself (Figure 17.1). Other critical anatomical and functional changes also occur. In this section, these various adaptations to first intrauterine and then extrauterine life are considered.

The cardiovascular system

Fetal cardiac output, peripheral resistance and blood pressure rise as parturition approaches and are underlain by rising cortisol levels (Figure 17.1). The fetal circulation differs from that in the adult because the **placenta, not the lung, is the organ of gaseous exchange**. Two fetal adaptations achieve this difference: (1) the **two fetal ventricles pump not in series but in parallel**; and (2) **three vascular shunts divert the fetal circulation away from the lungs and towards the placenta**. The effect of these adaptations is such that conversion to the adult form of circulation is initiated instantaneously at the first breath taken by the newborn. The fetal circulation is shown in Figure 17.2a.

Oxygenated blood returns from the placenta and is carried into two channels. The larger of these is the **ductus venosus**, a fetal shunt that bypasses the hepatic circulation and delivers blood directly into the inferior vena cava. The smaller channel perfuses the liver and enters the inferior vena cava through the hepatic veins. The inferior vena cava carries blood to the right atrium where it is split into two streams by the **crista dividens**, the free edge of the interatrial septum, which projects from the foramen ovale. The larger stream passes through the foramen ovale, another fetal shunt, into the left atrium, thereby avoiding

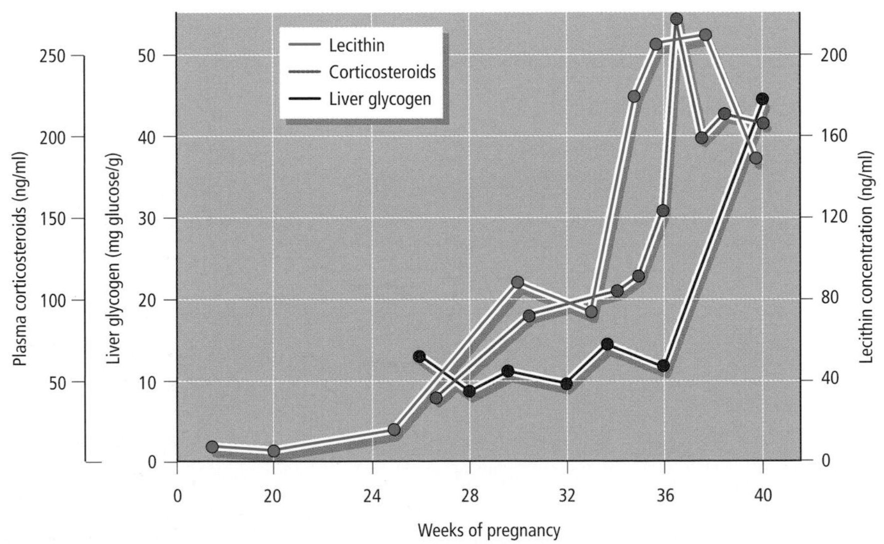

Figure 17.1 Time course of concentrations of corticosteroids in the plasma of the umbilical cord, of lecithin in amniotic fluid and of glycogen in the fetal liver. Lecithin may be used as an indicator of surfactant production in the fetal lung. The rise in lecithin content of human amniotic fluid just before birth is reflected in the lecithin:sphingomyelin ratio test for monitoring fetal development and well-being. (Sphingomyelin is a phospholipid, the concentration of which does not change near term and serves therefore as a baseline against which to measure lecithin.) A ratio of >2 indicates surfactant production is normal, and of <2 indicates the fetus may have respiratory distress syndrome. The close temporal relationship between corticosteroid concentration in the blood and amniotic lecithin levels indicates the causal relationship between the two. The increase in liver glycogen near term is also the result of the rise in corticosteroids, which induces the late cluster of fetal liver enzymes, including those required for glycogen synthesis.

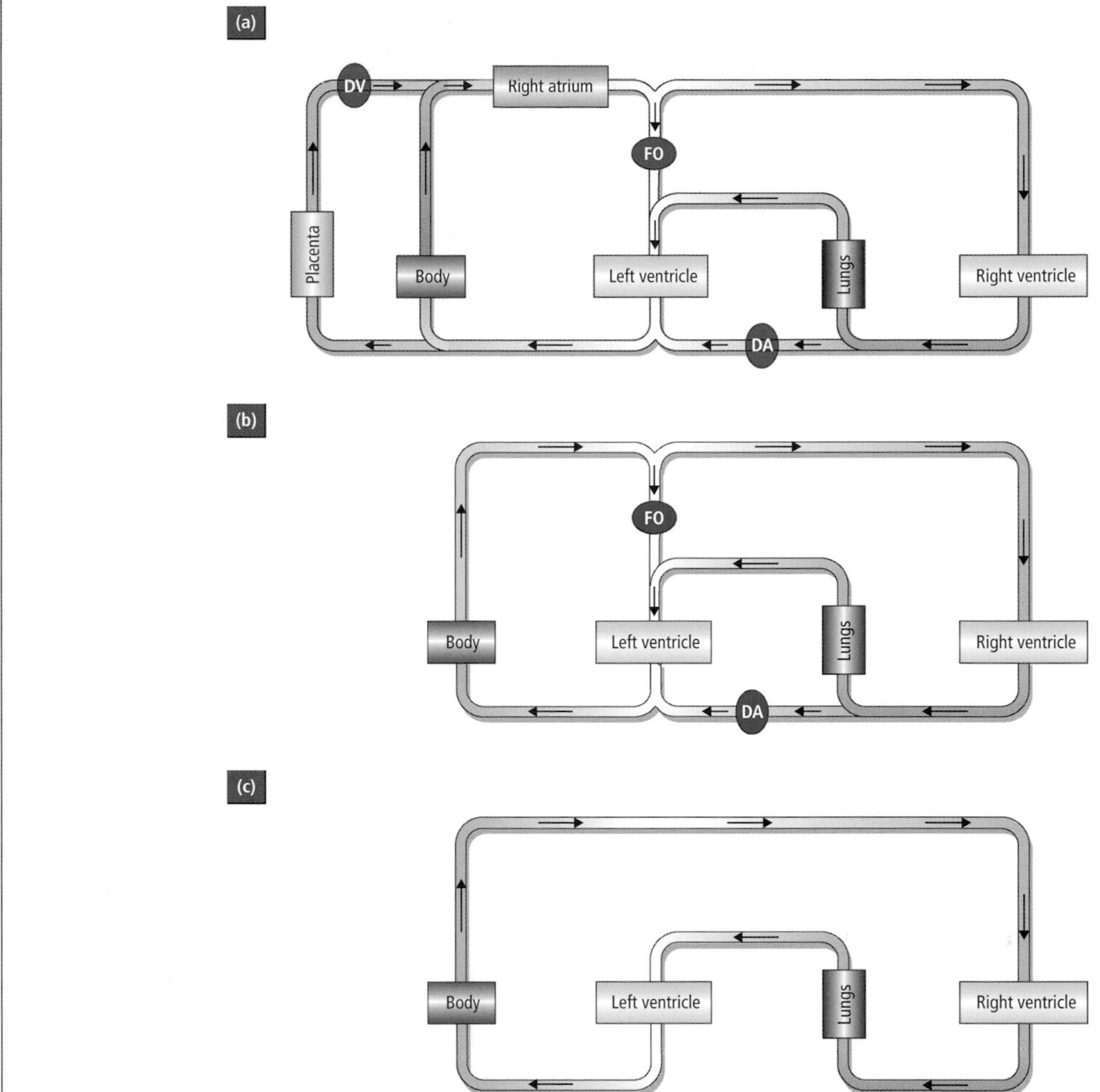

Figure 17.2 Highly schematic representation of the circulations of: (a) the fetus; (b) the neonate; and (c) the adult. The transitory form in (b) occurs through the occasional opening of fetal shunts prior to their complete, anatomical closure. DV, ductus venosus; DA, ductus arteriosus; FO, foramen ovale. (Source: Biggers J.D. (1979) *Medical Physiology* (ed. V. Mountcastle). Mosby, St Louis. Reproduced with permission from Elsevier.)

the pulmonary circulation. The smaller stream continues through the right atrium, as does coronary blood and blood returning from the head region via the superior vena cava. This blood flows into the right ventricle and out through the pulmonary artery. Thereafter, it also is split into two channels: the largest passing through yet a third fetal shunt, the **ductus arteriosus**, which carries blood to the aorta; while the smaller channel conveys blood to the fetal lungs. The small amount of poorly oxygenated blood passing through the pulmonary circulation returns to the left side of the heart.

The combined fetal cardiac output consists of about **one-third from the left and two-thirds from the right ventricle**. The three fetal vascular shunts combine in function to ensure the optimal distribution of oxygenated blood to the head and body. Thus, blood leaving the placenta in the umbilical vein is 90% saturated with O_2. Most of it is shunted past the liver to join with poorly oxygenated (20%) blood in the inferior vena cava, the resultant mix reaching the heart being 67% saturated. The blood shunted through the foramen ovale to the left atrium is joined by O_2-poor blood returning from the lungs. The resulting 62% saturated blood leaves the heart via the brachiocephalic artery to supply mainly the head region. The remainder of the O_2-rich blood from the inferior vena cava enters the right atrium to mix with poorly oxygenated blood (31% saturated) returning from the head region via the superior vena cava. The blood thereby entering the pulmonary artery via the right ventricle is 52% saturated with O_2, and the largest proportion of it is shunted through the ductus arteriosus to join O_2-rich blood in the aorta to yield an O_2 saturation of 58% in the descending aorta. The effectiveness of the ductus arteriosus shunt results largely from the high pulmonary vascular resistance due to constriction of pulmonary arterioles in response to the low fetal O_2 tension (20–25 mmHg, compared with 80–100 mmHg in the adult).

The **changes in the fetal circulation at birth involve closure of the three fetal shunts** to replace the placental circulation with a pulmonary circulation (Figure 17.2b,c). With the obliteration of the umbilical circulation, the ductus venosus ceases to carry blood to the heart. At the same time, a dramatic fall in pulmonary vascular resistance occurs as a result of inflation of the lungs with the first breath and the rise in pulmonary Po_2. Thus, overall, there is a net drop in pressure on the right side of the heart (with a loss of umbilical input and rise in pulmonary outflow) and a rise in pressure on the left side (with a return of pulmonary venous blood). This pressure imbalance leads to a brief reversal of blood flow through the ductus arteriosus, the muscular wall of which responds to the elevated Po_2 of neonatal blood by contracting. **Prostaglandin E_2 (PGE_2) prevents this contraction and a drop in its levels at birth is involved in closure,** which fails in about 6 in 10 000 term neonates, but more frequently with prematurity. Indomethacin can be used to treat a patent ductus arteriosus by inhibiting PGE_2. The foramen ovale has a flap valve over it in the left atrial chamber. In the fetus, this is maintained open by the stream of blood from the right atrium, but with the reversal of interatrial pressure, the flap is pressed against the interatrial wall, thereby separating the two sides of the heart **to yield two pumps working in series**. There is a recognizable, transitory form of circulation in the newborn that is the result of functional, rather than anatomical, closure of the foramen ovale and ductus arteriosus, and these shunts are able to reopen from time to time. The ductus venosus is closed permanently in most individuals within 3 months of birth and the ductus arteriosus by 1 year; the foramen ovale obliterates very slowly and it is found to have failed to close in about 10% of adults, in whom a probe may be passed through it.

The respiratory system

The fetus spends at least 1–4 hours each day making rapid respiratory movements, which are irregular in amplitude and frequency and generate negative pressures of 25 mmHg or more in the chest. These occur in episodes of up to 30 minutes during rapid eye movement (REM) sleep, but not during wakefulness or slow-wave sleep. They are purely diaphragmatic, moving amniotic fluid in and out of the lungs. Their functions may include an element of 'practice' of the reflex neuromuscular activities to be initiated in breathing at birth, and also promotion of the growth that follows lung distension. Prevention of fetal breathing, as occurs in congenital disorders of the nervous system or diaphragm, retards lung development to an extent that precludes support of extrauterine life.

The fetal lungs undergo major structural changes during pregnancy, and especially near parturition. Primitive air sacs are apparent in the lung mesenchyme from about week 20 and blood vessels appear around week 28. The pressure required to expand the fetal lungs decreases as the time of birth approaches, largely due to the appearance of **surfactant**, which reduces the surface tension of pulmonary fluid. The surfactant is a phospholipid (a di-saturated lecithin, mainly dipalmitoyl lecithin) attached to an apoprotein. It is synthesized by enzymes in the human fetal lung from weeks 18–20, but especially in the 2 months before birth (Figure 17.1). **Its synthesis is promoted by fetal corticosteroids**, which rise preterm. They also stimulate conversion of noradrenaline to adrenaline by activation of **phenylethanolamine N-methyl transferase** (PNMT) in both the adrenal medulla and, locally, in the lungs. **Catecholamine receptor numbers** in the lungs are also increased. Within the lungs, both water resorption and surfactant production rise. Failure to produce sufficient surfactant has serious consequences for lung expansion, as is seen in so-called **idiopathic respiratory distress syndrome** (or **hyaline membrane disease**). Lung maturation can be accelerated by injecting corticosteroids into the mother. Such treatment is highly effective for infants about to be born prematurely (see page 299).

To initiate normal, continuous breathing at birth, the first breath must overcome not only the viscosity and surface tension of fluid in the airways, but also the resistance of lung tissues. **Removal of fluids from the respiratory pathways** occurs through the mouth during vaginal delivery as a result of the rise in intrathoracic pressure, but this does not occur during Caesarean delivery. The remaining fluid is resorbed through the agency of pulmonary lymphatics and capillaries, helped by the cortisol/adrenaline stimulation. Prenatal episodic breathing is replaced rapidly with normal, continuous postnatal breathing, and the gaseous exchange function of the lungs is established within 15 minutes or so of birth. The **mechanisms bringing about the pronounced inspiratory effort at birth are varied.** Cold exposure and tactile, gravitational, auditory and noxious stimuli may all enhance respiration at birth, but are not absolutely necessary for the initiation of continuous postnatal breathing. Whichever factors are important, they operate on a newborn in whom the neuromuscular activities of respiration and swallowing (see below), including the ejection of foreign bodies from the pharynx and trachea, have been rehearsed, and **brainstem medullary respiratory rhythmicity has been established.** Pulmonary stretch receptors and arterial and central chemoreceptor mechanisms are also functional by this time. Again, corticosteroids seem to be involved in their activation.

Finally, during the last trimester, the spleen and bone marrow take over erythropoiesis. This latter transition is also regulated by the rising levels of fetal cortisol towards term, as is the switch from fetal to adult haemoglobins (see pages 261–2).

The gastrointestinal system

The human fetus swallows large volumes of amniotic fluid daily, which pass through the stomach to the large bowel. Water is absorbed readily through the small bowel as are electrolytes and other small molecules, such as glucose. Debris from fetal skin and larger molecules in amniotic fluid accumulate in the large bowel, together with sloughed cells from the small intestine and bile pigments, to form a green faecal mass, **meconium**. Defecation *in utero* does not normally occur.

The **endocrine pancreas** is active early in pregnancy, **glucagon (α-cells)** and **somatostatin (δ-cells)** predominating at first, followed by **insulin (β-cells)**, which is clearly present by 10 weeks in humans. Development of β cell function appears to depend on the activity of **fetal anterior pituitary growth hormone and ACTH/corticosteroid activity**. Initially, each type of cell is clustered separately, and only later do β-cells become surrounded by α- and δ-cells. Abnormalities in early pancreatic development tend to affect α- and δ-cells preferentially, leading to relatively **uncontrolled β-cell activity, hyperinsulinaemia and hypoglycaemia (nesidioblastosis syndrome)**. Insulin secretion from mid-pregnancy onwards responds positively to amino acids, glucose and short-chain fatty acids, and negatively to catecholamines. Details of how insulin acts in the fetus were discussed in Chapter 15 (see page 264).

A progressive increase in the activity of the fetal adrenal cortex near term is especially important in promoting the **deposition of liver glycogen** (Figure 17.1). Indeed, fetal adrenal hypoactivity is associated with major reductions in liver glycogen stores, a situation reversed by corticosteroid replacement, while **exogenous corticosteroids enhance liver glycogen content in normal fetuses**.

The renal system

Although the placenta is the main organ of excretion, the fetal kidneys are functional during pregnancy and produce substantial quantities of hypotonic urine, the tubules being inefficient at sodium reabsorption. Thus, **fetal aldosterone** levels rise only late in pregnancy and do not become responsive to a reduction in blood volume or nephrectomy until after birth. **Fetal atrial natriuretic factor (ANF)** is also produced prenatally, perhaps under corticosteroid control, and is able to regulate Na^+ excretion. Fetal urine contributes to total amniotic fluid volume to the extent of about half a litre per day. **Renal agenesis (Potter's syndrome)** is associated with a marked reduction of amniotic fluid volume; however, affected fetuses survive to term and are born alive despite various growth and developmental defects. At parturition, renal function must undergo a quite radical alteration, as the constant supply of water, sodium and other electrolytes through the placenta will be lost. Soon after birth, urine flow occurs at a high rate while sodium reabsorption is rather low, but urine flow is reduced over the following hours, rising again after a week. The neonatal glomerular filtration rate is only about one-third that expected for body size, and does not achieve maturity until 1.5–2 years of age. The newborn is in danger of hyponatraemia, as the ability to retain Na^+ is poor. Prematurity represents an important risk in this context, as the immature kidney leaks up to three times more Na^+ than that seen in babies born at term. The improvement seen in term babies reflects the action of rising fetal cortisol on Na^+/K^+-ATPase activity in the cortical tubules. The embryonic changes in the genital tract and external genitalia during sex differentiation were discussed in Chapter 3.

The nervous system

Given its complexity, the human nervous system develops more slowly than most other mammals, a process that continues after birth. Functionally, fetal movements occur early in pregnancy (from about 8 weeks), and these can be felt readily by mothers by week 14 (**quickening**). The function of the movements is uncertain, but 'exercise', which will contribute to both muscle growth and limb development, is likely to be among them. During the long human gestation period, the innervation of muscles by motor nerves and partial maturation of both ascending sensory and descending motor systems in the central nervous system (CNS) mean that some well-coordinated movements become possible by late in pregnancy (week 24 and onwards). A number of simple postural and other stereotyped reflexes is also apparent in the fetus from a relatively young gestational age. However, **complete myelination of the long motor pathways**, e.g. the corticospinal tract, does not occur until after birth, which is why fine movements of the fingers, e.g. apposition of fingers and thumb, are not possible until then. The fetus is also clearly capable of responding to extraneous stimuli. Loud noises and intense light, noxious stimulation of the skin and rapid decreases in the temperature of its fluid environment will result not only in movement but also in autonomic responses, such as acceleration of heart rate. However, presumably the level of stimulation in these sensory modalities is normally rather low and unvarying in the fluid-cushioned, constant temperature, light- and sound-attenuated chamber in which the fetus grows.

An important scientific, clinical and ethical issue concerning the development of sensory systems in the fetal brain is associated with the question, **When does a fetus feel pain?** Medical procedures during pregnancy (including amniocentesis), surgical procedures on the fetus itself and also the termination of pregnancy all may involve exposing the fetus to noxious stimuli and so an answer to the question is important. Clearly, an unambiguous definition of pain is required and this is obviously problematic in individuals who cannot speak and therefore cannot express what they are feeling. The International Association for the Study of Pain defines pain as '**an unpleasant sensory and emotional experience associated with actual or potential tissue damage, or described in terms of such damage … pain is always subjective. Each individual learns the application of the word through experiences related to injury in early life.**' This definition emphasizes the importance of cognitive factors such as prior experience of past injury. Thus, an important aspect of pain is its unpleasant emotional component: take that away and a noxious stimulus, which can still damage tissue, is not necessarily 'painful'.

Given the obvious complexity of the neural processes subserving pain, focusing on neural maturation of a specific part of the brain's nociceptive system will not necessarily help the determination of when a fetus feels pain. Much importance is often ascribed to the fact that **fetal skin has nocioreceptors** by 10 weeks (but later for deeper tissues) and that fetuses show clear **nociceptive withdrawal reflexes** by 15 weeks. But these generally reflect the fact that the spinal mechanisms mediating these simple reflexes are established at that time, occurring as they do in anencephalic fetuses. They cannot be taken to indicate that the fetus feels pain. **Stress responses** in connection with reflex withdrawals are not seen earlier than 18 weeks, implying that only by then are spinal and brainstem connections with the mid-brain established. But there is an emerging consensus among developmental neurobiologists that it is the establishment of **thalamocortical connections** (the pathway by which peripheral sensory information arrives at the cortex, where conscious sensation and feelings are processed) that must be a critical event. The penetration of thalamic fibres into the developing cortex occurs between weeks 12 and 18, but cortical in-growth cannot be observed until 24 weeks, by when **functional connections between the thalamus and cortex cannot reliably be detected**. Such data have led to the suggestion that a fetus cannot 'feel' pain before the cerebral cortex is able to process incoming sensory, including noxious, information and that this is therefore most unlikely to be the case before about week 24 of gestation. Of course, even then it is impossible to know whether the fetus is consciously feeling pain since we have few data on the development or neural basis of consciousness. Cortical responses to pain may be necessary but not sufficient for the experiencing of pain. The multidimensionality of pain perception, involving sensory, emotional and cognitive factors, may in itself be the basis of a conscious, painful experience, but it will remain difficult to attribute this to a fetus at any particular developmental age.

Indeed, data from fetal sheep and baboons suggest that **conscious wakefulness** is not a feature of fetal life, both showing sleep-like EEG pattern, as though sedated, perhaps due the effects of adenosine. The human fetus also seems to alternate between periods of **slow-wave and REM sleep** for most of the day, spending a much greater proportion of this time in REM sleep than an adult does. Postnatally, this sleeping time progressively decreases over the first 2 years of life to about 12 hours each day, and the proportion of REM sleep decreases to about one-quarter compared with slow-wave sleep. Such a conclusion is not incompatible with the evidence of fetal learning or conditioning to sensory stimuli, as wakefulness is not necessary for such conditioning to occur.

In light of these observations, the question arises whether some form of pain control should be considered for medical procedures occurring particularly during the third trimester, in order to minimize any possible adverse effects on the fetus. More research and many more data are needed to clarify these important issues, especially if rational approaches to **pain control in neonatal medicine** are to be developed (see Box 17.1 for a summary of a report on this topic).

A final matter of clinical and sociological significance is the impact of drugs from the maternal circulation on the developing brain of the fetus. Many are lipid-soluble and have no barrier to their free diffusion across the placenta and therefore into the fetal brain. The immature status of the blood–brain barrier also ensures that other chemical agents may gain access to the

Box 17.1 Fetal awareness

In reviewing the neuroanatomical and physiological evidence in the fetus, it was apparent that connections from the periphery to the cortex are not intact before 24 weeks of gestation and, as most neuroscientists believe that the cortex is necessary for pain perception, it can be concluded that the fetus cannot experience pain in any sense prior to this gestation. After 24 weeks there is continuing development and elaboration of intracortical networks such that noxious stimuli in newborn preterm infants produce cortical responses. Such connections to the cortex are necessary for pain experience but not sufficient, as experience of external stimuli requires consciousness. Furthermore, there is increasing evidence that the fetus never experiences a state of true wakefulness *in utero* and is kept, by the presence of its chemical environment, in a continuous sleep-like unconsciousness or sedation. This state can suppress higher cortical activation in the presence of intrusive external stimuli. This observation highlights the important differences between fetal and neonatal life, and the difficulties of extrapolating from observations made in newborn preterm infants to the fetus. The implications of these scientific observations for clinical practice are such that the need for analgesia prior to intrauterine intervention, for diagnostic or therapeutic reasons, becomes much less compelling. Indeed, in the light of current evidence, the Working Party concluded that the use of analgesia provided no clear benefit to the fetus. Furthermore, because of possible risks and difficulties in administration, fetal analgesia should not be employed where the only consideration is concern about fetal awareness or pain. Similarly, there appeared to be no clear benefit in considering the need for fetal analgesia prior to termination of pregnancy, even after 24 weeks, in cases of fetal abnormality.

(Summary of the report of a Working Party of the Royal College of Obstetricians and Gynaecologists entitled: Fetal Awareness. Review of Research and Recommendations for Practice, March 2010)

fetal brain in a way not seen in the adult. Drugs such as opiates (heroin, morphine) and nicotine, if taken by pregnant women, can produce dependence in their babies, including the appearance of withdrawal symptoms after birth. Pain-killing drugs, which are sometimes given during labour, may depress the behavioural repertoire of the newborn; e.g. sucking reflexes may be impaired. This may have adverse consequences for both lactation and mother–infant interaction (see Chapter 19).

Summary

Above we have examined how selected fetal systems operate during intrauterine life and prepare the fetus for parturition and extrauterine survival. The mechanisms involved vary from the instantaneous (cardiovascular system) to the gradual (urinary system), with some occurring late in pregnancy (gastrointestinal and respiratory systems) and many regulated, like parturition itself (see Chapter 18), by endocrine events involving the fetal adrenal cortex. In the next section, we examine aspects of fetal endocrine regulation more closely.

Fetal hormones orchestrate development and preparations for birth

It has already been emphasized that the fetoplacental endocrine system exerts important influences on maternal physiology (see Chapter 15). In contrast, the maternal endocrine system does not generally influence the fetus directly, except in the first trimester and in pathological circumstances, and few maternal hormones other than unconjugated steroids and releasing hormones cross the placenta readily.

In general, the fetal endocrine organs initiate function by the end of the first trimester. Most of the **fetal hormones do not cross the placenta to the mother**. The fetal endocrine system has a number of unique functions not apparent in the adult, e.g. in the differentiation of the reproductive tract, lungs, gastrointestinal system and even the brain itself.

The anterior pituitary

The anterior pituitary functions through most of fetal life from week 7 onwards, but is not fully functional, receiving pulses of hypothalamic RH only by 18 weeks. Growth hormone (GH) is produced by the fetal pituitary, but any role in the fetus is unclear as GH receptors are deficient and its role is probably taken by the IGFs (see page 259). Prolactin is present in amniotic fluid in concentrations 100-fold greater than those in either maternal or fetal circulations, but most of this prolactin is of decidual origin. After week 30 or so, the low levels of prolactin in fetal plasma rise markedly until term, presumably as a result of fetal pituitary activity, but then decline neonatally after a brief postpartum rise. Its functions are unknown, although decidual prolactin may have a role in regulating the permeability of the chorion and amnion to water. It may also be an important co-hormone in facilitating the effects of corticosteroids on the production of lung surfactant. The pituitary's thyroid and adrenal regulating activities are discussed below, and its gonadal regulating activity was considered in Chapter 3.

The thyroid gland

During early pregnancy, oestrogen stimulates increased production of **thyroxine-binding globulin (TBG)** and elevated albumin, both of which temporarily reduce free thyroxine. The resulting negative feedback provokes an increase in secretion of **thyroid stimulating hormone (TSH)** and hyperstimulation of the thyroid. This effect is enhanced by the fact that **hCG directly stimulates the maternal thyroid**, due to similarities between its β-chain and that of TSH. In 0.2% of women, these changes can lead to **maternal hyperthyroidism**, most commonly seen in women already at risk for this condition or with trophoblastic tumours or if carrying a multiple pregnancy. In those women with Graves' disease, their levels of anti-TSH antibodies tend to fall after the first trimester, meaning that the risk to the fetus from high levels of T_4 also falls then. Women who conceive despite poorly controlled thyrotoxicosis have a higher incidence of miscarriage, intra-uterine growth restriction, and perinatal mortality. Also, when untreated maternal antibodies cross the placenta in the third trimester, thyrotoxicosis in the late fetus and neonate can occur. Conversely, about 2% of women show symptoms of hypothroidism, usually due to a pre-existing condition (whether or not diagnosed), and for these women supplementation with T_4 may be required for at least the first trimester (see below).

Thyroxine (T_4) is essential for the normal development of the fetus, and an increased output of up to 30% is required during pregnancy. Maternal T_4 supports the needs of the conceptus through the first trimester, but has more limited transplacental access to the fetal circulation later due to placental de-ionidase activity, and in amounts insufficient to meet fetal needs. Fetal TRH and TSH are produced from 7–8 weeks, and the fetal thyroid is functional in producing T_4 by 11–12 weeks, but even at term up to 30% of fetal blood thyroid hormones are of maternal origin. By mid gestation, **maternal iodine** is required for the secretion of T_4 by the fetal thyroid under the influence of fetal TSH to yield circulating T_4 levels that exceed those in the mother by term. Some of this iodine comes from **placental de-iodinase activity**. **Fetal hypothyroidism** is associated with a bone age far behind chronological age, deficiency in body hair and, most importantly, behavioural retardation, since T_4 is essential for the normal differentiation of the CNS. At birth, TSH, T4, and T3 are abruptly released in the neonate to create a relatively hyper-thyroid state, considered facilitatory of the thermoregulatory adjustment to extra-uterine life.

Mineral-regulating hormones

Fetal levels of calcium, phosphorus and magnesium in late gestation are all higher than in maternal blood, reflecting their requirement for bone formation. The fetal parathyroid glands secrete two calcium-regulatory hormones from about 10 weeks of gestation: **parathormone (PTH)** and **parathormone-related protein (PTHrP)**. PTH's main function is to respond to lowered fetal blood levels of calcium, sensed via the **parathyroidal calcium sensing receptor (CaSR)**. PTH is present at rather low levels in the fetus, where it acts on target cells via the **PTH receptor (PPR)** to mobilize calcium. Its levels peak in fetal blood in mid-gestation, falling towards term, probably because of an increasing influx of maternal calcium, which thus becomes the main determinant of ossification rates. In contrast, although PTHrP similarly regulates fetal blood calcium, it is present at 15-fold higher levels at term, reflecting its additional production by the placenta and fetal liver, skeletal growth plate and amnion. Mouse embryos lacking PTHrP show abnormal ossification, hypocalcaemia and defective transplacental calcium transport, suggesting a paracrine role in the regulation of the latter. However, the identity of the placental calcium-sensitive receptor regulating PTHrP is unknown, as is the PPR-equivalent receptor by which it stimulates transport. Through their joint action, **calcium inflow to the fetus maintains fetal blood calcium even at the expense of hypocalcaemic mothers.** In mice genetically lacking parathyroids, the absence of PTH is compensated by increased secretion of PTHrP. In contrast, mice genetically null for *Prhrp* die at birth as a result of defective calcium transport and attendant mineralization problems.

Calcitonin is produced from thyroid parafollicular cells from 15 weeks, but is also produced by the trophoblast and is present at twice the maternal concentration in fetal plasma, where it helps maintain ossification. Vitamin D is transported across the placenta in its 25-hydroxylated form, 25OHD, where it is converted to its active form by the fetal kidneys and the placenta. On the maternal side, calcitonin and Vitamin D levels are elevated, whilst PTH is suppressed and PTHrH is not detectable. It may be necessary to provide the mother with extra Vitamin D to increase her dietary absorption of calcium thereby avoiding undue decalcification of her own skeleton.

The adrenal glucocorticoids

We have already encountered the importance of adrenal **glucocorticoid activity** for the fetus (see pages 282, 283 and Chapter 16) and for the endocrine function of the placenta (see page 251), and will do so again in Chapter 18 in the context of parturition. **Fetal pituitary secretion of ACTH** is under the control of the fetal **hypothalamic corticotrophin-releasing hormone (CRH)** and changes in the content of CRH mRNA precede alterations in the circulating levels of ACTH and cortisol. Fetal serum ACTH concentrations start to rise from 9 weeks, are quite high during weeks 12–19 of gestation, and gradually decline by around week 40. DHEA stimulation by hCG is also observed during the first trimester, and placental CRH also stimulates production of placental ACTH and so of fetal cortisol, especially as parturition approaches in the term fetus (see pages 294–5). Cortisol is found in fetal blood by week 10 and levels increase as gestation proceeds (Figure 17.1), to be especially high during labour.

After birth, the adrenal gland shows major structural changes. The fetal zone, which is responsible for the synthesis of dehydroepiandrosterone (DHEA; substrate for placental aromatization to oestrogen; see page 252), and which makes up some 80% of the adrenal volume in the fetus, regresses during the first neonatal month, while the cortex proper differentiates into the distinctive three zones characteristic of the adult gland. Table 17.1 summarizes the actions of fetal corticosteroids, especially as parturition approaches, and emphasizes its coordinating role during the transition from fetal to neonatal life.

Table 17.1 Summary of functions of the glucocorticosteroids secreted by the fetal adrenal cortex towards term (for details see text)

Function	Mechanism and/or relevant section
Circulatory system	Increase in fetal cardiac output, peripheral resistance and blood pressure (Chapter 16)
Lung maturation	Induction of enzymes necessary for surfactant synthesis, stimulation of alveolar water resorption and central and local respiratory mechanisms
Haematopoiesis	Promotes ‘switch’ in production of fetal to adult haemoglobin and shift of haematopoiesis to bone marrow and spleen
Glucose storage and gluconeogenesis	Induction of enzyme systems in liver and myocardium
Insulin secretion	Regulates maturation of fetal pancreatic islets (see ‘Glucose and carbohydrate’ in Chapter 15)
Synthesis of adrenaline	Induction of phenylethanolamine-N-methyl transferase in adrenal medulla
Production of thyroxine	May promote conversion of T_3 to T_4
Maturation of salt:water ratio control	Activation of atrial natriuretic factor? Stimulation of glomerular filtration rate and regulation of reabsorption of Na^+, DHEA
Lactogenesis	Ductular–lobular–alveolar growth in pregnancy (see Chapter 18)
Parturition	Induction of placental oestrogen-synthesizing enzymes. Increase in oestradiol precursor (dihydroepiandrosterone) concentrations (a function of fetal zone; see Chapter 18)

Conclusions

The fetus has to balance life in the womb with anticipation of leaving it. Corticosteroids seem to play key roles in managing this transition, but we have incomplete information as to how their prenatal rise is regulated and timed. This is unfortunate, as the transition to independent life is hazardous. If maturity comes too late, or if birth is premature, then neonatal survival and health are threatened. However, nature has evolved a mechanism for avoiding this potential lack of synchrony by tying birth itself to the corticosteroid rise – in most cases! We will see how it does so, and what happens when it does not, in the next chapter.

Key learning points

- Fetal and neonatal neuroendocrine systems, and especially the rise in fetal corticosteroids towards term, coordinate many aspects of later fetal development and preparations for birth.
- Fetal cardiac output, peripheral resistance and blood pressure rise as parturition approaches and are underlain by rising cortisol levels.
- The fetal circulation differs in two ways from that in the adult because the placenta, not the lungs, is the organ of gaseous exchange.
- The two fetal ventricles pump in parallel, not in series as in the adult, and three vascular shunts divert the fetal circulation away from the lungs and towards the placenta.
- At birth, closure of the three fetal shunts occurs to replace the placental circulation with a pulmonary circulation.
- The pressure required to expand the fetal lungs decreases as the time of birth approaches, because surfactant reduces the surface tension of pulmonary fluid.
- Synthesis of surfactant and resorption of water and salts from the respiratory tract are promoted by fetal corticosteroids.
- Failure to produce sufficient surfactant results in respiratory distress syndrome.
- The switch from fetal to adult haemoglobin and the relocation of erythropoiesis to the spleen and bone marrow are stimulated by corticosteroids.
- The endocrine pancreas is active early in pregnancy, glucagon and somatostatin predominating initially followed by insulin, which is clearly present by week 10 in humans.
- Liver glycogen deposition in late pregnancy is stimulated by fetal corticosteroids.
- Renal agenesis results in a marked reduction of amniotic fluid volume.
- Fetal movements or 'quickening' can be felt from week 14 onwards.
- It is unclear whether, and if so by when, pain can be felt by fetuses, but it is unlikely to be before 24 weeks at the earliest.
- Fetal reflex responses occur much earlier.
- Fetuses and neonates spend most of the day asleep and experience decreasing proportions of time in REM sleep.
- Thyroxine is essential for the normal development of the fetus, including the brain, and is of maternal origin in the first trimester, but mostly fetally derived thereafter, although it requires a supply of maternally-derived iodine.
- Fetal hypothyroidism is associated with a bone age far behind chronological age, deficiency in body hair and behavioural retardation.
- Fetal parathormone-related protein is required for adequate calcium transport into the fetus and ossification.

Clinical vignette

Antenatal steroids

A 23-year-old woman, Ms D, presented to her local maternity unit at 26 weeks in her first ongoing pregnancy with fever, nausea and lower abdominal pain. She had felt feverish three days previously and had attended her GP complaining of pain in passing urine. She was given oral antibiotics for infection, but felt progressively worse over the following days. After initial assessment at her local maternity unit, a diagnosis was made of ascending urinary tract infection and subsequent threatened preterm labour. On speculum examination her cervix was closed, but she was having regular uterine tightenings. She was admitted for observation and intravenous antibiotics to treat the urinary tract infection. Her obstetrician recommended a course of intramuscular dexamethasone, as she felt that Ms D was at high risk of a premature delivery. The following morning, Ms D's tightenings had increased in both strength and frequency. A speculum examination revealed that the cervix was 4 cm dilated and that the amniotic sac was bulging into the vagina. Magnesium sulphate was commenced for fetal neuroprotection and the neonatal team placed on stand-by. Two hours later, Ms D spontaneously delivered an extremely preterm but vigorous male infant. The neonatal team intubated the baby shortly after delivery and were able to ventilate the lungs at reasonable pressures. Ms D's infant had a complicated postnatal course, but was eventually discharged home 10 weeks after delivery with no requirement for any supplementary oxygen.

Synthetic glucocorticoids are regularly administered to pregnant women when there is a high chance of a spontaneous or medically indicated preterm delivery. Administration of steroids mimics the endogenous surge in fetal glucocorticoid production that matures the term fetus in preparation for delivery. The primary aim of steroid therapy is to promote fetal lung maturity by stimulating type 2 pneumocytes to produce surfactant (surface-active phospholipoprotein complex). Surfactant is critical for inflation of the lungs around the time of delivery because it reduces surface tension within the alveoli and thus reduces the initial expansion pressures required for breathing. Following administration of antenatal steroids, the risk of respiratory distress syndrome in the preterm babies is reduced by approximately 45% compared with babies who did not receive steroids. Steroid therapy is also of benefit for maturing many other organ systems. A major cause of morbidity in the severely preterm infant is from intraventricular haemorrhage in the developing brain. Antenatal steroids reduce the chance of intraventricular haemorrhage by 46% and are effective at protecting significant numbers of very premature babies against this potentially fatal complication. Furthermore, the risk of neonatal death from all causes is reduced following steroid administration.

FURTHER READING

General reading

Feldt-Rasmussen U, Mathiesen ER (2011) Endocrine disorders in pregnancy: physiological and hormonal aspects of pregnancy. *Best Practice & Research Clinical Endocrinology & Metabolism* **25**, 875–884.

Girling J, Martineau M (2010) Thyroid and other endocrine disorders in pregnancy. *Obstetrics, Gynaecology and Reproductive Medicine* **20**, 265–271.

Ohata Y, Ozono K, Michigami T (2016) Current concepts in perinatal mineral metabolism. *Clinical Pediatric Endocrinology* **25**, 9–17.

Owen D, Andrews MH, Matthews SG (2005) Maternal adversity, glucocorticoids and programming of neuroendocrine function and behaviour. *Neuroscience and Biobehavioral Reviews* **29**, 209–226.

Sack J (2003) Thyroid function in pregnancy – maternal–fetal relationship in health and disease. *Pediatric Endocrinological Reviews* **1** (Suppl. 2), 170–176.

Weinstock M (2005) The potential influence of maternal stress hormones on development and mental health of offspring. *Brain, Behavior and Immunity* **19**, 296–308.

More advanced reading

Bass JK, Chan GM (2006) Calcium nutrition and metabolism during infancy. *Nutrition* **22**, 1057–1066.

British Medical Journal (1996) Do fetuses feel pain?: For debate. *British Medical Journal* **313**, 795–798.

Ciba Foundation Symposium 86 (1981) *The Fetus and Independent Life*. Excerpta Medica, Amsterdam.

Coceani E, Olley PM (1973) The response of the ductus arteriosus to prostaglandins. *Canadian Journal of Physiology and Pharmacology* **51**, 220–225.

Fitzgerald M (1994) Neurobiology of fetal and neonatal pain. In: *Textbook of Pain* (ed. P Wall, R Melzak), pp. 153–163. Churchill Livingstone, Edinburgh.

Jones CT (ed.) (1988) *Research in Perinatal Medicine* (VII): *Fetal and Neonatal Development*. Perinatology Press, Ithaca, NY.

Merskey H (1991) The definition of pain. *European Psychiatry* **6**, 153–159.

Rodien P, Jordan N, Lefèvre A *et al.* (2004) Abnormal stimulation of the thyrotrophin receptor during gestation. *Human Reproduction Update* **10**, 95–105.

Sandman CA, Wadhwa PD, Dunkel-Schetter C *et al.* (1994) Psychobiological influences of stress and HPA regulation on the human fetus and infant birth outcomes. *Annals of the New York Academy of Sciences* **739**, 198–210.

Tobias JH, Cooper C (2004) PTH/PTHrP activity and the programming of skeletal development in utero. *Journal of Bone and Mineral Research* **19**, 177–182.

Weinstock M (1997) Does prenatal stress impair coping and regulation of hypothalamic–pituitary–adrenal axis? *Neuroscience and Biobehavioural Reviews* **21**, 1–10.

CHAPTER 18

Giving birth

Essential Reproduction, Eighth Edition. Martin H. Johnson.

Companion website: www.essentialreproduction.com

In the previous chapters, we described the development of a mature fetus and how it prepares itself for birth. In this chapter, we consider the birth process itself – **parturition**. Pregnancy has a **median duration of 280 days (40 weeks)** from the last menstrual period, with a **range from 259 to 294 days (37–42 weeks)**. Birth prior to 259 days is considered **preterm**, and later than 294 days **post-term**. Birth is a time of particular hazard for both mother and baby, indeed, a major source of evolutionary selective pressure comes at this time (see Figures 15.1 and 16.5). Thus, if the baby is born prematurely or too small, problems of neonatal viability and health arise (see Chapter 15). In contrast, if birth is delayed, the baby may become too large to deliver non-surgically. Even an 'on time' delivery can be problematic, given the risk of infection and the relative sizes of the neonatal head and maternal pelvic opening (Figure 18.1). Thus, **maternal mortality ratios (MMR**s; expressed as deaths/100 000 live births) vary geographically with MMRs of 450 versus 9 for developing and developed regions, 85% of all maternal mortality occurring in Sub-Saharan Africa and South Asia (see Figure 16.3). However, maternal mortality was high historically in the developed world until the mid-1930s (Figure 18.2). Indeed, a recent trend towards higher maternal mortality in the developed world has been noted, raising the question as to whether modern medicine might be making matters worse rather than better (see Box 18.1).

Complications during delivery are also associated with a significant impact on the neonate, resulting in poor pregnancy outcomes. In developing countries, only half of births are attended by a trained attendant and most occur at home. In these countries, **early neonatal deaths** (0–27 days) account for 41% – or a massive 3575 million babies – of all deaths in children younger than 5 years, most occurring at home. The reasons for this are many, but do indicate the importance of adequate health care. For example, in comparison, in 2008 in the UK, there were 3617 stillbirths plus 1729 deaths within 28 days of birth from a total of 708,000 births, giving a perinatal death rate of 0.75%. Moreover, many of these deaths were accounted for by multiple birth losses. Whilst any neonatal loss is a tragedy, the retreat from the massive losses of former times has come from a variety of medical, obstetric and social advances. We return to this theme at the end of this chapter. First, however, we consider what we know about the process of birth itself.

Parturition

In Chapter 17, we hinted that the fetus prepares not only itself but also its mother for birth. However, the factors affecting the onset and mechanism of parturition in women remain incompletely understood, despite intensive study. A lot of our knowledge on human parturition has come from pharmacological interventions in attempts to facilitate overdue or difficult births or to delay or prevent premature ones. Whether these interventions help us to understand endogenous controls or mislead us is unclear. More rigorous scientific data have been obtained from studies on other species, especially sheep and rodents. However, these species differ from the human in both the source and profile of their pregnancy hormones (see Chapter 14) and in many aspects of parturition, so at best they give us clues as to what questions to ask of the human situation – and then it is often unclear whether they are the appropriate questions! Given how species-specific parturition is, more recent studies in non-human primates have shed considerable light into the physiology underlying our birth process.

Labour

Before we discuss those factors that determine the onset of **labour**, the process by which the mother expels the fetus, we first describe some of its key features. The process of labour is divided into three stages. The **first stage** begins with its onset (regular painful contractions, and dilation and shortening of the cervix) and ends when the **uterine cervix is fully dilated**. It may further be divided into a **latent phase**, when the cervix slowly dilates to about 3 cm, and an **active phase** thereafter when the dilation of the cervix occurs more rapidly. The **second stage** of labour begins at full dilation of the cervix and ends with complete **delivery** of the fetus. The **third stage** begins with completion of fetal expulsion and ends with **delivery of the placenta**.

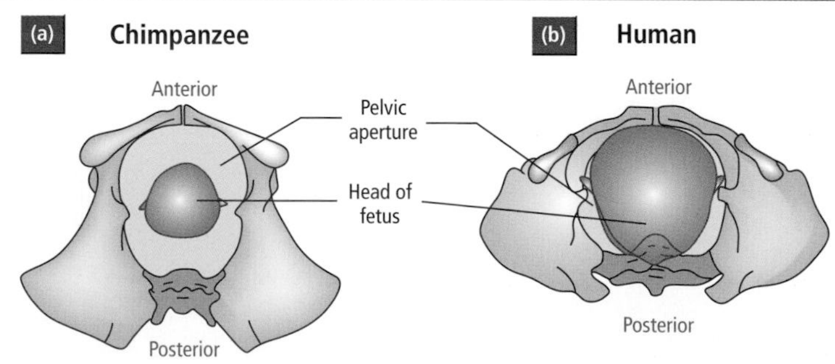

Figure 18.1 The chimpanzee and human pelvises compared. Note that the large aperture of the chimpanzee pelvis (a) permits easy passage of the relatively small head of the fetus in an anterior–posterior orientation. In contrast, the human pelvis (b) has an aperture just large enough to allow passage of the head of the fetus in a similar orientation.

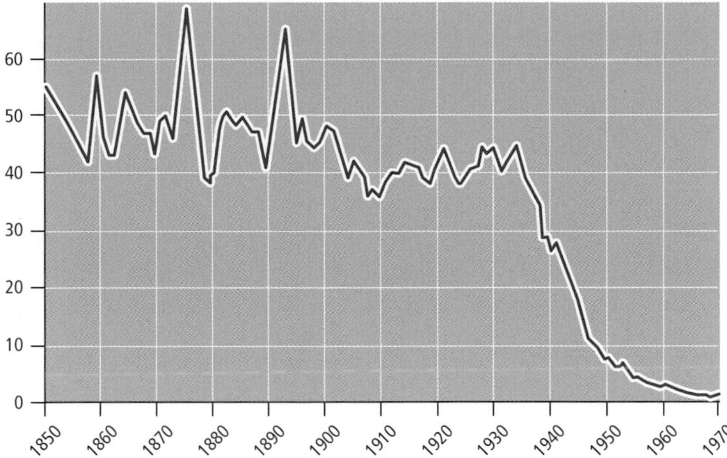

Figure 18.2 Annual death rate per 1000 total births from maternal mortality in England and Wales (1850–1970). Maternal death is defined by the International Classification of Diseases, Injuries and Causes of Death as: 'the death of a woman while pregnant or within 42 days of termination of pregnancy, from any cause related to or aggravated by the pregnancy or its management but not from accidental or incidental causes.' Note the drop in mortality from the 1930s, which can be attributed to the use of antibiotics (penicillin and sulphonamides) for infection control, and more effective obstetric care for all, enhanced by the introduction of the National Health Service. Currently in the less-developed world, almost 80% of maternal deaths are due to direct obstetric causes similar to those seen in the UK in the nineteenth century and up to the 1930s, including haemorrhage, infection, complications from unsafe abortion, eclamptic convulsions, and obstructed labour. (Source: Registrar General Reports.)

With the onset of labour, large contractions of the uterine musculature occur. These are regular, occur at shorter and shorter intervals and result in intrauterine pressures of 50–100 mmHg, compared with about 10 mmHg between contractions. One of the functions of these contractions is to retract the lower uterine segment and cervix upwards to allow the vagina and uterus to become one **continuous birth canal**, through which fetal expulsion occurs. This phenomenon of **brachystasis** reflects the properties of myometrial cells. Thus, shortening of each muscle cell during contraction is followed during relaxation by a failure to regain its initial length. With each subsequent contraction, further shortening of the cell occurs and so, eventually, each myometrial cell becomes shorter and broader, the fundal musculature becomes thicker and uterine volume decreases. The **lower uterine segment does not take part in these contractions** and remains quite passive during labour. As a result of this muscular phenomenon, the lower segment moves upwards, and is therefore retracted. This event may be palpated abdominally because the junction between the two uterine segments (the physiological **retraction ring**, marked because of the contrast between the thick myometrium above and the thin lower uterine segment below) gradually moves upwards. During this time the cervix becomes increasingly dilated and can no longer be pulled upwards because of its attachment to uterine and uterosacral ligaments and pubocervical fascia.

Moving into the second stage of labour, the **fully dilated cervix** is drawn up to just below the level of the pelvic inlet. Subsequent uterine contractions, and the resultant decrease in uterine volume, push the fetus downwards and through the pelvis (see Figure 18.3 for summary). This whole process of labour varies in duration between individuals but usually

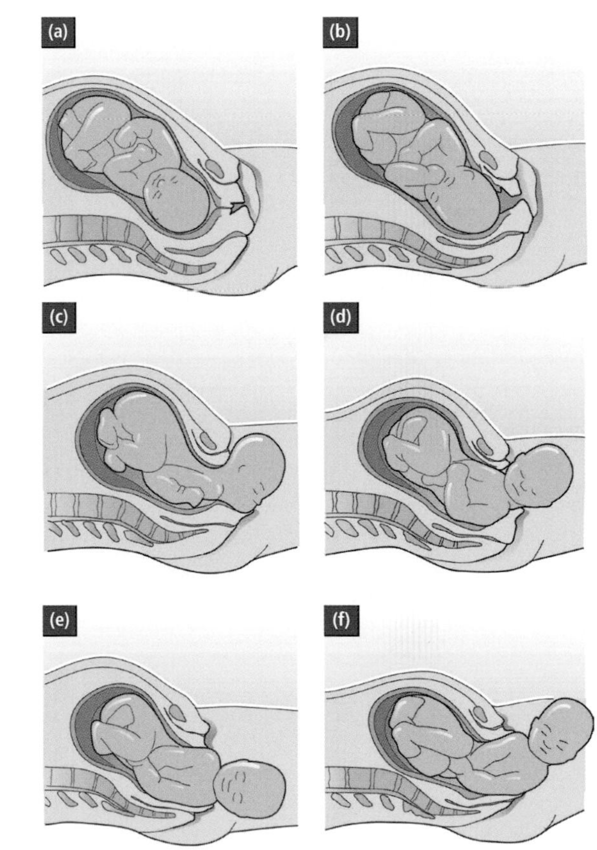

Figure 18.3 Normal labour showing: (a) engagement and flexion of the head; (b) internal rotation; (c) delivery by extension of the head after dilation of the cervix; (d–f) sequential delivery of the shoulders. (Source: Courtesy of K.R. Niswander.)

Box 18.1 Is modern medicine increasing maternal mortality?

Because doctors desire to improve maternal and neonatal survival, many in Europe and North America are cautious about natural home births, far from hospital facilities. However, they often also are oblivious to adverse outcomes that may arise from their interventions. Examination of recent UK maternal mortality rates (MMRs) shows a slight increase (Figure 18.4). Indirect deaths from underlying medical and psychiatric causes now exceed direct deaths. This slow rise is also seen in the Netherlands, Denmark, Austria, Canada, Norway and the USA, where it has been explained as being due to changes in data reporting and processing. However, contributions to MMR resulting from deferred child bearing, infertility treatment leading to multiple pregnancy and an increased choice of caesarian delivery may also be occurring. Might, paradoxically, the achievements of modern medicine now be generating their own contribution to MMR?

What is clear, however, is that, compared with the less-developed world, the appropriate use of caesarian deliveries is saving much maternal and fetal loss. Thus, it has been calculated from epidemiological data that maternofetal deaths were inversely correlated with a caesarian delivery rate up to 19%, and that in those countries that performed fewer caesarian deliveries, providing the high quality infrastructure to increasing them to 19% would save as many as 165,500 maternal and 803,100 neonatal lives!

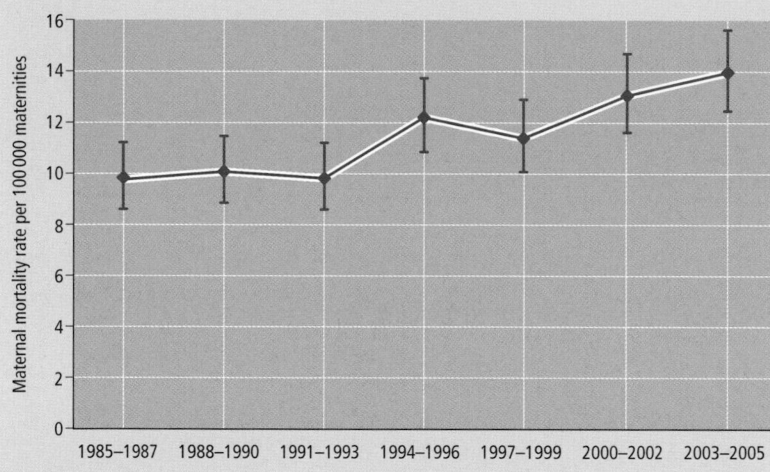

Figure 18.4 Plot of recent rise in maternal mortality rates in the UK. (Adapted from Lewis G, Macfarlane A (2007) Which mothers died, and why. In: *Saving Mothers' Lives (2003–2005)*, Seventh Report of the Confidential Enquiries into Maternal Deaths in the United Kingdom, p. 22. Public Health Agency, Belfast.)

Further reading

Bewley S, Foo L (2011) Are doctors still improving childbirth? In: *Birth Rites and Birth Rights* (eds. F Ebtehaj, J Herring, MH Johnson, M Richards), pp. 51–76. Hart Publishing, Oxford.

Braude PR, El-Toukey T (2011) Too late or too many – dilemmas facing the modern woman. In: *Birth Rites and Birth Rights* (eds. F Ebtehaj, J Herring, MH Johnson, M Richards), pp. 255–270. Hart Publishing, Oxford.

Jackson E, Abdalla H (2011) IVF birth data presentation: its impact on clinical practice. In: *Birth Rites and Birth Rights* (eds. F Ebtehaj, J Herring, MH Johnson, M Richards), pp. 270–284. Hart Publishing, Oxford.

Molina G, Esquivel MM, Uribe-Leitz T *et al.* (2015) Avoidable maternal and neonatal deaths associated with improving access to caesarean delivery in countries with low caesarean delivery rates: an ecological modelling analysis. *The Lancet*, 27 April, 33.

takes **less than 8 hours in multiparus** and **14 hours in primiparus women**. The first stage occupies much of this time and the second stage should generally last less than 1 hour. A few minutes after delivery of the fetus and **clamping of the umbilical cord**, the placenta becomes detached from the uterine wall as the result of a myometrial contraction. Within a short time, the placenta will be completely expelled by uterine contractions, a process often aided by the midwife or obstetrician pharmacologically and by steadily pulling on the umbilical cord (**active management** of the third stage of labour).

Parturition is a time of vulnerability for the fetus as it undergoes the transition to neonatal life. The necessary adjustments that it must make to its own cardiovascular and respiratory systems (see Chapters 16 and 17) are preceded by a period of dwindling maternal support as labour progresses. If this period is unduly protracted, the effectiveness of metabolic exchange can decline such as to cause **fetal distress, hypoxia and asphyxia**. The traditional obstetric approach has been to monitor the fetal heart rate by intermittent auscultation with a Pinard obstetric stethoscope, and to resort to **caesarean section** if necessary. Continuous electronic monitoring of fetal heart rate and/or sampling of fetal scalp blood pH is now more widely used in making the decision as to whether the fetus is genuinely hypoxic. Remember that periods of acute fetal hypoxia, as may occur intra-partum, lead to a fall in fetal heart rate, part of the fetal cardiovascular defence responses to periods of oxygen deprivation (pages 279–81). In obstetric practice, these decelerations in fetal heart rate are now used routinely to monitor fetal well-being. However, although introduction of this technique has improved survival rates in difficult cases, its effectiveness in routine cases is debatable. Some studies report a tendency to resort to caesarean section more readily, thus increasing the use of this delivery procedure disproportionately.

The cervix and myometrium are critically involved in parturition

It is clear from the foregoing account that both the cervix and myometrium are critically involved in parturition. Before the onset of parturition, the fetus lies within its fetal membranes in the uterus and is retained there by the cervix (Figure 18.5). The successful maintenance of the pregnancy requires that spontaneous phasic myometrial contractions are suppressed and the cervical canal is firmly closed so as to provide physical support for the growing fetus. Thus, at term, **two major physiological changes are necessary for the expulsion of the fetus** to proceed smoothly. First, the cervix must undergo a series of structural changes called **cervical remodelling**, such that it becomes sufficiently compliant to allow the expulsion of the neonate: a change in role **from fetal support to birth canal**. Second, and appropriately coordinated with the cervical changes, myometrial tone must change to allow **contractions of the body of the myometrium** (assisted later in labour by contractions of striated muscle in the abdominal wall and elsewhere) to increase intrauterine pressure. Before

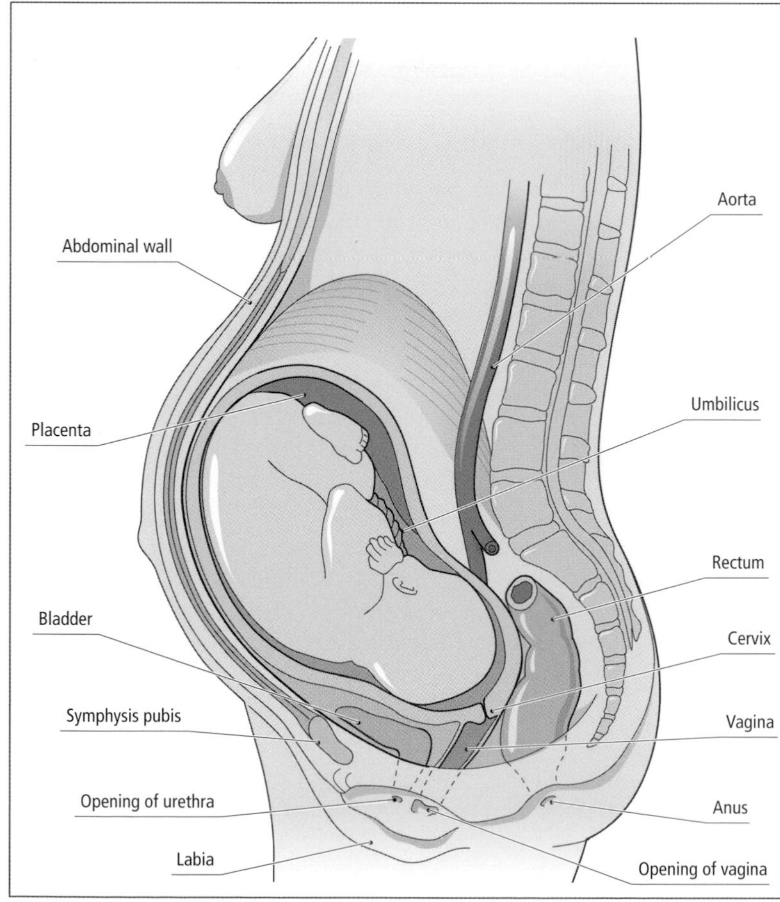

Figure 18.5 Diagram of a sagittal section through the body of a pre-term pregnant woman. Note particularly the size and position of the uterus, and how much of the abdomen it occupies.

examining how each of these changes is physiologically regulated, we first consider the nature of changes that occur in each tissue.

Cervical remodelling

During pregnancy, the cervix is of major importance in retaining the fetus in the uterus. This function is reflected in its initial high connective tissue content, which gives it rigidity and helps it resist stretch, the cervix forming a rigid structure with a firmly closed cervical **os**. The connective tissue is derived from collagen fibre bundles embedded in a proteoglycan matrix. For birth to occur, cervical remodelling occurs and is divided into distinct but overlapping phases: **softening, ripening, dilation** and **postpartum repair**. **Softening**, which begins in the first trimester of pregnancy in women and proceeds slowly thereafter, provides the first measurable decline in tissue compliance, but nonetheless maintains cervical competence. Softening involves gradual changes in the intercellular matrix: **collagen fibres** constitute a major functional component of the cervical tissue and structural changes to their bundling seem to underlie softening, but exactly how is not understood. However, collagen structure and packing are influenced by **glycosaminoglycans** (GAGs), the total content of which increases as pregnancy progresses, accompanied by a change in GAG composition.

Cervical **ripening** occurs more rapidly in the weeks or days preceding birth, and is characterized by maximal loss of tissue compliance and integrity. Cervical ripening has many of the features of a **pro-inflammatory reaction**, with increased vascularization, an influx of monocytes and raised levels of inter-leukins 6 and 8 (IL6 and IL8). The high molecular weight form of the GAG, **hyaluronan** (HA), increases during ripening but then, as dilation gets underway and hyaluronidase levels are observed to rise, is broken down into a lower molecular weight form. This latter transition seems to contribute to the increased viscoelasticity, tissue distensibility, hydration and disorganization of collagen matrix observed during dilation. The elevated activity of metalloproteinases also contributes to collagen breakdown. Together these changes result in a softened consistency, a shortened distance from fetal membranes to *os*, and a greater pliability of the **shortened cervical canal** to distension, essential as uterine contractions are initiated, allowing the ripened cervix to **dilate** sufficiently to allow passage of a term fetus. Finally, **postpartum repair** recovers tissue integrity and competency. The maternal amniotic and chorionic membranes also **rupture** during labour, and the pro-inflammatory reaction in the cervix may be paralleled by similar events in these membranes, and lead to their thinning during labour.

What regulates cervical remodelling? Clinical trials have demonstrated that both **prostaglandins** $\mathbf{E_2}$ (PGE_2) and $\mathbf{F_{2\alpha}}$ ($PGF_{2\alpha}$) increase the compliance of the cervix when given intravaginally or intracervically. Moreover, they have been shown to affect collagen bundle content and associations in cervical biopsies. PGE_2 also induces a leucocytic migration into the cervix by inducing the release of IL8. The use of PGs clinically facilitates delivery during the induction of birth or the evacuation of a late abortus. All these observations suggest a role for PGs in ripening at least. However, their role in cervical softening during normal delivery is less certain, and a role for the peptide hormone **relaxin** has been proposed, its receptors being detectable in cervical tissue (see Box 18.2). Recently, **NO** has been proposed as another possible physiological ripener of the cervix. Indeed, in animal studies, the pharmacological inhibition of iNOS prevents ripening and, moreover, NO has been shown to stimulate local release of PGs. It is possible therefore that **both NO and PGs** are involved in ripening.

Myometrial contractility: contractures vs. contractions

The myometrium consists of bundles of non-striated muscle fibres, intermixed with areolar tissue, nerves, blood and lymph vessels. In non-humans, the myometrium consists of distinct

Box 18.2 Relaxin is a pregnancy hormone that may influence parturition

The existence of relaxin was first postulated in the 1920s to explain the phenomenon in some species of prenatal separation of the **maternal pubic symphysis** caused by relaxation of the interpubic ligament; hence the name relaxin and its implied role as an aid to parturition. Relaxin is a cytokine related to **insulin**, comprising A and B peptide chains linked by two disulphide bonds There is considerable interspecies variation (>50%) in its amino acid composition and peptide chain lengths. Substantial species differences also exist in the source of relaxin; e.g. in the guinea pig, it is produced mainly in the uterus, whereas in the rabbit and horse, it is produced in the placenta. The major source of relaxin in the human (and pig, cow, sheep, rat and mouse) is the corpus luteum of pregnancy from the granulosa-derived large lutein cells, but it is also present in the placenta and decidua. In pregnant women, relaxin is detectable in the blood as early as weeks 7–10, is secreted in relatively small amounts during most of pregnancy, but rises to **maximum plasma concentrations during weeks 38–42**. Removal of the corpus luteum causes systemic levels of relaxin to fall, and women with multiple corpora lutea of pregnancy after super-ovulation show hyper-relaxinaemia and are at risk of premature birth. Indeed, elevated blood relaxin levels at 30 weeks are associated with premature birth. Relaxin may play a role in cervical softening (as definitively observed in rats and pigs), as the human cervix does express relaxin receptors.

outer longitudinal and inner circular layers, but in humans the outer longitudinal layer is less distinct and is organized as intertwined muscle bundles surrounding blood vessels: perhaps important for **haemostasis** following delivery of the haemochorial placenta. The inner subendometrial myometrium blends into the endometrial stroma and is composed of short muscle bundles arranged in a circular pattern. During pregnancy, oestrogens stimulate an **increase in myometrial bulk**, initially by increasing numbers of myocytes, but primarily by increasing myocyte size, from about 50 to 500 μm (hypertrophy), and glycogen deposition. Functionally, this system of muscle cells behaves as a **syncytium**, cells being coupled electrically via specialized gap junctions or nexuses. These allow coordination of the spread of current and contraction through the myometrium. Uterine activity can be recorded during pregnancy and, in all mammals studied to date some interesting patterns of myometrial activity begin to emerge. Throughout most of pregnancy, uterine activity is low but still active, composed predominantly of the **low amplitude, irregular, long-lasting epochs of the uterine activity**, which are known as '**contractures**'. At term, there is a marked switch in the pattern of these contractures to the typical **short-lived, high amplitude myometrial 'contractions'** normally associated with labour (see Figure 18.6). In sheep, the switch from contractures to contractions is gradual, taking place over several hours. Once the switch has occurred, myometrial contraction activity increases steadily and delivery of the newborn lamb generally occurs within hours. However, in primates the switch from myometrial contractures to contractions is a very dramatic event, occurring in the space of a few minutes, and having three very distinct characteristics. First, **the switch is nocturnal, occurring around the onset of darkness**. Second, and amazingly, **it is reversible, uterine contractions reverting back to contractures** by the following morning! Finally, **it is progressive**, augmenting on itself. So, a certain number of nights before birth the pregnant primate will have her first

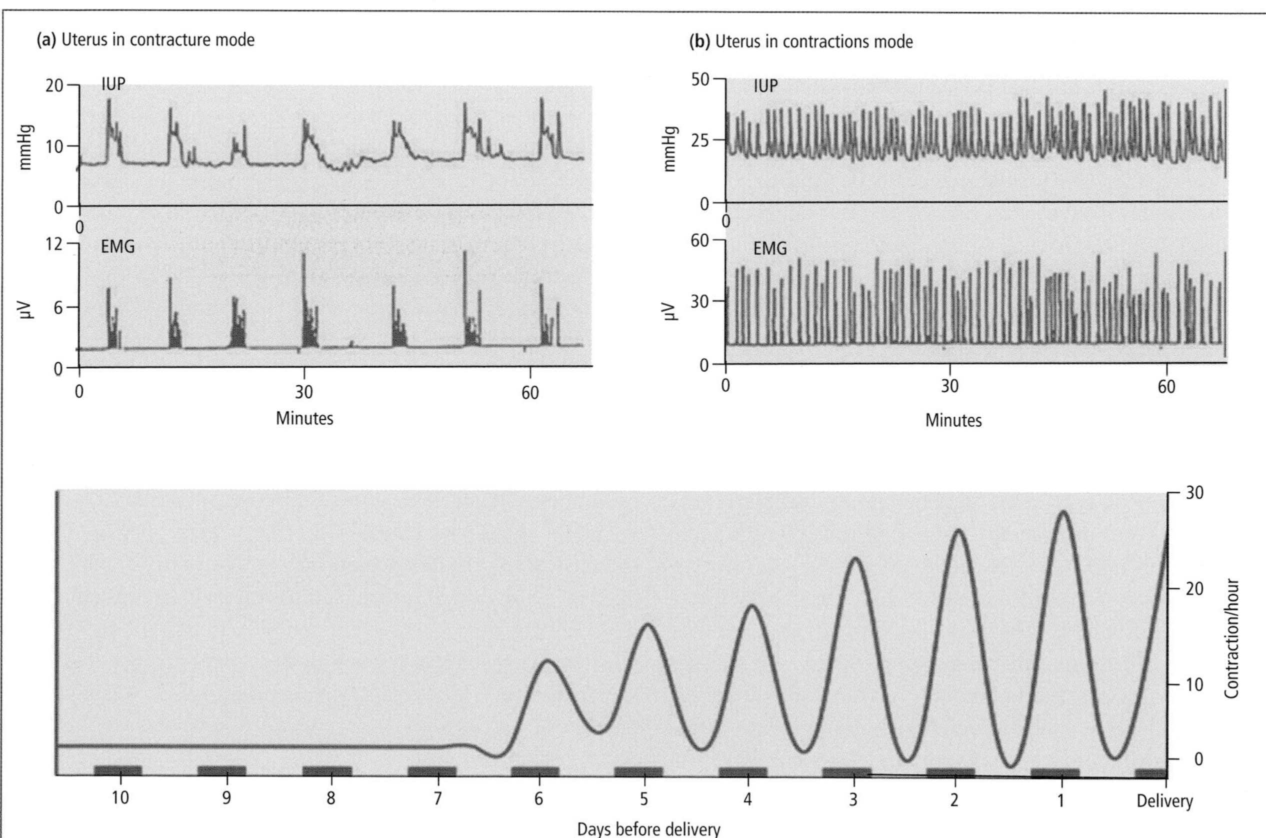

Figure 18.6 Top: myometrial activity recorded via a catheter measuring intrauterine pressure (IUP) or an electrode picking up uterine activity (myometrial electromyogram, EMG) recorded in a pregnant rhesus monkey. Note the difference between the low amplitude, irregular, long-lasting epochs of myometrial contractures, when compared to the short-lived, high amplitude myometrial contractions normally associated with labour. Bottom: diagrammatic representation of the terminal sequence of nocturnal, reversible and progressive nature of the switch in myometrial activity patterns from contractures to contractions in the primate. (Source: Nathanielsz PW *et al.* (1995) *Reproduction, Fertility and Development* **7**, 595–602, with permission.)

switch from myometrial contractures to contractions. By morning, uterine activity will revert back to contractures. The following night, myometrial activity will again switch to contractions; each contraction being greater and perhaps lasting a little longer and then reverting back to contractures by the second morning. And so on and so forth until delivery finally ensues (see Figure 18.6). In order to understand how these changes in myometrial patterns of contractility are regulated, we first consider briefly the physiology of myometrial contractility itself.

The contraction of myometrial cells depends on a rise in intracellular calcium concentration, both by liberation from intracellular stores and by entry into the muscle cells from the extracellular fluid. The calcium then binds to regulatory sites on the contractile proteins, actin and myosin, to allow expression of ATPase activity, and hence contraction. This release of calcium is stimulated by the presence of action potentials within the muscle cell. In the myometrium, spontaneous depolarizing pacemaker potentials occur. If the magnitude of such potentials exceeds a critical threshold, a burst of action potentials is superimposed on the pacemaker, a sharp increase in intracellular calcium occurs and a contraction follows. Calcium is then pumped back into intracellular stores and out of the cell, and the muscle relaxes. Contractility can therefore be modulated by **changing** (1) **the pacemaker potentials** or the relationship between these potentials and the **threshold for spiking**; (2) **the effect of spiking on calcium release**, thereby affecting the **contractility of the myofibrils**; and/or (3) the **interconnectivity of myocytes**, thereby facilitating the **spread of contractions**. In fact, all three mechanisms contribute to the increased myometrial activity as parturition approaches, although the precise way which contractures switch to contractions is not understood. Each of the mechanisms does so through the activity, directly or indirectly, of **PGs** and **oxytocin**. Thus, **PGF$_{2\alpha}$** and **oxytocin destabilize** the membrane potential, reducing the critical threshold for spiking; **PGE$_2$ and PGF$_{2\alpha}$** also enhance calcium entry into cells, whilst **oxytocin** acts mainly by enhancing the liberation of calcium from intracellular stores, both thereby increasing muscle contractility; **oxytocin** and **PGF$_{2\alpha}$** also increase connectivity among myocytes through increased connexin links, thereby facilitating the spread of contractile waves. In contrast, another PG, **PGI$_2$**, operating through a different receptor subtype, relaxes smooth muscle including the myometrium. It is possible that PGI$_2$ ensures that waves of myometrial relaxation are interspersed with contractility to maintain the blood supply to the fetus. In addition, PGI$_2$ may also play a role in relaxing the lower uterus.

Summary

During pregnancy, a dynamic balance must exist between forces that cause uterine quiescence and keep the cervix closed and those that produce coordinated uterine contractility and ripen the cervix and allow it to dilate. For delivery to occur, both balances must favour active uterine emptying. We have seen that oxytocin and PGs are critical players in both these physiological transitions. Next we consider how their activities are regulated. We will see that underlying all these changes at parturition is a functional shift from progesterone to oestrogen dominance.

Steroid hormones, prostaglandins and oxytocin

First, we review briefly the physiology of prostaglandin and oxytocin production and release, and then consider how these two processes are influenced by steroid hormones.

Prostaglandins and steroids

As discussed in Chapter 9, PGs are biologically active lipids synthesized in most body tissues (see Figure 9.7). They are essentially local hormones acting at or near their site of synthesis and are inactivated in the lung during one circulation in the bloodstream. The main prostaglandins produced by uteroplacental tissues are PGE$_2$, PGF$_{2\alpha}$ and PGI$_2$. Factors that increase PG synthesis are believed to act primarily by altering the stability of membranes binding phospholipase A$_2$, thereby liberating the active enzyme (see Figure 9.7). Those factors decreasing PG synthesis probably do the converse and stabilize membranes. It is interesting to note therefore that **steroid hormones exert opposing effects on phospholipase A$_2$**: **oestrogens activate** it, while **progesterone stabilizes** it. Thus, the high progesterone dominance during pregnancy results in low PG levels to provide a relatively quiescent uterine environment, and a **rise in the oestrogen:progesterone ratio** would result in increased production of arachidonic acid, and hence PG synthesis (see Figure 9.7).

The oestrogen:progesterone ratio also affects **the release of PGs**, but does so indirectly **via effects on oxytocin receptors**. Oxytocin then interacts with its receptors to stimulate the release of endometrial PGs directly. Thus, progesterone has been shown to decrease the number of myometrial oxytocin receptors, while **oestradiol** has the reverse effect, thereby **increasing the tissue's sensitivity to oxytocin**. Thus, a rising oestradiol:progesterone ratio can stimulate PG release independently of any alteration in circulating maternal oxytocin levels. This example of **receptor regulation** plays an important role in parturition. Thus, two routes to increased PG availability at parturition exist, and both can be induced by an increase in the oestrogen:progesterone ratio.

Oxytocin: sources and control

Oxytocin is a nonapeptide first identified as being synthesized by magnocellular neurons in the supraoptic and paraventricular nuclei of the hypothalamus and transported axonally to the posterior lobe of the pituitary for release into the blood (see Figure 2.12). However, oxytocin is also synthesized in several other tissues (e.g. the testis and elsewhere in the

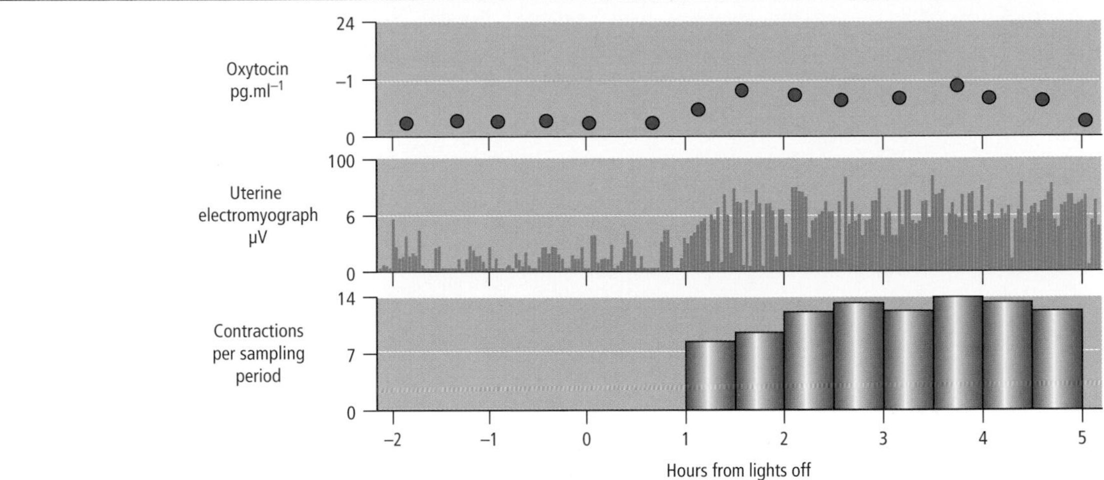

Figure 18.7 Diurnal changes in maternal plasma oxytocin concentrations, peaking around the onset of darkness, measured in a pregnant rhesus monkey during late gestation. Note how the peak concentrations of maternal plasma oxytocin correlate to the nocturnal switch in myometrial activity patterns from contractures to contractions recorded via a uterine electrode. (Source: Giussani D.A. *et al.* (1996) *American Journal of Obstetrics and Gynecology* **174**, 1050–1055, with permission.)

brain; see page 129 and pages 332–4), including the decidual tissue, but also lesser amounts in the amnion and chorion. The decidual oxytocin mRNA content rises markedly at labour onset. Oestrogens are the main up-regulators of oxytocin synthesis by uteroplacental tissues as labour progresses. Thus, two potential sources of oxytocin exist during parturition: the **endocrine posterior pituitary** and the **paracrine (and endocrine?) uteroplacental tissues**.

Endocrine control of myometrial activity

We now understand that the control of **the switch in myometrial activity patterns from contractures to contractions is under endocrine control**. Hormones involved include oestrogen, progesterone and oxytocin and they all have a clear diurnal rhythm, having peaks and troughs in maternal plasma. Further, subtle changes in the timing of the peak concentrations of one hormone relative to the other can influence the switch in myometrial activity from contractures to contractions. Blood levels of oxytocin in pregnant primates, including women, are normally low until term approaches, when plasma oxytocin levels begin to vary diurnally, with a nocturnal peak corresponding to a peak of uterine spontaneous activity (see Figure 18.7). Observations on pregnant monkeys also indicate that blood oestrogen levels correlate with this circadian uterine activity, and that suppression of oestrogen production blocks nocturnal myometrial activity altogether, suggesting a feto-placental origin for the diurnal changes in blood oxytocin.

Observations in baboon pregnancy have now revealed that maternal oestrogen and progesterone also vary diurnally during late gestation, peaking at night time. However, as term approaches, there is **a forward shift in the night-time peak of oestrogen relative to progesterone**, thereby creating a window of time during which there is a high oestrogen:progesterone ratio in maternal plasma. Interestingly, this window of time in **high oestrogen:progesterone correlates with the nocturnal switch in myometrial activity patterns from contractures to contractions** (see Figure 18.8). Therefore, subtle changes in the time of onset of the night-time peak of oestrogen relative to progesterone in maternal plasma can help us understand the nocturnal, reversible and progressive nature of the switch in myometrial activity patterns from contractures to contractions in primate species. Why the switch in myometrial activity patterns from contractures to contractions is so dramatically different between sheep and primates is a matter of ongoing discussion. Several agree that being born at night-time is an ancestral trait we have maintained to deliver our young around the safest hours of the day, away from predators. Certainly, most human births occur in the early hours of the morning. Why the switch is reversible and progressive in primates is more difficult to explain. Some have suggested that this is a type of 'practice' for mother and fetus, a dress rehearsal, a gentler way of making sure that all systems are aligned little by little for the big day.

Myometrial activity patterns may also be modified via endocrine and physical factors during the actual process of active labour, when the changes move from subtle to overt. During active labour, short-duration blood-borne additional pulses of oxytocin of increasing frequency and magnitude occur, peaking during the expulsive stage 2 of birth, and these seem to originate, at least in part, from the posterior pituitary. Release occurs in response to the distension of the uterine cervix via a **neuroendocrine reflex** (often called the **Ferguson reflex**; Figure 18.9) linking the sensory nerves from the vagina and cervix to the pituitary release of oxytocin for blood-borne carriage to the uterus. Since at the same time

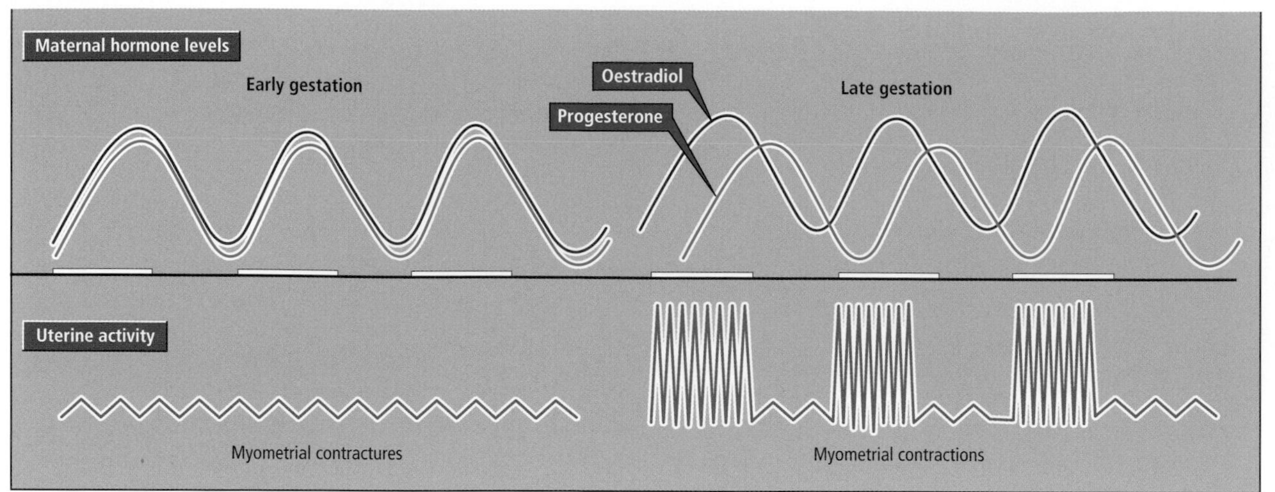

Figure 18.8 Diagrammatic representation of the diurnal changes in maternal plasma oestradiol and progesterone concentrations measured in pregnant baboons. Early in gestation the diurnal peaks of both steroids are in phase. Later in gestation, there is a forward shift in the night-time peak of oestradiol relative to progesterone, thereby creating a window of time of high oestrogen:progesterone ratio in maternal blood. This high ratio coincides with a switch in myometrial activity patterns from contractures to contractions. Maternal treatment with androgens in preterm pregnancy promotes a forward shift in night-time plasma oestrogen relative to progesterone, thereby accelerating the switch in myometrial activity patterns and inducing preterm labour. (Source: Giussani D.A. *et al*. (2000) *Endocrinology* **141**, 3296–3303, with permission.)

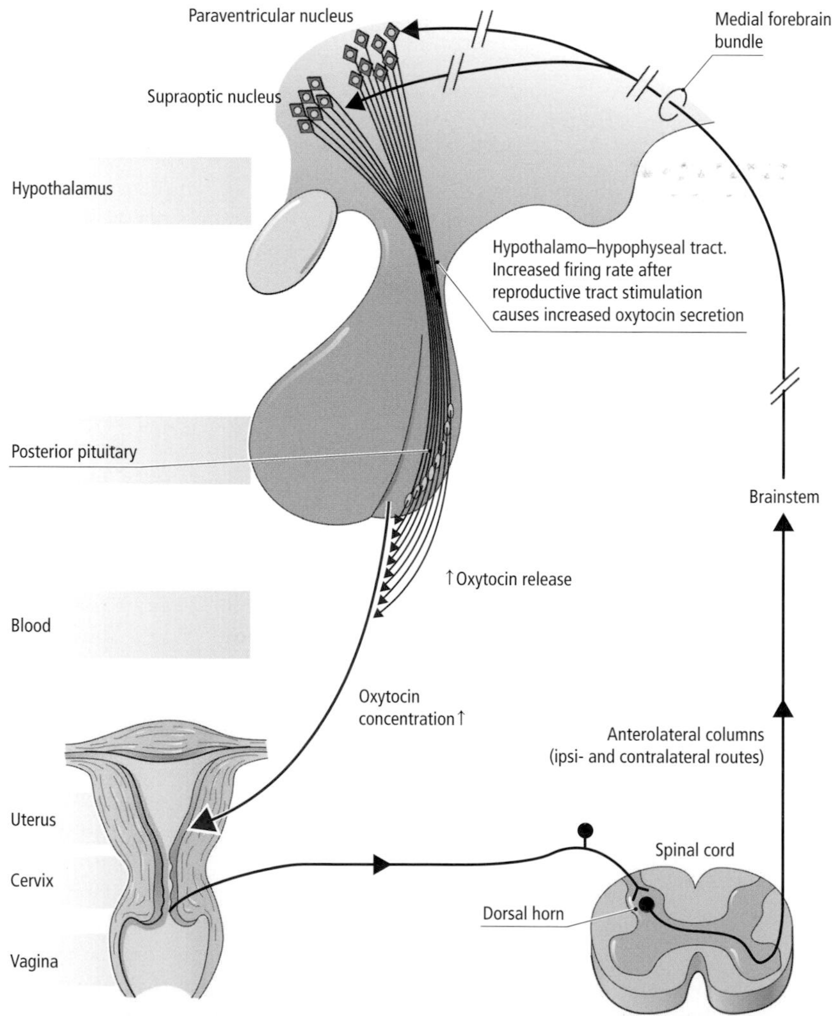

Figure 18.9 The neuroendocrine reflex (Ferguson reflex) underlying pituitary oxytocin synthesis and secretion. Stretching of the cervix (black line) activates the afferent limbs of the reflex which comprise: (1) the sensory nerves from the vagina and cervix; (2) the ascending somatosensory pathways in the spinal cord (the anterolateral columns); and (3) an incompletely described projection through the brainstem and medial forebrain bundle that ultimately reaches the hypothalamic magnocellular nuclei. The efferent limb is indicated by oxytocin release (red lines).

as oxytocin levels are peaking, an oestrogen-induced increase in the **density of myometrial oxytocin receptors is also occurring**, a positive feedback system operates to promote rapid expulsion. In addition, there is evidence that **uterine stretching** itself stimulates the appearance of receptors for PGs and oxytocin, but again only under low progesterone conditions. This stretch-induced response will thus further promote the transition to delivery.

Summary

During late gestation, uterine activity changes from mild contractures to more forceful myometrial contractions. In primates, the switch in uterine activity from contractures to contractions is nocturnal, reversible and progressive in preparation for actual labour. At the end of a pregnancy, during active labour, expulsion of the fetus is rapid to minimize the dangers to it from the intermittent fetal hypoxia (see pages 279–82) associated with uterine contractions. From a slow build, we have seen how the combined effects of PGs, oxytocin and their receptors generate a positive feedback chain of events to prompt both cervix and myometrium into acute labour. A rising maternal oestrogen:progesterone ratio appears to be critical in promoting the actions of PGs and oxytocin. So, in order to understand how the timing of parturition is controlled, we need therefore to understand whether and how such a rising ratio occurs in the first place.

Endocrine control of parturition

The processes by which the oestradiol:progesterone ratio altered differ among the species that have been studied, again mostly comparing sheep and non-human primates. Interestingly, it is **the differences rather than the similarities between species** that have shed more light on the processes mediating parturition in the higher primates, including women.

Studies in sheep

Combined, seminal studies by Liggins and by McDonald and Nathanielsz have provided significant evidence to support the idea that in ovine pregnancy it is the **fetus** itself that determines the timing of parturition through **maturational changes in the fetal hypothalamic–pituitary–adrenal axis.** Therefore, activation of the fetal hypothalamic paraventricular nuclei leads to an increase in the secretion of CRH. Hypothalamic CRH activates the synthesis of ACTH in the fetal pituitary and its release into the fetal systemic circulation. Fetal plasma ACTH arriving at the fetal adrenal cortex promotes an increase in fetal plasma cortisol. This is responsible for the exponential rise in fetal plasma cortisol concentrations close to term, known as **the fetal *prepartum* cortisol surge** (see Table 17.1). Thus, lesions of the fetal paraventricular nuclei in the hypothalamus (source of fetal CRH), fetal hypophysectomy or stalk section (interruption of source of fetal ACTH) or bilateral adrenalectomy (source of cortisol) each **prolongs pregnancy indefinitely**. In contrast, administration to the sheep fetus of sufficient ACTH or the synthetic glucocorticoid, dexamethasone, induces parturition prematurely.

The increase in fetal plasma cortisol is key because it increases the expression of placental 17α-hydroxylase that switches the steroidogenic synthesis cascade from making progesterone to making oestrogen in the ovine placenta. Consequently, maternal plasma progesterone concentrations fall and maternal plasma oestrogen increases, leading to the rising maternal oestrogen:progesterone ratio in maternal blood. Progesterone is the pregnancy-maintaining hormone and its fall in concentration in maternal blood is known as the **maternal progesterone withdrawal**. Conversely, oestrogen is **the parturition-activating hormone** and its increase leads to many events that lead to parturition, such as PG synthesis by the placenta, oxytocin synthesis by the maternal pituitary and the fetal membranes, and an increase in oxytocin receptor number and the expression of gap junction proteins in the myometrium. Combined, sustained increased PGs and oxytocin therefore lead to cervical ripening and its dilation, uterine contractions, active labour and delivery of the newborn lamb (see Figure 18.10). In many species, as we have seen before (Table 17.1), this surge in fetal plasma cortisol also serves to mature a number of fetal tissues and systems in preparation for the successful transition of the infant to neonatal life. Therefore, in sheep, the maturational effects of fetal plasma cortisol go hand-in-hand with those triggering parturition. Quite an intelligent dual effect of fetal cortisol, ensuring only fetuses mature enough to sustain postnatal life are actually delivered!

Studies in non-human primates

The problem with the parturitional system described above in sheep when applied to primates is that no primate placenta, including human, expresses the enzyme 17alpha hydroxylase – it is just not there! Therefore, while the fetal plasma *prepartum* cortisol surge still occurs in primates to mature the fetus in preparation for postnatal life, this increase in fetal plasma cortisol cannot stimulate an increase in the ratio of oestrogen:progesterone in maternal plasma and the consequent changes in PGs and oxytocin synthesis. Luckily for us, the primate placenta expresses another enzyme known as P_{450} aromatase, which is able to aromatize androgens to oestrogens. There is now significant evidence to suggest that in primates, including humans, a rise in fetal plasma androgen occurs towards term and it is this rise that itself serves as the local precursor for oestrogen synthesis within the primate placenta. There is also significant evidence to suggest that the fetal hypothalamus and pituitary drive the increase in fetal adrenal androgen production in primates. The primate adrenal gland cortex expresses receptors for ACTH and CRH. Further, treatment of the baboon fetus with ACTH led to a clear increase in fetal androgen production. Thus, in **both sheep**

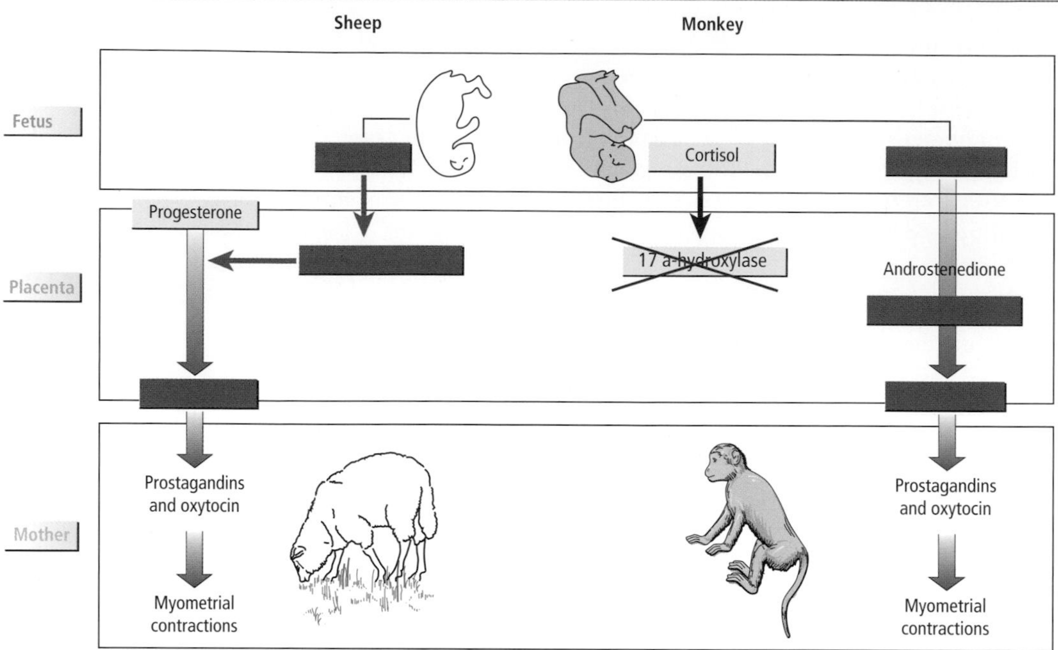

Figure 18.10 Comparison of the endocrine control of parturition between sheep and primates. In sheep the source of high maternal oestrogen concentrations in maternal plasma is an increase in fetal plasma glucocorticoid due to increased activity of the fetal hypothalamo-pituitary-adrenal axis. In primates, the placenta does not express 17 alpha hydroxylase but it expresses P_{450} aromatase. Therefore, the source of the oestrogenic drive in primates is fetal adrenal androgen, serving as the precursor, rather than fetal adrenal glucocorticoid. Downstream from the placenta, there is a commonality of mechanisms between species, leading to increased prostaglandins, oxytocin, uterine contractions, cervical ripening and dilation, culminating in delivery. (Source: Redrawn from Giussani D.A. Nathanielsz P.W. (1999) In: Cosmi E.V., Anceschi M.M., Di Iorio R., Marinoni M. (eds) *New Technologies in Reproductive Medicine, Neonatology and Gynecology*, Moduzzi Editore, Porto Conte, Italy. pp. 387–393.)

and primates the **fetal pituitary–adrenal axis plays the dominant role in timing the onset of parturition**.

Moreover, downstream from the placenta, there is a **commonality of mechanisms** promoting parturition in sheep and primates, including PG and oxytocin synthesis, myometrial contractions, cervical ripening and dilation, leading to active labour and delivery. However, it is **the source of the oestrogenic drive** that differs between the species – being **cortisol-driven in fetal sheep** and **androgen-driven in fetal primates**. The main fetal androgen is dehydroepiandrosterone sulphate (DHEA-S) produced by the fetal adrenals. We now know that there is a *prepartum* surge in fetal plasma DHEA-S in rhesus monkeys that closely resembles the *prepartum* surge in fetal plasma cortisol. Further, studies in rhesus monkeys in which the placental presentation of androgens was prematurely increased by maternal intravenous treatment with exogenous androgen prior to term reproduces all features associated with normal spontaneous term labour in this species. Namely, there is a premature increase in basal maternal plasma oestrogen (without a fall in basal maternal plasma progesterone) concentrations. There is also a premature shift in the diurnal night-time peak of maternal oestrogen relative to the night-time peak of maternal progesterone, creating a window of time of a higher oestrogen:progesterone ratio in maternal blood. This coincides with a premature increase in basal maternal plasma oxytocin and higher night-time peak levels, a premature switch in the final sequence of myometrial activity patterns from contractures to contractions, premature cervical ripening and dilation, all leading to preterm birth. Conversely, simultaneous treatment of pregnant monkeys with androgens and an aromatase inhibitor prior to term inhibits the androgen-induced preterm birth (Figure 18.10).

Studies in women

Despite the background of animal research summarized above, for women, the timing mechanism for parturition is not yet fully agreed. Plasma cortisol concentrations in the umbilical circulation of the fetus and in the amniotic fluid do rise during the last few weeks of normal human pregnancy. For example, corticosteroid sulphate concentrations in the amniotic fluid rise after week 30 of gestation, and this is paralleled by alterations in the palmitate:stearate ratio that provides an index of lung maturation (see Chapter 17). Fetal adrenal hypoplasia may be associated with post-maturity, but not necessarily so, and gestation and parturition within the normal range of variation have been seen in fetuses with congenital absence of adrenal glands. Maternal infusions of adrenocorticotrophic hormone (ACTH) or synthetic glucocorticoids, such as dexamethasone or betamethasone, do not induce parturition in women. This evidence that cortisol is an unlikely trigger of human parturition has a positive outcome, since otherwise, antenatal glucocorticoid therapy to accelerate fetal lung maturation in women threatened with preterm birth would actually trigger preterm birth (see later). Luckily for us, we are not like sheep!

In contrast to sheep, the human and non-human primate placenta is able to synthesize **corticotrophin-releasing hormone** (CRH). Placental (pCRH) synthesis increases with advancing gestation, increasing the circulating plasma concentrations of CRH in maternal plasma exponentially towards term. Studies in human pregnancy have shown that this ontogenic increase in maternal plasma CRH is associated with the timing of birth, being left-shifted in preterm birth and right-shifted in post-term birth. Consequently, it has been suggested that **placental CRH may act as a 'placental clock'**. Placental CRH may trigger the onset of parturition in human pregnancy, bypassing the fetal brain, to act directly at the level of the fetal pituitary and/or adrenal cortex. The cortisol then stimulates further pCRH production and so the makings of a positive feedback system are in place. Placental CRH also significantly stimulates fetal adrenal output of dehydroepiandrosterone sulphate (DHEA-S), the substrate for increasing oestrogen biosynthesis in primates. A possible placental-pituitary-adrenal axis mediating the onset of parturition in humans would also explain why anencephalic fetuses are also born at term. An alternative view is that in higher primates there are several redundant mechanisms, and that this is an example of an ancillary mechanism recruited to trigger the onset of parturition should the normal process involving activation of the fetal hypothalamo-pituitary-axis be unable to act.

Again, unlike sheep, maternal plasma progesterone concentrations do not generally fall towards term in pregnant women, such that there is no circulating progesterone withdrawal. However, use of antiprogestins, such as RU486, initiates parturition in women, which argues for the functional significance of progesterone being strongly involved in the maintenance of pregnancy in humans. It is possible that **a local rise in the effective oestrogen:progesterone ratio**, without detectable changes in the circulating plasma levels, are more important. Thus, changes in the **local metabolic stability or interconversion** of steroids have been proposed. Perhaps more significant, **the ratio of progesterone receptor B (activating) to receptor variants A and C (acting as dominant repressors)** is found to decline towards term in pregnant women. Additionally, during labour, steroid 5β-reductase expression and activity fall, which reduces levels of a potent relaxation-inducing metabolite of progesterone, **5β-dihydroprogesterone**. Such changes, when coupled with a decline in progesterone receptor co-activators, increasingly **blind the uterus to some of the actions of progesterone**. One consequence of progesterone action is the suppression of oestrogen α-receptor synthesis, so reducing the effectiveness of progesterone action leads to a **rise in uterine expression of the oestrogen receptor** as parturition approaches. Such changes locally amount to the uterus 'seeing' a shifting steroid ratio away from progesterone and towards oestrogen. However, the nature of the stimulus for these tissue changes and how they might be linked to the local cortisol and/or CRH changes observed is unclear.

How is the parturitional cascade unleashed?

While activation of the fetal hypothalamo-pituitary-adrenal axis appears essential for triggering the parturitional cascade in primates and sheep, for **most of gestation the axis needs to be inhibited** to maintain pregnancy and prevent the induction of preterm birth. Therefore, a question that has fascinated scientists for decades is what governs the switch in the fetal hypothalamo-pituitary-adrenal axis from being quiescent to being active at term. This is an area of ongoing research and one important hypothesis concerns the **relative bioactivity of glucocorticoids** arriving from the maternal to the fetal circulation. The fetal hypothalamo-pituitary adrenal axis expresses the glucocorticoid receptor at every level. Early in gestation, basal maternal cortisol crosses the placenta and acts on these glucocorticoid receptors in the fetal hypothalamus, pituitary and adrenal to suppress activation of the axis via negative feedback. Late in gestation, the primate placenta increases the expression of the enzyme **11-β-steroid-dehydrogenase**. This enzyme **converts cortisol to cortisone**, which is bio-inactive relative to cortisol. Thus, late in gestation, maternal cortisol arriving at the fetal circulation becomes inactivated, thereby disinhibiting its interaction with glucocorticoid receptors at every level of the fetal hypothalamo-pituitary-adrenal axis and unleashing the parturitional cascade of events.

Summary

In most mammals, the onset of parturition is timed primarily by the fetus via secretions of the adrenal cortex. Although the exact consequences of this increased secretion of fetal adrenocortical output may vary slightly in different species, the general result downstream from the placenta is an increasing oestrogen:progesterone ratio, which stimulates the synthesis and release of PGs and oxytocin. PGs are important players in the mechanical events of parturition, myometrial contractions and cervical ripening. Oxytocin also plays a key role, initially through an increased sensitivity as a result of its receptors increasing, but later as the positive feedback system clicks into action through increased secretion too, enhancing the synthesis and release of PGs as parturition proceeds. In women, some of these changes are observed but not the rising oestrogen:progesterone ratio in the circulation, which may, however, increase locally within the uterine tissues.

Atypical births

At the outset of this chapter, we described how birth is a time of particular hazard for both mother and baby; for a baby born prematurely or too small, problems of neonatal viability and health arise, whilst a post-mature birth risks the baby being too large to deliver non-surgically. For these reasons, ways to clinically control birth times have been sought – a search handicapped by our lack of understanding about how human birth is timed.

Inducing labour

Labour induction should not be attempted unless the cervix is sufficiently dilated (a so-called **Bishop score** of >6). Prostaglandins infused directly via the vagina aid ripening. Oxytocin is a potent uterotonic agent used clinically to induce labour, and an infusion is often given to women delivering naturally to ensure powerful myometrial contractions and thereby **reduce postpartum haemorrhage**. In addition, oxytocin also binds to receptors in the decidua and cervix to activate PG release. Plasma relaxin levels are not elevated in women during labour induced with either $PGF_{2\alpha}$ or oxytocin, but it is unclear why and with what consequences if any. RU486 to reduce progesterone is also used in induction.

Caesarian section

Caesarian section is one way of achieving a post-maturity delivery. The percentage of births being delivered in this way – either planned or emergency has increased from 9% in 1980 to 25% in 2009 in the UK. Such deliveries may occur through patient choice, and are also indicated clinically in the event of **prior uterine surgery**, **fetal compromise**, **breech presentation** or **failure to progress through labour**. There are clinical risks and benefits from undergoing caesarian compared with vaginal delivery. Here, however, we focus on the physiological differences between each, aside from those that are surgically associated. Thus, decidual oxytocin levels are four-fold higher after spontaneous labour than after delivery by caesarian section, presumably due to the lack of an acute phase of cervical distension and myometrial contraction. Given that the actions of oxytocin include the facilitation of complex social and bonding behaviours during care of the offspring (see Chapter 20) and preparation for lactation (see Chapter 19), it has been suggested but not shown that **maternal deficits can result from caesarian delivery**. Likewise, some aspects of lung development are impaired in babies born by elective caesarian section, because there is less activation of stress hormones, such as glucocorticoids and adrenaline, in the neonate.

Delaying births

Whilst clinical approaches to management of post-term delivery are reasonably well established, delaying premature births poses more problems obstetrically. Some 7–8% of singleton births in Europe and North America are preterm, but over 50% of multiple births are. Indeed, the incidence of prematurity has risen in recent years in association with increased obstetric intervention and a greater number of teenage mothers, older mothers, assisted conceptions and multiple births. Preterm birth can be classified as **spontaneous or induced**, the latter due to preterm premature rupture of the membranes or maternal or fetal conditions such as **pre-eclampsia** or **fetal growth restriction**. Amongst spontaneous preterm births, uterine haemorrhage is associated with preterm birth at any time, whilst infection and inflammation are major causes of early preterm births, and premature maternal or fetal hypothalamic–pituitary–adrenal activation is commonly associated with later (post 32 weeks) preterm births (see also Box 18.2 for discussion on hyper-relaxinaemia).

The fact that circulating oxytocin concentrations in women in preterm labour are often raised compared with those in women with normal pregnancies of similar gestational age (25–36 weeks) provides only a weak diagnostic predicator of prematurity. The ability to **diagnose preterm labour more reliably** would be invaluable. Thus, awareness of the importance of the preterm rise in corticosteroid levels for the terminal maturation of fetal systems has led to the use of synthetic analogues in preterm babies or in women at risk of preterm delivery. Synthetic analogues are used because, unlike natural corticosteroids, they can cross the placenta without being metabolized. Their use is not, however, without its dangers. Thus, acute exposure of the developing brain to high doses of corticosteroids can modify the brain biochemistry in ways that may lead to enduring changes in the function of the neonate's hypothalamic–pituitary–adrenal axis and to adverse impacts on mental health. Because our ability to predict which women threatening to deliver prematurely will actually do so is very poor, most of them treated with corticosteroids do not do so. This has led some doctors to treat women with multiple doses of corticosteroids, the effects of which, as shown by studies on neonatal animals and babies, are known to be adverse, and no obvious benefit to neonates was observed over a single injection. For this reason, **only a single maternal injection of corticosteroid is recommended** with largely beneficial effects on the survival and health of premature neonates. Likewise, for premature neonates, their postnatal treatment with a corticosteroid injection is on balance not recommended.

Conclusions

Once born, inspected and sexed, the fruit of the past 17 chapters enters the world. By simply passing down the birth canal to life outside the mother, the neonate acquires the **full legal status of personhood** and the rights of an individual. However, its capacity to exercise those rights is constrained by its immaturity. Indeed, to survive it must pass into a period of prolonged parental care during which, early on, its nutritional requirements are usually provided by the lactating mother. These topics are the subject of the next two chapters.

Key learning points

- Maternal and neonatal mortality has declined in developed countries since the 1930s due to improved medical care and use of antibiotics.
- Term labour occurs between 37 and 42 weeks of gestation with a median duration of 40 weeks.
- Preterm labour occurs between 24 and 37 weeks of gestation.
- Post-term labour occurs after 42 weeks of gestation.
- There are three stages of labour.
- Fetal monitoring can reveal fetal distress during parturition.
- The relatively firm non-pregnant uterine cervix must remodel by first softening as labour approaches and then ripening at parturition to allow dilation and passage of the fetus through the birth canal.
- These changes are thought to be mediated by prostaglandins (PGs) and possibly relaxin and nitric oxide.
- Myometrial contractility has distinct patterns known as contractures and contractions.
- In all mammals there is a switch from contractures to contractions during labour. In primates, the switch in myometrial activity patterns is nocturnal, reversible and progressive.
- In most species, a rising oestrogen:progesterone ratio late in pregnancy causes changes in PG synthesis and release.
- Part of this effect is mediated by increasing the levels of oxytocin receptors, enhancing the effectiveness by which oxytocin induces PG release.
- Mechanical stimulation of the cervix and contractions of the myometrium cause a reflex release of oxytocin from the posterior pituitary and a high oestrogen:progesterone ratio facilitates this reflex.
- Oxytocin is also produced in the decidua and placenta and a high oestrogen:progesterone ratio facilitates its production.
- Timing of the onset of parturition in most animals is determined largely by the fetus via increased secretion of CRH, ACTH and glucocorticoids.
- Fetal glucocorticoids or androgens bring about changes in the oestrogen:progesterone ratio via different mechanisms in different species.
- In sheep, increased fetal glucocorticoid secretion causes a rise in placental PGE_2 output which in turn activates enzymes that convert placental progesterone to oestrogen, thereby increasing the oestrogen:progesterone ratio.
- In primates, increased fetal adrenal androgen production serves as the precursor for placental oestrogen synthesis via P450 aromatase.
- In humans and great apes, there is also a source of placental CRH production.
- In primates, as term approaches, a forward shift in the night-time peak of oestrogen relative to progesterone creates a window of time of high oestrogen:progesterone ratio in maternal blood, coinciding with a switch in myometrial activity patterns from contractures to contractions.
- In primates including women, there may also be changes in the ratio of tissue receptors for progesterone and oestrogen such that the tissue 'sees' less progesterone and more oestrogen, but what causes this receptor change is not clear.
- Corticosteroid treatment of mothers threatening to deliver prematurely can save many premature babies from death by giving them an artificially boosted early maturation of fetal organs.
- However, most mothers treated in this way do not deliver prematurely, potentially exposing their fetuses to higher levels of corticosteroids than may be desirable.

Clinical vignette

Augmentation of labour

A 36-year-old woman, Ms P, was in early labour after an uncomplicated first pregnancy. She had attended for all routine antenatal care and had reached a gestation of 40 weeks and 2 days. The previous day, her membranes had ruptured with clear liquor and her baby remained active. Since rupturing her membranes, she had experienced irregular contractions, which had gradually become stronger and more regular. When Ms P felt unable to cope with the labour at home, she and her partner attended their local maternity unit. On arrival she was contracting 4 times in every 10 minutes. Vaginal assessment was performed to determine cervical dilation, and she was found to be 4 cm dilated. Ms P requested epidural analgesia to help with the pain and continuous fetal monitoring was commenced. On reassessment 4 hours later, however, the cervical dilation was unchanged and the contractions had spaced out to 2 in 10 minutes. The obstetrician on duty reviewed the case, and recommended that an intravenous infusion of syntocinon (a synthetic oxytocin analogue) be started. The syntocinon was titrated upwards in small increments until 4 regular contractions in every 10 minutes was achieved. The fetal heart rate was continuously monitored and showed no signs of distress. Ms P remained comfortable with the epidural analgesia. 4 hours after commencing the syntocinon drip a further cervical assessment was performed. At this stage the dilation was 8 cm and the fetal head had descended well. Two hours later Ms P's cervix was fully dilated (10 cm) and after a further 45 minutes of active pushing she gave birth to a healthy baby girl. Ms P and her baby went home well the following day.

Birth can be divided into three basic processes that are required for successful delivery: (1) the remodelling of the cervix; (2) contractions and consequent cervical dilation; and (3) the passage of the baby through the pelvic inlet and outlet. Oxytocin is crucial for the second of these steps. Oxytocin receptors are present on the gravid uterus during late pregnancy, and increase in number leading up to labour. Oxytocin is a small nonapeptide produced from the posterior pituitary gland and reproductive tract tissues. Pulses of oxytocin act on the myometrial receptors to cause rhythmic uterine contraction, which progressively shorten the myocytes in the uterine lower segment and bring about cervical dilation. The normal positive feedback loop of oxytocin release in labour can. however, be disrupted by a number of factors, including stress, dehydration, build-up of lactic acid (which prevents the uterine muscle from contracting in a coordinated fashion), or a malposition of the fetal head (which then fails to put sufficient mechanical pressure on the cervix). In this case, the obstetrician often resorts to by-passing the physiological positive feedback loop and administering synthetic oxytocin intravenously to keep the labour progressing. This therapy very often succeeds in replicating the normal process and vaginal delivery results. Artificial oxytocin must be carefully titrated to the uterine contractions, as it is possible to cause uterine hyper-stimulation, which can lead to fetal distress or uterine rupture.

FURTHER READING

General reading

Behrman RE, Butler AS (2007) *Preterm Birth: Causes, Consequences and Prevention*. National Academies Press, Institute of Medicine, Washington, DC.

Bewles S, Foo L (2011) Are doctors still improving childbirth? In: *Birth Rites and Birth Rights* (eds. F Ebtehaj, J Herring, MH Johnson, M Richards), pp. 51–76. Hart Publishing, Oxford.

Black RE, Cousens S, Johnson HL *et al.* (2010) Global, regional, and national causes of child mortality in 2008: a systematic analysis. *Lancet* **375**, 1969–1987.

Brown AG, Leite RS, Strauss JF 3rd (2004) Mechanisms underlying 'functional' progesterone withdrawal at parturition. *Annals of the New York Academy of Sciences* **1034**, 36–49.

Bulletti C *et al.* (1997) The uterus: endometrium and myometrium. *Annals of the New York Academy of Sciences* **828** (especially Part VIII, pp. 230–284).

Challis JRG (2002) Prostaglandins and mechanisms of preterm birth. *Reproduction* **124**, 1–17.

Crowther CA, Harding, JE (2007) Repeat doses of prenatal corticosteroids for women at risk of preterm birth for preventing neonatal respiratory disease. *Cochrane Database of Systematic Reviews*, CD003935.

Dalziel SR, Lim VK, Lambert A *et al.* (2005) Antenatal exposure to betamethasone: psychological functioning and health-related quality of life 31 years after inclusion in randomised controlled trial. *British Medical Journal* **331**, 665–676.

Dalziel SR, Rea HH, Walker NK *et al.* (2006) Long term effects of antenatal betamethasone on lung function: 30 year follow up of a randomised controlled trial. *Thorax* **61**, 678–683.

Dalziel SR, Walker NK, Parag V *et al.* (2005) Cardiovascular risk factors after antenatal exposure to betamethasone: 30-year follow-up of a randomised controlled trial. *Lancet* **365**, 1856–1862.

Forhead AJ, Fowden AL (2011) The consequences for preterm infants of antenatal glucocorticoid treatment. In: *Birth Rites and Birth Rights* (eds. F Ebtehaj, J Herring, MH Johnson, M Richards), pp.129–150. Hart Publishing, Oxford.

Giussani DA, Jenkins SL, Mecenas CA *et al.* (1996) Daily and hourly temporal association between delta4-androstenedione-induced preterm myometrial contractions and maternal plasma estradiol and oxytocin concentrations in the 0.8 gestation rhesus monkey. *American Journal of Obstetrics & Gynecology* **174**, 1050–1055.

Giussani DA, Jenkins SL, Winter JA, *et al.* (2000) Androstenedione treatment of pregnant baboons at 0.7–0.8 of gestation promotes a premature forward shift in the nocturnal maternal plasma estradiol surge relative to progesterone and increases myometrial contraction activity. *Endocrinology* **141**, 3296–3303.

Hertelendy F, Zakar T (2004) Prostaglandins and the myometrium and cervix. *Prostaglandins, Leukotrienes and Essential Fatty Acids* **70**, 207–222.

Larsen B, Hwang J (2011) Progesterone interactions with the cervix: translational implications for term and preterm birth. *Infectious Diseases in Obstetrics and Gynecology* **2011**, 1–13.

Loudon I (2000) Maternal mortality in the past and its relevance to developing countries today. *American Journal of Clinical Nutrition* **72** (Suppl.), 241S–246S.

McDonald TJ, Nathanielsz PW (1991) Bilateral destruction of the fetal paraventricular nuclei prolongs gestation in sheep. *American Journal of Obstetrics and Gynecology* **165**, 764–770.

Mecenas CA, Giussani DA, Owiny JR *et al.* (1996) Production of premature delivery in pregnant rhesus monkeys by androstenedione infusion. *Nature Medicine* **2**, 443–448.

Mesiano S (2004) Myometrial progesterone responsiveness and the control of human parturition. *Journal of the Society for Gynecological Investigation* **11**, 193–202.

Nathanielsz PW, Jenkins SL, Tame JD *et al.* (1998). Local paracrine effects of estradiol are central to parturition in the rhesus monkey. *Nature Medicine* **4**, 456–459.

Pepe GJ, Albrecht ED (1998). Central integrative role of oestrogen in the regulation of placental steroidogenic maturation and the development of the fetal pituitary-adrenocortical axis in the baboon. *Human Reproduction Update* **4**, 406–419.

Pepe GJ, Waddell BJ, Albrecht ED (1988) The effects of adrenocorticotropin and prolactin on adrenal dehydroepiandrosterone secretion in the baboon fetus. *Endocrinology* **122**, 646–650.

Smith R (2007) Parturition. *New England Journal of Medicine* **356**, 271–283.

Thijssen JHH (2005) Progesterone receptors in the human uterus and their possible role in parturition. *Journal of Steroid Biochemistry and Molecular Biology* **97**, 397–400.

Timmons B, Akins M, Mahendroo M (2010) Cervical remodeling during pregnancy and parturition. *Trends in Endocrinology and Metabolism* **21**, 353–361.

More advanced reading

Aguilar HN, Mitchell BF (2010) Physiological pathways and molecular mechanisms regulating uterine contractility. *Human Reproduction Update* **16**, 725–744.

Catalano RD, Lannagan TRM, Gorowiec M *et al.* (2010) Prokineticins: novel mediators of inflammatory and contractile pathways at parturition? *Molecular Human Reproduction* **16**, 311–319.

Challis JR, Bloomfield FH, Bocking AD *et al.* (2005) Fetal signals and parturition. *Journal of Obstetrical and Gynaecological Research* **31**, 492–499.

Giussani DA, Jenkins SL, Mecenas CA *et al.* (1996) The oxytocin antagonist atosiban prevents androstenedione-induced myometrial contractions in the chronically instrumented, pregnant rhesus monkey. *Endocrinology* **137**, 3302–3307.

Gomez-Lopez N (2010) The role of chemokines in term and premature rupture of the fetal membranes: a review. *Biology of Reproduction* **82**, 809–814.

Halliday HL, Ehrenkranz RA, Doyle LW (2009a) Late (>7 days) postnatal corticosteroids for chronic lung disease in preterm infants. *The Cochrane Collaboration*. John Wiley & Sons, Ltd, Chichester.

Halliday HL, Ehrenkranz RA, Doyle LW (2009b) Early (<8 days) postnatal corticosteroids for preventing chronic lung disease in preterm infants. *The Cochrane Collaboration*. John Wiley & Sons, Ltd, Chichester.

Hertelendy F, Zakar T (2004) Regulation of myometrial smooth muscle functions. *Current Pharmaceutical Design* **10**, 2499–2517.

Jayaraman P, Haigwood NL (2006) Animal models for perinatal transmission of HIV-1. *Frontiers in Bioscience* **11**, 2828–2844.

Lassi ZS, Haider BA, Bhutta ZA (2010) Community-based intervention packages for reducing maternal and neonatal morbidity and mortality and improving neonatal outcomes. *Cochrane Database of Systematic Reviews* Issue **11**, CD007754.

Nathanielsz PW, Giussani DA, Mecenas CA *et al.* (1995). Regulation of the switch from myometrial contractures to contractions in late pregnancy: studies in the pregnant sheep and monkey. *Reproduction, Fertility and Development* **7**, 595–602. Review.

Office of National Statistics (UK) (2010) Childhood, infant and perinatal mortality in England and Wales, 2008, *Statistical Bulletin*. Office of National Statistics, London.

Petraglia F, Imperatore A, Challis JR (2010) Neuroendocrine mechanisms in pregnancy and parturition. *Endocrine Reviews* **31**, 783–816.

Rezapour M, Bäckström T, Lindblom B, Ulmsten U (1997) Sex steroid receptors and human parturition. *Obstetrics and Gynaecology* **89**, 918–924.

Sherwood OD (2004) Relaxin's physiological roles and other diverse actions. *Endocrine Reviews* **25**, 205–234.

Weems CW, Weems YS, Randel RD (2006) Prostaglandins and reproduction in female farm animals. *Veterinary Journal* **171**, 206–228.

WHO (1996) Essential newborn care: report of a technical Working Group. WHO/FRH/MSM/96. World Health Organization, Geneva.

WHO (2000) *The World Health Report 2005: Make Every Mother or Child Count*. World Health Organization, Geneva.

Wood CE (2005) Estrogen/hypothalamus–pituitary– adrenal axis interactions in the fetus: the interplay between placenta and fetal brain. *Journal of Obstetrical and Gynecological Research* **12**, 67–76.

CHAPTER 19 Lactation

Essential Reproduction, Eighth Edition. Martin H. Johnson.

Companion website: www.essentialreproduction.com

In this chapter and the next, we consider the early postnatal events ensuring the survival of the newborn, including the provision of milk and associated nursing, and the relatively extended period of parental care that most mammals provide – a protective environment in which the young can grow and gradually attain independence. All mammals are characterized by lactational support for their young. Thus, postnatally, all female mammals have a critical role for the survival of their young. However, the duration and nature of the maternal dependency of the young vary from species to species, as does the involvement of more than the mother in neonatal care. However, in this chapter we are concerned with the provision of milk, an exclusively maternal role – for most mammals!

Lactation

Among the many changes occurring in the mother during pregnancy are those that involve the breast. In most mammals, this process is as vital to the success of reproduction as gamete production, fertilization and pregnancy, since the failure to lactate results in early postnatal death of the young. Humans have freed themselves from this absolute dependence on mothers by use of wet nurses or milk from cows or commercial formulae. However, it is becoming increasingly clear that babies who are not suckled on human breast milk, although not dying, may be disadvantaged (see Box 19.1). So, we start by describing the induction, control and regulation of human breast development, lactation and milk removal by the young, referring to other species only when data in the human are lacking, or when important differences between species are apparent.

Breast development from birth to sexual maturity

The human mammary gland consists of 15–20 **lobes** comprising glandular (or parenchymatous) tissue, with fibrous and adipose tissue between them. Each lobe is made up of **lobules of alveoli**, blood vessels and **lactiferous ducts**. The basic pattern of breast structure shown in Figure 19.1 is common to all species, but there are differences in details (see Figure 19.2). The alveolar walls are formed by a single layer of cuboidal to columnar epithelial cells, their shape depending on the fullness or emptiness of the alveolar lumen (Figure 19.3). It is these cells that are responsible for milk synthesis and secretion during lactation. The **myoepithelial cells** situated between the epithelial cells and the basement membrane have a contractile function, and are important for moving milk from the alveoli into the ducts before it is ejected (see page 303).

At birth, the mammary gland consists almost entirely of lactiferous ducts with few, if any, alveoli. Apart from a little branching, the breast remains in this state until puberty (see Chapter 4). Then, under the action primarily of **oestrogens**, the lactiferous ducts sprout and branch, and their ends form small, solid, spheroidal masses of granular polyhedral cells, which later develop into true alveoli. As menstrual cycles establish themselves, successive exposure of mammary tissue to oestrogen and progesterone induces additional, if limited, ductal–lobular–alveolar growth, and the breasts increase in size as a result of the deposition of fat and growth of connective tissue (see Figure 4.5). Adrenal corticosteroids may also contribute a supporting role to duct development at this time.

Cyclical changes to the breast occur in non-pregnant women and are especially evident premenstrually, when there

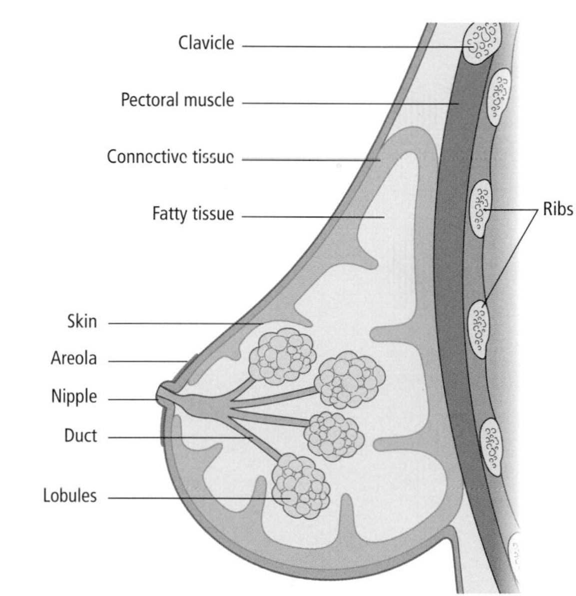

Figure 19.1 Dissection of the lateral half of the right breast of a pregnant woman. When fully developed, the lobules consist of clusters of rounded alveoli, which open into the smallest branches of milk-collecting ducts. These, in turn, unite to form larger, **lactiferous ducts**, each draining a lobe of the gland. The lactiferous ducts converge towards the **areola** of the **nipple (or papilla)** beneath which they form dilations, or **lactiferous sinuses**, that serve as small reservoirs for milk. After narrowing in diameter, each lactiferous sinus runs separately through the nipple to open directly upon its surface. The skin of the areola and nipple is pigmented brown and wrinkled. Also opening onto the peripheral area of the areola are small ducts from the **Montgomery glands**, which are large sebaceous glands whose secretions probably have a lubricative function during suckling. (Source: Peate I, Nair M. *Anatomy and Physiology for Nurses at a Glance*, First Edition. 2015 John Wiley & Sons, Ltd. Reproduced with permission of John Wiley & Sons.)

Box 19.1 Benefits claimed for exclusive breast-feeding

For the infant

- Reduced incidence and duration of diarrhoeal illnesses, respiratory infection, and occurrence of otitis media and recurrent otitis media
- Reduced risk of developing an allergy to cow's milk
- Improved visual acuity and psychomotor development, which may be caused by polyunsaturated fatty acids in the milk, particularly docosahexaenoic acid
- Higher IQ scores, which may be the result of factors present in milk or to greater maternal stimulation of infant during contact
- Reduced malocclusion due to better jaw shape and development
- Possible protection against neonatal necrotizing enterocolitis, bacteraemia, meningitis, botulism and urinary tract infection
- Possible reduced risk of autoimmune disease, such as diabetes mellitus type 1 and inflammatory bowel disease
- Possible reduced risk of sudden infant death syndrome and of adiposity later in childhood

For the mother

- Early initiation of breast-feeding after birth promotes maternal recovery from childbirth; accelerates uterine involution and reduces the risk of haemorrhaging, thereby reducing maternal mortality; and preserves maternal haemoglobin stores through reduced blood loss, leading to improved iron status
- Prolonged period of postpartum infertility, leading to increased spacing between successive pregnancies if no contraceptives are used
- Possible accelerated weight loss and return to pre-pregnancy body weight
- Reduced risk of premenopausal breast cancer
- Possible reduced risk of ovarian cancer
- Possible improved bone mineralization and thereby decreased risk of postmenopausal hip fracture
- Reduced risk of coronary heart disease

Duration?

Research has shown that there is no risk from exclusive breast-feeding for at least 6 months. WHO, the UK Department of Health and UNICEF all recommend 6 months' exclusive breast-feeding as optimal for health, followed by some breast-feeding for up to two years post partum. However, the individual circumstances of the mother and infant need to be taken into account when assessing whether this period of breast-feeding can be achieved.

Further reading

Kramer MS, Kakuma R (2002) Optimal duration of exclusive breastfeeding. *Cochrane Database of Systematic Reviews* Issue **1**, CD003517.

Michaelsen KF *et al*. (2003) *Feeding and Nutrition of Infants and Young Children: Guidelines for WHO European Region with Emphasis on Former Soviet Countries*. WHO Regional Publications, European Series, No. 87. World Health Organization, Geneva.

WHO (2015) 10 facts on breastfeeding. http://www.who.int/features/factfiles/breastfeeding/en/

may be an appreciable increase in breast volume and tenderness (see page 177). In addition, some secretory activity may occur in the alveoli and small amounts of secretory material can be expressed from the non-pregnant breast during the premenstrual period. Compared with other mammals, including non-human primates, mammary development in non-pregnant women is extensive. In most mammals, appreciable mammary growth is not achieved until the middle or end of pregnancy. Thus, in humans, **breasts are sexually dimorphic** and have assumed cultural significance in many societies.

Breast changes during pregnancy

In light of the uniquely developed nature of the non-pregnant breast in humans, it is not surprising that the hormonal requirements for human breast development during pregnancy differ from those in other mammals, in which a complex of sex steroids, adrenal steroids, growth hormone, prolactin and placental lactogen combine to induce mammary growth as pregnancy progresses. In women, neither placental lactogen nor growth hormone is essential. Rather, during early pregnancy, and under the influence of oestradiol and progesterone, the previously developed ductular–lobular–alveolar

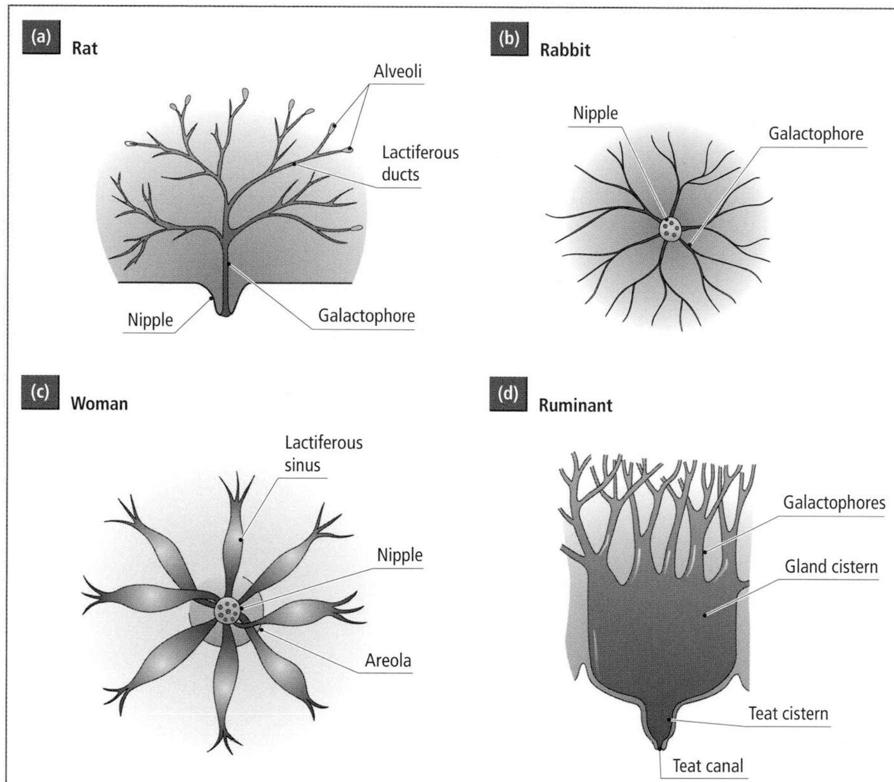

Figure 19.2 The number of mammary glands, their size, location and shape vary greatly among different mammals. The sow, for example, has up to 18 mammary glands (nine pairs), while the pairs of glands in the cow and goat (two and one, respectively) are closely apposed within an abdominal udder. Humans usually have only two breasts, although supernumerary nipples are not uncommon. Here we illustrate the organization of the ducts in mammary glands of four mammals. (a) The rat, in which the lactiferous ducts unite to form a single **galactophore**, which opens at the nipple. (b) The rabbit, in which a number of lactiferous ducts unite to form several galactophores. (c) The human female, in which one lactiferous duct drains each of 15–20 mammary lobes, dilating as a lactiferous sinus before emerging at the nipple. (d) The ruminant, in which, in the **udder**, the galactophores open into a large reservoir or **gland cistern**, which opens into a smaller **teat cistern** and thence to the surface via a **teat canal**.

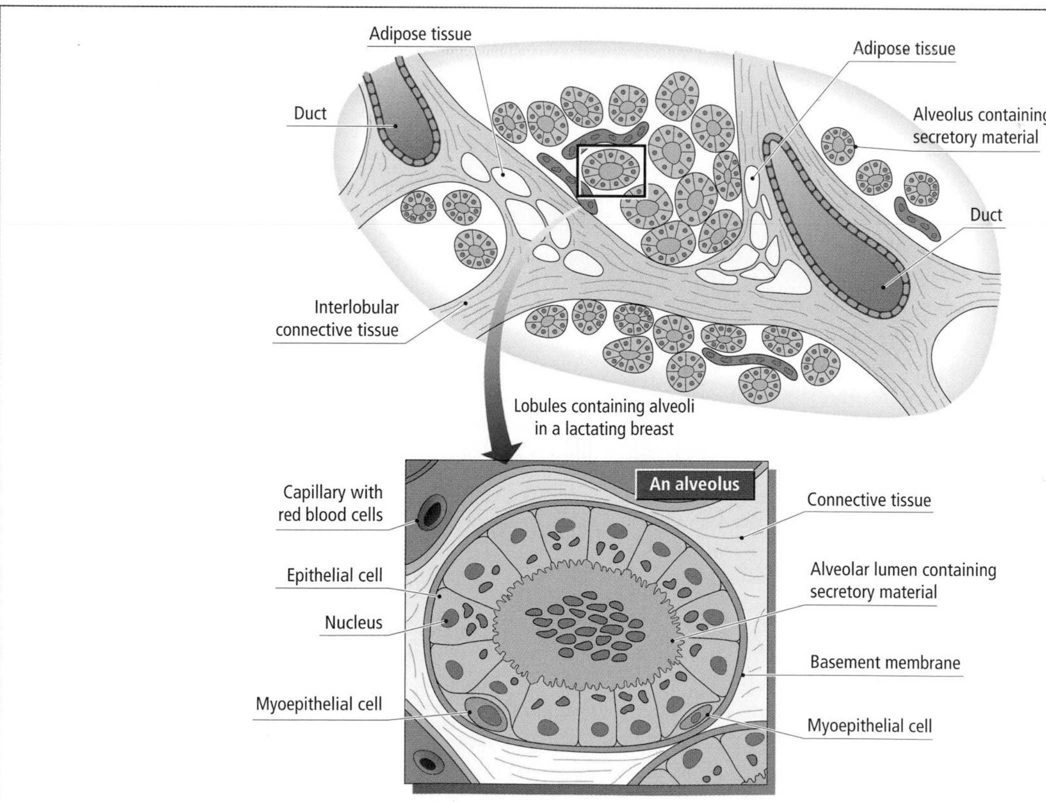

Figure 19.3 Microscopic structure of breast tissue. Microstructure of (above) lobules in a lactating mammary gland and (below) a high-power view of an alveolus. Note the rich vascular supply from which the single layer of secretory epithelial cells draws precursors used in the synthesis of milk. The myoepithelial cells situated between the basement membrane and epithelial cells form a contractile basket around each alveolus.

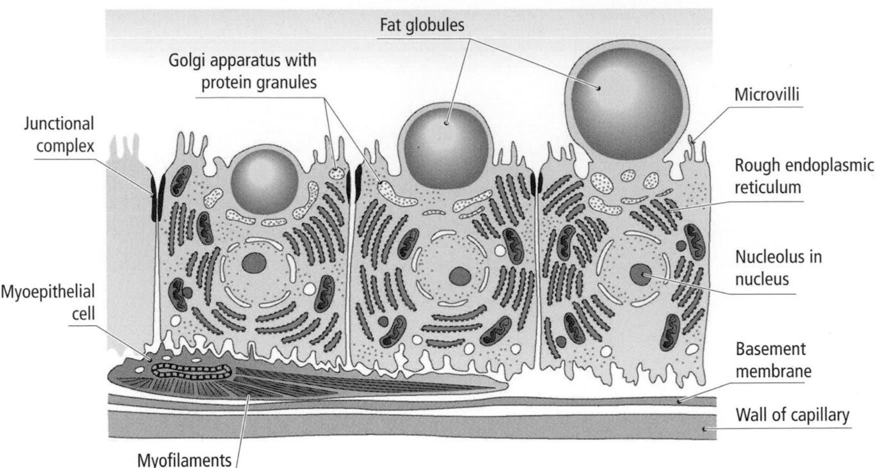

Figure 19.4 Schematic drawing of the ultrastructure of secretory alveolar epithelial cells. The position of the myoepithelial cell can be seen (compare with Figure 19.3). Adjoining alveolar cells are connected by junctional complexes near their luminal surfaces, which themselves bear numerous microvilli. The cell cytoplasm is rich in rough endoplasmic reticulum, particularly in the basal part of the cell. Many mitochondria are present and the large Golgi apparatus is situated nearer the luminal surface and close to the cell nucleus. The fat globules and protein granules are the cellular secretory products destined for the alveolar lumen. Their manner of extrusion from the cell is also indicated. (Source: Cowie A.T. (1972) *Reproduction in Mammals* (eds C.R. Austin, R.V. Short). Reproduced with permission from Cambridge University Press.)

system undergoes further hypertrophy. Under the regulatory control of these mammogenic hormones, **epidermal growth factor (EGF)** and **transforming growth factor α (TGFα)** stimulate the growth of normal mammary cells *in vivo* and *in vitro*, and have been localized to, or are synthesized in, mammary tissues.

Under this hormonal regimen, prominent lobules form in the breast, and the lumina of the alveoli become dilated. By mid-pregnancy, when duct and lobule proliferation has largely ended, differentiation of the alveolar cells to the form shown in Figure 19.4 occurs. The epithelial cells contain substantial amounts of secretory material from the end of the fourth month of human pregnancy, and the mammary gland is fully developed for lactation, awaiting only the endocrine changes described below for full activation.

Initiation and maintenance of milk secretion

Although the human breasts are sufficiently developed and the alveoli adequately differentiated to begin milk secretion by month 4 of pregnancy, copious milk secretion characterizing full lactation does not occur until after parturition. Why? The reduced levels of oestrogen and progesterone in the maternal circulation occurring at or soon after parturition hold the key to the initiation of lactation or **lactogenesis**. Thus, these hormones **inhibit milk secretion** by acting directly on mammary tissue, probably on the alveolar cells. They do so by rendering the breast **unresponsive** to **prolactin**, which increases in plasma concentration throughout pregnancy to reach a maximum at term (Figure 19.5).

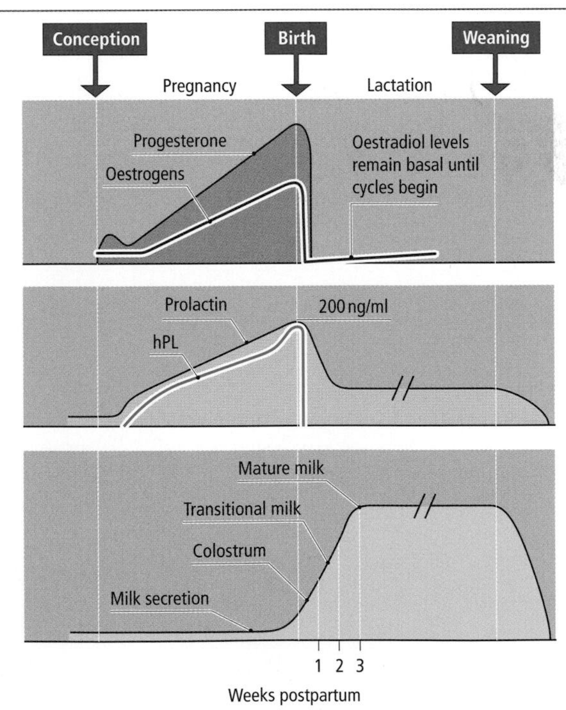

Figure 19.5 The sequence of hormone changes in the maternal circulation, which underlies the onset of lactation in women. Withdrawal of oestrogen and progesterone at parturition critically removes a block to prolactin-induced milk secretion in the gland. The pregnancy prolactin is partly decidual and partly pituitary in origin, hence the drop in its levels at parturition. (Source: Courtesy of A. McNeilly.)

After parturition, when **steroid levels, particularly progesterone, fall** precipitously, prolactin levels fall much more slowly. In the absence of suckling, the newly initiated milk secretion will last, albeit scantily, for 3 or 4 weeks, during which period blood prolactin concentrations remain well above normal non-pregnant levels. However, if prolactin levels are to remain elevated and full lactation is to continue with copious **lactopoiesis**, or milk secretion, **nipple stimulation by suckling** is essential. Suckling achieves this postpartum release of prolactin from the anterior lobe of the pituitary via a **neuroendocrine reflex** (Figure 19.6). The amount of prolactin released is determined by the strength and duration of nipple stimulation during suckling. Suckling at both breasts simultaneously, when feeding twins, for example, induces a greater release of prolactin than occurs during stimulation of one breast.

The circulating plasma concentration of prolactin during lactation appears, more than any other factor, to determine the amount of milk secreted by women. Thus, declining milk secretion in women can be boosted, with consequent breast engorgement, by treatments that release prolactin. Clearly, nipple stimulation during suckling fulfils a most important function in lactopoiesis. In other species, additional hormones (growth hormone, insulin and adrenal steroids among them) may be essential for the successful maintenance of lactation. The fact that nipple stimulation during a feed induces prolactin release, which subsequently induces further milk secretion, suggests **that the baby actually orders its next meal during its current one**.

Summary

We have seen that development of the ductular–lobular–alveolar system and milk-secreting capacity of the mammary glands during pregnancy is under the influence of sex steroids. The initiation of milk secretion depends on the presence of high levels of prolactin and withdrawal of oestrogen and progesterone. The maintenance of milk secretion then depends, in women, solely on the continued secretion of prolactin, maintained by nipple stimulation during suckling. These events are summarized in Table 19.1. Having produced milk, the mother

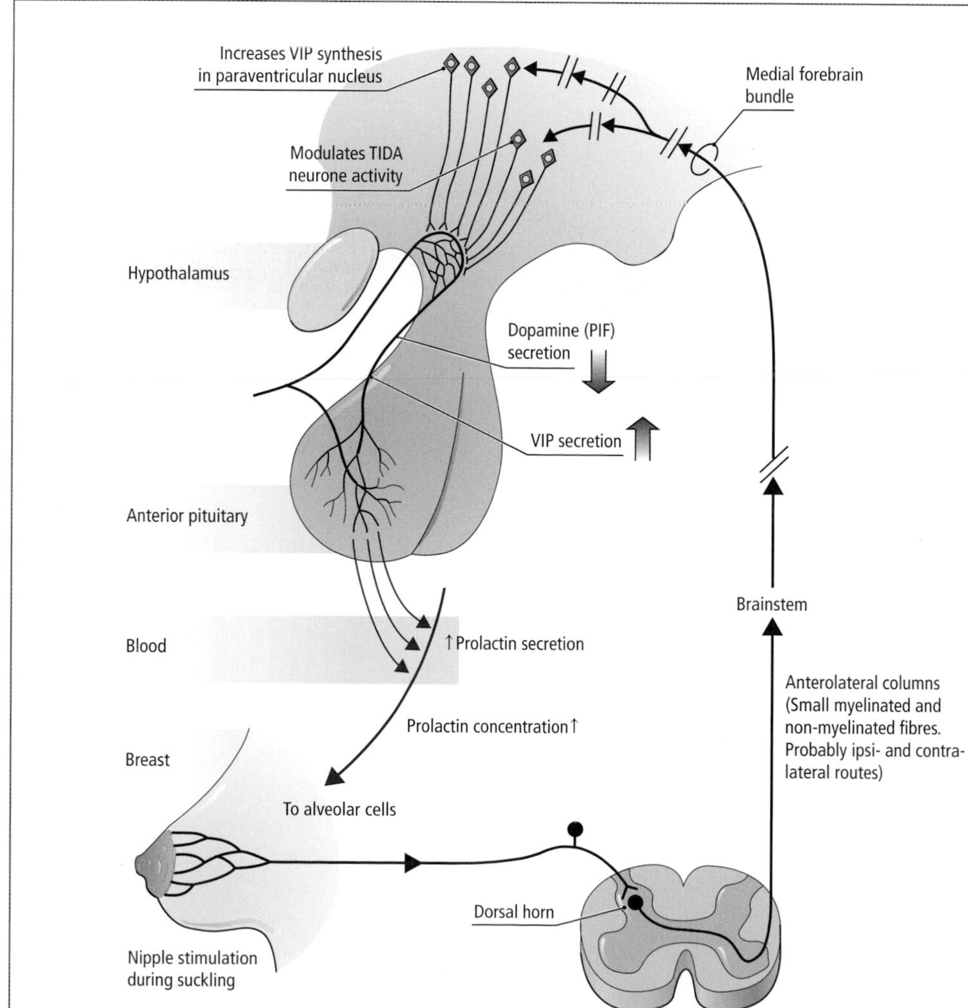

Figure 19.6 Somatosensory pathways in the suckling-induced reflex release of prolactin. The afferent limb of this reflex consists of neural pathways conveying sensory information from the nipples via the anterolateral columns in the spinal cord through the brainstem to the hypothalamus. Denervation of the nipple prevents prolactin release in response to nipple stimulation. Tuberoinfundibular dopamine (TIDA) neuron activity is modulated as a result of the arrival of this somatosensory-derived input to reduce secretion of dopamine (prolactin inhibitory factor, PIF) into the portal vessels during nipple stimulation. In addition, experiments in suckling rats have revealed a marked increase in the secretion of vasoactive intestinal polypeptide (VIP) into the portal vessels, which may facilitate prolactin secretion in the absence of dopamine, which sensitizes the lactotrophs to VIP, indicating that the two mechanisms may interact positively.

Table 19.1 Summary of events leading to full lactation in the human

Hormone or activity	Effect
Early to mid-pregnancy	
Oestrogen, progesterone and corticosteroids	Ductular–lobular–alveolar growth. Considerable branching of the duct systems until mid-pregnancy. Followed by considerable differentiation of epithelial stem cells into a true alveolar secretory epithelium
Oestrogen and progesterone (hPL?)	Little or no milk secretion occurs due to the inhibitory effects of these hormones on prolactin stimulation of alveolar cells. *Continues until ...*
Late pregnancy and term	
Oestrogen and progesterone	Pronounced alveolar epithelial cell differentiation, myoepithelial oxytocin receptors increase
Steroid and prolactin levels high	Colostrum secretion. *Steroids begin to fall at ...*
Parturition	
Oestrogen and progesterone fall precipitously Prolactin levels decline but basal concentrations remain high	Stimulation of active secretion of colostrum and, over 20 days or so, secretion of mature milk. Full lactation initiated
Suckling	
Induces episodic prolactin release at each feed	Maintains milk secretion by promoting synthesis of lipids, milk (particularly α-lactalbumin) and lactose

must deliver it to the neonate. In the next section, we describe how the suckling infant removes milk from the breast, and the **milk ejection reflex (MER),** which subserves this task.

The milk ejection reflex (MER)

Milk removal involves transport of milk from the alveolar lumina to the nipple (or teat) where it becomes available to, and is removed by, the suckling infant. The MER, which underlies this function, has much in common with the reflexly-induced release of prolactin described above.

The MER is a neurosecretory mechanism engaged by suckling

Stimulation of the nipple during suckling probably represents the most potent stimulus to milk ejection. The sensory information so generated travels via the spinal cord and brainstem to activate **oxytocin neurons** in the paraventricular and supraoptic nuclei in the hypothalamus (Figure 19.7). This input boosts not only the synthesis of oxytocin but also its release from the posterior pituitary into the bloodstream. On reaching the mammary gland, oxytocin causes contraction of the myoepithelial cells surrounding the alveoli to induce the expulsion of milk into the ducts and a build-up of intramammary pressure (**milk let-down**), which may cause milk to spurt from the nipple or teat. High-affinity oxytocin receptors are present on alveolar myoepithelial cells and show a gradual gestationally-dependent increase that reaches a maximum in the first week postpartum.

Although touch and pressure at the nipple are very potent stimuli to oxytocin release and milk let-down, the MER can also be **conditioned** to occur in response to other stimuli, such as a baby's hungry cry or, in cows, the rattling of a milk bucket. Such conditioned hormone secretion does not seem to occur in the case of prolactin, where nipple stimulation seems to be the only effective inducer. In Chapter 18 (see page 290), we described how stimulation of the female's reproductive tract, particularly vaginal and cervical dilation, also induces the reflex release of oxytocin. The oxytocin release so induced explains the phenomenon of milk ejection during coitus in lactating women and the rather ancient, but otherwise puzzling practice of blowing air into a cow's vagina to induce milk draught! The MER is particularly susceptible to **inhibition by physical and psychological stresses.** For example, discomfort immediately after parturition or worry and uncertainty about breast-feeding are potentially important inhibitors of the successful initiation and early maintenance of lactation. The way in which 'stress' inhibits the MER is not clear, but may involve inhibition of oxytocin release, and/or the release of catecholamines, such as adrenaline, and activation of the sympathetic nervous system. Constriction of mammary blood vessels induced by adrenergic stimulation might limit access of oxytocin to the myoepithelial cells.

Babies express maternal milk from the nipple during suckling

The mechanics of suckling involve the **expression of milk** from the nipple, aided by sucking. Thus, in the human, the nipple and areola are drawn out to form a teat, which is compressed between the infant's tongue and hard palate. The milk is then **stripped out of this 'teat'** by the tongue compressing the nipple from base to apex against the hard palate. Pressure on the base of the teat is then released, allowing its rapid refill with milk due to the oxytocin-induced increase in intramammary pressure that subserves let-down. There are undoubtedly species' differences in this process, and young also readily adapt to alternative means of obtaining milk. Bottle-feeding using stiff teats, for example, requires more sucking than stripping, and human infants or calves learn this skill rapidly.

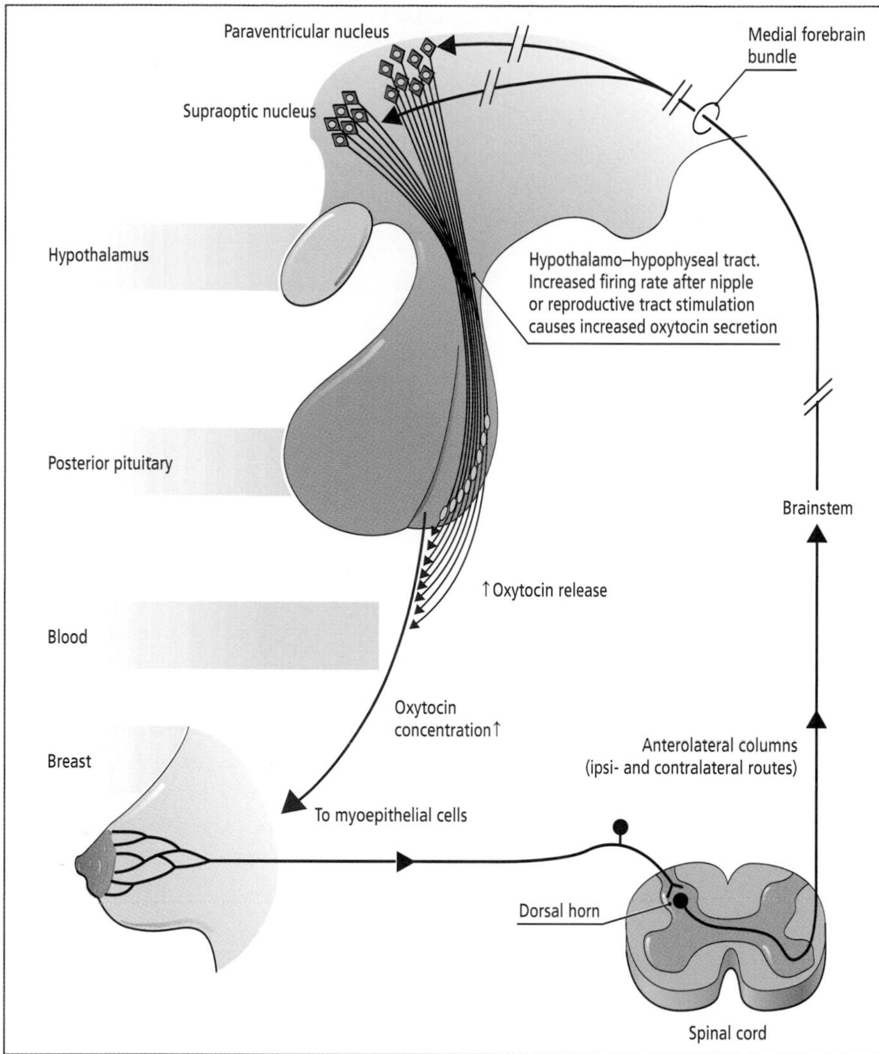

Figure 19.7 Somatosensory pathways in the suckling-induced reflex release of oxytocin. The afferent route taken by sensory information arising at the nipple is well established in rodents. It involves the peripheral sensory nerves, which enter the spinal cord via the dorsal roots, and a number of synaptic relays in the dorsal horn before transmission via the anterolateral columns. These pathways contain fibres destined for various sites in the thalamus and spinal cord and, in particular, the brainstem reticular formation. The brainstem–hypothalamic route taken by the sensory information probably involves an important relay in the midbrain peripeduncular nucleus before travelling in the medial forebrain bundle to reach the hypothalamic, magnocellular paraventricular and supraoptic nuclei.

Summary

Removal of milk from the breast is dependent on the suckling-induced MER. By this means, stimulation of the nipple and other cues associated with nursing induce the release of oxytocin from the neurohypophysis. This hormone stimulates contraction of the myoepithelial cells, which surround each alveolus, forcing milk out of the alveoli into the smaller lactiferous ducts. The resulting increase in intramammary pressure results in milk let-down, causing milk to spurt from the nipple and to be removed from the breast by the suckling infant. Psychological factors may be associated with the induction or inhibition of the MER.

Fertility is reduced during lactation

Lactation can continue for months. During this period, menstruation and ovulation return more slowly than in non-lactating women. Rarely will either occur before 6 weeks postpartum, normal reproductive cycles usually becoming reestablished by 3–6 months. However, menstruation itself is a poor indicator of fertility during this period, and conception often occurs in lactating women without an intervening menstruation. Approximately half of all contraceptively unprotected nursing mothers become pregnant during 9 months of lactation. This variable period of **postpartum lactational amenorrhoea and anovulation** is probably mediated primarily by **prolactin**, which can suppress the initiation of cyclical release of gonadotrophins, as discussed more fully in the contexts of **hyperprolactinaemia** (see page 372).

Cessation of lactation

Lactation may cease spontaneously as suckling declines or may be blocked pharmacologically.

The breast involutes when lactation ceases

When the suckling stimulus is discontinued, milk accumulates in the alveoli and small lactiferous ducts, causing distension and mechanical atrophy of the epithelial structures, rupture of the alveolar walls and the attendant formation of large hollow spaces in the mammary tissue. The alveolar distension also compresses capillaries, resulting in alveolar hypoxia and reduction in nutrient supply. **Milk secretion is therefore suppressed as a consequence of the effects of local mechanical factors**, rather than as the result of declining plasma prolactin (as suckling frequency diminishes). As desquamated alveolar cells and glandular debris become phagocytosed, the lobular–acinar structures become smaller and fewer, and the ductular system in the breast again begins to predominate. The alveolar lumina decrease in size and may eventually disappear, and their lining changes from the secretory single-layered to the non-secretory, two-layered type of epithelium, previously seen before pregnancy. This whole **involutional process takes about 3 months**, but is more intense if nursing is stopped suddenly rather than reduced in frequency as during gradual weaning.

Although these changes in the mammary gland are pronounced, they differ markedly from those in postmenopausal women. In the latter, there is clear structural atrophy of the breast rather than a transition to a period of inactivity. The breasts invariably remain larger after lactation than they were before pregnancy because of the increased deposition of fat and connective tissue that occurs between these two time points.

Lactation can be suppressed pharmacologically

It may be desirable to suppress lactation in women for a variety of reasons. Nursing may be contraindicated for clinical reasons, e.g. the mother may be HIV antibody positive (see page 327) or may simply prefer not to breast-feed her baby. Stillbirth or abortion after 4 months of pregnancy will be followed by unnecessary and often distressing lactation. A number of traditional methods for suppressing lactation may still be employed today, including breast-binding, application of ice packs or treatment with sex steroids to antagonize the effects of prolactin. However, the most widely used lactation suppressants today are dopamine D_2 receptor agonists, such as bromocriptine or cabergoline, which markedly depress prolactin and hence milk secretion (see page 372).

Milk

The composition of milk varies with time *postpartum*. Up to 40 ml/day of a yellowish, sticky secretion called **colostrum** is secreted during the first week postpartum. It contains lesser amounts of water-soluble vitamins (B complex, C), fat and lactose than mature milk, but greater amounts of proteins, some minerals and fat-soluble vitamins (A, D, E and K), and immunoglobulins (IgGs). During a transitional phase of 2–3 weeks, the concentrations of IgGs and total proteins decline, while lactose, fat and the total calorific value of the breast milk increase to yield mature milk, the contents of which are summarized in Table 19.2.

One or two features of human milk are emphasized here. **Lactose** (milk sugar) is the predominant carbohydrate in milk.

Table 19.2 Some contents of human mature milk

Water	Approximately 90 g%
Lactose	Approximately 7 g%
Fat	Including essential fatty acids, saturated fatty acids, unsaturated fatty acids
Amino acids	Including essential amino acids
Protein	Including lactalbumin and lactoglobulin
Minerals	Including calcium, iron, magnesium, potassium, sodium, phosphorus, sulphur
Vitamins	Including A, B1, B2, B12, C, D, E, K
pH	7.0
Energy value	65 kcal/100 ml

It is less sweet than common sugars and is important for promoting intestinal growth of *Lactobacillus bifidus* flora (lactic acid-producing), as well as providing an essential component (**galactose**) for myelin formation in nervous tissue. Lactose is formed within the Golgi apparatus of alveolar cells and is dependent on a combination of **α-lactalbumin** (the whey protein) and the enzyme **galactosyltransferase**, which together form **lactose synthetase**. The sugar passes to the alveolar lumen with the protein granules. **Galactosyltransferase activity is stimulated by prolactin.**

The **main energy source** in milk is fat, which is almost completely digestible, partly because it is present as small, well-emulsified fat globules. **Milk fat** is also an important carrier for vitamins A and D. The milk fat is synthesized in the smooth endoplasmic reticulum of the alveolar epithelial cells and passes in a membrane-bound droplet of increasing size towards the luminal surface of the cell (Figure 19.4). The droplet then pushes against the cell membrane, causing it to bulge and lose its microvilli. Gradually the cell membrane behind the lipid droplet constricts to form a 'neck' of cytoplasm, which ultimately is pinched off, releasing membrane-enclosed lipid into the alveolar lumen (Figure 19.4). **Milk protein** passes through the Golgi apparatus into vacuoles, and is released by exocytosis (Figure 19.4). Both fat and protein release processes depend on activation of the **prolactin receptors** present on the alveolar cells.

Breast or bottle milk?

The newborn gut is a sterile environment and thus the first influx of breast milk provides an acute dose of richly varied antigenic stimulation, including bacterial and viral antigens. The gut epithelium and its underlying and intraepithelial immunocompetent cells provide the major line of first defence via local innate and adaptive pro-inflammatory responses. These responses must be balanced to ensure that potential **infections can be neutralized** but without **excessive inflammation**, with its potential for tissue-damaging consequences. **Breast-feeding is associated with a lower incidence of gastrointestinal**

infection and allergic disease. This protective effect of breast milk comes about because it contains a number of agents that modulate the inflammatory response of the gut. Some, such as immunoglobulins, lysozyme and lactoferrin, may help to process or neutralize antigens, and others may act to influence the innate immune response itself by selective activation of receptors (such as the Toll-like receptor or TLR family) that lead to controlled pro-inflammatory responses.

In addition to its protective effects, breast milk also has a higher nutritional value than artificial or cow's milk. These and other clearly documented benefits of breast-feeding are described in Box 19.1. Moreover, newborn infants seem to prefer breast to artificial milk if given the choice. Increasingly, for premature babies, who are at particular risk of infection and malnutrition, it is recommended that colostrum or breast milk be included as part of their diet, with beneficial outcomes. Contra-indications to breast-feeding may include the risk of serious infection transmission (see below). Currently, a 6-month exclusive breast-feeding period is recommended as optimal for an infant's lifetime health (see Box 19.1). However, observed rates of initiation and especially of maintenance of breast-feeding in American–European societies fall far short of this ideal. For example, in the UK only around 70% of mothers initiated and only 30% had maintained any breast-feeding by 6 months, most of these non-exclusively. Among factors reported by women to reduce breast-feeding are nipple soreness and difficulties in milk production, lack of good cultural and/or family support for breast-feeding, midwife behaviour (especially undue pressure to breast-feed), inconvenience (especially in working mothers) and lack of maternal self-confidence. Most mothers are well aware of the benefits of breast-feeding, and not surprisingly therefore health campaigns based simply on the provision of information have not prevented premature termination of breast-feeding. Indeed, public health campaigns have generally not been very successful, although antenatal and postnatal support that boosts maternal confidence and assertiveness and builds peer support networks has helped increase the duration of breast-feeding.

HIV transmission

Numerous studies that directly test neonatal blood for HIV show unambiguously that HIV can be transmitted from mother to baby (**vertical transmission**). Without medical intervention, the proportions of babies born to HIV-positive mothers that carry the virus vary in different populations, being highest in East African populations (c. 25–35%), intermediate in the USA (c. 15–25%) and lowest in Western Europe (c. 15–20%). The reasons for this variation are uncertain, but may include the effects of co-factors such as maternal diet, vitamin A deficiency, co-infections and co-inflammations, intravenous drug abuse, general health and socio-economic status, as well as the virulence of the viral subtype, maternal immune status and viral load during and after pregnancy. The question is: **how has the virus arrived in the baby?** In principle, three routes are available: **transplacental, parturitional and lactational**.

It is clear from direct analysis for virus in fetal and placental tissues that infection can occur *in utero*, and that, in principle, both free virus and cells infected with virus can cross the placenta. However, the **placenta is a low-frequency route** of transmission of HIV, and indeed many other viruses, although not for viruses such as zika virus, cytomegalovirus and during early pregnancy, hepatitis B virus, which crosses via transcytosis through the undifferentiated trophoblast. There is little doubt from the results of controlled studies, in which HIV-positive mothers using exclusively either breast- or bottle-feeding are compared, that HIV transmission can be halved with bottle-feeding, clearly implicating milk during **lactation as a route of infective transmission**. However, it is important to note that abandonment of breast-feeding can, in poorer communities, have its own severe health costs, affecting the dietary intake and level of other infections in infants, which can erode any clear advantage of reduced HIV transmission. There is also evidence that maternal bleeding at parturition can expose neonates to HIV, and that delivery is a time of high risk for infectious transmission to the baby. With the increasing availability of cheaper antiretroviral drugs throughout the world, continuing HIV vertical transmission is in principle avoidable. However, this avoidability depends on women knowing they are HIV positive, having access to drugs and an environment in which they can take them safely, and on the absence of side effects that discourage continuation. Sadly, this is not the case for many women and their babies.

If neonates born to mothers who are HIV antibody positive are tested for the presence of anti-HIV antibodies, 100% of them test positive. This result is not surprising, as we saw in Chapter 15 (see page 267) that maternal IgG antibodies can cross the human placenta and provide a source of **passive immune protection** against infection. If the baby is not breast-fed, then these antibodies decline over a 6-month period, and if after this time babies are retested for the presence of antibodies to HIV, only a proportion of them remains positive. These antibodies are not of maternal origin, but reflect the baby's own **active immune response to HIV**, presumably largely due to viral exposure at parturition.

Conclusions

The provision of milk is a critical role for all female mammals, without which the survival of their young is in jeopardy. During pregnancy the hormonal environment prepares the breasts for lactation, but also applies the brake on its initiation. The fall in steroids at parturition releases that brake, and the stimulus of suckling ensures, reflexly, that both the making and the releasing of milk are attuned to the neonatal demand. Humans, however, have lost this absolute dependency on provision of maternal milk, but it is increasingly recognized that such provision is preferable to that provided by cow or formula milk. This is so not only for reasons of infant health, but also because maternal milk provision forms a key element in establishing mother–infant bonds. It is this parental care that is the subject of the next chapter.

Key learning points

- Lactation provides the primary source of nutrition for the newborn.
- The breast develops during pregnancy primarily under the influence of the sex steroids.
- Alveoli are grouped together in lobules and these are responsible for secreting milk, which is conveyed to the nipple via lactiferous ducts.
- Falling levels of progesterone late in pregnancy and high circulating levels of prolactin are key events in postpartum lactogenesis – the initiation of lactation.
- Maintenance of milk secretion, lactopoiesis, depends on sustained elevation of prolactin secretion, which is achieved through nipple stimulation by the suckling infant.
- Stimulation of the nipple by suckling is the primary stimulus to the milk ejection reflex, which is mediated by the release of oxytocin from the posterior pituitary and causes contractions of alveolar smooth muscle cells.
- Suckling involves primarily expression of milk from the teat, with a secondary role for sucking.
- Fertility is suppressed by lactation, mediated by the high levels of prolactin which suppress cyclical gonadotrophin secretion.
- Lactation can be suppressed by treatment with dopamine D_2 receptor agonists, which prevent prolactin secretion and hence lactopoiesis.
- When lactation ceases, the breast gradually involutes.
- Milk contains water, lactose, fats, amino acids, proteins, minerals and vitamins with an energy value of 65 kcal/100 ml.
- Milk, and especially colostrum, contain a range of immunoprotective agents that assists the sterile gut of the newborn to mount a controlled innate immune response to the antigenic load experienced with the first milk intake.
- There are clear medical advantages to breast-feeding, including bonding and secure attachment, immune protection and healthier nutrition. These may outweigh risks of maternal infection transmission.
- HIV may be transmitted vertically between mother and fetus *in utero*; this is a low-frequency route of viral transmission.
- HIV may be transmitted vertically between mother and fetus at parturition; this route is important.
- HIV may also be transmitted via breast-feeding; bottle-feeding can reduce vertical transmission.
- Treatment of pregnant and parturient women with anti-HIV drugs can reduce vertical transmission.

Clinical vignette

Breast milk for the preterm infant

A 25-year-old woman, Ms L, had an uncomplicated first pregnancy up until 27 weeks gestation, with low-risk screening results and a normal anatomy scan. At 27 weeks, she attended her midwife complaining of a headache and flashing lights in her vision. Her midwife measured her blood pressure, which was significantly elevated at 150/95 and found protein excreted in the urine. She was quickly referred to the local maternity unit, where antihypertensive medications were commenced and her blood pressure controlled. A diagnosis of early onset pre-eclampsia was made by the obstetric team. Over the next three weeks, Ms L received intensive monitoring and treatment to help prolong the pregnancy in the fetal interest. Eventually at 29 weeks and 5 days gestation, Ms L's blood pressure was no longer adequately controlled even on high doses of medication and her liver function was significantly compromised by pre-eclampsia. A preterm delivery was mandated in maternal interest, and this was balanced against the risk of prematurity to the fetus. A caesarean section was performed, and a live-born female infant delivered. Baby L was intubated and taken to the neonatal intensive care unit. At 30 weeks, Baby L was too premature to feed adequately and for the first two days of life was sustained on intravenous nutrition only. However, Ms L was encouraged to express breast milk to be stored for her baby. On the third postnatal day, the neonatal team considered Baby L well enough to start having small amounts of expressed breast milk every few hours, while maintaining the balance of nutrition with intravenous feeding. Gradually, over the next few weeks, the amount of breast milk was increased and intravenous nutrition decreased until Baby L was fully established on oral nutrition. Baby L was discharged well, close to her original due date.

While breast-feeding is an excellent feeding strategy for most babies, it is particularly important in the context of prematurity. While premature babies do not have a well enough developed gastrointestinal system to meet all or even most of their calorie requirements via trophic feeding, studies show that they benefit from having small amounts of breast milk introduced early in life. This stimulates the still-developing gut without overloading it, and reduces the amount of time that parenteral (intravenous) nutrition is required. Evidence suggests that small amounts of breast milk introduced into the gut is more beneficial than formula-feeding in reducing the risk of feed intolerance and necrotizing enterocolitis. Necrotizing enterocolitis is a common and serious complication of prematurity, where the delicate gut blood supply is compromised. This leads to bowel ischaemia and often perforation. Necrotizing enterocolitis can require extensive surgical excision of lengths of the intestines, including stoma formation in many cases. For a delicate premature baby, necrotizing enterocolitis is often a life-threatening event, and is a major cause of perinatal death in babies who require intensive care. Having small amounts of breast milk early on stimulates blood flow to the gut and encourages normal bowel function. It is also possible to introduce small amounts of formula milk where breast milk is not available, but this strategy has the disadvantage of lacking important antibodies and other molecules that promote the health of the newborn in the long term.

FURTHER READING

General reading

Al-husaini AM (2009) Role of placenta in the vertical transmission of human immunodeficiency virus. *Journal of Perinatology* **29**, 331–336.

Anderson J (2012) Women and HIV: motherhood and more. *Current Opinion in Infectious Diseases* **25**, 58–65.

Howie PW, Forsyth JS, Ogston SA, *et al.* (1990) Protective effect of breast-feeding against infection. *British Medical Journal* **300**, 11–16.

Insel TR, Young L, Wang Z (1997) Central oxytocin and reproductive behaviours. *Reviews of Reproduction* **2**, 28–37.

Jonas W, Woodside B (2016) Physiological mechanisms, behavioral and psychological factors influencing the transfer of milk from mothers to their young. *Hormones and Behavior* **77**, 167–181.

Senise JF, Castelo A, Martínez M (2011) Current treatment strategies, complications and considerations for the use of HIV antiretroviral therapy during pregnancy. *AIDS Reviews* **13**, 198–213.

More advanced reading (see also Box 19.1)

Kools EJ, Thijs C, Kester AD *et al.* (2005) A breast-feeding promotion and support programme: a randomised trial in the Netherlands. *Preventive Medicine* **40**, 60–70.

LeBouder E, Rey-Nores JE, Raby AC *et al.* (2006) Modulation of neonatal microbial recognition: TLR-mediated innate immune responses are specifically and differentially modulated by human milk. *Journal of Immunology* **176**, 3742–3752.

Marlier L, Schaal B (2005) Human newborns prefer human milk: conspecific milk odor is attractive without postnatal exposure. *Child Development* **76**, 155–168.

Reeve JR, Gull SE, Johnson MH *et al.* (2004) A preliminary study on the use of experiential learning to support women's choices about infant feeding. *European Journal of Obstetrics and Gynaecology and Reproductive Biology* **113**, 199–203.

Southgate DAT, Barrett IM (1966) The intake and excretion of calorific constituents of milk by babies. *British Journal of Nutrition* **20**, 363–372.

CHAPTER 20

Postnatal care

Essential Reproduction, Eighth Edition. Martin H. Johnson.

Companion website: www.essentialreproduction.com

In the previous chapter, we explored the postnatal lactational role that all female mammals have and which is critical for the survival of the young in most. However, the duration and nature of the maternal dependency of the young vary from species to species, as does the involvement of more than the mother in neonatal care. This variation depends on how developmentally advanced the young are at birth. Among **marsupials**, in which the young are very immature, mothers show little change in their behaviour at parturition. The young crawls into the mother's pouch and attaches to a teat, and the mother, apart from cleaning her pouch more often, shows little additional maternal behaviour, apparently not recognizing her own as distinct from other young at this time. Young that are born naked and blind (e.g. mice, rats, rabbits, ferrets and bears), so-called **altricial** young, have food and warmth provided by their nursing mothers in nests. Pup retrieval forms an important element of behaviour in these species, often elicited and directed by ultrasonic calls from misplaced young. **Semi-altricial** young are those born with hair and sight (e.g. carnivores, primates), but have poorly developed motor skills and may need to be carried by their parents. The parents may have nests or dens in which the female stays with the young most of the time (e.g. the canids), the males returning periodically to the nest to regurgitate food, initially for the non-excursive mothers, but later for the young as well. Primates do not normally leave their young in nests, but carry them around, the infants having well-developed clinging reflexes. Males often help in this task. **Precocious** young can move about well and, to some extent, fend for themselves (e.g. ungulates, guinea pigs and aquatic mammals). The parent–infant bond in these species seems to reflect more a need for contact and protection rather than nourishment; mothers appear to recognize their own young quickly, and thereafter will feed only them.

It will be clear from the foregoing that there is a variety of roles for both mother and father. Humans fall into the general primate pattern, but with an extraordinarily extended and varied period of care, through **infancy** (from birth to weaning), **childhood** (a period of dependence on adults for survival) and the **juvenile** stage (when survival without adults is possible) to the **adolescent** stage. In this extended period of care, not just the mother, but also the father, members of the extended family and others in the community play a key role. Not surprisingly, therefore, human postnatal care is strongly influenced by culture. This very varied pattern of neonatal care reflects the **evolved socialization** and **extended period of brain and behavioural maturation** in humans. Because of this, study of the more stereotypic neonatal care patterns observed in most other mammals does not always provide a good guide to our understanding of human postnatal care. You will recall we hit a similar problem when discussing gendered and sexual behaviours (see Chapters 5 and 6). There, we described how important the influence of behavioural interactions between a growing infant and its parents and social peers could be in shaping sexual and gender identities. These examples illustrate the vital role that parents play in ensuring not only the survival of their offspring but also their social development. Early on, the mother is of special importance and she displays a range of interrelated patterns of maternal behaviour so that her offspring is given protection, warmth, food and affection. The infant, however, is not merely the passive receiver of all this attention, but an active participant in a two-way interaction, eliciting by its own actions appropriate responses from its mother.

However, our understanding of how parental care operates effectively, particularly in humans, is far from complete. Thus, whilst observational study of human parental behaviour has become increasingly sophisticated, experimental studies are not readily performed. This problem can be partially overcome by studying primate behaviour, but that brings its own ethical dilemmas. So, much of what follows is based on animal studies, which provide the background for human observations. Here, we examine some of the important features in a comparative setting in order to reveal some of the general principles involved.

Patterns of maternal behaviour vary amongst mammals

Maternal/parental behaviour may be considered to occur in three sequential stages: (1) behaviour preparatory to arrival of the young displayed during gestation, e.g. **nest building**; (2) behaviour concerned with the **care and protection** of the young early after parturition and associated with lactation; and (3) behaviour associated with the progressive independence of the young and **associated with weaning**.

During gestation many mammals enter a phase of nest-building activity that is highly characteristic of each species. Very often at this time, the female may show a marked increase in aggression and defend the area in which the nest has been made. Females of very few species, human females among them, will accept the sexual advances of males as gestation proceeds. After birth, the females of all mammalian species are critical for the survival of their young, because of their lactational role. The other roles of the mother, and father or others, during the early postnatal period vary from species to species depending, in part, on how developmentally advanced the young are at birth (see earlier) and in part on the social structures of the species.

As infants grow, there is a gradual change in the behaviours of both them and their mother, which ultimately results in independence. During this time, infants tend to move away from the mother more often and to greater distances. Mothers, on the other hand, retrieve them less and encourage this exploration by rejecting them more often. Suckling occurs less frequently, lactation ceases and the infant is **weaned** and, in the human, enters childhood. In solitary species, the young move away from their mother quite rapidly, the females stop maintaining the nest and the temporary family aggregation disintegrates. In more social, group-living species, the young become independent of, and are rejected by, their mothers, but are progressively integrated within the group. In humans, childhood is an extended phase in which parental support remains integral to flourishing and survival.

Maternal behaviour therefore comprises a complex and variable pattern of interaction between mother and young, adapted to the developmental, social and environmental context in which it occurs. So what is known about the mechanisms underlying the onset and maintenance of the mother–infant bond?

Hormonal influences on maternal behaviour in non-primates

There is clear evidence that the distinctive profile of pregnancy hormones (elevated progesterone, oestradiol and prolactin/placental lactogen) are involved in nest building before birth, and in the prompt appearance of maternal behaviour directed so selectively towards the mother's own young after birth. However, some evidence appears to challenge this statement. Thus, ovariectomized, hypophysectomized female rats and even castrated or intact male rats, will all develop elements of the 'maternal' behaviour pattern when exposed to pups for a period of about 4–7 days. This **pup stimulation**, in which attributes of the pups evoke behavioural changes in the adults, is often called **sensitization** and can occur independently of any hormonal factor. It is also generally accepted that continuing maternal behaviour in the normal, postpartum lactating female is not dependent on hormones, as postpartum ovariectomy and hypophysectomy are not followed by any decline in behaviour (other than via endocrine effects on lactation, which must be taken into account). Thus, hormones do not seem to be required for either the **initiation or the maintenance of maternal behaviour**. Neither do they seem to influence the naturally occurring withdrawal of maternal behaviour, as both sensitized and lactating 'mother rats' show a similar decline in their maternal care over a period equivalent to 10–20 days postpartum, avoiding nursing and increasing their rejection of pups.

However, non-postpartum females (or males) require several days of exposure to pups before they display such behaviour, whereas postpartum females display all elements of maternal behaviour immediately after delivery. The hormones of gestation, and particularly the changes occurring before parturition, are important **determinants of the prompt onset of maternal behaviour**. Sequences of injections of oestradiol, progesterone and prolactin, to mimic the changes seen during pregnancy, induce a more rapid onset of maternal behaviour in virgin females presented with pups. Oestradiol and prolactin/placental lactogen (PL) appear to both be important hormones in this regard, supported by high progesterone that falls just before birth. However, although eliciting a more rapid response, this treatment has never completely mimicked the **immediacy of maternal care** following normal delivery. Clearly, there is something special about parturition itself that renders the mother uniquely sensitive to the newborn. Experiments on sheep have shown that this is indeed the case.

Parturition and maternal care

Non-parturient ewes, or parturient ewes 2 hours or more after having given birth, will not accept or nurse an orphaned lamb even if they are oestrogen primed. They can be induced to do so with 50% or so success by being made temporarily anosmic by use of a nasal spray, or alternatively by draping the pelt of the ewe's own dead lamb over an orphaned lamb that requires fostering. Olfactory cues from the lamb apparently, then, prevent it from being nursed by any mother, other than its own. The parturient ewe will accept an alien lamb and nurse it along with her own, **provided it is presented within 2 hours or so of the ewe having herself given birth**. The reason for this seems to be that the ewe has an altered olfactory responsiveness to the lamb during the immediate postpartum period. How is this olfactory mechanism brought into play?

In an ingenious experiment, stimulation of the cervix and distension of the vagina (using a vibrator or the bladder of a rugby football, respectively) in oestrogen-primed non-parturient ewes caused the immediate (within minutes) display of maternal behaviour towards alien lambs (Figure 20.1). Thus, mechanical stimulation of the reproductive tract, especially the cervix, during parturition is a critical trigger for the neural mechanisms underlying the rapid induction of maternal behaviour.

Mechanisms underlying parturient maternal behaviour change

Experimental data from ewes and rats suggest involvement of a **central oxytocinergic** neural system, activated by vaginocervical distension in parallel with the peripheral oxytocin hormonal system at parturition, as described in Chapter 18 (see Figure 18.9). Oxytocin synthesis is not restricted to the magnocellular neurons of the paraventricular and supraoptic nuclei, but also occurs in the parvocellular neurons of the medial paraventricular nucleus, which project to diverse areas of the brain, including the olfactory bulb, septal nuclei, amygdala and autonomic regions of the brainstem and spinal cord. Synthesis and release of oxytocin are elevated in the parvocellular paraventricular nucleus at parturition under high oestrogen, a time when oxytocin receptors are also increased in the ventromedial hypothalamus and bed nucleus of the stria terminalis. Indeed, at parturition, cerebrospinal fluid concentrations of oxytocin may reach those found in the plasma, but the oxytocin must be made locally as oxytocin does not readily cross the blood–brain barrier. It is not therefore surprising that peripheral injections

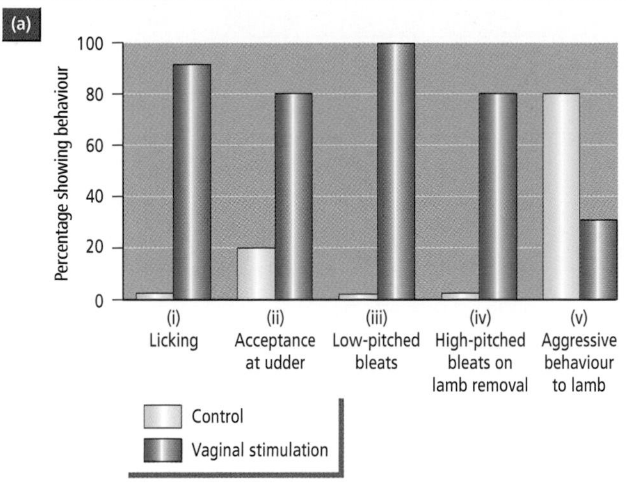

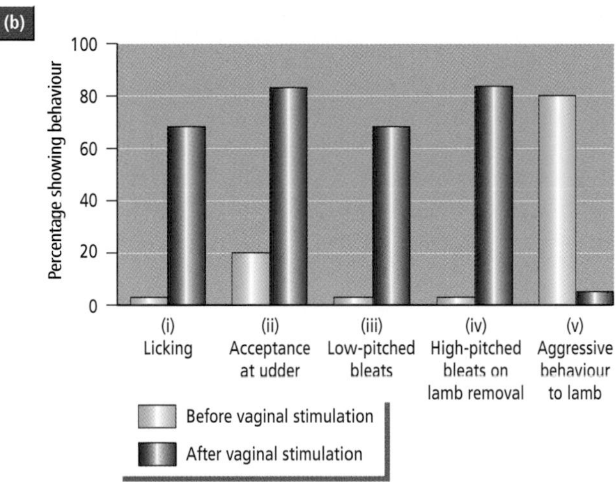

Figure 20.1 (a) The effects of vaginal stimulation on the maternal behaviour of non-pregnant ewes. After stimulation, ewes: (i) lick the alien lambs; (ii) allow them to suckle at the udder; (iii) emit low-pitched bleats characteristic of a maternal ewe; or (iv) emit high-pitched bleats, indicating distress, if the lamb is removed; and (v) exhibit a marked decrease in aggression towards the lamb. (b) Here, the controls in (a), who showed little or no maternal behaviour, were subjected to vaginal stimulation at the end of the observation period. As can be seen, their change in behaviour towards the alien lamb immediately afterwards is dramatic. The same immediate response has now been seen to occur in rats: cervical stimulation of virgin, oestrogen-treated females caused maternal behaviour within minutes of exposure to pups.

of oxytocin are without behavioural effect. In contrast, its central administration to virgin female rats induces full maternal behaviour within minutes, as long as the females have been primed with gonadal steroids. Treated females switch from having no interest in pups to building nests, and retrieving and licking pups. Finally, maternal behaviour at parturition can be prevented by prior treatment with oxytocin receptor antagonists, but these antagonists do not have any effect once maternal behaviour is established.

The oxytocin seems to act by temporarily suppressing the aversive olfactory effect of a lamb on the ewe, perhaps by an action at the olfactory bulb or amygdala, or at the medial preoptic area (mPOA; see next section), to which they project. Thus, in rats and sheep, **oxytocin released centrally at parturition induces maternal behaviour** in a way that is coordinated with its peripheral effects in the uterus during labour and on mammary tissue for milk ejection. These data suggest that central oxytocin may be a common mediator of the induction of maternal behaviour in many species, as summarized in Figure 20.2.

The neural basis of maternal behaviour

The neural basis of maternal behaviour has been studied mainly in rats and sheep, and in these animals, as in most others studied, **the medial preoptic area** (**mPOA**) appears to be the primary site for integrating parental behaviour. Thus, bilateral lesioning of this area severely impairs pup retrieval, nest building and nursing responses in postpartum female rats; oestradiol implants in this area facilitate maternal behaviour in hypophysectomized or ovariectomized female rats; and a correlation exists between quality of maternal care and higher expression levels of oestrogen α-receptor in this area; the area possesses, in addition, receptors for prolactin/PL and oxytocin, of which numbers increase toward the end of pregnancy and during the early postpartum period, and prolactin has been shown to act via this area. The mPOA is also a potential target for the rise in oxytocin activity at parturition, in addition to acting on the inputs to it. Thus, the mPOA receives converging information from every relevant offspring-related sensory input (including auditory, olfactory, tactile, temperature and visual offspring cues), and the hormonal events of late pregnancy and parturition combine with the sensory input from the newborn, to induce altered gene expression and protein synthesis patterns that result in changed electrophysiological activity in mPOA neurons. The mPOA then has efferent connections to those brain regions that influence the three aspects of the maternal response to her young via the midbrain, namely, to the nucleus accumbens for motivational aspects, to the amygdala, bed nucleus of the stria terminalis, and prefrontal cortex for emotional aspects and to the nucleus accumbens and prefrontal cortex for cognitive aspects. The change in mPOA activity is matched by structural changes, and marks the initiation of a series of structural and functional reorganizations in which the rat/sheep mPOA changes its role from facilitatory during the onset and early postpartum period of maternal behavior to inhibitory later postpartum, according to the changing needs of her young. Comparable changes in other brain nuclei emphasize the important role of neuroplasticity as the mother responds to the changing needs of her offspring. Once a female has had a litter, a memory trace is laid down that facilitates subsequent maternal behaviour, regardless of whether the females have given birth on exposure to a new litter, in other words a **heightened sensitivity** is evident. Two neural

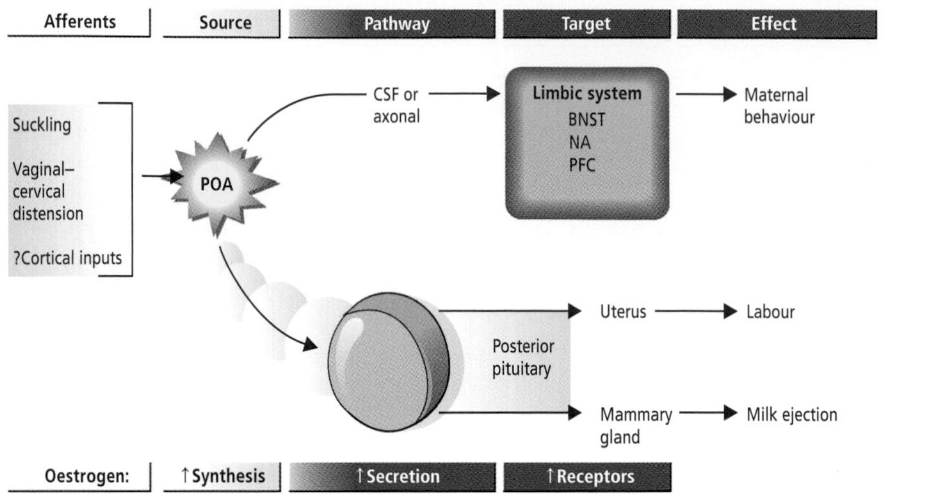

Figure 20.2 Schematic summary of central and peripheral oxytocin pathways activated in parallel at parturition. Oxytocin synthesized in supraoptic nuclei (SON) is transported to and released from the neurohypophysis into the peripheral circulation to affect the contractility of the uterus during parturition and the mammary alveoli to influence milk let-down. Oxytocin is also released either into the cerebrospinal fluid (CSF) or from the terminals of parvocellular oxytocin neurons in various structures such as the bed nucleus of the stria terminalis (BNST), nucleus accumbens (NA) and prefrontal cortex (PFC) where it may be involved with the initiation of maternal behaviour. Oestrogen increases the synthesis, release and binding of oxytocin to its receptors, but not necessarily by direct actions. (Source: Insel T.R., Young L., Wang Z. (1997) *Reviews of Reproduction* **2**, 28–37. Reproduced with permission from BioScientifica Ltd.)

sites involved in the establishment of this maternal memory have been identified, namely, the shell region of the nucleus accumbens and the medial amygdala, and at both locations an increased number of astrocytes was found.

Similar evidence of neuroplasticity has been found from scanning human brains, the volume of grey matter in the mPOA, amygdala and prefrontal cortex increasing over the first 3-4 months postnatally, and the mother's positive feelings towards her baby at the first month postpartum being correlated with a larger increase in grey matter volume over the subsequent 2–3 months.

Summary

The sequence of events influencing the display of maternal behaviour in non-primates appears to be: (1) exposure to hormones, particularly oestradiol and prolactin/PL, during late pregnancy; (2) parturition or cervical stimulation itself; (3) the release of oxytocin within key areas of the forebrain; and (4) continuing exposure to the newborn and growing infant(s). Maternal behaviour will be displayed in the absence of (1)–(3), but with reduced success and increased latency. The realization that cervical stimulation has such dramatic effects has already had an impact on sheep farming. It has also provided a stimulus to studies of maternal behaviour in women, where implications for the success of mother–infant bonding following non-vaginal deliveries are obvious. Indeed, mother–infant bonding after caesarean delivery may differ from that seen after vaginal delivery, but it is unclear whether this is related to differences in cervico-vaginal stimulation. We will now examine in more detail the extent to which our understanding of non-primate maternal behaviour has helped studies in humans and other primates.

Parent–infant interactions in primates

Baby monkeys spend most of their first few months (or years in apes) of life clinging tenaciously to their mothers, and occasionally fathers, aunts and juveniles, in characteristic ventro-ventral, back-riding or arm-cradled positions. This **clinging reflex** requires a well-developed motor capability. Similarly, a well-developed **rooting reflex**, also seen in human babies, by which the head turns towards a tactile stimulus around the mouth, particularly the cheeks, ensures that the infant gains the nipple for suckling. Clinging and rooting reflexes are present immediately after birth, and are elicited by cues from the mother. **Contact comfort** is also an important determinant of an infant monkey's early goal-directed movements, as experiments using surrogate mothers have shown. Thus, a 'cuddly' towelling-covered wire model is much preferred to a bare wire surrogate, even if the latter can provide milk. Similarly, a warm, moving, milk-providing surrogate is preferred to one lacking any one of these attributes. It is inferred that babies prefer such cues from the feelings of security and comfort generated. Human infant behaviour towards a warm, rocking mother or soft blanket is comparable.

Clinging, contact comfort and rooting behaviours emphasize the baby's contribution to the interactions at the earliest moments after birth, and help establish the bond with the mother, as well as securing warmth, contact and food with only minimal help from her. The mother must also keep her infant clean and protect it. Such behaviour is elicited quite specifically by the infant, which emphasizes that the **close contact during suckling** forms the focal point around which subsequent patterns of maternal behaviour develop. Although the baby initiates contact by clinging, the mother clearly derives considerable rewards too, as evidenced by the fact that a monkey mother will carry her dead baby around for some days and show distress when it is removed.

Initially, an infant does not respond to its mother as an individual. Its responses can be elicited by a wide range of stimuli (fur, nipples, etc.), usually, but not necessarily, associated with its own mother's body. Eventually the range of stimuli eliciting responses in a baby becomes narrowed to those from its mother alone. Similar processes occur in human babies. Olfactory cues seem to be of particular importance in the mechanisms by which babies recognize their mothers and vice versa. Communication between mother and infant takes several forms, and these reinforce both mutual recognition and the mother–infant bond. Thus, **baby monkeys make characteristic vocalizations**, e.g. 'whoos' when separated from their mothers and 'geckers' when frightened or denied access to the nipple. The mother usually responds rapidly to such vocalizations with physical contact, giving an immediate soothing effect. There are obvious parallels with human babies who have characteristic cry patterns when in pain, frightened, hungry or in a temper, and who are soothed by close contact with their mothers. Mothers can distinguish the different types of crying and learn to respond appropriately. Undoubtedly, there are many more subtle and, as yet, poorly understood means of communication between mother and infant, e.g. facial expressions such as 'grins' in monkeys and smiles in human babies, which contribute to the mother–infant bond.

As an infant grows and begins to move away from its mother, it is essential that it understands and complies with its mother's signals concerning potential hazards, e.g. proximity of a predator. Thus, communication between mother and infant changes in parallel with, and often as a consequence of, an infant's cognitive and motor development. However, the gradual process of gaining independence from the mother and exploring the environment, including play with other infants, in turn influences cognitive development. Extreme fear of strangers and novel surroundings could easily result in an infant never leaving its mother and thus failing to gain experience of the wider world in which it must live. Indeed, monkeys reared in isolation show great fear of novel stimuli, probably because their cognitive development has been restricted and impaired. It is in this context, then, that the delicate balance between a mother's rejection of her infant and the latter's curiosity and exploratory tendencies assumes importance. The mother's proximity and availability allow the infant to resolve the conflict between **explore–retreat** tendencies and so increase its familiarity with strange objects while assessing their safety or hostility (Figure 20.3). If, during this time, a more prolonged separation of the infant from its mother is imposed by her

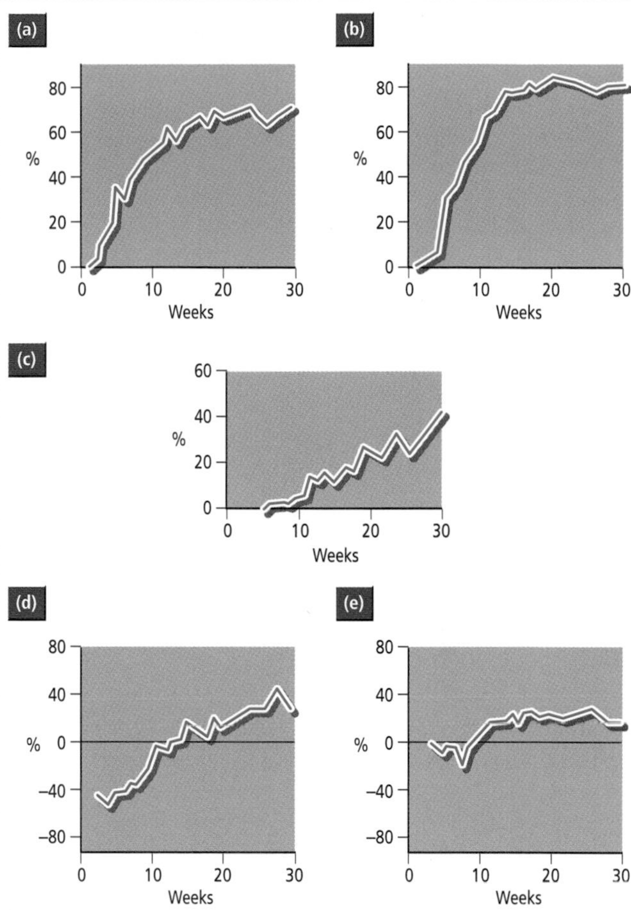

Figure 20.3 The course over the first 30 weeks of life of mother–infant interaction patterns in small captive groups of rhesus monkeys. (a) Total time infant spent off mother as a percentage of total time watched. (b) Time spent out of arm's reach (>60 cm) of mother. (c) Relative frequency of rejections (ratio of numbers of occasions on which infant attempted to gain ventroventral contact and was rejected, to number of occasions on which it made contact on mother's initiative, on its own initiative or attempted unsuccessfully to gain contact). (d) Infant's role in ventroventral contacts (number of contacts made on infant's initiative, as a percentage of total number made, minus number of contacts broken by infant, as a percentage of total number broken). (e) Infant's role in the maintenance of proximity. Note that, initially, the infant stays on the mother all the time, clasping her ventroventrally, but gradually it spends more time off her, both at hand and also out of arm's reach. Early on, the mother is primarily responsible for the close contact, restricting the infant's sorties by hanging on to a tail or foot. During this time, therefore, the infant is responsible for breaking, and the mother for making, contact. Later, however, the infant becomes primarily responsible for making contact, as the mother rejects its approaches more often and initiates contact less often. (Source: Adapted from Hinde R.A. (1974) *Biological Bases of Human Social Behaviour*, McGraw-Hill Inc., New York.)

removal, devastating effects can follow. The baby shows considerable distress, withdraws into a hunched, depressed posture and decreases its motor activity. Reuniting the pair is followed by a period of intense contact, but with eventual and gradual rejection of the infant by the mother to re-establish pre-separation patterns and levels of interaction. The longer the separation the more severe and potentially enduring the effects, and the mother's behaviour is then critical in restoring the infant's security. It is important to emphasize, however, that because mother–infant separation can have long-term behavioural effects, it need not necessarily have them and many factors may influence the final outcome.

Mechanisms?

In social primates, young females have plenty of opportunity to learn and 'rehearse' skills that they will subsequently need as mothers. They will watch other monkeys, particularly their mothers, holding infants and may even 'practise' by holding siblings themselves. Experiments demonstrate that females reared in a socially deprived environment make poor, aggressive and rejecting mothers. There may be important parallels in disturbances of parental care in humans. However, these studies do not mean that maternal behaviour is simply learnt, as a deprived upbringing could have elicited negative social behavioural patterns generally. Indeed, studies have shown that hormones can and do influence behaviours of both parents, notwithstanding the important role that learning by example plays. Thus, both correlational and experimental studies suggest that high circulating oestrogen concentrations during pregnancy can increase maternal responsiveness to other young during pregnancy. However, whether pre- or post-partum steroids affect subsequent post-birth patterns of maternal behaviour is less clear.

Is there also a role for oxytocin in primate maternal behaviour? A correlation between plasma oxytocin concentrations in lactating females, with various measures of maternal care has been reported – but peripheral oxytocin is no guide to possible central effects. More convincingly, injection of oxytocin centrally into adult nulliparous, female rhesus monkeys did increase their interest in infants, and this interest increase was prevented by use of an oxytocin receptor antagonist. These few studies suggest tentatively that there may be **a similar role for oxytocin within the primate brain** as was seen in non-primates. There is also some evidence that the CRH–ACTH–cortisol system in primates can act in opposition to the oxytocin system. Thus, central CRH release also occurs in stressed females in parallel with the elevation of peripheral CRH–ACTH–cortisol, and this is associated with adverse maternal behaviour patterns. Moreover, experimentally-induced acute elevations of CRH within the brain seem to reduce some aspects of maternal behaviour.

Paternal care amongst primates

Primate paternal care takes many forms, ranging from indirect protection against predators, through directly carrying and even feeding the young. Amongst those primates most studied are marmosets, in which paternal investment is high, because they have multiple offspring and the female has an early post-partum oestrus, and is thus pregnant and lactating at the same time, so requiring more support. However, amongst our closest relatives, the great apes, paternal care is almost non-existent, largely because the apes are promiscuous or polygynous, whereas humans have evolved pair-bonding. What about hormonal and physical changes in males in association with pregnancy? Male marmosets show progressive weight gain during pregnancy, implying a closely attuned responsiveness to their mate. They also show hormonal changes that accompany their partner's pregnancy, with elevated estradiol, prolactin and testosterone concentrations that peak in the last month of gestation. The elevated paternal prolactin prenatally seems to correlate with the father's parental experience, and postnatally with their responsiveness to, and carrying of, infants. However, pharmacological reduction of prolactin levels rather inconsistently reduced infant carrying, perhaps suggesting that the correlation between prolactin and infant care in these primates reflects the infant stimulation of prolactin secretion, rather than activation of infant care by prolactin. In contrast, testosterone concentrations fall postnatally, being lower in males that are caring for their offspring and lowest in males with most paternal experience or that carry young more frequently. These responses come largely from infant odours, as experienced marmoset fathers display decreasing serum testosterone, and increasing oestrogens, within 20–30 minutes of contact with a scent from their own infant, but not from alien infants.

In summary, the few experiments that have been done on primates are consistent with the endocrine data from animals, but with attenuation, social and cognitive factors playing a larger role. What about the situation in humans?

Human attachment behaviour

It is beyond the scope of this book to do credit to the wealth of data on mother–infant interaction in humans. However, one or two aspects are presented as they amplify some of the comparative data described above.

Despite the common assertion that newborn babies cannot see, they clearly can, and do direct their attentive responses selectively to specific visual stimuli. Stimuli associated with the human face seem especially important and there is good evidence for facial mimicry in babies just a few hours old. Observations of early maternal behaviour after home deliveries show that immediately postpartum, mothers pick up their infant, stroke its face and start breast-feeding while gazing intently into its face. In hospital deliveries, the pattern differs only slightly, mothers particularly exploring their baby's extremities, but still with considerable emphasis on eye-to-eye contact. This early period of intense eye-to-eye contact and physical exploration may be very important. But the interaction is not one-way; the baby emits signals to the mother that evoke her maternal responses and willingness to nurse. Thus, a

reciprocal mother–infant interaction establishes their bond, leading subsequently to mutual recognition.

In addition to the stimuli associated with the mother's face being important for eventual maternal recognition by her infant, **olfaction is used by human neonates** to differentiate between their own and another mother. In experiments in which a 6-day-old infant was presented with breast pads (which had absorbed milk) from its own mother or from an 'alien' lactating female, significantly more time was spent turning towards its own mother's pad. Babies, particularly when several weeks old, attune to their mother's facial expressions when they talk and coo. In one study, a 4-week-old infant, with a blind mother who had never been sighted and displayed a mask-like face during speech, tended to avert his eyes and face from his mother when she leaned over to talk to him. When interacting with other, sighted, individuals, however, this was not the case. The normal interaction had therefore been distorted, but not completely so, as other modes of communication (verbal–auditory in particular) were used successfully to overcome this interaction deficit.

The essential contribution of the baby to the bond between it and its mother is highlighted when babies display behaviour that disrupts the relationship. A baby who cries and shows avoidance responses when picked up may very easily induce feelings of frustration, confusion and anxiety in the parents. They may, in fact, feel rejected by the infant, quite the opposite to what is usually encountered when examining the occurrence of rejection in a mother–infant dyad. The behaviour displayed subsequently by the 'rejected' parents will affect their infant's developing behaviour and so the path is potentially set for an unsatisfactory and enduring pattern of interaction between them.

The **effects of early separation** on subsequent mother–infant interaction in rhesus monkeys may be paralleled in humans. Enforced early separation of women from their newborn babies for periods of up to 3 weeks, as might occur after premature delivery, can be associated with differences in **attachment behaviour** (bonding) when compared with mothers similarly separated from their babies, but allowed additional contact during the first few days after birth. Modern paediatric practice takes account of such findings, and where possible ensures as much contact as is practicable between mothers and their babies. Again, it must be emphasized that although postnatal separation of mother and baby can have delayed and long-lasting effects, it need not necessarily do so. Maternal attitudes, psychophysiology and behaviour do change during pregnancy so that even mothers with initially negative attitudes towards being pregnant generally become more positive by about 5 months. Whether the increasing concentrations of oestrogens and progesterone are involved is unclear. During the postpartum period and the fall in hormone levels, 40–80% of women experience mood changes, the intensity of which is thought to arise from the endocrine change, and the type of which (depression, elation) to depend on circumstances. Moreover, an effect of the oestrogen:progesterone ratio on maternal behaviour is suggested by the observation that mothers who during pregnancy experienced a high oestradiol:progesterone ratio exhibited more depression and anxiety, whereas postpartum mothers with low oestrogen:progesterone ratios during pregnancy exhibited greater attachment feelings. In particular, mothers exhibiting an increase in the oestrogen:progesterone ratio from the 5th and 7th months to the 9th month of pregnancy had higher attachment feelings than those with smaller or no changes in ratios. Evidence for the role of **oxytocin in the initiation of human parental bonding also remains both circumstantial and controversial**. Thus, plasma oxytocin levels provide an unreliable indicator of central oxytocin levels, and not surprisingly show variable correlations with maternal behaviours. Notwithstanding, mothers displaying high levels of affectionate contact towards their babies showed an oxytocin increase following mother–infant interactions, no increase being detected in mothers showing low levels of affectionate contact.

Whereas strong social bonding outside of the mother–infant dyad is atypical for most mammals, social behaviour in humans typically includes engagement, empathy and mutual attachment. Indeed, Hrdy has suggested that the prolonged and energy-costly bonding between human parents and offspring was a key trait in the **evolution of human sociability**, as the provision of social parenting support allowed more extended development. In humans, **oxytocin** has been shown though its effects on the CNS **to play a key role in decreasing anxiety, reducing cortisol reactivity, and increasing communicative behaviour in couples**. This finding resonates with studies on those mammalian species that do show strong pair bonding, as well as bonding between young and both parents (not just the mother), such as some types of vole, in which oxytocin is also implicated in pair bonding. Functional magnetic resonance imaging has suggested a correlation between the distribution of oxytocin receptors and patterns of brain activation while viewing partners or, for mothers, when viewing their infant. Some studies claim that oxytocin administration by intranasal infusion enhances sociability and increases mutual trust. Indeed, it has been suggested that a functional deficit in CNS oxytocin activity may underlie some forms of autism. These claims need further testing, but do offer a plausible mechanism for enhancing mother–father–neonate relationships, as well as for wider sociability, given that central oxytocin levels increase maternally at birth and during breastfeeding, in other mammals.

As far as men are concerned, the endocrine findings largely resemble those for male marmosets. Thus, testosterone levels drop after birth from levels that are lower in previous fathers than non-fathers, and decrease further when the fathers provide effective care. Likewise, prolactin levels pre-birth tend to be higher in existing than in new fathers, and rise in experienced fathers when they hear cries or hold their baby. Also, oxytocin levels are higher in fathers than in nonfathers pre-partum and the postpartum use of intranasal sprays of oxytocin, which facilitates passage to the brain, leads to increased interactions between fathers and babies. However, more human research is needed.

Summary

There is still much to learn about the mechanisms regulating the onset, course and maintenance of human parental care. In as much as there is a message, it is a now familiar one, namely, that humans show an emancipation from the effects of hormones, but not a total escape. Cultural and cognitive factors seem to dominate. The brief account above is not intended to represent the 'way' a mother or father should behave towards their baby, or to say that adverse consequences will necessarily result if they do not behave in this way. However, by studying the behaviour of both human and non-human primate mother–infant pairs, we are beginning to understand the factors contributing to the success and richness of the parent–infant bond. Equally, we may discover what contributes to its breakdown, and how failure to establish an adequate bond at an appropriate time leads to disturbed behaviour in the parents or the infant, or both, later on. Given the pervasive effects of good or bad parental care on subsequent social and, indeed, parental behaviour, the importance of such research is obvious. Recent research in genetics is beginning to shed some light on this subject.

Genetics and maternal care

We have seen how aspects of parental care can influence the long-term behaviour of the offspring. However, one reliable but puzzling feature of this work is the **variability in the response** among different offspring exposed to an apparently similar level and quality of parental care. In addition, we do not yet have a clear understanding of exactly how these early events produce such lasting responses – the underlying causal mechanism(s). Might there be useful genetic explanations? The sequencing of the genomes of various species provides a powerful first step along a path that should lead to explanations as to how environmental signals interact with genes to produce distinctive phenotypes. Here we give just two sorts of examples relevant to maternal care of how this work is progressing.

The impact of maternal care can depend on the genetic make-up of the offspring

An explanation for at least part of the variability in offspring responses to specific patterns of maternal care may lie in the subtle influences exerted by different genetic polymorphisms. A polymorphism is a sequence variation within the gene or its regulatory regions that does not prevent the expression of a functional protein (thus the allele is still considered to be 'wild-type') but may affect the properties of the mRNA and/or the protein in subtle ways that are revealed only in particular circumstances, e.g. exposure to certain infectious agents. In the context of maternal behaviour, prospective studies on children from birth to adulthood have been informative.

In one type of study, it was found that **patterns of antisocial behaviour** were higher in young men who were maltreated as children. However, not all maltreated young men showed this response to the same extent (Figure 20.4a). The highest levels of antisocial behaviour were associated with the presence of a genetic polymorphism in the promoter of the gene encoding **monoamine oxidase A (MAOA)** – but this relationship only emerged as significant when the boys had been maltreated. This polymorphism resulted in lower levels of gene expression and enzyme activity in the brain. Now, MAOA is an enzyme that controls the destruction of monaminergic neurotransmitters, such as noradrenaline, serotonin (5-HT) and dopamine, reducing their levels. Both MAOA deficit and elevated levels of these neurotransmitters are associated with increased aggression.

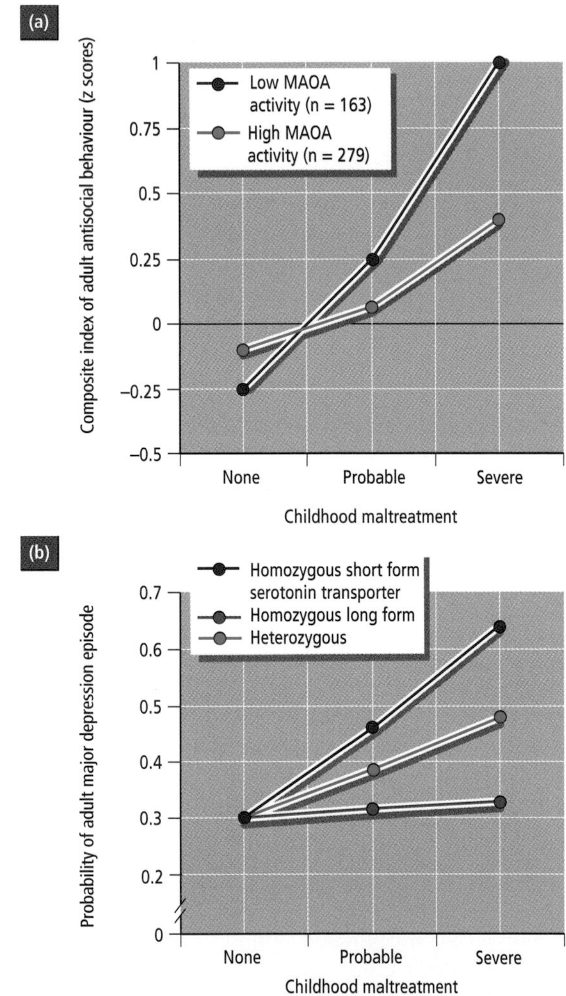

Figure 20.4 (a) Summary plot of the relationship between antisocial behaviour as adult men and the maltreatment of boys in childhood (classified as None, Probable and Severe at 3 and 11 years) who either had high or low monoamine oxidase A (MAOA) activity. Note that severe maltreatment as boys is strongly associated with high levels of antisocial behaviour in boys with low MAOA activity, but much less so for boys with high MAOA activity. (b) A similar plot for adult depressive episodes (18–26 years) in adults homozygous or heterozygous for the 5-HT transporter. (Source: Adapted from Caspi *et al.* (2002, 2003).)

So, one interpretation of these results is that childhood abuse interacts with MAOA activity to influence the susceptibility of the maltreated child to expression of antisocial behaviour.

A second example from the same prospective study examined how childhood abuse might interact with genetic polymorphisms to influence patterns of depression later in life. This time the gene examined encoded the **serotonin transporter**, a gene pertinent for depression because current pharmaco-therapies target selective serotonin reuptake. The polymorphism studied was in the gene promoter region and was characterized as being long or short, the latter polymorphism showing lower transcriptional efficiency. When childhood maltreatment had occurred, children with one or two short alleles had a significantly elevated probability of a major depressive episode compared with children having two long alleles (Figure 20.4b). Again, this significant difference in depression only emerged with childhood maltreatment.

These studies focused on genes with a strong probability of involvement in the psychiatric behaviour under study. It is not claimed that these polymorphic variations can provide the sole explanation for population variations in responses to behaviours experienced in childhood. The value of these studies is that they offer us paradigms for studying how parental behaviours might interact with genetic susceptibilities to produce different behaviour patterns later in life. Of course, the outstanding question then becomes: just how might their interaction function to create this enduring change in behaviour? One answer to that question may lie in another interesting study that again involves genetic expression. In this case, however, **we are not dealing with polymorphic variants but with epigenetic variants**.

The pattern of maternal care may induce heritable epigenetic changes in the chromatin of the newborn

Studies in rats have shown that one consequence of high-quality postpartum maternal care is a lowered sensitivity of the offspring to stressors later in life – **greater resilience**. In contrast, offspring that experience low-quality care become more sensitive to stressors and show higher acute corticosteroid responses – they **stress easily**. Indeed, stable lines of rats have been bred in which the mothers consistently show low-quality or high-quality mothering with reliable effects on their offspring. However, if the offspring of low-quality mothers are cross-fostered at birth to high-quality mothers, then they grow up with relatively robust reactions to stressors, and vice versa. This experiment shows that the mothering seems to have a causal effect on the subsequent stress behaviour.

Where and how is the memory of maternal care quality laid down? It is found that the **offspring experiencing high-quality care** showed **increased sensitivity of the hypothalamic–pituitary–adrenal (HPA) axis to negative feedback by glucocorticoids** and a damped anxiety/stress response compared to offspring that had experienced low-quality care. The increased glucocorticoid feedback sensitivity in the latter rats was associated with a greater expression of glucocorticoid receptors in the hippocampus. In addition, the adult offspring that had received high quality maternal care displayed increased estrogen sensitivity in the MPOA together with increased levels of maternal behaviors toward their own pups following birth. Moreover, as pups, these offspring had increased mRNA and protein of the estrogen receptor alpha in the MPOA than that found in the low maternal care offspring. These findings appeared to locate at least two sites of enduring maternal effect, but what caused this difference in expression? The answer found was surprising, and involved epigenetic changes to chromatin.

You will recall that there are several mechanisms by which epigenetic imprinting can be achieved. It can involve variation in the histone subtypes making up the chromatin, various post-translational modifications to histones, and also direct methylation of certain cytosine residues in the DNA itself (see Figure 1.7). It is the latter sort of epigenetic imprint that was described here. Thus, when the promoter regions of the hippocampal glucocorticoid receptor and the mPOA oestrogen receptor genes were analysed for their cytosine methylation pattern, it was found that **those offspring that had experienced high-quality care had much lower levels of cytosine methylation** at certain key bases in the promoter sequences. Moreover, this lower level of methylation was found regardless of whether the neonates had been born of high-care mothers or had been cross-fostered to them from low-care mothers immediately after birth. This latter control is very important, because had they been left with their birth mothers, they would have had high methylation levels as adults. This result implies that the lower methylation levels were determined by high-quality maternal care and were in turn (at least in part) responsible for offspring anxiety behaviour – a chain of causality. This conclusion was further strengthened by the experimentally induced reduction in the methylation levels of the glucocorticoid receptor promoter in the adult hippocampus, which converted easily stressed animals to more stressor-resilient ones. Conversely, increasing the methylation levels in the adult offspring of good mothers rendered them less resilient to stressors.

A small post-mortem study on the brains of human suicides has also suggested that **a similar imprinting mechanism may operate in humans**. Thus, it is known that childhood abuse alters adult HPA stress responses and is associated with an increased risk of suicide. When the hippocampal activity of a neuron-specific glucocorticoid receptor gene promoter was compared among brains from suicides with and without a history of childhood abuse and those from controls, decreased expression was found in those from abused suicides. In addition, the promoter was more methylated (Figure 20.5). Consistent with these data are those showing that **glucocorticoid responses are blunted in maltreated or bullied 12-year-olds** who show social and behavioural problems compared with non-bullied peers.

These studies leave many questions of mechanism unanswered. However, the studies offer examples of how a complex

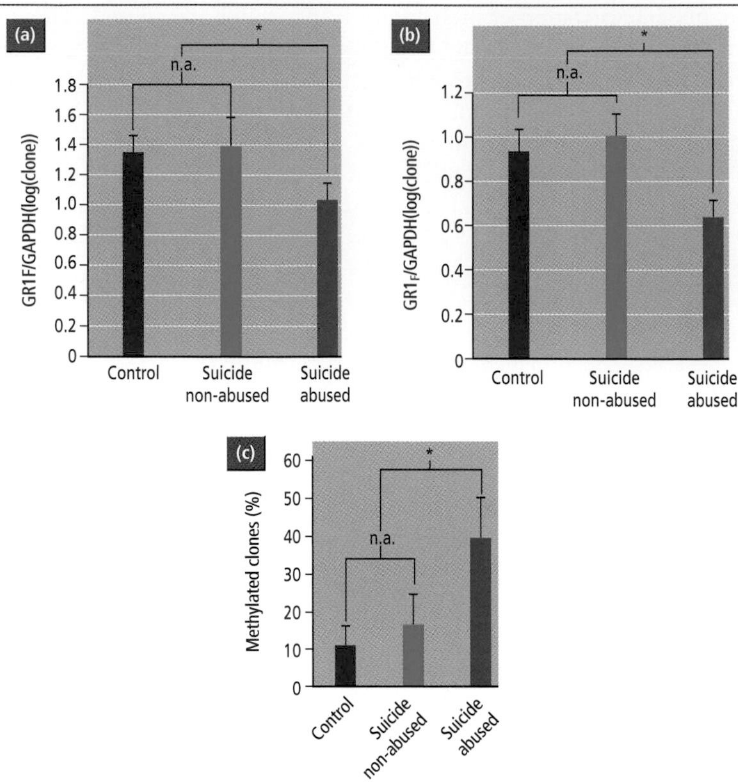

Figure 20.5 Human data on hippocampal glucocorticoid receptor expression levels in postmortem brains from suicides with a history of childhood abuse, non-abused suicides victims and controls (n = 12 for each group). Expression levels of (a) total glucocorticoid receptor (GR) mRNA and (b) of a glucocorticoid receptor splice variant mRNA (GR1F). (c) Cytosine methylation of the glucocorticoid receptor promoter (called NR3C1) in the hippocampus expressed as a percentage of methylated clones found for each group. * $P > 0.05$. N/A, not significantly different. (Source: McGowan P.O. *et al.* (2009) *Nature Neuroscience* **12**, 342–348. Reproduced with permission from *Nature*.)

environmental input (maternal care) can operate epigenetically to influence chromatin organization and thereby patterns or levels of gene expression in ways that are heritable across cell generations and have a subtle effect on a complex behavioural phenotype. If these sorts of epigenetically-induced change were applied to genes with polymorphic variations, then it might enhance further an underlying difference in expressivity that could also explain the results with MAOA and 5-HT.

One question that hangs in the air from these findings is why such a system might have evolved. It is interesting to speculate that we might be seeing a behavioural version of the maternal metabolic programming described in Chapter 16 (see pages 282–4). Thus, in a high-stressor environment, mothers may become more blunted in their responsiveness and so give less good care to their offspring. Their offspring then pick up this signal through epigenetic modifications so that in adulthood their own responsiveness is blunted. In consequence, their chances of survival to sexual maturity, and thus reproduction, might increase, but this survival may be at the expense of their longer-term health and welfare.

Implications?

Our adult behavioural phenotype results from complex interactions between **genotype and environment**. The question has always been: how does this interaction function? What is now becoming clear is that at least one mechanism is through **durable epigenetic remodelling** of our genome: **our epigenome**. The role of early parental care in this remodelling seems to be highly significant. It is unlikely that the quality of parental care will only affect delinquent and depressive behaviours. Thus, many of the gendered and sexual differences between men and women described in Chapters 5 and 6 may also prove to be epigenetic in origin.

However, the studies described above also emphasize that the genetic substrate on which epigenetic marks can be imposed is also important. **Humans are uniquely varied in their range of genetic isotypes**, probably an evolutionary consequence of their wide geographic dispersal and climatic penetration. The realization that certain isotypes are associated with a tendency towards certain behaviours brings both **risks and benefits**. It may help early diagnosis and so facilitate early behavioural interventions. It could also lead to the tailoring of drug treatments for some severe psychiatric conditions to the individual genotype of the patient concerned. However, the temptation to extend the variety of human behaviours viewed as being 'pathological' is of enduring concern, given the cultural, geographic and historical variation in perceptions of **what is 'normal' versus 'deviant'**. Overdiagnosis and premature intervention are likewise a constant temptation for doctors with new (and often incomplete) knowledge and technology. A clear conversation between the medical profession and society on how best to deploy this emerging knowledge is essential.

Conclusions

In humans, the period of parental care is prolonged and complex. The mechanisms that induce maternal responses and maintain this long period of parental care are not clearly understood. They undoubtedly also depend increasingly on cognitive processes, in addition to any neuroendocrine influences around the time of parturition. It is also clear that the quality of care, whether parental or social, can have profound influences on subsequent well-being in ways that depend upon the genotype and epigenotype of the neonate. This clarity of understanding has the potential to provide benefits individually, but also poses risks to society. The increasing capability of medicine to intervene in these early life events has for some years now been present and exercised in respect of the stages of reproduction that lead up to neonatality. The final section explores how these interventions in the control of fertility and the treatment of infertility occur and with what consequences.

Key learning points

- The rapid induction of maternal behaviour at parturition is critical for survival of the young.
- Stimuli from the newborn of many species are critical in inducing maternal responses; they are able to do so even in nulliparous females, but with a longer latency.
- In most mammals, exposure to sex steroids and prolactin during pregnancy greatly hastens the onset of maternal behaviour in response to cues from the newborn.
- Parturition itself and the associated stimulation of the cervix are also a potent stimulus to the onset of maternal behaviour in sheep and rats.
- The central release of oxytocin from neurons within the brain appears to be a common neural mechanism underlying maternal behaviour in most mammals.
- This oxytocin release is stimulated by the vaginocervical dilation at parturition.
- Infusions of oxytocin can induce maternal behaviour within seconds in females who have been exposed previously to oestrogens.
- Oxytocin receptor antagonists prevent the induction of maternal behaviour at parturition.
- Very little is known of the neuroendocrine mechanisms underlying maternal behaviour in non-human primates and in women.
- Mother–infant interaction soon after parturition supports the formation of a close bond between mother and infant that is the focal point for maternal behaviour.
- In monkeys, infants gradually gain independence from their mothers through a dynamic interaction which involves the infant leaving the mother progressively more often to explore its environment and the gradual rejection of the infant's approaches to the mother for contact.
- In humans, the period of maternal care is prolonged and complex.
- Human bonding depends increasingly on cognitive processes, in addition to any possible neuroendocrine influences around the time of parturition.
- The quality of parental care can impact differently on offspring depending on their genetic make-up, such that some offspring may be genetically more susceptible to enduring effects of parental care quality.
- Parental care may also impose epigenetic imprints on genes in the neonate that heritably affect the gene expressivity later in life with consequences for subsequent behaviour.

Clinical vignette

Neonatal jaundice

A 3560 g male baby was born by spontaneous vaginal delivery at 37 weeks and 3 days. He is the first child born to fit and well parents. The pregnancy and delivery were uncomplicated. He was discharged from hospital along with his mother on day 1 of life, following a routine 'baby check', at which no abnormalities were detected. His mother was attempting to establish breast-feeding. The community midwife visited the family on day 3 of life. She noticed that the baby's skin looked yellow, although he was otherwise well hydrated and handled normally. She referred the baby to hospital, where a non-invasive check of his bilirubin level was carried out. The resulting value was 350 μM/L, indicating moderately severe jaundice requiring treatment. The baby was admitted to hospital and underwent phototherapy. His mother was given additional help in breast-feeding and the jaundice resolved over a two-day period. The baby was discharged well on day 5 of life.

In preparation for birth at full term, most fetal physiological systems switch from a growth and differentiation phase to a maturation phase under the influence of rising levels of glucocorticoids. However, a counter-example is the bilirubin-conjugating system, which is not matured until postnatal life. The reason for this is that during gestation, bilirubin (which is an important by-product of red blood cell metabolism) is excreted from the fetus via the placenta. In order for this to happen successfully, the bilirubin must remain unconjugated, otherwise it cannot be excreted via the transplacental route and will build up to dangerously high levels within the fetus. After delivery, however, bilirubin is very poorly excreted from the body unless it is water-soluble, i.e. conjugated. In order for bilirubin to be excreted postnatally, it must thus be conjugated by glucuronosyltransferase. Because the induction of these conjugating enzymes only occurs after delivery, the pace of bilirubin conjugation often cannot keep pace with the breakdown of red blood cells in the early neonatal period. Newborns produce bilirubin at more than twice the adult rate because of a relative surplus of red blood cells and increased turnover. Hence bilirubin levels rise and give rise to the yellow skin pigmentation known as jaundice. A build-up of high levels of bilirubin is dangerous in the fetus or neonate because it is toxic to developing neurons in the brainstem. It is thus important to recognize and treat jaundice in the neonatal period. Phototherapy is a common, non-invasive treatment, which uses light to promote photo-oxidation and excretion of bilirubin. In most cases, physiological neonatal jaundice spontaneously resolves as the initial excess of red blood cells is broken down.

FURTHER READING

General reading

Bainham A *et al.* (1999) *What Is a Parent?* Hart Publishing, Oxford.

Bridges RS (2015) Neuroendocrine regulation of maternal behavior. *Frontiers in Neuroendocrinology.* 0:178–196. doi:10.1016/j.yfrne.2014.11.007.

Churchland PS, Winkielman P (2012) Modulating social behavior with oxytocin: how does it work? What does it mean? *Hormones and Behavior* **61**, 392–399.

Gabory A, Attig L, Junien C (2009) Sexual dimorphism in environmental epigenetic programming. *Molecular and Cellular Endocrinology* **304**, 8–18.

Hammock EAD, Young LJ (2006) Oxytocin, vasopressin and pair bonding: implications for autism. *Philosophical Transactions of the Royal Society B* **361**, 2187–2198.

Hrdy SB (2009) *Mothers and Others: The Evolutionary Origins of Mutual Understanding.* Harvard University Press, Cambridge, MA.

Insel TR, Young LJ (2001) The neurobiology of attachment. *Nature Reviews of Neuroscience* **2**, 129–136.

Insel, TR, Young L, & Wang Z (1997) Central oxytocin and reproductive behaviours. *Reviews of Reproduction* **2**, 28–37.

Nowak R, Porter RH, Lévy F *et al.* (2000) Role of mother–young interactions in the survival of offspring in domestic mammals. *Reviews of Reproduction* **5**, 153–163.

Owen D, Andrews MH, Matthews SG (2005) Maternal adversity, glucocorticoids and programming of neuroendocrine function and behaviour. *Neuroscience and Biobehavioral Reviews* **29**, 209–226.

Pereira M, Ferreira A (2016) Neuroanatomical and neurochemical basis of parenting: dynamic coordination of motivational, affective and cognitive processes. *Hormones and Behavior* **77**, 72–85.

Rutter M (2006) *Genes and Behaviour Nature–Nurture Interplay Explained.* Blackwell Publishing, Oxford.

Saltzman W, Maestripieri D (2011) The neuroendocrinology of primate maternal behavior. *Progress in Neuro-Psychopharmacology & Biological Psychiatry* **35**, 1192–1204.

Singh I, Rose N (2009) Biomarkers in psychiatry. *Nature* **460**, 202–207.

Storey AE, Ziegler TE (2016) Primate paternal care: interactions between biology and social experience. *Hormones and Behavior* **77**, 260–271.

Weaver ICG, Meaney MJ, Szyf M (2006) Maternal care effects on the hippocampal transcriptome and anxiety-mediated behaviors in the offspring that are reversible in adulthood. *Proceedings of the National Academy of Sciences of the United States of America* **103**, 3480–3485.

Zhang T-Z, Bagot R, Parent C *et al.* (2006) Maternal programming of defensive responses through sustained effects on gene expression. *Biological Psychology* **73**, 72–89.

More advanced reading

Agrati D, Lonstein JS (2016) Affective changes during the postpartum period: influences of genetic and experiential factors. *Hormones and Behavior* **77**, 141–152.

Bowlby J (1973) *Attachment and Loss*, Vol. **2**. Hogarth, London.

Bowlby J (1980) *Attachment and Loss*, Vol. **3**. Hogarth, London.

Bowlby J (1982) *Attachment and Loss*, Vol. **1**, 2nd edn. Hogarth, London.

Caspi A, McClay J, Moffitt TE *et al.* (2002) Role of genotype in the cycle of violence in maltreated children. *Science* **297**, 851–854.

Caspi A, Sugden K, Moffitt TE *et al.* (2003) Influence of life stress on depression: moderation by a polymorphism in the 5-HTT gene. *Science* **301**, 386–389.

Champagne FA, Weaver IC, Diorio J *et al.* (2006) Maternal care associated with methylation of the estrogen receptor-1b promoter and estrogen receptor-expression in the medial preoptic area of female offspring. *Endocrinology* **147**, 2909–2915.

Champagne F, Weaver ICG, Diorio J *et al.* (2003) Natural variations in maternal care are associated with estrogen receptor α expression and estrogen sensitivity in the medial preoptic area. *Endocrinology* **144**, 4720–4724.

Ciba Foundation Symposium 33 (1975) (New Series) *Parent–Infant Interaction*. Elsevier, Amsterdam.

Corter CM, Fleming AS (1995) Psychobiology of maternal behaviour in human beings. In: *Handbook of Parenting*, Vol. 2 (ed. MH Bornstein), pp. 87–115. Lawrence Erlbaum Associates, Mahwah, NJ.

Harlow HF, Suomi SJ (1970) Nature of love—simplified. *American Journal of Psychology* **25**, 161–168.

Harlow HF, Zimmerman RR (1959) Affectional responses in the infant monkey. *Science* **130**, 421–432.

Hinde RA (1974) *Biological Bases of Human Social Behaviour*. McGraw-Hill Inc., New York.

Keverne EB, Levy F, Poindron P, Lindsay DR (1983) Vaginal stimulation: an important determinant of maternal bonding in sheep. *Science* **219**, 81–83.

Marlier L, Schaal B (2005) Human newborns prefer human milk: conspecific milk odor is attractive without postnatal exposure. *Child Development* **76**, 155–168.

McCormack K, Newman TK, Higley JD et al. (2009) Serotonin transporter gene variation, infant abuse, and responsiveness to stress in rhesus macaque mothers and infants. *Hormones and Behavior* **55**, 538–547.

McGowan PO, Sasaki A, D'Alessio AC *et al.* (2009) Epigenetic regulation of the glucocorticoid receptor in human brain associates with childhood abuse. *Nature Neuroscience* **12**, 342–348.

Ouellet-Morin I (2011) Blunted cortisol responses to stress signal social and behavioral problems among maltreated/bullied 12-year-old children. *Biological Psychiatry* **70**, 1016–1023.

Weaver IC, Cervoni N, Champagne FA *et al.* (2004) Epigenetic programming by maternal behavior. *Nature Neuroscience* **7**, 847–854.

Weaver ICG, Champagne FA, Brown SE *et al.* (2005) Reversal of maternal programming of stress responses in adult offspring through methyl supplementation: altering epigenetic marking later in life. *Journal of Neuroscience* **25**, 11045–11054.

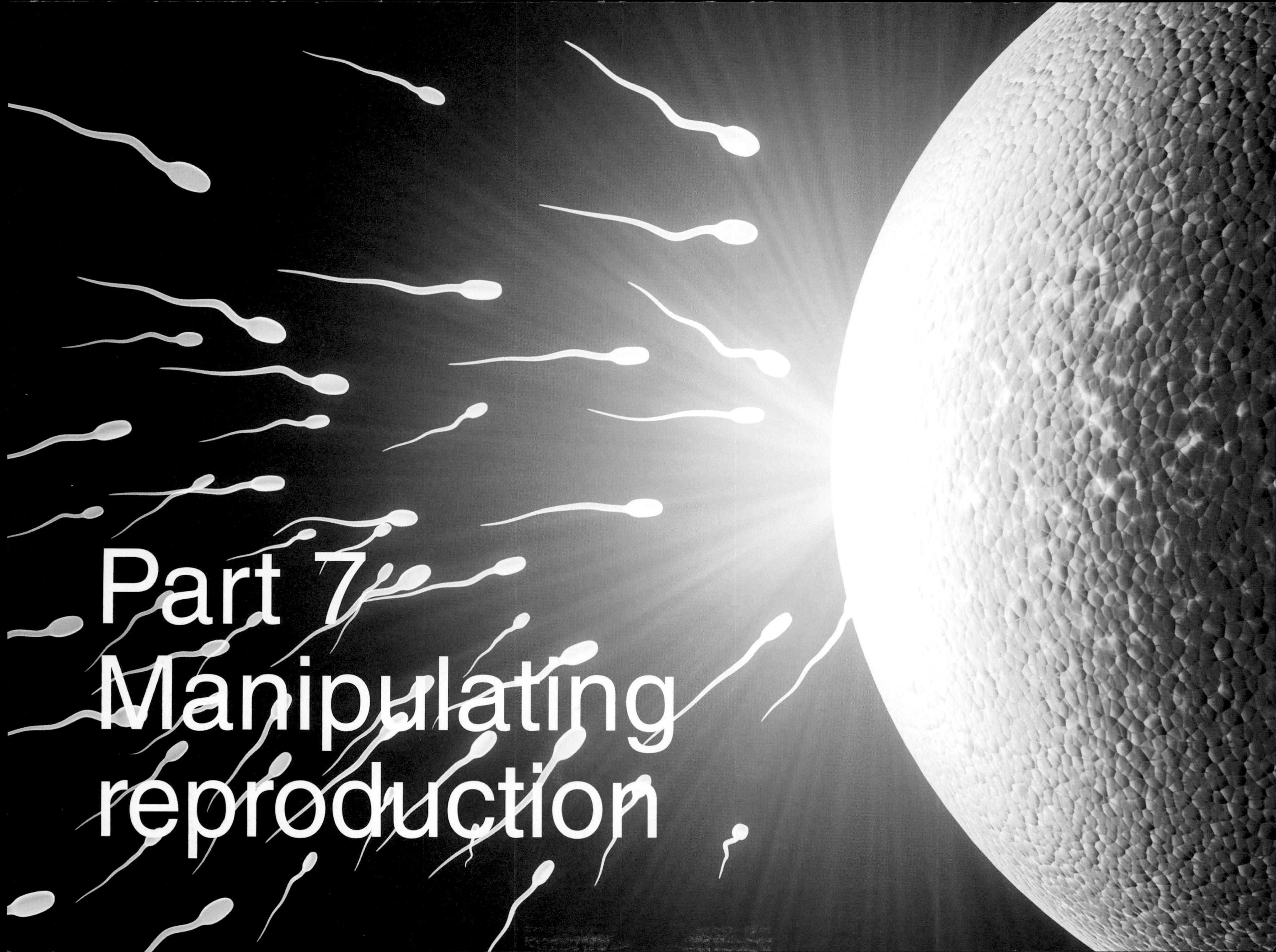
Part 7
Manipulating
reproduction

CHAPTER 21
Controlling fertility

Essential Reproduction, Eighth Edition. Martin H. Johnson.

Companion website: www.essentialreproduction.com

As we have seen throughout this book, humankind, whilst grounded in evolved biology, has achieved remarkable plasticity in many aspects of its reproduction. It has also been remarkably successful reproductively, as witnessed by the exponential increase in world population (Figure 21.1). This rise has come about largely as the result of our increasing ability to exert control over our environment, thereby to extend our reproductive capacity by allowing more of us to reach reproductive age. This largely scientific success has brought its own problems, notably over-extended resources and climate change. Three reactions to these problems are (1) to ignore them; (2) to preach a despairing apocalyptic vision; or (3) to rely on science to solve them. Here we are concerned to enquire whether and how science can help with the population problem itself. To begin with, some definitions are necessary.

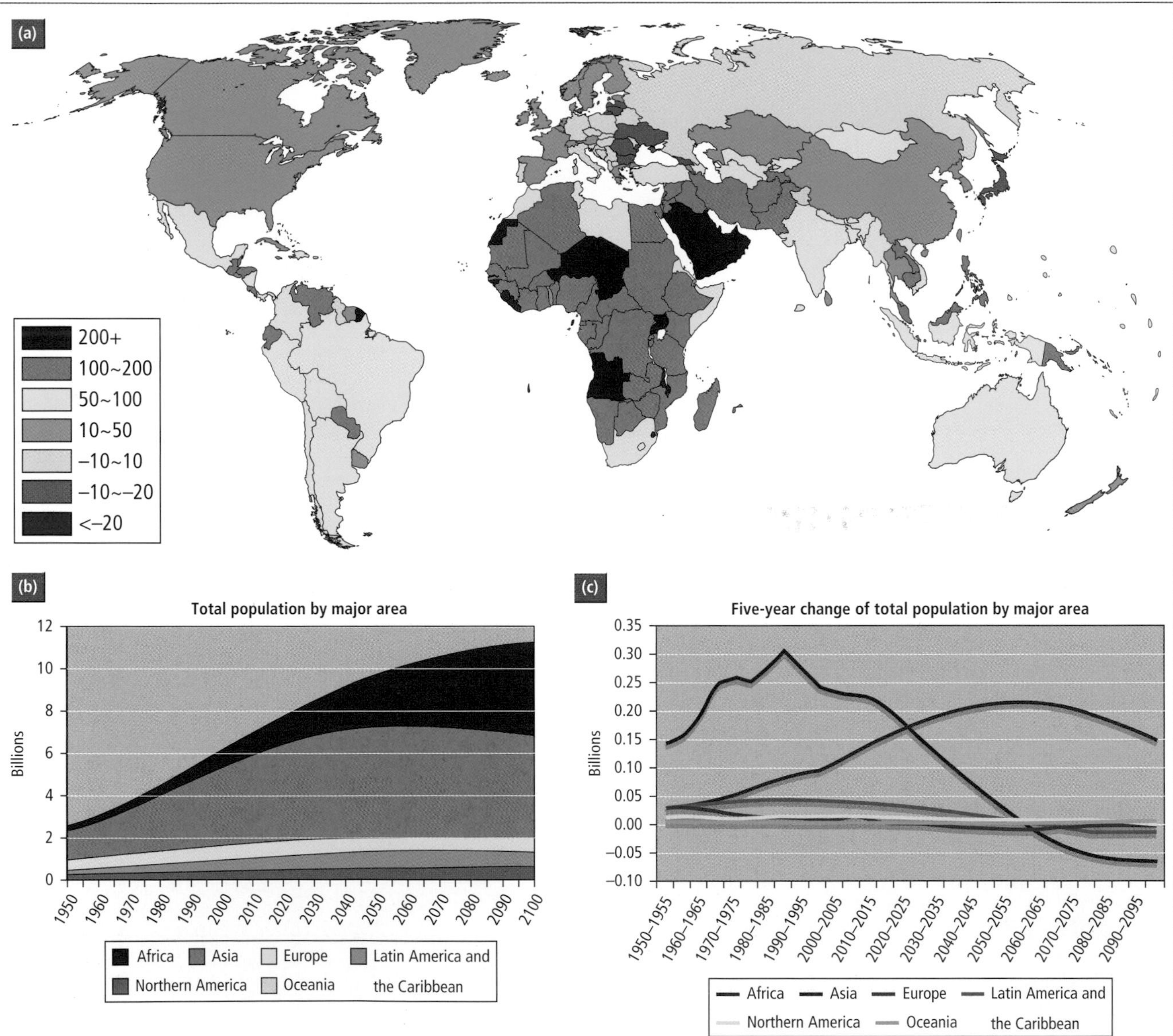

Figure 21.1 (a) A world map showing estimated population growth from 1980 to 2015: note the hotspots in Africa – and the largely successful control of population growth in Asia. (b) Plot of estimated total population by region between 1950 and 2100: note the initial exponential growth, largely from the developing world, but that growth will plateau and decline in Asia as the reduced numbers of reproductivelv active women feed through. (c) A plot of the estimated five-year change in total population by major region between 1950 and 2100, to illustrate the declining Asian population and the projected delayed decline in African population. (Source: Department of Economic and Social Affairs, Population Division, United Nations. *World Population Prospects: The 2015 Revision*.)

Fertility, fecundibility and fecundity

The **fertility** of an individual, a couple or a population refers to the number of children born, and the **fertility rate** of a population is the number of births in a defined period of time divided by the number of women of reproductive age. Women rather than men are used in this calculation because their reproductive potential is limiting. Thus, fertility is a measure of the **actual outcome of the reproductive process**. In contrast, **fecundibility** is defined as the probability of conceiving in a menstrual cycle in a woman who has regular periods and engages in regular unprotected sex, whether or not conception goes to term. Conception is indicated by either a biochemical pregnancy (positive human chorionic gonadotrophin test) and/or a clinical pregnancy (from an ultrasound scan: see Figure 16.1). Finally, **fecundity** is a measure of the capacity to both conceive and produce a live birth under the same circumstances. Both fecundibility and fecundity are therefore measures of the **reproductive potential** of women.

Reliable estimates of human fecundibility have come recently from prospective population-based Chinese and European studies. These recruited couples, mostly with women under 35, using no contraception and trained to detect the most fertile 6 days around ovulation and to try for pregnancy every cycle during this period. Cumulative conception rates of around 50% after two cycles and 85% by 6 months were observed. Approximately half the remaining couples conceived within a year, leaving a residual 5–7% of couples as **clinically subfertile**. Comparable data from prospective studies of fecundity are limited, but, together with data from reliable retrospective studies, suggest that 10–15% of all clinically diagnosed pregnancies end in spontaneous miscarriage. Thus, human fecundibility and fecundity rates of 25% and 22% per cycle seem reasonable estimates for women of peak reproductive age. We should remember that, for women, puberty and menopause mark the limits of this period. In reality, the fertility of a population or individual will rarely reach its theoretical maximum (fecundity), because of constraints placed on reproductive efficiency. Traditionally three factors have limited their fertility: (1) the use of contraception and induced pregnancy termination; (2) pathology (infertility or subfertility); and (3) social or religious practices and traditions. Each of these constraints is considered in turn in this and the ensuing two chapters in this final part of the book

Artificial control of fertility

The three socially accepted artificial controls over fertility available to varying extents in different societies are (1) **sterilization**; (2) **contraception**; and (3) induced **pregnancy termination**. Not all of these controls are 100% effective and, thus, some of them should be seen as regulating fertility by delaying births or increasing the interval between births. Their relative cost-effectiveness for both the population as a whole and the individual in particular is, of course, of the greatest importance. The main methods of fertility regulation used worldwide are shown in Table 21.1, and in the UK in 2008–2009 by age of

Table 21.1 Reported pregnancy rates and patterns of usage for various forms of contraception

Method	Pregnancy rate*	Use world-wide†	Region in which the method is most used (%)/least used (%)†
Oral contraception	9	8.8	Western Europe (37.5)/Middle Africa (2)
Progesterone-only pill		n.a.	
Implant	0.5		
Injectable	6	4.6	Southern africa (29)/Southern Europe (0.2)
IUCD (copper)	0.8	13.7	Central Asia (41) /Middle Africa (0.3)
Progesterone-releasing IUCD	0.2	n.a.	
Male condom	18	7.7	Eastern Europe (24)/Northern Africa (0.8)
Rhythm	24	2.6	Middle Africa (7)/Southern Africa (0.3)
Withdrawal	22	3.1	Western Asia (15)/Western Europe (0.6)
Cap/diaphragm/female condom	9/12/21	0.1	Eastern Europe (1.5)/Africa, South America, Polynesia (0)
Sterilization (female)	0.15	19.2	Central America (33)/West Africa (0.4)
Sterilization (male)	0.5	2.4	North Europe (16.4)/Africa, West Asia (0)
None	85	36.4	

*Expressed as number per 100 woman years of exposure (i.e. 100 women for 1 fertile year) note that these data are based on reported unintended pregnancies within the first year of use. Depending upon the user's motivation, better or worse outcomes might be experienced. (Source: CDC, 2011.)
†Expressed as percentage of women aged 18-49 in a married or potentially fertile relationship. (Source: United Nations data, 2015.)

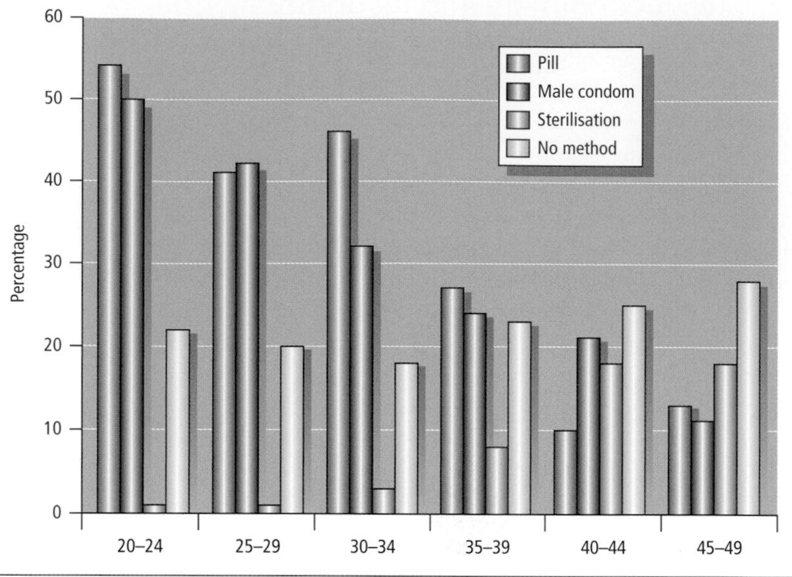

Figure 21.2 The major fertility control methods used by UK women aged 20–49 in 2008-2009. (Source: based on the Office of National Statistics Report).

woman in Figure 21.2. However, global figures mask the wide variation with local traditions, some of which will be captured below. These variations emphasize that fertility control methods interact with cultural and economic factors to determine usage patterns. Special issues arise when considering fertility control for young people (see Box 21.1).

Sterilization

Sterilization should be entered into as an irreversible procedure. It is therefore the method of fertility limitation selected by individuals or couples who have achieved their desired family size or who, for personal or health reasons, want to avoid reproduction. If there is any question of the person requesting sterilization not having the mental capacity to consent, the case should be referred for legal advice to establish whether the procedure is in the best interests of the person concerned and not for the convenience of others.

Vasectomy

In men, the recommended method of sterilization involves local diathermic electro-coagulation followed by division of the *vas*, with or without excision of a small segment, and the interpositioning of some fascia between the cut ends to minimize spontaneous *vas deferens* re-anastomosis. The *vas* is accessed through a small incision in the scrotum preferably under local anaesthesia as an outpatient procedure. Aspermic ejaculates result within 3 months. Between 0.2 and 0.4% of men father a child after vasectomy performed in this way, half of these because they cease contraceptive use too early (i.e. prior to repeated ejaculates containing less than 100,000 non-motile spermatozoa/ml after 12 weeks from the operation, and preferably having had 11–20 ejaculations), but in the other 50% due to spontaneous recanalization. The failure rate amongst men who have been declared aspermic at 12 weeks is 1 in 2300, Spermatozoal production is unaffected, and so the spermatozoa and the small volume of fluid that continues to pass out of the epididymides can build up and lead to a local granuloma in some 3% of men. Certain consequences follow: (1) there is about twice the incidence of chronic or intermittent tenderness in the scrotal region, sufficient to cause some distress in about 6% of cases, and severe distress in 0.3%; (2) in 50% of men the leakage of spermatozoal debris into the systemic circulation from the site of inflammation induces an immune response to their own spermatozoa, but there is no evidence that this has ill-effects; (3) in several species of experimental animal (but not in men), a progressive decline in spermatogenic output occurs.

Some specialist centres claim very high rates of success with the micro-surgical reversal of vasectomy, but dependent on the method of sterilization, and the length of time after the procedure. Patency rates of 97% can be achieved with average pregnancy rates around 75%, if reversal is undertaken within 3 years of the vasectomy, falling to 40% and <10% after >19 years. Developments in the technology of assisted conception (TESE and PESA followed by ICSI) now make the infertility resulting from vasectomy potentially fully reversible regardless of the time since vasectomy (see page 369), although the relative expense of repeated ICSI cycles versus surgery might be a significant consideration. Around 15% of men request reversal for a range of reasons, including psychological factors, particularly arising from the erroneous equation of **fecundity** with **potency**, which may result in anxiety that leads to erectile dysfunction after vasectomy in some men. This fear also prevents many men from accepting the procedure. Sexual arousal and ejaculation are, of course, independent of spermatozoal release. The evidence shows that there has been a decline in the cumulative total of vasectomies performed between 2000 and 2010 in England from 41,100 to 18,000.

Box 21.1 Fertility control for young people

The legal definition of 'young person'

This varies with different jurisdictions, as do the legal requirements for doctors when dealing with young people. In the UK, for example, a child is a person who has not reached the age of 18 in England, Wales and Northern Ireland, but a person below the age of 16 in Scotland. The age at which young people can legally consent to sexual intercourse also varies from 14 to 18 years across Europe, sometimes depending on whether partners of the same sex or the opposite sex are involved and often with extra protection for 'children' between 14 and 18 from exploitation. It is essential that medical practitioners understand the local legal situation, but are also aware that people from overseas may not know of possible conflicts with the legal situation in their own country of residence.

Unwanted teenage pregnancy

This (together with genitourinary disease, especially HIV and hepatitis infection, but also chlamydial infection with its attendant infertility consequences) provides a major stimulus to sexual health and relationship education for young people. Table 21.2 shows the wide variation in teenage pregnancies across the world. In much of Africa and the Indian subcontinent, the high birth rates reflect the early age of marriage. In the developed world there is clear evidence that effective education works to delay first sexual experience, reduce unwanted pregnancy rates, and enhance self-esteem and relationship skills. The political will to implement national effective educational schemes is very variable across different countries, with corresponding differences in the effectiveness of programmes and the prevalence of unwanted

Table 21.2 The wide variation in teenage pregnancies across the world (%)

Country	1995–2000	2000–2005	2005–2010	Country	1995–2000	2000–2005	2005–2010
Australia	19.2	16.7	16.5	Japan	4.4	5.8	5.0
Austria (%)	14.6	13.5	12.8	Malawi	160.9	159.1	119.2
Bangladesh	130.5	105.9	78.9	Malaysia	15.0	15.2	14.2
Belgium	22.7	16.1	14.2	Netherlands	6.2	7.1	5.1
Bulgaria	50.6	44.8	42.8	Nigeria	135.2	127.2	118.3
Canada	20.1	15.0	14.0	Pakistan	60.0	41.3	31.6
China	6.8	8.2	8.4	Poland	20.4	14.9	14.8
Cyprus	13.1	7.2	6.6	Portugal	20.7	20.2	16.8
Czech Republic	16.7	11.4	11.1	Romania	41.9	35.0	32.0
Denmark	8.0	6.6	6.0	Russian Federation	35.7	27.9	30.0
Finland	9.6	13.1	9.3	Sierra Leone	150.9	143.4	143.7
France	7.2	8.0	7.2	Somalia	72.5	71.9	70.1
Gambia	130.9	102.8	76.6	South Africa	80.6	70.7	59.2
Germany	12.8	10.4	7.9	Spain	8.1	10.3	12.7
Ghana	90.1	79.9	71.1	Sri Lanka	28.0	27.9	23.6
Greece	11.7	10.7	11.6	Sweden	8.9	8.1	6.0
Hungary	27.0	18.5	16.5	Switzerland	5.7	5.3	4.6
India	116.1	98.6	86.3	Turkey	52.0	42.8	39.2
Ireland	19.0	19.0	17.5	Uganda	191.0	172.5	149.9
Israel	17.7	16.0	14.0	United Kingdom	31.0	27.0	29.6
Italy	7.0	7.0	6.7	United States of America	50.5	43.1	41.2
Jamaica	93.5	85.4	77.3	**World**	**66.8**	**60.4**	**55.7**

pregnancies. Recently, it has become clear that the emphasis on preventing pregnancy has perhaps led the young men and women to under-estimate the adverse effects of maternal age on fertility later in life, and has led to calls to include education about the limits of fertility in young people's sexual education.

Professional clinical practice

This requires that everyone, regardless of age, should be viewed as potentially autonomous and competent to give informed consent, and all decisions must be taken within local laws for the young person's benefit, and not for the convenience of other parties. Clinicians must therefore assess competence. However, children should not be encouraged to have intercourse below the legal age of consent, should have all pertinent health and legal issues of doing so explained to them, and should be encouraged to include their parents or carers in any decision about contraception or pregnancy termination, and be supported in doing so. The confidentiality of the consultation and its limits (where coercion, maltreatment or exploitation is suspected) should be made clear.

Further reading

Bastable R, Sheather J (2005) Mandatory reporting to the police of all sexually active under-13s: new protocols may undermine confidential sexual health services for young people. *British Medical Journal* **331**, 918–919.

Faculty of Family Planning and Reproductive Health Care Clinical Effectiveness Unit (2004) Contraceptive choices for young people. *Journal of Family Planning and Reproductive Health Care* **30**, 237–251.

Herring J (2006) *Medical Law and Ethics* pp. 208–210. Oxford University Press, Oxford.

Ingham R (2004) Sexual health and young people: the contribution and role of psychology. In: *Sexuality Repositioned: Diversity and the Law* (ed. B. Brooks-Gordon *et al.*), pp. 235–260. Hart Publishing, Oxford.

Kirby D (1999) Sexuality and sex education at home and school. *Adolescent Medicine* **10**, 195–209.

O'Brien SH, Kaizarc EE, Goldd MA, Kellehera KJ (2008) Trends in prescribing patterns of hormonal contraceptives for adolescents. *Contraception* **77**, 264–269.

O'Sullivan I *et al.* (2005) *Contraception and Sexual Health, 2004/05*. Office for National Statistics, London.

Tubal occlusion

Sterilization of women is more invasive, expensive and risky and less effective than male sterilization. It involves **oviducal ligation**, usually by clipping with a clip or ring (safest methods), or less frequently, because of increased complications and difficulties of reversal, electrocoagulation by **diathermy**. The procedure, viewed by **laparoscope** sub-umbilically is generally undertaken under general anaesthesia as a day-case procedure. A small suprapubic incision allows insertion of the ring or loop applicator. In some cases, a larger incision (**mini-laparotomy)** may be needed to access, ligate and section the oviduct for women who are obese or have had previous abdominal surgery with adhesion formation. **Hysteroscopic sterilization** is a new procedure still undergoing safety and efficacy evaluation. It involves the transcervical insertion of a titanium microcoil (Essure®) into the ostia of each oviduct, which induces scar formation to block the tubes. Overall, sterilization has a failure rate of around 2–3 in 100 women becoming pregnant over 10 years, with an increased risk of tubal pregnancy and potential for perforation of the fallopian tubes. Although not easily reversed surgically, the new techniques of assisted reproduction (see pages 369–71) allow the infertility to be circumvented.

Unwanted effects of sterilization can include complications at the time of or following the procedure including haemorrhage (2 in 1000), persistent pain at the site of the ligation, risk of ectopic pregnancy (partial occlusion), a very slight risk (1 in 12,000) of a surgically related death, and psychological disturbance associated with loss of fecundity and the perception that femininity or womanhood may also have been damaged. As for vasectomy, the evidence shows that there has been a decline in the cumulative total of females aged 15–49 who were sterilized in England between 2000 and 2010 from 35,300 to 9700.

Contraception

Contraception differs formally from sterilization only in its potential or actual ease of reversibility, and thereby in the control that the individual exerts over its use. It is, thus, the artificial control of choice for those who wish to delay the expression of their fecundity or to express it with discrimination. The methods currently available are listed in Table 21.1, together with estimates of their efficiency, expressed in terms of their failure rates. The reported rates vary widely, depending on the education, motivation and experience of the users, and their differential access to technology, good hygiene, education and medical follow-up. A breakdown of the main methods used to regulate fertility in the UK by women aged 16–49 is given in Figure 21.2. Note the dominance of oral contraception, condom use and sterilization, and that the balance of each varies with female age, a marked shift to sterilization occurring from 34 years, reflecting the differing objectives of fertility limitation. Even the most cursory glance through these data makes evident the complexity of contraceptive use, reflecting in turn the variety of needs, resources and beliefs available to those exercising usage.

Natural methods

Contraceptive approaches that rely on coital technique, rather than the use of technical or pharmaceutical aids, straddle the boundary between social and artificial approaches to fertility control. In consequence, such approaches seem to pose particular problems for theologians and lawyers when prescribing or proscribing sexual behaviours. For example, of the techniques listed below, only the rhythm method is sanctioned by the Roman Catholic Church on the grounds that the other approaches are 'unnatural'. All the methods described below occur naturally in the biological, if not in the theological, sense.

The rhythm method. For many centuries common belief wrongly equated menstruation with the period of maximum fertility. In fact, of course, ovulation occurs approximately mid-cycle. However, given the potential life of spermatozoa in the cervix of several days and a 24-hour fertilizable life of the oocyte, coition should be avoided through much of the first part of the cycle (Figure 21.3). Moreover, the timing for this method is **counted back** from the small postovulatory rise in temperature of 0.2–0.6 °C (Figure 21.3). Thus, the approach is retrospective and will only be safe if the woman's cycle is sufficiently constant, which it often is not, especially given its susceptibility to emotional or stressful disturbance. These factors, and the high motivation required by the sexual partners, make for **a high failure rate** (Table 21.1), this method being some 20-fold less effective than oral contraceptives at their worst. Additionally, there are (disputed) claims of a significantly increased frequency of pregnancy loss and genetic abnormalities in conceptuses derived by fertilization of aged eggs occurring in couples using this method of fertility control (see page 208). Only around 1% of women now use this method in the UK, but it contributes significantly to population control in many countries (see Table 21.1).

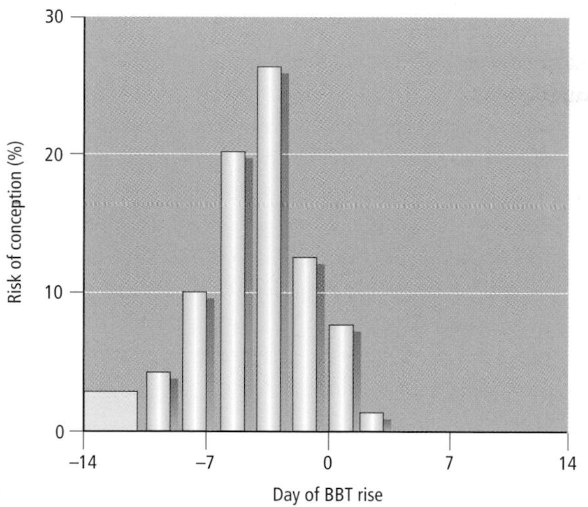

Figure 21.3 The percentage risk of conception at different times during the unprotected cycle of women using the rhythm method, plotted as days before or after the day of rise in basal body temperature (BBT) (day 0). Coital interactions during much of the first part of the cycle carry a substantial risk of pregnancy. Only the period after the BBT rise until the end of the period of menstrual flow is truly safe. This extended period of risk is due to the potential longevity of spermatozoa in the female tract and variability in the length of the follicular phase in some women. Pink area represents menstrual flow. (Source: Guillebaud J. (1993) *Contraception*. Churchill Livingstone, Edinburgh. Reproduced with permission from Elsevier.)

Coitus interruptus. The withdrawal of the penis from the vagina during copulation but before ejaculation has been, and remains, one of the most frequently used forms of contraception, and the technique is credited (together with abortion) as being responsible for much of the decline in birth rate at the time of the Industrial Revolution in Europe. Failures occur due both to lack of adequate control by men at the moment of ejaculation and to insemination with spermatozoa that leak from the urethra before ejaculation or persist from a prior ejaculation. It is difficult to estimate the true usage prevalence, but high usage in western Asia, including the Middle East, are claimed (see Table 21.1).

Masturbation and other forms of sexual interaction. Mutual masturbation is one commonly employed means of reducing the incidence of pregnancy, and is highly effective as such. It is also freer of risks of transmission of human immunodeficiency virus (HIV) as long as fresh semen or vaginal fluids are not subsequently rubbed manually into the vagina, anus or skin lesions. **Oral and anal sex** is used commonly in many societies to gain sexual pleasure without risking fertility. Although the evidence linking oral sex to HIV transmission is weak and disputed, other genital infections, such as herpes, hepatitis and gonorrhoea are readily spread by this sexual practice. Oral sex (unprotected by use of a condom) is, however, clearly lower risk for HIV transmission in either direction when compared with unprotected vaginal or anal sex. In the latter case, the epithelia lining the anus and rectum are more easily damaged than the vaginal epithelium, making bleeding more frequent and infection transmission to both the penetrated and penetrating partners more common. High-strength condoms substantially reduce the risk of transmission of infection anally. There is now good evidence that circumcision reduces the transmission of HIV to men.

Caps, diaphragms and spermicidal foams, jellies, creams and sponges

The combination of both a physical seal at the cervix between the vagina and the uterus, and a **spermicide** at the vaginal site of seminal deposition, has been used for over a century. The method became popular early in the twentieth century as, for the first time, it gave women some control over the use of

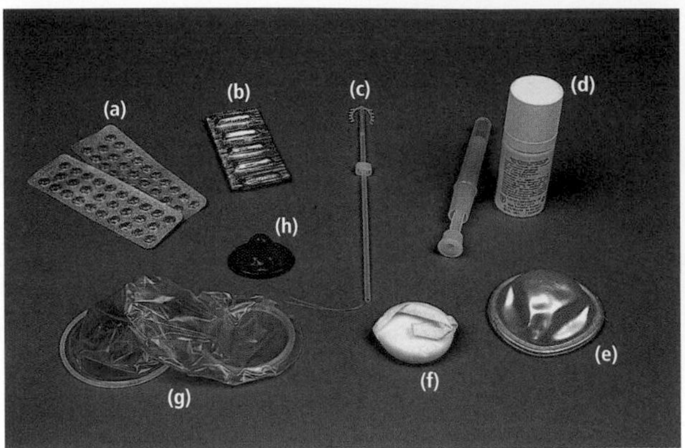

Figure 21.4 Contraceptives. (a) Two packs of oral contraceptives. (b) Pack of vaginal spermicidal suppositories (no longer available in the UK). (c) Intrauterine contraceptive device (IUCD) (MultiLoad 375) emerging from its insertion catheter. Note that the IUCD is not entirely within the catheter; the two serrated arms remain exposed as the cannula is inserted through the cervix into the uterine lumen until the arms reach the fundus. When so located there, the insertion catheter is withdrawn, leaving the IUCD in place with the thread tail on its end passing through the cervix. (d) Spermicidal contraceptive foam with its applicator for insertion into the vagina (no longer available in the UK). (e) A diaphragm. (f) Vaginal contraceptive sponge impregnated with spermicide (no longer available in the UK). (g) Female condom (Femidom). (h) Male condom.

contraception. However, the method has decreased in popularity with the development of more effective and convenient methods of contraception, and spermicidal sponges and pessaries are now no longer available in the UK. This is because spermicides contain **nonoxinol-9**, a non-toxic detergent that destroys the sperm membrane. However, when used frequently, it can cause irritation and in some women epithelial disruption and thereby increases the risk of HIV transmission.

The diaphragm (Figure 21.4e), via its sprung margin, occludes the top of the vagina including the cervix. The smaller '**Dutch**' or **cervical cap** fits directly over and around the cervical os, where it is held by suction. These devices can be fitted well in advance of intercourse, and should be left in place for at least 6 hours after coitus. Neither of them gives a perfect barrier, but do reduce the chances of spermatozoal passage deeper into the genital tract.

These contraceptive approaches give more control to the female partner and their use is relatively independent of intercourse compared with the condom (see below). They also offer protection from pelvic inflammatory disease. Their disadvantages are the requirements for careful fitting and training in use, high motivation, good hygiene and a willingness of women to touch their own internal genitalia. When used by motivated women, they can be very effective, but otherwise can be of relatively low efficiency, clinical trials suggesting six-cycle pregnancy rates of around 9–21% (see Table 21.1).

Table 21.3 Condom use

Do	Don't
Use kite-marked only Use water-based lubricants (KY jelly, glycerol) Use only once. Take care not to tear or snag condom Put on penis before any genital/anal contact Use condoms if risk of HIV, hepatitis, herpes, gonorrhoea, chlamydia or cervical dysplasia	Don't use if 5 or more years old or exposed to excessive heat or UV light Don't use oil-based lubricants (Vaseline, baby oil, suntan lotions) Don't allow penis to become flaccid inside partner Don't have genital/anal contact after condom removal, unless penis is washed Don't use ordinary strength condoms if having anal intercourse

Condoms

Condoms (Figure 21.4h) are the commonest form of mechanical contraceptive in use (see Figure 21.2), particularly since their promotion with the appearance of HIV. The improved strength, lubrication and design of modern condoms have considerably enhanced their efficiency and durability. Condoms are cheap, readily available and relatively easy to use, and also give protection against venereal diseases. Good use is crucial to their success (Table 21.3).

The **female condom** has been marketed (under the brand name **Femidom**). It resembles an extra-large lubricated condom with a rimmed structure at the closed end, akin to the rim of a diaphragm, which fits into the vagina, thus protecting the cervical os in the vaginal vault (Figure 21.4g). It combines the protective advantages of the penile condom with many of the advantages of the diaphragm. Unlike the diaphragm, it does not require fitting or complex training in its use. Its user-friendliness and effectiveness remain the subject of research, but it is much less effective than a properly used male condom, and is not popular (see Table 21.1).

Steroidal contraceptives for women

Since the initial development of the female 'pill' in the 1960s, a range of steroid-based contraceptives has been developed (Table 21.4 and Figure 21.4a), the oral pill being the most popular in the UK (Figure 21.2). Synthetic steroids, as described in Box 2.1, are used in these preparations as their half-lives in the body are longer and their effects therefore sustained. The general underlying principle is the suppression of ovulation by the negative and antipositive feedback effects of progesterone (with or without oestrogen) on the pituitary and hypothalamus (see pages 162–5). Additionally, progesterone can exert direct antifertility effects on the female genital tract to suppress sperm penetration through cervical mucus (see pages 174–5) and endometrial receptivity (see page 226).

Table 21.4 Steroid-based contraceptives for women

	Oral COC*	Oral POP†	Injectables	Implants	LNG IUS‡
Administration					
Frequency	Daily	Daily	2- to 3-monthly	5-yearly	3 months
Relative progestagen dose	Low	Ultra-low	High	Ultra-low	Ultra-low
Blood levels	Rapidly fluctuating	Rapidly fluctuating	Initial peak then decline	Constant	Low constant
How does it work?					
Ovary: ovulation suppressed§	+++	+	++	++	–
Cervical mucus: sperm penetrability down	Yes	Yes	Yes	Yes	Yes
Endometrium: receptivity to blastocyst down	Yes	Yes	Yes	Yes	Yes
User failure rates (%)	0.2–3	3–5	0.5–1	0.1	0.1
Menstrual pattern	Regular	Often irregular	Irregular	Irregular	Irregular
Amenorrhoea during use	Rare	Occasional	Common	Common	Usual
Reversibility					
Immediate termination possible?	Yes	Yes	No	Yes	Yes
By woman herself at any time?	Yes	Yes	No	No	No
Time to first likely conception from first omitted dose/removal	3 months	c. 1 month	3–6 months	c. 1 month	c. 1 month
Usage rates in the UK (%)	18	5	3		1

*COC, combined (oestrogen and progestagen) oral contraceptive.
†POP, progesterone-only pill.
‡LNG IUS, levonorgestrel-releasing intrauterine system.
§By two mechanisms – no preovulatory follicles formed and/or no LH surges occur.
Source: Data adapted from Guillebaud J. (2016) *Contraception Today*. 8th edition. CRC Press, Taylor and Francis Group, London

When considering these methods, it is important to remember that, 'naturally', female mammals living with males will not experience a series of oestrous cycles but will be pregnant or nursing for much of their fertile lives, and so will experience extended exposure to higher levels of oestrogens and progestagens. Humans, in contrast, are **atypical in being reproductively cyclical**, especially in modern developed societies. The steroidal contraceptives, in essence, mimic the continuous exposure to steroids experienced during pregnancy, during which the hypothalamic–pituitary axis is suppressed, and so, paradoxically, may take the endocrine status of women closer to that of most other female mammals.

The variables to consider in any steroidal contraceptive are: the **nature of the steroidal components** (oestrogens + progestagens, or progestagens only?); the **potency and dose** of the synthetic steroid(s) used; and the **duration of exposure** to steroids of different types or doses. A **maximal contraceptive effect** and complete continuous amenorrhoea are provided by high doses of both types of steroid continuously present. However, against this optimal contraceptive regimen must be balanced: the **side effects of the steroids**; a **woman's perception and concern** about what is happening to her reproductive system when it is closed down completely such that menstruation ceases; and the requirement that **reproductive capacity be recovered** reasonably rapidly when contraceptive practice ceases. How can each of the potential variables be adjusted to maintain effective contraception while minimizing or eliminating the influence of these non-contraceptive concerns?

Combined oestrogen/progestagen contraception. Progesterone alone can suppress ovulation, but it does so with variable efficiency unless present constantly (see later). Oestrogen addition exerts an additional negative feedback effect of its own and also promotes the development of progesterone receptors, which renders the progestagens in the contraceptive more effective. Thus, the **combined oral contraceptive (COC)** uses, as its name implies, both types of steroid (Table 21.4). This contraceptive is usually taken for 21 days, during which time the output of endogenous gonadotrophins, follicular growth and thus ovarian oestrogen are completely suppressed (Figure 21.5). The exogenous hormones develop and maintain the endometrial lining. This period of suppression is then followed by a 7-day break, with either no pills or placebo pills. During this period, the endometrium breaks down, which leads to a **withdrawal bleeding** that **simulates**

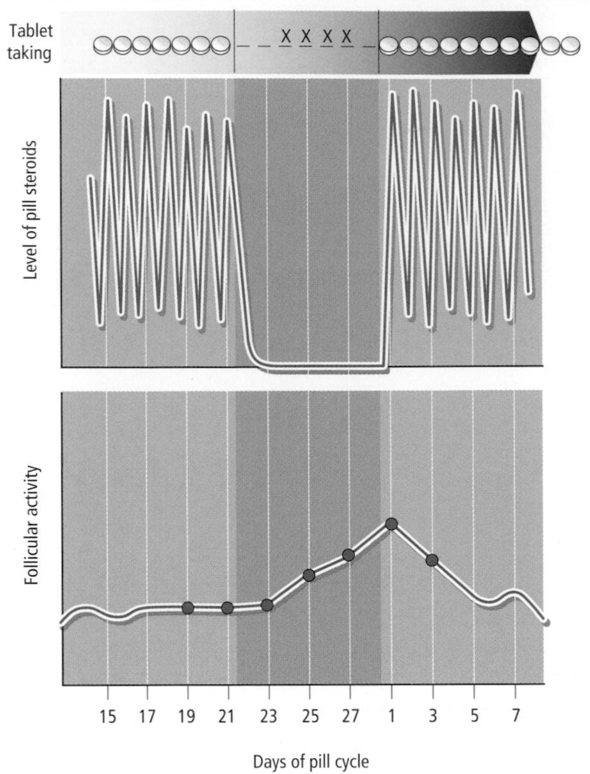

Figure 21.5 Steroid levels and follicular activity during monocyclic use of the combined oral contraceptive. The days of pill-taking are indicated at the top by small pill symbols, while the pill-free days are dashed. 'X' indicates the days of withdrawal bleeding. The profile of synthetic steroids in the blood (upper panel) reflects the episodic (daily) taking of the pill. Levels fall rapidly when pill-taking ceases. The lower panel shows follicular activity measurable by ultrasound scan (see Figure 22.7) and reflected in the output of endogenous steroids. Activity is low while the pill is taken, but note how rapidly follicles develop when pill-taking ceases, so reducing negative feedback and leading to rising gonadotrophin output. Resumption of pill-taking leads to immediate suppression. Note that a failure to resume pill-taking on time leaves a woman potentially vulnerable to ovulation and pregnancy.

menstruation. This withdrawal bleeding is important to many women psychologically, as it 'confirms' their cyclicity and lack of pregnancy (although, of course, it does not reflect a true ovarian cycle). Without it, some women will discontinue contraceptive use, and so it is of importance for **compliance**, an essential element of a contraceptive's effectiveness. However, the absence of exogenous steroids for 7 days results in a reawakening of the woman's own hypothalamic–pituitary axis, a rising output of gonadotrophins and follicular development (Figure 21.5). Indeed, towards the end of the 7-day pill-free period, some women can come dangerously close to ovulating. Thus failure to take the first active pill of the new series, or interference with its effectiveness (e.g. vomiting), can lead to contraceptive failure, leading to the suggestion that 21 days on:4 days off may be safer. This regimen of 21 days on, 4–7 days off is called **monocyclic**.

For more assured contraceptive efficiency, and for women who experience severe withdrawal symptoms (quasi-menstrual tension) or for whom a regular bleed is socially or medically difficult, a recommended alternative is the use of the pill on a **bi- or tri-cyclic** regimen, in which the relative durations of active pill:placebo are 42:7 or 63:7 days, respectively. Such regimens, of course, lose the 'regular cycle' appearance, important for some pill users. (*Note:* these regimens should not be confused with bi- and tri-phasic pills, see below.)

It will be clear from this discussion that the regimen to be used is best tailored to the psychological, physical and social condition and needs of each woman. This personal tailoring also extends to the potency and dose of the constituent steroids used. Individual women react differently to the same preparation, some being more sensitive than others and so requiring lower doses or less potent preparations. The level and balance of oestrogens (now usually 20–35 mg of **ethinyl oestradiol**) and one of several progestagens (usually **norethisterone** or **levonorgestrel**; see Box 2.1) vary in different preparations, thereby offering a range of contraceptives to suit women with different clinical histories, physiologies and perceptions of risk. As a rule of thumb, it is best to go for the lowest dose of steroids possible, and this may be estimated as being just above the dose that gives **breakthrough bleeding**, an endometrial blood loss occurring **while the woman is taking an active preparation**. Breakthrough bleeding reflects inadequate support for the endometrium by the exogenous contraceptive steroids, and provides an indication of the woman's sensitivity to their action.

Other variants of the oral contraceptive include changing the ratio of oestrogen to progesterone in different pills during the monocyclic 21-day cycle. The intention here is to mimic the natural cycle more closely. The now discontinued **sequential oral contraceptives** used oestrogens only for 7–14 days and a combined oestrogen–progestagen mix for 14 or 7 days, followed by 7 pill-free days. However, these pills required higher doses of oestrogen and were less efficient contraceptively. In contrast, **bi-phasic and tri-phasic oral contraceptives** combine the mimicry of a normal cycle attempted in sequential preparations with the contraceptive efficiency of monocyclic COCs. Thus, in biphasic preparations, both oestrogen and progestagen are given for the first half of the cycle, but the progestagen dose is stepped up at mid-cycle. The contraceptive efficiency is as good as in the conventional combined tablets, but the doses of steroid used are lower. In tri-phasic preparations, 5 or 6 days have low oestrogen and low progestagen, a further 5 or 6 days have both oestrogen and progestagen slightly elevated, and the remaining 10 days have low oestrogen and doubled progestagen. These preparations place increasing emphasis on the capacity of progestagens to block the positive feedback effect of oestrogen and less on direct negative feedback effects (see pages 161–3). In consequence, oestrogen levels are lower and a more normal 'cycle' is achieved.

A **transdermal patch contraceptive** combining norelgestromin and ethinyl oestradiol (Evra) has been licensed. Each patch is applied for 7 days and then replaced by a new patch. After three patches, a 4–7-day patch-free period allows breakthrough bleeding. The contraceptive effectiveness is similar to that of tri-phasics, and self-reported compliance was higher. Side effects were more marked than for oral contraceptives for the first two to three cycles of use, but not thereafter.

The transvaginal ring (NuvaRing) is a variant of the patch, in which a combined oestrogen and progesterone release occurs from a ring inserted into the vagina for 21 days and removed for 4–7 days as for the monocyclic COC.

Progestagen-only contraception. These contraceptives may be taken daily by an oral route or given by injection subcutaneously for continuous release over a period of 8 weeks, or as implants for up to 5 years. In addition, progesterone-impregnated IUCDs are available (see below and Table 21.4). **Progesterone-only pills** (**POPs**) are taken continuously, contain low doses of progestagen, and work primarily by effects on cervical mucus and perhaps the endometrium. However, in a significant proportion of users (around 20%), ovulation is suppressed, and, in around 40%, follicular–luteal activity is abnormal. In all users, irregular bleeding may occur. Because the dose of progesterone used is small and the pill is taken daily, side effects are relatively few compared with COCs, although weight gain can be problematic. However, the effect of the progesterone on cervical mucus lasts for only 22–26 hours. Therefore, if the woman is significantly delayed in taking her daily pill, fertility returns. This requirement for a highly organized lifestyle and level of commitment is reflected in the much more variable contraceptive effectiveness, and so is less popular.

Subdermally implanted depot progestagens, in contrast, are extremely effective contraceptives and used by millions of women worldwide, although less so in more developed countries (Table 21.1). Implants lasting 5 years (levonorgestrel), 3 years (etono-gestrel; the only one available in UK, formerly marketed as Implanon, now as Nexplanon) or 2 years (nestorone) are available commercially in silicon capsules or rods (crystalline steroid). Follicular growth and ovulation are disrupted, leading to inadequate luteal phase function, the primary mechanism of contraceptive action, any effect on mucus and the endometrium being secondary. No major adverse effects have been reported, the method is fully reversible and, despite some irregular bleeding, there appears to be high user satisfaction.

Injectable progestagens lie between orals and implants, and include depot medroxyprogesterone acetate (Depo-Provera or DMPA; renew every 13 weeks i.m.) and norethisterone enantate (NET-EN; renew every 8 weeks; not licensed in the UK). Injectables are delivered intramuscularly and provide faster release systems that generate relatively high, ovulation-suppressing levels of steroid initially, which fall to mucus-affecting levels as the time for the next injection approaches. Their use is often associated with irregular bleeding and weight gain, with the positive side effect of reduced iron loss and anaemia. **Postcoital emergency contraception** was discussed earlier (see Box 10.5) and antiprogestagens below (see page 358).

Side effects of steroidal contraceptives

We have stressed the balance between contraceptive efficiency and unwanted side effects. Some of the unwanted side effects are social or psychological, such as the need either to experience a regular bleed or no bleed at all, or a requirement for highly organized self-administration or the availability of good medical care. Other minor side effects can include weight gain, headaches and acne, each of which will be specific for an individual woman and will vary with different preparations. Claims of reduced libido, because of reduced androgens, have not proved sustainable. All of these factors are extremely important, as marrying the physiology, psychology and social condition of the woman to her contraceptive is the key element in providing effective and **acceptable** contraception for her.

However, beyond these elements, there has also been widespread concern and discussion about more severe life-threatening side effects of steroidal contraceptives. There are two general points that need to be made emphatically. First, the life-threatening risk associated with modern steroidal contraception is four-fold lower than that from pregnancy, childbirth and pregnancy termination, and much less than that associated with driving a car. Second, there is little evidence that steroidal contraceptives **cause** life-threatening conditions, but they may be co-associated with other causal agents to **promote their effects**. Because these co-associated factors are known, the risk can be quantified for each woman.

There is evidence that the oestrogen in COCs **increases clotting factors and blood coagulability** (see page 177), although they also increase fibrinolytic activity but only in the absence of smoking. The current increased risk of venous thromboembolism in women taking low-dose oestrogen monophasic COCs is estimated as a three- to four-fold relative risk compared with women not taking the pill, but the **absolute risk is small**, increasing from 5 to 15 or 20 per 100 000 woman-years of use, and there is no increase in mortality. The companion progestagen in the COC influences relative risk, desogestrel and gestodene (relative risk: 3.8) counteracting the prothrombotic effects of ethinyl oestradiol less than do levonorgestrel (relative risk: 2.8) or norethisterone (relative risk: 3.2). None of the progesterone-only contraceptives increases the risk of venous thromboembolism. Thus, two main general factors **contraindicate** combined steroidal contraception: **smoking and obesity** (body mass index >30), especially in **older women** or women with a **history of thromboembolic or cardiovascular disease**. Additionally, the presence of **migraine, breast and liver tumours, and severe cirrhosis** make COCs unacceptable. Finally, women should discontinue COC usage at least 4 weeks **before major elective surgery or immobilization**. For all of these cases, progesterone-only contraception can be considered (see below).

The broad effects of modern steroidal contraceptives on cancers are neutral. Benign breast cancer and three very serious conditions, **carcinomas of the endometrium, ovary and colon/rectum**, are **reduced** in pill users. A slight increase in hepatic carcinoma and breast cancer occurs among women who have taken the pill for several years before the age of 25, an outcome perhaps related to the effects of progestagens on cell division in the postpubertal, developing breast. Increased cervical dysplasia is also reported, but may be secondary to increased coital activity without condoms rather than a direct effect of steroids. These observations emphasize the value of regular cervical smear and breast examinations in women on steroidal contraceptives.

The **latent diabetes** revealed in some women during pregnancy (see page 264) can also become evident in users of steroidal contraceptives, who should therefore only use low-dose oestrogen preparations under careful supervision.

A plus for most women on steroidal contraception is the reduced incidence of pelvic inflammatory disease, a significant cause of infertility (see later). It is not clear why this is so, as genitourinary infections of the vagina are, if anything, increased. One idea is that infectious agents hitch a ride on spermatozoa, but the hostility of cervical mucus prevents their passage further into the female genital tract.

Antiprogestagens

Antiprogestagens such as **mifepristone** (also called **RU486**, see Box 2.1), act as a luteal phase contraceptive and early abortifacient. They function by occupying and blocking progesterone receptors but do not block follicular growth and oestrogen output. A putative advantage of mifepristone is that it might only be needed once a month to target particular points in the cycle. Trials to test this possibility show that administration early in the luteal phase (as emergency contraception within 120 hours after coitus) prevented most pregnancies. When taken at the mid or late luteal phase, by which time implantation has occurred, RU486 fails to induce menses in a significant portion of cases. It is thus not as useful as modern COCs and progesterone-only contraceptives.

Steroidal contraceptives for men

Just as in the female, the output of gonadotrophin-releasing hormone (GnRH) and/or gonadotrophins can be suppressed in the male via negative feedback using androgens, progesterone or the continuous administration of a GnRH analogue (see Box 2.1). In the latter two cases, endogenous androgens are depressed, reducing masculinizing stimulation and libido, both unacceptable side effects. Thus, all contraceptive preparations for men include synthetic androgen itself. However, trials have been disappointing. Trials using weekly injections of testosterone enanthate produced azoospermia in two-thirds of men and oligospermia (as few as 3×10^6/ml of semen) in the rest. However, androgen levels were very high, and there were large individual and ethnic variations in response. Combination of implants of testosterone with oral, injected or implanted progestagen has proved more promising, with severe oligospermia ($<1 \times 10^6$/ml) in most men. Side effects include acne, weight gain and oily skin, which may affect long-term compliance. There is also some concern that treatment with the high doses of testosterone required may exacerbate the risk of prostatic hyperplasia, which is quite common in older men. It seems unlikely that hormonal contraception for men will have a major impact on fertility control.

Intrauterine contraceptive devices (IUCD or IUD)

Modern IUCDs, used by around 5% of UK women, are made of copper, the older purely plastic models being no longer in use. **T380A** (T Safe 380A in the UK) is the gold standard among available models, and is approved for use for 10 years after insertion. The IUCD functions as a foreign body within the uterus to produce a low-grade, local, chronic inflammatory response, the composition of the uterine luminal fluid resembling a serum transudate containing large numbers of invading leucocytes. The use of copper enhances contraceptive effectiveness because, in addition to copper's inflammatory effects, it also has specific spermotoxic and embryotoxic effects. There is a broad correlation between the capacity of the IUCD to induce inflammation and its efficiency as a contraceptive. Such a uterine environment reduces transport of viable sperm to the oviduct and thereby fertilization, is toxic for oocytes and any fertilized eggs that do form, and impairs implantation and decidualization. In addition, luteal life may be abbreviated because of premature prostaglandin release in some species (see page 142). Overall, most but not all the contraceptive effect is thought to be prefertilization. These multiple sites of action make the IUCD highly effective, the copper T380A being considerably more effective than the combined oral contraceptive (Table 21.1).

For obvious reasons, contraindications to IUCD insertion include a history of pelvic inflammatory disease, heavy or painful periods and anatomical abnormality of the uterus. Possible complications following IUCD insertion are: heavy menstrual or irregular uterine blood loss (**dysmenorrhoea**); uterine pain or muscular spasm; and rarely (<0.01%) uterine perforation. Earlier suggestions to avoid IUCD insertion in nulliparous women seem to be without foundation, probably because copper IUCDs are smaller than plastic ones and produce fewer side effects and spontaneous expulsions. Immediately after insertion, but not thereafter, there is an elevated risk of pelvic inflammatory disease. Unnoticed expulsion of the IUCD can also occur (up to 1 in 20 in the first year of use), as can rare pregnancies with the IUCD still in place. There is no evidence of any effect of IUCD use on subsequent fertility or increased risk of reproductive tract cancer. The skills required for insertion, and the desirability of regular monitoring of patients, mean that trained and available medical or paramedical staff are needed. The IUCD is a widely, if variably, used and very effective contraceptive device, and is also highly effective as a postcoital contraceptive (Box 10.5).

A novel advance to the contraceptive action of the IUCD is to combine a plastic T-shaped IUCD with steroidal contraception by using the **progesterone-impregnated IUCD** (**levonorgestrel-releasing intrauterine system or LNG IUS** or **progestasert**; approved for use for 5 years after insertion). The IUS delivers progesterone directly to its major local and peripheral sites of action: thickening the cervical mucus and thinning the endometrium, leading to amenorrhoea, and so it is also used to regulate severe bleeding (**menorrhagia**; see Table 21.4).

The contraceptive future

A range of variably effective sexual techniques, devices and pharmaceutical preparations now exists to limit reproduction. The use of these, tailored to the needs of the individual or couple concerned, should be adequate for effective control of fertility, high acceptability and good compliance. The problem is not with the technology but with its use, issues of education, support and motivation being paramount. Given the restrictions on clinical trials imposed in recent years, it is unlikely that a totally new range of contraceptive approaches will be available in the near future. Most advances have come from modifying and refining existing methods in the light of evidence-based practice.

Considerable research into **immunological contraceptive approaches** is ongoing. For example, active immunity to the specific β-chain terminal sequence of human chorionic gonadotrophin (hCG) neutralizes the embryonic hormone but leaves LH unaffected, thereby giving normal (or slightly lengthened) cycles and protection against pregnancy (see Figure 13.12). Results of advanced trials using this approach have been encouraging. No side effects have been noted, and the immunity declines after about 6 months in the absence of a booster injection, so the approach should be reversible. Alternatively, **immunity to antigens on spermatozoa, in semen or on the *zona pellucida*** is known, from clinical and experimental studies on primates, to be associated with infertility. Such an approach has been harnessed for contraceptive use, using immunization to recombinant proteins of the *zona pellucida* and an epididymidal protein called **eppin** (see pages 185 and 199). However, there is a reasonable reluctance to tinker with the body's immune system until more is understood about its natural regulation. Uncontrollable side effects, such as a wider autoimmune response, might develop after extended use of a 'contraceptive vaccine', particularly one that utilizes cellular or cell-associated antigens. Moreover, there is wide variation among individuals in immune responsiveness, which makes for a corresponding and unacceptable variation in contraceptive effectiveness.

More speculative are approaches to the control of implantation deriving from our increasing understanding of the cytokines involved. The potent anti-pregnancy effects of knocking out genes for some of these cytokines have encouraged speculation that **anticytokines** might provide a good route to contraception. However, this possibility seems to be some way off.

Pregnancy termination

Clinical termination of pregnancy (abortion) is appropriate for women whose mental or physical health would be put at risk by a continuing pregnancy. Pregnancy termination is also, whether performed legally or illegally, an approach to fertility control in many communities, especially where contraception is not readily available or is unreliable. In the UK, it is estimated that at least one-third of women will have had a pregnancy termination by the age of 45. The legal position concerning medical termination is complex and details vary widely in different jurisdictions. Where **fetal abnormality** is suspected (through age of mother, a parent with a balanced translocation, exposure of the mother to potential teratogens, prior evidence of familial risk or as a result of a suspicious ultrasound scan), confirmatory tests may be recommended clinically, which may include chorionic villous sampling and/or amniotic fluid sampling (amniocentesis). See Box 21.2 for more details.

About 40% of UK terminations are carried out in the first trimester (or 3 months) of pregnancy. Over this period of pregnancy, medical terminations in the UK are recommended using **antiprogestagens** (such as mifepristone/RU486; see above) plus a **prostaglandin E_1** analogue (**misoprostol**) or **menstrual extraction** by **vacuum**. Failure rates using these approaches are low (<1%), as are complications, such as major blood loss, incomplete aspiration or damage to the cervix or uterus. The side effects of postoperative infection and damage to the uterus or cervix may affect subsequent fertility. Between 8 and 16 weeks, surgical termination by **dilatation and physical evacuation** is recommended. Thereafter, the rare terminations use **oral prostaglandin**, followed by oxytocin once membranes have broken.

Conclusions

An individual ideally will have control over their fertility that is 100% effective at each sexual encounter, with zero risk of side effects. In practice, this requirement is not always achievable using a single existing method of fertility control, so resorting to multiple levels of fertility control tends to occur. Social regulation of sexual encounters is followed by use of natural or barrier techniques. Failure of these approaches leads to use of postcoital contraception or to menstrual regulation. There should be little need for unwanted pregnancy in more sophisticated societies, given the cumulative contraceptive efficiency of these various approaches.

With the use of some contraceptive approaches there may be attendant risks to health, well-being or even life. However,

Box 21.2 Chorionic villous sampling (CVS), amniocentesis and maternal blood sampling

Some 5% of pregnant women are offered one of these invasive prenatal diagnostic tests, usually for chromosomal analysis of the conceptus. There are small associated risks that are related strongly to operator experience, which should ideally exceed a minimum of 10 procedures per year and be audited.

Amniocentesis is more commonly offered, and mostly undertaken from 15 weeks of pregnancy (advised only exceptionally at earlier times as pregnancy losses are higher). An ultrasound-guided 0.9-mm (20 gauge) needle tip is inserted transabdominally into the amniotic sac from which fluid containing desquamated fetal skin cells is obtained. Ultrasound guidance reduces maternal blood contamination, and damage to maternal and fetal organs. The aspirated fluid sample is spun and divided into amniocyte cells, which are considered fetal in origin, and the overlying liquid. The cells are cultured *in vitro* and then examined cytogenetically or using modern molecular techniques, and the fluid can be examined (especially in the third trimester) for evidence of rhesus disease, lung maturity, infection and insulin levels. Women who are HIV positive should avoid amniocentesis if possible as the chance of maternofetal transmission increases. There is an associated miscarriage rate of around 1% in amniocentesis carried out later than 15 weeks of pregnancy.

Before 15 weeks of pregnancy, **chorionic villus sampling** (**CVS**) **is safer than amniocentesis**. It is usually undertaken between 10 and 13 weeks of gestation, and is considered unsafe earlier. Overall, CVS is usually regarded as less safe than amniocentesis, but the evidence is variable, the only direct randomized comparison of trans-abdominal CVS with second-trimester amniocentesis reporting similar total pregnancy loss in the two groups (6.3% compared with 7%). CVS involves ultrasound-guided aspiration of placental tissue, usually trans-cervically but sometimes (especially if done later than 13 weeks) via a percutaneous trans-abdominal route. The placental/trophoblastic tissue sample may be taken by aspiration or microforceps.

Recently, the knowledge that trophoblastic cells can be detected in the maternal circulation has led to the development of **maternal blood tests** to detect genetic problems in these cells of conceptus origin.

Further reading

RCOG (2010) *Amniocentesis and Chorionic Villus Sampling*. Green top Guideline No.8. Royal College of Obstetricians and Gynaecologists, RCOG Press, London. https://www.rcog.org.uk/en/guidelines-research-services/guidelines/gtg8/

although these risks should not be dismissed, their impact is often somewhat exaggerated. For example, many of the contraindications to the use of steroidal contraceptives, such as thromboembolic episodes, cardiovascular problems or latent diabetes, also apply to pregnancy or the surgical procedures required for sterilization. Among groups not at risk from steroidal contraceptives, pregnancy itself and abortion constitute much greater risks to health and life. Thus, the balance of risks must be evaluated for each individual. The optimal situation is a combination of well-educated general practitioners, a readily available range of contraceptives, and easy and rapid access to early clinical pregnancy termination. Sadly, this combination is rare anywhere, and in most of the world it is simply not available.

Population control programmes ideally match their objectives to the individual requirements through education and provision of a family planning service. However, state intrusion into personal reproductive decisions has been tried in a number of countries with variable success. Conversely, the fact of overpopulation has been used as an excuse to deny fertility treatments to those experiencing difficulty conceiving. We return to that issue in Chapter 23. First, however, we explore in Chapter 22 the biology of infertility.

Key learning points

- Fertility, fecundibility and fecundity are measures of the actual and potential reproductive outputs of females.
- Sterilization involves vasectomy in men and tubal occlusion or section in women.
- Contraception can be achieved without mechanical or chemical intervention by behavioural approaches.
- Physical barriers to fertility (caps, diaphragms, condoms) are more effective and provide better protection against genitourinary infection.
- Some steroidal contraceptives combine synthetic oestrogens and progestagens (COCs), and may be administered orally or transdermally.
- Other steroidal contraceptives use progestagens alone, and may be administered orally, by injection, by implant or by an impregnated IUCD.
- Most oral contraceptives are relatively safe and effective if taken responsibly and with medical advice.

- Oral contraceptives work by suppressing ovulation and/or sperm and embryo transport in the female tract.
- Steroidal contraceptives for men are still in clinical trials and use synthetic androgens with or without progestagens to suppress FSH output, but are not promising.
- Clinical termination of pregnancy may be used to control fertility where contraception is lacking or ineffective. Early (medical or surgical) termination is preferable to late (surgical) termination on maternal health grounds.
- Termination of pregnancy may be offered for reasons other than fertility control, e.g. to avoid birth of an afflicted child or in the interests of maternal health.
- A contraceptive strategy needs to be tailored to the needs and circumstances of each individual, which is reflected in the wide variation by age and geography in use of different contraceptive approaches.

Clinical vignette

Contraception

A 16-year-old girl, Ms T, began to have sexual intercourse with her first boyfriend, who is aged 19. She knew that her parents disapproved of the relationship and would have been very angry if they were aware that she was sexually active. She wished to continue the relationship, but was very concerned about becoming pregnant. She attended a local family planning clinic for advice on contraception, but specified that her mother must not be able to find any evidence of contraceptive use at home. She was not able to get to the clinic on a regular basis, as she had no independent means of transport, but had managed to get to the appointment as she is currently staying with a friend for the weekend. She stated that both she and her partner were virgins when they began the relationship and hence she was not concerned about sexually transmitted infections. She was in good physical health, although she used an inhaler for mild asthma, and had bad migraines with visual aura around the time of exams each year. After some discussion, she opted to have an implant placed in the upper arm, which would provide her with contraception for 3 years. She understood that this could cause some bruising when first placed, but this would fade over a couple of days and thereafter be undetectable unless the arm was specifically examined. Ms T's periods became somewhat irregular for several months after placement with some intermittent spotting, but other than this there were no side effects. After moving out of home to attend university two years later she decided to continue with the implant as it was highly effective and had given her peace of mind with regard to contraception.

There are many options for contraception available, and facilitating the process of each woman picking the best option for her individual situation can be complex. The range of factors that should be taken into account includes medical safety, effectiveness, social factors, and personal preference. In the case outlined above, all of these must be considered. In terms of medical safety, Ms T suffers from migraine with aura, which is a contraindication to prescribing contraception containing oestrogen (such as a combined pill, patch or vaginal ring) because they increase the risk of a stroke. Effectiveness is also an important concern here because an unintended pregnancy would present difficulties for Ms T, either in terms of accessing termination care or of continuing a pregnancy. However, Ms T's primary concern is about having a contraceptive method that cannot easily be detected and does not require frequent trips to a family planning clinic. This means that progesterone-only pills or injectable progesterone (which has to be given every 3 months) are not suitable here. At 16, Ms T is likely to wish to preserve her options for fertility in the future. Her best options are therefore to have a long-acting reversible method, such as an implant or intrauterine device fitted. In the UK, parental consent to medical treatment (including, but not limited to, contraceptive provision) is not required for 16-year-olds, provided that they are judged to be competent to understand and consent to treatment of their own accord.

FURTHER READING

General reading

Archer DF, Lasa IL (2011) Tailoring combination oral contraceptives to the individual woman. *Journal of Women's Health* **20**, 879–891.

Attia MM, Ibrahim AM, Abou-Setta AM (2013) Role of the levonorgestrel intrauterine system in effective contraception. *Patient Preference and Adherence*, **7**, 777–785.

Bancroft J (2005) The endocrinology of sexual arousal. *Journal of Endocrinology* **186**, 411–427.

Buhling KJ, Zite NB, Lotke P, Black K for the INTRA Writing Group (2014) Worldwide use of intrauterine contraception: a review. *Contraception* **89**, 162–173.

Croxatto HB (2003) Mifepristone for luteal phase contraception. *Contraception* **68**, 483–488.

Dunson DB, Colombo B, Baird DD (2002) Changes with age in the level and duration of fertility in the menstrual cycle. *Human Reproduction* **17**, 1399–1403.

ESHRE Capri Workshop Group (2003) Hormonal contraception without estrogens. *Human Reproduction Update* **9**, 373–386.

Frye CA (2006) An overview of oral contraceptives: mechanism of action and clinical use. *Neurology* **66** (6, Suppl. 3), S29–S36.

FSRH Guidance (September 2014) *Male and Female Sterilisation.* RCOG Press, London.

Guillebaud J (2016) *Contraception Today*, 8th edn. CRC Press, Taylor and Francis, London.

Jensen JT (2011) The future of contraception: innovations in contraceptive agents: tomorrow's hormonal contraceptive agents and their clinical implications. *American Journal of Obstetrics & Gynecology* **205**, S21–S25.

Kamischke A, Nieschlag E (2004) Progress towards hormonal male contraception. *Trends in Pharmacological Sciences* **25**, 49–57.

Meirik O, Fraser IS, d'Arcangues C; WHO Consultation on Implantable Contraceptives for Women (2003) Implantable contraceptives for women. *Human Reproduction Update* **9**, 49–59.

National Institute for Health and Clinical Excellence (2005) *Long-acting Reversible Contraception* (2005) NICE, London.

O'Sullivan I *et al.* (2005) *Contraception and Sexual Health, 2004/05.* Office for National Statistics, London.

RCOG (2004a) *Male and Female Sterilisation.* Evidence-based Clinical Guideline Number 4. Royal College of Obstetricians and Gynaecologists, RCOG Press, London.

RCOG (2004b) *The Care of Women Requesting Induced Abortion.* Evidence-based Clinical Guideline Number 7. Royal College of Obstetricians and Gynaecologists, RCOG Press, London.

Short RV (2009) Population growth in retrospect and prospect. *Philosophical Transactions of the Royal Society B* **364**, 2971–2974.

Simon C (1996) Potential molecular mechanisms for the contraceptive control of implantation. *Molecular Human Reproduction* **2**, 475–480.

More advanced reading (see also Boxes 21.1 and 21.2)

Dennis J, Hampton N (2002) *IUDs: Which Device?* RCOG Press, London.

Faculty of Family Planning and Reproductive Health Care Clinical Effectiveness Unit (2003a) *New Product Review: Norelgestromin/ Ethinyl Oestradiol Transdermal Contraceptive System (Evra).* RCOG, London.

Faculty of Family Planning and Reproductive Health Care Clinical Effectiveness Unit (2003b) First prescription of combined oral contraception. *Journal of Family Planning and Reproductive Health Care* **29**, 209–223.

Faculty of Family Planning and Reproductive Health Care Clinical Effectiveness Unit (2004a) The copper intrauterine device as long-term contraception. *Journal of Family Planning and Reproductive Health Care* **30**, 29–42.

Faculty of Family Planning and Reproductive Health Care Clinical Effectiveness Unit (2004b) The levonorgestrel-releasing intrauterine system (LNG-IUS) in contraception and reproductive health. *Journal of Family Planning and Reproductive Health Care* **30**, 99–109.

Gnoth C, Godehardt D, Godehardt E, Frank-Herrmann P, Freundl G (2003) Time to pregnancy: results of the German prospective study and impact on the management of infertility. *Human Reproduction* **18**, 1959–1966.

Liu PY, Handelsman DJ (2003) The present and future state of hormonal treatment for male infertility. *Human Reproduction Update* **9**, 9–23.

Naz RK (2011) Antisperm contraceptive vaccines: where we are and where we are going? *American Journal of Reproductive Immunology* **66**, 5–12.

O'Rand MR, Widgren EE, Hamil KG, Silva EJ, Richardson RT (2011) Epididymal protein targets: a review brief history of the development of epididymal protease inhibitor as a contraceptive. *Journal of Andrology* **32**, 698–704.

RCOG (2004) *Venous Thromboembolism and Hormonal Contraception.* Guideline No. 40. Royal College of Obstetricians and Gynaecologists, London.

Schaffir JA, Isley MM, Woodward M (2010) Oral contraceptives vs injectable progestin in their effect on sexual behavior. *American Journal of Obstetrics and Gynecology* **203**, 545.e1–5.

Stanford JB, Mikolajczyk RT (2002) Mechanisms of action of intrauterine devices: update and estimation of postfertilization effects. *American Journal of Obstetrics and Gynecology* **187**, 1699–1708.

United Nations Department of Economic and Social Affairs Population Division (2001) *World Population Prospects: The 2000 Revision.* ST/ESA/SER.A/204. Sales No. E.01. XIII.12. United Nations, New York.

CHAPTER 22

Restoring fertility

Essential Reproduction, Eighth Edition. Martin H. Johnson.

Companion website: www.essentialreproduction.com

In Chapter 21, we defined clinical subfertility in couples or individuals as occurring when they fail to conceive within the time scale expected from the fecundibility level for their age group (see page 349). Conventionally, couples having frequent, unprotected intercourse are not defined as being clinically subfertile until they have **failed to conceive within 1 year**. Subfertility may be contributed to by either or both partners, a relatively minor problem for each sometimes becoming a more serious problem for both. Conventionally, a clinical investigation attempts to locate the subfertility to the male, the female or the couple, and to categorize it as **primary or secondary subfertility**. If neither partner has conceived or fathered a pregnancy before, the subfertility of each and of the couple is said to be primary. If either or both have parented a pregnancy with another partner previously, then subfertility is primary for the couple only, and secondary for each individual. Where the couple has had a previous pregnancy (whether or not it went to term). but now are having problems conceiving, their subfertility is described as secondary.

Estimates of the prevalence of subfertility are rough and ready, as many factors, especially the age distribution of the population, impinge on it. Conventionally, 10–15% of couples are said to experience difficulty with conception and successful delivery of a healthy child, but the prospective studies on women mostly under 35 years suggest only 5%. The higher conventional estimate may reflect the increased proportion of older women trying to conceive.

Causes of subfertility

The causes of subfertility vary greatly with socio-economic and geographical factors. The male and female partners contribute more or less equally to the problem, but for many couples the diagnosis is 'unexplained subfertility', a catch-all diagnostic category that in part reveals our ignorance (Figure 22.1). Psychological factors causing sexual dysfunction (such as erectile dysfunction, vaginal spasm, premature ejaculation and disorders of sexual identity) are not uncommon among infertile patients, but they may often be responses to, rather than the primary causes of, the infertile state. Eating disorders (obesity and anorexia) contribute significantly to subfertility.

Three major classes of disorder account for about 75–80% of all explicable cases of infertility (Figure 22.1): disorders of the female tract (in particular blocked or damaged Fallopian tubes); disorders of ovulation; and poor quality of spermatozoa. In addition, there is evidence that human fertility patterns are influenced by the substantial (20–25%) loss of clinically diagnosed pregnancies in both fertile and sub-fertile couples. Finally, and of increasing significance for many societies as couples delay their decision to have a family, the question of age hovers over the whole question of subfertility.

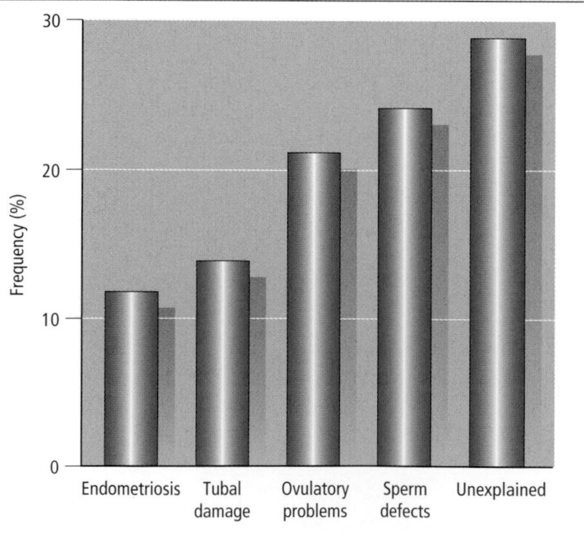

Figure 22.1 Causes of subfertility in UK couples, expressed as a percentage frequency distribution.

Age, senescence and reproductive capacity

Ageing should be distinguished from **senescence**. Ageing begins at conception and is a description of what happens with the passage of time. Senescence describes deterioration related to dysfunction and disease. Typically, reproductive senescence begins during middle age, but its onset and time course vary widely among individuals of the same age. Both men and women experience a number of senescent changes to the central nervous system (CNS), which, as we have seen, has an important regulatory role in reproductive and sexual function. There is evidence that the weak androgenic hormone dehydroepiandrosterone (DHEA), produced in large amounts by the adrenal, may have neuroprotective effects which might be reduced in some elderly people and thus might account for a reduced capacity to repair and sustain neurons, thereby contributing to the senescence of CNS function. In women, a major change occurs with the **menopause**, defined as the last menstrual cycle, which results from the exhaustion of functional ovarian follicles and is thus primarily part of the ageing process. However, senescent changes in the reproductive potential of women may precede this landmark. Most men, in contrast, do not experience gametic exhaustion or a sudden fall-off in fertility as part of the ageing process, there being no evidence for a male equivalent of the menopause, despite popular claims to the contrary. They do, however, experience senescent changes, which, as for women, show a great variation among individuals.

Age and reproductive capacity in women

The fecundity of a woman varies with her age. Menstrual cycles begin during puberty, and menarche marks the earliest expression of the potential fertility of the female with its implication that a full ovarian and uterine cycle has been achieved. Failure of menarche indicates a diagnosis of **primary amenorrhoea.** This may result from a failure or delay of normal maturation of the underlying neuroendocrine mechanisms (see pages 61–5); from primary defects in the gonad, such **as ovarian dysgenesis or agenesis** due to chromosomal abnormality or mutations (e.g. Turner syndrome and true hermaphroditism (see pages 39 and 43); or from primary defects in the genital tract, such as lack of patency between the uterus and vagina (presenting as **cryptomenorrhoea** or hidden menses) or, indeed, the absence of internal genitalia as seen in androgen insensitivity syndrome (see page 47). However, even in most women, early menstrual cycles are rarely regular. Some cycles are anovulatory and may lack, or have abbreviated, luteal phases (Figure 22.2). In general, the follicular phase tends to become shorter with age and the luteal phase tends to lengthen, for reasons that are unclear.

The climacteric and menopause

Fertility is highest in women in their 20s and declines thereafter (Figure 22.3). This decline could, in principle, be due to a reduction in fecundity or to changes in sexual behaviour resulting in older women having a reduced chance of, or inclination for, unprotected sex. Measures of the risk of childlessness in women of different ages at marriage, who then try actively to have offspring, are also shown in Figure 22.3, and suggest a real decrease in fecundity. It is clear that the ability to sustain a pregnancy through to successful parturition declines slowly to the age of 35 years, but rapidly thereafter when increasing frequencies of failed ovulation, early pregnancy loss, perinatal or neonatal mortality, low birth-weights, maternal hypertension and congenital malformations especially, but not exclusively, due to fetal chromosomal imbalance are encountered. Maternal oocytes, stored in second meiotic metaphase since fetal life, show evidence of increased meiotic spindle instability and chromosomal dispersal, and are the major contributor to the early phases of this decline in fecundity.

Later, eggs actually start to run low as well as being less robust, and so follicular development is impaired. Accordingly, it is not surprising that irregular menstrual cycles may begin to reappear in some women in their early 40s and mark the onset of the **climacteric**, a period of reproductive change that may last for up to 10 years before the last menstrual cycle (the **menopause**). This **secondary amenorrhoea** occurs at a mean age of 52 years in the USA. Symptoms associated with the climacteric can include mood changes, irritability, loss of libido and hot flushes. The climacteric reflects declining numbers of ovarian follicles and their reduced responsiveness to gonadotrophins, and is a direct consequence of the fixed number of oocytes a woman has arising from the early termination of mitotic proliferation in female germ cells (see page 8).

The final cessation of reproductive life is, thus, a function of ovarian failure, although senescent changes to hypothalamic and pituitary cells may also make a contribution to the dysfunctional uterine bleeding often encountered, and often a stimulus for clinical intervention, including the vogue for (unnecessary) hysterectomies. Premature loss of oocytes, and

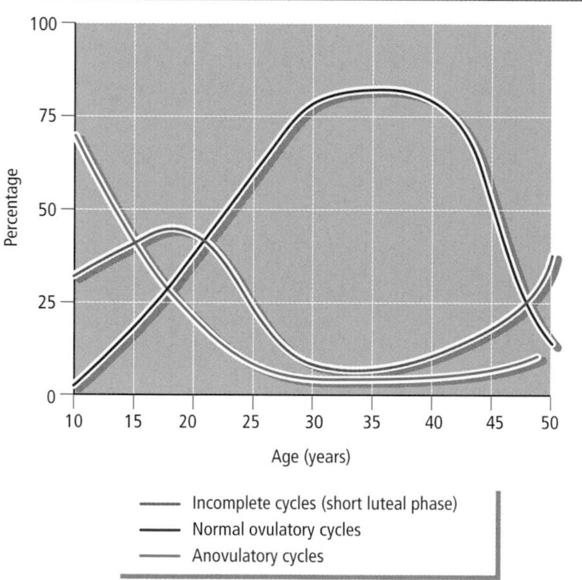

Figure 22.2 Relative incidence of three types of menstrual cycle with age of woman.

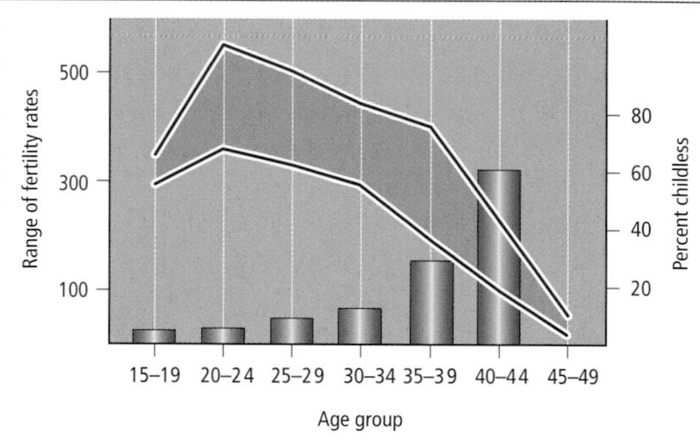

Figure 22.3 Rates of fertility (red range) and childlessness (blue bars) by age of woman. The fertility rate data were collected from populations of married women in whom, it is alleged, no efforts were made to limit reproduction. The range reflects the different circumstances of the populations, drawn from different parts of the world. These data approach a measure of fecundity by age in humans. The histograms show the proportions of women remaining childless after first marriage at the ages indicated despite continuing attempts to deliver a child. Note the sharp rise above 35 years, implying a fall in fecundity from this time onwards.

Table 22.1 Steroidogenesis in the postmenopausal woman. Change from resting plasma concentrations (%)

Hormone	Dexamethasone suppression (% decrease)*	hCG stimulation (% increase)†
Oestradiol	>50	0
Oestrone	>75	10
Progesterone	>75	10
17α-OH progesterone	53	42*
Dehydroepiandrosterone	65	60*
Androstenedione	70	10
Testosterone	40	60*
Dihydrotestosterone	>30	—

* Dexamethasone inhibits adrenal cortical steroidogenesis by depressing the output of ACTH, so the degree of reduction in plasma steroid levels shown suggests the degree of adrenal contribution, which is high.
† hCG provides a gonadotrophic stimulus to the ovaries, so the degree of increase in plasma steroid levels shown suggests the degree to which the ovaries are able normally to secrete the hormones. In the second column, there are only three significant changes from baseline (asterisked).

thus **premature menopause**, occur in about 2% of women, in some as early as their late teens and early 20s. The causes of premature menopause are unclear, although in some women a familial element exists – an important fact to elicit as it will be relevant to decisions about timing of childbearing. With the decreasing numbers of small preantral/early antral follicles, the secretion of follicular inhibin B declines early in the climacteric, elevating FSH levels because of the reduced negative feedback. These elevated FSH levels can then temporarily salvage a greater proportion of the small follicles and can shorten the follicular phase. During this period, enough antral follicles still survive for adequate levels of inhibin A and oestrogen to rise. Only in the later part of the climacteric, as menopause approaches, do inhibin A and oestrogen also fall, and then LH rises too. Paradoxically, androgen levels may rise after the menopause, as a result of adrenal synthesis combined with increasing LH-responsive ovarian interstitial cell synthesis, and this may lead to **hirsutism** and further exacerbate the adverse effects of lowered oestrogen on the cardiovascular system. Postmenopausal hormone levels and responsiveness are shown in Figure 10.3b and Table 22.1.

The consequences of the menopause

The physical, functional and emotional changes of the climacteric and menopause are driven directly or indirectly by these ovarian changes. Oestrogen withdrawal is responsible for: vasomotor changes, such as '**hot flushes**' and '**night sweats**'; changes in ratios of blood lipids, associated with an increased risk of coronary thrombosis; reduction in size of the uterus and breasts; and a reduction in vaginal lubrication and a rise in the pH of vaginal fluids, in consequence of which discomfort during intercourse (**dyspareunia**) and vaginal skin thinning (**atrophic vaginitis**) may occur. The antiparathormone activity of oestrogen is also lost at this time, resulting in increased bone catabolism, **osteoporosis** and rising incidence of fractures. These symptoms of the menopause may be prevented or reduced by oestrogen treatment (**hormone replacement therapy, HRT**) usually combined with progesterone therapy to avoid endometrial hyperplasia, which could predispose to endometrial cancer. However, the results of recent trials have contraindicated HRT as a routine treatment, as it can be associated with increased risks of breast cancer, cardiac disease and cerebrovascular accident that more than offset the advantages of reduced osteoporotic fractures. Whether or not to receive HRT and for how long is a matter for consultation between patient and doctor, when individual risk and circumstances can be discussed.

A number of behavioural changes may occur during the climacteric, e.g. depression, tension, anxiety and mental confusion. However, these changes may not be related solely to effects of steroid withdrawal on the brain, but rather may result secondarily from difficulties in psychological adjustment to a changing role and status ('empty nest syndrome'), and in part from insomnia due to night sweats. Loss of libido is common and may be related to dyspareunia due to vaginal dryness. In Chapter 6 (see page 98), we discussed the relationship between androgens and libido: given the rising postmenopausal androgens, libido might be expected to rise in at least some women, but findings are not consistent.

Fertility and the postmenopausal woman

Regardless of the time of the menopause, women can nonetheless, if provided with a 'young' oocyte (or a conceptus derived from one), carry a pregnancy to term successfully. The early part of this pregnancy requires administration of exogenous hormones: first, to build up the regressed reproductive tract (see pages 172–4); and, second, to mimic luteal support until the fetoplacental unit takes over endocrine control (see pages 230–1). The capacity of women in their 40s to 60s to give birth after receiving younger oocytes emphasizes that the primary reproductive ageing process is oocyte and follicle loss. Senescent changes in other tissues, such as the uterus, may reduce the likelihood of a successful pregnancy outcome in some women, although the effects are statistically marginal, as is attested to by the high multiple pregnancy rate seen in this group after receiving multiple embryos. A case-by-case approach to these patients is best adopted.

The issue then is: from where do the younger oocytes come? Currently, most oocytes are **donated** or **sold** by younger women. However, the increasingly successful use of cryopreservation for ovulated oocytes or for ovarian strips means that a woman could in principle cryo-bank some of her own ovarian germ cells when young for use later. Indeed, such **cryo-banking** is already being used successfully for younger female cancer patients undergoing chemotherapy or radiotherapy. It is important that younger women, currently making the decision to delay having a family for personal and/or professional reasons, realize that whilst their **social clock** may be telling them that it is not yet timely to reproduce, their **biological clock** is still ticking. They do have options of having a genetically related child now, or of waiting and risking a high technology intervention using donated eggs, or risking a third option of an even more invasive procedure to cryopreserve their own eggs or ovarian issue.

Age and reproductive capacity in men

There is a reproductive deterioration with age in men, although much of it may be due to senescence. Thus, fertile spermatozoa can be produced well beyond the age of 40 years, and in most men throughout life. Nonetheless, semen volume and the motility, quantity and quality of spermatozoa decline steadily throughout adult male life from the early 20s, but with no obvious sudden change equivalent to the menopause in women (Figure 22.4a). Studies of sperm fertilizing capacity in IVF procedures, in which the oocyte providers were under 30 years, show little evidence of age-related functional decline in spermatozoa until after the age of 40 years. However, we must be cautious when extrapolating this to the general population who are not experiencing problems conceiving. There is some evidence to suggest that the incidence of certain congenital diseases increase with paternal age (e.g. achondroplasia and Apert's syndrome: Figure 22.4c), although the effect seems to be smaller than with maternal age. This increase may be due to the much higher levels of spontaneous mutation observed in spermatogenic cells from ageing men. In addition, loss of libido, erectile dysfunction and failure to achieve orgasm occur with higher frequency from 30–40 years onwards (Figure 22.4b), mostly resulting from senescent or iatrogenic factors, such as diabetes, pharmacological control of high blood pressure and neurodegeneration (see pages 188–9 for the biology of erectile dysfunction and its treatment).

There is little evidence for a fall with age in testosterone to below the threshold levels required for behavioural effects (see pages 94–5 and Figure 22.4b). However, as men age beyond 60 years, atypical patterns of LH and testosterone pulses and a changed responsivity of LH to GnRH occur with higher frequency, and feedback control by testosterone seems more erratic. Some (disputed) evidence for a fall in the proportion of free/bound testosterone has been proposed to lead to a decline

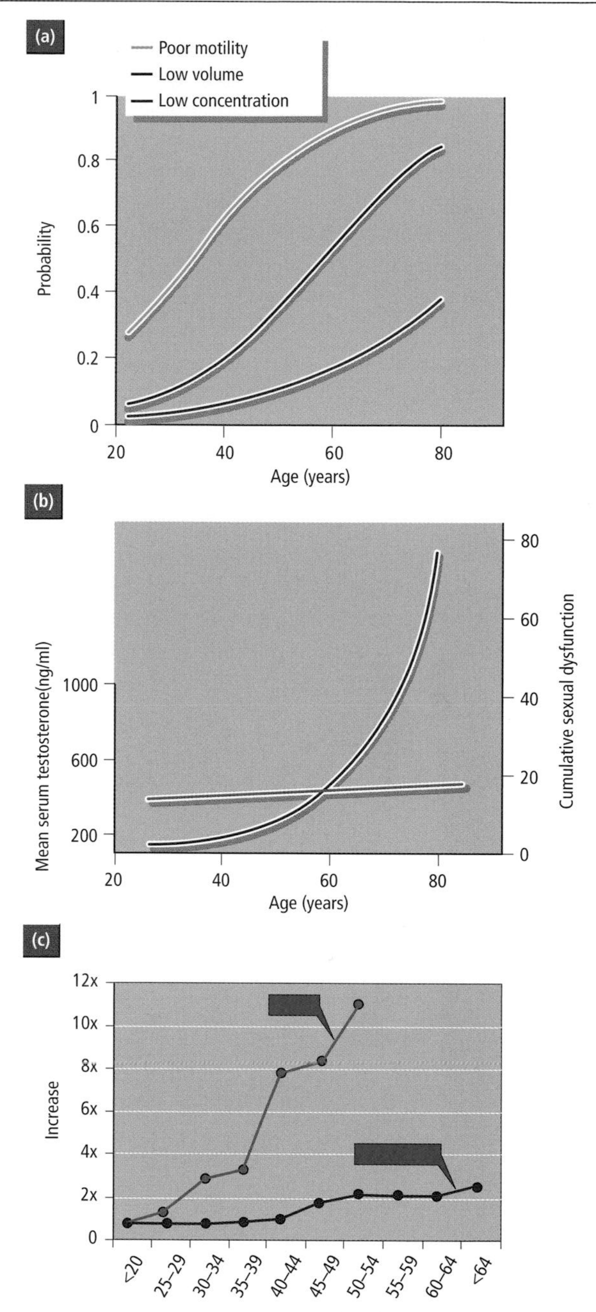

Figure 22.4 Reproduction and ageing in males. (a) Probability of having various semen parameters by age (sperm motility <50%; semen volume <2 ml; concentration $<20\times10^6$); (b) Note that although there is no obvious fall in testosterone with age, the cumulative incidence of sexual problems in men rises steadily and becomes marked after 40 years. (Source: Eskenazi *et al.* (2003) *Human Reproduction* **18**, 447–454. Reproduced with permission from Oxford University Press.) (c) The incidence of cases of spontaneous achondroplasia and of mutations in the FGFR genes in the DNA from men by age. (Source: Tiemann-Boege *et al.* (2002) *Proceedings of the National Academy of Sciences of the United States of America* **99**, 14952–14957.)

in **available androgens**, and a reduced responsivity of androgen receptors has been proposed to contribute to impaired androgen efficacy. All of these changes are thought to be primarily senescent. The muscle wasting and increased adiposity seen with age are plausibly related to a reduced anabolic effectiveness of androgens. Given the uncertainty about the exact underlying causes of male reproductive dysfunction with age, there is currently little evidence to support **androgen therapy**, the effectiveness of which remains unproven, especially when balanced against the increased risk of prostatic cancer.

Approaches to subfertility treatment

For a couple wishing to have a child, a diagnosis of subfertility can be protracted, stressful and socially isolating, as are many of the therapeutic options available. Moreover, for many couples, the biological clock is ticking and so pressure for early results builds. The increasing desperation of the couple combined with an eagerness of clinical staff to appear helpful and media hype about new developments can inflate expectations unrealistically. Although great strides have been made in treating subfertility, it is important for couples to realize at the outset that many treatment attempts end in failure, and that recourse to multiple lines of treatment, often through a hierarchy of different approaches, may be used. Given that most couples fund their treatment, wholly or in part, the **stresses of uncertainty** and of the clock ticking are often compounded by **financial stress**.

The emphasis that clinics and the media place on so-called **success rates** in treating subfertility implies for many of those couples who leave treatment childless that they have failed. For this reason, it is useful to talk of **outcome or live birth rates**, and to build into the consultations from the outset both realism and serious discussion of options, including **acceptance of the state of childlessness**. The best clinics will view a successful treatment as one in which the total experience of the couple has been positive, regardless of the precise outcome. Fortunately, more clinics are now adopting this approach, being attentive to the psychosocial support and honest informational needs of the couple and the attitudes underlying the actions and words of the whole subfertility team.

Certain general factors predispose to reduced fertility and need to be explored sensitively, with support and advice, early on. Key adverse factors to consider are poor communication or conflicting desires between the partners being treated, lack of knowledge about how and when best to conceive, obesity (especially in the woman), low-body weight and excessive exercise, smoking, moderate to excessive alcoholic consumption, drug use, excessive caffeine intake, poor diet and stress (occupational or social). Studies show that simply attending to adverse factors can result in spontaneous pregnancies without technical intervention, as well as enhancing the chance of pregnancy after more technological treatments. A general outline of approaches to treatment is given below for different conditions, emphasizing the underlying biology explored in earlier chapters.

Disorders of the female tract

A range of disorders of the female tract exists, some treatable, some manageable. Here we focus on three of the main disorders.

Tubal damage

Tubal damage is usually a consequence of pelvic infection, being associated with sexually transmitted disease such as chlamydia, untreated frank or asymptomatic gonorrhea, or, tuberculosis, or sepsis following a termination or a completed pregnancy. Tubal infection leads to loss of cilia on the intraluminal cells, causing impaired oocyte and spermatozoal transport, and to extraoviducal scarring, leading to adhesions that restrict oviducal movement and fimbrial oocyte pick-up, or may result in physical blockage of the fimbrial ends of the tubes (**phimosis**) (Figure 22.5a).

(a)

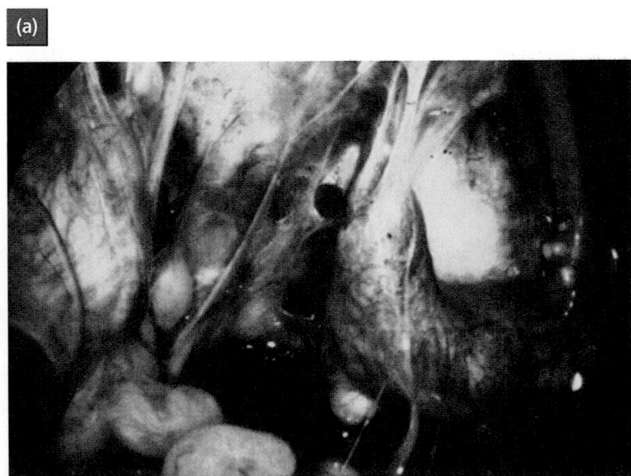

(b)

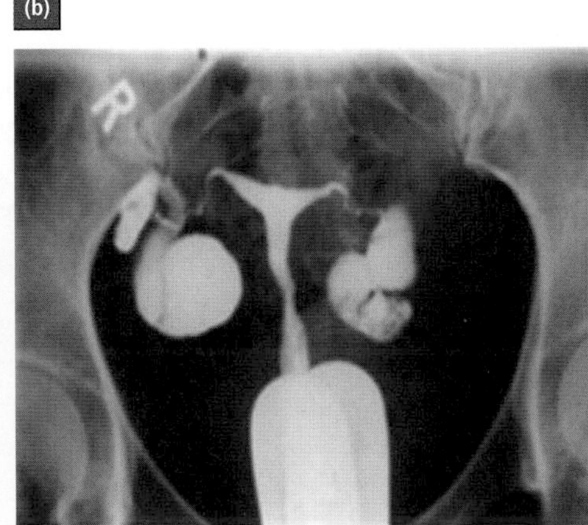

Figure 22.5 (a) A pelvis with severe post-infection adhesions that have resulted in phimosis (occlusion) and dilation of the left Fallopian tube. (b) Hysterosalpingogram (HSG) of a post-inflammatory pelvis where, despite a normal-shaped uterus, both tubes have been left occluded and dilated in the classical retort shape. Such badly damaged tubes are not amenable to restorative surgery. IVF is a better option.

The diagnosis of **tubal obstruction** is preferably made either by X-ray (**hysterosalpingogram**) using a radio-opaque dye (Figure 22.5b) or using ultrasound after instilling a small volume of fluid into the cavity of the uterus (**sonohysterography**). Both are less invasive than visual assessment of the intra-abdominal pelvic organs with a laparoscope and insufflation of dye from the cervix through the tubes under direct vision, which may be advised if predisposing co-morbidities are present. Microsurgical treatment for this large group of patients has limited success for mild disease only, and requires a centre with high expertise for good outcomes. More effective therapy is provided by aspiration of oocytes from the ovary and ***in vitro* fertilization** (**IVF**) (Figure 22.6), followed by placement of the conceptuses into the uterus, thereby bypassing the damaged tubes. When IVF technology is to be used, follicular growth is usually controlled exogenously by: (1) shutting down the woman's own hypothalamic activity via several days' administration of a GnRH analogue (such as buserelin; see Box 2.1 and Chapter 4); (2) then administering a recombinant FSH preparation; and (3) monitoring follicular growth by ultrasound scanning of the ovary (Figure 22.7), and (4) injecting hCG to stimulate the completion of oocyte maturation. Intrafollicular oocytes are recovered when almost fully matured using a needle inserted through the vault of the vagina under continuous monitoring by trans-vaginal ultrasound scanning. The needle is inserted into each follicle, and the oocyte collected by aspiration with its cumulus cells into a sterile receptacle (Figure 22.8). Spermatozoa are provided from the male partner by masturbation or, if obstructive azoospermia is present, by sampling the epididymis (percutaneous epididymal sperm aspiration; PESA) or isolating spermatozoa from a biopsy of the testis (testicular sperm extraction; TESE: see later), following which intra cytoplasmic sperm injection (ICSI) is used instead of IVF (Figure 22.6).

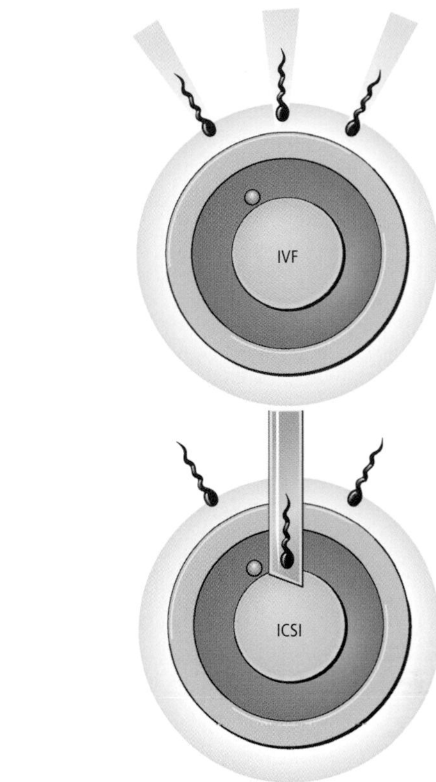

Figure 22.6 Assisted conception techniques used to overcome infertility. In IVF, oocytes (shown as pink) are recovered from the female partner's ovary (or by donation from another female), usually after hormonally induced stimulation, and spermatozoa are recovered from the male partner (or by donation from another male). The two are then mixed *in vitro* for 24–48 hours; the oocytes that have been fertilized are identified by the appearance of pronuclei and passage through cleavage to two to four cells, and up to two fertilized conceptuses are placed in the uterus via a trans-cervical catheter. Any fertilized oocytes remaining may be frozen with reasonable success for later use, should this be necessary. After IVF, average pregnancy rates per treatment cycle initiated at established clinics in the UK are about 25%. Alternatively, if numbers of spermatozoa are very low, or where there are no motile spermatozoa or even only epididymidal or testicular spermatozoa or spermatids (e.g. in obstructive azoospermia), intracytoplasmic injection (ICSI) can be used. ICSI involves use of the micropipette to inject a single spermatozoon into the ooplasm.

Endometriosis

Endometriosis is a condition in which endometrial-like tissue grows inappropriately in **ectopic sites** such as the oviduct, ovary or peritoneal cavity (Figure 22.9), and which may cause a severe reaction to which the body responds by scarring and adhesion formation. It is often associated with **dyspareunia** and **dysmenorrhoea**, and its origins and pathogenesis remain unclear. Endometriotic deposits and symptoms can be controlled or ameliorated by suppressing estrogen secretion with the use of GnRH analogues or the oral contraceptive pill, thereby removing the stimulus to endometrial growth. However, these treatments prevent pregnancy and are unhelpful in these circumstances, but may provide a period of relief before instigation of other forms of fertility treatment, such as the removal of deposits or ovarian endometriotic cysts (endometriomas) by **laparoscopic ablation**. Even when the endometriosis is minimal, treatment almost halves the time to conception.

Uterine absence or malformation

Uterine absence or malformation, or health problems in the female partner that make carrying a pregnancy unwise, can be circumvented by the use of **surrogacy**. Surrogacy is also used when two men wish to start a family together. In **partial surrogacy**, a male partner provides sperm to inseminate the surrogate mother, who provides the oocyte. Partial surrogacy would be used where ovarian damage or absence has occurred in

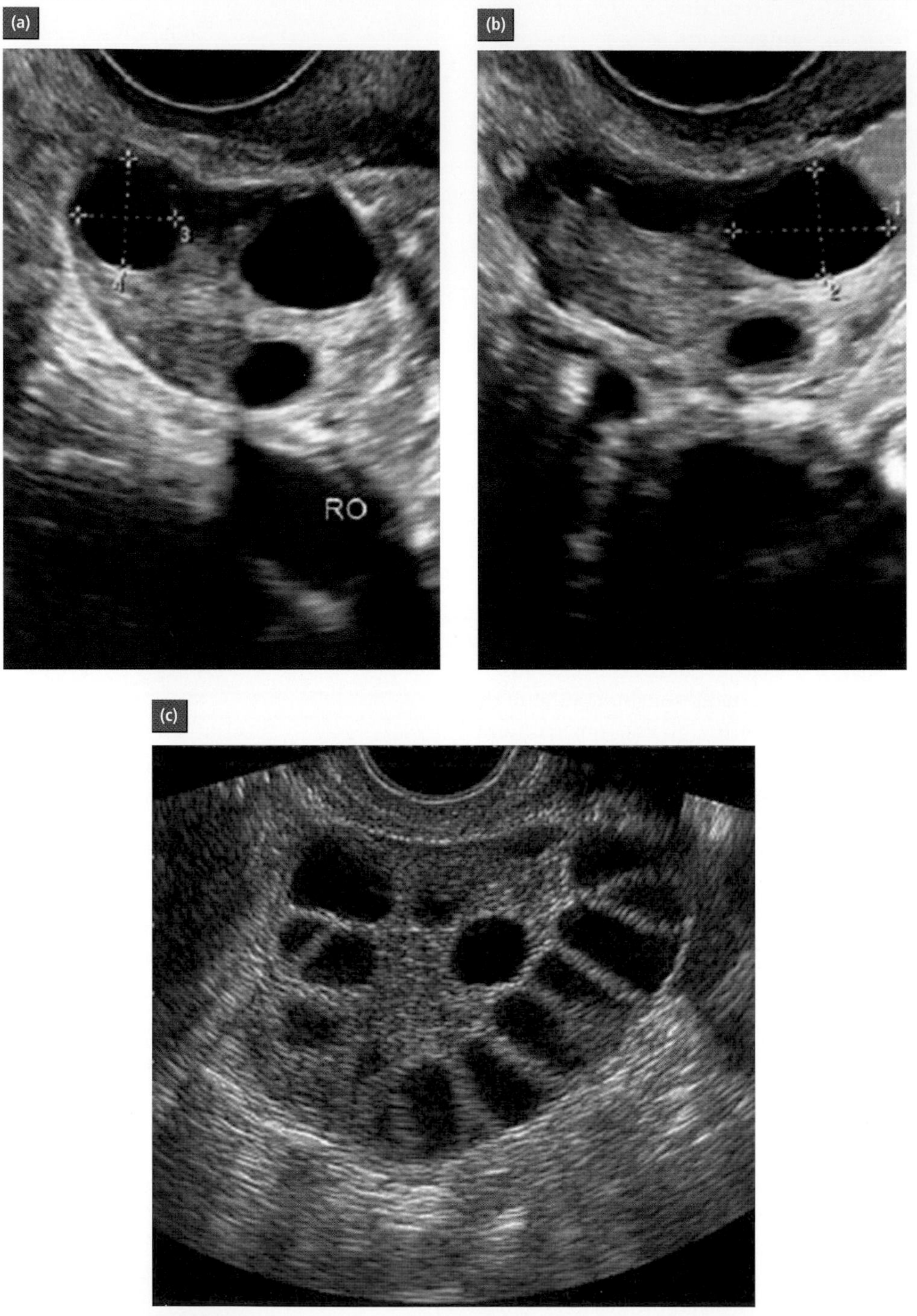

Figure 22.7 (a,b) Two transvaginal ultrasound scans in different planes of a human ovary showing two developing follicles, the largest of which has an average diameter of about 16 mm. (c) Ultrasound scan of a hyperstimulated ovary, which, if not controlled by not administering hCG (LH), will probably result in OHSS. (Source: Courtesy of Professor Yacoub Khalaf.)

the female partner, and does not **require** medical intervention, but is most safely undertaken with medical assistance. **Full or complete surrogacy** involves use of IVF with the gametes of the **commissioning couple** to produce a conceptus for transfer to the surrogate woman's uterus. Legal regulation of surrogacy arrangements varies among different jurisdictions (See https://www.surrogacyuk.org/legalitie and https://www.gov.uk/government/uploads/system/uploads/attachment_data/file/261435/Intercountry-surrogacy-leaflet.pdf, for details for the UK).

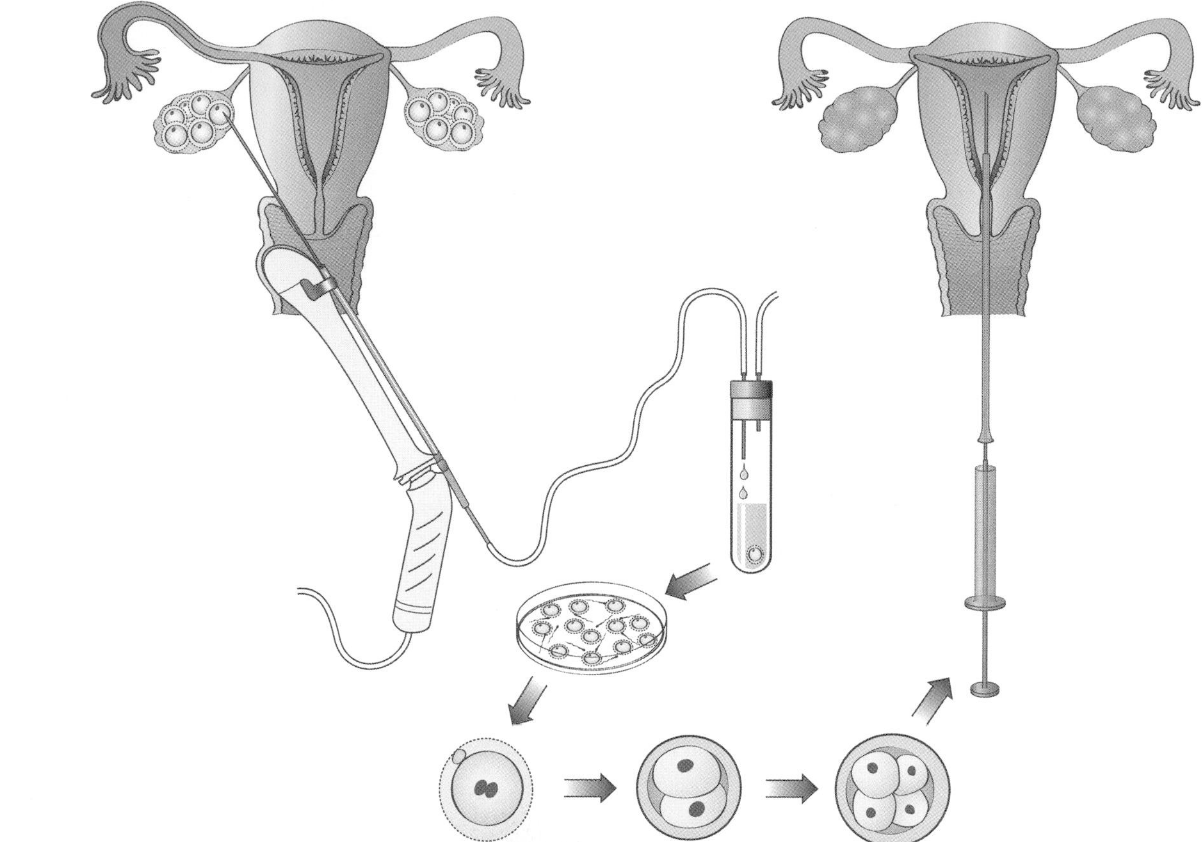

Figure 22.8 Summary of IVF procedure. (a) Egg recovery is achieved laparascopically by inserting needle under ultrasound guidance into each follicle, which is then flushed with warm culture medium, and the oocyte collected by aspiration with its cumulus cells into a sterile receptacle. It is then transferred to a drop of capacitated sperm for fertilization and early cleavage after which it is transferred at the 8-cell or blastocyst stage to the uterus via a cannula inserted through the vagina and cervix (b).

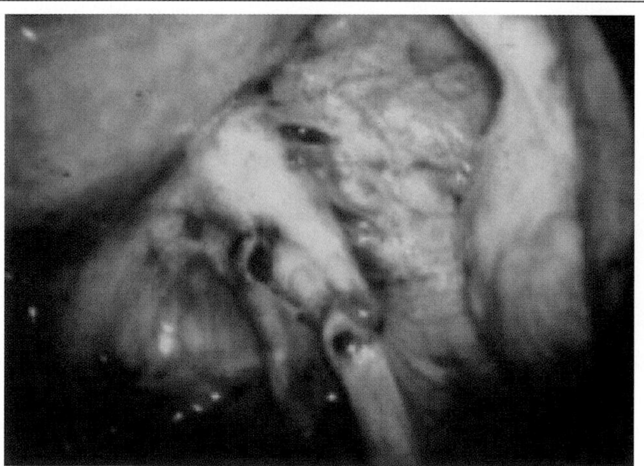

Figure 22.9 Laparoscopic image of the right side of the pouch of Douglas showing the uterosacral ligament with implants of endometriosis, which are likely to be a cause of dysmenorrhoea and dyspareunia.

Disorders of ovulation

This general classification covers a range of disorders. Primary amenorrhoea was discussed earlier, as was premature menopause through early ovarian failure (see pages 365–6). Here we consider only disorders relating to malfunction of the matured reproductive system: namely, absent menses (**secondary amenorrhoea**) and irregular menses (**oligomenorrhoea**), both of which are indicative of **anovulatory cycles**. As the hypothalamus plays such a key role in regulating ovarian function via the pituitary, it is not surprising that these conditions are often associated with stress, eating disorders (obesity, anorexia nervosa), strenuous exercise or use of various drugs, such as neuroleptics or tranquillizers, and may resolve if the primary cause is removed or its effects alleviated. Indeed, in one study, merely taking women into clinical care and giving placebo treatments resulted in 30% of cases having successful pregnancies! Many other patients, although classified as oligomenorrhoeic or secondarily amenorrhoeic, have never had entirely normal cycles and could therefore represent failures of terminal maturation of the neuroendocrine system at puberty.

The endocrine features of a normal menstrual cycle and the associated cyclical changes in the woman's anatomy and physiology are described in Figure 10.15 and see pages 168–77. An ideal clinical investigation would examine all of these through at least one cycle, but such a procedure would be time-consuming, costly and inconvenient. In practice, therefore, a more limited range of preliminary tests is applied in an attempt to locate the endocrine defect. In the patient with cycles, a **mid-luteal phase progesterone measurement** is the simplest screening test to give an indication as to whether or not she is ovulating, but the timing of this test is critical. Several underlying causes of ovulation disorders can be identified, and have been classified by the World Health Organization (WHO) into three groups:

- Group I: Hypothalamic pituitary disorders and hyperprolactinaemia
- Group II: Idiopathic anovulation
- Group III: Ovarian failure, Polycystic Ovary Syndrome (PCOS: hypothalamic-pituitary-ovarian dysfunction) and luteal phase defect.

Group I

Hypothalamic pituitary disorders

Deficiency of GnRH in reaching the pituitary results in depressed gonadotrophin and oestrogen levels and failure of ovulation with oligomenorrhoea or amenorrhoea (**hypogonadotrophic hypogonadism**). Ultrasonically, quiescent ovaries and a thin endometrium are identified. Women of low body weight (BMI < 19Kg/m^2) who can gain a little weight, and those who reduce over-exercising, may achieve natural ovulation. Patients who require therapy can be treated with exogenous gonadotrophins, carefully used and monitored to achieve unifollicular ovulation. The use of a pump to deliver pulsatile infusions of GnRH subcutaneously, as described for initiating puberty in Chapter 4 (see pages 63–4), is a more physiological approach and does not carry the same risk of multiple pregnancy, but the medication is no longer available in some countries.

Hyperprolactinaemia

Hyperprolactinaemia is a cause of subfertility in both women and men, in the latter associated with decreased circulating levels of testosterone, impotence and loss of libido. The endocrine profile of women with hyperprolactinaemia is characterized by an absence of pulsatile LH secretion, a reduced pituitary LH response to injected GnRH, a failure of positive feedback, and hence chronic anovulation and amenorrhoea. Ovarian responsiveness to FSH and LH is not necessarily impaired; indeed, normal cyclicity can be maintained in the presence of high prolactin levels if exogenous gonadotrophins are administered. Exactly how elevated prolactin levels cause these changes is unclear. The decrease in pituitary sensitivity to GnRH might reflect an indirect effect of prolactin via decreased ovarian oestradiol secretion, rather than an effect exerted directly on the pituitary.

Hyperprolactinaemia has various causes, including physiological (e.g. pregnancy and the first few months of breast-feeding), iatrogenic (e.g. psychiatric use of dopamine receptor-blocking neuroleptic drugs that increase prolactin secretion, as can oestrogens in some oral contraceptives), or an underlying pathology (e.g. prolactin-secreting pituitary tumours – so-called prolactinomas, some 20% of postmortem samples showing evidence of microadenomas).

Treatment is effective and simple. Dopamine D2 receptor agonists, such as bromocriptine and cabergoline, are used to lower serum prolactin concentrations immediately, and daily treatment results in the return of ovulation and cyclicity in the vast majority of women within 2 months. In the case of prolactin-secreting tumours, dopamine receptor agonists have an antimitotic action and, in addition to lowering plasma prolactin concentrations, reduce tumour size. Surgical removal of a macroprolactinoma is reserved for cases of intolerance to side effects or drug resistance.

Group II

Idiopathic anovulation

In this condition, gonadotrophin secretion seems to be occurring within the normal range, but is insufficient to support a normal cycle, probably as a result of ovarian insensitivity. In consequence, oestrogen levels fail to rise appropriately and ultrasonography of the ovary reveals antral follicles that fail to mature fully. Most cases will respond to therapy with exogenous gonadotrophins to recruit or maintain follicular growth and oestrogen output. The endogenous LH surge is usually attenuated and, hence, may need to be supplemented or replaced by an injection of hCG (to mimic the ovulatory effect of LH). However, the appropriate doses of gonadotrophins can be difficult to gauge and slight overdosing can lead to multiple ovulations and implantations or, in some cases, to **ovarian hyperstimulation syndrome (OHSS)**. Their use should be accompanied by ovarian ultrasound monitoring to measure follicle size and number, and to adjust gonadotrophin doses accordingly. For the future, a real advance in avoiding ovarian hyperstimulation is on the horizon as we learn more about small inter-individual variations in FSH receptor gene structure. Thus, single nucleotide polymorphism analysis has revealed at least four variant haplotypes, each having slightly different FSH-binding properties. These then reflect the differing sensitivities of the women to FSH. This understanding opens the possibility of haplotyping women in advance and then tailoring the stimulation regimen to their haplotype.

Because of these difficulties and the cost of using gonadotrophins, a simpler regimen using **anti-oestrogens**, such as oral **clomiphene citrate** or **tamoxifen**, tends to be used initially. Anti-oestrogens act on the hypothalamus to compete with endogenous oestrogen and thereby reduce its negative feedback, resulting in elevated gonadotrophins. Clomiphene also stimulates aromatase activity in the ovary, and the ovarian oestrogen produced then stimulates granulosa cell proliferation and development of LH receptors (see page 147), and thereby increases ovarian responsiveness. Clomiphene tablets are administered for 5 days early in the follicular phase, usually on days 2–6 of the cycle, and yield positive ovulatory responses in around 70% of anovulatory women (especially when they are not obese). If the patient is amenorrhoeic, pregnancy needs to be excluded before treatment. A withdrawal bleed can then be induced with a progestagen (e.g. medroxyprogesterone acetate) to initiate the cycle.

Group III

Ovarian failure

Secondary amenorrhoea with elevated serum FSH and LH **(hypergonadotrophic hypogonadism)** with a low estradiol confirms the presence of ovarian failure, due to lack of inducible ovarian follicles (**premature menopause**). There is no rational treatment for this condition, although some women who are recently amenorrhoeic may have ovarian insufficiency with an occasional ovulation which allows a surprise pregnancy. Once established this type of infertility can only be overcome by the use of egg donation. (See section 'Fertility and the postmenopausal woman' above). These women will require long-term hormone replacement therapy to prevent the deleterious effects of hypo-oestrogenism on their bones.

Polycystic Ovary Syndrome (PCOS: hypothalamic-pituitary-ovarian dysfunction)

PCOS is a distinct and common syndrome and accounts for 80–90% of anovulatory infertility. In addition, the syndrome generally includes hyperandrogenism and secondary insulin resistance and obesity. PCOS is characterized by two to three times the normal number of preantral growing follicles that then arrest at the early antral stage, no dominant follicles being selected (Figure 22.10a). Hypersecretion of LH is found in about 40% of women with PCOS. The finding of a tonically elevated LH but not FSH distinguishes this syndrome from secondary ovarian failure and (early) menopausal onset. The primary cause of ovarian failure of this kind is not known, but is thought to be a thecal cell defect in androgen biosynthesis. Significant weight reduction in the obese (BMI >30Kg/m^2) and the use of metformin (the anti-diabetic medication) may be sufficient to reinstate regular cycles and ovulation in some women with PCOS.

In the past, treatment by surgical removal of part of the ovary (**wedge resection**) was recommended. This procedure apparently produced an acute reduction in the prevailing high level of follicular circulating androgens, perhaps re-instating more normal feedback relationships and the possibility of an LH surge and ovulation. However, the precise sequence of events is far from clear, and the procedure has fallen out of favour, because of the removal of significant amounts of oocyte containing cortical tissues and possible associated damage to the oviducts and post-operative formation of adhesions. A variant procedure, involving the drilling of small holes in the ovary using diathermy or a laser, has achieved good results in some studies (Figure 22.10b,c).

Nowadays, treatment with anti-oestrogens is often used (see preceding section). However, PCOS ovaries are often exquisitely sensitive to gonadotrophins and clomiphene, and ovarian hyperstimulation and multiple pregnancies are a particular risk. Lower initial doses of clomiphene may be needed in PCOS to achieve unifollicular ovulation.

Luteal phase defect

Finally, some women with evidence of ovulation nonetheless show slow or reduced rises in progesterone, associated with impaired fertility arising from an abbreviated luteal phase. This pattern is observed more frequently in women who have undergone gonadotrophin therapy, e.g. during IVF treatment. Whether it is due to a deficiency in the maturation of granulosa cells leading to poor luteinization, or is a primary defect, e.g. in the development of LH or prolactin receptors, is unclear. However, as it is not yet clear whether, in women, either LH or prolactin are luteotrophic, a deficiency in these cannot be invoked by way of explanation. Treatment with progesterone or hCG during the luteal phase is often helpful (see also Box 14.1).

Oligozoospermia

Oligozoospermia (strictly meaning too few spermatozoa) is a term usually expanded to include a wide range of defects in semen quality, such as **asthenozoospermia** (reduced motility) and **teratozoospermia** (abnormal morphology). Although each of these deficiencies may occur individually, they are more usually found together (**oligo–astheno–terato–zoospermia**), reflecting a general deficiency in spermatogenesis (Table 22.2). A systematic and quantitative assessment of semen quality from the ejaculate is an essential part of an infertility examination and should be repeated at least once after 2–3 months. However, the data in Table 22.2 should not be regarded as representing a cut-off at or below which men will inevitably be infertile, but more a guide to suggest that there may be a male component to the infertility. Attempts to improve the reliability of semen analysis have included assessment of the integrity of the sperm DNA but with mixed outcomes that do not currently warrant inclusion of these analyses in standard semen testing. Total absence of spermatozoa in the ejaculate (**azoospermia**) may be due to deficient production (**aspermatogenesis**) or deficient transport through the excurrent ducts (**obstructive azoospermia**). Deficiencies in the seminal plasma volume or composition usually reflect disease or malfunction of the accessory glands, such as the prostate or seminal vesicles.

It has been proposed recently that sperm counts in men from several developed countries may have declined during the

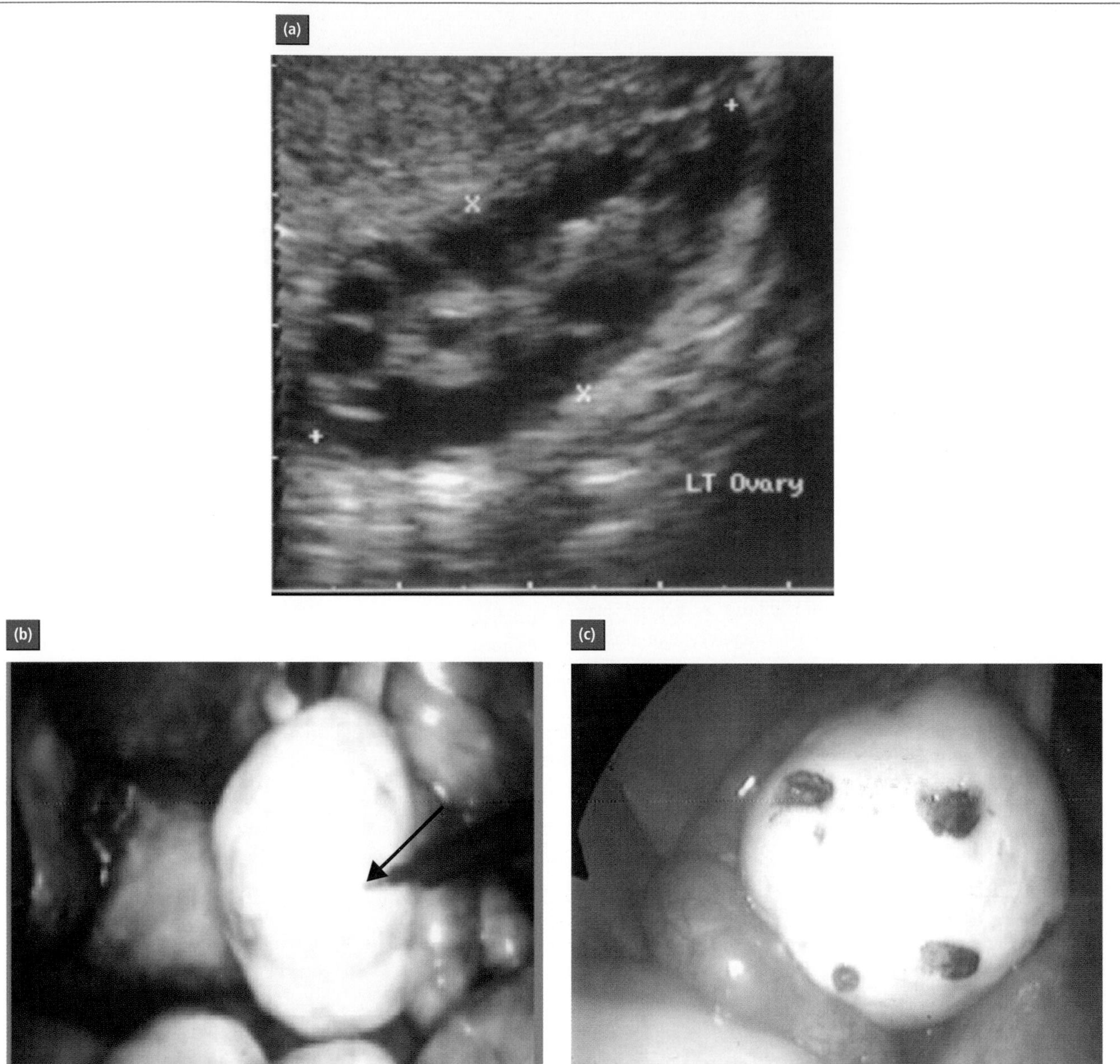

Figure 22.10 (A) Ultrasound picture of polycystic ovary showing multiple small follicles (dark areas) in a thick walled ovary (Source: Courtesy of Adam Balen); (B) Laparoscopic image of pearly white thick walled polycystic ovary with diathermy 'drilling' needle (arrowed) right (Source: Courtesy of PR Braude); (C) An example of a polycystic ovary that has been treated by 'drilling' four holes laparoscopically (Source: Courtesy of Adam Ballen).

last few decades, the suggestion being that environmental toxins, including some that appear to be oestrogenic metabolites, might have a causal role.

Causes

Characteristically, hypospermatogenesis is associated with smaller testes (<20 ml volume) of softer consistency and can result from dietary deficiency, X-irradiation, heating of the testis, exposure to a range of chemicals (notably cadmium, antimitotic drugs used in tumour therapy, insecticides and antiparasitic drugs), and excessive alcohol intake. Removal of the offending agent may restore spermatogenesis from the stem cell population of spermatogonia, if this is undamaged (see Box 7.1). Irreversible forms of hypospermatogenesis include: cryptorchid testes (see page 54); genetic abnormalities such as XXY, XYY, some autosomal translocations and an increasing number of identified deletions on the Y chromosome; and germ cell aplasia of unknown cause, often with hyalinization of tubules or as a sequel to severe orchitis or to prolonged and

Table 22.2 WHO (2013) characteristics associated with 'normal' and 'subfertile' semen

Criterion	Normal	Subfertile
Volume (ml)	2–5	<1.5
Sperm concentration (number/ml)	$50–150 \times 10^6$	$<15 \times 10^6$
Total sperm number per ejaculate	$100–700 \times 10^6$	$<40 \times 10^6$
Spermatozal vitality	>75% live	<58% live
Spermatozoa swimming forward vigorously (progressive motility)	>50%	<32%
Abnormal spermatozoa	>30%	>4%
Viscosity after 60 minutes (liquefaction)	Low	High
White blood cells (number/ml sperm cells)	Low	$>10^6$

intense drug therapy for tumours. In all these patients (except those with Klinefelter syndrome), testosterone and LH levels may well be in the normal range. FSH levels, however, tend to be elevated, probably because of the absence of inhibin production (see page 130). In general, elevated FSH is associated with a poor prognosis, and treatment endocrinologically is not recommended. Hypospermatogenesis due to a primary neuroendocrine deficit is relatively rare in men and is easily recognized. Treatment may be attempted with exogenous GnRH or gonadotrophins and is often successful.

Obstructive azoospermia is not associated with obvious endocrine disorder, and testes are of normal size and consistency. Obstruction to spermatozoal transport usually occurs in the epididymides, secondary to infection with, for example, gonorrhoea or tuberculosis, and surgical correction can be effective. Men with cystic fibrosis (CF) will have congenital bilateral absence of the vas deferens (CBAVD). Conversely, men who have CBAVD (albeit that they do not manifest any other signs of CF) are likely to have mutations in the CF gene and/or its non-coding region In either case, usually spermatogenesis is normal, hence their sperm can be retrieved using PESA or TESE (see below) for assisted conception using ICSI. However partners of CBAVD men should be screened for CF mutations as their children may be at risk, which can be avoided by preimplantation genetic diagnosis (PGD see page 377).

Abnormal, slow-swimming or dead spermatozoa in the ejaculate, as distinct from low numbers, might also result from suboptimal spermatogenesis, a condition that is certainly increased in cases of **varicocoele** (varicosity of the spermatic vein), genetic abnormality, maintained elevated scrotal temperatures, deficiencies in spermatozoal maturation, genital tract infections, and cytotoxic factors or anti-sperm antibodies in the fluids of the accessory glands. Recently, it has been suggested that aberrant reprogramming of the sperm epigenome may be the basis for some forms of male subfertility (see Box 7.3).

Treatments

Couples with mild male factor problems can be offered six cycles of intra-**uterine insemination** at mid-cycle (with or without exogenous ovarian stimulation, depending on the female partner). **Donor sperm insemination** can be offered in the complete absence of spermatozoa, or where the male partner has genetic reasons for not reproducing.

IVF and variants of it have provided useful routes for circumventing hypospermatogenesis by avoiding the dilution of spermatozoa that would occur during their passage through the female genital tract (see pages 192–3). In IVF, small numbers of recovered viable spermatozoa are concentrated around the oocyte(s) to promote fertilization. IVF can also be used where the male partner carries an infection potentially transmissible to the female partner, such as HIV. In such cases, the careful washing of sperm has been required, together with post-wash testing for clearance, although with more effective antiviral drug use, this is often no longer considered essential.

A variant of IVF allows successful pregnancies in cases where there are very few or no spermatozoa in the ejaculate, or where all or most of the spermatozoa in the ejaculate are immotile or clumped. A single spermatozoon, or even spermatid, is picked up in a micropipette and injected directly into the ooplasm (**intracytoplasmic sperm injection; ICSI**). Sometimes sperm must be recovered from the epididymis (**percutaneous epididymidal sperm aspiration; PESA**) or testis (**testicular sperm extraction; TESE**). This approach is summarized in Figure 22.11. Its use is not without controversy. For those oligospermic conditions with a genetic component, transmission to any male offspring will occur unless only female conceptuses are selected for transfer. In addition, about 50% of obstructive azoospermia cases are associated with heterozygosity for cystic fibrosis (CF), which raises the possibility of transmission of CF to offspring. Men with obstructive azoospermia can also be offered microsurgical correction to restore patency, although with variable outcome, reflecting the important role that the epididymis plays in spermatozoal maturation (see pages 184–6).

Miscarriage

Loss of pregnancy, especially in the first trimester, is common (around 30% of conceptions) and was described in detail in Chapter 16 (see pages 271–3). Rarer recurrent miscarriage was discussed in Box 16.2.

Assisted reproductive technologies (ART)

Over the past 30 years, the plight of the infertile has been taken much more seriously by the medical profession. There is little doubt that this change of attitude has stemmed from the pioneering work of Bob Edwards, Patrick Steptoe and Jean Purdy

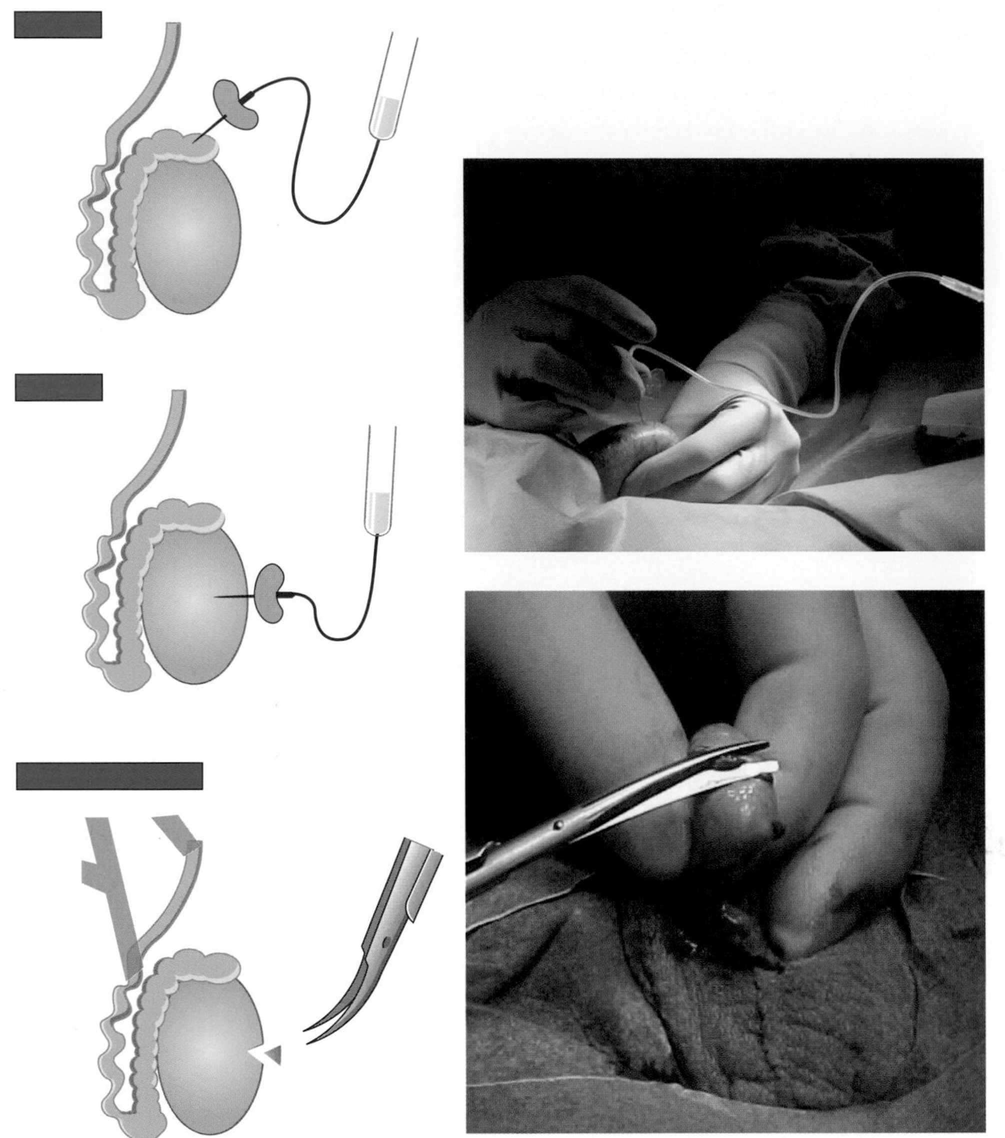

Figure 22.11 Diagrammatic illustration to show differences between PESA, TESE and testicular biopsy [left]. On the right are two photographs of (upper) TESE and (lower) testicular biopsy.

in the development of IVF technology that led to the first IVF birth of Louise Brown in 1978 (who is now a mother herself!). In 2010, the Nobel Foundation recognized these achievements through the award of the Nobel Prize for Physiology or Medicine to Bob Edwards (Patrick having died in 1995 and Jean in 1985). Over 5 million people worldwide exist as a result of the reproductive technologies that have subsequently been refined or developed. Those ARTs mentioned thus far in this book include IVF, ICSI, TESE and PESA, donor insemination (DI), oocyte donation and surrogacy. This capacity to help infertile people has also been a major stimulus to clinical research, as a result of which we now know much more about the basic science underlying human reproduction, and we are less reliant on animal models, essential though these remain. As a result of this research, new technologies related to reproduction have been and are being developed. Here we examine five of these. Now we can:

- Freeze and thaw eggs, sperm and conceptuses, with retention of their viability (**cryopreservation**).
- Remove individual blastomeres from early eight-cell stages and test their DNA for the presence of faulty genes (**preimplantation genetic diagnosis** or **PGD**) or screen

for sporadic chromosome anomalies (**preimplantation genetic screening** or **PGS**), thereby offering the initiation of an unaffected pregnancy to affected couples.

- Mix together cytoplasm and chromosomes from different eggs or embryos to produce at least some viable conceptuses.
- Developing the conceptus *in vitro* to peri- and post-implantation stages.
- Persuade the pluriblast cells from the blastocyst to proliferate *in vitro* as **embryonic stem cells (ES cells)**.

Cryopreservation

Cryopreservation at −196 °C, by immersion in liquid nitrogen (liquid phase storage) or in the vapour above it (gas phase storage), allows long-term storage of gametes or conceptuses with minimal degradation. There are two types of cryopreservation: conventional **slow freezing** or ultra-rapid cooling to achieve **vitrification**. In both cases, ice crystal formation is prevented by use of **cryoprotectants** – a form of 'antifreeze'.

Cryopreserved tissues can, in principle, be stored indefinitely, although storage is usually limited to 5–10 years. Thawed material may be used reproductively by the person(s) from whom it was obtained or for donation to others. Cryopreservation is used for various reasons: for convenience; for quarantining the material in case it carries infectious agents such as HIV, hepatitis viruses or cytomegalovirus; to provide the opportunity for repeated or multiple use of the same donor; and in advance of cancer (and other) therapies that use potentially damaging drugs, radiation or surgery.

Cryopreservation followed by pregnancies after thawing has been offered clinically for spermatozoa since the 1950s, for cleavage or blastocyst stages since the 1980s, but for oocytes only recently. Cryopreservation of strips of ovarian tissue with subsequent autografting and successful pregnancies has also been reported recently. Studies of the obstetric outcome of the use of cryopreserved gametes or conceptuses do not indicate any major problems, although data are thin for oocytes and long-term follow-up studies of the resultant children are lacking.

Preimplantation diagnosis (PGD)

PGD was first demonstrated in animals in the 1960s, but its clinical application leading to pregnancies were first reported in 1990, and only since 2000 has PGD been widely practised clinically. The principle is simple: take a biopsy from the cleavage or blastocyst stages, test it genetically and decide whether to transfer the biopsied conceptus. Broadly, this approach is now used in four types of clinical application:

- To detect any one of a growing number of (mostly rare) **single gene defects**, or recurrent chromosome aberrations (such as translocations) so that an at-risk couple can start a pregnancy confident that a disease-free child will be born.
- To screen for sporadically occurring chromosomal anomalies, such as errors of somy, ploidy (often called **preimplantation genetic screening or PGS**) usually in older women, at-risk couples, or couples/women experiencing recurrent miscarriage, despite the fact that there is little evidence that this use is helpful.
- For **sexing of conceptuses** to avoid transmission of sex-liked genetic disease or for family balancing – selecting one gender over another.
- To produce so-called '**saviour siblings**', in which a conceptus is typed for histocompatibility with an existing sibling who has a disease requiring a tissue graft – to be provided by the umbilical cord blood cells from the 'saviour sibling'.

Each of these applications of PGD raises ethical issues, and the ethico-legal response to consideration of these issues has been very varied (see pages 386–90). Technically, PGD is performed by one of three approaches to biopsy. Post-fertilization, either a **trophoblast biopsy** from a day 5 blastocyst is used, or, more commonly until recently, one or two **blastomeres are biopsied** from a day 2–3 cleaving conceptus. Pre-fertilization sampling of the **polar body** can be used to identify genetic problems in eggs but only reflects maternal information. Each sampling method has advantages and disadvantages in terms of sensitivity and reliability. The technology used for the genetic diagnosis of a biopsied sample is tailored to the purpose for which the test is used. Originally, **fluorescent *in-situ* hybridization** (**FISH**) was used to detect abnormal karyotypes, structural chromosomal anomalies or embryo sex, and the **polymerization chain reaction** (**PCR**) to amplify the tiny amounts of DNA before probing it in various ways was used for single-gene detection or sexing. Recent improvements in **whole genome amplification** (**WGA**) and **array technologies** promise to revolutionize detection methodology. Thus, WGA is being used in conjunction with two types of array: **array comparative genomic hybridization** (aCGH) and **single nucleotide polymorphism (SNP)**-based arrays. Both array types can be used to determine anomalous chromosome numbers, but only SNP-based arrays can also be used for haplotyping the sample. More recently testing has been extended to whole genome sequencing (NGS – next generation sequencing).

Egg reconstruction

We encountered one form of egg reconstruction in Chapter 1 (see page 9) when the issue of **somatic cell nuclear transfer (SCNT)** was raised. This technique, when applied to produce live young is known as '**reproductive cloning**' and gave rise to the famous sheep, Dolly, who shared all her nuclear genes and chromosomes with the donor (Figure 12.6). Reproductive cloning has now been achieved in many species, including monkeys, so there is every reason to believe it would work in humans too. However, internationally there are legal prohibitions on human reproductive cloning, but not on its experimental variant of **therapeutic cloning**. Therapeutic cloning, which is only allowed under licence in some jurisdictions, involves exactly the same SCNT procedures except that placement of the created clone in a uterus is not permitted. The reason for permitting what is in essence a research technology is to help us to understand better the early epigenetic changes in human development.

A second form of egg reconstruction has, however, been attempted clinically, but its focus is on **aberrant ooplasmic function**, rather than on nuclear function. Thus, we know that during oocyte maturation both chromosomal and cytoplasmic maturation must occur for the production of a fully fecund egg (see pages 206–7). It was suggested that in some women experiencing subfertility, an unidentified ooplasmic deficit might be responsible and so the transfer of a small amount of ooplasm (<10% volume) from a healthy egg was attempted. However, this **ooplasmic transfer** was discontinued in the USA in 2001, but continued sporadically in other countries for the treatment of unexplained cleavage arrest. A newer and more specific form of ooplasmic replacement has been proposed and extensively investigated as a means of overcoming certain **mitochondrial diseases**. A small number of individuals manifest serious genetically transmitted diseases, which can be very debilitating and even lethal, involving defects in energy production and including deafness, blindness, diabetes, loss of motor skills and/or heart and liver failure. You will recall that all mitochondria are derived from your mother (see pages 200–6), and it turns out that many of these diseases result from the inheritance of genetically mutated mitochondria. The mitochondrial genome (**mtDNA**) contains 37 genes, each present in multiple copies – at least one set per mitochondrion. The ratio of genetically aberrant to normal mitochondria can vary (**heteroplasmy**), and only people with a large proportion of faulty mitochondria may develop a mtDNA disease. Given that there are no treatments to cure these disorders, avoidance reproductively is a possible solution. One option is **egg donation** and IVF, but this approach breaks the genetic link to the mother. Alternatively, prenatal diagnosis plus termination of pregnancy or preimplantation genetic diagnosis could be offered. However, none of the embryos may be guaranteed to be mutant-free. Hence the suggestion that defective maternal mtDNA be removed by transferring the nuclear genome of the prospective mother to a donated enucleated egg (Figure 22.12). This can either be achieved by transferring the couple's pronuclei after fertilization to a zygote from which its pronuclei have been removed (pronuclear transfer; PNT), or prior to fertilization transferring the maternal spindle to a donor egg from which its spindle has been removed (maternal spindle transfer; MST) and then fertilizing with the intended father's sperm. This approach should result in healthy offspring carrying the nuclear genome of the prospective parents and the healthy mitochondrial genome from the donor. Extensive preclinical investigations suggest that very low levels of mutant mtDNA are carried over during this process, but there is still a theoretical risk that even that small amount may be selected for or amplified during development negating the intended benefit. In 2016, the UK approved in principle the use of this technique in very specific cases as a risk-reduction technology and is likely to be put into practice soon. Indeed, a reported birth using spindle transfer has been described recently in Mexico.

Developing the conceptus in vitro to peri- and post-implantation stages

Recently, the mouse and human conceptus has been successfully cultured *in vitro* over the period of implantation, when remarkably the whole structure resembled that observed *in vivo.*

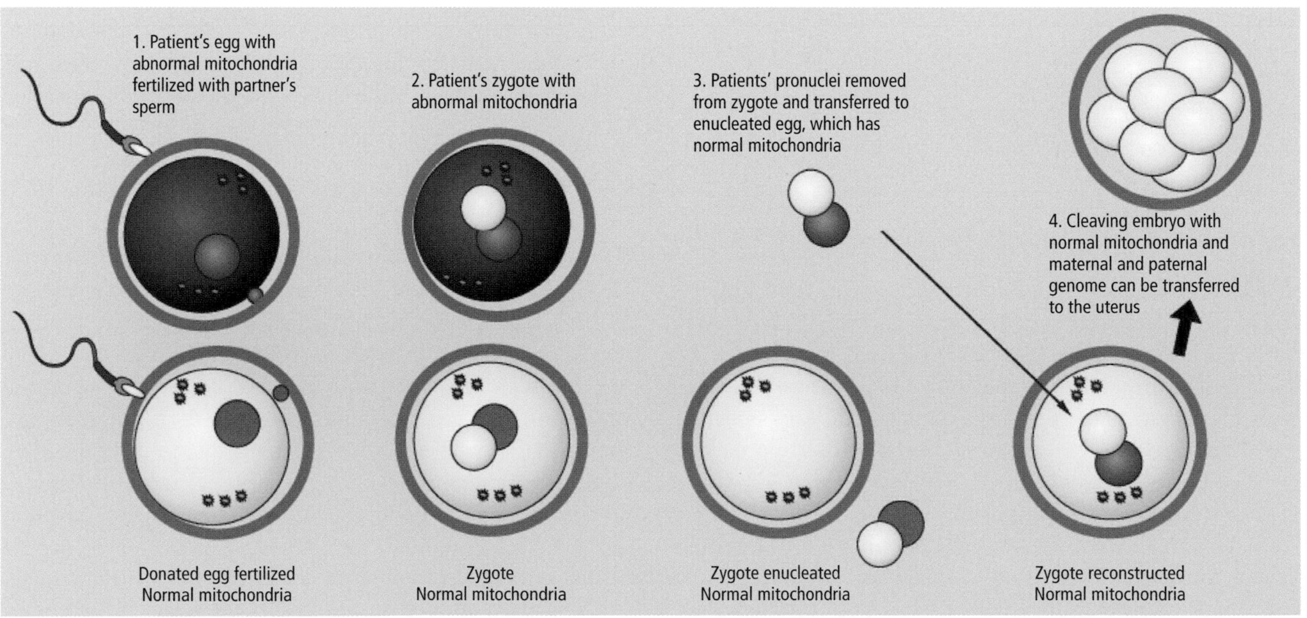

Figure 22.12 Outline of a procedure for reconstituting a zygote to rescue it from mitochondrial disease. (Source: Bredenoord A, Braude P (2010) *British Medical Journal* **342**, 87–89. Adapted with permission from the BMJ.)

The conclusion drawn was that the conceptus has within it the complete programme for the construction of both the extra-embryonic and embryonic structures, and that the uterus does not provide any essential instructional elements. For the human conceptus, the culture had to cease for legal reasons before 14 days and thus before the germ cells could be identified. It will be of interest to see whether such cultures beyond 14 days will become legally permitted or will occur in non-restrictive jurisdictions, and if so, what more can be learnt about these hitherto inaccessible phases of early human development and frequent embryo loss.

Embryonic stem cells

IVF has also permitted the production of embryonic stem cells (ESCs). Stem cells are pluripotent, self-renewing cells that arise during normal development (see page 8). ESCs are one early type derived from the ICM of the blastocyst or its derivative epiblast. ESCs are in principle fully competent and able to generate all the tissues required to make a complete individual. In combination with somatic cell nuclear transfer (SCNT), the production of pluripotent cell lines matched genetically to the nuclear donor offers one approach that opens the possibility for use in **tissue repair**. Also, there have been several claims to have developed gamete-like cells *in vitro* from human and animal stem cells, some of which, in animals, have been used successfully in fertilization to produce live young – albeit with low success rates and limited long-term survival. These procedures are highly experimental, but were they to improve for humans, they might offer a route to **infertility alleviation**. Such a scenario is unlikely to work for some genetic forms of infertility, but would be of use for those who have lost functional gametes through accident, disease or age. It is, however, unlikely to work for **gay or lesbian couples** hoping to make 'opposite-sex' gametes from ESCs of one of the partners. This is because, in lesbian couples, the presence of two X chromosomes seems to be incompatible with sperm formation, whilst for gay couples, the absence of two X chromosomes seems to be incompatible with egg cell development (see page 44). Recently, a combination of mouse ESCs with mouse trophoblastic stem cells has been cultured under the conditions used to culture intact conceptuses over the implantation period (see above), when a conceptus-like structure developed, reinforcing the idea that ESCs are equivalent to epiblast cells and providing a way of potentially circumventing the legal ban on 14-day cultures.

Conclusions

Much of our understanding of the early stages of human pregnancy and development, and our ability to treat or manage subfertility, have come, and are continuing to come, from the development of IVF, one of the landmark contributions to human medicine and health in the 20th century. It is a tribute to the vision of Bob Edwards (the dedicatee of this book) that he predicted some 30–40 years ago many of the major advances in treatment described here.

Key learning points

- Approximately 10–15% of couples experience clinical subfertility.
- Clinical subfertility is defined as failure to conceive after 12 months of unprotected intercourse.
- Men and women contribute more or less equally to subfertility problems.
- Female age contributes to lower fertility, reproductive fecundity being time limited and optimal between 15 and 35 years.
- The menopause marks the end of fecundity in women and is preceded by the climacteric, a time of declining fecundity.
- Loss of female fecundity is due to abnormalities in and then exhaustion of oocytes.
- In men, the decline in reproductive function is progressive and more extended, and is largely the result of senescent changes, not loss of germ cells.
- Major causes of subfertility are blocked oviducts, ovulatory disorders and aspermia or oligozoospermia, but around 25% of cases are unexplained.
- Spontaneous miscarriage occurs in about 10–15% of clinically diagnosed pregnancies.
- A range of techniques is now available to help the infertile including induced ovulation, IVF, ICSI, TESE, PESA, DI, egg/embryo donation and surrogacy.
- New technologies associated with these technologies include cryopreservation, preimplantation genetic diagnosis, stem cell derivation, implantation *in vitro* and cell nuclear transfer.

Clinical vignette

Uterine transplantation

A 28-year-old woman, Ms V, was diagnosed in her teens with complete uterine agenesis (Mayer-Rokitansky-Küster-Hauser Syndrome). At the time of diagnosis she received counselling and was informed that in the absence of a uterus, she would be unable to become pregnant and deliver children. In her late twenties. however, she and her partner wished to revisit options that might be available to them for reproduction. They understood the options of gestational surrogacy and adoption, but Ms V felt strongly that she wished to pursue any possibility of carrying a baby herself. The couple was able to enrol in a trial of uterine transplantation, and Ms V's older sister (who had been sterilized following the birth of her third child) offered to donate her uterus. Prior to the uterine transplant procedure, Ms V underwent an egg retrieval and IVF procedure, from which five frozen blastocysts were obtained. For the uterine transplant procedure itself donor surgery and recipient surgery took place concurrently in order to minimize the ischaemic time of the uterus. Ms V then commenced an anti-rejection regimen of medications following the surgery. Ms V experienced her first-ever menstrual period 4 days after the new graft was placed. Ten months following the transplant she had two frozen blastocysts replaced in the donor uterus. An ultrasound scan performed 6 weeks later showed a single viable intrauterine pregnancy. Ms V continued on her anti-rejection medication during pregnancy. The pregnancy was uncomplicated up to 32 weeks, when Ms V developed pre-eclampsia. She underwent caesarean delivery of a healthy baby girl at 34 weeks. Following pregnancy, Ms V underwent an uncomplicated hysterectomy to remove the transplanted uterus.

Uterine transplantation is a new procedure, currently offered only within a research trial context. However, within specialist centres there have been some successful live-births reported. The procedure is difficult to perform and only suitable for a small sub-set of patients whose infertility is related to uterine factors. In common with other solid organ transplant procedures, early data suggest that live donor transplants are more successful than those from deceased donors. This may be partly because of the increased ability to minimize ischaemic time from a live donor and also because of demographic differences between deceased and living donors. The most complicated part of the surgery is preserving the complex vascular connections of the uterus during retrieval. It is imperative that these are suitable to form anastomoses with the recipient's pelvic vasculature, which may be unusual due to under-lying syndromes such as Mayer-Rokitansky-Küster-Hauser Syndrome (complete uterine agenesis). The number of live-births that have occurred following uterine transplant is currently small, but as international experience grows, the success rate is likely to increase.

FURTHER READING

General reading

Balen AH (2017) Polycystic ovary syndrome (PCOS). *The Obstetrician & Gynaecologist* **19**, 119–129. DOI:10.1111/tog.12345.

Bewley S, Davies M, Braude P (2005) Which career first? The most secure age for childbearing remains 20–35. *BMJ* **331**, 588–589.

Bredenoord A, Braude P (2010) Ethics of mitochondrial gene replacement: from bench to bedside. *British Medical Journal* **342**, 87–89.

Brinsden PR (2003) Gestational surrogacy. *Human Reproduction Update* **5**, 483–491.

Brosens I, Gordts S, Valkenburg M *et al.* (2004) Investigation of the infertile couple: when is the appropriate time to explore female infertility? *Human Reproduction* **19**, 1689–1692.

Cunningham GR, Toma SM (2011) Why is androgen replacement in males controversial? *Journal of Clinical Endocrinology & Metabolism* **96**, 38–52.

De Jonge C (2012) Semen analysis: looking for an upgrade in class. *Fertility and Sterility* **97**, 260–266.

Dunson DB, Colombo B, Baird DD (2002) Changes with age in the level and duration of fertility in the menstrual cycle. *Human Reproduction* **17**, 1399–1403.

Farquhar C, Rishworth JR, Brown J *et al.* (2015) Assisted reproductive technology: an overview of Cochrane Reviews. *Cochrane Database of Systematic Reviews* Issue 7. Art. No.: CD010537. DOI: 10.1002/14651858.CD010537.pub4.

Handyside AH (2010) Preimplantation genetic diagnosis after 20 years. *Reproductive BioMedicine Online* **21**, 280–282.

Handyside AH (2011) PGD and aneuploidy screening for 24 chromosomes by genome-wide SNP analysis: seeing the wood and the trees. *Reproductive BioMedicine Online* **23**, 686–691.

HFEA (1997) *Sixth Annual Report*. HFEA, London. http://www.hfea. gov.uk

HFEA (2016) Scientific review of the safety and efficacy of methods to avoid mitochondrial disease through assisted conception: 2016. update. http://www.hfea.gov.uk/docs/Third_Mitochondrial replacement_scientific_review.pdf.

Johnson MH (2011) Robert Edwards: the path to IVF. *Reproductive BioMedicine Online* **23**, 245–262.

Johnson MH, Elder K (2015) The Oldham Notebooks: an analysis of the development of IVF 1969–1978. V. The role of Jean Purdy reassessed. *Reproductive BioMedicine and Society Online* **1**, 46–57.

Klein J, Sauer MV (2002) Oocyte donation: best practice and research. *Clinical Obstetrics and Gynaecology* **16**, 277–291.

Kuhnert B, Nieschlag E (2004) Reproductive functions of the ageing male. *Human Reproduction Update* **10**, 327–339.

Macklon NS, Geraedts JP, Fauser BC (2002) Conception to ongoing pregnancy: the 'black box' of early pregnancy loss. *Human Reproduction Update* **8**, 333–343.

National Collaborating Centre for Women's and Children's Health (2004) *Fertility Assessment and Treatment for People with Fertility Problems*. RCOG Press, London.

Nuffield (2012) *Novel Techniques for the Prevention of Mitochondrial DNA Disorders*. The Nuffield Foundation, London. https://nuffieldbioethics.org/wp-content/uploads/2014/06/Novel_techniques_for_the_prevention_of_mitochondrial_DNA_disorders_compressed.pdf

Pitkin J, Rees MC, Gray S *et al.* (2005) Managing the menopause: British Menopause Society Council consensus statement on hormone replacement therapy. *Journal of the British Menopause Society* **11**, 152–156.

Santoro N (2005) The menopausal transition. *American Journal of Medicine* **118** (12B), 8S–13S.

Sigman M (2014) Beyond bulk semen parameters. *Fertility and Sterility* **102**, 1501.

Silber S, Pineda J, Lenahan K *et al.* (2015) Fresh and cryopreserved ovary transplantation and resting follicle recruitment. *Reproductive Biomedicine Online* **30**, 643–650.

Thacker PD (2004) Biological clock ticks for men, too: genetic defects linked to sperm of older fathers. *Journal of the American Medical Association* **291**, 1683–1685.

Theodosiou AA, Johnson MH (2011) The politics of human embryo research and the motivation to achieve PGD. *Reproductive BioMedicine Online* **22**, 457– 471.

van der Steeg JW, Steures P *et al.* (2011) Role of semen analysis in subfertile couples. *Fertility and Sterility* **95**, 1013–1019.

Veldhuis JD, Keenan DM, Iranmanesh A (2005) Mechanisms of ensemble failure of the male gonadal axis in aging. *Journal of Endocrinological Investigation* **28** (Suppl. 3), 8–13.

Zhang J, Liu H, Luo S *et al.* (2017) Live birth derived from oocyte spindle transfer to prevent mitochondrial disease. *Reproductive Biomedicine Online* **34**, 25–32.

More advanced reading

Arce J-C, Nyboe Andersen A, Collins J (2005) Resolving methodological and clinical issues in the design of efficacy trials in assisted reproductive technologies: a mini-review. *Human Reproduction* **20**, 1757–1771.

Barritt JA, Brenner CA, Malter HE, Cohen J (2001) Mitochondria in human offspring derived from ooplasmic transplantation. *Human Reproduction* **16**, 513–516.

Battaglia DE, Goodwin P, Klein NA, Soules MR (1996) Influence of maternal age on spindle assembly in oocytes from naturally cycling women. *Human Reproduction* **11**, 2217–2222.

Chian RC, Huang J.Y, Tan SL *et al.* (2008) Obstetric and perinatal outcome in 200 infants conceived from vitrified oocytes. *Reproductive Biomedicine Online* **16**, 608–610.

Cook R, Day Sclater S (eds) (2003) *Surrogate Motherhood*. Hart Publishing, Oxford.

de la Rochebrochard E, Thonneau P (2002) Paternal age and maternal age are risk factors for miscarriage; results of a multicentre European study. *Human Reproduction* **17**, 1649–1656.

Eskenazi B, Wyrobek AJ, Sloter E *et al.* (2003) The association of age and semen quality in healthy men. *Human Reproduction* **18**, 447–454.

Gnoth C, Godehardt D, Godehardt E *et al.* (2003) Time to pregnancy: results of the German prospective study and impact on the management of infertility. *Human Reproduction* **18**, 1959–1966.

Hunt PA, Hassold TJ (2002) Sex matters in meiosis. *Science* **296**, 2181–2183.

Liverman CT, Blazer DG (eds) (2004) *Testosterone and Aging: Clinical Research Directions*. IOM Committee on Assessing the Need for Clinical Trials of Testosterone Replacement Therapy, Institute of Medicine. The National Academies Press, Washington, DC.

McElreavey K, Ravel C, Chantot-Bastaraud S, Siffroi JP (2006) Y chromosome variants and male reproductive function. *International Journal of Andrology* **29**, 298–303.

Norman RJ, Noakes M, Wu R *et al.* (2004) Improving reproductive performance in overweight/obese women with effective weight management. *Human Reproduction Update* **10**, 267–280.

RCOG (2011) *Venous Thromboembolism and Hormone Replacement Therapy*. Green-top Guideline No. 19 3rd edition. Royal College of Obstetricians and Gynaecologists, London.

Sammartino A, Cirillo D, Mandato VD *et al.* (2005) Osteoporosis and cardiovascular disease: benefit–risk of hormone replacement therapy. *Journal of Endocrinological Investigation* **28** (Suppl. 10), 80–84.

Sauer MV, Kavic SM (2006) Oocyte and embryo donation 2006: reviewing two decades of innovation and controversy. *Reproductive BioMedicine* **12**, 153–162.

Schiavi RC, Schreiner-Engel P, Mandeli J *et al.* (1990) Healthy aging and male sexual function. *American Journal of Psychiatry* **147**, 766–771.

Steptoe PC, Edwards RG (1978) Birth after the reimplantation of a human embryo. *Lancet* **2**, 366.

Tiemann-Boege I, Navidi W, Grewal R *et al.* (2002) The observed human sperm mutation frequency cannot explain the achondroplasia paternal age effect. *Proceedings of the National Academy of Sciences of the United States of America* **99**, 14952–14957.

van der Linden M, Buckingham K, Farquhar C *et al.* (2011) Luteal phase support for assisted reproduction cycles. *Cochrane Database of Systematic Reviews* Issue **10**, CD009154.

Wang X, Chen C, Wang L *et al.* (2003) Conception, early pregnancy loss, and time to clinical pregnancy: a population-based prospective study. *Fertility and Sterilility* **79**, 577–584.

Wennerholm U-B, Söderström-Anttila V, Bergh C *et al.* (2009) Children born after cryopreservation of embryos or oocytes: a systematic review of outcome data. *Human Reproduction* **24**, 2158–2172.

Yin W, Gore AC (2006) Neuroendocrine control of reproductive aging: roles of GnRH neurones. *Reproduction* **131**, 403–414.

Zhou GB, Meng QG, Li N (2010) In vitro derivation of germ cells from embryonic stem cells in mammals. *Molecular Reproduction and Development* 77, 586–594.

Zhu JL, Madsen KM, Vestergaard M *et al.* (2005) Paternal age and congenital malformations. *Human Reproduction* **20**, 3173–3177.

CHAPTER 23
Society and reproduction

Essential Reproduction, Eighth Edition. Martin H. Johnson.

Companion website: www.essentialreproduction.com

In the preceding chapters, we have attempted to describe the important mechanisms that underlie the establishment of sex, the attainment of sexual maturity, the production and successful interaction of male and female gametes, the initiation and maintenance of pregnancy, and the production and care of the newborn. These processes absorb much of our physical, physiological and behavioural energies. Sexual reproduction and its associated activities permeate all aspects of our lives. This ramification of sex, in turn, makes it highly susceptible to social and environmental influences, and we are beginning to understand some of the mechanisms by which these influences interact with the developmental, genetic and behavioural components of sexual reproduction. We have also seen plausible ways in which these social and environmental influences might be of adaptive value to us. However, although the adaptive value of sensitivity to environmental and social signals seems undeniable for the reproductive efficiency of the species as a whole, it is not necessarily so for the individual. In contemporary human society, the increasing emphasis on individual rights and welfare has prompted a more sympathetic attitude towards the fertility of the individual, both its limitation and its encouragement. Moreover, sexual interaction among humans is not related directly and exclusively to fertility, but rather has evolved a wider social role. Sexuality and its expression have, in their own right and independent of any reproductive considerations, become an important element within human society (see Chapter 6). These contemporary considerations raise social and political issues and bring us into the domains of ethics and law.

Social constraints on fertility

There is a rich literature on the sociology, anthropology and history of sex and reproduction, which is too large for this book to cover comprehensively. It is an important body of knowledge for public health management, as all studies show that the simple availability of educational information about scientific and medical technology is not in itself sufficient to change sexual and reproductive behaviours. A further element, namely the **motivation** required for a change in behaviour to occur, is critically dependent on an individual's social network, values, beliefs and self-esteem, and the peer pressures that reinforce these. Among the important social variables that may influence reproduction are the following:

- Accepted social roles of men and women (their **gender stereotypes**; see pages 74–6), the perceived and legal balance of power between the genders, and, in particular, the extent to which women are educated and have economic independence.
- These in turn relate to the age of a woman at marriage and at birth of her first child (see pages 351–2), and the maternal mortality rate (see pages 273–5).
- Accepted **size of the family**, the desirability of spacing children, the perceived economic advantages of a given family size, which in turn may be related to the anticipated child mortality pattern, the availability of child care, and the preferred sex ratio, which may also be influenced by inheritance patterns.
- Knowledge or beliefs about the required or expected **frequency of intercourse** in relation to the point in the menstrual cycle, the time of year or religious calendar, age of partners, social conditions and opportunities, and the delivery or suckling of children.
- Extent to which maternal lactation (with its consequent hyperprolactinaemia and depression of fertility; see page 325) is replaced by use of milk substitutes or wet nurses.
- Acceptability of sexual interactions outside (or to the exclusion of) the usual (normative) framework in which successful pregnancy might result, emphasizing the sexual rather than the reproductive function of coition (e.g. paid-for sex, natural methods of contraception, atypical patterns of sexual expression; see pages 86–93 and 353).
- Social role and status of **celibacy.**
- Social, ethical and legal acceptability of reproduction outside of a tightly controlled social structure, such as the nuclear or extended family (single or unmarried mothers, lesbian and gay couples, use of gametes from deceased partners, and mothering/fathering by the very elderly).
- Incidence of divorce and the attendant delay before remarriage;
- Extent to which **eugenic considerations** (whether ethnic, caste, ability, economic status; see pages 101–3) are an integral part of social fabric and determine with whom a person can reproduce or who can be born.
- Strength with which **religious or other social beliefs** are held by, or imposed on, the individual, which may also affect access to education, information, contraception, pregnancy termination and assisted reproductive technologies (ART).

Each of these factors will affect, to varying degrees, the overall sexual and reproductive expectations and strategy of the individual woman, couple or population, and the various factors will interact with each other. For example, high economic expectations coupled with low infant mortality and gender equality tend to reduce fertility by both delaying birth of the first child and reducing overall family size. These in turn tend to lead to a greater individualization of society, and to more freedom of sexual expression and greater reproductive choice.

Perhaps the single largest change that has occurred in most societies over the past century is the fall in expected infant mortality arising from improved diet and hygiene, and the introduction of antibiotics and prophylactic immunization. However, this fall has led to a corresponding rise in the numbers of women surviving beyond puberty and, thus, to an increased fecundity of the population as a whole. It is the time lag between the decline in infant mortality and the reproductive adjustment to it, via the compensatory social factors listed above, that has led to a world population explosion. Indeed, it is not clear that social adjustment alone is adequate to correct this imbalance. The increased use of artificial constraints on fertility in various societies and, most notably in China, the recourse to state control over reproduction, have become an essential part of the social response to population growth, so as to regulate individual fertility according to desired or imposed social patterns (see Chapter 21). Often, the desirability of controlling the profligate reproduction of humanity has resulted in adverse reactions against some sections of society – in particular, and paradoxically, the infertile.

Biological constraints on fertility

Good preconception health is important for achieving pregnancy, whether by natural means or through the use of ARTs. Smoking, excessive alcohol use and extreme changes in weight are among factors known to contribute to a decline in fertility among both men and women. Besides these, age is among the most influential factors contributing to a decline in fertility. As already discussed, this is increasingly pertinent, as more women and couples are waiting until their 30s and 40s before deciding to have children, for a range of social reasons, that include the absence of an identified partner or of sufficient income to support a child, or for reasons relating to career progression. When these reasons are coupled with press reports of famous women having babies by IVF late in life (often unreported as being via oocyte donation), a false belief in the power of IVF to treat older women can take hold. Indeed, the demographic trend towards delaying childbearing among women increased between 1970 and 2000, the percentage of first births in women older than 30 increasing 6-fold. The proportion of first births to women above the age of 30 has continued to rise since then and it is believed that around 20% of women in the western hemisphere now have their first child after the age of 35. This trend has two important implications.

First, as we have already discussed (pages 364–7), human fertility declines with age. Women's fertility normally starts to decline in their early thirties, with the decline speeding up after the age of 37. Accordingly, age has a large influence on how rapidly female patients in particular are referred to reproductive specialists for ART treatments. One response to the realization of declining fertility with age has been an increased interest in egg freezing by younger women.

Second, and relatedly, increasing age at first birth may reveal inherited factors related to reproductive health that may not have manifested themselves in older relatives who mostly conceived at an earlier age. This could mean that reproductive specialists, treating an increasingly older population, turn to diagnostic metrics and criteria other than age in order to diagnose and treat infertility. Thus, the age-related decrease in women's fertility is concurrent with, and mechanistically linked to, gradual changes in levels of reproductive hormones such as follicle stimulating hormone, estradiol and anti-mullerian hormone (AMH). Reproductive specialists assess the levels of these hormones in their evaluation of infertile patients. However, there is now a large body of evidence demonstrating the association, if not the mechanistic role, of further, subclinical factors that may contribute to the aetiology of infertility and various reproductive conditions. For example, causes of female infertility include metabolic disorders, chromosomal abnormalities, endometriosis, pelvic adhesions, and endocrine disorders, such as hyperprolactinaemia (see Chapter 22). Moreover, there is growing evidence that genetic abnormalities are present in as many as 10% of infertile females and 15% of infertile males. These genetic abnormalities can range from chromosomal changes (of ploidy) to single-gene and single-nucleotide alterations.

For many conditions resulting from chromosomal changes, infertility is noted as just one manifestation of a more complex phenotype. For example, Turner syndrome occurs in women when one of the X chromosomes is either missing or structurally altered on a large scale. Infertility is just one aspect of Turner syndrome, stemming from an increased susceptibility to primary ovarian insufficiency due to congenital hypogonadotropic hypogonadism. Infertility can also be noted as a 'minor' manifestation in monogenic conditions including galactosaemia, caused by mutations in the *GALT* gene, and in cystic fibrosis, which is caused by mutations in the *CFTR* gene and associated with infertility in almost all male sufferers due to congenital bilateral absence or blockage of the *vas deferens*.

In addition to these single-gene defects, additional rare genetic abnormalities have been identified in infertile men and women. These include mutations in genes like gonadotropin-releasing hormone 1 (*GnRH1*) in association with Kalmann syndrome, a condition characterized by a failure to start or complete puberty Extending these more established associations, both targeted and genome-wide association studies have aimed to explore the genetic underpinnings of a number of different reproductive conditions (Table 23.1) However, while they have revealed plausible candidate genes, such association studies are unlikely to reveal the full genetic landscape, perhaps most fundamentally because fertility is a complex, rather than monogenic, trait. Thus, changes at multiple, rather than single, genetic loci are likely to contribute to extended periods of reduced fecundity.

Fertility as a complex genetic trait

Most human genetic traits can be classified as either monogenic or complex. Monogenic traits are influenced primarily by variation within a single gene and typically conform to classic patterns of inheritance within families. However, while monogenic traits are the basis of 'classical' genetics, it has become clear that conditions, the inheritance of which strictly conform to Mendelian principles, are relatively rare, at least in humans. Complex genetic traits are caused by a combination of genetic,

Table 23.1 Proposed genetic markers for some of the most common and high impact reproductive conditions to affect men and women

Reproductive condition	Proposed genetic markers	References
Early menopause/primary ovarian insufficiency (POI)	*FMR1/2, BMP15, FSHR, FIGLA, NOBOX, POF1B, MCM8, SYCP2L, FOXL2*	Pelosi *et al*., 2015
Endometriosis	*CYP19A1, GSTM1, ID4, VEGFA, WNT4, GREB1, VEZT, FN1*	Rahmioglu *et al*., 2015
Uterine fibroids	*FH, FASN, COL4A5/6, HMGA2, SLK, MED12*	Levy *et al*., 2012 Ordulu, 2016
Polycystic ovary syndrome (PCOS)	*FSHR, LHCGR, INSR, DENND1A, THADA, RAB5B, YAP1, CAPN10, FBN3*	De Leo *et al*., 2016 McAllister *et al*., 2015
Recurrent pregnancy loss (RPL)	*HLA-G, P53, P73, NLRP7, VEGFA, F5, F2, ACE, IFN-g, TNF-a, NOS3, SERPINE1, Il6, IL10*	Hyde and Schust, 2015 Tur-Torres *et al*., 2017
Spermatogenic failure	*DAZ, CDY1, HSFY, USP9Y, EIF1AY, TEX11, MCM8, KLHL10, SYCP3, RBMY*	Mitchell *et al*., 2017 Miyamato *et al*., 2015
Obstructive azoospermia	*CFTR, ADGRG2*	Mitchell *et al*., 2017

behavioural and environmental factors. Thus, while complex traits like infertility can have a genetic component, equally changes in behaviour or in the environment, as we have already seen, play a considerable part.

The path to reproductive success involves the coordination of processes such as folliculogenesis, oogenesis, spermatogenesis and hormone regulation. Complex networks of genes and functional non-coding elements regulate each of these processes. These have evolved to include elements of pleiotropy and redundancy that phenotypically buffer genomic mutations and safeguard mechanisms of reproduction. On a population level, this means that infertility and reproductive disorders are therefore likely to comprise of multifactorial defects driven by several predisposing genetic alleles. Thus, there is often inconsistent predictive value in genetic markers identified by studies of particular fertility phenotypes based on single or small numbers of genetic variants. Discoveries from *a priori* genome-wide association studies have undoubtedly revealed more of the genetic landscape underlying particular fertility disorders. However, in most cases the link between genetic pathology and medical intervention is yet to be made. Making such a link is essential if genomic markers are ever going to have an impact in an ART setting. Thus, though alterations in the genes that drive many reproductive processes have been well documented across thousands of studies to date, they still largely present medical practitioners with more questions than answers.

This raises the question of whether practitioners should begin to incorporate particular kinds of genetic information into reproductive medicine. Currently, the most common indications for genetic counseling include advanced maternal age, family history of a genetic condition, and/or suspected diagnosis of a genetic condition. In addition, the use of chromosomal screening, in the form of parental and preimplantation genetic screening (PGS) is gradually becoming more refined and the utility (or otherwise) of such tools is becoming clearer. What about genetic markers that are not strictly diagnostic or completely predictive? Such markers might confirm a suspected diagnosis, and may be useful in guiding diagnosis and treatment in cases of borderline clinical findings. Identification of pathogenic variants can also be useful in counseling the patient and their families, and can help to identify family members at increased risk for the disorder, as well as those family members who are not at an increased risk. However, should genetic information about patients and their family members be obtained and communicated if no treatment or intervention exists? Could the results of genetic testing create a false sense of security or undue panic? In addition to potential psychological harm, could the results of the genetic test be used to limit access or rights and privileges to such things as insurance or employment? Though it seems inevitable that genomics will increasingly impact on reproductive medicine, such questions dictate caution before it becomes clearer that such technology can have a positive impact on patient care.

Society and the infertile

The infertile have been the butt of social disapproval, from textual assertions that barrenness in a marriage is a punishment (often from God) to infertility being grounds for annulment, divorce or polygamy. Two features recurred traditionally in these expressions of disapproval: the **woman was responsible** and **inheritance** (through the male line) was paramount. The socially gendered values system underlying these negative attitudes has persisted until very recently. Indeed, medical research on male reproduction is relatively new compared with that on female reproduction, and even now prostatic cancer screening is way behind that for female reproductive cancers and in the UK only girls are being immunized against HPV not boys! To this gendered pattern of negative attitudes to infertility has been added some contemporary twists. Thus, the argument is often heard, and acted on, that infertility treatment and research should not be funded by the state, reasons given including: infertility being due to 'misbehaviour' (i.e. venereal disease!), there being a population problem, and women should reproduce when young enough not to need help. There remains in these contemporary attitudes, elements of punishment, blame and scapegoating.

Aside from these atavistic critiques of infertility research and treatment can be added some with more potential legitimacy. These include concerns about the **efficacy of treatments**. Thus, despite the large numbers of healthy babies born through IVF, there nonetheless remain some lingering concerns about the longer-term safety of these relatively novel technologies. These concerns have been fuelled by the increased understanding we now have of how the early developmental environment can influence patterns of health and behaviour later in life and even transgenerationally, and the role that epigenetics can play in mediating these influences (see pages 282–4 and 339–40). These concerns emerge into the media from time to time, often in alarmist terms and out of context. What is the evidence that actually bears on them? We find that social rather than medical issues lurk in many of the fears.

Multiple births

There is one clearly established and highly significant adverse clinical outcome of ART, namely an increase in multiple births, which can constitute 25% or more of ART outcomes. Pregnancies carrying more than one conceptus show markedly **increased maternal and child morbidity** (Figure 23.1 and Table 23.2). This increase in multiple births in recent years comes from two sources: (1) induced ovulation for ovulatory

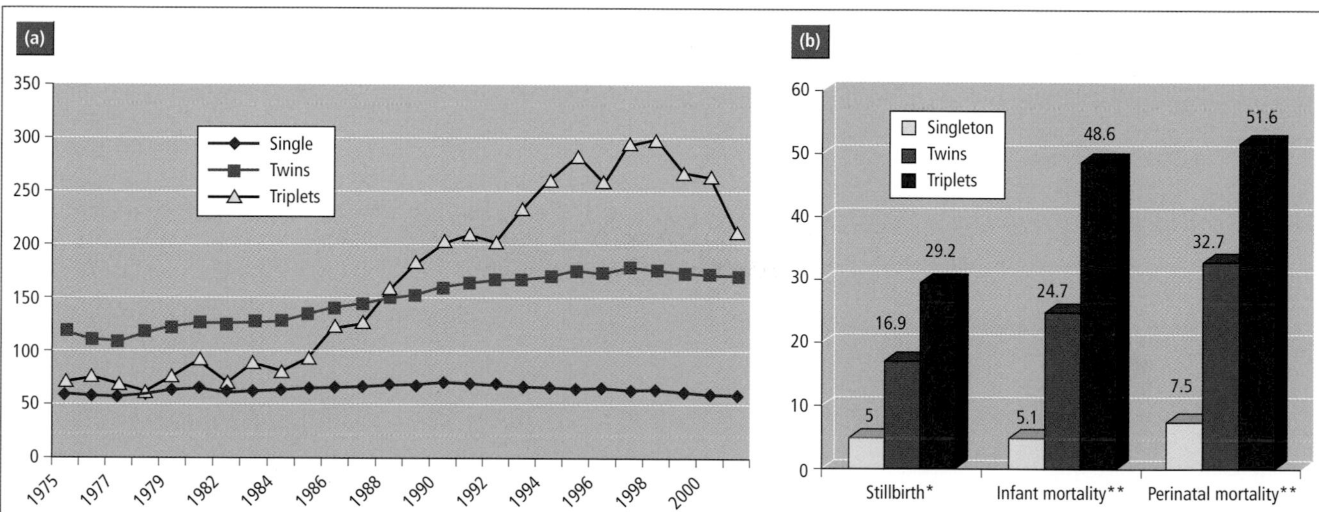

Figure 23.1 (a) A plot of the relative numbers of singleton and multiple births in England and Wales. Note the rise coinciding with the onset of infertility treatment and the dip in triplets following changes in regulatory advice concerning numbers of embryos to be transferred. (b) Histograms to show adverse neonatal outcome resulting from multiple births. *per 1000 births; **per 1000 live births. (Source: El-Toukhy *et al.* (2006) *American Journal of Obstetrics and Gynecology* **194**, 322–331. Reproduced with permission from Elsevier.)

Table 23.2 Data from a randomized control trial comparing elective single transfer (eSET) followed where unsuccessful by a frozen embryo transfer (FrET) versus double (DET) embryo transfer

Number of pregnancies	eSET			DET		
	Pregnancy rate (fetal heart+sac) (%)	Delivery rate (%)	Twin rate (%)	Pregnancy rate (%)	Delivery rate (%)	Twin rate (%)
906	31	28	2	48	42	35
	eSET+FrET (1+1)			**DET (2+0)**		
661	40	39	0.3	44	43	33

Source: Bergh (2005) *Human Reproduction* **20**, 323–327. Reproduced with permission of Oxford University Press.

disorders; and (2) placement of more than one conceptus into the uterus after IVF/ICSI to increase the chance of a pregnancy occurring. Multiple births occur through a **social collusion** between patients anxious to get pregnant (almost at any cost) and doctors anxious to get good pregnancy rates for their clinics. Multiple births are largely avoidable, and only a single conceptus should be transferred in most cases. Indeed, many countries or states are now incentivizing or imposing **transfer of reduced embryo numbers**.

The mental and emotional well-being of the offspring

It is often asserted that the children of 'unnatural' families will have psychological problems about their origins. However, ongoing prospective studies on the psychological adjustment and emotional well-being of children born through IVF reveal that, if anything, they score higher than control babies conceived naturally. Plausibly this may be related to the intense effort and commitment of their parents in producing them in the first place. Similarly, prospective studies on the children of lesbian parents do not reveal major adjustment problems. Studies on children of gay parenting or on saviour siblings are not advanced enough to provide clear data, but seem to be moving in the same direction. All that we know about child development leads us to expect these positive outcomes. Thus, **a child who is loved and offered secure attachments is likely to thrive**. However, what is clear from studies on children born through donation is that any dishonesty during childhood about their origins can, if discovered later in life, be very traumatic. It is better for the **child to absorb the origins of their identity** as part of their growing up.

Congenital abnormalities

A significant increase in prevalence of congenital abnormalities in IVF babies has been reported in a meta-analysis of various retrospective comparative population studies. The data equate to one additional congenital abnormality over normal for every 60–250 ART babies born. This increased risk does not appear to be related to multiple births, and is similar for both minor and severe defects. Does ART cause these defects, or are they a consequence of the parental genetic status that may underlie or predispose to the infertility? Since male genital tract defects are commonly reported in IVF babies, at least part of the increased incidence may relate to the parental infertility itself. In retrospective analyses, the prevalence of **defects in parental imprinting** is higher among IVF babies than in the general population. However, the absolute prevalence is extremely low so numbers are very small. These studies are retrospective and thus not optimally controlled, and it is also unclear, should the difference turn out to be real, what the explanation might be. Did ART cause the imprinting errors? Or are certain types of infertility associated with a higher probability of imprinting error? Current clinical advice is that there is no proven reason for concern. Thus, overall, there appears to be a small but significantly increased risk of problematic outcome amongst ART-produced babies, aside from the well-established substantial risk increase arising from multiple births. There are certainly features of ART procedures that give cause for concern that the procedures themselves might be responsible for at least some of this increased risk. Thus, the freezing and thawing of embryos, any delay in the fertilization of oocytes, the culture media composition and the risk of exposure to reactive oxygen species, variations in laboratory temperature and its effects on maternal meiotic spindle stability, and the medications used to induce ovulation could each impact adversely on outcome if not carefully and skilfully managed and monitored. It is also possible that some of the increased risk may arise from the underlying infertility of the couples seeking treatment, rather than the treatments themselves. Perhaps examining outcome data more carefully to discriminate between different types of infertility would be useful, since tubal damage secondary to infection is a very different sort of pathology from azoospermia. Where does this leave us when talking with prospective patients for ART therapy?

What is reasonably clear is that doctors should properly discuss the increased risks from whatever source. However, it is a matter of concern that patients remain willing to accept the established high risks arising from multiple births in order to increase their chances of conceiving at all. This suggests that among at least some patients, their desperation for a child is leading them to rash decisions, and so it seems unlikely that the other smaller and less well understood risks will deter them from undergoing ART at all. The second major conclusion from the evidence base is that much better quality analysis is needed, both from existing retrospective data and ideally from high-quality prospective matched cohort studies. The continuing and careful monitoring of children conceived after ART for any long-term health impacts is essential if we are to improve the quality of our evidence base from which to offer advice.

Commodification?

The advent of IVF and related therapies, especially when aligned with private medicine – as is the case in most of the USA and much of the UK – has raised the fear that people or embryos are becoming commodified. Examples abound: surrogate mothers and egg donors might be induced to 'rent' their wombs or to 'sell' their eggs; frozen eggs, embryos or sperm might be sold for commercial profit; saviour siblings might be produced for purposes other than to be loved for themselves – to provide 'spare parts' for a sibling or parent; embryo sexing may be requested and performed because a boy is wanted and a girl would be aborted. Undoubtedly all these fears are being realized somewhere by some unethical doctors. Should the fact that a new technology can be abused by some mean that its use should be prohibited for all? The answer has generally been No – it is **better to manage the risks than ban the benefits**. The form that this management takes, however, varies with different jurisdictions. Thus, some countries rely on professional self-regulation, others on state regulation and yet others on the free market plus recourse to the courts. All these systems should be underpinned by clearly articulated guidance on medical ethics and its enforcement professionally (see pages 390–1). This latter point assumes particular significance given the global market in reproductive technology. Thus, patients, doctors, embryos, gametes and uteri are moving around the globe for a range of reasons all under the broad heading of **cross-border reproductive care**. Unless doctors have a strong ethical framework to guide their own actions and to police those of their colleagues, abuse of patients will undoubtedly occur.

Personhood?

Considerable controversy has surrounded the ethics of producing conceptuses *in vitro*, as well as freezing them, discarding some of them or undertaking research on them. In particular, the decisive role that fusion of an oocyte and a spermatozoon might play in the establishment of an individual human life has been stressed. It is important to be aware that scientific evidence does not support the view that one single event is decisive for the creation of an individual life. The process is a continuum which starts with the growth of the oocyte, synthesis of the maternal cytoplasmic inheritance, and distinctive epigenetic changes to the sperm and oocyte genomes, and continues with the formation and development of the genetically unique zygote, the expression of these unique genes to replace the maternally inherited gene products, the signalling by the conceptus to the mother of its presence, the segregation of the embryonic from the extraembryonic lineages, the point at which homozygous twinning can no longer occur, the subsequent gradual acquisition of a distinctive human form capable ultimately of independent existence, and the pre- and post-natal epigenetic modifications wrought by environmental signals. The natural losses in mammals are massive over the early events in this sequence, particularly from oocyte atresia, sperm wastage and preimplantation death. Human preimplantation losses are increased further by our changed social and sexual habits, and the use of various forms of contraceptive. So, biology cannot provide the answer to the question 'When does a human life begin?' because the question is based on false premises. Biologically, life is a continuum. Ethical decisions cannot therefore be taken on the basis of biological observation alone. However, biological observations should inform the ethical conclusions reached. Broadly, biology tells us that we acquire progressively and continuously those features that make us individual up to the time of our death. This has led to the adoption in many societies of a **gradualist view** of the degree of ethico-legal protection and care due to developing humans. Where in that sequence full legal personhood is granted depends on arbitrary criteria. In most jurisdictions, it is the **birth of a viable baby**, as, prior to that, the rights of the carrying mother must trump those of the fetus.

Eugenics

Eugenics means literally 'generating fine offspring', but has come to carry connotations of hereditary improvement by controlled selective breeding. Whilst we started this book with natural and sexual selection (see page 4), we are ending it with a consideration of **selection by human intervention**. In fact, examples of such interventions have featured throughout this book – not least the role played by society in the selection of sexual partners (see pages 101–3) and the many ways in which medicine has prevented the deaths of fetuses, neonates and mothers. However, these selection processes are usually regarded as 'morally' (if not practically) neutral, whereas eugenic selection carries overtones of moral reprehension. Here we will consider two types of selection.

In Boxes 13.1 and 13.2, we considered the problem of how to select which of a number of IVF-generated conceptuses for transfer to the uterus. Three issues were identified: (1) how many to transfer? (2) what stage to transfer them – cleavage or blastocyst? and (3) what criteria to use? In essence, these decisions are eugenic in that they involve choosing among conceptuses to give the best chance of pregnancy. Most of those not selected will likely perish. However, usually these decisions are not seen as being eugenic, a term reserved for selection that has a genetic basis.

Genetic selection has been exercised through abortion for several decades since the advent of prenatal diagnosis (PND) by amniocentesis and chorionic villus sampling (Box 21.2). However, two newer forms of testing technologies have expanded the possibilities for eugenic selection: preimplantation genetic diagnosis (PGD) (see page 377) and the detection in maternal blood of trophoblastic cells (see Box 21.2), often called **non-invasive testing** to distinguish it from amniocentesis and CVS. PGD is expensive and involves IVF, so will not be undertaken lightly, but the increasing sensitivity and discriminatory power of non-invasive testing are bringing forward the time when testing can be achieved reliably; soon it will be routinely possible well within the first trimester, when an early drug-induced termination is possible. These techniques open up the possibility of eugenic selection. Such selection can be negative or positive.

Negative eugenic selection avoids certain traits being passed on. Typically, it has been used in at-risk families who are known to carry mutations that can lead to severely debilitating or lethal conditions, especially those that affect the baby neonatally or early in childhood, such as haemoglobinopathies like **thalassaemias**. However, selection to prevent the transmission of cancer predisposition genes (e.g. the ***BRCA1* and *2*** mutations predisposing to breast cancer) or those impacting seriously on health later in life, such as **Huntington disease**, is also being offered via PGD. For some genetic conditions that are **sex-linked**, such as **haemophilia** or **Duchenne muscular dystrophy**, sex selection can be offered. In addition to avoiding the transmission of these mutations in single genes, the use of **preimplantation genetic screening** (PGS) to screen for preimplantation conceptuses free of chromosomal anomalies is also offered by some clinics to older women (see page 377). In general, there is a social acceptance that where the birth of a severely handicapped baby can be avoided, the potential parents, who would have to bring the child up, should have the choice as to whether they do so. The ethical issue becomes one of how severe should a condition be for them to be offered the choice? Different societies handle this question in different ways. Some offer no choice, others offer a free choice, others exercise statutory or professional regulatory systems.

These ethical questions become even more vexed when **positive eugenics** – the active selection for certain characteristics – is concerned. The active selection of **histocompatible conceptuses** to provide a tissue donor for an existing sibling or relative (see page 377) is one example of positive eugenics. Another is positive selection of a child of a specific sex. Sex selection for non-medical purposes has been advocated for **family balancing**, where a family has two or three sons and wants a daughter, for example. Such a use seems reasonable and would not unduly **unbalance the sex ratio of a population**, unlike sex selection for a single child, for which there is already evidence of distorted sex ratios in certain countries. Even more contentious is selection for a specific trait such as blue eyes or blond hair, or for deafness or dwarfism by parents who are deaf or dwarfs themselves. The recent development of reliable technologies for modifying genes in the zygote brings the potential power of genetic enhancement even closer. Should societies regulate these choices? If so, how? Given the ease of international travel, moving between different jurisdictions for reproductive treatments already occurs. Moreover, non-invasive testing, if commercialized, would remove much of the high technology input required for PGD.

Reproduction, sexuality and ethics

Humans, like many other primates and some cetaceans and insectivores, but unlike most other mammals, do not confine their sexual activity to one narrow phase of the cycle at or around the time of ovulation. Neither do humans confine their sexual activity to vaginal penetrative sex with a member of the 'opposite' gender. Thus, for humans, coition and the expression of sexuality have a significant function over and above that of reproduction. The nature of that function in evolutionary terms remains a matter for debate, but the clear biological separability of coital sex from procreation is undeniable. Moreover, through humankind's own talents and capabilities, whether viewed as evolved or God-given or both, this separability has been further enhanced. Thus, the development of methods for reducing fertility or overcoming and circumventing infertility has further divorced reproduction from the sexual act. Reproduction may be commenced outside of the body, doctors and scientists becoming essential agents in the process. Oocytes, spermatozoa or embryos may be donated by individuals other than the partners being treated, leading to unusual patterns of 'parenting'; indeed, the very definition of a parent is being challenged. Thus, a woman may carry a fetus derived from gametes neither or only one of which came from her or her partner (**gamete/embryo donation**). A woman may carry a fetus derived from IVF with the purpose of giving it at delivery to two other parents, one or both of whom may be genetically related to the child (**surrogacy**). Parents may both be women or both men, only one of them being related genetically to the child. A woman may carry a pregnancy well after her menopause. Genetic parenthood after death by use of frozen gametes or embryos is real. Reproductive cloning has approached feasibility.

This blurring of boundaries between the expression of sexuality and procreation, between 'genetic', 'uterine' and 'social' parents, between conventional and novel patterns of parenting is confusing for many, and challenging for traditional religions and social values. This challenge derives from the fact that religious, social and legal control of sexual and reproductive activities has been a feature of all human societies, as these activities bear heavily on patterns of inheritance, power and individual freedom of expression, as well as on individual and social health. Thus, these new possibilities appear to, and do, threaten many established aspects of social structure. Clear ethical thinking about their impact is required, and not simply reactive assertion from tradition, if sensible legal controls on their use are to be enacted. Such legal responses are already occurring with variably sensible outcomes in many countries. What are the key ethical issues that have influenced these discussions and decisions?

There is insufficient space here to do justice to the arguments, but some of the main points are summarized, not all of which are universally accepted, but all of which can form the basis for reflection and discussion – especially by doctors.

- It is not legitimate to resort to ethical arguments that equate traditional with natural, nor of novel with unnatural. What is possible biologically is natural and what is happening is therefore natural.
- There is a presumption that individual liberty to choose how to act sexually and reproductively should be protected, as long as others are not affected adversely.
- With this liberty goes the moral responsibility to protect the welfare of the parties practising sexual and reproductive interactions, and also of any children that might be born as a result. This means that the adult parties should give adequate and informed consent to their mutual activity, and should not be placed under duress. The interests and welfare of any child born, and of the 'siblings' of any child born (or not born), must be taken into account. Such a process should consider the prevailing social values and how these will affect both society's perception of the child and the ability of the child to form a clear and positive identity. It will also take into account the health of the child.
- There is a general acceptance, based on biological and ethical arguments, that humans acquire an individual status of personhood progressively, not suddenly. There is no moment at which a human being exists where one did not before, just as there was no moment in evolution at which humankind existed where no humans had existed before.
- Arising from this consideration, the status of the human gametes, preimplantation conceptus, embryo and early fetus differs from that of both existing humans and non-humans. Human material with the potential for development into an

individual demands respect for that potential, while not having the moral or legal rights accorded to an existing human; where conflict arises between the needs of a potential mother and those of her developing embryo/fetus, there is a presumption in favour of the mother's rights until such time that the fetus becomes capable of an independent existence.

- There is a presumed right to parenthood, although not at any cost.
- There is a presumption that individuals should be treated confidentially and should have control over the reproductive information held about them.
- There is also a presumption that it is better not to withhold from individuals information about their origins, e.g. if they are conceived by techniques of assisted conception; and that genetic information should be stored securely and made available so that genetic incest can be avoided.
- There should be a limit on the numbers of children produced from donated gametes and embryos from one individual, so as to limit the chances of genetic incest occurring subsequently.
- There is a presumption that it is better to exist than not to exist.

In the UK, in which much of the pioneering medical and scientific work that has led to this reproductive revolution has occurred, ethical debate and legal action also came early – indeed, it was initiated by Bob Edwards himself in the 1970s. There is currently in place, through the Human Fertilization and Embryology Act of 2008, a Human Fertilization and Embryology Authority (HFEA) set up to regulate the generation, use and storage of human embryos *in vitro*, the storage and use of human gametes *in vitro* for later therapeutic use, and the use of human embryos and of human fertilization *in vitro* in research. Many of the other issues raised above are also being brought to the HFEA. This body represents an attempt to protect the human individual, society and conceptus from exploitation and excess, by applying ethical principles to biological and medical problems through legal processes. It is being observed with interest by many other states as a possible model of enlightened regulation to copy or react against.

Conclusions

Our failure to control fertility adequately, concomitant with our success in decreasing neonatal and infant mortality, has, for much of humanity, replaced one tragedy by another. The spectre of deprivation, starvation, mass migration and war is still all too real for much of the world's population. One of the greatest indictments of medical science has been its relative neglect of, and belated initiation of, research in human reproduction, human sexuality and the associated clinical disciplines. The social and religious conservatism that has delayed or prevented analysis of these subjects means that much of our knowledge is recent or incomplete. Indeed, many of the pioneers in this area of scientific study are still alive. Echoes of this conservatism recur in debates on the ethics of the so-called 'test-tube babies', abortion, PGD, contraception, attitudes to sexual behaviour and the response to HIV and AIDS. More important, however, are the continuing effects of social and religious attitudes on the effective dissemination and application of the knowledge that has accrued from such a relatively brief period of rigorous scientific and medical study. Deliberate policies to promote reproduction in some countries fearful of racial imbalance, and the frustration of family planning programmes and sexual health education programmes in others, are sadly as much a feature of the so-called civilized world as of countries struggling to improve the economic circumstances of their inhabitants. Throughout this book we have refrained from imposing a personal view on the account of the science of reproduction. However, we hope that study of this book will help students training to take a place in the medical and scientific professions to realize just how pervasive the reproductive process is, both for the individual and through society at large, and how an adequate understanding and control of it represent a crucial element in the survival and well-being of both.

Key learning points

- Social and legal constraints on fertility and the expression of sexuality exist in most human communities, but differ in nature among them.
- Societies tend to blame the infertile for their state.
- Most genetic influences on fertility are multi-genic.
- Multiple births are the one major adverse outcome of assisted conception.
- There is little evidence of assisted conception in itself adversely affecting the health and well-being of the offspring produced.
- There is evidence of the commodification of some aspects of assisted reproduction in some jurisdictions.
- The concept of personhood cannot be applied usefully to eggs, embryos or fetuses prior to their birth and survival.
- Both negative and positive eugenics are practised using a range of reproductive technologies, and are regulated variously in different countries.
- Cross-border reproductive travel for reproductive services is growing and makes local regulatory practices more difficult to enforce.
- Clear ethical reasoning is required to chart a moral path through the new assisted reproductive technologies and the new family patterns that are being generated.

Clinical vignette

Genetic testing of the fetus

A 28-year-old woman, Ms H, was 7 weeks into her first pregnancy. Ms H's father was diagnosed with Huntington disease at the age of 55, and she had opted for genetic testing to assess her risk of the disease. Ms H's genetic analysis indicated that she had an expanded number of trinucleotide repeats and hence a high chance of developing the condition later in life. She attended a fetal medicine clinic wishing to have a genetic test to find out if her unborn baby might be similarly affected. The fetal medicine consultant referred the case to the local ethics committee for an opinion on whether or not it would be ethical to perform the test. As Ms H was of the strong opinion that she would terminate the pregnancy if the fetus were affected by Huntington disease, the ethics committee considered that it would be ethical to go ahead with genetic testing. An uncomplicated chorionic villus sampling procedure (sampling of cells from the placenta with a needle through the mother's abdomen) was performed. After 10 days, the results of the genetic testing showed that the fetus had a normal number of trinucleotide repeats and was therefore unlikely to be affected by Huntington disease. Ms H continued with the pregnancy and gave birth to a healthy baby boy at 38 weeks.

In this case, the ethics committee judged that it would only be ethical to perform the genetic testing if Ms H intended to terminate the pregnancy in the event of a positive result. This is because it can be viewed as unethical to determine information about the genetics of a living child who is unable to consent. If Ms H was aware that her child was destined to develop Huntington disease later in life, it might influence the child's upbringing and opportunities in life even if this information was not shared with the child. For this reason, it is usually viewed as unethical for parents to obtain genetic information about their child in cases where the knowledge would not have medical or psychological benefits until adulthood. This is the case in Huntington disease, where testing could be safely deferred until the child is old enough to understand the implications of testing and consent for themselves. However, it can be argued that the ethic's committee's decision conflicts with Ms H's right to all available information before deciding whether to continue with a potentially affected pregnancy. Termination of wanted pregnancies affected by genetic diseases is a very complex decision for many women, and may be considered unfair to ask women to make decisions about what to do in the abstract, prior to testing being carried out. Indeed, Ms H could have changed her mind after an adverse diagnosis.

FURTHER READING

General reading

Bergh C (2005) Single embryo transfer. *Human Reproduction* **20**, 323–327.

Dunstan GR (1990) *The Human Embryo: Aristotle and the Arabic and European Traditions*. University of Exeter Press, Exeter.

Inhorn MC, Gurtin ZB (2012) Cross-border reproductive care: a future research agenda. *Reproductive BioMedicine Online* **23**, 665–676.

Johnson MH (1998) Should the use of assisted reproduction techniques be deregulated? The UK experience: options for change. *Human Reproduction* **13**, 1769–1776.

Johnson MH (2001) The developmental basis of identity. *Studies in the History and Philosophy of Biology and Biomedical Sciences* **32**, 601–617.

Johnson MH (2006) Escaping the tyranny of the embryo? A new approach to ART regulation based on UK and Australian experiences. *Human Reproduction* **21**, 2756–2765.

RCOG (2011a) *Multiple Pregnancy Following Assisted Reproduction*. Royal College of Obstetricians and Gynaecologists Scientific Advisory Committee Opinion Paper **22**, 1–5.

RCOG (2011b) *Reproductive Ageing*. Royal College of Obstetricians and Gynaecologists Scientific Advisory Committee Opinion Paper **24**, 1–5.

Spar DL (2006) *The Baby Business: How Money, Science and Politics Drive the Commerce of Conception*. Harvard Business Press, Boston.

More advanced reading

Adashi EY, Barri PN, Berkowitz R *et al.* (2003) Infertility therapy-associated multiple pregnancies (birth): an ongoing epidemic. *Reproductive BioMedicine* 7, 515–542.

Arce J-C, Nyboe Andersen A, Collins J (2005) Resolving methodological and clinical issues in the design of efficacy trials in assisted reproductive technologies: a mini-review. *Human Reproduction* **20**, 1757–1771.

Bainham A *et al.* (eds) (1999) *What Is a Parent? A Socio-legal Analysis*. Hart Publishing, Oxford.

Bonduelle M, Wennerholm UB, Loft A *et al.* (2005) A multi-centre cohort study of the physical health of 5-year-old children conceived after intracytoplasmic sperm injection, in vitro fertilization and natural conception. *Human Reproduction* **20**, 413–419.

Bouligand J, Ghervan C, Trabado S *et al.* (2010) Genetic defects in GNRH1: a paradigm of hypothalamic congenital gonadotropin deficiency. *Brain Research* **1364**, 3–9.

Cook R, Day Sclater S (eds) (2003) *Surrogate Motherhood.* Hart Publishing, Oxford.

De Leo V, Mussacchio MC, Cappelli V, *et al.* (2016) Genetic, hormonal and metabolic aspects of PCOS: an update. *Reproductive Biology and Endocrinology* **14**, 38.

De Rycke M, Liebaers I, Van Steirteghem A (2002) Epigenetic risks related to assisted reproductive technologies: risk analysis and epigenetic inheritance. *Human Reproduction* **17**, 2487–2494.

Golombok S, Brewaeys A, Giavazzi MT *et al.* (2002) The European study of assisted reproduction families: the transition to adolescence. *Human Reproduction* **17**, 830–840.

Hansen M, Bower C, Milne E *et al.* (2005) Assisted reproduction and the risk of birth defects – a systematic review. *Human Reproduction* **20**, 328–338.

Hyde KJ, Schust DJ (2015) Genetic considerations in Recurrent Pregnancy Loss. *Cold Spring Harbor Perspectives in Medicine* **5**, a023119.

Levy G, Hill MJ, Beall, S *et al.* (2012) Leiomyoma: genetics, assisted reproduction, pregnancy and therapeutic advances. *Journal of Assisted Reproduction and Genetics* **29**, 703–712.

McAllister JM, Legro RS, Modi BP, Strauss JF (2015) Functional genomics of PCOS: from GWAS to molecular mechanisms. *Trends in Endocrinology and Metabolism* **26**, 118–124.

Mitchell MJ, Metzier-Guillemain C *et al.* (2017) Single gene defects leading to sperm quantitative anomalies. *Clinical Genetics* **91**, 208–216.

Miyamato T, Minase G, Okabe K *et al.* (2015) Male infertility and its genetic causes. *Journal of Obstetrics and Gynaecology Research* **41**, 1501–1505.

Ordulu Z (2016) Fibroids: genotype and phenotype. *Clinical Obstetrics and Gynecology* **59**, 25–29.

Pelosi E, Simonsick E, Forabosco A *et al.* (2015) Dynamics of the ovarian reserve and impact of genetic and epidemiological factors on age of menopause. *Biology of Reproduction* **92**, 130.

Rahmioglu N, Montgomery GW, Zondervan KT (2015) Genetics of endometriosis. *Women's Health* **11**, 577–586.

Tur-Torres MH, Garrido-Gimenez C, Alijotas-Reig J (2017) Genetics of recurrent miscarriage and fetal loss. *Best Practice and Research: Clinical Obstetrics and Gynecology* pii: **S1521-6934**(17)30054–8.

Index

Page numbers in italics refer to Figures, Tables or Boxes

Essential Reproduction, Eighth Edition. Martin H. Johnson.

Companion website: www.essentialreproduction.com